Climate of the Past, Present and Future

Climate of the Past, Present and Future
A Scientific Debate, 2nd ed.

Javier Vinós

Critical Science Press

Madrid

2022

Cover image: Spilhaus continuous ocean projection displaying 1955-2012 average sea surface temperature at 2 °C contour interval after Locarnini RA, Mishonov AV, Antonov JI et al (2013) World Ocean Atlas 2013. Volume 1: Temperature. In: Levitus S & Mishonov A (eds) NOAA Atlas NESDIS 73. Global annual afternoon cloudiness derived from observations from the Aqua satellite (NASA). ©Author 2022. Some rights reserved. CC BY-NC 4.0

Published by Critical Science Press
ISBN: 978-84-125867-0-1 (Hardcover)

*To my parents, José Antonio and María del Carmen, and to my children, Blanca and Miguel.
May the torch of knowledge and progress be handed to the next generation.*

CONTENTS

LIST OF FIGURES AND TABLES

PREFACE

Towards the end of the summer of 2014, I walked alone the "Camino de Santiago" (Way of Saint James) from the Pyrenees to Santiago de Compostela, near the Atlantic coast. It is an ancient pilgrimage route that had its heyday during the Medieval Warm Period and decayed with the Black Plague, but has seen a modern revival as a spiritual and cultural European route that is now an UNESCO World Heritage Site. I walked 750 km in a month visiting from the Atapuerca archaeological site, famous for its *Homo antecessor* and *neanderthalensis* remains, to the medieval architecture of northern Spain. I had a lot to think after the recent death of my parents in less than two years. What kind of world are we leaving to our children and their children? In the long days at "El Camino" I had plenty of time to deeply think about the passage of time and the changing of the world and its people through pre-history and history. A testimony I could see before my eyes. As a biologist (of the laboratory type) I was familiar with the effects of global warming. Not only I can remember the colder winters of my childhood in the early 1970s, I can also attest to the lengthening of the growing season, the earlier appearance of insects over the years, or the recent decision by some migratory birds to remain in Spain through the winter instead of migrating to Africa.

One of my decisions was to start a blog to explore the risks of global warming in the fall of 2014. It is easier to research and learn things when one has to explain them to others. As a scientist, when I need information, I don't rely on second-hand opinions. I go directly to the evidence and the scientific literature. But in my carefully laid out plan of warning the world about the dangers of climate change I found a problem. The evidence that the planet was warming was clear (I already knew that), although no warming had taken place for over a decade. The evidence that we have greatly increased the atmospheric levels of carbon dioxide was clear (I also knew that). What wasn't clear at all was the evidence that the carbon dioxide was causing the warming. Clearly the warming had started long before the fast increase in carbon dioxide.

The more I researched climate change the less certain I was about the IPCC conclusions about anthropogenic warming. Particularly troublesome was the treatment of skeptical scientists. In science strong evidence defends itself. When Albert Einstein was told of the publication of the 1931 book *"A hundred authors against Einstein,"* he is credited with saying *"Why 100 authors? If I were wrong, then one would have been enough!"* I decided to go deeper and learn what was known about how climate changed when humans could not have affected it. Paleoclimatology articles are rife with claims for a stronger role from solar variability than is currently accepted by the IPCC and coded into climate models.

By 2015 I had made my transition from accepting IPCC claims at face value to being very skeptical that we had sufficient knowledge and understanding of climate change to support them. I don't really understand why it was decided that global warming should be fully blamed on us. I know most scientists that hold that belief are sincere, but how many of them have looked at the evidence critically as I have been doing for the past seven years, free from assumptions and group-think? Before 2014 I had never looked at the evidence and I would have defended the official position as I would have found unthinkable that the extraordinary evidence to support those extraordinary claims wasn't there. I am sure many scientists concentrated on their own narrow subject assume the evidence is there and are too busy to check by themselves. It is also not a wise career movement to frontally oppose the climate academic status quo. As a non-climatologist I am also free from that pressure.

One of the dangers of being an outsider to a field is being unable to judge the quality and solidity of one's work. Was I just overestimating the importance of the arguments I was rising? Perhaps everything I was finding was of a trivial nature and already dealt with scientifically long ago. I didn't think so because I was reading several articles every day and the count was already in the thousands. If my doubts had been solved it should be reflected in the scientific literature. Quite the contrary I was finding authors subtly expressing similar doubts between lines in their articles. I decided that I should find an expert with an open mind that could judge my work and tell me if it had any value. Judith Curry, then a professor at the Georgia Institute of Technology, was the perfect person. She is an expert in climate and atmospheric sciences and besides being the president of the climate forecast company CFAN, she runs a blog where high quality collaborations are welcome.

I sent my first article to Judith Curry in May 2016. She sent it out for review by an outside expert and came back with a lot of requests for changes including a change in focus. I rewrote it into what was essentially a draft version of chapter 6 and resubmitted it in September, when it was published at Judith Curry's blog "Climate.Etc" I was very fortunate then in knowing Andy May through the comments of a blog. He is a petrophysicist from Texas and also a researcher of climate change that kindly offered to correct the English language in my web articles. He actually did a lot more and over the past five years he has contributed with valuable opinions and together we have written some web articles, developing a friendship. He is also the author of two popular climate books for general readership, *"Climate Catastrophe! Science or Science Fiction?"* on the practical aspects of climate change and how it affects our lives, and *"Politics & Climate Change: A History"* on how climate change became a politically-loaded issue.

In October 2016 I had already written an article on the climate of the Pleistocene and I sent it to Dr. Willie Soon, at the Harvard-Smithsonian Center for Astrophysics, requesting his opinion on it. He was so kind as to read it and tell me that in his opinion it had sufficient quality for publication. With that reassurance, over the next three years the book took form with drafts for most chapters appearing on Judith Curry's blog, where many readers contributed with valuable opinions that improved them. Publication of a climate book that is not skeptical of climate change, but is skeptical of its causes proved difficult. Some reviewers frontally opposed publishing a book that contradicted IPCC conclusions. But the 2-year delay due to rejections was fruitful, as the book kept improving. I had been a little unsatisfied because I did not have a good answer to how the climate changes and how the Sun affects the climate, although I found consolation in thinking that nobody did. Then a warm night in late spring, while walking along a Mediterranean seafront promenade eating an ice-cream, I let my thoughts wander on the Early-Eocene low gradient paradox. How could the poles have been so warm then if much less energy could be transported through such low gradient? It then struck me that the answer required just to invert the question. The Early-Eocene poles were so warm because much less energy was transported to them. Transporting more energy to the poles is how the planet cools. Time will tell if I was correct, but I have been able to provide an answer to my questions, developing what I named the Winter Gatekeeper hypothesis.

Over the past six years I have put more dedication, effort and time in researching climate change than many people dedicate to obtain their university degree. No doubt sufficient effort to have obtained a second doctorate if I have focused it into a sufficiently narrow aspect. But my goal was not a title, yet contributing to the most interesting and important scientific question of our times. Without question science historians will have a feast in the future with the climate change scare, that Michael Crichton termed *"State of fear."* I want to be in the right side of it and for that I only have to follow the evidence wherever it takes me. I became a scientist to look for answers to important questions. The quest is what makes the effort valuable to me.

This book would not have been possible without the support and diffusion given to my work by Judith Curry, who had to endure my assault on her blog with articles several times longer than prudence recommends for the web. Publishing at her blog has given me a motivation for doing my best to deserve such distinction. Andy May has accompanied me in this trip, being the first to read the material and improving it in multiple aspects. If the book can be understood it is thanks to his unpaid generosity, and all the mistakes are mine only. I also thank Willie Soon for his encouragement and for interesting and educative exchanges.

In the best spirit of science, many scientists have shared their data and figures with me even when disagreeing with my interpretation of the evidence. They have contributed to make the book better. They are: Jean-François Berger, Maxime Debret, Sarah Doherty, Trond Dokken, William Fletcher, Jacques Giraudeau, Rüdiger Hass, Andrea Kern, Thomas Marchitto, Paul Mayewski, Adriano Mazzarella, Nick McCave, Kerim Nisancioglu, Olga Solomina, Christopher Scotese, Frank Sirocko, Willie Soon, Ilya Usoskin, Heinz Wanner and Bernie Weninger. I am grateful to all of them. I am also grateful to all the commenters of my web articles and the reviewers of the book. They have also made the book better.

Finally, for enduring all the time and dedication that I have taken away from more important things, and for all the support she has given me through the years, I am deeply indebted and grateful to my companion Mar Lagunas.

Javier Vinós
Madrid, December 27, 2021.

FOREWORD

In May 2016, I received an email from Javier Vinós—someone who was unknown to me at the time—proposing a guest post for my blog Climate Etc. (judithcurry.com) on the role of solar variability on climate. I jumped at the opportunity, since this was a topic I knew very little about. This post kicked off a 10 part series by Javier entitled 'Nature Unbound' that was published on my blog over the course of two years -- this series provided the nucleus for *"Climate of the Past, Present and Future."*

As a climate researcher myself, I learned an enormous amount from Javier's blog posts. Like a majority of climate researchers, my expertise is on recent climate variability that is studied primarily in the context of elucidating manmade global warming. Most climate researchers focus on the period since 1950; I have been somewhat of a maverick in the climate research community by drawing attention to climate variations over the past several hundred years and also to natural climate variability.

Given the 'consensus enforcement' surrounding the issue of manmade climate change, there has been little incentive for an academic climate scientist to develop an alternative but comprehensive narrative of climate change. Javier Vinós, an academic researcher from outside the traditional fields from which climate scientists are drawn, has taken an independent look at climate variations and their causes – past, present and future. This is an enormous undertaking for a single scientist. However, reasoning by single intellect about all of the relevant processes is a very much needed complement to the fragmented top-down consensus seeking approach employed the Intergovernmental Panel on Climate Change (IPCC), that is focused on 'dangerous anthropogenic climate change.'

In the heavily politicized debate surrounding climate change, this book returns the debate to a rational, scientific one. Rather than starting from the assumption that recent warming is caused by manmade emissions of greenhouse gases, Javier Vinós examines how climate has changed naturally and then assesses how it is different from what is happening now.

This book reminds us that climate 'is' climate change, with change being intrinsic to a very complex, strongly-regulated dynamical system. The book provides strong support for the idea that the belief in a stable benign climate suddenly thrown out of equilibrium by human actions is, in all probability, wrong. It raises the suspicion that anthropogenic forcing of climate change has been seriously overestimated.

The first half of the book is a trip backwards in time – the past 800,000 years. Throughout the book, a sense of the history of scientific debate on these topics is provided, including the current uncertainties.

The book highlights the importance of solar variations in controlling the Earth's climate. The extraordinary coincidence of Grand Solar Maximum in the late 20th century with a period of warming should raise all type of questions. Instead solar variability is assigned no role in Modern Global Warming by the IPCC. A great deal of the science discussed in this book suggests that the climatic effect of solar variability has been significantly underestimated, out of ignorance and neglect. This underestimation of solar forcing has the inevitable consequence of an overestimation of the role of CO_2 and to the incorrect hypothesis of CO_2 as the control knob on climate.

In pondering the climate of the 21st century, Vinós readily acknowledges that we are dealing with a situation without precedent and so the answers that we can obtain from science carry a huge uncertainty that cannot be properly constrained by evidence. His forecast for a stabilization of the current warming does not depend on any change in policies or heroic reductions in emissions. He expects that atmospheric CO_2 levels should reach 500 ppm but might stabilize soon afterwards. Afterwards global warming could end, with temperatures stabilized around +1.5 °C above pre-industrial, followed by a slow decline.

After reading this book, I am perhaps more concerned about a coming ice age in several thousand years time than I am about the possibility of catastrophic warming from greenhouse gas emissions on the timescale of the 21st century. If Vinós' analysis is correct, thinking that we can control the Earth's climate by reducing CO_2 emissions may turn out to be the greatest folly of the 21st century. This is a debate that we need to have.

Judith Curry
President, Climate Forecast Applications Network
Professor Emerita, Georgia Institute of Technology
Reno, NV USA.
5 March 2019

ABBREVIATIONS

21stC-SGM: Mid-21st century solar grand minimum
97CS: 1997–1998 climate shift
$\delta^{18}O$: Change in oxygen isotope 18, expressed as ‰
δD: Change in deuterium (hydrogen isotope 3), expressed as ‰
µm: One millionth of a meter

- A -
a: Anni. Years taking 1950 as the reference present date
Aa index: The antipodal amplitude geomagnetic index
AABW: Antarctic bottom water
AAM: Atmospheric angular momentum
ACE: Abrupt climatic event
AD: Anno Domini. Number of years since the beginning of the Christian era in the Gregorian calendar
AGW: Anthropogenic global warming.
AIM: Antarctic isotope maxima.
AMO: Atlantic Multidecadal Oscillation.
AMOC: Atlantic Meridional Overturning Circulation.
AO: Arctic oscillation
AOO: Arctic Ocean Oscillation index
AP: After present. Number of years after 1950 in the Gregorian calendar
AR: Assessment report published by the IPCC
ASR: Absorbed short-wave radiation.

- B -
B2K: Before 2000. Number of years before the year 2000 in the Gregorian calendar
BA: Bølling–Allerød Period
BC: Before Christ. Label to indicate a number of years before the beginning of the Christian era in the Gregorian calendar
BDC: Brewer–Dobson circulation
BO: Biennial Oscillation of the polar vortex
BP: Before present. Number of years before 1950 in the Gregorian calendar

- C -
c.: Circa, approximately.
Cal (years): Calibrated years, also calendar years. Dating obtained from converting radiocarbon years to calendar years
CE: Christian Era
CFC: Chlorofluorocarbon
CH_4: Methane
CMIP: Coupled model intercomparison project
CO_2: Carbon dioxide

COVID-19: Coronavirus disease 2019

- D -
D: Deuterium (hydrogen isotope 3)
D–O: Dansgaard–Oeschger
DACP: Dark Ages Cold Period
DLR: Downward longwave radiation
DNA: Deoxyribonucleic acid

- E -
ECMWF: European Center for Medium-range Weather Forecast
ECS: Equilibrium climate sensitivity.
EDC3: EPICA Dome–C deuterium age scale
ELA: Equilibrium line altitude, a glaciological term
ENSO: El Niño/Southern Oscillation
EPICA: European Project for Ice Coring in Antarctica
ERA: European Reanalysis
EROEI: Energy return on energy invested
ETCW: Early Twentieth Century warming
EUMETSAT: European Organisation for the Exploitation of Meteorological Satellites

- G -
Ga: Giga anni. Number of 10^9 years before the present
GCM: General circulation model
GHE: Greenhouse effect
GHG: Greenhouse gas
GI: Greenland interstadial
GICC05: Greenland Ice Core Chronology 2005
GISP2: Greenland Ice Sheet Project 2
GRIP: Greenland Ice Core Project
GS: Greenland stadial
GSAT: Global surface average temperature
GSL: Geological Society of London
Gt: Gigatonnes
GtC: Gigatonnes of carbon
Gyr: Giga years, billions (10^9) of years

- H -
HadCRUT: Hadley Climate Research Unit temperature
HadSST: Hadley sea-surface temperature
HCO: Holocene Climatic Optimum
HE: Heinrich event

- I -
IACP: Intra-Allerød Cold Period
IERS: International Earth Rotation and Reference Systems Service

IPCC: Intergovernmental Panel on Climate Change
IPWP: Indo–Pacific Warm Pool
IR: Infrared radiation
IRD: Ice-rafted debris
ISGI: International Service of Geomagnetic Indices
ITCZ: Intertropical Convergence Zone

- K -

ka: Kilo anni. Number of 10^3 years from the present
KNMI: Koninklijk Nederlands Meteorologisch Instituut
kyr: Kilo years, thousands of years

- L -

LBK: Linearbandkeramik (German), Linear Pottery culture
LGM: Last Glacial Maximum
LIA: Little Ice Age
LIG: Latitudinal insolation gradient
LOD: Length of day.
LTCW: Late Twentieth Century warming
LTG: Latitudinal temperature gradient
LWR: Longwave radiation

- M -

Ma: Mega anni. Number of 10^6 years before the present
MGW: Modern Global Warming
MHT: Mid-Holocene Transition
MIS: Marine Isotope Stage
MSM: Modern Solar Maximum
MPT: Mid-Pleistocene Transition
MT: meridional transport
mtDNA: mitochondrial deoxyribonucleic acid
MWP: Medieval Warm Period
Myr: Mega years, millions of years

- N -

NAC: North Atlantic Current
NADW: North Atlantic Deep Water
NAO: North Atlantic Oscillation
NASA: National Aeronautics and Space Administration
NCEI PDO index: Pacific Decadal Oscillation index produced by the National Centers for Environmental Information from NOAA
NCEP/NCAR: National Center for Environmental Prediction/National Center for Atmospheric Research reanalysis
NEEM: North Greenland Eemian Ice Drilling Project
NGRIP: North Greenland Ice Core Project
NH: Northern Hemisphere
NOAA: National Oceanic and Atmospheric Administration
NOAA/ESRL: National Oceanic and Atmospheric Administration/Earth System Research Laboratories
NSIDC: National Snow and Ice Data Center

- O -

OHC: Ocean heat content

OLR: Outgoing long-wave radiation
ONI: Oceanic Niño Index

- P -

PCI: Polar Circulation Index
PDO: Pacific Decadal Oscillation
PETM: Paleocene–Eocene Thermal Maximum
PV: Polar vortex

- Q -

QBO: Quasi-biennial oscillation
QBOe: Easterly orientation of the quasi-biennial oscillation
QBOw: Westerly orientation of the quasi-biennial oscillation

- R -

RCP: Representative concentration pathway
REE: Rare-Earth Elements
RF: Radiative forcing
RSR: Reflected shortwave radiation
RWP: Roman Warm Period

- S -

SC: solar cycle, referred only to specific numbered 11-year sunspot cycle occurrences
SGM: Solar grand minimum/minima
SH: Southern Hemisphere
SLR: Sea level rise
SN: International sunspot number
SPECMAP: Spectral Mapping Project
SPM: Summary for policymakers
SR: Short-wave radiation
SST: Sea-surface temperature
SSW: Sudden stratospheric warming

- T -

TSI: Total solar irradiance
ToA: top of the atmosphere
TOR: Total outgoing radiation

- U -

UN: United Nations
UNEP: United Nations Environment Programme
UV: Ultraviolet

- W -

WACC: warm Arctic/cold continents
WDC–SILSO: World Data Center–Sunspot Index and Long-term Solar Observations
WGK-h: Winter Gatekeeper hypothesis
WMO. World Meteorological Organization
WWV: Warm water volume

- Y -

YD: Younger Dryas
yr: Years

INTRODUCTION

"Whenever a theory appears to you as the only possible one, take this as a sign that you have neither understood the theory nor the problem, which it was intended to solve."

Karl R. Popper (1972)

Outstanding questions in climate change

One of the main global themes of the past decades is the global climate debate across science, policy, and advocacy. It has its root in the 1988 Toronto Conference on the Changing Atmosphere. Consensus had been building among climate scientists since 1980 that the increase in carbon dioxide (CO_2) during the previous decades was finally having an effect on surface temperatures. In 1988 that concern became global, and since then it has been a central issue to the UN and many nations. The Intergovernmental Panel on Climate Change (IPCC), created in 1988, has become the world reference in climate science through its vast assessment reports on the scientific knowledge about climate change.

As the most authoritative voice of climate science, the IPCC has a great influence in shaping the political discourse. Through the development of future scenarios, the IPCC has decisively contributed to an *"emergency discourse"* supported by three powerful, yet unproven, scientific concepts (Asayama 2021):

- The idea of a temperature threshold, a safe limit beyond which irreversible climate damage will take place. Initially set at +2 °C, this arbitrary threshold was moved to +1.5 °C in 2018 (IPCC Special Report on Global Warming of 1.5°C; SR15).
- The idea of an allowable carbon budget to stay below a given temperature, despite the lack of an established and accepted value of CO_2 climate sensitivity. This allows the translation of temperature limits into policies to restrict greenhouse gases (GHGs) emissions.
- The idea of a climate deadline, derived from the calculation of when some temperature threshold will be crossed for a given level of emissions. This idea has captured the attention of climate activists and politicians. It can be illustrated by the phrase *"we only have 12 years to save the planet,"* (The Guardian 2018) extracted from the 2018 SR15 report that states emissions must decline by about 45% from 2010 levels by 2030 to avoid surpassing the temperature threshold.

The UN secretary general, António Guterres, urged all countries to declare a state of climate emergency until the world has reached net zero CO_2 emissions (The Guardian 2020). At least 15 countries and over 2,100 local governments in 39 countries have done so, covering over 1 billion people. According to most scientists, politicians, international organizations, and the media, there is no greater pressing problem faced by humankind that only the reduction of our CO_2 emissions will allow avoidance of the clear and impending danger of an irreversible climate catastrophe.

Given the stakes, it is our scientific duty to make sure we properly understand the totality of climate change. After all, climate change has always happened. In the words of Freeman Dyson: *"Climate change is part of the normal order of things, and we know it was happening before humans came."* (Roychoudhuri 2007). It is clear that unusual changes are taking place in the climate system of the planet. The global glacier retreat that has been occurring since c. 1850 has no precedent for several thousand years (Solomina et al. 2015). This is happening at a time when the Milankovitch glacial cycle theory indicates glaciers should be growing, something that happened for most of the time during the last five thousand years until the 19th century. The cryosphere retreat for the past two centuries has all the marks of a strong anthropogenic contribution. However, the claim by the IPCC that the climate change since pre-industrial times shows no evidence of a significant natural contribution should be met with a dose of skepticism. After all, there is no generally agreed cause for the Little Ice Age (LIA), the early 14th century to mid-19th century cold period, well registered in human history (Zhang et al. 2007; Parker 2012). If we are uncertain of what caused the LIA, how can we be so sure about the cause of what came afterwards?

This book is the result of a profound and detailed study on natural climate change, a relatively neglected part of today's climate science. It is not a review of what is known on natural climate change, as other excellent books accomplish that. Quite the contrary, the book is a detailed examination of unanswered natural climate change questions and problems, whose discussion is usually restricted to highly specialized scientific works. The relevant evidence to these outstanding climate questions, painstakingly gathered by climate researchers over the last decades, is displayed in a multitude of custom-made illustrations. The book discusses the evidence for the presence and causes of natural climate cycles, and other natural climate events, that have had a profound effect on the past climate of the planet, and their relevance to present climate change.

Chapter 2 examines the unsettled questions in the glacial–interglacial cycle of the Pleistocene, and standing problems with their most accepted explanation, the Milankovitch theory. The Mid-Pleistocene Transition changed the interglacial frequency from a 41-kyr obliquity cycle to a poorly defined 100-kyr cycle of uncertain origin. A drastic change with great repercussions for the climate of the planet for which no explanation can be found in a corresponding change in Milankovitch forcings, as they are currently understood. Trying to explain the 100-kyr frequency in terms of eccentricity leads to the problematic weakening in Pleistocene climate records of the 125 and 405-kyr periodicities in eccentricity (Nie 2018), and to the puzzling observation that for the past 5 Myr eccentricity and its supposed climatic effect display anti-correlation (Lisiecki 2010). Two hypotheses within Milankovitch theory try to explain interglacial determination. The best well-known one focuses on 65°N peak summer insolation, a property that depends on precession changes. A competing hypothesis, that traces its origin to Milutin Milanković himself, focuses on a caloric summer or summer energy integral that mostly depends on obliquity changes (Huybers 2006). This hypothesis, practically unknown outside specialized circles, is the one best supported by evidence, and readily solves many of the problems detected with Milankovitch theory, including that, for some interglacials, the effect appears to precede the cause.

Chapter 3 investigates the Dansgaard–Oeschger cycle found in proxy records of the last glacial period. These events served as blueprint for the definition of abrupt climate change. Their cause is unsettled, and different hypotheses put the focus on the Atlantic Meridional Overturning Circulation, meltwater events, sea-ice processes, or abrupt changes in thermohaline water stratification. The unresolved question of their periodic occurrence is revisited, producing an interesting twist: Evidence suggests that Dansgaard–Oeschger events are triggered according to two inter-related periodicities of a suggested tidal origin.

Chapter 4 analyzes the evidence for Holocene climatic variability. The Holocene was long considered a quite stable climate period, but evidence has been uncovered in the last few decades of the occurrence of over twenty centennial periods when the rate of climate change greatly surpassed the long-term average. The most intense and best studied of these periods is the 8.2-kyr event, but the existence of so many periods of abrupt climate change at times when GHGs hardly changed reveals that our understanding of natural climate change is inadequate. The Mid-Holocene Transition that separates the Holocene Climatic Optimum from the Neoglacial was accompanied by a change in climatic frequencies that has not been properly explained (Debret et al. 2009).

Does a 2500-yr climate cycle exist? It was first proposed by Scandinavian researchers in the early 20th century that high latitude Holocene climate was comprised of four botanical periods of c. 2500 years (the Blytt-Sernander sequence). Their relevance is not just regional, because the high latitudes are more sensitive to climate change, amplifying its variations (e.g., Arctic amplification) and more clearly display less prominent global changes. This climatic classification was popular among researchers before the 1970s. Chapter 5 investigates the related c. 2500-yr climate cycle first proposed by Roger Bray (1968), for which abundant proxy evidence exists. He proposed a solar cause for it, and cosmogenic isotope records show a remarkable degree of agreement, showing

the coincidence of Spörer-type solar grand minima with the most conspicuous periods of abrupt climate deterioration in the Holocene proxy climate evidence. The features of the proxy evidence that display this solar-climate correspondence for the Bray cycle suggest what aspects of the climate system are most affected by the persistent changes in solar activity when they last several decades.

Chapter 6 deals with the archaeological and historical evidence, in addition to climate proxy evidence, for the abrupt climatic events tied to the Bray cycle, and the mark they left in some human societies at the times they took place. There is a clear association between periods of profound climate worsening and periods of social crisis, that often coincide with important cultural transitions, lending support to the hypothesis that climate change acts as an engine for societal progress and adaptation (Roberts et al. 2011). Archaeological climate studies are increasingly important and both scientific disciplines can benefit from their interaction.

For the past two decades, since Gerard Bond et al. published their landmark article (2001), there has been a scientific debate over the existence and importance of a 1500-yr Holocene climate cycle. This unresolved discussion has greatly abated in the latest years due to contradictory evidence resulting in the cycle's waning acceptance. Chapter 7 takes a critical look at this question, and shows that when properly framed, the existence of the cycle is clearly supported by a particular subset of climate proxies. The nature of these proxies gives important clues regarding the 1500-yr cycle's possible mode of action. Unusual features displayed by some of these proxies raise the possibility that the cycle has a different cause than generally assumed.

Holocene solar activity, as inferred from cosmogenic isotope records, displays several periodicities in frequency analysis. These controversial quasi-cycles present a variability in the period and amplitude of their oscillations similar in proportion to the substantial variability in the accepted 11-yr solar cycle. Chapter 8 examines the evidence supporting their existence and their correspondence with quasi-cyclical climate changes. The very good phase agreement between solar oscillations and climate oscillations explains why this association is so pervasive in the paleoclimatological scientific literature. This association is not often discussed outside the discipline.

Chapter 9 looks at the climatic impacts resulting from changes in the greenhouse effect. It starts with the history of how it evolved to become at present the most accepted explanation for climate change, and then it focuses mainly on the role of GHG variations as an agent for natural climate change, a perspective less explored than its anthropogenic role. The unsettled Faint Sun Paradox and the possible factors proposed as an explanation are examined. The role of CO_2 in Phanerozoic climate evolution is controversial and, when closely examined, the quality of the data does little to resolve the debate. The usual explanation that long-term CO_2 changes precisely compensated for long-term changes in solar brightness leads directly to the anthropic principle, i.e., if it had not happened we would not be here. A unsatisfactory, unfalsifiable answer. The Cenozoic era, with better data, displays a puzzling lack of correspondence between climate changes and CO_2 changes that is seldom discussed. A role for the CO_2 changes of the past 70 years in modern global warming is generally accepted, but the recent proposition that CO_2 is the *"principal control knob governing Earth's temperature"* (Lacis et al. 2010), here referred to as the CO_2 hypothesis, lacks in supporting evidence.

Chapter 10 examines one of the most fundamental and least well-known properties of the climate system, the meridional transport of energy along the latitudinal temperature gradient. It involves the stratosphere, troposphere, and ocean in a not well understood coupling, that is variable in time and space. Most of the variability in the energy transported is seasonal, tied to the strengthening of the winter atmospheric circulation when the temperature gradient becomes steeper. Evidence is presented that relates different climate phenomena to changes in this transport, examining the role of the Quasi-Biennial Oscillation, El Niño/Southern Oscillation (ENSO), and the solar cycle in modulating meridional transport. A particularly neglected piece of evidence is crucial to revealing the solar modulation of meridional transport, as changes in atmospheric circulation can be related to the correlation between the solar cycle and changes on Earth's rate of rotation (Lambeck & Cazenave 1973; Le Mouël et al. 2010). The effect of changes in solar activity on ENSO is far from being accepted, despite an abundant bibliography. The possibility that ENSO acts as a tropical pump linked to meridional transport leads to new evidence of the solar modulation of ENSO presented in this book.

Chapter 11 is a continuation of chapter 10, reviewing the evidence that two drivers of climate change, volcanic forcing and multidecadal internal variability, induce changes in meridional transport. It is known that changes in multidecadal oscillations are associated with climate regime shifts (Tsonis et al. 2007). Evidence is presented in support of one such shift taking place in 1997–98, associated with a change in meridional transport, that altered the energy budget of the planet. This evidence leads to the intriguing possibility that meridional transport changes constitute an unrecognized climate change driver. An all-encompassing hypothesis is presented to propose that meridional transport is the principal driver of natural climate change. This hypothesis links all other natural causes, volcanic eruptions, multidecadal internal oscillations, ENSO, and solar activity, through their effects on the amount of energy directed to the two gigantic cooling radiators that the planet has at its poles in the current ice age. This hypothesis is named the Winter Gatekeeper because the variable amount of energy lost by the planet at the winter dark pole is proposed as the main climate effect mediator.

With the knowledge gained about natural climate change, modern global warming is examined in chapter 12. The recent warm period presents some highly unusual characteristics, like its non-cyclical cryosphere retreat that appears to have undone most of the Neoglacial advance, and its human-caused very high, and fast rising, CO_2 levels, higher than at any time during the Late Pleistocene. But unlike the anthropogenic forcing, the increase in temperature and sea level over the past 120 years shows little acceleration. According to the evidence the anthropogenic contribution is clear, but how much is human-caused and how much is natural is still an open question. The IPCC and most climate scientists are confident of the answers provided by climate models. Whether they deserve such confidence remains to be seen.

The IPCC has decisively contributed to the climate *"emergency discourse"* through the development of a series of gloomy future scenarios, not only in temperature but also in other climate phenomena, like sea-level or Arctic sea-ice. Leaving aside the debate about how unusual the IPCC's business-as-usual scenarios are, chapter 13 attempts to produce an alternative set of projections. Unlike IPCC projections, they consider fossil fuel supply side constraints and human population dynamics. They also take advantage of the advances made by systematic studies on forecasting (Armstrong et al. 2015). The golden rule of forecasting establishes the need to be conservative and adhering to cumulative knowledge about the subject. The set of conservative climate forecasts for the 21st century presented in chapter 13, is opposite to IPCC's wildly extremist projections. Time will be the final arbiter between this author's modest efforts and IPCC's multi-million-dollar bureaucracy's scientific projections.

The final chapter deals with the very distant future when the present interglacial should come to an end and the planet returns to the glacial conditions that have dominated the Pleistocene. The IPCC has reached the outlandish conclusion that a new glacial inception is not possible for the next 50 kyr if CO_2 levels remain above 300 ppm (Masson–Delmotte et al. 2013). The evidence presented in chapter 14 shows that glacial inceptions have taken place for the past two million years every time obliquity has gone below 23° during an interglacial. Interglacials are simply unsustainable under low obliquity conditions and there is no evidence that this time will be different. The long interglacial hypothesis rests on the wrong astronomical parameter, high equilibrium climate sensitivity to CO_2, and uncertain model predictions of very slow long-term CO_2 decay rates. The evidence supports a long delay between orbital forcing and its global ice-volume effect. If correct, the orbital threshold for glacial inception is crossed several millennia before glacial inception takes place. The orbital threshold calculated for the interglacials of the past 800,000 years supports that it was crossed for the Holocene 1400–2400 years ago. This interpretation of the evidence suggests that it is just a matter of 1500–4500 years before the next glacial inception takes place, putting an end to the Anthropocene.

The future is unknown, but unless we attempt to answer the outstanding questions about natural climate change reviewed in this book, the climate science of the future will not have a solid foundation. Science is about skepticism and debate. In the words of Richard Feynman:

"Once you start doubting, just like you're supposed to doubt. You ask me if the science is true and we say 'No, no, we don't know what's true, we're trying to find out, everything is possibly wrong' ... When you doubt and ask it gets a little harder to believe. I can live with doubt and uncertainty and not knowing. I think it's much more interesting to live not knowing, than to have answers which might be wrong." (Feynman 1981)

If we disallow the skepticism and the debate, we end with no science.

References

Armstrong JS, Green KC & Graefe A (2015) Golden rule of forecasting: Be conservative. Journal of Business Research 68 (8) 1717–1731

Asayama S (2021) Threshold, budget and deadline: beyond the discourse of climate scarcity and control. Climatic Change 167 (3) 1–16

Bond G, Kromer B, Beer J et al (2001) Persistent solar influence on North Atlantic climate during the Holocene. Science 294 (5549) 2130–2136

Bray JR (1968) Glaciation and solar activity since the fifth century BC and the solar cycle. Nature 220 (5168) 672–674

Debret M, Sebag D, Crosta X et al (2009) Evidence from wavelet analysis for a mid-Holocene transition in global climate forcing. Quaternary Science Reviews 28 (25-26) 2675–2688

Feynman R (1981) In: "Feynman: The Pleasure of Finding Things Out" BBC Horizon, Series 18, episode 9 (23 November 1981) https://vimeo.com/340695809

Huybers P (2006) Early Pleistocene glacial cycles and the integrated summer insolation forcing. Science 313 (5786) 508–511

IPCC (2018) Summary for Policymakers. In: Masson-Delmotte V, Zhai P, Pörtner H-O (eds) Global Warming of 1.5°C. An IPCC Special Report on the impacts of global warming of 1.5°C above pre-industrial levels and related global greenhouse gas emission pathways, in the context of strengthening the global response to the threat of climate change, sustainable development, and efforts to eradicate poverty. Cambridge University Press, Cambridge, p 3–24

Lacis AA, Schmidt GA, Rind D & Ruedy RA (2010) Atmospheric CO_2: Principal control knob governing Earth's temperature. Science 330 (6002) 356–359

Lambeck K & Cazenave A (1973) The Earth's rotation and atmospheric circulation—I Seasonal variations. Geophysical Journal International 32 (1) 79–93

Le Mouël JL, Blanter E, Shnirman M & Courtillot V (2010) Solar forcing of the semi-annual variation of length-of-day. Geophysical Research Letters 37 (15) L15307

Lisiecki LE (2010) Links between eccentricity forcing and the 100,000-year glacial cycle. Nature Geoscience 3 (5) 349–352

Masson–Delmotte V, Schulz M, Abe–Ouchi A et al (2013) Information from paleoclimate archives. In: Stocker TF, Qin D, Plattner G–K et al (eds) Climate Change 2013: The Physical Science Basis. Contribution of Working Group I to the Fifth Assessment Report of the Intergovernmental Panel on Climate Change. Cambridge University Press, Cambridge, p 383–464

Nie J (2018) The Plio-Pleistocene 405-kyr climate cycles. Palaeogeography, Palaeoclimatology, Palaeoecology 510 26–30

Parker G (2012) Global crisis. War, climate change and catastrophe in the seventeenth century. Yale University Press, New Haven.

Roberts N, Eastwood WJ, Kuzucuoğlu C et al (2011) Climatic vegetation and cultural change in the eastern Mediterranean during the mid-Holocene environmental transition. The Holocene 21 (1) 147–162

Roychoudhuri O (2007) Our rosy future, according to Freeman Dyson. Salon 29 Sep 2007. https://www.salon.com/2007/09/29/freeman_dyson/ Accessed Aug 13 2022

Solomina ON, Bradley RS, Hodgson DA et al (2015) Holocene glacier fluctuations. Quaternary Science Reviews 111 9–34

The Guardian (2018) We have 12 years to limit climate change catastrophe, warns UN. Watts J (writer) Oct 8 2018. https://www.theguardian.com/environment/2018/oct/08/global-warming-must-not-exceed-15c-warns-landmark-un-report Accessed Aug 13 2022

The Guardian (2020) UN secretary general urges all countries to declare climate emergencies. Harvey F (writer) Dec 12 2020. https://www.theguardian.com/environment/2020/dec/12/un-secretary-general-all-countries-declare-climate-emergencies-antonio-guterres-climate-ambition-summit Accessed Aug 13 2022

Tsonis AA, Swanson K & Kravtsov S (2007) A new dynamical mechanism for major climate shifts. Geophysical Research Letters 34 (13)

Zhang DD, Brecke P, Lee HF et al (2007) Global climate change, war, and population decline in recent human history. Proceedings of the National Academy of Sciences 104 (49) 19214–19219

THE GLACIAL CYCLE

2.1 Introduction

The Earth has spent most of its time during the past million years within the coldest 5% of the past 500 million years. It is locked in a very cold stage known as the Late Cenozoic Ice Age. The reasons for this are unknown. An ice age is defined as any period when there are extensive ice sheets over vast land regions, as we see now. Since the last four cold periods of the Phanerozoic Eon (three of them with ice age conditions; see Sect. 9.3.2 & Fig. 9.4) have taken place roughly 150 million years apart, some scientists favor an astronomical explanation (changes in the Sun, the orbit of the Earth, or passage of the solar system through more dense regions or galactic arms), while others prefer a terrestrial explanation (changes in the continental distribution, or in greenhouse gases concentration).

During the Quaternary or Pleistocene Glaciation (last 2.58 Myr), the Earth's climate alternates between very cold conditions during glacial stages and conditions similar to the present during interglacials. The glacial/interglacial alternation is known as the glacial cycle. For the past 800,000 years the Earth has spent roughly 82% of its time in glacial conditions and 18% of its time in interglacial conditions.

Milankovitch Theory on the effects of Earth's orbital variations on insolation remains the most accepted explanation for the glacial cycle since 1976 (Hays et al. 1976), when evidence was uncovered that the glacial cycle presents the same frequencies as Earth's orbital changes. According to its defenders, the main determinant of a glacial period termination is high 65°N summer insolation, and a 100-kyr cycle in eccentricity induces a non-linear response that determines the pacing of interglacials. Available evidence, however, supports that the pacing of interglacials is determined by obliquity, that the 100 kyr spacing of interglacials is not real, and that the orbital configuration and thermal evolution of the Holocene does not significantly depart from the average interglacial of the past 800,000 years.

So, we don't know why the Earth is in an ice age, but at least we think we know why 18% of the time the Earth gets a brief respite from predominantly glacial conditions and enters the milder conditions of an interglacial.

2.2 Milankovitch Theory

The currently favored theory on glacial–interglacial climate change was first proposed in 1864 by James Croll, a self-educated janitor at the Andersonian College in Scotland, who demonstrated that scientific advance can be produced by anybody with a good mind. He was offered a position in 1867, corresponded with Charles Lyell and Charles Darwin, and was awarded an honorary degree. But scientific knowledge at the time and his own limitations in mathematics and astronomy led to the final rejection of the theory. Croll wrongly concluded that orbital eccentricity and lack of winter insolation were responsible for glacial periods, and although he was the first to propose a positive ice-albedo feedback as a mechanism, his model called for asynchronous glaciations at the poles and timings for glaciations that were not supported by the evidence then available.

The Serbian genius Milutin Milanković was, in 1920, the first to undertake the work of calculating the intricacies of the Earth insolation at different latitudes due to orbital variations in a time without computers. He had adopted Joseph Murphy's (1876) view that long cool summers and short mild winters were the most favorable conditions for a glaciation, and he identified summer insolation as the key factor to explain the drastic climate changes of the past. His theory was not accepted until 1976, when geological evidence was found on multiple glacial–interglacial cycles, although their timing (100 kyr) was a bit off relative to Milankovitch Theory. Proper dating of glaciations during the past 2.6 million years showed that for the most part they have taken place at intervals of 41,000 years, a period more akin to orbital insolation forcing.

Milankovitch Theory is based on the complex changes that the orbit and orientation of the Earth suffers due to the slowly changing gravitational pull from other bodies in the Solar System. There are three types of orbital changes that affect Earth's insolation over the long term (Fig. 2.1).

2.2.1 Eccentricity

If the Solar system was only composed of the Sun and the Earth, Earth's elliptical orbit would always have the same eccentricity. However, the movements of the other planets, mainly the closest giant, Jupiter, and the closest planet, Venus, introduce gravitational perturbations that slightly change the eccentricity of the Earth's orbit. Eccentricity changes with a major beat of 405 kyr and two minor beats of 95 and 125 kyr. A change in eccentricity is the only

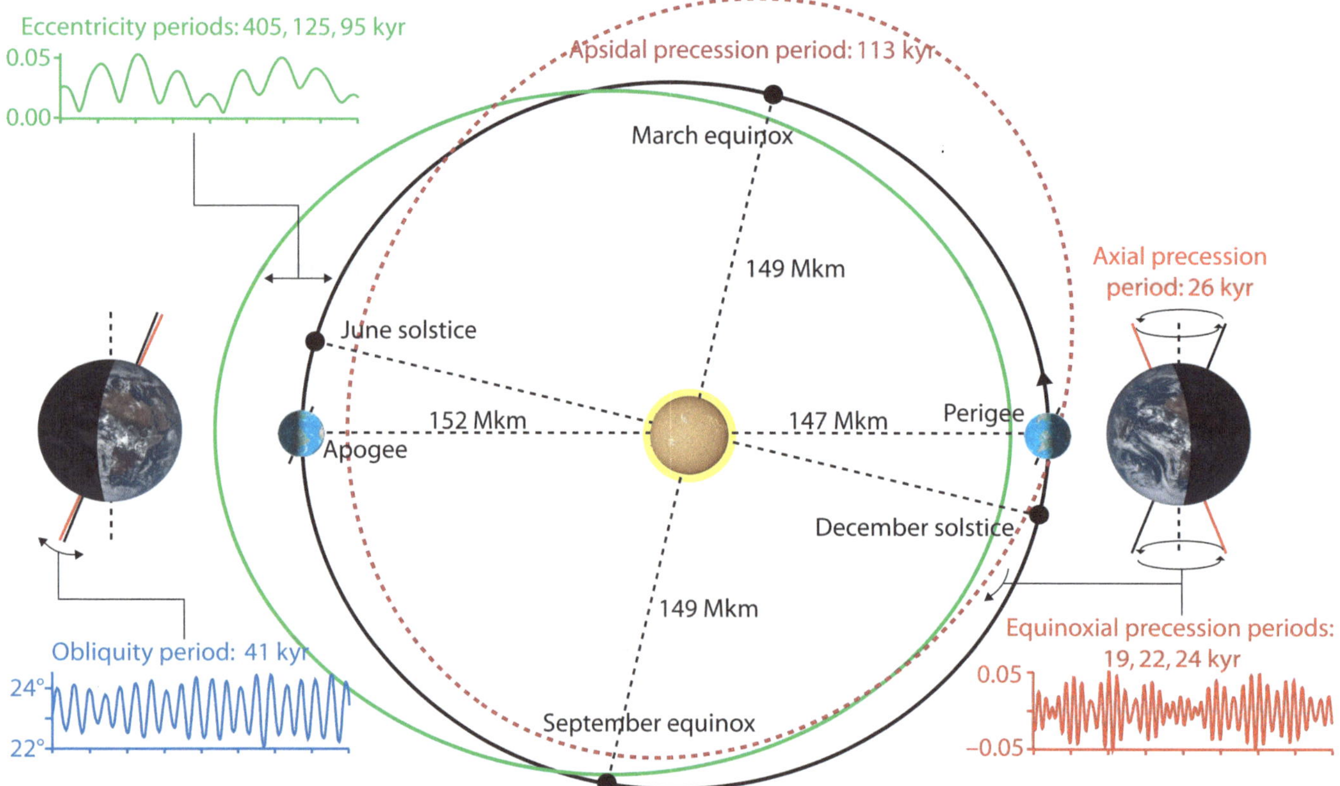

Fig. 2.1 Changes in Earth's orbit as the basis for Milankovitch theory
The orbital eccentricity variation produces changes in the shape of the Earth's orbit with periods of 405 kyr and 100 kyr. Axial tilt changes with obliquity periods of 41 kyr. The apsidal precession rotates the orbit around one of the elliptical foci, while the axial precession wobbles the Earth. Both together produce an average precession period of c. 23 kyr.

orbital modification that alters the amount of solar energy that the Earth receives as it changes its distance from the Sun. However, the yearly averaged insolation shows only a very small change that is due to the increase in Earth's average distance to the Sun with increasing eccentricity. This effect is caused by the planet spending more time farther from the Sun in a more eccentric orbit due to Kepler's second law. Since the Earth's orbit is always quite circular (eccentricity varies from 0.004 to 0.06) the change in insolation between perihelion and aphelion (now at January and July) is small, currently about 6.4% (eccentricity 0.016). The changes in eccentricity also produce a shortening and lengthening of the seasons as the Earth speeds at perihelion and slows at aphelion. Currently the Northern Hemisphere winter (at perihelion) is 4.6 days shorter than Southern Hemisphere winter (at aphelion). The important thing to remember in terms of climatic change is that due to the length of its main cycle, and the low eccentricity of Earth's orbit, the eccentricity cycle results in an exceedingly small forcing. Or in other words, the annually averaged insolation changes by less than 0.2% due to eccentricity. It is only through its effect on precession and obliquity that eccentricity becomes relevant. The combined effect makes eccentricity a very relevant climatic factor, and the 405 kyr precession cycle has been identified by its effect on strata over hundreds of millions of years (Kent et al. 2018).

2.2.2 Obliquity

This cycle is given by the changes in the inclination of Earth's axis, or axial tilt, with respect to Earth's orbital plane. This changes are caused by the torque exerted by the gravitational pull from the Sun and the Moon on the equatorial bulge of the Earth, with a minor contribution by the planets. The axial tilt varies between 22.1° and 24.3° over the course of a cycle, that takes 41 kyr. Currently the tilt is 23.44° and decreasing. The change in tilt changes the distribution of the solar energy between the seasons and through latitudes. The higher the obliquity, the more insolation in the poles during the summer and the less insolation in the poles during the winter and in the tropical areas all year. High obliquity promotes interglacials, while low obliquity is associated with glacial periods. Obliquity does not change the amount of insolation the Earth receives, but it changes the amount of insolation each latitude receives and the change is large at high latitudes. Surprisingly, obliquity has a bigger effect on climate than expected from the latitudinal distribution of its insolation changes. It should have little effect in the tropics, but there is a clear obliquity imprinting in West African monsoonal derived records. It is believed that the effect from obliquity changes is enhanced through changes in the latitudinal insolation gradient (Bosmans et al. 2015).

2.2.3 Precession

There are two precessional movements. The axial precession is the Earth's slow wobble as it spins on its axis due to the gravitational pull on its equatorial bulge mainly by the

Sun and the Moon. The Earth's axis then describes a circle against the fixed stars in 26 kyr, so if it is now pointing to Polaris, at 13 ka it was pointing to Vega. The apsidal (or elliptical) precession is the slow rotation of the elliptical orbit around the focus of the ellipse closest to the Sun in a period of 113 kyr (Fig. 2.1). It is the result of a combination of factors including the effects from general relativity and perturbations from other planets. The combined precession (of the equinoxes) from these two movements displaces progressively the seasons around the year and around the orbit, so that if now Northern Hemisphere winter takes place at perihelion (perigee closest to the Sun), in about 11,500 years it will be taking place at aphelion (apogee farthest from the Sun). Precession is therefore modulated by eccentricity as the precession angle would be irrelevant at zero eccentricity (circular orbit). It is important to note that precession doesn't change the amount of insolation that the Earth receives or the amount of insolation that each latitude receives during the year. Whatever insolation precession gives to one season, it takes it back from the other seasons, thus precession is an important contributor to seasonal insolation and to the latitudinal insolation gradient. The interaction of the various components of precession produce oscillations at 19, 22 and 24 kyr with a mean cycle period of roughly 23 kyr. Since the Northern Hemisphere summer now takes place at aphelion, we are at a minimum, in the precessional cycle, from the point of view of summer insolation at 65°N.

2.2.4 Modern interpretation of Milankovitch Theory

As currently viewed by followers of Milankovitch Theory, glacial inception takes place when summer insolation at 65°N allows more ice to survive the summer every year. This starts the buildup of the Laurentide, Fennoscandian and Siberian ice sheets. This process is fueled by ice-albedo and other feedbacks and progressively cools the Earth with a simultaneous drop in sea level. The glacial period survives several cycles of increased 65°N summer insolation and progressively gets colder with a lower sea level. The next eccentricity cycle, at 95 or 125 kyr later, induces a non-linear response on precession such that the next rise in 65°N summer insolation triggers a glacial termination. This is a much faster process than glaciation as is helped by feedback effects such as a reduction in ice-albedo or a buildup of greenhouse gases.

However, this was not the original view of Milanković. In 1869 Joseph Murphy debated the conditions that promoted glaciation with James Croll, and before reviewing the evidence he concluded: *"We have plenty of observed data; and I think I can show that they all go to prove a cool summer to be what most promotes glaciation, while a cold winter has, usually, no effect on it whatever"* (Murphy 1869). Had Croll taken Murphy's advise perhaps we would be studying Croll Theory of glaciations. Murphy's reasoning proved influential on the following decades, and Milanković adopted his view. Milanković did not consider peak summer insolation at 65°N, but took the effort to calculate the caloric half-year summer insolation at high latitude, an integral of the caloric power from insolation for the equinoxial half-year that contains the summer. This calculation is not only a much better representation of Murphy's proposal of cool summers than peak insolation on a certain date, but it also results in a different consideration of orbital parameters. Precession is mostly a seasonal factor, while obliquity is a semiannual factor. Therefore, peak insolation depends mainly on precession, while half-year summer insolation depends mainly on obliquity at high latitudes. Milutin Milanković proposed that high-latitude caloric half-year summer insolation determined the amount of snow cover that can survive summer melt, causing ice-sheet growth or retreat. In Milanković's caloric summer, despite not taking into account seasonal duration, precession is in control of the low latitudes while obliquity is in control of the high latitudes.

Milanković's concept of caloric summer was abandoned in favor of 21st June insolation after Berger (1978). 65°N summer insolation depends almost completely on precession, and was introduced into the first models (Kutzbach 1981). It was quickly adopted as the Milankovitch parameter of choice despite not being Milanković's proposal. This is an important mistake because precession as a glacial control has an Achille's heel in Kepler's second law. The closest the Earth is to the Sun during the Northern Hemisphere summer, the highest its velocity, and the shortest the summer. So, the number of days with enough insolation to melt the ice is lower, resulting in less melting. The SPECMAP project has literally carved in stone this mistake by tuning the oceanic isotopic records to precession, resulting in circular argumentation, as now precession affects climate as climate records are tuned to precession. But the imprinting of obliquity on climate is everywhere. Even in the tropics, where the effect of obliquity should be very low, there is a clear obliquity effect on climate proxies, like Mediterranean sapropel patterns resulting from the West African monsoon insolation-driven changes. An effect of the latitudinal insolation gradient, that depends mainly on obliquity, has been proposed as explanation (Bosmans et al. 2015).

The deviation of modern Milankovitch Theory from obliquity-linked half-year summer insolation to precession-linked peak summer insolation has important climatic repercussions. Milankovitch Theory was originally linked to a 41 kyr effect, while currently it struggles searching for a 23 kyr effect with most authors unaware of the source of the problems. Without having clarified this important issue that will be reviewed below, current discussions on Milankovitch Theory are about the fashionable role of CO_2 in glacial termination (Shakun et al. 2012), about a three stage model with interglacial, mild glacial and full glacial conditions (Paillard 1998), or about a sea-ice switch to explain why other peaks in 65°N summer insolation fail to get the world out of a glacial until the eccentricity cycle kicks in 100 kyr later (Gildor & Tziperman 2000).

2.3 Problems with Milankovitch Theory

The current theory, as presented in textbooks, explains glaciations through summer insolation at 65°N, paced by the 100-kyr eccentricity cycle, and is supported by the scientific consensus. But, it has some important problems that challenge its validity, and most of them can be traced to the adoption of 65°N peak summer insolation as the significant parameter.

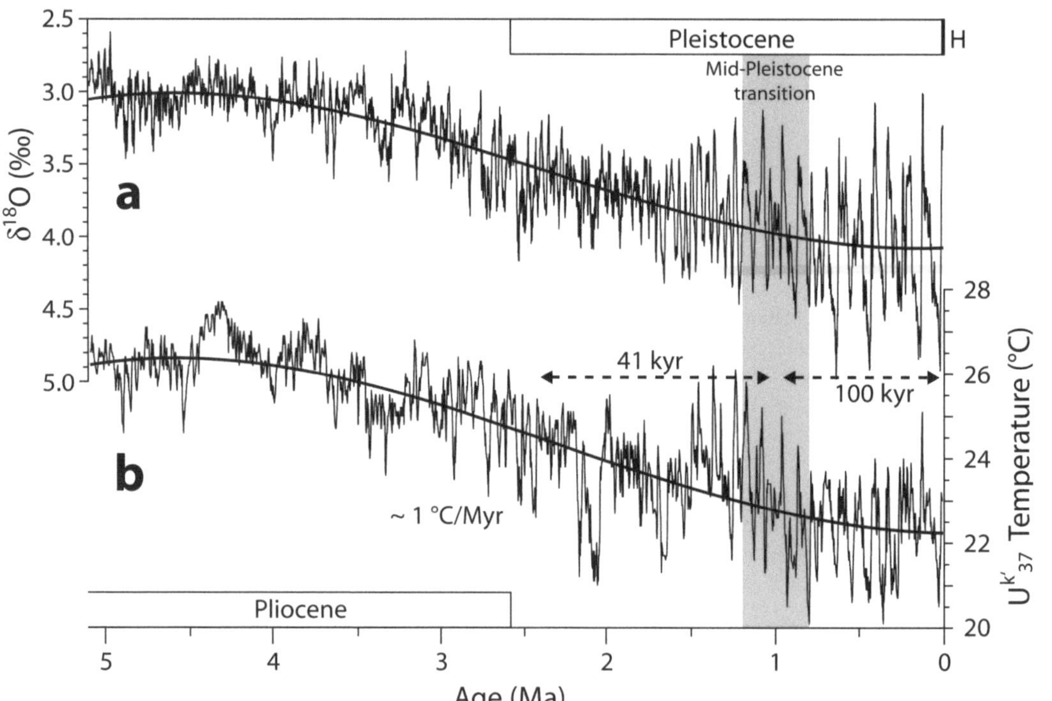

Age (Ma)

Fig. 2.2 The Mid-Pleistocene Transition Two different proxies for temperature, **a)** $\delta^{18}O$ isotope in benthic cores, and **b)** the alkenone $U^{K'}_{37}$ in marine sediments, show the progressive cooling of the Earth through the Pliocene. At the early-Pleistocene, glaciations took place at 41 kyr intervals. As the cooling progressed, this interval lengthened to 100 kyr in what is called the Mid-Pleistocene Transition. H, Holocene. After Zachos et al. (2001), and Lawrence et al. (2006).

2.3.1 The Mid-Pleistocene transition

Milutin Milanković's hypothesis, based on Joseph Murphy's cool summers, called for the 41-kyr obliquity cycle affecting the amount of energy received at high latitudes during caloric summer half-years, leading to glaciation or deglaciation as it went through the ice melting/accumulation threshold. When Hays et al (1976) analyzed benthic cores over the past 468 ka, they found the 41-kyr signal, but a stronger one at 100-kyr. Milankovitch Theory had no particular place for eccentricity, as its forcing is exceedingly small. The problem was compounded when more benthic cores extended the record over the past 5 Myr (Lisiecki & Raymo 2005). Before 1.2 Ma the record showed interglacials spaced on a 41-kyr cycle, but after 0.8 Ma on a 100-kyr cycle (Fig. 2.2). The period 1.2–0.8 Ma was named the Mid-Pleistocene Transition (MPT), and nobody knows what caused the change in glacial periodicity. There are no known orbital changes, no known solar activity changes and no known greenhouse gas changes that can explain a geologically abrupt transition from 41-kyr to 100-kyr interglacial spacing. Essentially scientists are caught in a double bind. An explanation of the 100-kyr glacial periodicity in terms of eccentricity and northern summer insolation leads to an unexplainable 41-kyr pre-MPT planet, while an explanation of the glacial cycle in terms of 41-kyr obliquity falls short of explaining the 100-kyr post-MPT planet. It is actually a false dilemma that can be easily explained in Occam's principle terms.

2.3.2 The 100-kyr problem

The most important problem with the modern interpretation of Milankovitch Theory is the 100-kyr problem. It comes from the need to explain the observed 100-kyr frequency in global ice-volume proxies in terms of orbital changes that do not match the observed climatic response to the calculated forcing. For the past one million years glacial ice-volume, measured by changes in ^{18}O isotope,

has oscillated with a main 97-kyr periodicity (Fig. 2.3). The only orbital cycle at that periodicity is the eccentricity cycle, but eccentricity is an almost negligible forcing on its own. So Hays, Imbrie, and Shackleton (1976), proposed that eccentricity was playing its role in a non-linear way. The problem is compounded because the main cycle of eccentricity is 405-kyr and that cycle is barely seen in the benthic record (Fig. 2.3). Nie (2018) showed that the 405-kyr periodicity in East Asian summer monsoon and global ice-volume intensified during the Pliocene and Early Pleistocene weakening significantly at the MPT, so, we are left with the conclusion that since then eccentricity produces a multiplicative effect during its minor 95 and 125-kyr cycles, yet no important effect from its major 405-kyr cycle. Maslin & Ridgwell (2005) call it *"the eccentricity myth."* In addition, the change from early-Pleistocene 41-kyr glaciations to late-Pleistocene 100-kyr glaciations was achieved without any known change in insolation, so, in its current incarnation, Milankovitch Theory is at odds to explain it.

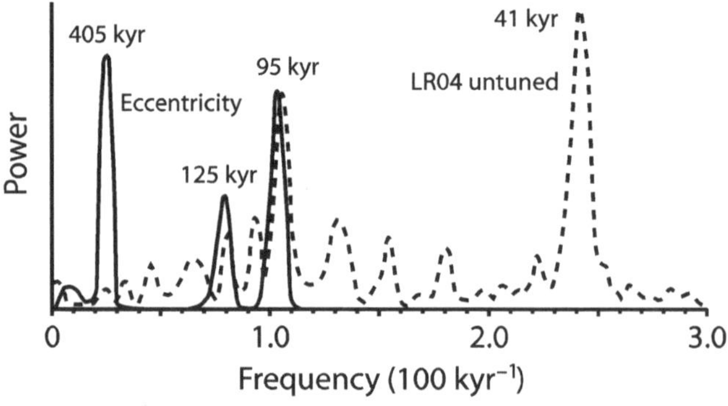

Fig. 2.3 Spectral differences between eccentricity and global ice-volume
Orbital eccentricity multitaper spectrum (solid line) shows three major peaks at 405, 125, and 95 kyr. Untuned LR04 benthic stack $\delta^{18}O$ isotope ice-volume and temperature proxy record shows a peak close to 100 kyr, but an unremarkable peak at 125 kyr and very little signal at 405 kyr. After Rial et al. (2013).

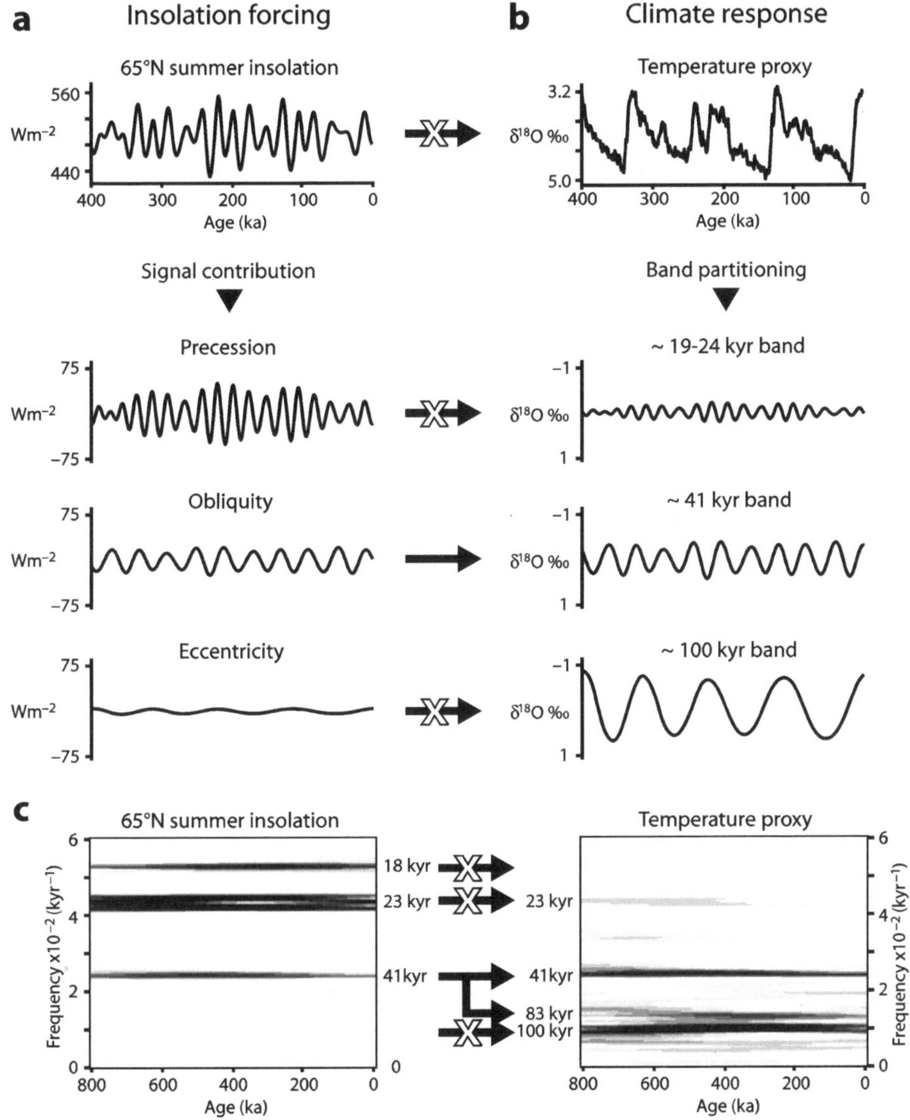

Fig. 2.4 The 100-kyr problem
Milankovitch theory, in its current consensus form, runs into problems explaining the disparity between predictions and observations. **a)** The calculation of 65°N summer insolation shows that the predicted range of 105 W/m² is mainly due to the contribution of precession, followed by obliquity with a similar magnitude. The contribution from eccentricity is one order of magnitude smaller. **b)** When the global ice-volume proxy spectra is analyzed, the main band is a 100-kyr band, followed in intensity by a 41-kyr band, while the 19–24-kyr band is barely detectable. So, the strongest contributor gives the weakest signal, while the strongest signal comes at a frequency of what should be a negligible contributor, and the only orbital change giving a commensurate response to its forcing is obliquity. After Imbrie et al. (1993). **c)** Gabor transform (windowed time-frequency Fourier analysis) of 65°N summer insolation calculations from the orbit of the Earth during the last 800 kyr (left panel). The main contributor to the signal is the 23-kyr period, followed by the 18-kyr period, both from precession cycles, with the less intense band from 41-kyr obliquity cycle next, as shown in a). Right panel, the same analysis performed over temperature data from observations (EPICA Dome C ice core deuterium record). The temperature proxy barely responds to precession, as the band at 23 kyr is very tenuous. Instead we see obliquity bands at 41 and 83 kyr (double harmonic) and the prominent band at 100 kyr, that cannot be the eccentricity, since it is missing what should be an even stronger band at 405 kyr. After Pollard & Baez (2013).

The 100-kyr problem is best illustrated in Fig. 2.4 where we analyze Milankovitch Theory, through the decomposition of 65°N peak summer insolation into its components: eccentricity, obliquity and precession (Fig. 2.4a); and compare it with evidence from temperature proxy records (Fig. 2.4b), analyzed in the frequency domain to reveal their main cyclic components. Note that you rarely see eccentricity plotted at its true comparative forcing. The disparity is so evident (Fig. 2.4c) that the current consensus glacial cycle hypothesis cannot be right. Except for the 41-kyr obliquity band there is a complete mismatch between the proposed forcing and the observed climatic effect. Precession contributes the majority of the energy variation in 65°N peak summer insolation, while its effect on ice-volume is small, and its frequency peaks (or bands in Fig. 2.4c) show little prominence. Eccentricity contributes one order of magnitude less energy than the other orbital components, yet it shows the biggest climatic response.

Muller & MacDonald (1997) showed that the 100-kyr spectral peak from climatic isotope data is too narrow to be the result of two components of 95 and 125 kyr, as astronomical data requires (Fig. 2.3). Different explanations have been proposed for the absence of 405 and 125-kyr frequency bands in proxy records, like the 100-kyr cycle being due to Earth's orbital inclination (Muller & MacDonald 1997), being paced by every fourth or fifth precession cycle (Ridgwell et al. 1999), or as an emergence of forced synchronization as an harmonic of the 405-kyr band (Rial et al. 2013). Another problem is that for the past 5 Myr there is a strong (−0.69) negative correlation between the 100-kyr power of eccentricity and the climatic proxy δ¹⁸O in benthic cores (Lisiecki 2010), so in essence the stronger the eccentricity forcing, the weaker the climatic response. This is a serious objection to the nonlinear response proposed by Hays et al. (1976), as it has to be simultaneously non-linear and inverse.

2.3.3 The causality problem

Second in importance is the causality problem, exemplified in "the stage 5 problem." Marine Isotope Stage 5e is used here as an alternative name for the previous interglacial, also known as Eemian. According to insolation, warming towards Eemian or Marine Isotope Stage (MIS) 5e should have started at the earliest at 135 ka, however data from crystals in a Nevada cave named Devils Hole in 1992 indicate that by that date Glacial Termination II was essentially finished (Winograd et al. 1992; Ludwig et al. 1992; Fig. 2.5). A great controversy erupted over that data in the literature and has not abated since (see

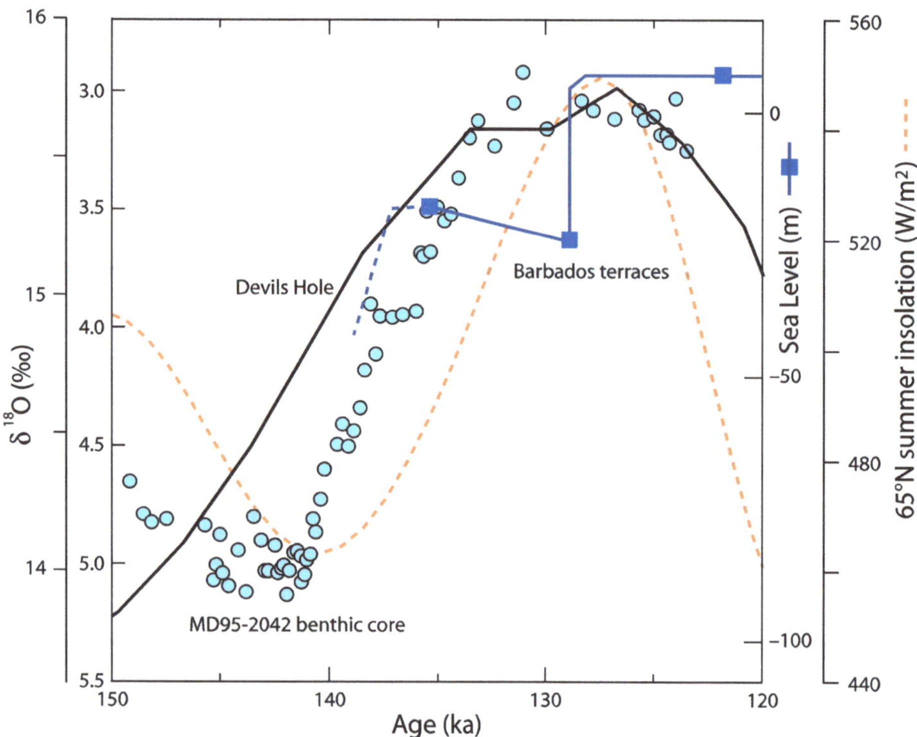

Fig. 2.5 The causality problem
According to Milankovitch theory, glacial termination II, leading to MI Stage 5e or the Eemian interglacial, could not have commenced earlier than 135 ka due to low 65°N summer insolation (dashed line). However, data from Devils Hole cave (black line, left outer scale; after Landwehr et al. 1997) indicates a much earlier start since deglaciation was already well under way at 140 ka. Data from Barbados coral reefs (squares, right inner scale; after Gallup et al. 2002)) supports the early start as by 136 ka, according to sea levels, termination II was already 80% complete. Iberian-margin benthic core MD95–2042 δ¹⁸O record supporting an early Termination-II (circles, left inner scale; after Drysdale et al. 2009). Obliquity (not shown) started increasing 10 kyr earlier than insolation, at 150 ka.

Moseley et al. 2016, and multiple comments to it in Science vol. 354, issue 6310, 2016). But Devils Hole data is not alone, as similar data has been uncovered from coral reefs in the Bahamas (Gallup et al. 2002), Barbados and Papua New Guinea, and from Iberian-margin sediments and Italian cave speleothems (Drysdale et al. 2009), and all of it indicates that termination was essentially completed by 135 ka. A date when 65°N summer insolation was still below the levels of 70% of the previous 100 kyr (Fig. 2.5). The issue remains unresolved, but additional data indicates that MIS 5e may not be the only glacial termination where the effect appears to precede the cause. MIS 15c shows the same situation. The problem is further complicated because summer insolation has been used as a defining criterion to date the start and end of glaciations in sediments in the official UN sponsored SPECMAP series. This results in circular reasoning since computed insolation is assumed to pace the glaciations and terminations and has been used to date them. See Puetz et al. (2016) for a strong criticism of the practice of record orbital tuning.

2.3.4 The asymmetry problem

A third issue is that glacial cycles are symmetric between the hemispheres, as both are warming or cooling simultaneously, whereas the seasonal precession forcing (and 65°N summer insolation) is anti-symmetric. That is, when one hemisphere warms the other cools. As CO_2 is a relatively well mixed gas in the global atmosphere, it is claimed that it can be also responsible for interhemispheric synchroneity (Lorius et al. 1985).

2.3.5 The 41-kyr problem

A fourth problem that is seldom discussed is the 41-kyr problem. If modern Milankovitch Theory struggles to explain the glacial cycle in the last 0.8 million years, it has no less problems to explain it between 3–0.8 million years ago. During that period temperatures and global ice-volume varied almost exclusively at the 41 kyr obliquity period, while high-latitude peak summer insolation is always dominated by precession. If we explain interglacials in terms of peak summer insolation as the driving mechanism, then the 41 kyr periodicity found in the Plio–Pleistocene record prior to the MPT becomes unexplained.

2.4 Evidence that interglacial pacing does not follow a 100-kyr cycle

The claim that interglacials follow a 100-kyr cycle is surprising. According to the LR04 marine sediment cores stack or EPICA Dome C Antarctic ice core no single interglacial of the past 800,000 years starts 100,000 years after the previous one. Eccentricity-linked 95 and 125 kyr periodicities are also not particularly favored intervals. Table 2.1 shows the distance from the start of every interglacial for the past 800 kyr to the start of the previous one, using the temperature at the start of the Holocene (0 °C anomaly in EPICA Dome C data) as the defining criterion for the start date (extrapolating the warming rate for interglacials that didn't reach 0 °C, see Fig. 14.7). It is also difficult to understand how a 100-kyr cycle hypothesis can be supported based upon 11 interglacials within the last 800 kyr that have an average spacing of 72.7 kyr, very far from 100 kyr.

To clarify this issue, I have plotted the interglacial start date versus distance from previous interglacial start date, after Mearns & Milne (2016). The result is given in Fig. 2.6. The data strongly indicates that the spacing of interglacials tends to fall on multiples of the 41,000-year obliquity cycle. There are two anomalous interglacials, MIS 11c that was unusually long, and MIS 7e that was unusually short. It is possible that MIS 11c was long because it started early, and MIS 7e was short because it started late. Correcting for this possibility sets also the next two interglacials interval as multiples of the obliquity

Interglacial	Start date	kyr from previous
MIS 19c	−787,000	77
MIS 17	−707,000	80
MIS 15c	−624,400	82.6
MIS 15a	−579,600	44.8
MIS 13a	−499,000	80.6
MIS 11c	−424,800	74.2
MIS 9e	−335,500	89.3
MIS 7e	−243,800	91.7
MIS 7c–a	−214,700	29.1
MIS 5e	−131,400	83.3
MIS 1	−11,700	119.7

Table 2.1 Interglacials of the past 800,000 years
Interglacial start date was determined directly from EPICA Dome C temperature data from δDeuterium isotopic changes (Jouzel et al. 2007). Temporal distance between interglacials was calculated between start times. Average distance is 72.7 kyr, while most frequent distance is c. 82 kyr.

frequency (Fig. 2.6, arrows) and shows every interglacial of the past 800 kyr spaced at multiples of obliquity.

Another observation is that at times of high eccentricity in the 405-kyr eccentricity cycle, as at 600 ka and 200 ka, interglacial spacing drops to c. 41-kyr (MIS 15c/MIS 15a, MIS 7e/MIS 7c–a; Fig. 2.6); so how is it possible that eccentricity promotes a 100-kyr interglacial periodicity if at the times when eccentricity is highest the frequency of interglacials increases to 41-kyr? This observation supports Lorraine Lisiecki finding of anti-correlation between eccentricity and climate in the Plio–Pleistocene (Lisiecki 2010).

The presence of two interglacials separated by only one obliquity cycle (41 kyr) at times of very high eccentricity (Fig. 2.6) suggests the existence of a repeating pattern following the 405-kyr eccentricity cycle where the length of a unit is given by the distance between MIS 15a and MIS 7c–a, 365,000 years, or nine obliquity cycles,

during which five interglacials take place, four of them separated by 82 kyr and one by 41 kyr. The average spacing of interglacials would then be 73 kyr, very close to the average value of 72.7 kyr for the entire series. Interglacials would take place every 1.8 obliquity cycles. However, the cycle is irregular, as the existence of short and long interglacials and the past glacial period lasting three obliquity cycles shows.

2.5 Evidence that obliquity, and not precession, sets the pacing of interglacials

The evidence that obliquity sets the pace of interglacials is so abundant and clear that it is surprising that the prevailing consensus view opposes it. In the early 1980s it was proposed that the pace of interglacials was set by changes in peak insolation forcing caused mainly by precessional variations. Since then, the belief in the climatic effect of summer insolation variations at 65°N has been deeply ingrained. But it is the evidence, that endures, and not the consensus of opinion, that changes over time, what determines the correctness of a hypothesis. Let's review the evidence in favor of obliquity:

2.5.1 Obliquity controlled glaciations before the Mid-Pleistocene Transition

Glacial cycles were indeed governed by the 41-kyr obliquity cycle for most of the Quaternary Ice Age prior to the MPT (Figs. 2.2 & 2.7), and the 23-kyr and 100-kyr cycles were nowhere to be seen in that period. The simplest "Occam's razor" explanation is that obliquity still does the job.

2.5.2 Interstadials are still under obliquity control

Throughout the Plio–Pleistocene, the Earth has been cooling progressively (Fig. 2.2). The cooling of the planet reached a point at around 1.5 Ma when some interglacials started to be affected and did not reach what we consider interglacial temperatures, so we do not consider them to be interglacials and do not assign them numbers in the MIS

Fig. 2.6 The 100-kyr Myth
Plot of Interglacial start date versus distance to the previous interglacial start date, after Mearns & Milne (2016). The spacing of interglacials shows a strong tendency to fall into multiples of the obliquity frequency (horizontal bands). The anomalous interglacials MIS 11c (white star) and MIS 7e (black star) abnormal length can be partly explained as being due to an early and late start respectively. If their length transgressions were accounted for and corrected in the graph, that would make the next interglacials (white and black circles) adjust themselves to multiples of the obliquity frequency (arrows), and every dot would be near the bands. Dotted curve, eccentricity. Note how high eccentricity coincides with a shorter spacing of interglacials, and in MIS 1 case, low eccentricity coincides with a longer spacing.

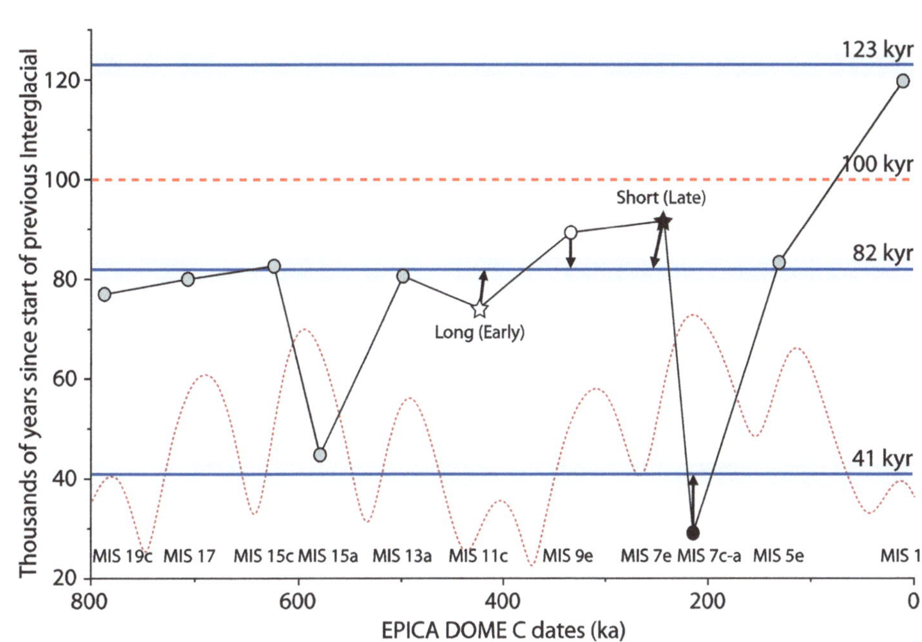

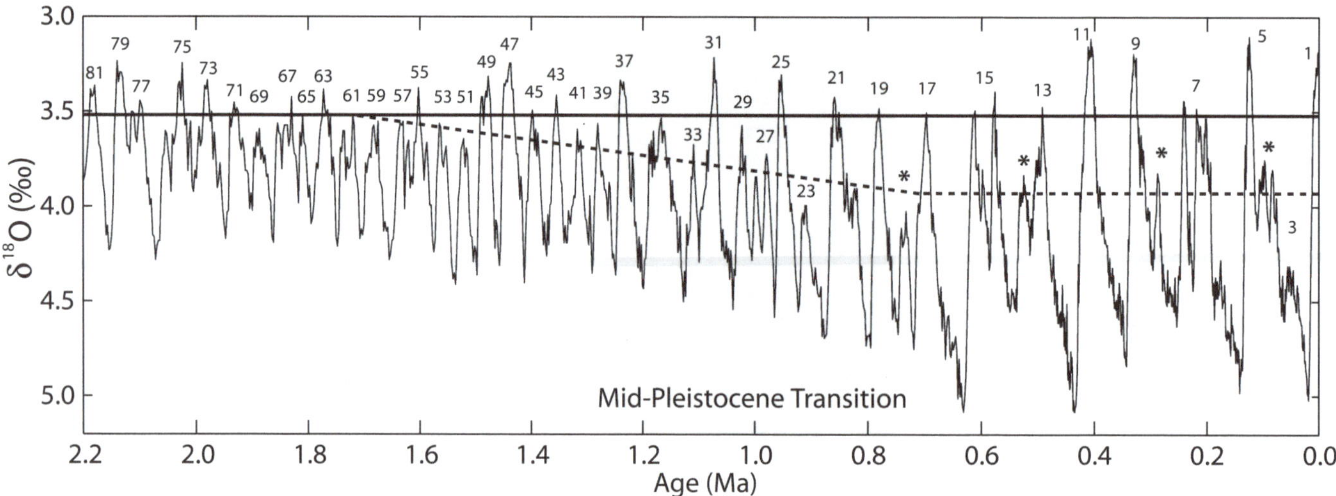

Fig. 2.7 Pleistocene temperature proxy record
δ¹⁸O isotopic record from LR04 stack of 53 benthic cores from all over the world (after Lisiecky & Raymo 2005) shows that from about 1.5 million years ago some interglacials continued reaching the previous interglacial average temperature (solid horizontal line), while others showed a decreasing trend in interglacial average temperature (dashed line), and are not considered interglacials. Periods of higher temperature more recent than MIS 23 that did not reach interglacial levels are usually not assigned an MIS number (asterisks). For the past 600 kyr the world has stopped or very much reduced its previous cooling rate.

sequence (Fig. 2.7, asterisks). We call them interstadials and are nothing but cool interglacials between cold stadials. The MPT did not involve any change in insolation, or orbital cycles, so proponents of the 100 kyr-Insolation Milankovitch hypothesis are at odds to explain how an obliquity cycle turned into an eccentricity cycle.

The most interesting question is not why some obliquity induced periods of warming fail to reach what we consider interglacial temperatures, but why some still manage to reach them given the cooling of the planet.

2.5.3 Temperature shows a clear response to obliquity-linked changes in 70–90° insolation

Although precessional changes greatly affect the amount of insolation during a three-month period, that change is quickly averaged over the following nine months, leaving total annual insolation unchanged. By contrast obliquity

changes add a significant amount of warming at high latitudes (70–90°) year after year over a period of thousands of years and can have an enormous cumulative effect (Fig. 2.8). The temperature proxy record clearly shows temperature decreasing during periods of obliquity-linked polar insolation deficit (dashed blue lines in Fig. 2.8), and increasing during periods of obliquity-linked polar insolation surplus (dashed red lines in Fig. 2.8).

2.5.4 Temperature responds poorly to precession-linked changes in insolation

Peak summer insolation is dominated by the 23-kyr precession cycle. When a frequency analysis is performed on insolation astronomical data and temperature proxy data, only a very small temperature response to the precession bands is detected (Fig. 2.4c right panel). The only consistent response between astronomical insolation and tem-

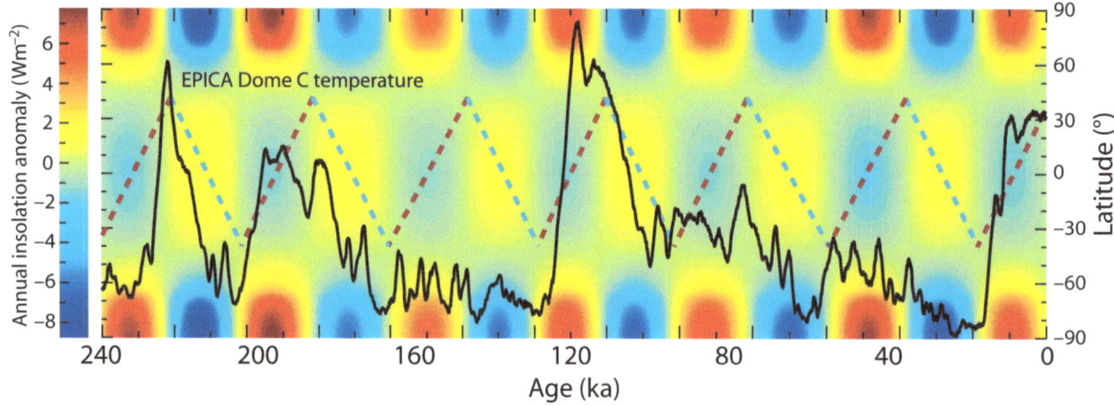

Fig. 2.8 Annual insolation changes at high latitudes and the symmetry problem
Changes in annual insolation by latitude and age are shown in a colored scale. They are essentially due to changes in obliquity, since changes in insolation by precession are averaged between seasons within the same year. The high latitude persistent changes in insolation last for thousands of years and correspond quite well to temperature changes in Antarctica, shown as a black line overlay (no scale). Glacial–interglacial cycles show symmetric temperature responses in both hemispheres. The obliquity cycle is schematically shown, shifted 90° (10.25 kyr) so ascending red dashed lines correspond to periods with polar insolation surplus, and descending blue dashed lines correspond to periods with polar insolation deficit. Antarctic temperature responds mainly to obliquity changes, not to high-latitude south summer insolation changes (not shown). The effect helps keep interhemispheric synchronicity during the glacial cycle.

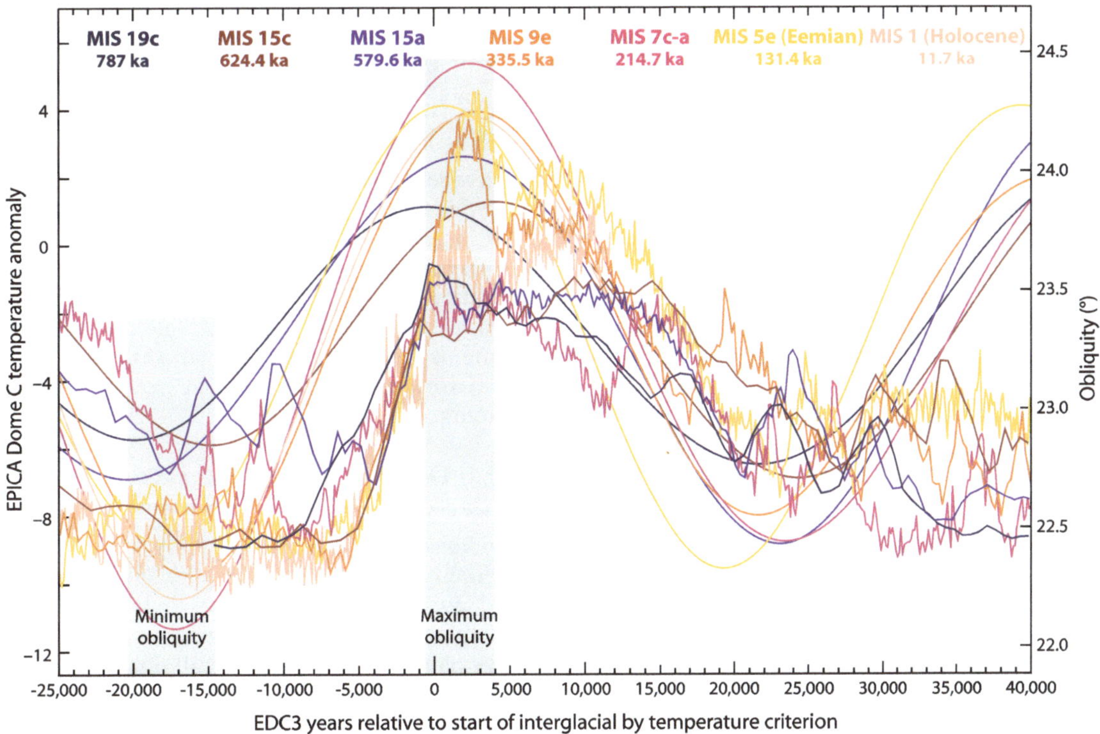

Fig. 2.9 Interglacial alignment with obliquity
Interglacials MIS 1, 5e, 7c–a, 9e, 15a, 15c, and 19c were aligned by temperature at their start. Their obliquities then display a significant degree of synchronization. Obliquity bottoms 20 to 15,000 years before the start of the interglacial. The warming in Antarctica starts about 10,000 years later, and proceeds so fast that interglacial average temperatures are reached by the time obliquity peaks about 19,000 years after it started rising. The interglacial comes to an end with a delay of about 5,000 years over the falling obliquity. EPICA Dome C data after Jouzel, et al. (2007). Astronomical data after Laskar et al. (2004). Some interglacials were not considered for their abnormal duration (MIS 11c, MIS 7e), or abnormal temperature profile (MIS 13a, MIS 17).

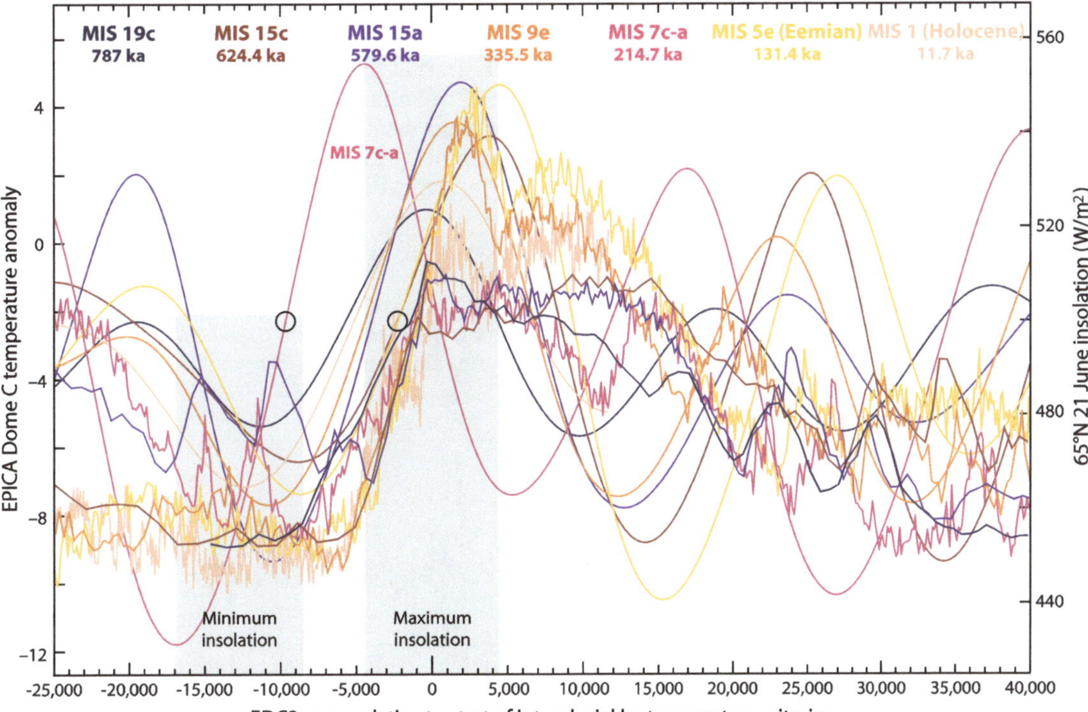

Fig. 2.10 Interglacial alignment with 65°N summer insolation
Same as Fig. 2.9 for 65° northern June 21st insolation. Although insolation also has a tendency to align indicating that interglacials cannot take place if insolation is working in the opposite direction, the spread is clearly higher in this case. Insolation for MIS 7c–a came too early and for MIS 15c and MIS 5 too late to be held responsible for driving interglacial warming. Black circles indicate 500 W/m² at 65°N on June 21st for these three interglacials. EPICA Dome C data after Jouzel, et al. (2007). Astronomical data after Laskar et al. (2004).

perature data is given by obliquity. Not only is there no significant signal for a 23-kyr cycle in the data, but if 65°N summer insolation is so important it becomes difficult to explain why it sometimes has a huge effect on temperatures and at other times it has almost no effect.

2.5.5 Temperature shows better phase agreement with obliquity

When seven interglacials of the past 800 kyr are aligned by temperature, their obliquity curves also align (Fig 2.9). Maximum obliquity takes place within only 4,000 years of the interglacial start, despite being a 41-kyr cycle. Such tight phase coherence between obliquity and temperature indicates that temperature is under obliquity control. Obliquity and temperature phases are shifted by 28.5° (c. 6,500 yr delay) indicating a lag in the climatic response to orbital forcing. A similar phenomenon is observed in present regional temperatures that show a delay of c. 1 month to insolation changes (warmest days are about one month after summer solstice). Milankovitch forcing is much smaller and requires a much longer time to effect much bigger changes.

The same temperature alignment including 65°N 21st June insolation curves (Fig. 2.10) shows much more variability (compare to Fig. 2.9). This variability underscores that for MIS 7c–a, MIS 5e, and MIS 15c insolation could not have driven glacial termination. In the first, interglacial insolation increase took place too early and in the last two it took place too late (Fig. 2.10, circles; see Sect. 2.3.3). This problem has been widely discussed for MIS 5e, know as the causality problem, as there is an abundance of data that shows Glacial Termination II taking place before insolation was sufficiently high to drive it (Fig. 2.5). MIS 7c–a and MIS 15c support that it is not an exception.

2.5.6 Temperature changes almost perfectly match obliquity changes

Allowing for the lag in the temperature response to the obliquity forcing, both show an almost perfect match in their temporal variability. As shown in Fig. 2.11, temperature rarely gets out of the obliquity envelope, although often is unable to show a strong response to obliquity increases. The match is so good that it leaves very little room for precession and eccentricity in determining temperature temporal variability, indicating they can only af-

fect the amplitude of temperature response to obliquity changes and therefore they must be a second order factor.

2.5.7 Interglacials show a duration consistent with obliquity cycles

Average duration of MIS 5e, 7c–a, 9e, 15a, 15c, and 19c interglacials, measured at the –3 °C anomaly in the EPICA data is c. 18,000 years. Average duration of the up swing of the obliquity cycle at 23.5° is c. 18,000 years. Average duration of the northern summer insolation cycle at 500 W/m² is c. 11,000 years. Interglacials tend to last the same as the obliquity cycle but shifted c. 6,500 years due to Earth's thermal inertia and lagged response to orbital forcing.

2.5.8 Obliquity-paced interglacials solve all Milankovitch Theory problems

Evidence from interglacial pacing, temperature response to obliquity, temperature-obliquity alignment, and interglacial average duration clearly indicates that, in general, interglacials respond primarily to the obliquity cycle as they have always done and still do. Despite a general consensus ignoring what the data clearly indicates, some authors have realized this fact and are proposing hypotheses where changes in insolation driven by obliquity are the main forcing responsible for the glacial cycle (Huybers & Wunch 2005; Huybers 2006; Liu et al. 2008; Tzedakis et al. 2017).

The hypothesis that obliquity drives the glacial cycle solves most of Milankovitch Theory problems. The 100-kyr problem is solved because there is no 100-kyr cycle, just a 41-kyr cycle that skips one or two beats. And it solves the 41-kyr problem for similar reasons. It solves the causality problem because now glacial terminations usually start at the bottom of the obliquity cycle and therefore MIS 5e termination is well underway at 135 ka when 65°N summer insolation is still too low. It also solves the lack of asymmetry in the polar response, as the obliquity cycle is symmetrical in both poles.

2.6 The 100-kyr ice cycle

Having showed that interglacials are paced by obliquity and not by eccentricity, and that they do not follow a 100-kyr cycle, it is time to revisit the actual 100-kyr cycle. The

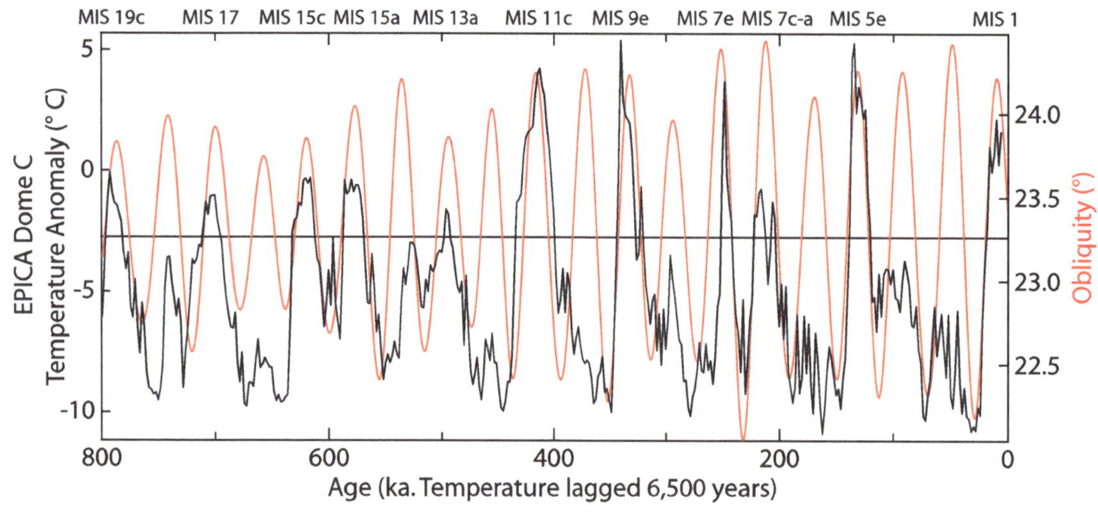

Fig. 2.11 Temperature changes due to axial tilt changes
Black curve, temperature anomaly at EPICA Dome C ice core for the past 800,000 years, lagged 6,500 years. Red curve, changes in obliquity of the planetary axis in degrees. The fall in obliquity always terminates interglacials. EPICA Dome C data after Jouzel et al. (2007). Astronomical data after Laskar et al. (2004).

evidence for the existence of a 100-kyr cycle in Pleistocene climate is overwhelming, starting with Hays et al. (1976) frequency analysis that provoked the reinstatement of Milankovitch theory. Since that data was measuring changes in ^{18}O isotope, it is clear that the nature of the 100-kyr cycle corresponds to an ice cycle. Extensive continental ice sheets expand and wane simultaneously at both hemispheres according to a 100-kyr periodicity. This periodicity obviously results from eccentricity, as it shows phase coherence with it. It can be concluded from the evidence that the 100-kyr cycle determines global ice build-up, while the 41-kyr cycle determines interglacials. They are two related but independently paced cycles and this is the source of most confusion in the field. In brief, low eccentricity promotes global ice build up, while high eccentricity promotes global ice melting, and obliquity is responsible for opening up windows during which an interglacial is possible, and closing them making an interglacial impossible. The decision whether there is an interglacial or not when obliquity opens the window of opportunity is taken by a combination of insolation forcing and the status of the 100-kyr ice cycle at the time (see next section). Figure 2.12 shows how the two cycles interplay constituting two sides of the same coin. The top side determines the frequency of interglacials (focus on Fig. 2.12a, that depends on obliquity), while the bottom side reflects the 100-kyr cycle determined by the ice proxy (focus on Fig. 2.12b, that depends on eccentricity). Frequency analysis does not differentiate between both halves of the data, producing the 100-kyr, 80-kyr and 41-kyr bands observed in Fig. 2.4c, right panel, without informing us that the bands come from different values in the data, the obliquity band from the low $\delta^{18}O$ values, and the 100-kyr band from the high $\delta^{18}O$ values.

With this information the 100-kyr periodicity stops being a mystery. It is linked to eccentricity but in the opposite way most authors consider. Eccentricity-derived precession-linked insolation forcing does its main role fighting the intrinsic accumulation of ice during glacial periods and has only a supporting role at terminations. At the MPT the world became so cold and the intrinsic ice build-up so extensive that the change in insolation caused by the increase in obliquity was insufficient to melt all the ice that accumulated during the previous low obliquity period. Some interglacials became colder, with more ice at high latitudes. When eccentricity is low, 23-kyr precession cycles produce low summer insolation at high latitudes, resulting in cool summers, with the ice growing through the precession cycle. Low eccentricity thus leads to ice accumulation due to low summer energy precession cycles, while high eccentricity leads to ice melting through high summer energy precession cycles. When an obliquity increase coincides with high eccentricity and high insolation from precession at the right time, it produces an interglacial, melting all the extrapolar ice and resetting the ice clock. Then, as eccentricity and obliquity decline, ice accumulates. 41-kyr after the interglacial, eccentricity has become low, and precession cycles have low insolation during the summer. Under those conditions high obliquity is unable to produce an interglacial on its own. Therefore more ice accumulates. By the time conditions are right again, and eccentricity is sufficiently high, four to five precession cycles had taken place (90–115 kyr) until insolation is again high enough to melt the ice and contribute to an obliquity paced interglacial. The larger the ice-volume accumulated over the 100-kyr period the more unstable the ice sheets and the stronger the feedbacks to drive fast melting dynamics (albedo, rising sea levels,

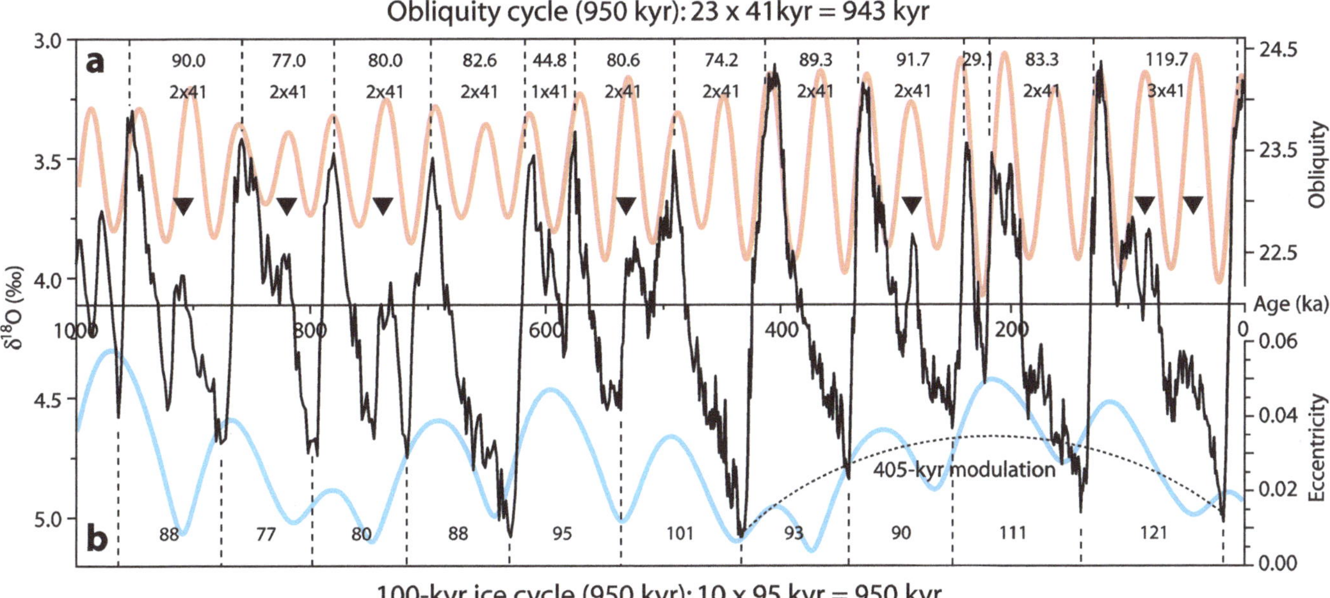

Fig. 2.12 Cycles of ice – cycles of warmth
LR04 temperature and ice proxy (black line) shows two different periodicities during periods of high and low global ice-volume. **a)** Periods of low ice-volume follow the obliquity cycle (orange curve). Time between interglacials (upper numbers) fits multiples of 41 kyr. Obliquity oscillations that do not produce an interglacial often coincide with periods of warming (arrowheads). Obliquity curve has been shifted to the right by 6,500 years to account for the lag in the response. Data from Lisiecky & Raymo 2005. **b)** Periods of high ice-volume follow the eccentricity cycle (blue curve). The amount of ice extends as an inverse of the eccentricity, and in the graph tends to reach the eccentricity curve as drawn. The average of the periods between ice maxima is close to 95 kyr (bottom numbers), with the latest 6 periods averaging 102 kyr. The 405-kyr cycle modulation is clearly observable in the last four periods (dotted line).

dust, CO_2, volcanic activity). The large ice-sheet instability hypothesis introduced by Paillard (1998) has received support from studies that show it could be mediated at least in part by the delayed effect of slow glacial isostasis adjustment maintaining ice-sheets at low altitude through the rapid melting process (Abe–Ouchi et al. 2013). Once the ice is melted in a new interglacial, the ice clock is reset. This explains why the 405-kyr periodicity is absent, as the ice cycle is re-initialized after every successful interglacial. However, the amplitude modulation by the 405-kyr eccentricity cycle is clearly visible in the high ice-volume during the Last Glacial Maximum (LGM), and the glacial maximum 415 kyr earlier, compared to the low ice-volume during the glacial maxima c. 200 ka ago (Fig. 2.12b, dotted line). It follows that since we are in a very low eccentricity situation and moving towards even lower, almost zero eccentricity, there will be considerable ice build up over the next three precession oscillations, preventing a new interglacial in the next obliquity cycle, and leading to high volume of continental ice (though not as high as at the LGM) in 70 kyr (see Chap. 14). So low eccentricity does not promote long interglacials. It does exactly the opposite. It promotes long harsh glacial periods.

2.7 Interglacial determination for the past million years

Obliquity determines when an interglacial might take place but, for the past million years, if the interglacial finally takes place or not does not depend on obliquity. If it did there would be an interglacial in every obliquity oscillation, as it happened in the Pliocene–Early Pleistocene. The 100-kyr ice cycle appears at the MPT, when the cooling of the planet stimulates ice growth and allows the survival of large continental ice sheets outside Greenland and Antarctica during certain obliquity oscillations. These ice sheets can grow for two or even three periods of low obliquity reaching sizes not seen in the planet for the last 250 million years.

The growth of these extrapolar ice sheets has two opposing effects. On one side they make the planet colder, making it more difficult to warm and leading to a decrease in the frequency of interglacials. On the other side when they become very large they reach lower latitudes and extend on continental platforms freed by lowering sea levels, and become increasingly unstable, prone to catastrophic break ups, releasing iceberg armadas, as during Heinrich

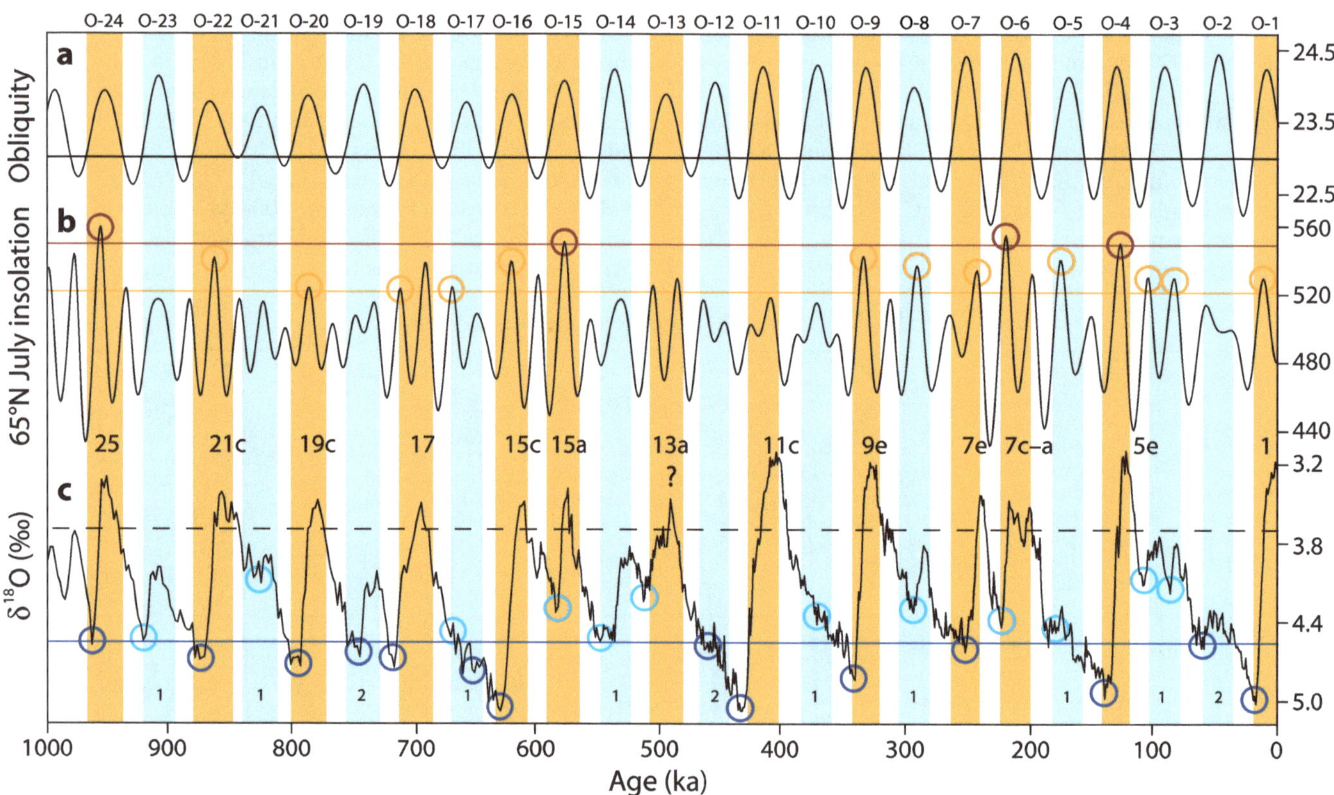

Fig. 2.13 Elements participating in interglacial determination
a) Interglacials are allowed when obliquity (black curve) is above 23° (black horizontal line), defining windows of opportunity marked with a colored bar, orange for interglacial positive and light blue for interglacial negative, and numbered on top from O–1 to O–24. **b)** 65°N summer insolation (black curve) favors interglacials when above 521 W/m² (orange horizontal line and circles) and hinders them when below that value. Insolation values above 549 W/m² (dark red horizontal line and circles) directly produce an interglacial. Astronomical data after Laskar et al. (2004). **c)** Global ice-volume (LR04 proxy, black curve; after Lisiecki & Raymo 2005) promotes interglacials when above (lower in the graph) a threshold value (4.56‰ δ18O dark blue horizontal line and circles) and hinders them when below that level (light blue circles). The graph reveals that after obliquity, global ice-volume is the most important determinant for interglacials (see O–11, for example), and most high obliquity periods without an interglacial are due to insufficient ice (small number 1 at bottom), while some are due to insufficient insolation (small number 2 at bottom). The odd MIS 13a interglacial (question mark), cannot be explained in these terms, but it is quite similar except for a temperature spike to MIS5c (O–3) that is not considered an interglacial. Observe that the numbering of obliquity cycles almost exactly coincides with Marine Isotope Stages numbering derived from data. It is another indication of the great importance of obliquity changes for climate.

events. They also result more susceptible to positive melting feedbacks that increase in strength proportionally to the previous growth of the ice sheets. The feedbacks include rising sea levels, decreasing albedo, warming, dust accumulation, and increased volcanic activity due to ice unloading, leading to increasing CO_2 levels. As a result both the chances of deglaciation and the speed of deglaciation when it finally takes place grow with the increase in global ice. The effect on interglacials is double. They changed their shape after the MPT with deglaciations becoming faster than glaciations, acquiring a sawtooth aspect. And their temperature amplitude increased. Some of the latest interglacials are the warmest in over 2 million years, while glacial maxima for the past 650 kyr are the coldest periods probably in 250 million years (Fig. 2.7). It is possible that the increase in amplitude of temperature oscillations since the MPT has contributed to stabilize the long-term average temperature of the planet, that showed a worrisome steep decreasing trend prior to the MPT (Fig. 2.2).

Obliquity opens windows of interglacial opportunity when it is above 23°, and there have been 24 such windows in the past million years (Fig. 2.13a, numbered colored bars). 13 of them produced what we consider an interglacial ($\delta^{18}O$ values of < 3.7‰; Fig. 2.13c, orange bars), while 11 of them did not (Fig. 2.13 blue bars). The average interglacial frequency is therefore one every 77 kyr. Interglacials cannot exist outside these windows and die a rather quick death when obliquity decreases below 23° (Fig. 2.13a, horizontal line). This is a rule without exceptions. It doesn't matter if 65°N July insolation is high, as it happened between O–22 and 21, or O–16 and 15, underscoring that precession-linked insolation is a secondary factor, as already established.

The second factor in importance is global ice-volume (Fig. 2.13c). It is found that global ice-volume strongly promotes interglacials at obliquity windows when benthic $\delta^{18}O$ is $\geq$ 4.56‰ (LR04 stack; Fig. 2.13c, blue line and dark blue circles), but not when it is below that value (Fig. 2.13c, light blue circles). We know global ice-volume is more important than precession-linked high latitude summer insolation because high insolation does not produce an interglacial at times when global ice-volume is low in O–17, O–8, O–5, and O–3, and because high global ice-volume can produce an interglacial at times when summer insolation is not particularly high like in O–11. Reaching ice-volume values higher than 4.55‰ benthic $\delta^{18}O$ takes longer than one obliquity window except when eccentricity is very low, and that is the main reason why obliquity oscillations fail to produce an interglacial. They lack sufficient extrapolar ice to recruit strong melting feedbacks that allow the melting of so much ice in so little time.

The third factor in importance is the one currently believed to be the most important and incorporated as such into climatic models, 65°N summer insolation. It is found that there are two critical levels in precession-linked insolation for interglacial determination. Values of 65°N July summer insolation above 549 W/m² result in an interglacial at the obliquity window regardless of global ice values (Fig. 2.13b, dark red line and circles). But such values only happen every c. 400 kyr when eccentricity is at the highest values of the 405-kyr cycle. Values above 521 W/m² (Fig. 2.13b, orange line and circles) promote interglacials, while values below that hinder them, as in O–23, O–12, and O–2, when an interglacial would have been very useful to Neanderthals, that were crippled by the increasingly harsher glaciation.

The most effective time for a maximum effect from global ice-volume and high-latitude summer insolation differs. Global ice-volume has a maximum effect when it is high at the start of the obliquity window. If it is insufficient to trigger a termination then, it will continue growing, and ice growth within the obliquity window reduces the chance of an interglacial by reducing the time to pro-

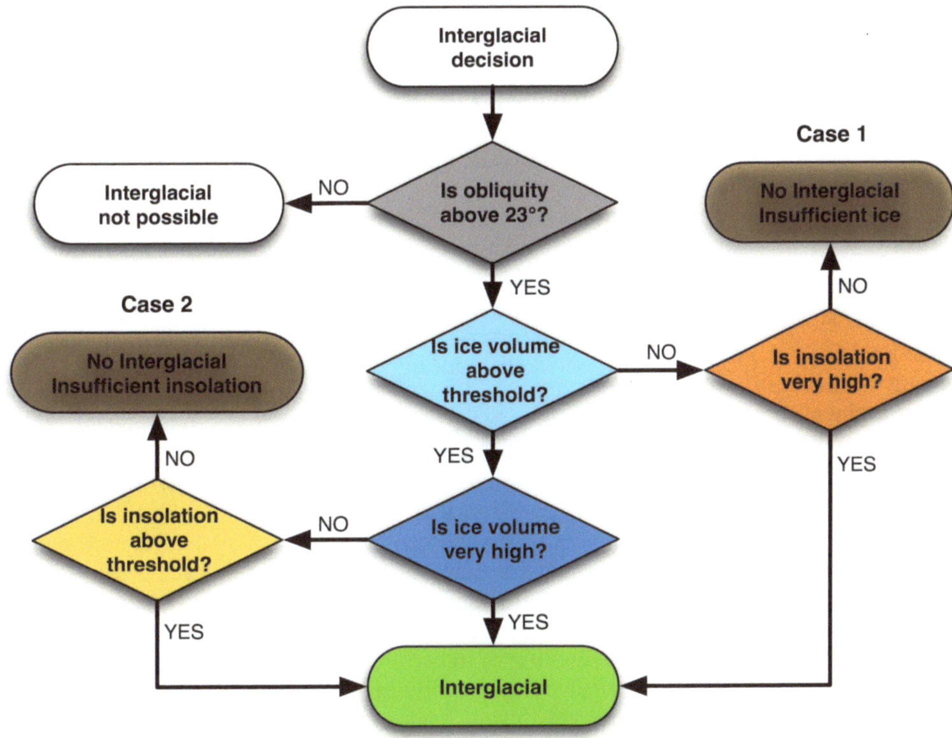

Fig. 2.14 Interglacial flow chart A simple flow chart incorporates the criteria deduced from the past 1 million years of glacial cycle. Interglacials do not take place when obliquity is below 23°. If obliquity is above 23°, interglacials require an ice-volume higher than the equivalent to 4.55‰ benthic $\delta^{18}O$ or a very high 65°N July summer insolation, above 549 W/m². If insolation is lower, interglacials require an insolation above 521 W/m², or very high ice-volume, around 4.90‰ $\delta^{18}O$. There are two scenarios when interglacials fail to take place. The most common is case 1, when ice-volume is not sufficiently high (8 cases). Case 2 is when there is sufficient ice but insolation is too low, associated to very low eccentricity (3 cases) .

duce a deglaciation before obliquity starts decreasing. In contrast, peak high-latitude summer insolation has a bigger effect when it happens close to the obliquity maximum, as then both act together. When two peaks in high-latitude summer insolation occur in the same obliquity window, as in O–18 and O–13, the obliquity peak coincides with an insolation trough, resulting in a cooler, oddly shaped, more symmetrical interglacial (MIS 17 and MIS 13a), similar to interglacials before the MPT.

One result from this analysis is that future interglacials should be predictable to a high degree (Fig. 2.14). Interglacials will skip one obliquity window to allow for sufficient ice build up, unless insolation is very high. The simple flow chart in Fig. 2.14 only fails to hindcast the highly unusual MIS 13a interglacial (that without its temperature spike it might not be considered an interglacial), but correctly hindcasts the outcome of the other 23 cases, including the very infrequent occurrence of two consecutive missing interglacials at O–2 (MIS 3). There was no interglacial at MIS 3 despite sufficient global ice-volume because high-latitude summer insolation was not high enough due to the low eccentricity at the time.

The global ice-volume requirement is surprising because global temperature correlates well with global ice-volume, and therefore the planet is in a colder state when there is a very high ice-volume, as it happened during the LGM. It is likely a proxy for strong feedback factors that operate more strongly when temperatures are very low and ice levels very high. Among the known factors are:

- Reduction of ice-albedo
- Increased melting of ice
- Rising sea levels
- Increase in dust
- Increased volcanism
- Increase in greenhouse gases

The effect of the temperature decrease during a glacial period prior to the next obliquity cycle has the effect of pulling a spring. The stronger it is pulled, the stronger and faster it will go in the opposite direction when released. This spring acts as a negative feedback to further cooling, and its existence can be inferred from the narrow thermal regulation of the planet during at least the past 540 million years (see Sect. 9.3.2). It is what allows interglacials to take place during this very cold period of the planet, as otherwise for the last 1.5 million years the planet would have been locked in a permanent glacial period only interrupted by interglacials every 400 kyr, at the peak of eccentricity. It is possible that there wouldn't be humans if that had happened as conditions are already too close to CO_2 starvation for plants during glacial maxima. Only the arrival of the occasional interglacial prevents further cooling.

When obliquity starts rising during a glacial period it starts moving energy little by little from tropical to polar areas. Its effect on the global average temperature is not noticeable for many thousands of years. If the planet is very cold, with a great portion of the water in huge ice sheets over continents and continental shelves then powerful feedbacks will start. Temperatures will rise after about ten thousand years of increasing energy transfer to higher latitudes and warming will accelerate. It is at about this time when rising precessional insolation during the summer in the Northern Hemisphere will start contributing to the undergoing melting of the northern ice sheets. The contribution of feedback factors and northern summer insolation is what allows the Earth, every 1.8 obliquity cycles on average, to overcome the cold inertia of the planet. It is an additive process where obliquity sets the pace, and is helped by feedback factors and northern summer insolation. If one of these two is strong enough the other might be dispensed. The result is that every interglacial is different. It is the response to forces that assemble and come apart at different times and with different intensities.

2.8 Summer energy as the relevant insolation forcing

Peter Huybers (2006) observed that the melting or growing of the ice-sheets must depend on the cumulative time spent at the ice-sheet border latitude above 0°C during a melting season. It is the same reasoning that led Milutin Milanković to propose his theory 86 years earlier, but in the meantime the time factor had been diluted in favor of the maximum intensity of the insolation responsible. Huybers' observation led to the proposal of a Milankovitch parameter that is close to caloric summer but accounts better for the different duration of summers. He called it summer energy and is calculated by adding the day-time insolation energy (in GJ/m^2) at 65°N for every day that was above a certain insolation threshold enough for ice-melting, that at 65°N was determined to be 275 W/m^2 (Huybers 2006).

Didier Paillard (1998) added the last piece of the puzzle when he proposed a simple model that reproduced the glacial cycle by introducing an ice-volume factor that was needed to transition from interglacial to mild-glacial state, and from mild-glacial to full-glacial state. The model forbids the reverse transitions. In essence Paillard's model introduced the brilliant concept that ice build-up made the transitions unidirectional towards full-glacial, and when ice-volume was very high, ice-sheet instability caused a glacial termination when enough summer energy was available.

Figure 2.15 explains how the glacial cycle responds to summer energy changes (mainly due to obliquity), and to ice-volume changes, and how ice-volume responds to eccentricity. Figure 2.15a shows the ice-volume proxy (LR04 benthic $\delta^{18}O$) for the past 340 kyr, overlain by the summer energy parameter that has been lagged by 6000 years, to account for the observed delay of the effect to the forcing (Huybers 2009; Donders et al. 2018). By plotting ice-volume versus lagged summer energy (Fig. 2.15b), it is observed that during the 41-kyr oscillations in summer energy, ice-volume starts and ends at repeatable states defined, following Paillard (1998), as interglacial, mild-glacial, full-glacial, and deep-glacial.

A simple excitation/relaxation model (Fig. 2.15c) explains the timing of glaciations. During the Early Pleistocene the situation can be described by a reversible oscillation both in summer energy and ice-volume (dashed bi-directional orange arrow) between mild glacial (D) and cool interglacial (D') at a 41-kyr frequency. At the MPT the cooling of the world caused the beginning of the build-up of extensive continental ice-sheets outside the polar regions during glaciations. Now glacial periods would

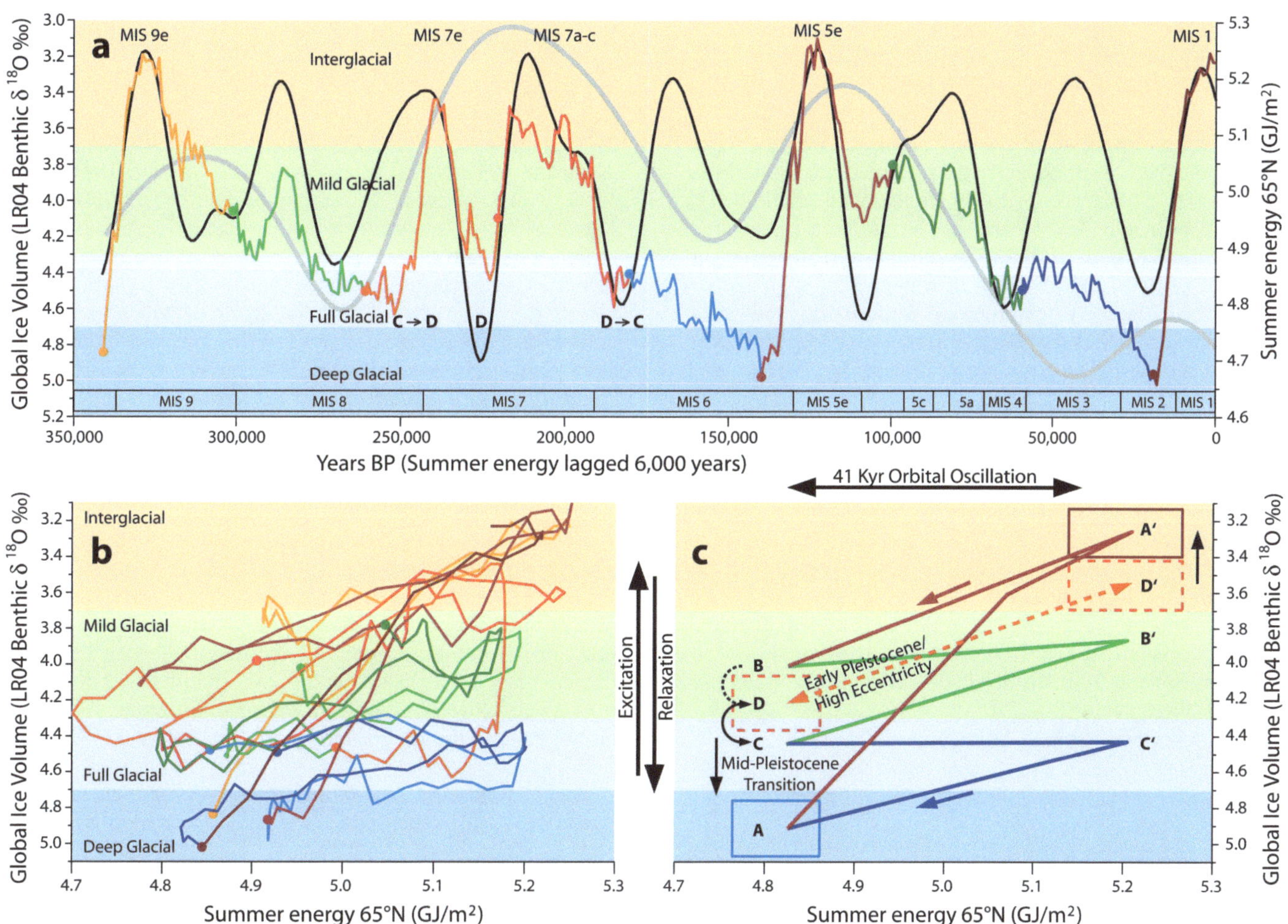

Fig. 2.15 The timing of Pleistocene glaciations as a function of summer energy, ice-volume and eccentricity
a) LR04 benthic stack as an ice proxy (multi-colored line; after Lisiecki & Raymo 2005) for the past 341 kyr. The line was colored in 40–41 Kyr segments with orange-red tones for low ice segments, green tones for intermediate ice segments and blue tones for high ice segments. Dots of the same color mark the start of each segment. Background color defines four different states, light orange for interglacial, light green for mild-glacial, light blue for full-glacial, and cyan for deep-glacial. Black curve is summer energy at 65°N with a 275 W/m^2 threshold (Huybers 2006), lagged by 6000 years to compensate the delay between forcing and effect. Thick grey curve is eccentricity (without scale; after Laskar et al. 2004). **b)** Plot of the multi-colored ice proxy curve versus summer energy. It is evident that the ice-volume at the start of the orbital 41-kyr oscillation in summer energy determines the subsequent ice-volume evolution during the oscillation and the possibility of an interglacial taking place in that oscillation. **c)** Simple excitation/relaxation multi-state model explains the timing of Pleistocene glaciations. Under Early Pleistocene or high eccentricity conditions climate operates as a simple oscillatory system represented by dashed lines, reversibly transitioning at 41-kyr frequency between mild-glacial (D) and cool interglacial (D'). The MPT when eccentricity is not high introduced an ice-volume requirement for excitation out of glacial conditions, represented by the downward arrow (A), and at the same time resulted in warmer interglacials (A'), as the upward arrow indicates. Depending on the speed of ice-volume accumulation, that is inversely correlated to eccentricity, the system must transition through usually one or exceptionally two oscillations (two represented, B' & C') in the relaxation process to reach the excitable state (A). Low eccentricity favors high ice accumulation accelerating the relaxation (only one oscillation required). Medium eccentricity delays the relaxation as ice accumulates more slowly (two oscillations required). High eccentricity bypasses the ice requirement, returning the system to Early Pleistocene conditions as it happened in the C → D transition 245 kyr BP when due to high eccentricity MIS 7e was produced despite low ice-volume and being very late in the summer energy oscillation. The system transitioned back to Mid-Pleistocene conditions after MIS 7a–c (D → C). The double dependency on ice-volume and insolation to produce interglacials at obliquity maxima (Fig. 2.14), results in the absence of a regular pattern. Interglacials are produced at 41, 82, or 123-kyr intervals. However, the ice-volume dependency on eccentricity results in a 100-kyr cycle on ice accumulation that is clearly appreciable in ice proxies.

transition towards more extreme conditions until ice-volume would be so high as to cause high ice-sheet instability (A). This ice-sheet instability is reflected in massive iceberg discharge when perturbed, resulting in Heinrich events. Also, the rebound effect from ice-sheet instability causes warmer interglacials (A'). Mid and Late Pleistocene are characterized by bigger temperature swings between deep glacial and warm interglacial.

When the interglacial ends, the system must relax back to the initial (A) state, but the ice-volume required is so high that it must transition through one or two oscillations (two shown in Fig. 2.15c) during which little ice is melted during the summer energy increase (B → B', C → C'), but considerable ice-sheet growth takes place during the summer energy decrease (B' → C, C' → A). This causes some glacial periods to last one or exceptionally two complete obliquity oscillations.

Global ice-volume is under control of eccentricity, because high eccentricity enhances the effect of precession and low eccentricity damps it. When eccentricity is very high its effect is like returning to the Early Pleistocene, facilitating an interglacial at every summer energy oscillation. Under very high eccentricity intermediate ice-volume glacials (B, C) behave as Early Pleistocene glacials (D). This can be seen very clearly at the MIS 7e interglacial 245 kyr ago (Fig. 2.15a). As the model indicates, the glacial state prior to MIS 7e was of full-glacial, with an ice-volume insufficient to produce an interglacial, thus little ice was melting despite high summer energy. However, when eccentricity became very high (Fig. 2.15a, grey curve), a late interglacial suddenly took place (C → D transition) with very little time left before low obliquity put an end to it. Then, as eccentricity continued being very high, a new interglacial was produced (MIS 7c–a). Both MIS 7 interglacials happened due to high eccentricity, and they were cool interglacials of the Early Pleistocene type. The effect that high eccentricity has in promoting interglacials and low eccentricity in inhibiting them results in more ice-volume accumulating at times of lower eccentricity. The consequence is that although interglacials do not follow a 100-kyr eccentricity cycle, ice-volume does present a 100-kyr cycle (Fig. 2.15a).

2.9 Interglacials of atypical duration

Six interglacials out of the past ten during the last 800 kyr (MIS 13a not considered for the reasons stated in Sect. 2.7), display a very similar temperature profile in EPICA Antarctic records (MIS 5e, 7c–a, 9e, 15a, 15c, 19c). They show a fast increase in temperatures for 5–7,000 years, followed by a temperature stabilization for about 5,000 more years, and then a slow temperature decline that accelerates over time for the next 10–12,000 years during which they lose two thirds or more of the temperature gained from the glacial maximum before the interglacial start. During the period of high temperature (above –2 °C anomaly), that lasts about 15,000 years, each interglacial presents a different temperature profile, highlighting interglacial uniqueness.

An average interglacial was constructed from those six interglacials by aligning them and obtaining their average and standard deviation temperature, average obliquity profile, and the average 65°N summer insolation profile for five of them. MIS 7c presents a very deviant insolation profile that would significantly alter the average of the rest, so its insolation was not included. We can compare this average interglacial to the two interglacials that display a very different duration, the short interglacial MIS 7e at 244 ka, and the long interglacial MIS 11c at 425 ka (Fig. 2.16).

MIS 7e started very late in the obliquity cycle because when obliquity increased above 23°, northern summer

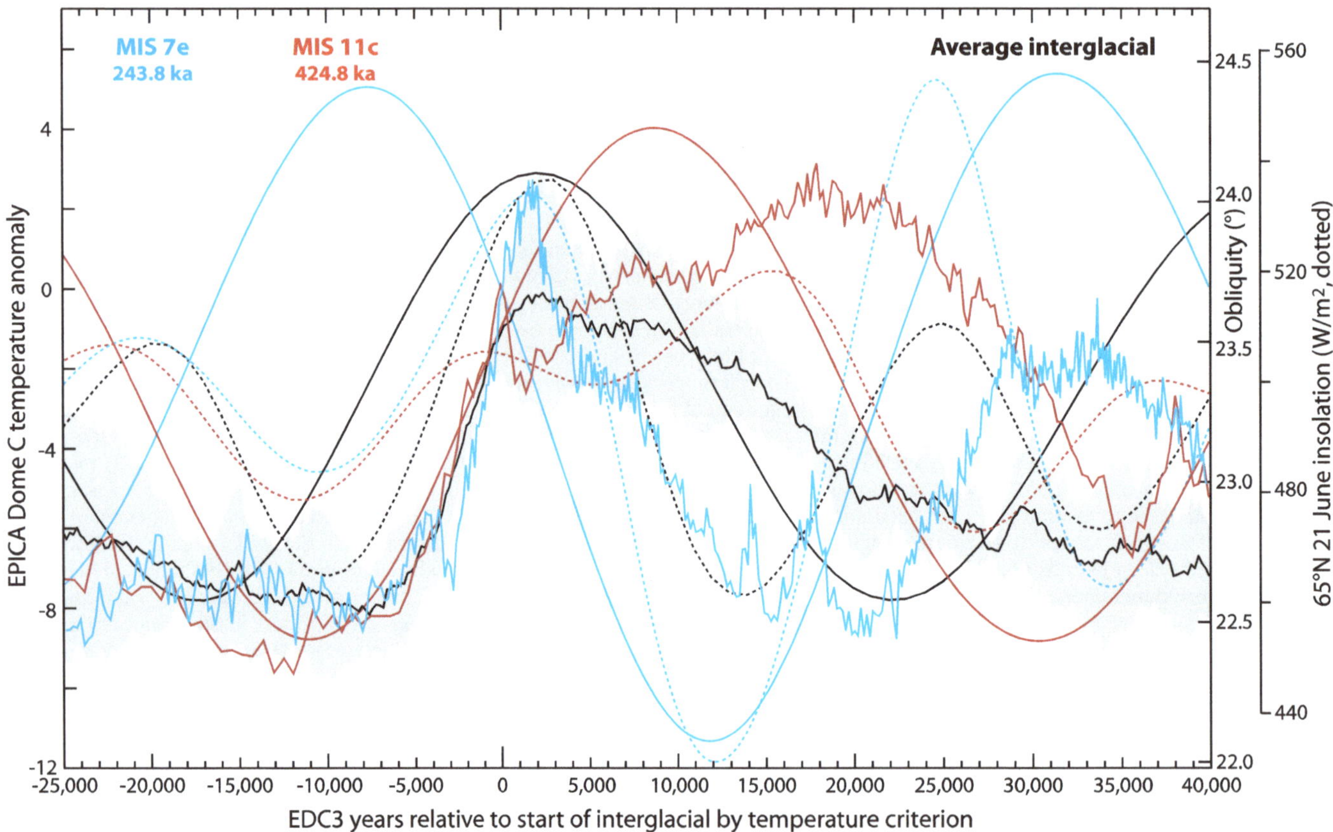

Fig. 2.16 Comparison of atypical interglacials to the average interglacial
An average interglacial (black curve and 1σ grey bands) was constructed from interglacials MIS 5e, 7c–a, 9e, 15a, 15c and 19c, after aligning them at the date specified in table 2.1. The obliquity for all of them (black sinusoid continuous line) and the insolation curves at 65°N 21st June for all but MIS 7c–a (black dotted line) were also averaged. MIS 7e temperature, obliquity and insolation data are similarly plotted in blue, and MIS 11 in dark red. Temperature data from EPICA Dome C, after Jouzel et al. (2007). Astronomical data after Laskar et al. (2004).

insolation was decreasing (Fig. 2.16), and ice-volume was below the threshold (Fig. 2.13). Under normal circumstances MIS 7e would have been a cycle without interglacial, however at 250 ka eccentricity was very high and rising quickly (Fig. 2.15), and ice-volume surpassed the threshold. At 248 ka, with obliquity still above 24°, and insolation at 515 W/m^2 and increasing rapidly, a delayed interglacial was triggered. But as soon as insolation peaked at 242 ka, the simultaneous falling of obliquity and insolation could not sustain the interglacial. MIS 7e started late because it was triggered by precessional-linked insolation due to high eccentricity, but ended on schedule for the obliquity cycle becoming a short interglacial.

MIS 11c is a very exceptional interglacial. As figure 2.13 shows it was triggered by its extraordinary ice-volume values under very modest insolation as soon as increasing obliquity and increasing insolation indicated the feedbacks the right direction. Ice-volume at 433 ka was the highest for the past 5.3 Myr (tied with 630 ka; Lisiecki & Raymo 2005). MIS 11c jumped the gun and obviated the c. 6 kyr delay between forcing and effect. It is unknown how it did it, but it got a 6 kyr lead over the rest of the interglacials. Then MIS 11c proceeded to increase its temperature in three steps. The first step, triggered by rising insolation and a strong feedback response, ended early when insolation peaked at 424 ka. But then rising obliquity provided the impetus for a second warming period (as insolation did not decrease much) that ended at 235 ka when obliquity peaked. Then a third warming step took place caused by a second insolation peak at 226 ka. The three warming steps responsible for the extraordinary duration of MIS 11c are clearly detected in the temperature record (see Figs. 2.11 and 2.16) and give MIS 11c the opposite temperature profile to most interglacials since it evolves from lower to higher temperature. It is the interglacial with highest temperature for the longest time despite occurring at a time of low eccentricity. Given the high increase in energy and the normal thermal inertia of

the planet, its decline was also a very long one, despite being more pronounced than the average decline (Fig. 2.16). Due to the confluence of all these unlikely circumstances, MIS 11c is simply unrepeatable. The Holocene has absolutely nothing in common with MIS 11c except taking place at a time of low eccentricity.

2.10 Role of obliquity in the glacial cycle

Most scholars publishing on the glacial cycle have focused on local conditions to try to explain terminations. Insolation, changes in albedo, and dust deposition are supposed to act maximally at a certain latitude at the edge of the ice sheet to melt it. However, the evidence that obliquity plays an important role poses a significant problem, as obliquity's effect on insolation decreases very fast as the distance from the pole increases (Fig. 2.8). Solving the glacial cycle, therefore, may require out of the box thinking. Such thinking has been provided by researchers focusing on a very little studied property of solar insolation, the latitudinal insolation gradient (LIG). Moisture delivery to Antarctica (Vimeux et al. 1999) and Greenland (Masson–Delmotte et al. 2005) has been shown to strongly correlate with obliquity and interpreted as resulting from changes in LIG affecting the hydrological cycle. LIG differences arise from differences in the angle of incidence of solar rays resulting from seasonal changes in the orientation of the Earth. At any time the amount of insolation at a certain latitude depends mainly on precession, but the summer pole points towards the Sun, and on summers the amount of insolation at high latitudes changes significantly with obliquity. When obliquity is high the summer LIG flattens and when it is low the summer LIG steepens. Changes in summer LIG follow almost exactly changes in obliquity at both hemispheres (Fig. 2.17). By contrast winter LIG changes do not depend on obliquity, as the winter pole is in the dark, and depends almost exclusively

Fig. 2.17 Changes in the summer latitudinal insolation gradient depend on obliquity
a) Mean summer insolation depends mainly on precession. Northern Hemisphere in dark red. 60°N (solid curve) and 30°N (dotted curve) are for the month of July (21st June–21st July). Southern Hemisphere in light blue 60°S (solid curve) and 30°S (dotted curve) are for the month of January (21st December–21st January). Observe how insolation in both hemispheres is in anti-phase, largely cancelling its net effect. **b)** The summer latitudinal insolation gradient is similar and in phase in both hemispheres, and depends mainly on the obliquity cycle. Dark red solid curve is the result of subtracting the 60°N minus the 30°N mean insolation. When obliquity is high both are similar and the gradient is flatter. When obliquity is low the gradient becomes steeper (more negative value). A steeper gradient drives more energy and moisture towards the poles, favoring planetary cooling and ice growth, leading to glacial inception. Light blue solid curve is the same for the Southern Hemisphere. Black dotted curve corresponds to obliquity. Data from Laskar et al. (2004) analyzed with AnalySeries (Paillard et al. 1996).

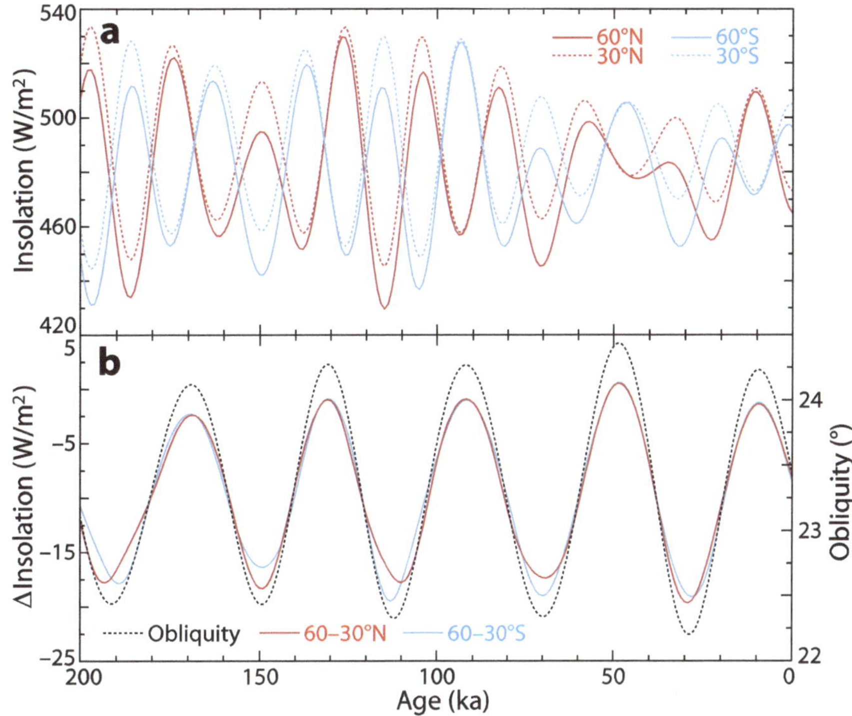

on precession. On summers polar energy budget depends mostly on insolation and is very much affected by obliquity, while on winters poles get most of their energy from winter energy transport from lower latitudes. But as we have seen it is the summer conditions that are critical for terminations and glacial inceptions.

The role of moisture transport in glaciations is rarely considered. In 1872 John Tyndall argued: *"So natural was the association of ice and cold that even celebrated men assumed that all that is needed to produce a great extension of our glaciers is a diminution of the sun's temperature. Had they gone through the foregoing reflections and calculations, they would probably have demanded more heat instead of less for the production of a 'glacial epoch'."* (cited by Kukla & Gavin 2004). Indeed such is the role of obliquity, which increases insolation in the tropics as it decreases it at the poles and at the same time steepens the LIG during the summers bringing the necessary moisture that will remain locked there as ice until the process reverses (Kukla & Gavin 2004). And it is not only moisture transport what depends on the LIG, the latitudinal temperature gradient (LTG) also depends on the LIG (Davis & Brewer 2009).

The LTG is a central property of Earth's climatic system at all time scales. It drives the atmospheric-oceanic circulation and helps explain the propagation of orbital signatures through the climatic system, including the Monsoon, Arctic Oscillation, and ocean circulation (Davis & Brewer 2009). Scotese (2016) has shown that one of the main differences between a hothouse and an icehouse planet is in the LTG. The same applies to the difference between glacial and interglacial periods.

Polar amplification of global average temperature changes is prominently displayed as one of the main features of Modern Global Warming, yet polar amplification is the result of changes in the LTG. Climate models treat LTG as an emergent feature and underestimate its sensitivity to changes in LIG, probably overestimating the role of non-condensing greenhouse gases in driving polar amplification (Davis & Brewer 2009).

The interpretation of the glacial cycle in terms of LTG emphasizes summer enthalpy and moisture transport from the tropics to the poles as the decisive factor under obliquity control. Within this hypothesis the tropics, with their huge thermal and moisture capacity, become principal agents in the formation and waning of ice sheets, orchestrated by obliquity changes, while local factors like latitudinal insolation, albedo, and dust are important secondary players that sometimes become decisive.

2.11 Role of CO_2 in the glacial cycle

There is no doubt that CO_2 is one of several feedbacks that must act on the glacial cycle, as CO_2 levels increase with glacial terminations and decrease with the cooling of glacial inceptions. Its exact role remains controversial. About half of the CO_2 increase at terminations comes from increased volcanic activity, probably stimulated by load changes effected by ice sheet melting (Maclennan et al. 2002; Huybers & Langmuir 2009), highlighting its feedback role. Being a positive feedback implies that the signal output is amplified, and it is generally accepted that the CO_2 increase must contribute to the warming at glacial terminations. There is an objection to a more substantive role for CO_2 on glacial terminations. We have reviewed in sections 2.7 to 2.9 (Figs. 2.12 to 2.15) the important role that global ice-volume plays in determining interglacial temperature. Interglacials preceded by very high global ice-volumes are warmer than interglacials preceded by lower global ice-volumes. It means that we can predict, to a certain point, how warm an interglacial is going to be from the amount of ice accumulated before it starts. And therefore we can predict how much CO_2 it is going to have (absent an anthropogenic contribution), given the clear correlation between CO_2 and temperature. Factors that can be predicted from causative agents in highly variable phenomena cannot be in control. The issue is even worse at glacial inceptions when the correlation between temperature and CO_2 is lost, and temperature decrease leads by several millennia the CO_2 decrease. At the end of MIS 5e temperature decreased by 4 °C in Antarctica in 8,000 years, without any decrease in CO_2 (Fig. 2.18). Further, the rate of temperature decrease between 120.5 and 111 ka was fairly uniform at the resolution of the EPICA Dome C

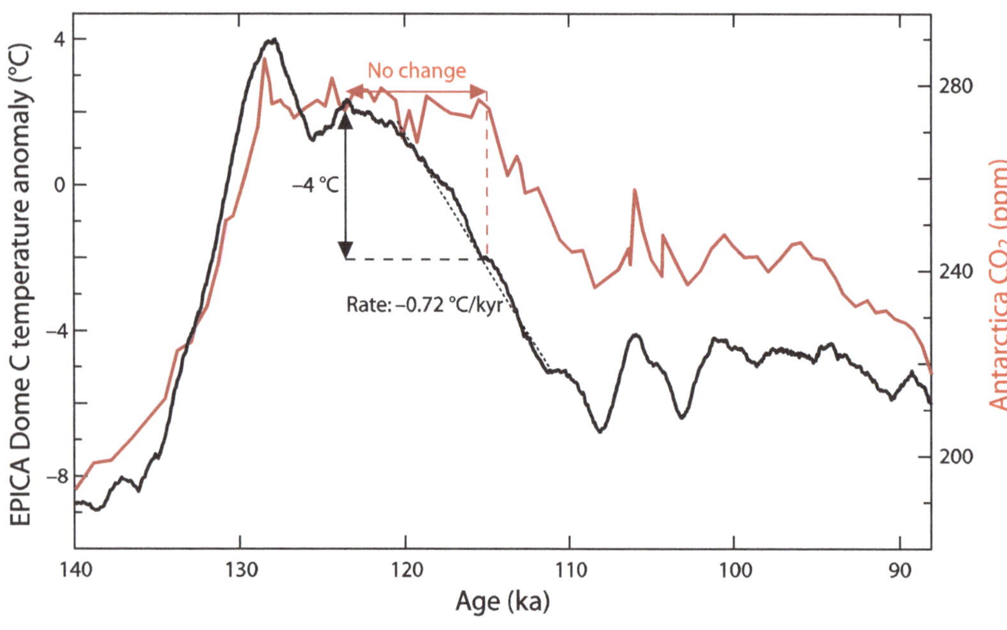

Fig. 2.18 No role for CO_2 at glacial inceptions
EPICA Dome C deuterium proxy for Antarctic temperature (black curve) shows a cooling of 4 °C between 123.5 and 115 ka (black arrow), when Antarctic CO_2 levels (red curve) showed no change (red arrow). The rate of cooling between 120.5 and 111 ka (black dotted line) is nearly linear, with a trend of approximately −0.72 °C/kyr, that shows no response to the important differences in CO_2 rate of change. Temperature data averaged with a 1,000-year running window. Epica Dome C temperature data after Jouzel et al. (2007); CO_2 data after Bereiter et al. (2015).

ice core, unaffected by CO_2 levels being constant until 115 ka and then falling by one third of the glacial-interglacial difference in the last 4 kyr (Fig. 2.18). It is hard to argue for an important role by CO_2 on glacial terminations without evidence, when it clearly doesn't have one at glacial inceptions.

We know from ice core measurements that glacial termination I (the closest to us, starting at 18 ka) involved a change in CO_2 atmospheric concentrations from 190 to 265 ppm, an increase of 75 ppm. Concurrently the temperature increased globally by an estimated 4–6 °C (Schneider von Deimling et al. 2006; Annan & Hargreaves 2013; Tierney et al. 2020). CO_2 role has been estimated at causing directly 30% of the warming, and indirectly through forcing interactions another 20% (Gregoire et al. 2015), so between half and one third of the warming (or about 2–3 °C) is attributed to the CO_2 increase.

A simple calculation tells us that the rise from 190 to 265 ppm is 48% of a doubling. This is true because we are dealing with a logarithmic scale, (ln(265)–ln(190))/(ln(190×2)–ln(190))=0.48. So 48% of a doubling is estimated to have produced c. 2–3 °C of warming between 18–10 ka. The rise from preindustrial to current levels of CO_2 (280 to 415 ppm, or 135 ppm) constitutes 57% of a doubling. That is (ln(415)–ln(280))/(ln(280×2)–ln(280))=0.57, so it should be similar in terms of warming effect. Yet, even if CO_2 is responsible for 100% of modern warming, as the IPCC claims (see Chap. 9), why has it produced only about 0.8 °C increase (HadCRUT4 1850–2020)? Something is not right. If our knowledge of past CO_2 levels is correct, and the hypothesis that CO_2 was responsible for one third to half of the warming at glacial termination is correct, at 18 ka CO_2 was two to three times more potent than now. If anthropogenic CO_2 is not responsible for all the warming observed since 1850, as it appears probable, then the situation is much worse. There is no way to reconcile the disparity that was already noticed by the late Marcel Leroux (2005). So we must accept, based on current data, that CO_2 had a very minor role during the glacial cycle, responsible for, at most, one sixth of the warming at terminations, and therefore conclude that CO_2 is not an important climate factor during the Plio–Pleistocene glacial–interglacial transitions.

A more pessimistic view is to consider that the recent CO_2 increase entails more than 1 °C of committed warming not yet manifested, implying an equilibrium climate sensitivity of 4 or more. There is no evidence to support this belief. In fact, there is ample evidence against it:

- The continued removal of anthropogenic CO_2 via increasingly robust carbon sinks. The more we produce, the more is removed from the atmosphere. An increasing removal rate works against a hypothesized high warming commitment from current CO_2 levels.
- The lack of evidence for a climate sensitivity of 4 or more. Most experimentally deduced values for equilibrium climate sensitivity are between 1.5 and 2.5 (see Sect. 9.5), half of the rate required for the claimed role of CO_2 in deglaciation.
- The lack of a significant increase in the rate of warming during the last century. If we had actually increased the committed warming significantly, the rate of warming should have increased proportionally, but that is not what has been observed (see Sect. 12.7).

- A great amount of committed warming should make periods of decades with little or no warming increasingly unlikely. Again this goes against observations.

The only reasonable way to reconcile the disparity in CO_2 increases and temperature increases between glacial termination I and Modern Global Warming is to conclude that CO_2 had a minor role in glacial termination. Further, it is reasonable to expect it will have a minor role in the next glacial inception.

2.12 Conclusions

2a. Obliquity is the main factor driving the glacial–interglacial cycle. Precession, eccentricity and 65°N summer insolation play a secondary role. The glacial cycle is the result of the interplay between two cycles operating simultaneously, the 41-kyr obliquity cycle and the global ice-volume 100-kyr cycle related to eccentricity.

2b. The Mid-to-Late Pleistocene pacing of interglacial periods is the consequence of the Earth being in a very cold state that prevents almost half of obliquity oscillations from successfully emerging from glacial conditions. The rate for the past million years has been 72.7 kyr/interglacial, or 1.8 obliquity oscillations between interglacials. This can be generally described as one interglacial every two obliquity oscillations except when close to the 405 kyr eccentricity peaks, when interglacials take place at every obliquity oscillation.

2c. Glacial terminations require, in addition to rising obliquity, the existence of very strong feedback factors recruited at very high global ice-volume levels. Increasing northern summer insolation once obliquity is rising is a positive factor, and if high enough during eccentricity peaks, it can drive the termination process on its own.

2d. The Mid-Pleistocene Transition was likely provoked by the continuous cooling of the planet causing too much accumulation of extrapolar ice. This change introduced the counter-intuitive requirement for large ice sheets build up to trigger an interglacial, starting the 100-kyr global ice-volume cycle and causing obliquity-driven interglacial generation to skip one oscillation unless high eccentricity removes the ice volume requirement.

2e. The most convincing hypothesis on the effect of obliquity is through changes to the summer latitudinal insolation gradient, that affects the latitudinal temperature gradient controlling energy and moisture transport to the poles.

2f. CO_2 can only produce a minor effect in glacial terminations since its measured change in concentration (roughly a third of a doubling, which represents half of the warming effect of a doubling) is too small to account for any important contribution to the large observed temperature changes.

References

Abe–Ouchi A, Saito F, Kawamura K et al (2013) Insolation-driven 100,000-year glacial cycles and hysteresis of ice-sheet volume. Nature 500 (7461) 190–193

Annan JD & Hargreaves JC (2013) A new global reconstruction of temperature changes at the Last Glacial Maximum. Climate of the Past 9 (1) 367–376

Bereiter B, Eggleston S, Schmitt J et al (2015) Revision of the EPICA Dome C CO_2 record from 800 to 600 kyr before present. Geophysical Research Letters 42 (2) 542–549

Berger A (1978) Long-term variations of daily insolation and Quaternary climatic changes. Journal of the Atmospheric Sciences 35 (12) 2362–2367

Bosmans JHC, Hilgen FJ, Tuenter E & Lourens LJ (2015) Obliquity forcing of low-latitude climate. Climate of the Past 11 (10) 1335–1346

Bryson WM (2003) A short history of nearly everything. Broadway Books. New York

Davis BA & Brewer S (2009) Orbital forcing and role of the latitudinal insolation/temperature gradient. Climate dynamics 32 (2–3) 143–165

Donders T, Van Helmond NA, Verreussel R et al (2018) Land-sea coupling of early Pleistocene glacial cycles in the southern North Sea exhibit dominant Northern Hemisphere forcing. Climate of the Past 14 (3) 397–411

Drysdale RN, Hellstrom JC, Zanchetta G et al (2009) Evidence for obliquity forcing of glacial termination II. Science 325 (5947) 1527–1531

Gallup CD, Cheng H, Taylor FW and Edwards RL (2002) Direct determination of the timing of sea level change during Termination II. Science 295 (5553) 310–313

Gildor H & Tziperman E (2000) Sea ice as the glacial cycles climate switch. Paleoceanography 15 605–615

Gregoire LJ, Valdes PJ and Payne AJ (2015) The relative contribution of orbital forcing and greenhouse gases to the North American deglaciation. Geophysical Research Letters 42 (22) 9970–9979

Hays JD, Imbrie JN & Shackleton J (1976) Variations in the Earth's Orbit: Pacemaker of the Ice Ages. Science 194 1121–1132

Huybers P (2006) Early Pleistocene glacial cycles and the integrated summer insolation forcing. Science 313 (5786) 508–511

Huybers PJ (2009) Pleistocene glacial variability as a chaotic response to obliquity forcing. Climate of the Past 5 481–488

Huybers P & Langmuir C (2009) Feedback between deglaciation, volcanism, and atmospheric CO_2. Earth and Planetary Science Letters 286 (3–4) 479–491

Huybers P & Wunsch C (2005) Obliquity pacing of the late Pleistocene glacial terminations. Nature 434 491–494

Imbrie J, Berger A, Boyle EA et al (1993) On the structure and origin of major glaciation cycles 2. The 100,000-year cycle. Paleoceanography 8 (6) 699–735

Jouzel J, Masson–Delmotte V, Cattani O et al (2007) Orbital and millennial Antarctic climate variability over the past 800,000 years. Science 317 (5839) 793–796

Kent DV, Olsen PE, Rasmussen C et al (2018) Empirical evidence for stability of the 405-kiloyear Jupiter–Venus eccentricity cycle over hundreds of millions of years. Proceedings of the National Academy of Sciences 115 (24) 6153–6158

Kukla G & Gavin J (2004) Milankovitch climate reinforcements. Global and Planetary Change 40 (1–2) 27–48

Kutzbach JE (1981) Monsoon climate of the early Holocene: Climate experiment with the earth's orbital parameters for 9000 years ago. Science 214 (4516) 59–61

Landwehr JM, Coplen TB, Ludwig KR et al (1997) Data from Devils Hole Core DH–11. US Geological Survey Open File Report 97–792 http://pubs.usgs.gov/of/1997/ofr97-792/ Accessed 13 Jun 2022

Laskar J, Robutel P, Joutel F et al (2004) A long-term numerical solution for the insolation quantities of the Earth. Astronomy & Astrophysics 428 (1) 261–285

Lawrence KT, Liu Z, & Herbert TD (2006) Evolution of the Eastern Tropical Pacific Through Plio–Pleistocene Glaciation. Science 312 79–83

Leroux M (2005) Global warming-myth or reality?: The erring ways of climatology. Springer, Berlin

Lisiecki LE (2010) Links between eccentricity forcing and the 100,000-year glacial cycle. Nature Geoscience 3 (5) 349–352

Lisiecki LE & Raymo ME (2005) A Pliocene-Pleistocene stack of 57 globally distributed benthic $\delta^{18}O$ records. Paleoceanography 20 1

Liu Z, Cleaveland LC & Herbert TD (2008) Early onset and origin of 100-kyr cycles in Pleistocene tropical SST records. Earth and Planetary Science Letters 265 (3–4) 703–715

Lorius C, Jouzel J, Ritz C et al (1985) A 150,000-year climatic record from Antarctic ice. Nature 316 (6029) 591–596

Ludwig KR, Simmons KR, Szabo BJ et al (1992) Mass-spectrometric ^{230}Th–^{234}U–^{238}U dating of the Devils Hole calcite vein. Science 258 (5080) 284–287

Maclennan J, Jull M, McKenzie D et al (2002) The link between volcanism and deglaciation in Iceland. Geochemistry, Geophysics, Geosystems 3 (11) 1–25

Maslin MA & Ridgwell AJ (2005) Mid-Pleistocene revolution and the 'eccentricity myth'. In: Head MJ and Gibbard PL (eds) Early-Middle Pleistocene transitions: The land-ocean evidence. Geological Society, London. Special Publications 247 19–34

Masson–Delmotte V, Jouzel J, Landais A et al (2005) GRIP deuterium excess reveals rapid and orbital-scale changes in Greenland moisture origin. Science 309 (5731) 118–121

Mearns E & Milne A (2016) The Shrinking Glacier Conundrum. The Alpine Journal 120 195–207

Moseley GE, Edwards RL, Wendt KA et al (2016) Reconciliation of the Devils Hole climate record with orbital forcing. Science 351 (6269) 165–168

Muller RA & MacDonald GJ (1997) Spectrum of 100-kyr glacial cycle: Orbital inclination, not eccentricity. Proceedings of the National Academy of Sciences 94 (16) 8329–8334

Murphy JJ (1869) On the nature and cause of the glacial climate. Quarterly Journal of the Geological Society 25 (1–2) 350–356

Nie J (2018) The Plio–Pleistocene 405-kyr climate cycles. Palaeogeography, palaeoclimatology, palaeoecology 510 26–30

Paillard D (1998) The timing of Pleistocene glaciations from a simple multiple-state climate model. Nature 391 378–381

Paillard D, Labeyrie L & Yiou P (1996) Macintosh program performs time-series analysis. Eos, Transactions American Geophysical Union 77 (39) 379–379

Pollard BS & Baez JC (2013) Milankovitch cycles and the Earth's climate. Climate Science Seminar at California State University. Northridge, April 26. https://math.ucr.edu/home/baez/glacial/glacial.pdf Accessed 29 Jun 2022

Puetz SJ, Prokoph A & Borchardt G (2016) Evaluating alternatives to the Milankovitch theory. Journal of Statistical Planning and Inference 170 158–165

Rial JA, Oh J & Reischmann E (2013) Synchronization of the climate system to eccentricity forcing and the 100,000-year problem. Nature Geoscience 6(4) 289–293

Ridgwell AJ, Watson AJ & Raymo ME (1999) Is the spectral signature of the 100 kyr glacial cycle consistent with a Milankovitch origin? Paleoceanography 14 (4) 437–440

Schneider von Deimling T, Ganopolski A, Held H & Rahmstorf S (2006) How cold was the last glacial maximum? Geophysical Research Letters 33 (14) L14709

Scotese CR (2016) Some thoughts on global climate change: The transition from icehouse to hothouse. In: Scotese CR (author) The Earth History: The evolution of the Earth System. PALEOMAP Project, Evanston, IL, https://www.researchgate.net/publication/275277369_Some_Thoughts_on_Global_Climate_Change_The_Transition_for_Icehouse_to_Hothouse_Conditions Accessed 18 Oct 2018

Shakun JD, Clark PU, He F et al (2012) Global warming preceded by increasing carbon dioxide concentrations during the last deglaciation. Nature 484 (7392) 49–54

Tierney JE, Zhu J, King J et al (2020) Glacial cooling and climate sensitivity revisited. Nature 584 (7822) 569–573

Tzedakis PC, Crucifix M, Mitsui T & Wolff EW (2017) A simple rule to determine which insolation cycles lead to interglacials. Nature 542 (7642) 427–432

Vimeux F, Masson V, Jouzel J et al (1999) Glacial–interglacial changes in ocean surface conditions in the Southern Hemisphere. Nature 398 (6726) 410–413

Winograd IJ, Coplen TB, Landwehr JM et al (1992) Continuous 500,000-year climate record from vein calcite in Devils Hole, Nevada. Science 258 (5080) 255–260

Zachos J, Pagani M, Sloan L et al (2001) Trends, rhythms, and aberrations in global climate 65 Ma to present. Science 292 (5517) 686–693

3

THE DANSGAARD–OESCHGER CYCLE

"Chaos is merely order waiting to be deciphered."
José Saramago (2002)

3.1 Introduction

It was already known by botanists early in the 20th century that peat bogs and lake sediments in Scandinavia showed quite abrupt climatic changes reflected in vegetation changes that indicated that the end of the last glacial period was marked by alternating cold (stadials) and warm (interstadials) periods. The last two stadials were named after a tundra flower whose pollen and leaves became abundant, *Dryas octopetala*, as the Older Dryas and the Younger Dryas (YD).

In 1972, after analyzing the isotopic composition of ice cores from Camp Century in Greenland, Willi Dansgaard reported that the last glacial period showed over 20 abrupt interstadials marked by very intense warming (Johnsen et al. 1972). The discovery was met with indifference by a scientific community still struggling to identify Milankovitch cycles in the data, because the new abrupt changes were not found in Antarctic records. Twelve years later Hans Oeschger reported that the abrupt changes were accompanied by sudden increases in CO_2 in the Greenland ice cores (Stauffer et al. 1984). From then on the abrupt changes were known as Dansgaard–Oesch-

ger (D–O) events. It was later found that Greenland elevated CO_2 records were the result of a chemical contamination, since they did not match Antarctic CO_2 records. Since then scientists have been struggling to find a cause for them, and there is controversy over being a cyclical manifestation or not. The controversy is relevant because some researchers believe D–O events might be related to the millennial climate variability taking place during the Holocene, and thus would constitute part of the background to present climate change.

3.2 Dansgaard–Oeschger oscillations

US Greenland Ice Sheet Project 2 (GISP2) took five years and over 3 km of drilling to reach bedrock in 1993, producing the GISP2 ice core. It has good temporal resolution allowing the counting of annual ice layers over tens of thousands of years. Analysis of D–O events in the GISP2 ice core, led to the discovery that they displayed a prominent frequency at 1470 years (Grootes & Stuiver 1997). The result was not so clear in similar ice cores, like the European GRIP (Greenland Ice Core Project) completed in 1992, but GISP2 was considered to have better resolution.

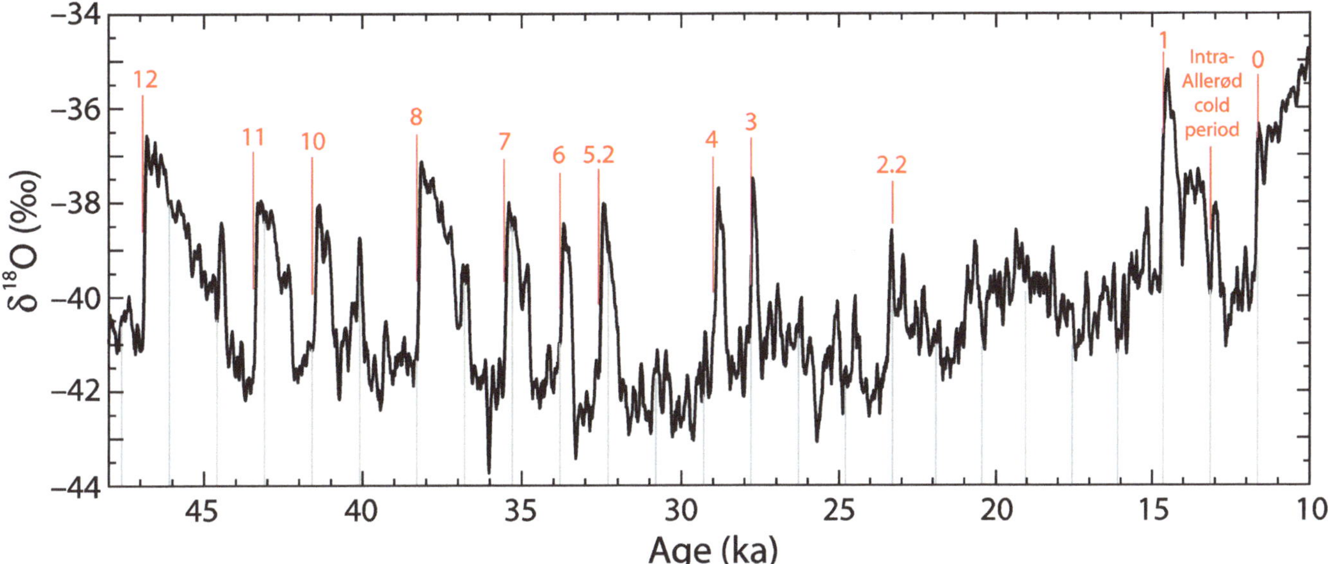

Fig. 3.1 The Dansgaard–Oeschger cycle
GISP2 record of $\delta^{18}O$ proxy for Greenland temperature. GISP2 has been placed on the new GICC05 timescale (Seierstad et al. 2014) Dividing the 48–10 kyr BP period in boxes of approximately 1500 years (grey lines) helps display the periodicity of the D–O cycle (Ramstorf 2003). D–O events are numbered according to the Greenland revised stratigraphical scheme (Rasmussen et al. 2014). The abrupt start of the Holocene in the North Atlantic is considered here the most recent D–O warming (number 0).

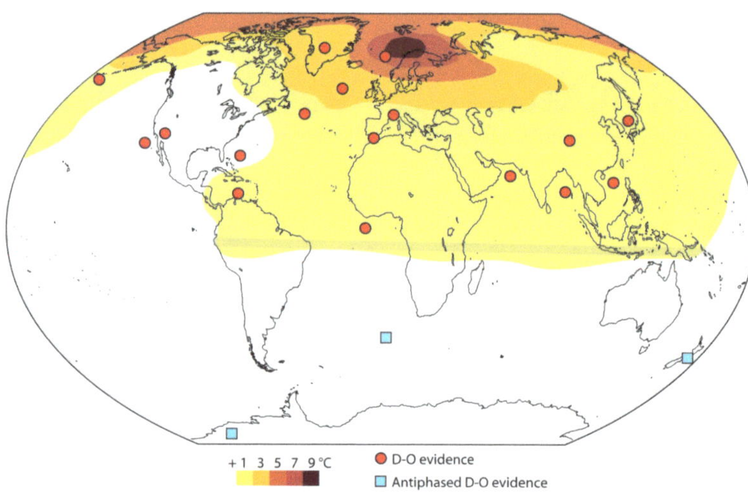

Fig. 3.2 Widespread effects of the Dansgaard–Oeschger cycle

Warming in Greenland is coincident (red dots) with warmer wetter conditions in Europe, higher sea-surface temperatures in the western Mediterranean, increased precipitation in the Venezuelan coast, enhanced summer monsoon in the Indian Ocean, aridity in the southwestern North America and China, changes in ocean ventilation in California, increased sea temperature and productivity in the Arabian sea. Warming in Greenland is also coincident (blue squares) with anti-phased cooling in Antarctica, and the circumpolar seas (See Broecker 1999 and Rohling et al. 2003 for references). Modeled changes in temperature (after Menviel et al. 2014) do not fully reproduce the extent of a D–O interstadial, even after doubling the temperature scale, as Greenland shows an 8–10 °C increase. D–O events are centered in the Nordic Seas, between Greenland, Iceland, the Faroe Islands, and Norway, south-west of what the model shows.

According to Stefan Rahmstorf (2003), the periodicity of the D–O cycle had less than 1% probability of being due to chance, and the deviations for the later, best dated oscillations from the period was of only 2% (Fig. 3.1; Rahmstorf 2003). In 2003 the European Consortium completed the NGRIP (North-GRIP) drilling in Central Greenland, obtaining a more undisturbed and in several aspects better ice core, that again did not support a clear 1470-yr periodicity.

In 2005 the Danish Center for Ice and Climate constructed a new timescale, the Greenland Ice Core Chronology 2005 (GICC05), by comparing three ice cores. It reached 60 ka, the earliest time when annual layers can still be counted. The GICC05 timescale has been extended back to the last interglacial with more ice cores, and using volcanic ash layers, ^{10}Be deposition rates, the global CH_4 record, and impurity patterns. Comparison of GISP2 with GICC05 showed significant discrepancies, and ash layers confirmed GISP2 was in need of correction. When GISP2 data was converted to the GICC05 timescale (Fig. 3.1; Seierstad et al. 2014), the power of the 1470-yr peak in frequency analyses not surprisingly diluted into several smaller peaks in the 1000–2000-yr range. It shows the dangers of relying on a single, supposedly superior, record. We must reject the notion that D–O events occur at 1470-yr intervals, but they still show a tendency to occur at multiples of c. 1500 years (Fig. 3.1). In fact the D–O event that put an end to the Oldest Dryas period and started the warmer Bølling period, and the D–O event that started the Holocene, took place at almost exactly 3000 years apart (c. 14,750 and 11,750 yr BP). And in the middle of that period, another abrupt warming put a sudden end to the intra-Allerød cold period (Fig. 3.1). The naming convention of the Greenland revised stratigraphical scheme by Rasmussen et al. (2014), that divides D–O oscillations into Greenland stadials (GS, cold phase), and interstadials (GI, warm phase), numbered within a sequence that also includes non D–O climatic changes, is followed in this work.

D–O oscillations are the most dramatic and frequent abrupt climatic change in the geological record. They helped define the concept of abrupt climate change, as prior to their discovery it was assumed that climate changed slowly over time from a human perspective. In Greenland, D–O oscillations are characterized by an abrupt warming of c. 9 °C in annual average temperature

from a cold stadial to a warm interstadial phase within a few decades, followed by slower gradual cooling before a more rapid return to stadial conditions. Initially they were thought to be regional phenomena, since they were not prominently displayed in Antarctic ice core records, however evidence uncovered since shows that they display a hemispheric-wide climate effect reaching also the Southern Hemisphere (Fig. 3.2).

D–O oscillations are not the only climate change taking place during the last glacial period. Temperature variability is very high (Fig. 3.3b & c), and the changes have different shapes, durations and spacing. They are sometimes separated by other intense climate changes of a different nature called Heinrich events (HEs). Lets describe those changes starting with Greenland.

With an irregular periodicity of c. 6,000 years (Fig. 3.3 vertical bars) 1–4 kyr long HEs took place in the northern Atlantic region, causing a drop of 1–2 °C from the already cold glacial climate. Sea surface temperatures in the North Atlantic fell to what are now Arctic conditions as far south as 45°N, and were probably covered by sea-ice during the winter. Global methane levels decreased during HEs (Fig. 3.3e). HEs are also characterized by a greatly enhanced iceberg production from the Laurentide ice sheet, or less often from the Fennoscandian one, carrying with them big amounts of eroded material that, when the icebergs melted, were deposited on the sea bed as ice-rafted debris (IRD). HEs are labeled H0 to H6 (Fig. 3.3a), with the most recent coinciding with the YD.

A HE is followed by the triggering of a D–O interstadial warming. Although only one in four D–O oscillations is preceded by a HE, the other three are preceded by a similar albeit reduced cooling and IRD deposition in North Atlantic marine sediments. Gerard Bond suggested that HEs are part of the D–O cycle (Bond et al. 1993). Since HEs involve more profound cooling and much more intense ice-shelf calving we can therefore distinguish between HE D–O oscillations (GI 1, 4, 8, 12, 14, 17.2) and non-HE D–O oscillations.

D–O oscillations are characterized by their asymmetric change in temperatures. They all display a very fast warming, with temperatures rising by about 7–13 °C (9 °C average, Central Greenland temperature) in less than seven decades, within the span of a human life (Fig. 3.4). The warming rate is an amazing 1 °C every seven years. This abrupt warming ceases synchronously for nearly all D–O

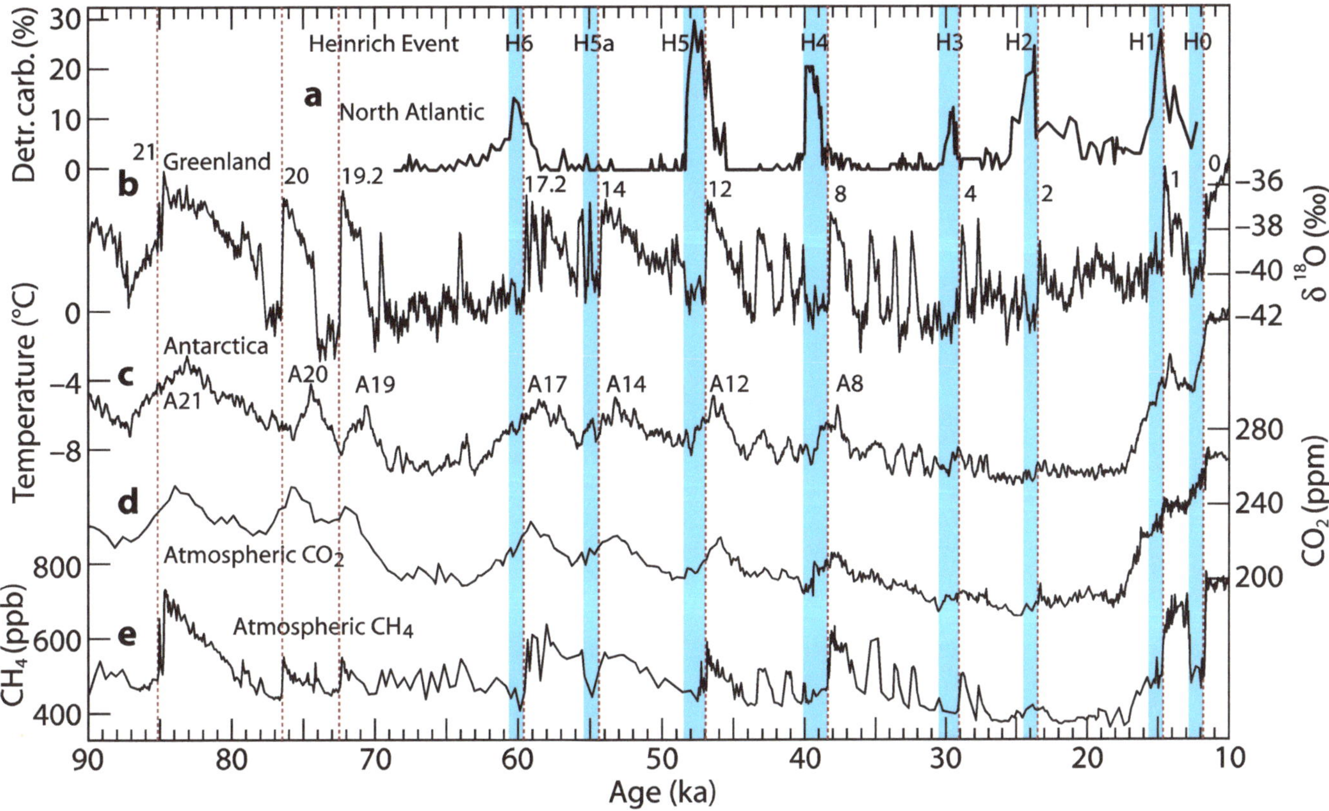

Fig. 3.3 Chronology of climatic events for the Last Glacial Period
a) Percentage of detrital carbonate in deep sea core DSDP 609 (after Bond et al. 1999), a proxy for iceberg activity in the North Atlantic. Heinrich events are defined on the basis of inferred iceberg activity. The amount of specific petrological tracers changes at different Heinrich events, reflecting differences in the origin of the icebergs. **b)** GISP2 Greenland $\delta^{18}O$ temperature proxy on GICC05 timescale (after Seierstad et al. 2014). Numbers denote D–O events. **c)** Antarctic EPICA Dome C temperature reconstruction from deuterium records (after Jouzel et al. 2007). Data averaged with a 200-yr running window. A8 to A21 are Antarctic warming events (Antarctic Isotope Maxima) numbered according to Greenland interstadials. **d)** Antarctic atmospheric CO_2 concentration from several ice cores (after Bereiter et al. 2015). **e)** Greenland CH_4 concentration from GISP2 ice core on GICC05 timescale (after Seierstad et al. 2014). Vertical blue bars, timing of Heinrich events (H3 to H6 from Ahn & Brook 2008; H0 to H2 from Hemming 2004). Vertical dotted lines, abrupt warming in Greenland after Heinrich events.

events (Fig. 3.4). Given that the rate and duration of the warming phase is very similar, despite very dissimilar previous conditions, it suggests that the source of heat probably has a constrained enthalpy capacity, and the amount of heat released is similar for all D–O events. The abrupt warming is followed by a period of cooling of c. 175 years when temperature drops by 2–4.5 °C. From this point D–O oscillations display great variability. Some D–O oscillations (GI 3, 4, 5.2, and 6) will loose all the temperature gained in just three centuries, showing cooling rates of − 0.3 °C/decade or even faster, and often reaching temperatures colder than prior to the D–O event (Fig. 3.4). The total span for these short D–O events is just 400–500 years. Other D–O oscillations will take 500 to 800 years to complete a more irregular descent, for a total span of 800–1000 years (GI 7, 10). Finally some D–O oscillations will take more time to go back to glacial stadial baseline than the length of a cycle oscillation (Fig. 3.4, GI 8). In such cases new D–O events are prevented from taking place until the cooling ends.

3.3 Dansgaard–Oeschger oscillations in the Antarctic record

When studying the D–O cycle in Antarctic records it became apparent that Greenland temperature changes matched methane level changes at a global scale (Fig. 3.3). Since methane levels were increasing simultaneously in both Greenland and Antarctic ice cores, this provides a precise way to align both records (WAIS Divide Project Members 2015). The alignment of Antarctic and Greenland records shows that there is an inverse temperature relation between both poles. During a D–O GS (cold phase) temperature rises in Antarctica. This rise is especially intense if the stadial is a HE. Temperature in Antarctica peaks c. 220 years after the D–O abrupt warming triggers in Greenland (WAIS Divide Project Members 2015; Fig. 3.5). This delay suggests an oceanic link between both poles is responsible. Afterwards temperature goes down simultaneously in both poles, but Antarctic temperature bottoms early and starts to raise again in preparation for the next oscillation.

How is this temperature connection achieved? The planet receives most of the energy from the Sun through the tropical areas where part of it is radiated back. The rest of the energy that has been converted to heat has to be

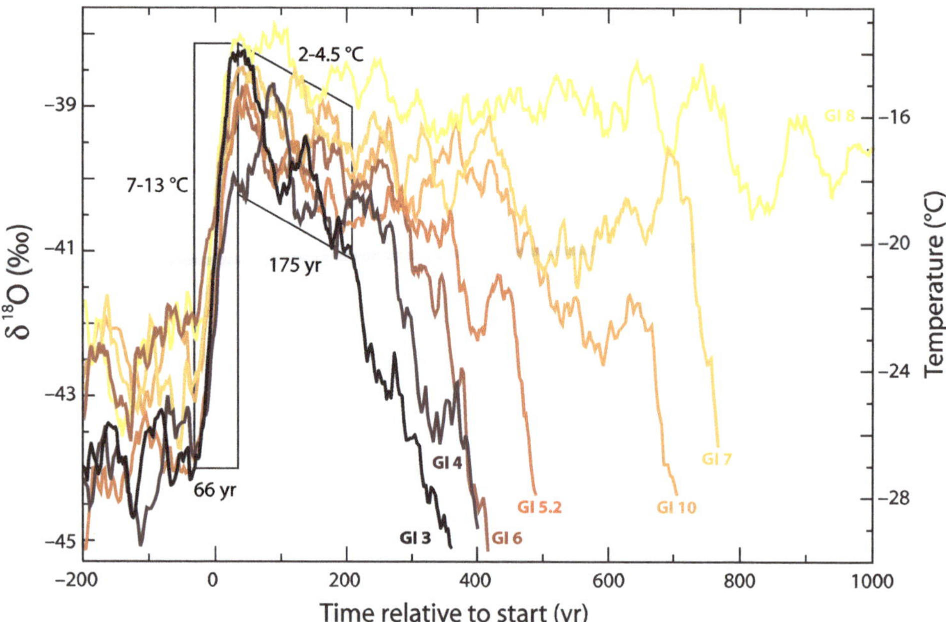

Fig. 3.4 Time evolution of recent D–O oscillations
NGRIP δ^{18}O temperature proxy (NGRIP members 2004) on GICC05 timescale, averaged with a 40-yr running window. D–O oscillations corresponding to Greenland interstadials (GI) 3 to 10, between 42–27 ka have been aligned according to their mid-warming time. GI 9 does not fit the D–O pattern and is not included. D–O oscillations show a very abrupt warming phase in a very narrow time-span of less than a century, followed by the loss of about one fourth of the temperature increase over the next two centuries. After that, D–O events stop behaving consistently and return to glacial conditions at different rates. After Ganopolski & Rahmstorf (2001).

directed towards the poles, where most of the surplus heat is radiated to space. This meridional heat transport is achieved in part through the atmosphere, that transports two thirds of the surplus heat towards higher latitudes, and in part through the oceans, where another third of the surplus heat is carried through the Meridional Overturning Circulation as warm subsurface currents, that return as cold deep currents once they ventilate the heat to the high latitudes atmosphere. What makes the meridional heat transport function is the temperature gradient between the tropics and the poles.

Pacific warm waters are mostly prevented from reaching the Arctic, so the only effective connection is through the North Atlantic between Greenland and Scotland. The closure of the Panamanian Pacific–Atlantic connection converted the North Atlantic Ocean into a cul-de-sac for meridional heat transport. During glacial periods the Bering Strait throughflow is stopped by ice and low sea levels. The South Atlantic is the only southern ocean that transports heat northward across the Equator. Therefore the Atlantic is an avenue for subsurface warm waters that originate in the Southern Ocean and go to the Arctic Ocean, returning as deep cold waters. This is the cause of the temperature connection between the poles. It has been proposed that when the Atlantic current is strong it cools the Antarctic and warms the Arctic by changing the energy budget in favor of the last, and when the Atlantic current is weak it warms the Antarctic and cools the Arctic through the opposite effect. This heat-piracy hypothesis is the basis of the bipolar see-saw model, that is supported by available evidence on changes of the Atlantic Meridional Overturning Circulation (AMOC; Stocker & Jonhsen 2003).

The explosive growth of methane during D–O oscillations raised fears that abrupt climatic changes had repeatedly triggered the hypothetical clathrate gun. Substantial methane deposits of several hundred Gt exist as solid hydrates (clathrates) at 300–500 m depths in the oceans, maintained as solid by low temperature and high pressure. A hypothesis was developed that a fast warming could trigger their release in the near future. However deuterium isotopic analysis of ice core methane showed that the increase in methane was accompanied by a depletion in deuterium (Bock et al. 2010; Fig. 3.6). This depletion indicates that its origin is in deuterium poor methane from boreal wetlands, one of the main natural sources of methane, and not from deuterium rich clathrate hydrates. The increase in temperature and precipitation associated with the D–O cycle (Fig. 3.2) would be responsible for boreal wetlands expansion and increased CH_4 emissions.

The decrease in methane deuterium starts during HE 4, several centuries before the abrupt D–O warming commences and CH_4 levels increase (Fig. 3.6; Bock et al. 2010). The authors explain it as the result of an AMOC intensification increasing precipitation at high latitudes

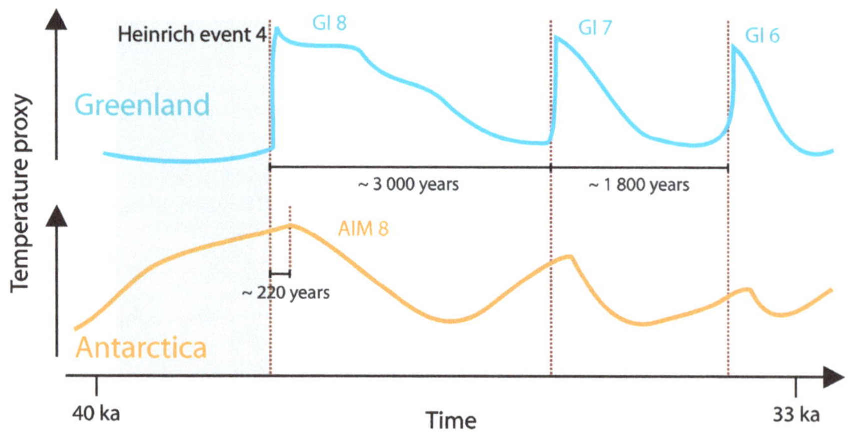

Fig. 3.5 Cartoon of the D–O interpolar phasing of temperatures
During a Heinrich event Greenland temperature (top) gets very cold, while it raises in Antarctica (bottom). Once the abrupt warming takes place in Greenland, temperature peaks in Antarctica on average 220 years later. If the previous Antarctic warming has been very intense, as in AIM 8 (Antarctic Isotope Maxima), the corresponding Arctic cooling, GI 8 (Greenland interstadial) can take much longer and then one or more cycle periods are skipped until temperature is cold enough. After WAIS Divide Project Members (2015) and van Ommen (2015).

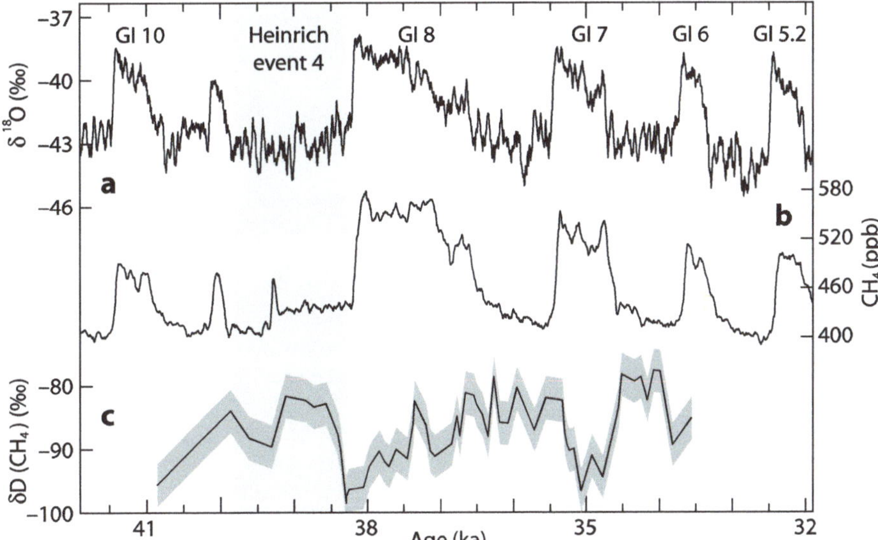

Fig. 3.6 Methane changes and origin during D–O events
a) NGRIP $\delta^{18}O$ Greenland temperature proxy (after NGRIP members 2004) on GICC05 timescale, averaged with a 40-yr running window. **b)** CH₄ record from West Antarctic Ice Sheet (WAIS) Divide ice core (after Rhodes et al. 2015), synchronized to GICC5 age model. **c)** NGRIP methane deuterium changes, $\delta D(CH_4)$, in per mil. Dark grey area represents 1σ uncertainty of measurement. The light grey bar indicates the Heinrich event 4. Changes in methane levels are inversely correlated to methane deuterium content, indicating a deuterium poor source, most likely boreal wetlands. After Bock et al. (2010).

(Bock et al. 2010). This is important because an AMOC intensification during HE is believed to be responsible for the extraordinary increase in North Atlantic iceberg activity when it causes widespread ice shelves collapse in the North Atlantic (Hulbe et al. 2004). D–O events not following an HE appear to be the response to similar but less intense bipolar see-saw changes, and thus a more intense AMOC could provide the energy required for D–O phenomena.

D–O events constitute the most clear example of abrupt climatic change other than asteroid impacts. They have taken placed repeatedly in the past at millennial frequency, changing the climate of an entire hemisphere and affecting the whole planet. They were capable of raising methane levels by 30% globally (Fig. 3.6), yet surprisingly they were neither cause nor consequence of significant CO_2 changes. Antarctic ice core records do not register any contribution or response from CO_2 to D–O events (Fig. 3.3d). CO_2 levels only increase during HEs. As we have seen HEs are associated with Antarctic warming while the North Atlantic region cools down and iceberg discharge is greatly enhanced. Since the Antarctic region is the only one warming during a HE, it is generally believed that the increase in CO_2 originated from enhanced CO_2 ventilation from a warming Southern Ocean (Ahn & Brook 2014).

Ahn and Brook (2014) investigated the relationship between CO_2 changes and temperature changes in Antarctica during the stadials that precede the abrupt D–O warming. They are stadials only in Greenland (GS), because as we have seen the cold period between D–O events is characterized in Antarctica by warming (Fig. 3.5). They found that GS display different CO_2 changes whether they coincide or not with a HE (Fig. 3.7; Ahn & Brook 2014). HE–GS display an increase in CO_2 while non-HE–GS do not show any increase at all in CO_2 levels (Fig. 3.7 box). This result suggests that it is the HE cooling the North Atlantic that is causing the Southern Ocean to warm and release CO_2, and not the Antarctic warming, that in all cases is unrelated to CO_2 levels.

We can conclude that according to available evidence, CO_2 plays no role at all in the most abrupt and frequent climate changes of which we have knowledge, the D–O cycle, and that the increase in CO_2 observed in Antarctica associated to HEs appears to be a consequence of Southern Ocean warming, and neither a cause or consequence of Antarctic warming. Furthermore, the increase in CO_2 during HEs (of about 10–15 ppm) does not appear to signifi-

Fig. 3.7 CO₂ and Antarctic temperature relationship during Greenland stadials
Relationship between changes in Antarctic temperature (expressed as changes in the difference in the ratio of deuterium to hydrogen respect to a standard), and changes in CO_2 (ppm) at the Siple Dome (Antarctica) during periods coincident with GS. Records are long-term detrended and three hundred-year running averaged (after Ahn & Brook 2014). GS are numbered according to the Greenland revised stratigraphical scheme (Rasmussen et al. 2014), and therefore are terminated by a D–O warming that initiates an interstadial numbered with the preceding number. Dashed lines, GS coinciding with Heinrich events. Continuous lines, GS not coinciding with a Heinrich event. Antarctic warming prior to D–O events takes place in the absence of CO_2 changes, except when associated to Heinrich events.

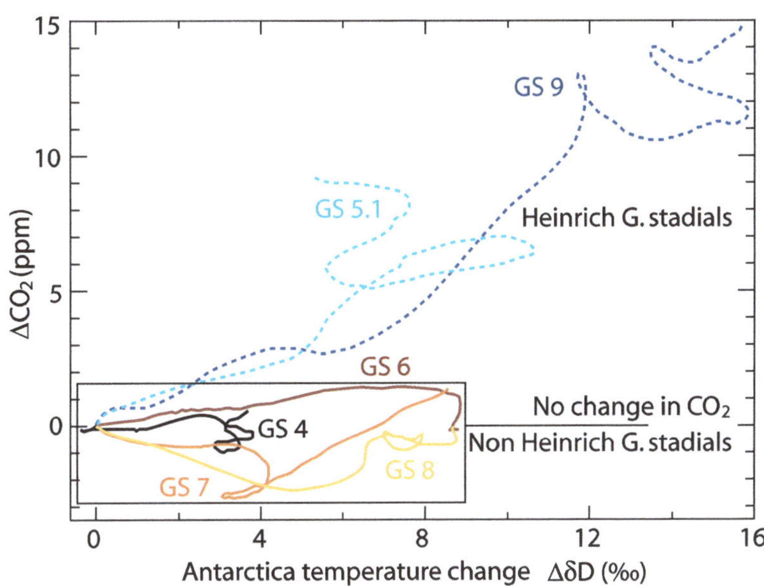

cantly alter the rate or magnitude of the warming during the subsequent D–O oscillation (see for example that GI 8 shows a similar warming to the rest, Figs. 3.4 & 3.6).

Now we can understand the difference between HE and non-HE D–O oscillations. HEs are likely caused by a reduction in inter-hemispheric heat transport by AMOC. As it happens at c. 6000 years intervals, we can speculate that it is due to an intensification of glacial conditions and/or Northern Hemisphere ice build up over time. As a consequence Antarctica warms and the Arctic and North Atlantic cool, clear evidence of a reduced AMOC. Warming at the Southern Ocean releases CO_2, while ice continues growing in the Northern Hemisphere. The increasing temperature difference between the Southern Ocean and the North Atlantic reactivates AMOC. This causes an increase in precipitation in boreal wetlands that release deuterium-poor methane, although methane levels are not changing yet. The strengthened AMOC should be responsible for the collapse of the extended northern ice shelves provoking an iceberg avalanche. Iceberg melting and meltwater production result in the formation of a fresh water lid that promotes North Atlantic and Nordic Seas water stratification. When the D–O warming takes place it puts a sudden end to the stadial. The increased warming and precipitation cause an increase in methane production from boreal wetlands, and globally methane levels increase by c. 30%. Interestingly, the warming recorded in Greenland is the same regardless of whether the stadial was under a HE or not (Fig. 3.6). This important fact suggests that D–O events do not depend on the intensity of AMOC, that should be higher at the end of HE due to a bigger temperature difference between the Southern Ocean and the North Atlantic, and places important restrictions on the cause of the D–O cycle. However the length of the GI after a D–O event does depend on the previous stadial situation. When the D–O abrupt warming occurs after an HE, the subsequent cooling usually takes much longer and normally other D–O events are prevented from taking place until the cooling is completed (Fig. 3.3b). It appears probable that the enhanced AMOC continues providing an increased amount of heat for 1.5 to 3 kyr, sustaining a slower cooling.

3.4 Does the Dansgaard–Oeschger cycle have a periodicity?

Some authors have disputed the existence of a D–O cycle on the basis that the oscillation distribution is not significantly different from random (Ditlevsen et al. 2007). There is obviously the difficulty in correctly dating with suffi-

cient precision oscillations that took place so long ago. The new Greenland GICC05 timescale strongly supports that if a D–O cycle exists, it is not regularly spaced, as it was once believed by some authors. Glacial climate records are very noisy because climate variability is enhanced during glacial periods (see Sect. 4.7), and mathematical analysis of the D–O frequency becomes more inconclusive due to it. D–O events need to be correctly identified, and for that they have to be correctly defined, as some researchers work with what might not be D–O events, but temperature spikes that could respond to different causes. A D–O oscillation requires several signature conditions. It is highly asymmetric with rapid warming in a few decades and slow cooling over at least 200 years followed by cooling over at least 200 more years for a minimum duration of 400 years (Fig. 3.4). It is matched by a similar peak of methane levels of similar duration (Fig. 3.6). And it is preceded by prior Antarctic warming that peaks about 220 years after the Greenland warming peak (Fig. 3.5). Temperature peaks that do not respond to this pattern, like GI 9, cannot be a priori identified as a D–O event, yet they do affect the frequency analysis. Nevertheless the identification of a 1470-yr periodicity in D–O events was always fraught with problems. It was shown that most of the power for the peak came from just 3 D–O events (GI 7, 6 & 5.2; Schultz 2002).

Does the new Greenland chronology mean that D–O events have a random distribution? Not necessarily. The new chronology provides an opportunity for discarding incorrect assumptions and improve our knowledge of the D–O cycle. Casual examination of the best dated glacial period in GISP2 on GICC05 timescale between 11.7–14.7 ka shows that between Greenland sudden warming at the start of the Holocene, and the sudden warming that ended the Oldest Dryas and started the warm Bølling–Allerød (BA) period, there was a time span of 3000 years. The same interval separates the start of GI 7 and GI 5.2 between 35.5 and 32.5 ka (Fig. 3.8a). These are some of the most recent and best dated D–O events, so it is interesting to examine if the 3 kyr interval is significant in D–O events frequency. Multitaper spectral analysis of GISP2 on GICC05 does indeed show a prominent peak at 3 kyr (Obrochta 2015; Fig. 3.8b). More information can be obtained if we focus our frequency distribution analysis strictly to the abrupt warming phase of *bona fide* D–O events, according to the criteria listed above. That way we eliminate the noise from other temperature changes. Table 3.1 lists the 19 D–O events in the last glacial period whose warming phase takes place at ≤ 5.5 kyr from another D–O event (GI 21 and above are excluded by this criterion). This list results in 24 time intervals of 5.5 kyr or less,

D–O	Date	D–O	Date	D–O	Date	D–O	Date
GI 0	11,700	GI 6	33,700	GI 12	46,850	GI 18	64,150
GI 1	14,700	GI 7	35,500	GI 13	49,300	GI 19.1	69,650
GI 3	27,800	GI 8	38,250	GI 14	54,250	GI 19.2	72,400
GI 4	28,900	GI 10	41,500	GI 15.2	55,800	GI 20	76,500
GI 5.2	32,500	GI 11	43,350	GI 17.1	59,100		

Table 3.1 Partial list of D–O events
19 Greenland Interstadials (GI) identified as D–O events and occurring at 5.5 kyr or less from another D–O event are used in the analysis, with the dates (BP) assigned to the fast-warming phase. They define the 24 intervals of ≤ 5.5 kyr shown in Fig. 3.8c as vertical bars.

Fig. 3.8 D–O events periodicity
a) GISP2 δ^{18}O Greenland temperature proxy on GICC05 timescale (after Seierstad et al. 2014), 60-yr averaged, for the periods 11.2–15.1 ka (top graph in a) and 32–35.9 ka (bottom graph in a). The common 3000-yr periodicity, and one instance of 1800-yr and 1200-yr spacing each, between D–O events are indicated by horizontal arrows. **b)** Multitaper spectral analysis of GISP2 data on original timescale (grey line), and GICC05 (black line) for 1–4 kyr periodicities (after Obrochta 2015). The strong 1470-yr peak (vertical grey line) in the original timescale is absent in GICC05. **c)** Distribution analysis of the interval between D–O fast-warming phases in the GISP2 record on GICC05 timescale. Over half of the intervals cluster at 3.0 ± 0.3 and 4.8 ± 0.3 kyr values.

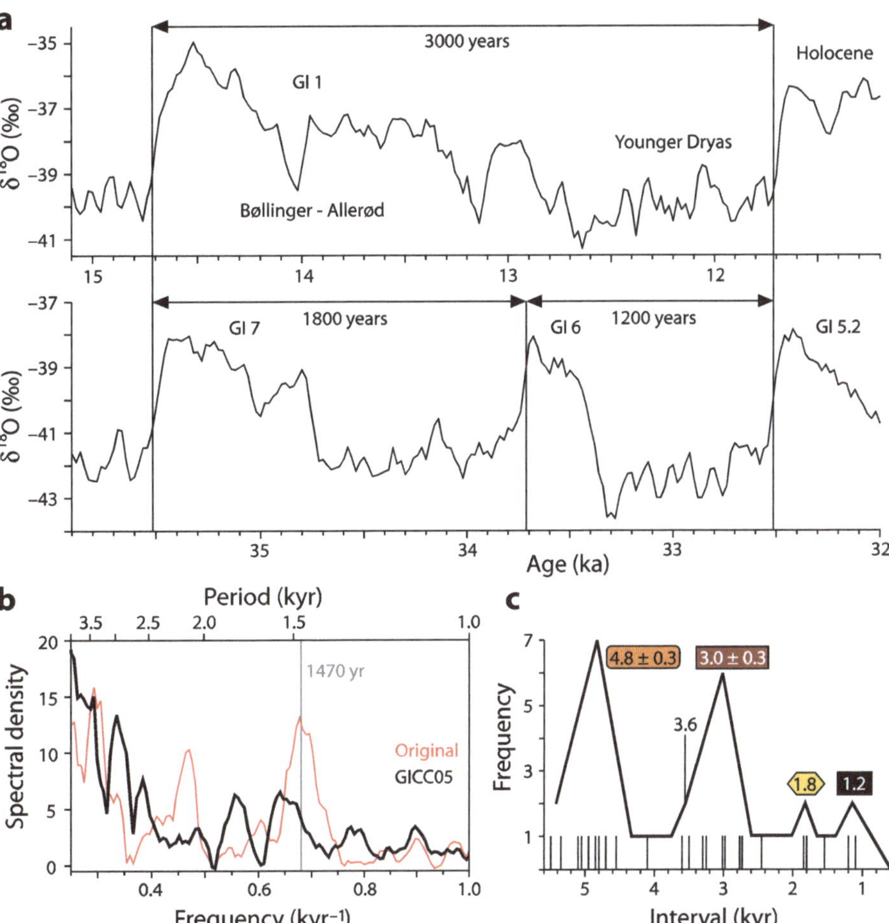

separating D–O events. When this sample is analyzed, over half of the intervals cluster at two lengths, 4.8 ± 0.3 kyr and 3.0 ± 0.3 kyr (Fig. 3.8c).

The occurrence of over half of the D–O events at preferred specific intervals could be due to chance, indicate the existence of intrinsic delays, or alternatively denote the presence of one or more external pacers that produce favorable conditions at different periodicities. When we analyze the temporal disposition of the observed intervals between D–O events, it becomes evident that it is too ordered to be due to chance (Fig. 3.9). So we can rule out that possibility, as highly unlikely. Furthermore, the best explanation for the observed intervals is that two different periodicities are taking place simultaneously in the D–O cycle. One periodicity with 4.8 ± 0.3-kyr period is represented by the sequence spanning GI 4-6-8-11, of c. 14,500 years in a set of three oscillations, and by the sequence spanning GI 13-14-17.1-18, of c. 14,800 years in another set of three oscillations (Fig. 3.9, orange rounded rectangles). The other periodicity with 3.0 ± 0.3-kyr period is represented by the sequence spanning GI 5.2-7-8-10, of c. 9,000 years in a set of three oscillations (Fig. 3.9, brown rectangles). Interestingly, the short periodicities of 1.2 and 1.8 kyr, are the residuals that fit both periodicities together (Fig. 3.9, yellow hexagons and black boxes). 1.8 is the difference between 4.8 and 3.0, and 1.2 is the difference between 3.0 and 1.8. 3.6 kyr, an interval that appears twice, is double the 1.8 number.

The surprising answer to the long-sought D–O cycle periodicity riddle is that it does not have one but two periodicities mixed together, a result difficult to extract from a noisy record through computer analysis. This extraordinary disposition of D–O events according to two slightly-irregular simultaneous periodicities is very difficult to explain in terms of intrinsic delays in meltwater events, salinity changes, or thermohaline circulation strength changes. The most parsimonious explanation is the existence of an external pacer that simultaneously cycles in two frequencies. The obvious relationship between both periodicities through the appearance of residuals that approach the difference between their periods (1.8 and 1.2 kyr) is a strong indication of a common cause for both periodicities. If we observe the pattern that appears from the analysis at Fig. 3.9, eleven of the nineteen D–O events are the consequence of belonging to two series of three 4.8 kyr units (GI 4, 6, 8, 11, 13, 14, 17.1, 18), and one series of three 3.0 kyr units (GI 5.2, 7, 8, 10). What these triplets have in common is that three 4.8 kyr units have the same duration as eight 1.8 kyr units (14.4 kyr), and three 3.0 kyr units have the same duration as five 1.8 kyr units (9 kyr). We must logically conclude that the two periodicities observed are a manifestation of a fundamental 1800-yr cycle. Such a fundamental cycle is already known to exist in nature and affect climate. It is the 1800-yr oceanic tidal cycle (Keeling & Whorf 2000).

The strongest possible tides on Earth take place when (1) there is a Sun–Earth–Moon syzygy (alignment), (2) when the Moon is closest to the Earth, at perigee, (3) when the Moon is located at one of the nodes of the Earth's ecliptic, and (4) when the Earth is closest to the Sun, at perihelion. These conditions are met simultaneously on average every 1800 years (1682, 1823, or 2045 years ± 18

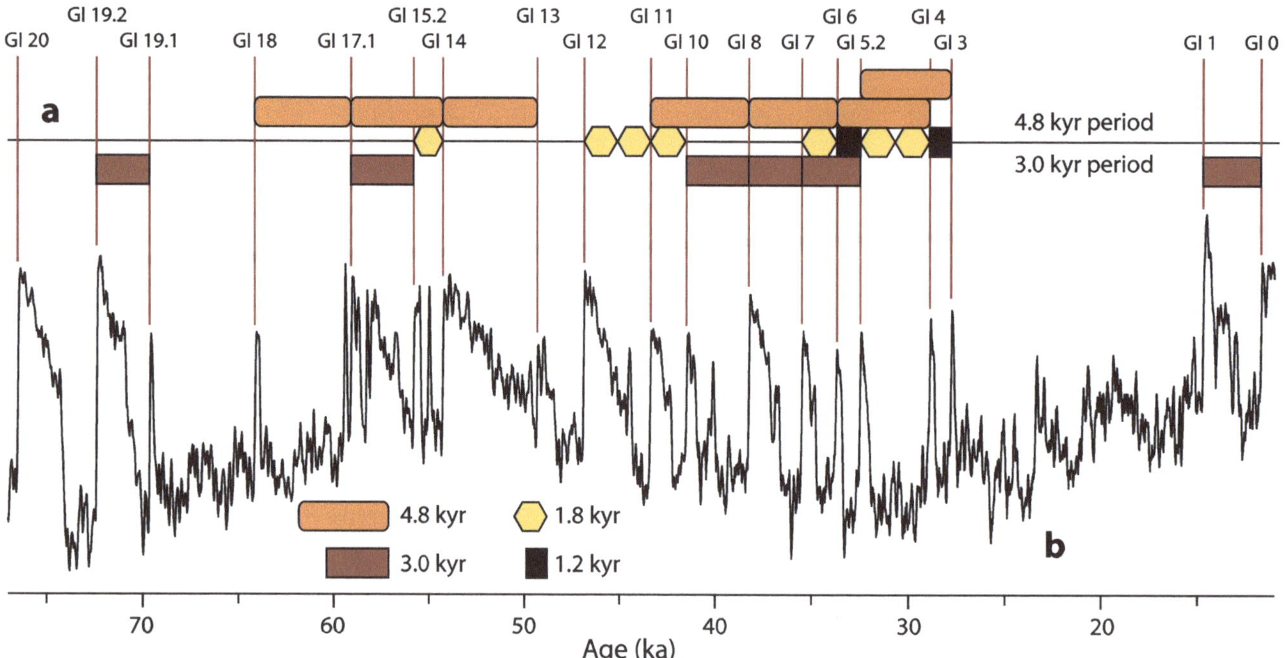

Fig. 3.9 The D–O cycle
a) Diagram showing the disposition of the 4.8-kyr intervals (orange rounded rectangles), 3-kyr intervals (brown rectangles), 1.8-kyr intervals (yellow hexagons), and 1.2-kyr intervals (black boxes) according to the 4.8 and 3.0-kyr observed periodicities. Greenland interstadials (GI) named according to Rasmussen et al. 2014. Note that all D–O events, except GI 20, participate in the periodicities. **b)** GISP2 $\delta^{18}O$ Greenland temperature proxy on GICC05 timescale (after Seierstad et al. 2014), 100-yr averaged. The 19 GIs showed have been selected because they fit the D–O signature (see main text). Other temperature spikes have not been considered even if they fit in the double D–O periodicity and might constitute unrecognized D–O events.

years). The 1800-yr lunar periodicity arises from small mismatches between syzygy and the lunar node that increase and decrease over time (Keeling & Whorf 2000). But the theoretical strength of the tides, measured as the angular velocity of the Moon respect to perigee, is not the same at every 1800-yr peak. The cycle is subjected to modulation by small differences (± 20 days) between syzygy and perihelion, resulting in a cycle modulation that currently has a periodicity of 4650 years, but could have been longer in the past due to secular changes in the periods of lunar months and anomalistic year produced among other things by changes in precession (Keeling & Whorf 2000). So there is a known explanation for the relationship between the 1.8 and 4.8-kyr periodicities in D–O events. At present we can only speculate on a lunar origin for the 3.0-kyr periodicity, but lunar cycles are intrinsically variable. The closeness in value between the rotation of the lunar nodes around the Earth, 18.61 years, and the precession of the lunar orbit, 18.03 years, produces a tidal cycle of 366 years (345, 363, and 405 years, plus occasionally other values) as the perigee and the lunar node take place closer in time or separate. This 366-yr cycle could produce a 3.0-kyr cycle at every eighth unit (and a 1.5-kyr cycle at every fourth). The 366-yr cycle shows modulation by the 1800-yr cycle, so it also becomes stronger at times when the 1800-yr cycle peaks. This could explain why the 3.0-kyr cycle synchronizes with both the 1.8-kyr and the 4.8-kyr cycles. Lunar cycles also account for a high degree of variability in the periodicity. Tidal intensity does not change smoothly along the 1800-yr cycle, but goes up and down continuously and it is almost as high at 1800 ± 90, ± 180, and ± 360 years, and so on, so a threshold value can be reached centuries before or after the cycle peaks, and thus the 10% variability observed (Fig. 3.8c) is easily ex-

plainable. Although far from conclusive, it is clear that nearly all the periodicities that the D–O cycle presents can be corresponded with calculated lunisolar tidal periodicities, centered on the 1800-yr cycle. Calculating a more precise correspondence is a complicated task, and has the problem that we don't know how lunar cycles have evolved over such a long time.

3.5 Conditions for the Dansgaard–Oeschger cycle

Since D–O oscillations, as previously defined, are a glacial feature, they appear to be influenced by global temperature and therefore by orbital changes. D–O events have a reduced frequency at warm times of maximal obliquity at 90, 50, and 10 kyr BP and at very cold times after minimal obliquity at 65 and 20 kyr BP (Fig. 3.10). As with the glacial cycle (see Chap. 2), the effect of obliquity lags the forcing by c. six millennia. So it appears that D–O abrupt changes cannot take place when the world is warm or very cold (deep glacial).

Schulz et al. (1999) investigated the irregular distribution of D–O oscillations during the past 100 kyr, and noticed a strong relationship between the D–O signal and variations in continental ice mass, as recorded in sea-level variations. Times of reduced D–O frequency coincide with inflection points in sea level variation, both at maxima and minima. And D–O events tend not to happen when sea levels are above −35 m or below −100 m from present level (Fig. 3.10d, e). Optimal conditions for D–O events must have been those between 60–49 ka and between 43–27 ka, i.e. sea levels between −40 and −100 m, and lagged-obliquity outside maxima or minima. It is suggested that

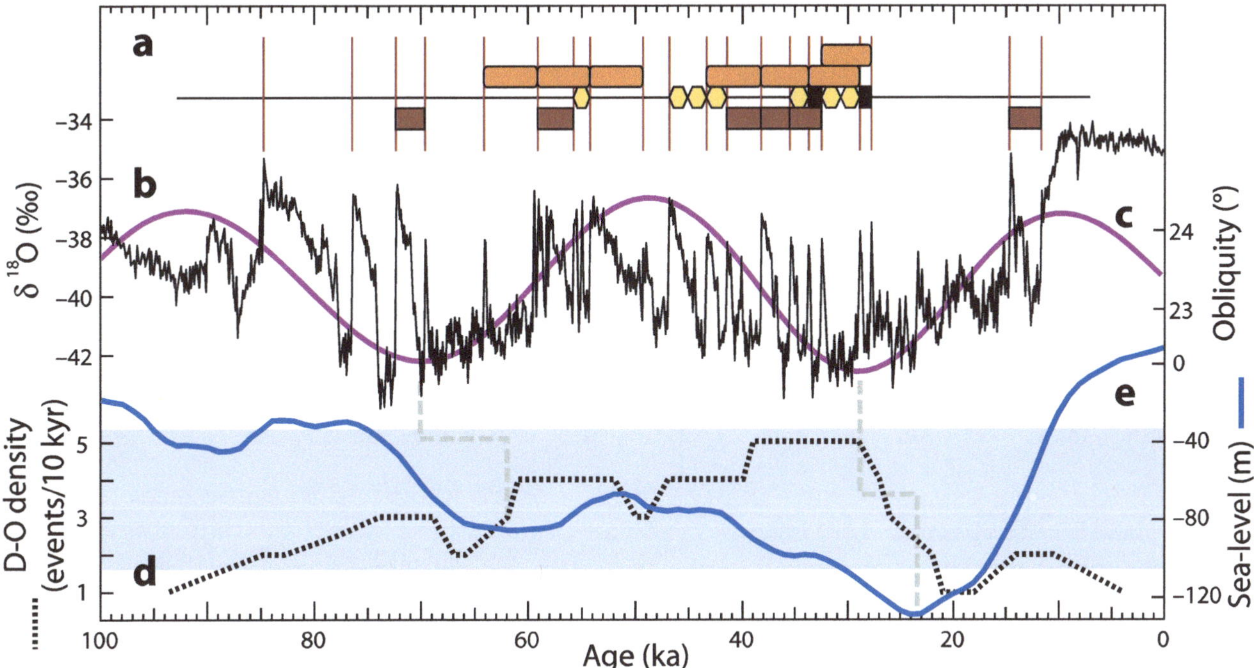

Fig. 3.10 D–O oscillations and changes in sea levels
a) Diagram showing the disposition of D–O events according to the observed periodicities: 4.8-kyr intervals (rounded rectangles), 3-kyr intervals (rectangles), 1.8-kyr intervals (hexagons), and 1.2-kyr intervals (black boxes). **b)** GISP2 $\delta^{18}O$ Greenland temperature proxy on GICC05 timescale (thin line, left scale; after Seierstad et al. 2014), 100-yr averaged. **c)** Earth's axial obliquity (thick purple line, right scale; after Laskar et al. 2004). **d)** D–O event density. Smoothed number of identified D–O events per 10 kyr (thick dotted line, left scale). **e)** 3000-yr averaged sea level in meters from present level (thick blue line, right scale; after Spratt & Lisiecki 2016). Dashed grey line indicates the lag in sea levels respect obliquity. Blue shaded area indicates sea level range when D–O events happen at regular intervals.

failure to meet these optimal conditions might be the reason why the D–O cycle periodicity is interrupted between 49–43 ka (Fig. 3.10a).

The right conditions for a D–O oscillation require the build up of extensive ice sheets over the northern continents that cause a drop in sea levels of at least 35 m. Once that happens, the bipolar see-saw must be set to warm Antarctica and cool the North polar regions. These conditions will extend the sea-ice cover over ample regions of the Nordic seas and North Atlantic and produce an increase in iceberg discharge. Then sufficient heat has to be accumulated to be suddenly released when the next tick of the pacer triggers a D–O cycle. Whenever those conditions are set a new D–O oscillation might be triggered. Warming from obliquity maxima will prevent the conditions from taking place, as will a profound cooling that reduces sea levels below –100 m, producing too much ice.

3.6 The Bølling–Allerød and Younger Dryas as part of the Dansgaard–Oeschger cycle

The abrupt climate change events that took place in the North Atlantic region at c. 14,700 and 11,700 years, initiating the BA GI and the Holocene respectively, have been considered D–O events in the previous analysis (Fig. 3.8 and table 3.1) because they not only display the correct temperature profile, magnitude, and GHG signature, but are also separated by the favored spacing of 3,000 years. However, the YD has been frequently considered a unique event, rather than part of a series of similar events. A glacial meltwater flood event into the North Atlantic (Broecker et al. 1989), or the Arctic (Murton et al. 2010), a

volcanic eruption (the Laacher–See eruption; Baldini et al. 2018), or a cosmic impact (Wolbach et al. 2018) have been proposed as specific YD causes. Less often the BA and YD are considered the last in the D–O series of events (Broecker et al. 2010; Rial 2012). Nye & Condron (2021) use a principal component outlier detection scheme and, after taking into consideration the warming at glacial termination, conclude that the BA/YD is not statistically different from the other D–O events in the Greenland record. If the BA and YD are part of the D–O series of events, as the evidence supports, then one-time events, like a cosmic impact or a volcanic eruption, become an unlikely explanation for the YD. What makes the BA/YD transition different from other D–O stadial transitions is that it takes place during a glacial termination and it is thus seen by many researchers as a reversal in the deglaciation process. This is not supported by continuing sea level rise through the YD, and by proxies from other parts of the world and other latitudes that show uninterrupted warming (Shakun et al. 2012). This uniqueness is also disputed by the existence of a possible analogous event during Glacial Termination III at 247 ka (Broecker et al. 2010).

The YD period can be considered a HE (H0; Fig. 3.3). It was characterized by cooling in the Northern Hemisphere. Benthic cores show the chronologically coherent deposition of an iceberg-rafted Heinrich layer in the North Atlantic, rich in detrital carbonate from the Hudson area (Andrews et al. 1995), basaltic glass, and hematite grains (Bond & Lotti 1995), like during other HEs. And it was abruptly terminated by the sudden warming accompanied by methane increase characteristic of D–O events, like all other HEs. Other factors might have participated in the YD cooling, but it appears probable that it took place an intensification of the bipolar see-saw with Antarctic/Southern

Ocean warming and Arctic/North Atlantic cooling due to an AMOC weakening, followed by AMOC intensification when the latitudinal temperature gradient became too steep, leading to an intense ice-shelf extension/collapse cycle, very much as in other HEs (Hulbe et al. 2004). The abrupt warming that terminated the YD in Greenland ice cores was of a magnitude comparable to other D–O events. The GS 1–GI 0 (Holocene) transition is indistinguishable from the GS 13–GI 12 transition at 46.8 ka by several criteria (Rasmussen et al. 2014), so we have dated the start of the Holocene according to the D–O chronology. The main difference is that after the abrupt warming of D–O 0 the subsequent cooling to return to glacial conditions did not take place, and interstadial temperatures continued increasing for c. 4000 years as the ice sheets melted.

3.7 Consensus Dansgaard–Oeschger cycle theory and challenges

The leading D–O cycle hypothesis was established by Wallace Broecker et al. (1990), and is defended by one of the foremost experts on abrupt climatic changes, Richard Alley (2007). It is known as the "Salt Oscillator" hypothesis, and it is based on oscillatory changes of the AMOC, in response to fresh-water pulses due to ice melting water (meltwater pulses, MWPs) being stored and released periodically from ice-sheets (Fig. 3.11).

According to this hypothesis, the AMOC is controlled by warm sea surface waters of its North Atlantic Current component (NAC) becoming saltier through evaporation that takes fresh water out of the Atlantic basin, and further becoming saltier and colder through evaporation in subarctic regions until they become dense enough to sink and then turn South to constitute the cold North Atlantic Deep

Water (NADW) component. In this hypothesis, the intensity of the NADW determines the state of the AMOC.

The NADW flows southward along the bottom of the Atlantic Ocean exporting with it the excess salt, resulting in a gradual reduction in North Atlantic surface salinity over time. Furthermore, tropical heat transferred to the high-latitude North Atlantic produces ice melting and MWPs that further reduce water salinity. The theory states that if surface waters at the sites of deep-water formation become too fresh, then AMOC weakens or shuts off because the surface waters are not dense enough to sink. Once AMOC weakens enough or even shuts down, salt starts to accumulate again in the North Atlantic due to the absence of NADW export. According to this theory, weak AMOC conditions are associated with cold stadials (Fig. 3.12a).

As salt continues to accumulate in the North Atlantic during periods of reduced NADW formation, eventually surface waters at key sites of deep-water formation would become salty and dense enough again to sink, thus restarting AMOC and causing an abrupt warming in the high-latitude North Atlantic, triggering the warm phase of a D–O cycle.

Several studies have suggested that it only takes a small reduction in sea surface salinity to alter the rate of NADW formation, to the point that some scientists, including late Wallace Broecker and Richard Alley became worried that an increase in the hydrological cycle due to current global warming could reduce North Atlantic salinity, leading to the shut down of the AMOC causing an abrupt cooling in the near future (Broecker 1999; Alley 2007). They do not take into consideration that before the Mid-Holocene transition, around 5000 yr BP, the northern Atlantic region was warmer and generally wetter than present, when the Sahara was a savanna type of environment,

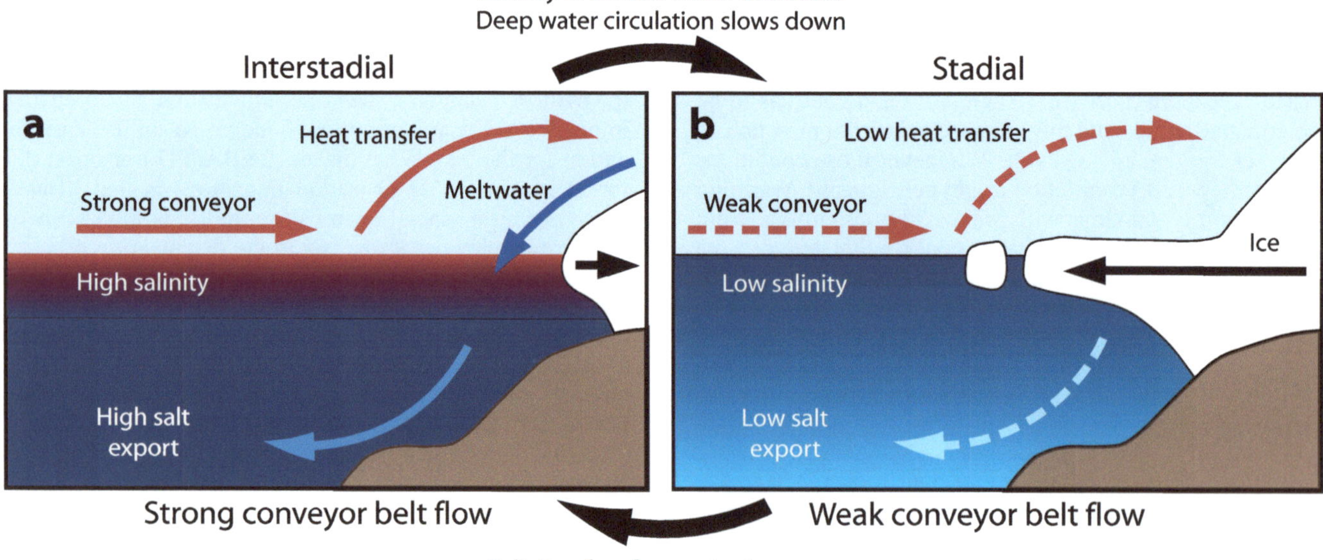

Fig. 3.11 The salt oscillator hypothesis
a) During warm interstadials, a strong AMOC transports heat northward causing the ice sheets around the North Atlantic to melt, gradually reducing surface water salinity until it no longer sinks and deep water formation ceases to form, stopping NADW. Eventually, surface salinity is reduced enough to weaken AMOC, shifting the climate into a cold stadial. **b)** During stadials, cooler conditions in the North Atlantic reduce meltwater input from the ice sheets, allowing an increase in surface salinity that eventually causes water to sink restarting NADW and causing AMOC to strengthen, returning the climate system to an interstadial. After Ruddiman (2001)

Fig. 3.12 Mechanism of the salt oscillator hypothesis
a) During warm interstadials, when AMOC is stronger, enhanced northward oceanic heat transport results in warmer conditions in the North Atlantic, causing ice sheets around the North Atlantic to produce meltwater, gradually reducing surface water salinity. Eventually, surface salinity is reduced enough to weaken AMOC, shifting the climate into a cold stadial. **b)** During stadials, cooler conditions in the North Atlantic reduce meltwater input from the ice sheets, allowing an increase in surface salinity that eventually causes AMOC to strengthen, returning the climate system to an interstadial. After Schmidt & Hertzberg (2011). **c)** Atlantic profile from 5°S to 80°N showing the underwater crest between Greenland and Scotland. Interstadial conditions (left) show a strong AMOC capable of crossing the crest. Stadial conditions show a weakened AMOC that turns further South (center). During Heinrich events the AMOC collapses (right; after Rahmstorf 2002). AMOC: Atlantic Meridional Overturning Circulation. NAC: North Atlantic Current. NADW: North Atlantic Deep Water. AABW: Antarctic Bottom Water.

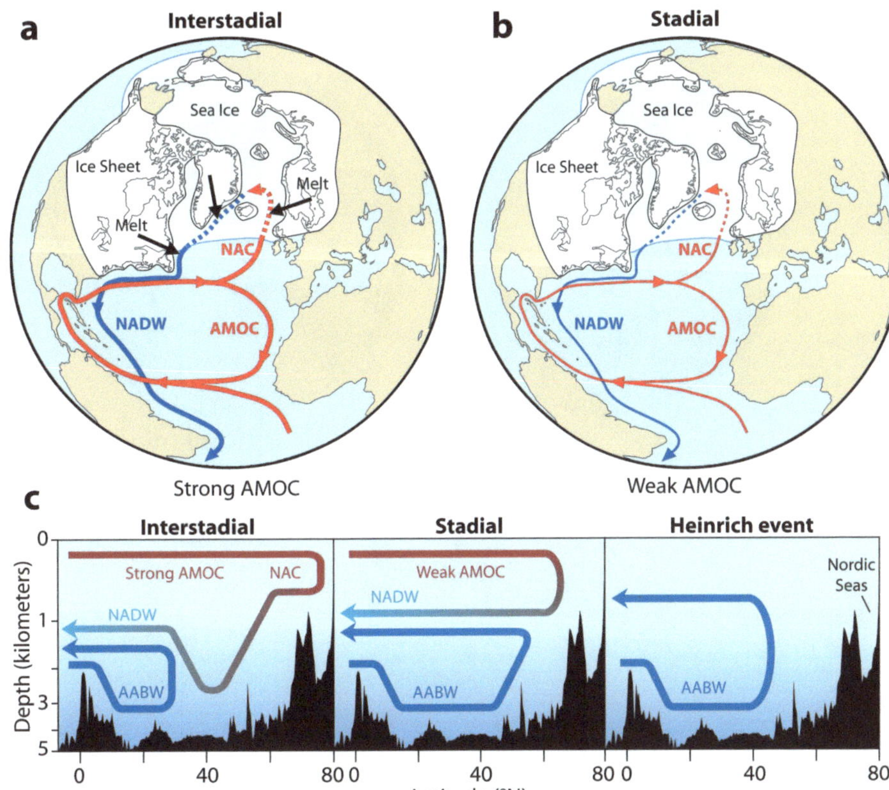

with the Intertropical Convergence Zone (ITCZ), a belt of daily rains reaching 30°N, and the AMOC did not shut down.

Evidence does not support that the cooling at the beginning of stadials is caused by meltwater and iceberg increase. Studies using the increase in abundance of the cold-loving left-coiling form of the foraminifer *N. pachyderma (s)* to track the southward displacement of the Arctic Front show that the North Atlantic cooling at the start of stadials preceded by several centuries to 2 kyr the increase in benthic rafted detritus associated to enhanced iceberg activity (Barker et al. 2015). It has been found that as the atmosphere cools down and seasonal sea-ice increases at the start of an stadial, North Atlantic surface and intermediate waters start warming (Rasmussen et al. 2016), and it is suggested that the ice increase coupled with water warming is responsible for increased iceberg activity and IRD deposition during HEs and GSs. The salt oscillator hypothesis has no particular explanation for the periodicities in the D–O cycle. The pacing must come from the intrinsic delays in the salinity and meltwater build up and depletion, and the oceanic currents response delay for the cycle to proceed. As simply put, the pacing of a pendulum depends on its length, but climate variability is far from the regularity of simple physics.

In recent years this consensus view of D–O cycle formation through salt-oscillation has come under assault on several fronts. While several studies have questioned the occurrence of MWPs at the expected time intervals, others indicate that AMOC is a lot more stable than required by the theory and even extreme MWPs would not be able to destabilize it persistently. The work of Rasmussen & Thomsen (2004), also confirmed by Dokken et al. (2013), and Ezat et al. (2014), and theoretically framed by Petersen et al. (2013), shows that during stadials the flux of warm water to the North Atlantic and Norwegian Sea does not cease. Instead during cold stadials warm waters enter the Arctic under the sea-ice at a subsurface level and thus instead of ceding the heat to the atmosphere they warm the subsurface waters below a double insulating cold water layer made of fresh superficial water and a cold and saline halocline. Thus while the atmosphere gets colder and sea-ice increases, the ocean heat accumulates at the subsurface level. The heat stored in this way must be limited by the Nordic Seas basin geometry. After the amount of warm water accumulated at the subsurface level of the basin reaches a certain point, since it is prevented from going up by the double insulating layer, and since it will not sink past certain point due to the presence of colder denser bottom water, additional warm water is likely to displace water already in the basin without increasing the heat stored in this way. This would answer one of the outstanding questions in D–O events. Regardless of the time passed since the previous event, they all seem to release a similar amount of heat, suggesting a physical constraint.

3.8 Mechanistic explanation of the Dansgaard–Oeschger cycle

After the "Ice Ages Mystery" was solved (Imbrie & Imbrie 1979), the discovery of *"the most dramatic, frequent, and wide-reaching abrupt climate changes in the geologic record"* (Dokken et al. 2013) became the most tantalizing, biggest mystery in paleoclimatology. The leading hypothesis of the salt-oscillator by Broecker present in textbooks (Fig. 3.11; Ruddiman 2001), is based on the surprising idea that deepwater formation drives the strength of surface currents, when it is known since antiquity that surface

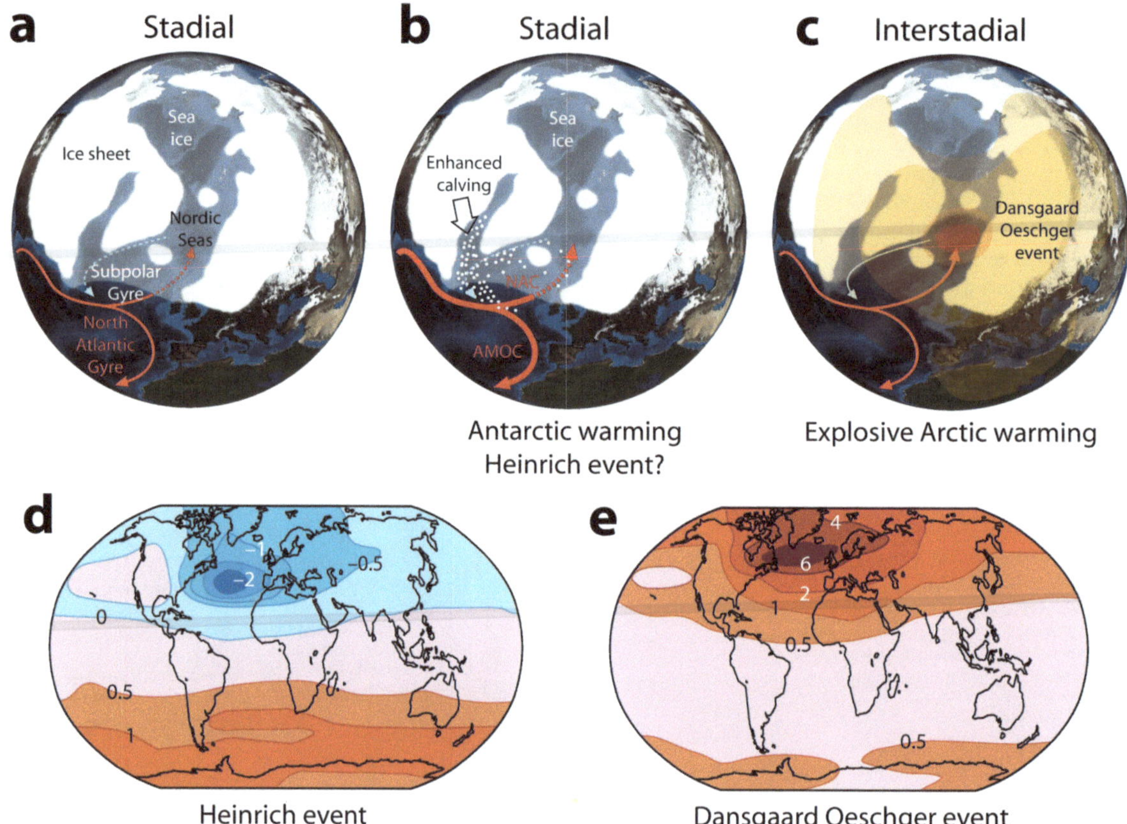

Fig. 3.13 Mechanism of the D–O cycle
a) At the beginning of a stadial the Arctic is very cold, sea-ice has grown to maximum extent, and AMOC strength (red arrow) is reduced. The North Atlantic Current (NAC) enters the Nordic Seas below the sea-ice (dotted red arrow), returning as cold water (dotted blue arrow). **b)** At the end of the stadial while Greenland is still cold, the North Atlantic Ocean has warmed due to the increase in the warm North Atlantic Current, producing an increase in iceberg discharge (white dots) carrying ice-rafted debris (IRD). At the Nordic Seas, warm water sinks below the ice preventing surface warming despite the increased heat transport. At certain times these conditions are enhanced producing a Heinrich event. **c)** An interstadial is abruptly produced when in an explosive manner the accumulated subsurface warm water rises and melts the sea-ice, transferring heat to the atmosphere. **d)** Modeled conditions during a Heinrich event, that supposes an increased cooling in the North Atlantic. **e)** Modeled conditions during a D–O abrupt warming. After Ganopolski & Rahmstorf (2001).

currents are driven by lower troposphere winds (Wunsch 2010).

We can start the cycle at the beginning of the stadial, that appears to be the point when North Atlantic sea surface is colder (Fig. 3.13a; Rasmussen et al. 2016). While northern sea-ice expands, the warming of Antarctica and the Southern Ocean start activating the bipolar see-saw, causing the AMOC to bring increasing amounts of warm water that initiate the warming of North Atlantic surface and intermediate waters. Water warming activates iceberg calving, increasing IRD deposition (Fig. 3.13b). The oceanic warming during the stadial contrasts with cold atmospheric conditions in Greenland, indicating limited moisture exchange. A possible explanation is provided by records that indicate generalized intermediate water warming through the North Atlantic and Nordic Seas, but restricted surface and subsurface waters warming (Rasmussen et al. 2016). Open marine waters with low iceberg contribution display gradual warming during stadials, while the Nordic Seas, the Greenland–Scotland ridge, coastal areas, and the IRD belt, maintain a quasi-stable stratification with a mass of warm subsurface water overlaid by a thin cold insulating double layer made by cold fresh water from melting and a cold halocline (Dokken et al. 2013; Rasmussen et al. 2016). This quasi-stable stratifi-

cation prevents the warming of higher latitudes because instead of venting the heat to the atmosphere it submerges below the sea-ice where it gets layered and insulated by the halocline (Fig. 3.13b). During Heinrich stadials, unusually large ice shelves prolong stadial conditions, by providing a bigger source of icebergs and meltwater to maintain the quasi-stable stratification, and their break up provides the huge amounts of icebergs that contribute to the IRD-heavy layers identified by Hartmut Heinrich (Fig. 3.13d; Heinrich 1988). Ice-shelf dynamics have been implicated in several hypotheses explaining HEs (Hemming 2004). Towards the end of a GS the decrease in IRD indicates a reduced meltwater input leading to increasing instability of the quasi-stable stratification maintained at coastal and shallow areas like the Greenland–Scotland ridge and the Nordic Seas.

After several centuries to a few millennia, the quasi-stable stratification, weakened by reduced meltwater input, breaks down, and the subsurface warm waters at high northern latitudes raise to the surface and abruptly warm the atmosphere (Fig. 3.13c, e), starting a GI. This warming alters the latitudinal temperature gradient and inverts the bipolar see-saw, so the Antarctic region starts to cool after about 200 years. As warm waters cool down they sink and form the NADW, so the higher northern latitudes also start

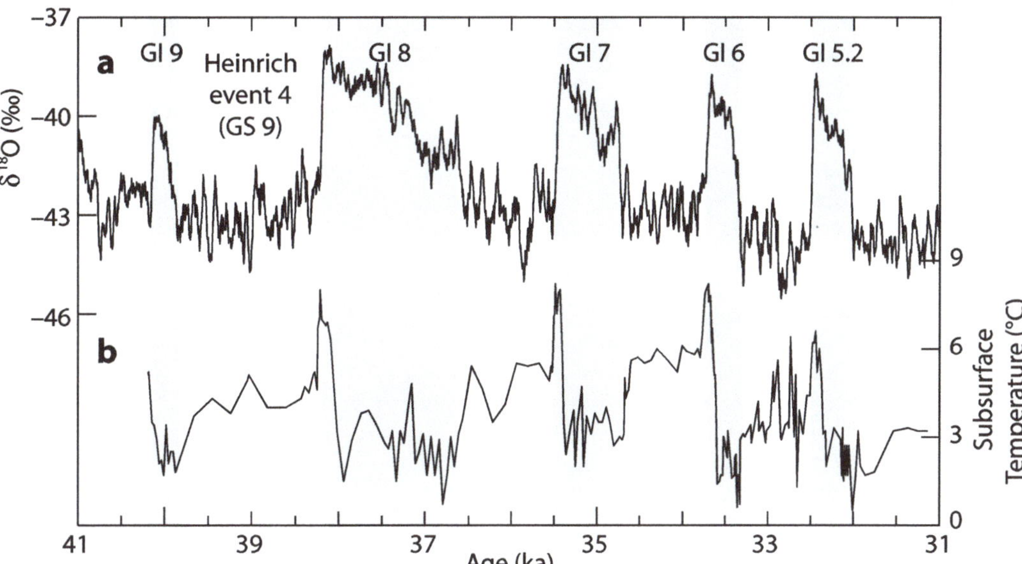

Fig. 3.14 Subsurface temperature abrupt changes in the Norwegian Sea
a) NGRIP δ^{18}O proxy for Greenland temperature on the GICC05 timescale (after Seierstad et al. 2014), for the period 41 to 31 ka. Greenland interstadials (GI, shaded) corresponding to D–O events and Greenland stadial 9 (GS-9) corresponding to Heinrich event 4, are indicated. **b)** Benthic core MD992284 Norwegian Sea sub-surface temperature reconstruction below the halocline based on planktonic foraminifera assemblages. After Dokken et al. (2013).

to cool. Once sea-ice re-growths and the halocline forms, it insulates again the warm waters from the atmosphere and the temperature drops putting an end to the interstadial. The deep cooling of the new stadial flips again the bipolar see-saw restarting the cycle. The length of a GI is determined by the previous warming in Antarctica. After Antarctic Isotope Maxima (AIM, Fig. 3.3c), when the amount of warming in Antarctica is substantial, D–O related GIs last much longer, indicating that the heat to keep them going is provided by the bipolar see-saw from the Southern Ocean.

There is evidence from Norwegian Sea sediments that have preserved the temperature stratification of the sea that changes in sea temperature and stratification precede the abrupt atmospheric changes (Dokken et al. 2013). During the stadial phase, the planktonic foraminifera are mainly recording the temperature of cold water within, or just below, the halocline. As the stadial phase progresses, the planktonic foraminifera show an increase in temperature (Fig. 3.14b) consistent with the continuous arrival of relatively warm and salty Atlantic water below the halocline. With no possibility of venting heat to the atmosphere due to the sea-ice cover, the decrease of subsurface waters density weakens the stratification that allows the halocline and sea-ice cover to exist. The thermodynamics of the system favors an abrupt transition if the stratification is perturbed beyond a certain tipping point. The instability of the reverse-temperature stratification increases as the meltwater input decreases towards the end of the stadial. The transition to a warm GI occurs when the stratification collapses, at which point heat from the subsurface layer is rapidly mixed up to the surface, melting the sea-ice cover (Fig. 3.15). This sudden mix-up is seen in the planktonic foraminifera proxy record as an abrupt sea temperature warming that precedes the atmospheric warming (Fig. 3.14b).

Sea-ice biomarker abundance at a high resolution Nordic Seas core supports that the stratification collapse is preceded by several centuries of decrease in sea-ice cover (Sadatzki et al. 2019). The D–O abrupt warming appears to be primed by a reduction in the ratio of cold fresh meltwater input to warm saline Atlantic Current advection that would increase the instability of the reverse vertical

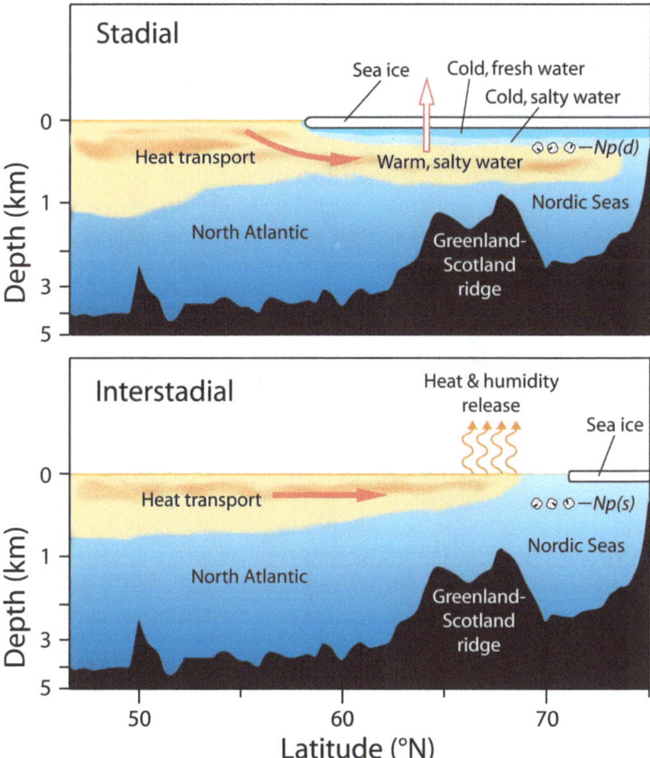

Fig. 3.15 North Atlantic–Nordic Seas vertical reorganization model
Schematic representation showing conditions in the North Atlantic and Nordic Seas during typical (top) cold stadial periods and (bottom) warm interstadial periods of the D–O cycle. During stadial periods warm northward Atlantic water is stratified below insulating layers of sea-ice, cold fresh water and cold saltier water (top). Despite atmosphere and surface cold temperature, foraminifera reflect subsurface warming, as indicated by the presence of the warm-loving dextral (right-coiling) *Neogloboquadrina pachyderma* (*Np(d)*). At interstadial periods the stratification collapses and the warm Atlantic water reaches the surface after melting the ice, warming the atmosphere. Foraminifera reflect an abrupt subsurface cooling as the assemblage becomes dominated by the cold-loving sinistral (left-coiling) *N. pachyderma* (*Np(s)*). Abrupt subsurface cooling in the Nordic Seas coincides with atmospheric warming in Greenland. After Rasmussen & Thomsen (2004), and Dokken et al. (2013).

stratification, manifested as a reduction in the sea-ice required for its maintenance. The growing instability of the marine stratification towards the end of the stadial is manifested by growing temperature fluctuations that act as early warning signals (Rypdal 2016). Only in this unstable state reached towards the end of the GS would the North Atlantic–Nordic Seas system respond to the next triggering of the D–O pacemaker. Interestingly, Sadatzki et al. (2019) detect an abrupt increase in Nordic Seas sea-ice at 1–4 centuries into the GI that coincides with the time when the D–O events end their common slow cooling phase and enter their highly variable terminal cooling phase (Fig. 3.4). This sea-ice spike suggests both cooling phases are driven by heat of different origin, the first one probably from decreasing accumulated subsurface heat that once depleted causes the sea-ice spike, and the second one from heat provided by AMOC that is dependent on the magnitude of the prior Antarctic warming.

Much less attention has been dedicated to the study of the GI/GS transitions, with the notable exception of the BA/YD transition at c. 12,900 BP. Unlike the very homogeneous GS/GI transitions (Fig. 3.4), GI/GS transitions are highly heterogeneous. Some of the transitions are very abrupt (GI/GS 4, 6, 7), spanning only a few decades, while some are very gradual, extending over several centuries (GI/GS 8, 10, 11, 12). The BA/YD transition is intermediate lasting around a century (Rasmussen et al. 2014). This heterogeneity added to the intrinsic instability of GIs, that require a continuous supply of exogenous heat to be maintained during a glacial period, question the need for an event to explain the GI/GS transitions versus a deterioration of the conditions that make a GI possible until a tipping point is reached. HEs, that are characterized by a much bigger Antarctic warming, are followed by longer GIs, indicating that increased heat supply to the North Atlantic by the bipolar see-saw stabilizes GIs. This is surprising because the leading hypothesis for GI/GS transitions is that an increase in meltwater input to the Arctic/North Atlantic reduces deep-water formation leading to a slowing of the AMOC and a reduction in the heat supply, yet meltwater production is the result of GI conditions. Stadials are destabilized by insufficient fresh meltwater production and stabilized by very cold conditions (AMOC strongly reduced during HEs), while interstadials are destabilized by increased meltwater production and stabilized by warm conditions (enhanced AMOC after HEs). The balance between meltwater production and AMOC strength appears to determine the stability of stadials and interstadials producing the conditions for an abrupt GS/GI transition and for a GI/GS transition that can be gradual or abrupt if a catastrophic flood event at the right time and place precipitates the transition. Under this paradigm freshwater-sensitive sea-ice dynamics capable of strong feedback responses provide the climate forcing that drives the transitions (Petersen et al. 2013).

3.9 Tidal cycles as an explanation for Dansgaard–Oeschger triggering mechanism

D–O events are paced at 1 to 3 kyr intervals at the period when they show a highest frequency (44–32 ka, Fig. 3.9). Proposed explanations for their triggering fall into two classes: internal variability due to oscillations in ocean circulation or ice-shelf dynamics, and external forcings, like variations in the Sun. But each explanation has shortcomings. Internal variability hypotheses have a problem explaining how the precise double periodicity observed (3.0 and 4.8 kyr; Fig. 3.9) can be achieved given the great intrinsic variability of the involved phenomena and given the variability in the duration of GIs and GSs. Solar cycles of the required periodicities are unknown. More promising as a pacing candidate is the c. 1800-yr lunar cycle. As shown above (Figs. 3.8 & 3.9), 1.8 kyr is the common relationship between the observed triple repetitions of 4.8 and 3.0 periodicities, the difference between 4.8 and 3.0, the difference between 3.0 and 1.2, and half the twice observed 3.6 spacing, besides appearing also twice as a spacing between D–O events.

There are very few scientists defending a tidal origin to the pacing of D–O cycles, and curiously, Charles Keeling, the father of global CO_2 measurements since 1956 at Mauna Loa, dedicated his later years to find a tidal origin to temperature changes (Keeling & Whorf 1997; 2000). Yet as outlandish as the tidal hypothesis sounds initially, it is uniquely capable of explaining some mechanistic features of the D–O cycle supported by evidence. As with any suspect, we have to analyze if it has the means and the opportunity. Are tides capable of producing the required

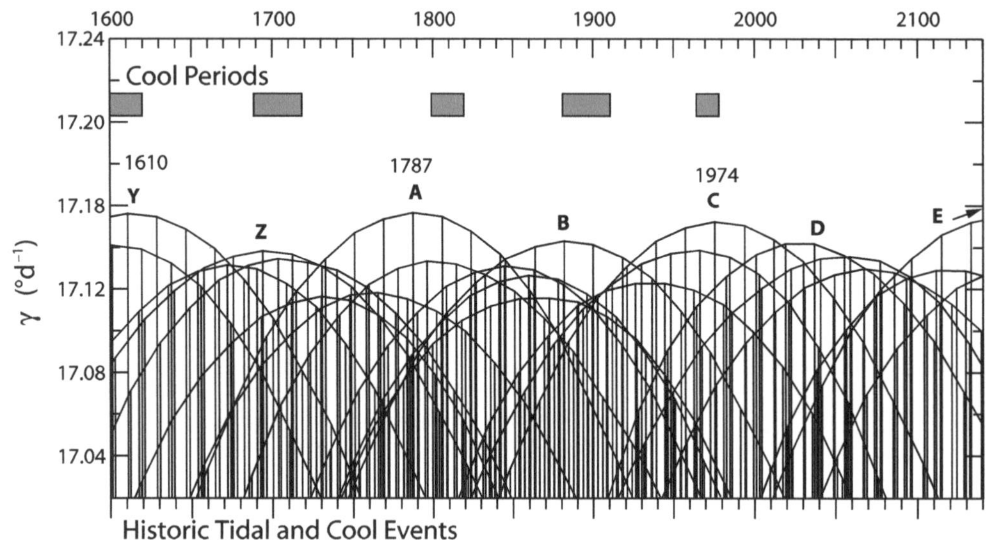

Fig. 3.16 Timing of lunisolar tidal forcing from AD 1600 Each event, shown by a vertical line, gives a measure of the forcing in terms of the angular velocity of the moon, in arc degrees per day, at the time of the event. Arcs connect events of each prominent 18.03-year tidal sequence. Also plotted are times of cool episodes seen in climate data. Centennial maxima are labeled with letters. Climatic events (boxes) of the dominant tidal sequences (letters) are at about 90-year intervals. After Keeling & Whorf (1997).

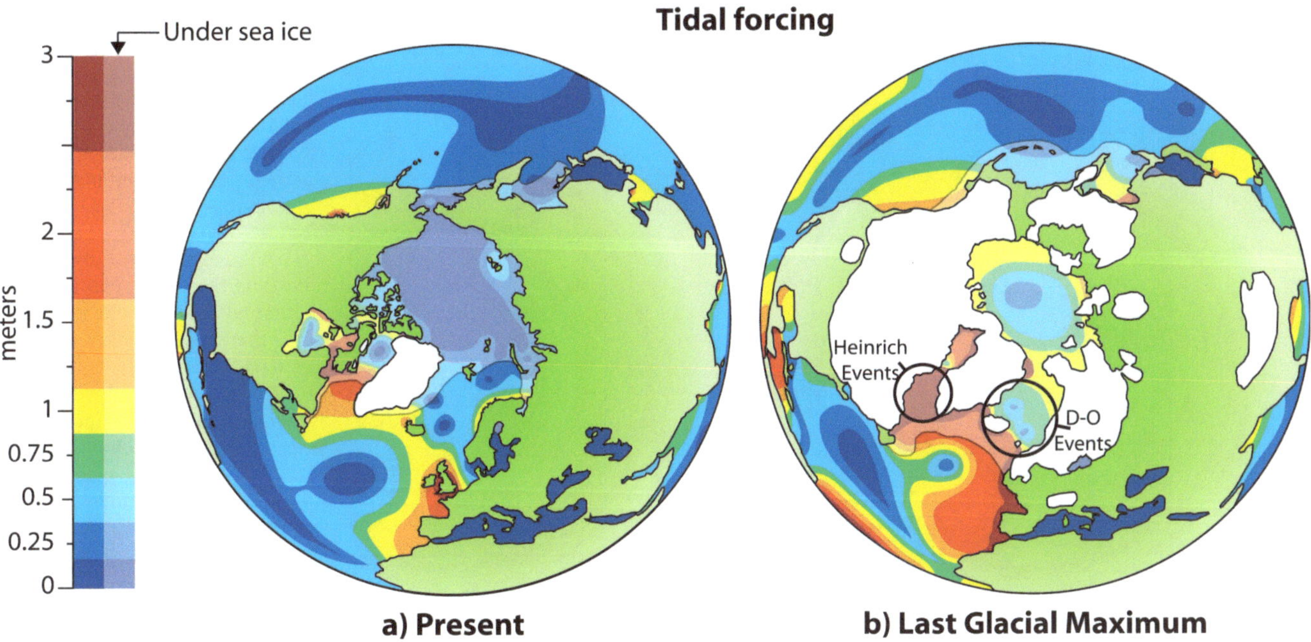

Fig. 3.17 Ice age tidal amplitude
Tidal amplitude (m) of the principal lunar semidiurnal tide M2 at **(a)** present time and **(b)** 23 kyr BP in a hydro-dynamical model coupled to a gravitationally self-consistent (hence geographically variable) prediction of sea-level change. Among the areas with stronger tides are those producing iceberg discharges during Heinrich events, and Nordic Seas area where D–O abrupt warming originates (black circles). After Griffiths & Peltier (2008).

effect? Regular tides already have a strong effect on ocean water vertical mixing. The vertical mixing effect of tides is calculated annually at 4 terawatts, versus 2 terawatts for the wind (Keeling & Whorf 1997). Since ocean waters are temperature stratified, vertical mixing is one of the main factors in ocean temperature changes. Moreover, tides also take place below sea-ice, where they are the only factor affecting vertical mixing.

Tides also increase their power in a non-linear way according to cycles, the main one being the 18.6 years nodal cycle. Since the orbit of the Moon has an inclination of c. 5° with respect to the orbit of the Earth, the nodes are the points at which the Moon crosses the ecliptic plane, and the line that joins both nodes produces a full rotation every 18.6 years. This produces alignment cycles with different periodicities, that occur when the Earth is at perigee, and the Moon is at apogee or perigee at the time of being at one of the nodes where the Moon orbit crosses the ecliptic, and with Earth the Moon and the Sun being in syzygy. Even more important that these alignment cycles, tidal strength varies with harmonic beats of tidal frequencies at longer cycles. This cycles act on a centennial scale and unlike the alignment cycles produce very high tides over a period of months or years. They have been associated with cool periods of historic times (Keeling & Whorf 1997; Fig. 3.16).

The strongest dominant tidal sequence of the last 200 years took place on January, 8th, 1974 (Fig. 3.16). Therefore we can check if anything unusual happened with tides around that date. According to historical records unusually high tides affected the western coasts of US and Europe on January 1974. In Western Europe the tides coupled with storms caused severe flooding in Ireland, where the severity of the damage on the 11–12th January flooding was higher than a previous hurricane "Debbie", causing the worst disaster in history for the Electricity Supply Board

of Ireland (Keane & Sheahan 1974). In the US, Fergus Wood, a researcher for the National Oceanic and Atmospheric Administration, gave a public warning on December 26, 1973, on the upcoming very close perigee–syzygy alignment of January 8, 1974, and coastal damage was prevented by sandbagging, backfilling, and other precautionary measures. The Los Angeles Times reported on Wed. Jan. 9, 1974 (CC Ed. Part I, Page 1, Cols. 2, 3) *"Giant waves pound Southland coast, undermine beach homes. Sandbag barriers erected to ward off tidal assault."* (Wood 1978). The next alignment on February 9 also caused a tidal flooding along the southern coast of England. At Fort Denison, Sidney Harbour, Australia, analysis of water levels since 1914 to 2009 shows that the largest tidal anomaly was recorded on 26 May 1974 during the most significant oceanic storm event on the historical record. Over this timeframe some 96.8% of the measured anomalies fall within the bandwidth between –10 cm and +20 cm. The anomaly of 1974 measured 59 cm (Watson & Frazer 2009). Ocean tides beneath the Ross Ice Shelf in Antarctica were measured between December 1973 and February 1974 by Robinson et al. (1974), where they detected tides of about 2 meters at that time underneath the ice shelf by gravimetry.

So it is clear that unusually strong tides take place with centennial periodicities capable of exerting powerful vertical mixing even below the sea-ice, thus providing a mechanism for triggering a D–O abrupt interstadial warming. Tides were already demonstrated to enhance iceberg calving by Otto Petterssen in 1914, but tides are also sensitive to sea levels and so some researchers are showing through models that reproduce current tides, that with glacial conditions of low sea level, much bigger tides would be produced at certain areas of the world (Arbic et al. 2004; Griffiths & Peltier 2008). These areas are located mainly in the North Atlantic region (Fig. 3.17), so the

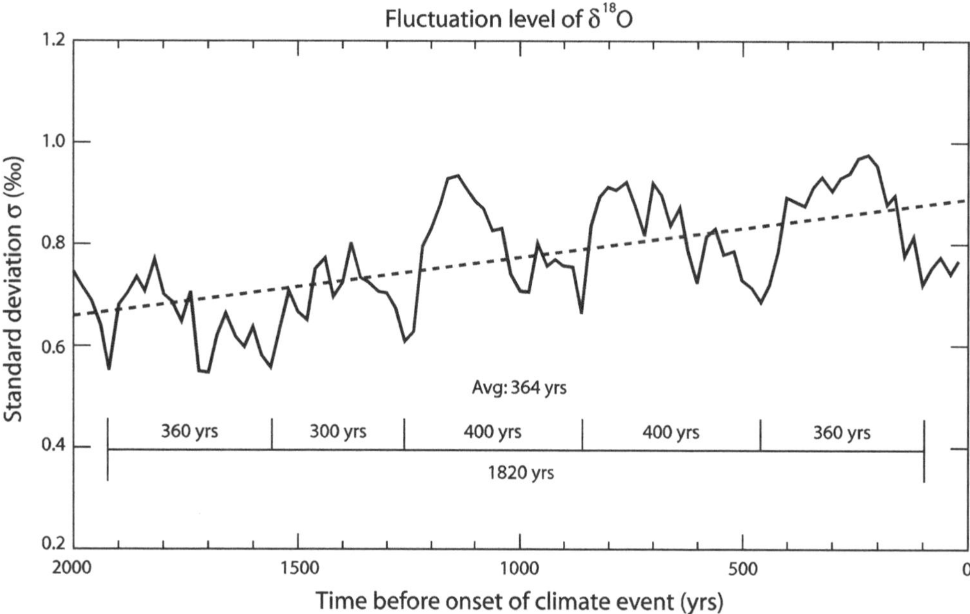

Fig. 3.18 Fluctuations in the temperature signal during stadials display lunisolar frequencies The NGRIP $\delta^{18}O$ time series from 18 stadials filtered by subtracting a 100-year moving average had their standard deviation computed in running 100-yr windows. These standard deviations from the ensemble are averaged at time points preceding the onset of an interstadial period producing an estimate of the fluctuation level in the $\delta^{18}O$ signal as a function of time. The dotted line is a linear fit showing an increase in thermal fluctuations as the stadial approaches its ending. Interestingly these early warning signs of the Greenland stadial termination fluctuate at the 366-yr and 1800-yr lunisolar periodicities. After Rypdal (2016).

authors propose a tidal origin for HEs. As tidal waves propagate, these mega tides of the glacial period would have affected the North Atlantic–Nordic Seas area where D–O cycles abrupt warming took place.

We have seen that lunisolar tidal cycles have had the capability to have produced megatides during the glacial period, strong enough to produce intense vertical water mixing, and thus capable of explaining the triggering of D–O cycles. The tidal hypothesis is very consistent with the known requirements for D–O cycles: cold conditions that favor extensive sea-ice cover, water temperature stratification of enough differential, sea levels low enough for huge tides to be produced, but not too low as the ice cover can be too thick and stable so the tide effect is not strong enough.

The tidal hypothesis appears to have the means, does it have the opportunity? Lunar cycles display a high intrinsic degree of variability that together with variable geographical conditions result in very complicated local tidal patterns. But tidal forcing displays two main long periodicities simultaneously. One is the c. 1800-yr cycle resulting from mismatches between syzygy and nodal periodicities, modulated by c. 4.7 kyr mismatches between syzygy and perihelion periodicities. The other is the c. 366-yr periodicity resulting from the closeness between lunar nodal and precessional periodicities, modulated by the 1800-yr cycle that could result in c. 1.5 and 3.0-kyr periodicities. This periodicity was already proposed by Berger et al. (2002) to be responsible for the D–O cycle. They suggested that seasonal constrains could explain why the 366-yr harmonic beat manifests as a c. 1500-yr periodicity in D–O events.

Further evidence that lunisolar tidal cycles are responsible for triggering the abrupt D–O GS/GI transitions comes from the growing instability towards the end of stadials manifested in the temperature fluctuations detected as early warning signs of an approaching abrupt transition (Rypdal 2016). The standard deviation of these temperature fluctuations, an indication of their intensity, waxes and wanes over time presenting the basic 366 and 1800-yr periods that constitute the basis of the 3.0 and 4.8-kyr periodicities displayed by D–O events (Fig. 3.18).

As should be expected if the lunisolar tidal hypothesis is correct, the times at which tides are stronger following the 366 and 1800-yr lunar cycles should provide time points when the stadial becomes significantly more unstable, manifested in larger temperature fluctuations and eventually an abrupt GS/GI transition.

A lunisolar tidal cycle of millennial scale is therefore a promising hypothesis, despite lack of clear evidence, for the pacing of the D–O cycles abrupt warming. During the Holocene the D–O cycle disappears, as it is based on low enough sea levels, extensive sea-ice and inverse temperature-stratified waters, with the possible participation of enhanced glacial tides. However tidal cycles continue exerting a climate effect during the Holocene, and a c. 1500-yr oceanic and atmospheric signal resonates over the Late Holocene (Neoglacial) climate. The 1500-yr climate periodicity during the Holocene is reviewed in chapter 7. The general features of this signal also agree well with what could be expected from a lunisolar tidal cycle during a warm interglacial: increased storminess and decreased sea-surface temperature.

3.10 Conclusions

3a. Between 85 and 12 thousand years ago Greenland proxy temperature records display 20 abrupt and intense climatic events known as Dansgaard–Oeschger events that constitute a pseudocycle irregularly paced according to a double 3.0 and 4.8-kyr periodicity.

3b. Each D–O oscillation is preceded by North Atlantic cooling and iceberg discharges that when intense and prolonged constitute a Heinrich event.

3c. D–O oscillations present an asymmetric change in temperature with warming of 8–10 °C in a few decades followed by a cooling in stages from a few centuries to a few millennia.

3d. Prior to the Greenland abrupt warming, temperature increases in Antarctica until about 220 years after the start of the Greenland warming.

3e. The abrupt Northern Hemisphere warming increases global methane levels from boreal wetlands due to increased temperature and precipitation.

3f. CO_2 has no role during D–O cycles, and its levels are neither cause nor consequence of the most frequent most abrupt climate changes of the past. The increase in CO_2 levels during Heinrich events does not significantly alter the rate or magnitude of the warming during the subsequent D–O oscillation.

3g. D–O cycles require sea levels between 35 and 100 m below present, and appear to be inhibited by maximal obliquity.

3h. The leading "salt oscillator hypothesis," has no explanation for D–O spacing and relies on unproven meltwater pulses and a contrary to evidence shut down of the Atlantic Meridional Overturning Circulation.

3i. Challenger D–O hypothesis proposes the stratification of warm subsurface waters below the halocline and the sea-ice in the North Atlantic and Norwegian Sea, with the abrupt warming taking place due to the collapse of this stratification.

3j. Lunisolar tidal periodicities based on the 1800-yr lunar cycle provide a promising hypothesis for D–O oscillations pacing and triggering mechanism.

References

Ahn J & Brook EJ (2008) Atmospheric CO_2 and Climate on Millennial Time Scales During the Last Glacial Period. Science 322 (5898) 83–85

Ahn J & Brook EJ (2014) Siple Dome ice reveals two modes of millennial CO_2 change during the last ice age. Nature Communications 5 3723

Alley RB (2007) Wally was right: Predictive ability of the North Atlantic "conveyor belt" hypothesis for abrupt climate change. Annu Rev Earth Planet Sci 35 241–272

Andrews JT, Jennings AE, Kerwin M et al (1995) A Heinrich-like event H-0 (DC-0): Source (s) for detrital carbonate in the North Atlantic during the Younger Dryas chronozone. Paleoceanography and Paleoclimatology 10 (5) 943–952

Arbic BK, MacAyeal DR, Mitrovica JX & Milne GA (2004) Palaeoclimate: Ocean tides and Heinrich events. Nature 432 (7016) 460

Baldini JU, Brown RJ & Mawdsley N (2018) Evaluating the link between the sulphur-rich Laacher See volcanic eruption and the Younger Dryas climate anomaly. Climate of the Past 14 (7) 969–990

Barker S, Chen J, Gong X et al (2015) Icebergs not the trigger for North Atlantic cold events. Nature 520 (7547) 333–336

Bereiter B, Eggleston S, Schmitt J et al (2015) Revision of the EPICA Dome C CO_2 record from 800 to 600 kyr before present. Geophysical Research Letters 42 (2) 542–549

Berger WH, Pätzold J & Wefer G (2002) A case for climate cycles: Orbit Sun and Moon. In: Wefer G, Berger WH, Behre KE and Jansen E (eds) Climate development and history of the North Atlantic realm. Springer, Berlin, pp 101–123

Bock M, Schmitt J, Möller L et al (2010) Hydrogen isotopes preclude marine hydrate CH_4 emissions at the onset of Dansgaard–Oeschger events. Science 328 (5986) 1686–1689

Bond GC & Lotti R (1995) Iceberg discharges into the North Atlantic on millennial time scales during the last glaciation. Science 267 (5200) 1005–1010

Bond G, Broecker W, Johnsen S et al (1993) Correlations between climate records from North Atlantic sediments and Greenland ice. Nature 365 (6442) 143–147

Bond GC, Showers W, Elliot M et al (1999) The North Atlantic's 1-2 kyr climate rhythm: relation to Heinrich events, Dansgaard/Oeschger cycles and the Little Ice Age. In: Clark PU, Webb RS and Keigwin LD (eds) Mechanisms of global climate change at millennial time scales. Geophysical Monograph Series Vol 112. American Geophysical Union, Washington, DC, p35–58

Broecker WS (1999) What if the conveyor were to shut down? Reflections on a possible outcome of the great global experiment. GSA Today 9 (1) 1–7

Broecker WS, Bond G, Klas M et al (1990) A salt oscillator in the glacial Atlantic? 1 The concept. Paleoceanography 5 (4) 469–477

Broecker WS, Denton GH, Edwards RL et al (2010) Putting the Younger Dryas cold event into context. Quaternary Science Reviews 29 (9–10) 1078–1081

Broecker WS, Kennett JP, Flower BP et al (1989) Routing of meltwater from the Laurentide Ice Sheet during the Younger Dryas cold episode. Nature 341 (6240) 318–321

Ditlevsen PD, Andersen KK & Svensson A (2007) The DO-climate events are probably noise induced: statistical investigation of the claimed 1470 years cycle. Climate of the Past 3 (1) 129–134

Dokken TM, Nisancioglu KH, Li C et al (2013) Dansgaard–Oeschger cycles: Interactions between ocean and sea ice intrinsic to the Nordic seas. Paleoceanography and Paleoclimatology 28 (3) 491–502

Ezat MM, Rasmussen TL & Groeneveld J (2014) Persistent intermediate water warming during cold stadials in the southeastern Nordic seas during the past 65 ky. Geology 42 (8) 663–666

Ganopolski A & Rahmstorf S (2001) Rapid changes of glacial climate simulated in a coupled climate model. Nature 409 (6817) 153–158

Griffiths SD & Peltier WR (2008) Megatides in the Arctic Ocean under glacial conditions. Geophysical Research Letters 35 (8) L08605

Grootes PM & Stuiver M (1997) Oxygen 18/16 variability in Greenland snow and ice with 10− 3-to 105-year time resolution. Journal of Geophysical Research: Oceans 102 (C12) 26455–26470

Heinrich H (1988) Origin and consequences of cyclic ice rafting in the northeast Atlantic Ocean during the past 130,000 years. Quaternary research 29 (2) 142–152

Hemming SR (2004) Heinrich events: Massive late Pleistocene detritus layers of the North Atlantic and their global climate imprint. Reviews of Geophysics 42 (1) RG1005

Hulbe CL, MacAyeal DR, Denton GH et al (2004) Catastrophic ice shelf breakup as the source of Heinrich event icebergs. Paleoceanography 19 (1) PA1004

Imbrie J & Imbrie KP (1979) Ice ages: solving the mystery. Enslow Publishers, Short Hills

Johnsen SJ, Dansgaard W & Clausen HB (1972) Oxygen isotope profiles through the Antarctic and Greenland ice sheets. Nature 235 (5339) 429–434

Jouzel J, Masson–Delmotte V, Cattani O et al (2007) Orbital and millennial Antarctic climate variability over the past 800,000 years. Science 317 (5839) 793–796

Keane T & Sheahan MF (1974) Storms of January 1974. (Report) Met Éireann Internal Memorandum 79/74 http://edepositireland.ie/handle/2262/73584 Accessed 29 Jun 2022

Keeling CD & Whorf TP (1997) Possible forcing of global temperature by the oceanic tides. Proceedings of the National Academy of Sciences 94 (16) 8321–8328

Keeling CD & Whorf TP (2000) The 1800-year oceanic tidal cycle: A possible cause of rapid climate change. Proceedings of the National Academy of Sciences 97 (8) 3814–3819

Laskar J, Robutel P, Joutel F et al (2004) A long-term numerical solution for the insolation quantities of the Earth. Astronomy & Astrophysics 428 (1) 261–285

Menviel L, Timmermann A, Friedrich T & England MH (2014) Hindcasting the continuum of Dansgaard–Oeschger variability: mechanisms, patterns and timing. Climate of the Past 10 (1) 63–77

Murton JB, Bateman MD, Dallimore SR et al (2010) Identification of Younger Dryas outburst flood path from Lake Agassiz to the Arctic Ocean. Nature 464 (7289) 740–743

North Greenland Ice Core Project members (2004) High-resolution record of Northern Hemisphere climate extending into the last interglacial period. Nature 431 (7005) 147–151

Nye H & Condron A (2021) Assessing the statistical uniqueness of the Younger Dryas: a robust multivariate analysis. Climate of the Past 17 (3) 1409–1421

Obrochta S (2015) Comment. Climate of the Past Discussions 11 C2548–C2554 https://cp.copernicus.org/preprints/11/C2548/2015/cpd-11-C2548-2015.pdf Accessed 15 Jun 2022

Petersen SV, Schrag DP & Clark PU (2013) A new mechanism for Dansgaard-Oeschger cycles. Paleoceanography 28 (1) 24–30

Rahmstorf S (2002) Ocean circulation and climate during the past 120,000 years. Nature 419 (6903) 207–214

Rahmstorf S (2003) Timing of abrupt climate change: A precise clock. Geophysical Research Letters 30 (10) 1510–1514

Rasmussen SO, Bigler M, Blockley SP et al (2014) A stratigraphic framework for abrupt climatic changes during the Last Glacial period based on three synchronized Greenland ice-core records: refining and extending the INTIMATE event stratigraphy. Quaternary Science Reviews 106 14–28

Rasmussen TL & Thomsen E (2004) The role of the North Atlantic Drift in the millennial timescale glacial climate fluctuations. Palaeogeography Palaeoclimatology Palaeoecology 210 (1) 101–116

Rasmussen TL, Thomsen E & Moros M (2016) North Atlantic warming during Dansgaard–Oeschger events synchronous with Antarctic warming and out-of-phase with Greenland climate. Scientific reports 6 20535

Rhodes RH, Brook EJ, Chiang JC et al (2015) Enhanced tropical methane production in response to iceberg discharge in the North Atlantic Science 348 (6238) 1016–1019

Rial JA (2012) Synchronization of polar climate variability over the last ice age: in search of simple rules at the heart of climate's complexity. American Journal of Science 312 (4) 417–448

Robinson ES, Neuburg HAC & Williams R (1974) Ocean tides beneath the Ross Ice Shelf. Antarctic Journal of the United States 9 (4) 162–164

Rohling E, Mayewski P & Challenor P (2003) On the timing and mechanism of millennial-scale climate variability during the last glacial cycle. Climate Dynamics 20 (2–3) 257–267

Ruddiman WF (2001) Earth's Climate: Past and Future, 1st edn. WH Freeman and Co, New York, p345–348

Rypdal M (2016) Early-warning signals for the onsets of Greenland interstadials and the Younger Dryas–Preboreal transition. Journal of Climate 29 (11) 4047–4056

Sadatzki H, Dokken TM, Berben SM et al (2019) Sea ice variability in the southern Norwegian Sea during glacial Dansgaard–Oeschger climate cycles. Science Advances 5 (3) eaau6174

Saramago J (2002) O Homem Duplicado. The Double. English 2004 ed. Houghton Mifflin Harcourt. Boston, Massachusetts

Schmidt MW & Hertzberg JE (2011) Abrupt Climate Change During the Last Ice Age. Nature Education Knowledge 3 (10) 11

Schulz M (2002) On the 1470-year pacing of Dansgaard-Oeschger warm events. Paleoceanography 17 (2) 4–1

Schulz M, Berger WH, Sarnthein M & Grootes PM (1999) Amplitude variations of 1470-year climate oscillations during the last 100,000 years linked to fluctuations of continental ice mass. Geophysical Research Letters 26 (22) 3385–3388

Seierstad IK, Abbott PM, Bigler M et al (2014) Consistently dated records from the Greenland GRIP, GISP2 and NGRIP ice cores for the past 104 ka reveal regional millennial-scale $\delta^{18}O$ gradients with possible Heinrich event imprint. Quaternary Science Reviews 106 29–46

Shakun JD, Clark PU, He F et al (2012) Global warming preceded by increasing carbon dioxide concentrations during the last deglaciation. Nature 484 (7392) 49–54

Spratt RM & Lisiecki LE (2016) A Late Pleistocene sea level stack. Climate of the Past 12 (4) 1079–1092

Stauffer B, Hofer H, Oeschger H et al (1984) Atmospheric CO_2 concentration during the last glaciation. Annals of Glaciology 5 160–164

Stocker TF & Johnsen SJ (2003) A minimum thermodynamic model for the bipolar seesaw. Paleoceanography 18(4) 1087–1093

van Ommen T (2015) Palaeoclimate: Northern push for the bipolar see-saw. Nature 520 (7549) 630–631

WAIS Divide Project Members (2015) Precise interpolar phasing of abrupt climate change during the last ice age. Nature 520 (7549) 661–665

Watson P & Frazer A (2009) NSW King Tide Photo Event Summary 2009; Appendix B4. Paper presented at the 18th NSW Coastal Conference Ballina Australia 3–6 November 2009 https://coastalconference.com/2009/papers2009/Phil%20Watson%20Full%20paper%202.pdf Accessed 29 Jun 2022

Wolbach WS, Ballard JP, Mayewski PA et al (2018) Extraordinary biomass-burning episode and impact winter triggered by the Younger Dryas cosmic impact ~ 12,800 years ago. 1. Ice cores and glaciers. The Journal of Geology 126 (2) 165–184

Wood FJ (1978) Perigean spring tides: A potential threat toward coastal flooding disaster. Preface to: Wood FJ (author) The strategic role of perigean spring tides in nautical history and North American coastal flooding 1635–1976. US Dept of Commerce, National Oceanic and Atmospheric Administration, Washington, p V–XIII

Wunsch C (2010) Towards understanding the Paleocean. Quaternary Science Reviews 29 (17–18) 1960–1967

HOLOCENE CLIMATIC VARIABILITY

"Extended into the future, the Holocene pattern of climatic change implies that the Little Ice Age, if it is not already over, will be succeeded by a climate regime similar to that of the Roman Empire and Middle Ages. This suggests that predictions of an inminent new ice age may be premature."
George H. Denton and Wibjörn Karlén (1973)

4.1 Introduction

Botanists studying peat stratigraphy in the late 19th and early 20th century were among the first to notice abrupt climate changes reflected in peat layers. These sudden transitions were later confirmed by changes in sediment pollen composition. Scandinavian palynologists established the Blytt–Sernander sequence that divided the Scandinavian Holocene into five periods. They used the terms Boreal for drier, and Atlantic for wetter (Fig. 4.1).

The Blytt–Sernander sequence fell out of fashion in the 1970s when new techniques allowed a more quantitative reconstruction of past climates. However, it captures the essence of Holocene climate as four periods of roughly 2500 years each. Every period shows a characteristic vegetation pattern in Scandinavia, indicative of relatively stable climatic conditions, separated from other periods by rapid vegetation changes suggestive of abrupt climatic changes. The dates and conditions generally accepted (Ammann & Fyfe 2014) are:

- Pre-Boreal, 11,500–10,500 yr BP. Cool and sub-arctic
- Boreal, 10,500–7,800 yr BP. Warm and dry
- Atlantic, 7,800–5,700 yr BP. Warmest and wet
- Sub-Boreal, 5,700–2,600 yr BP. Warm and dry
- Sub-Atlantic, 2,600–0 yr BP. Cool and wet

The transition from Sub-Boreal to Sub-Atlantic took place in Scandinavia at the end of the Bronze Age. Rutger Sernander proposed that this climatic change was abrupt, even a catastrophe that he identified with the Fimbulwinter, or great winter of the Sagas. At the time other scientists believed in a more gradual climatic change, but recent studies on the 2.8 kyr abrupt climatic event (ACE; Kobashi et al. 2013) agree with Sernander.

Another classification divides the Holocene climatically into two periods: the Holocene Climatic Optimum (HCO, also known as Hypsithermal, Altithermal or Holocene Thermal Maximum), between 9,000 and 5,500 yr BP (although some authors only consider it from 7,500 yr BP after the 8.2 kyr ACE), and the Neoglacial period, between 5,000 and 100 yr BP, separated by the Mid-Holocene Transition (MHT) that roughly coincides with the start of the Bronze Age. They would be preceded by a warming phase at the early Holocene (Anathermal).

The most popular subdivision of the Holocene is in three periods. The Early Holocene, up to the 8.2 kyr ACE,

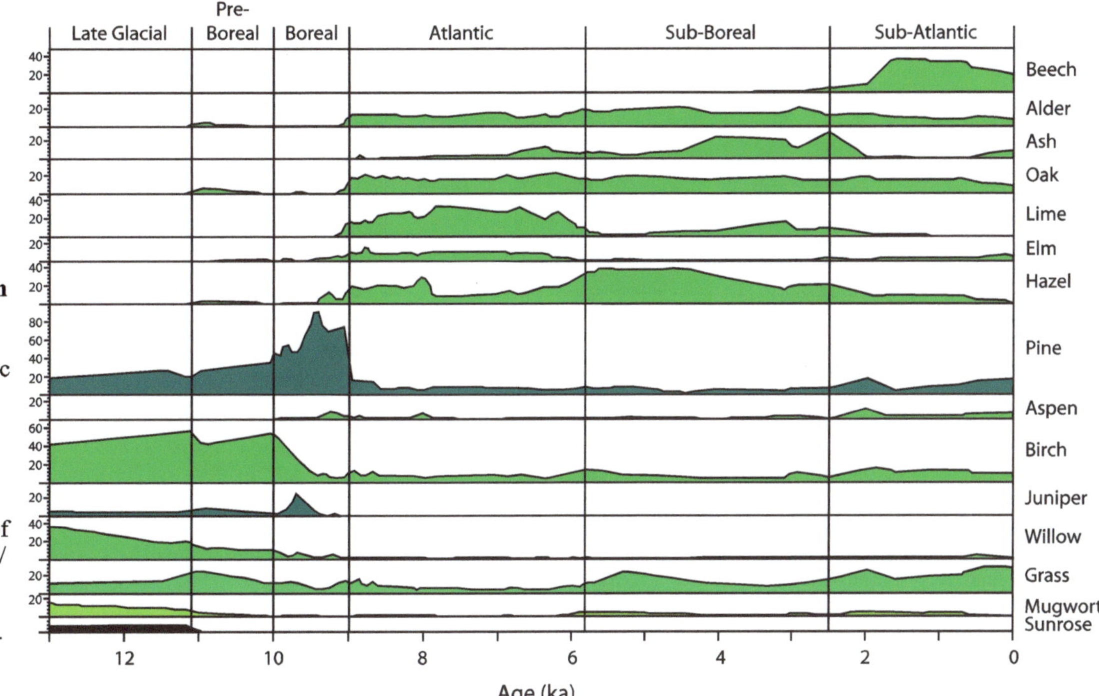

Fig. 4.1 Pollen diagram at Roskilde Fjord An example of the Blytt–Sernander climatic zones established with the traditional pollen indicators, with the distinct elm-fall at the Atlantic/Sub-Boreal transition, and the rise of beech at the Sub-Boreal/Sub-Atlantic transition. Period dates might change at different locations. After Schrøder et al. (2004).

the Middle Holocene, between the 8.2 and the 4.2 kyr ACEs, and the Late Holocene since the 4.2 kyr ACE. In 2018 the International Union of Geological Sciences ratified the stratigraphic division of the Holocene into three stages or ages: Greenlandian between 11,700–8,326 B2K (before 2000), Northgrippian between 8,326–4,250 B2K, and Meghalayan from 4,250 B2K to the present. Although this popular subdivision has been made official, in the author's opinion it fails to properly capture the Holocene climatic trends, as the 4.2 kyr ACE is climatically irrelevant for the Holocene climate evolution. A millennium earlier the 5.2 kyr ACE that took place during the MHT fundamentally altered the climate of the Holocene, initiating the Neoglacial cooling.

4.2 Holocene general climate trend

Broadly speaking the Holocene had an abrupt start at 11,700 yr BP, after the Younger Dryas cold relapse, and reached maximal temperatures in about 2,000 years. Since about 9,500 yr BP, a time that coincides with maximal obliquity of the Earth axis, the climate of the Holocene stopped warming and a few thousand years later started a progressive cooling.

By far the main factor driving Holocene climatic change are the insolation changes due to the orbital variations of the Earth (Fig. 4.2). These changes are of two types that produce two different effects not always properly differentiated by Holocene climate researchers. Changes due to precession (modulated by eccentricity) have the effect of redistributing insolation between the different seasons of the year by latitude. The c. 23-kyr precession cycle determines axial orientation as the Earth orbits the Sun, and thus the amount of insolation received by each hemisphere at any point of the orbit. Insolation changes due to precession are represented in Fig. 4.2 with

three month insolation curves for a North and South latitude, relative to present values. These changes increase or decrease seasonality or the difference between summer and winter. So, Northern Hemisphere seasonality was minimal at the Last Glacial Maximum, and maximal at the start of the Holocene, 10,500 yr BP, and will become minimal again in a thousand years.

Precession changes do not alter the annual amount of insolation at any latitude, since whatever insolation they take from one month at a particular location, they give back in another month within the same year. Precession changes are also asymmetrical, as their effect is opposite in each hemisphere, so the Northern Hemisphere summer (June–August, N–JJA thick red line in Fig. 4.2) has become progressively cooler during most of the Holocene, while the Southern Hemisphere summer (December–February, S–DJF thick blue line in Fig. 4.2) has become progressively warmer during most of the Holocene. Precession changes are responsible for sea surface temperature (SST) patterns, and thus oceanic currents. North–South differences set the position of the ITCZ (Intertropical Convergence Zone or the climatic equator). Therefore they are responsible for the African Humid Period, monsoon patterns and the important MHT, that changed the climate mode of the Holocene globally.

Changes due to obliquity have the effect of redistributing insolation between different latitudes following an obliquity cycle of c. 41 kyr. When obliquity was maximal 9.5 kyr BP, both poles received more insolation due to obliquity, while the tropics received less. Obliquity also affects seasonality; at maximal axial tilt, there is an increased difference between summer and winter at high latitudes. But unlike precession changes, obliquity alters the amount of annual insolation at different latitudes in a 41-kyr cycle. This is represented by the background color of figure 4.2, that shows how the polar regions received

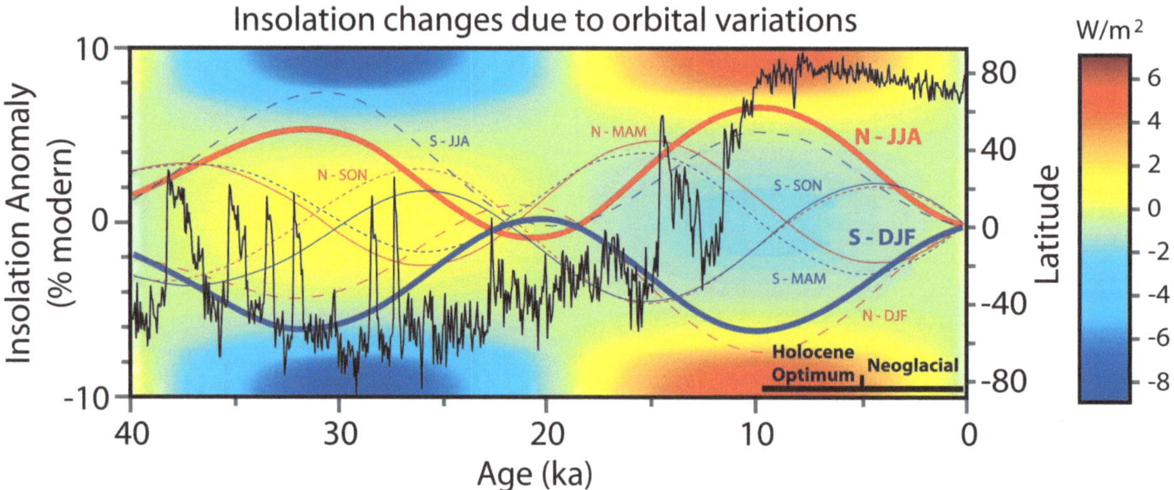

Fig. 4.2 Insolation changes due to orbital variations of the Earth
Black temperature-proxy curve represents δ¹⁸O isotope changes from NGRIP Greenland ice core (without scale). The insolation curves are presented as the insolation anomaly for summer, winter, spring, and fall. N (red) or S (blue) are the northern or Southern Hemisphere and the three letters are the month initials. Northern and southern summer insolation represented with thick curves. Background color represents changes in annual insolation by latitude and time due to changes in the Earth's axial tilt (obliquity), shown in a colored scale. This figure essentially shows how global temperature changes respond mainly to persistent changes in insolation caused by changes in obliquity that are symmetrical for both poles. Changes in seasonal insolation caused by the precession cycle (modified by eccentricity) are asymmetric and less important for the global response, although they cause profound changes in regional climatic differences. The Holocene Climatic Optimum corresponds to high insolation surplus in polar latitudes (red area), while Neoglacial conditions represent the first 5,000 years of a 10,000 year drop into a high glacial insolation deficit in polar latitudes (blue area). Insolation curves after Polissar et al. (2013). NGRIP δ¹⁸O isotope curve after NGRIP members (2004).

increasing insolation from 30 to 9.5 kyr BP, coinciding with the period of warming. Since then, and for the next 11.5 kyr, the poles will be receiving decreasing insolation. Unlike precessional insolation changes, obliquity changes are symmetrical. Although the annual insolation change is not very large, it accumulates over tens of thousands of years and the total change is staggering, creating a huge insolation deficit or surplus. This changes the latitudinal (equator-to-pole) temperature gradient, and is largely responsible for entering and exiting glacial periods (Tzedakis et al. 2017) and for the general evolution of global temperatures and climate during the Holocene. Obliquity changes contribute to the lack of warming of Antarctica during the Holocene, despite increasing Southern Hemisphere summer insolation. Ultimately obliquity changes will be responsible for the glacial inception that will put an end to the Holocene interglacial in the distant future (see Chap. 14).

In the Holocene, the precession cycle and the obliquity cycle are almost aligned so that maximal obliquity and maximal northern summer insolation were almost coincident at the beginning of the interglacial about 10 kyr ago. See in figure 4.2 how the thick red curve representing northern summer insolation reaches maximal values 10 kyr BP, almost coinciding with the center of the background polar red color, representing highest warming from maximal obliquity about 9.5 kyr BP. However this has an interesting consequence. 19,000 years ago obliquity was the same as it is now (only increasing), and the precession cycle was at the same position as it is now (same 65 °N summer insolation; Fig. 4.2). The Earth was receiving the same energy from the Sun, and the orbital configuration was distributing it over the planet in the same way during the Last Glacial Maximum as today. Why is the climate so different for the same energy input? One possible answer is the huge thermal inertia of the planet due mainly to its water content. 21 kyr ago the increasing obliquity had been adding energy to the poles for 10,000 years, reducing the latitudinal insolation gradient, and adding energy to the summers (Huybers 2006; Tzedakis et al. 2017), and was on its way to overcome the huge cold inertia with the help of precession changes that were about to take place. In the present, decreasing obliquity has been taking energy from the poles for 10,000 years, increasing the latitudinal insolation gradient that favors energy loss and increased polar precipitation, and reducing energy during summers. These changes will also overcome the huge warm inertia even against precession changes, but will do so progressively over many thousands of years. A comparison between temperature and obliquity over the past 800 kyr shows that while variable, the thermal inertia of the planet delays the temperature response to obliquity changes by an average of 6,500 years (see Fig. 2.11).

On a multi-millennial scale, global average temperature follows mainly the 41-kyr obliquity cycle with a lag of several thousand years. Holocene temperatures are no exception, and a few thousand years after the peak in obliquity (9,500 years ago), temperature started to decline. This general pattern of Holocene temperature was already known by the late 1950s from a variety of proxy records from different disciplines (Lamb 1977; Fig. 4.3a). Greenland ice cores confirmed this pattern, when corrected for uplift (Vinther et al. 2009), and greatly improved the dating of temperature changes (Fig. 4.3b).

Proxies do not record temperatures but physical, chemical, or biological processes that are affected by temperature. They suffer from known but difficult to estimate uncertainties, like dating uncertainties, uncertainties about

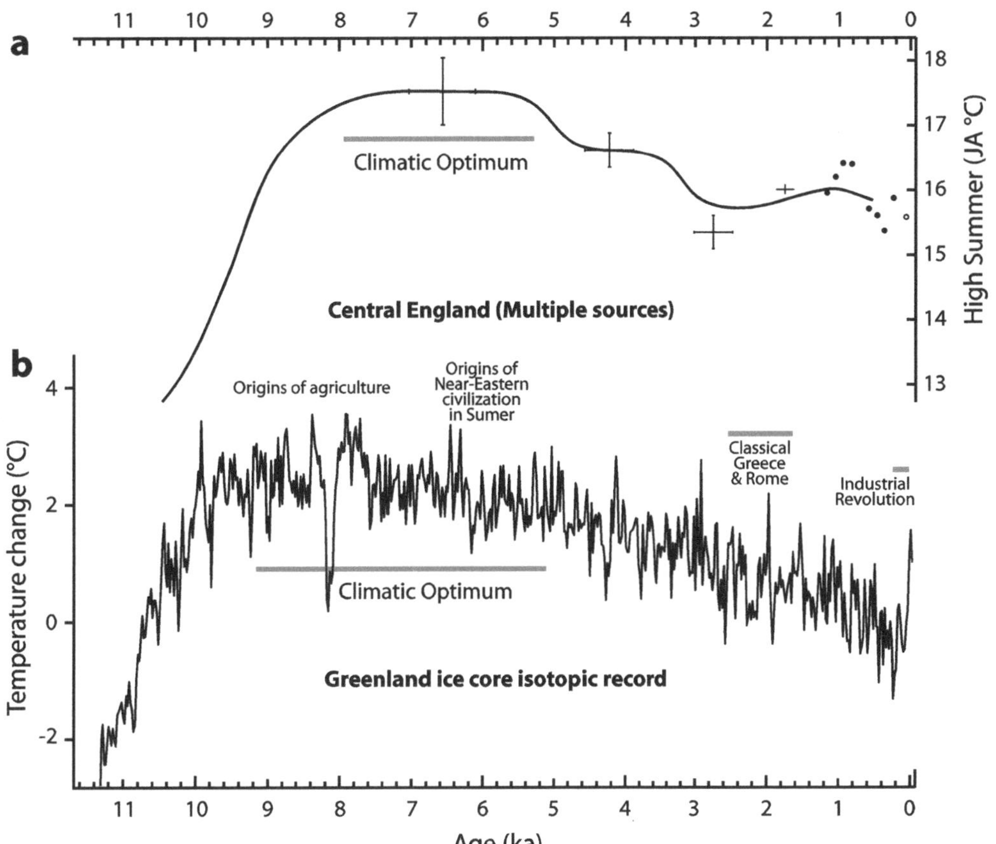

Fig. 4.3 Holocene temperature profile

a) Summer (July–August) Central England temperature reconstruction from multiple proxies and sources by HH Lamb. Crosses represent dating and temperature uncertainty. Black dots are centennial averages. White dot is 1900–1965 average. After Lamb (1977). **b)** Greenland temperature reconstruction based on an average of uplift corrected $\delta^{18}O$ isotopic data from Agassiz and Renland ice cores. This average has been corrected for changes in the $\delta^{18}O$ of seawater and calibrated to borehole temperature records. Some historical periods are indicated. After Vinther et al. (2009).

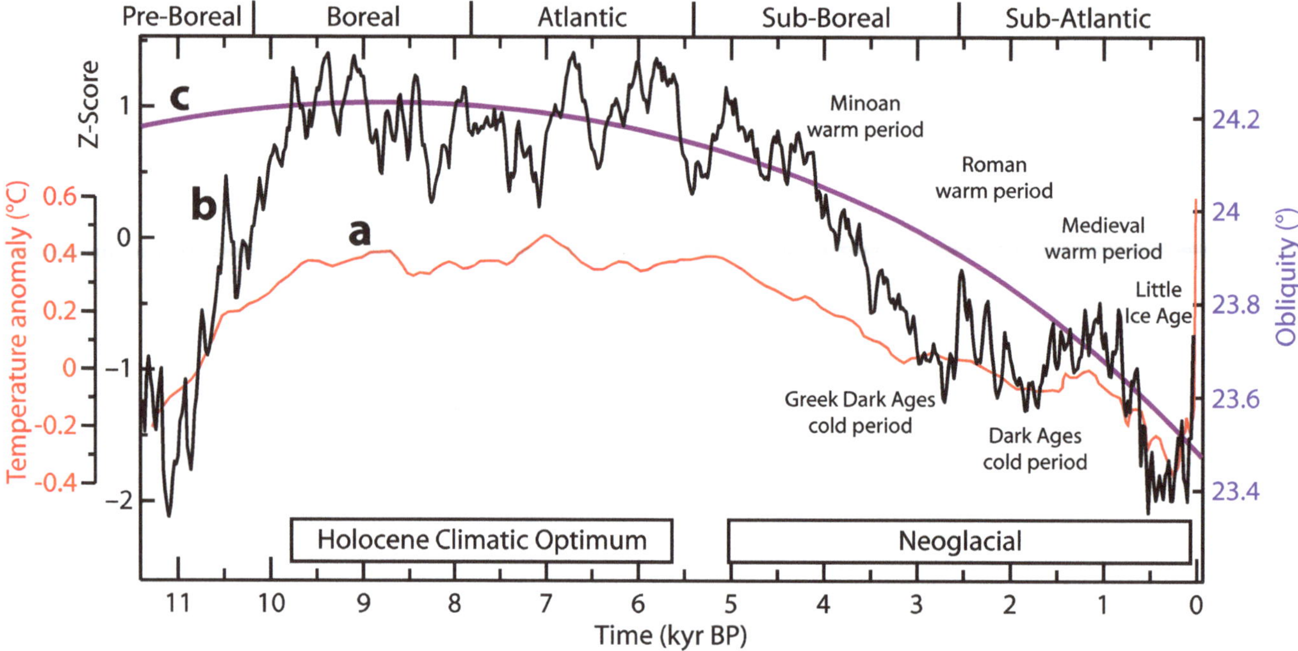

Fig. 4.4 Holocene global temperature reconstruction
a) Red curve (left scale), global average temperature reconstruction from Marcott et al. (2013). The averaging method does not correct for proxy drop out which produces an artificially enhanced terminal spike, while the Monte Carlo smoothing eliminates most variability information. **b)** Black curve (left scale), global average temperature reconstruction after Marcott et al. (2013), using proxy published dates, and differencing average. Temperature anomaly is expressed as Z-score, or distance to the mean in standard deviation units. **c)** Purple thick curve (right scale), Earth's axis obliquity is shown to display a similar trend to Holocene temperature.

uncontrolled environmental effects (e.g. precipitation), uncertainties about non-linear responses to temperature changes, or seasonal response uncertainties. They are also likely affected by unknown inhomogeneous factors. Mixing land and ocean proxies adds another source of uncertainty, as the result is very dependent on the proxies included. Global reconstructions add the uncertainty from estimating the temperature of the entire planet from just a few hundred of low-precision uncalibrated temperature-affected factors (not thermometers) that provide a data point once a decade at best. Although little confidence can be placed on the result of global proxy reconstructions, that are very dependent on researchers choices, they are useful to study Holocene climate evolution. One Holocene temperature reconstruction available (Marcott et al. 2013; Fig. 4.4a) presents several problems regarding proxy re-dating, lack of sufficient proxies in the last centuries producing an artifactual deviation, and heavy smoothing. To correct some of the problems it presents, that reconstruction has been repeated here the with the same set of proxies, each averaged to their own anomaly, without any smoothing, and with the original published dates for the proxies (Fig. 4.4b). The temperature changes have been expressed as Z-score (distance to the mean), to avoid making inferences about temperatures of the past that we cannot possibly know. Nevertheless, there is a vast literature and consilience of evidence from different fields that indicates that the HCO was on average between 1 and 2 °C warmer than the Little Ice Age (LIA). The resulting temperature curve is extraordinarily similar to Hubert Lamb regional reconstruction from the 1970s (Fig. 4.3a), with significant temperature drops at 5.5, 3, and 0.5 kyr BP.

4.3 The controversial role of greenhouse gases during the Holocene

What role, if any, have greenhouse gases (GHG) played in Holocene climate change? Available data indicates that despite significant changes in GHG concentration in the atmosphere during the period of 10,000 to 600 yr BP, their contribution to temperature changes cannot have been important. According to Monnin et al. (2004), CO_2 concentrations measured in Antarctic ice cores decreased from 267 to 258 ppm between 10,000 and 6,800 yr BP, and afterwards increased more or less linearly to 283 ppm by 600 yr BP, just prior to the LIA (Fig. 4.5c). This increase of 25 ppm represents about 10% of a doubling. Considering the period from the Last Glacial Maximum (20 kyr BP) to the HCO, atmospheric CO_2 increased by 70 ppm or 36% of a doubling. We can see that the CO_2 increase between 6.8 and 0.6 kyr BP constitutes almost one third of the CO_2 increase from the coldest point of the last glacial period to the warmest point of the present interglacial. Almost a third of the glacial-interglacial span cannot be considered insignificant for the increase in CO_2 that took place between 6.8 and 0.6 kyr BP. If CO_2 is as potent warming agent as purported in some hypotheses and models, one should expect some warming coming out of this CO_2 increase, especially because from 5,000 yr BP it was accompanied by an increase in atmospheric CH_4 concentrations (Kobashi et al. 2007; Fig. 4.5d). But instead of an increase in temperatures, what we find is a progressive decrease from the HCO to the LIA driven by changes in insolation.

Climate models adjusted to explain present global warming do not reproduce the Holocene climate. The mean temperature of an ensemble of three models

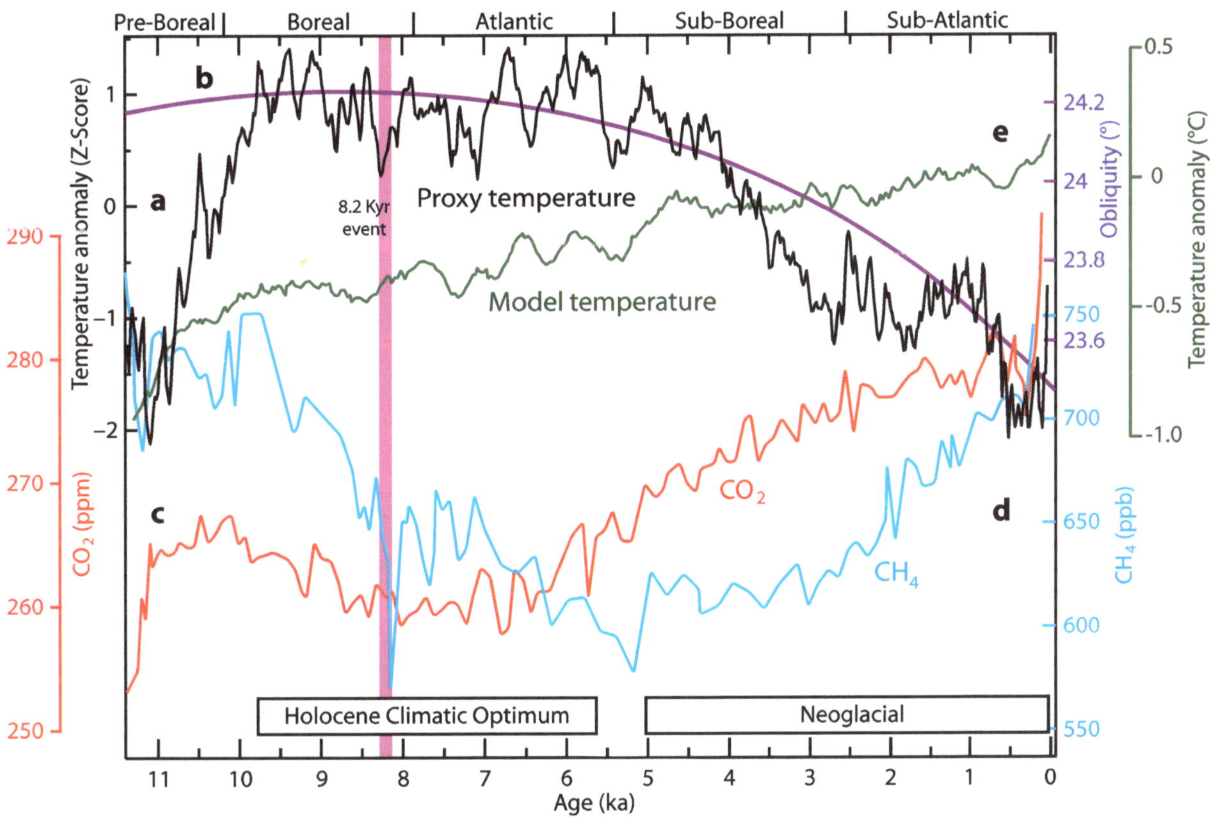

Fig. 4.5 Temperature and greenhouse gases during the Holocene
a) Black curve, global temperature reconstruction as in Fig. 4.4b. **b)** Purple curve, Earth's axis obliquity cycle. **c)** Red curve, CO_2 levels as measured in Epica Dome C (Antarctica) ice core, reported in Monnin et al. (2004). **d)** Blue curve, methane levels as measured in GRIP, GISP2, and NEEM (Greenland) ice cores as reported by Kobashi et al. (2007). Notice the great effect of the 8.2 kyr event on methane concentrations. **e)** Green curve, simulated global temperature from an ensemble of three models (CCSM3, FAMOUS, and LOVECLIM) from Liu et al. (2014), show the inability of general climate models to replicate the Holocene general temperature downward trend. Vertical bar, 8.2 kyr BP ACE. Major Holocene climatic periods are indicated.

(CCSM3, FAMOUS, and LOVECLIM; Liu et al. 2014; Fig. 4.5e) shows a constant increase in temperature during the entire Holocene, driven by the increase in GHG. This disagreement between models and data-derived reconstructions of Holocene climate has been termed by the authors the Holocene temperature conundrum (Liu et al. 2014).

Climate modelers should take the opportunity to adjust their models to Holocene conditions. It is clear that the main driver of Holocene climate has been changes in insolation due to orbital variation. Changes in GHG concentrations appear to have had only a minor effect.

4.4 The Holocene Climatic Optimum

The issue of Holocene temperature has become controversial. While the Holocene Altithermal or Climatic Optimum (c. 9800–5700 BP) is well characterized in the Northern Hemisphere as 1–5 °C warmer than the bottom of the LIA depending on latitude, much less information exists regarding the tropical and Southern areas. Marcott et al. (2013), take the view that, globally, the HCO was 0.7 °C warmer than the bottom of the LIA. Such low temperature variability for the Holocene rests on tropical warming of 0.4 °C during the HCO, and Southern area HCO cooling of 0.4 °C.

At the core of the issue is the question if current temperature is outside the registered bounds for Holocene

temperatures. The cryosphere clearly shows that glaciers all over the world were significantly more reduced during the HCO than at present (Koch et al. 2014). The biosphere generally agrees since the extension of species such as the water chestnut and the pond turtle were then north of their present European climatic limits and the treeline has not reached its HCO maximum latitude or altitude in Sweden (Kullman 2001), Canada (Pisaric et al. 2003), Russia (MacDonald et al. 2000), the Alps (Tinner et al. 1996), or Colombia (Thouret et al. 1996). The marine biosphere agrees as current levels of coccolithophores in the tropical oceans are lower than during the HCO (Werne et al. 2000), which is another indication that the oceans are not as warm as then.

In contrast to Marcott et al. (2013), the non-tropical Southern Hemisphere post-HCO Neoglacial cooling is well documented in the many glaciers from the Southern Andes and New Zealand, reviewed by Porter (2000). Their data demonstrates that Southern Hemisphere glaciers were smaller during the HCO, and that the early Neoglacial advance began between 5400–4900 BP. In southern Africa, Holmgren et al. (2003) have shown persistent Holocene cooling since 10,000 yr BP. In Antarctica Masson et al. (2000), identify an early Holocene optimum at 11,500–9,000 BP followed by a second optimum at 7,000–5,000 BP. Shevenell et al. (2011), show that the Southern Ocean has cooled by 2–4 °C at several locations in the past 10–12 kyr. The Holocene cooling of just 0.4 °C proposed by

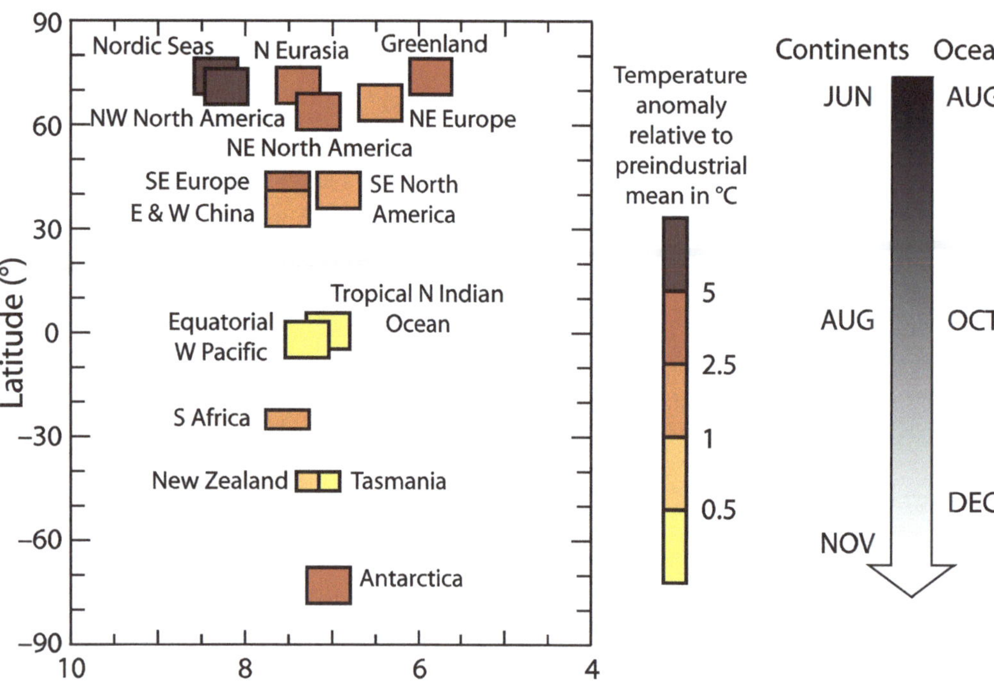

Fig. 4.6 Model characterization of the Holocene Climatic Optimum Global atmosphere-ocean-vegetation model summary of Holocene timing, intensity and seasonal timing of the simulated maximum positive temperature deviation from pre-industrial levels, shown as a function of latitude (vertical axis) and time (horizontal axis). According to the authors, the timing and magnitude of the HCO in their model results is generally consistent with global proxy evidence. The arrow at the right indicates the seasonal timing of the maximum temperature deviation. After Renssen et al. (2012).

Marcott et al. (2013) for the southern 30–90°S region appears to be an underestimation. At southern latitudes, the HCO cannot be explained by precessional summer insolation changes, and large-scale reorganization of latitudinal heat transport has instead been invoked. Decreasing obliquity should also be considered a cause.

However, it is in the tropical areas where Marcott et al. (2013) becomes more controversial. The fossil coral Sr/Ca record at the Great Barrier Reef, Australia, supports a mean SST c. 5350 BP 1.2 °C warmer than the mean SST for the early 1990s (Gagan et al. 1998). At the Indo–Pacific Warm Pool, the warmest ocean region in the world, Stott et al. (2004) find that SST has decreased by c. 0.5 °C in the last 10,000 years, a finding confirmed by Rosenthal et al. (2013), who show a decrease of 1.5–2 °C for intermediate waters. East African lakes show temperatures peaking toward the end of the HCO, followed by a general decrease of 2–3 °C to the LIA (Berke et al. 2012). Tropical glaciers at Peru (Huascarán) and Tanzania (Kilimanjaro) display their highest $\delta^{18}O$ values (warmest) at the HCO, followed by a general decline afterwards (Thompson et al. 2006). The position that the tropics have experienced warming since the HCO appears to be based, in large part, on marine alkenone proxies. However, many alkenone records are from upwelling areas that have high sedimentation rates, but often display inverted temperature trends. Even worse, they generally do not agree with Mg/Ca proxies. Leduc et al. (2010) attempt to resolve the discrepancy between these two paleo-thermometry methods and note that none of the seven Mg/Ca records available for the East Equatorial Pacific have exhibited monotonous warming during the Holocene. They attribute the discrepancy with the alkenone annual temperature signal, by suggesting that it only captures the winter season and thus responds mainly to changes in insolation during that season. This explanation brings the divergent alkenone records in agreement with the rest of the marine and land tropical records that display a tropical cooling since the HCO. If we estimate this cooling in the 0.5–1 °C range it is clear

that Marcott et al. (2013) are underestimating global Holocene cooling and therefore HCO global temperatures.

An estimate of c. 1.2 °C global temperature decrease between average HCO temperatures and the bottom of the LIA is therefore consistent with global proxies, glaciological changes, and biological evidence, and might even be a conservative estimate. Model reconstructions (Renssen et al. 2012; Fig. 4.6) also disagree with Marcott et al. 2013 since they show warming in the tropical areas during the HCO with respect to pre-industrial temperatures. The modeled result cannot be reconciled with Marcott et al. proposal of only 0.7 °C cooling from the HCO to the bottom of the LIA. As the modeled result by Renssen et al. is in better agreement with the abundant biological and glaciological evidence mentioned above, Marcott et al. proposal is difficult to defend, and is probably in error due to proxy bias.

4.5 The Mid-Holocene Transition and the end of the African Humid Period

The MHT is a period of time between 6,000 and 4,800 yr BP when a global climatic shift took place at a time of significant human cultural changes associated with the transition from the Neolithic to the Bronze Age. The MHT separates the HCO from the Neoglaciaciation, and is characterized by periods of global glacier advance interrupted by periods of partial recovery. The principal cause of this global climatic shift was the redistribution of solar energy as the northern summer insolation decrease reached its maximum rate. This redistribution of solar energy, due to orbital forcing, produced a progressive southward shift of the Northern Hemisphere summer position of the ITCZ. This displacement was accompanied by a pronounced weakening of summer monsoons in Africa and Asia and increased dryness and desertification at around 30°N latitude in South America, Africa and Asia. The associated summertime cooling of the NH, combined with changing

Fig. 4.7 Climate pattern change at the Mid-Holocene Transition The shift from the Holocene Climate Optimum to the Neoglaciation involved a complete reorganization of the Earth's climate, mainly directed by the southward migration of the Intertropical Convergence Zone (ITCZ) and the weakening of the African, Indian and Asian summer monsoons. **a)** During the Holocene Climatic Optimum the summer ITCZ reached a higher northern latitude producing a wetter (green horizontal hatching) 30°N tropical band. Northern high latitudes were warmer (red right-slanted hatching), while southern high latitudes were cooler (blue left-slanted hatching). The North American West and Southern Hemisphere mid-latitudes were drier (brown vertical hatching). The Indo–Pacific Warm Pool was warmer (red area), while the equatorial Atlantic and Pacific were cooler (blue areas) and under a weaker El Niño/Southern Oscillation (ENSO) regime. **b)** In the Neoglacial the southern displacement of the ITCZ created the 30°N arid tropical band, and many of the mentioned patterns inverted. The North Atlantic Oscillation (NAO) acquired a more negative character. ENSO became progressively stronger as the planet cooled. After Wanner & Brönnimann (2012).

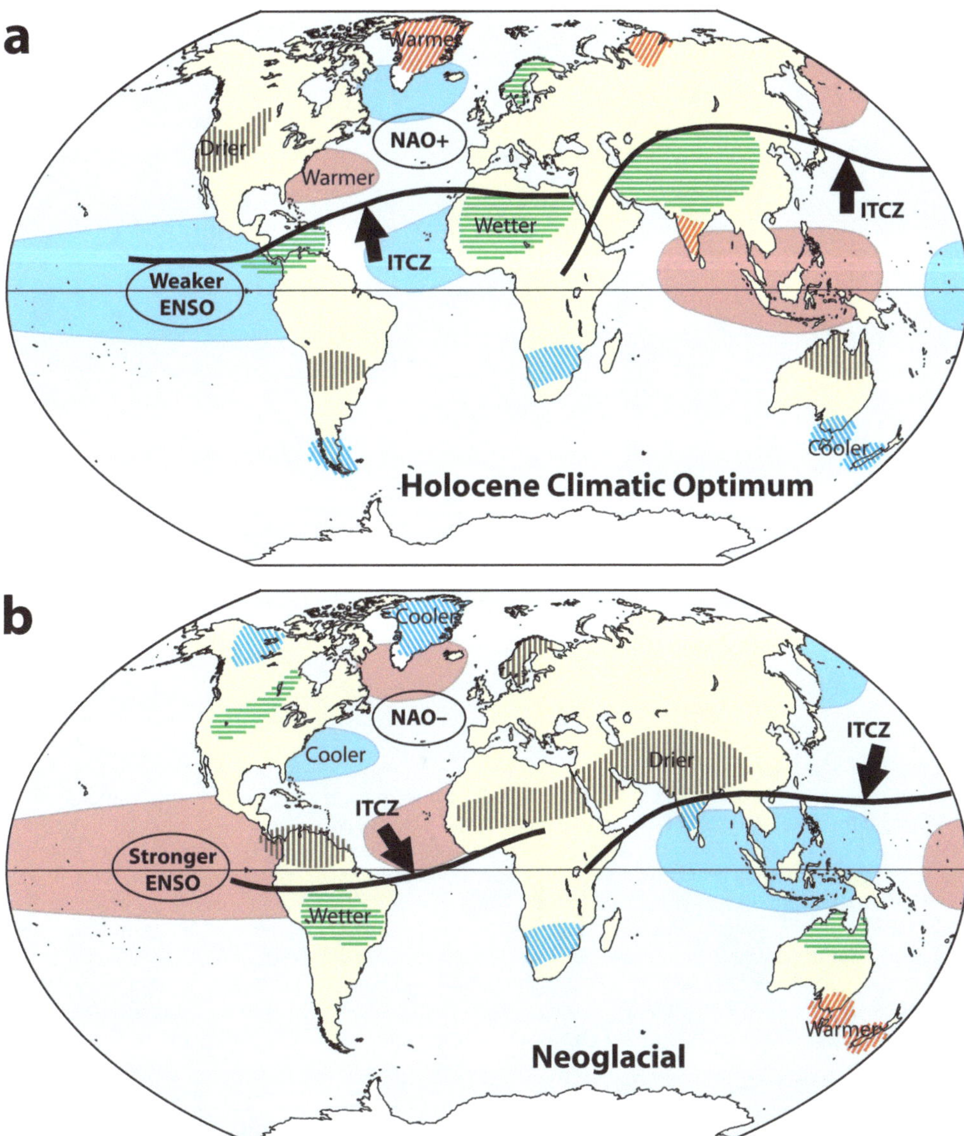

latitudinal temperature gradients in the world oceans, likely led to an increase in the amplitude of the El Niño/Southern Oscillation (ENSO). The effect of these changes had worldwide repercussions in temperature and precipitation patterns (Fig. 4.7).

The 23-kyr precession cycle is the main force behind the seasonal changes in the North–South temperature gradient that are so important for the climate in general and the precipitation regime of the c. 30°N tropical band. The orbital monsoon hypothesis was first proposed by Kutzbatch (1981) and is supported by current evidence. Rossignol–Strick (1985) demonstrated that the dark organic-rich layers in Mediterranean sediments, known as sapropels, represent periods of intense African monsoon every 23 kyr, modulated by the precession and eccentricity cycles. These sapropel formation periods also correspond to African Humid Periods, when the African monsoon produces enough precipitation over the Sahara to sustain a savanna type ecosystem. The last such period started about 15 kyr BP, but had a dry interval during the Younger Dryas at 12.5 kyr BP. As the Sahara became covered in vegetation, with large rivers and lakes, and populated by large mammals, it became inhabited by humans (Fig. 4.8). The Green Sahara entered a dryness crisis around 5.8 kyr BP

and became a desert in just 500 years when the ecosystem collapsed and its human population crashed. Climate refugees from the Sahara greatly increased the population in the Nile valley and shortly after 5500 BP Egyptian society began to grow and advance rapidly towards refined civilization. Extensive use of copper became common during this time (Chalcolithic Period). The process culminated 5100 yr BP with the unification of Egypt under the first pharaoh in one of the first complex civilizations.

The MHT (6 to 4.8 kyr BP), in addition to a global climatic pattern change, also underwent a significant change in the principal climate forcings (Fig. 4.9). This change is reflected in multiple proxies, like North Atlantic ice-rafted debris (Bond et al. 2001), ENSO-linked sedimentation (Moy et al. 2002), or temperate Mediterranean forest pollen abundance (Fletcher et al. 2013), as a change in frequency at the MHT. Several authors have noticed this change in the principal climate forcings at the MHT. Debret et al. (2009) after studying 15 climate proxies from marine sediments (North Atlantic, West Africa and Antarctic), ice-core records (South America and Antarctica), cave speleothems (Ireland) and lacustrine sediments (Ecuador) by wavelet analysis, concluded that the first part of the Holocene was characterized by frequencies typical of solar

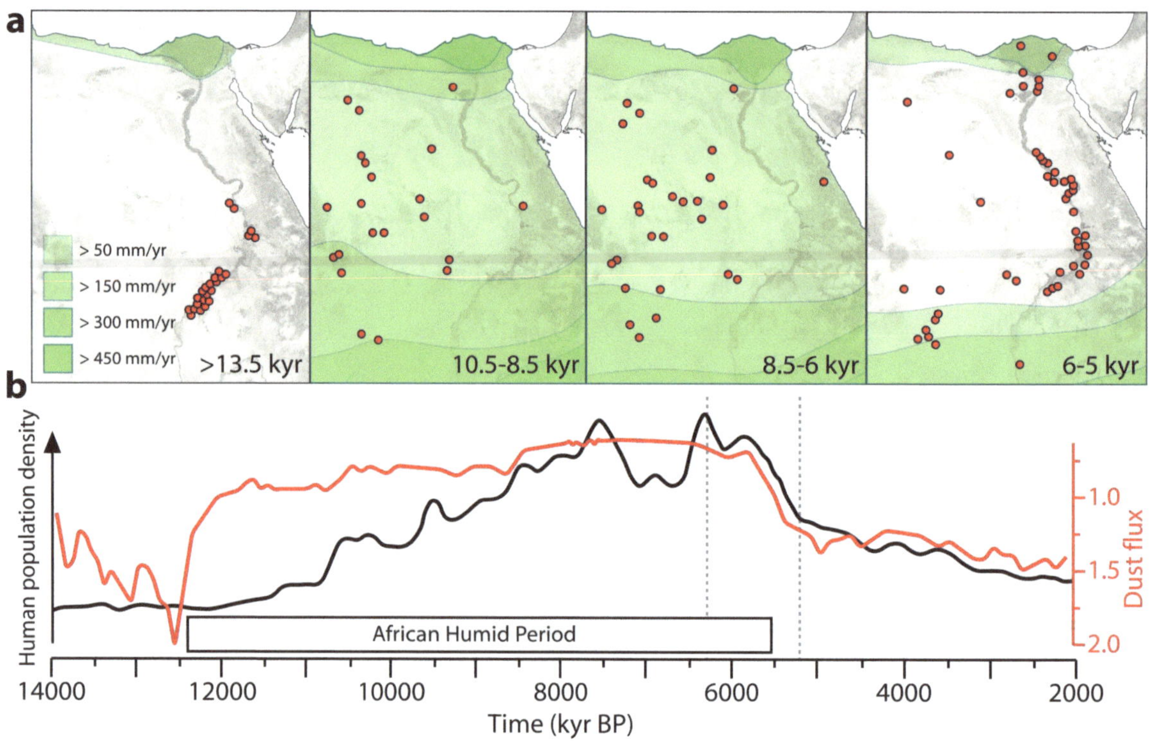

Fig. 4.8 The African Humid Period

a) Climate-controlled occupation in the Eastern Sahara during the main phases of the Holocene. Red dots indicate major occupation areas. During the Last Glacial Maximum and the terminal Pleistocene (20,000 to 12,500 BP), the Saharan desert was void of any settlement outside of the Nile valley. With the abrupt arrival of monsoon rains at 12,500 BP, the desert was replaced by savannah-like environments and became inhabited by prehistoric settlers. After 9,000 BP, human settlements became well established all over the Eastern Sahara, fostering the development of cattle pastoralism. Retreating monsoonal rains caused the onset of desiccation of the Egyptian Sahara at 6,300 BP. Prehistoric populations were forced to the Nile valley. The return of full desert conditions all over Egypt at about 5,500 BP coincided with the initial stages of pharaonic civilization in the Nile valley. After Kuper & Kröpelin (2006). **b)** Dust flux (aridity proxy, red curve, inverted) record from the NW African coast 658C core related to a population proxy (black curve) based in the summed probability distribution of 3287 calibrated ^{14}C ages from 1011 archeological sites between 14,000–2,000 years BP. Dotted lines indicate the concurrent end of the African Humid Period and population collapse. After Manning & Timpson (2014).

variability at 1000 and 2500 years, with a shift to 1200–1800-year oceanic-atmospheric frequencies in the last half of the Holocene (Fig. 4.9f). Some of the proxies go from displaying a c. 1000-yr solar frequency during the HCO to displaying a c. 1500–1800-yr oceanic-atmospheric frequency during the Neoglacial (Fig. 4.9e). Even more proxies start showing a c. 1500-yr frequency at the MHT they did not show in the early Holocene (Debret et al. 2007). Several causes can contribute to these changes. For most of the Holocene, Northern Hemisphere summer insolation had been decreasing and winter insolation increasing. As a result, the difference in insolation between summer and winter had decreased by about 55 W/m^2 at 45°N (Fig. 4.9a, grey area). This decrease in seasonality was more pronounced during the MHT, and must have had the effect of reducing seasonal differences and climate response to seasonal differences in insolation. At the MHT Earth's axial obliquity transitioned from high to an increasingly faster decrease (Fig. 4.9b). The decrease in obliquity reduced insolation at the poles and increased it at the tropics. This had the effect of increasing the latitudinal insolation gradient and the latitudinal temperature gradient, activating atmospheric and oceanic circulation resulting in an enhanced heat meridional transport, particularly during the winter. Another possible cause is the emergence of the modern oceanic circulation with the stabilization of the thermohaline circulation c. 5 kyr BP (Fig. 4.9c), after sea

level stopped increasing (Fig. 4.9d), and ocean salinity stopped decreasing (Fig. 4.9c), with the end of the ice-sheets melting. Finally, changes in solar periodicities might be responsible for some of the observed changes (Fig. 4.9f). The c. 2500-yr periodicity is nearly continuous through the Holocene, with a missing oscillation at 7.7 kyr BP (see Chap. 5), but the c. 1000-yr periodicity, very pronounced until 5.2 kyr BP, becomes undetectable during most of the Neoglacial, until near the end of the Roman Warm Period (see Chap. 8), when it becomes noticeable again.

A bimodal climate pattern was also identified during the Holocene by Moy et al. (2002) in the Laguna Pallcacocha sediment proxy for ENSO activity for the last 12 kyr (Fig. 4.10). They found that around 5 kyr BP ENSO variance shifted from a c. 1000-yr period in the HCO to a c. 2500-yr period in the Neoglacial. ENSO activity was very reduced during most of the HCO increasing from around 7 kyr BP, and more prominently after 5.6 kyr BP, displaying many very strong peaks of activity during the Neoglacial period. Moy et al. also observe that periods of high North Atlantic iceberg activity, indicative of significant cooling (Bond events) are flanked by periods of high ENSO activity, but coincide with periods of marked decrease in ENSO activity (Fig. 4.10). This relation suggests that some link may exist between the two systems (Moy et al. 2002). It is proposed in chapters 10 and 11 that ENSO

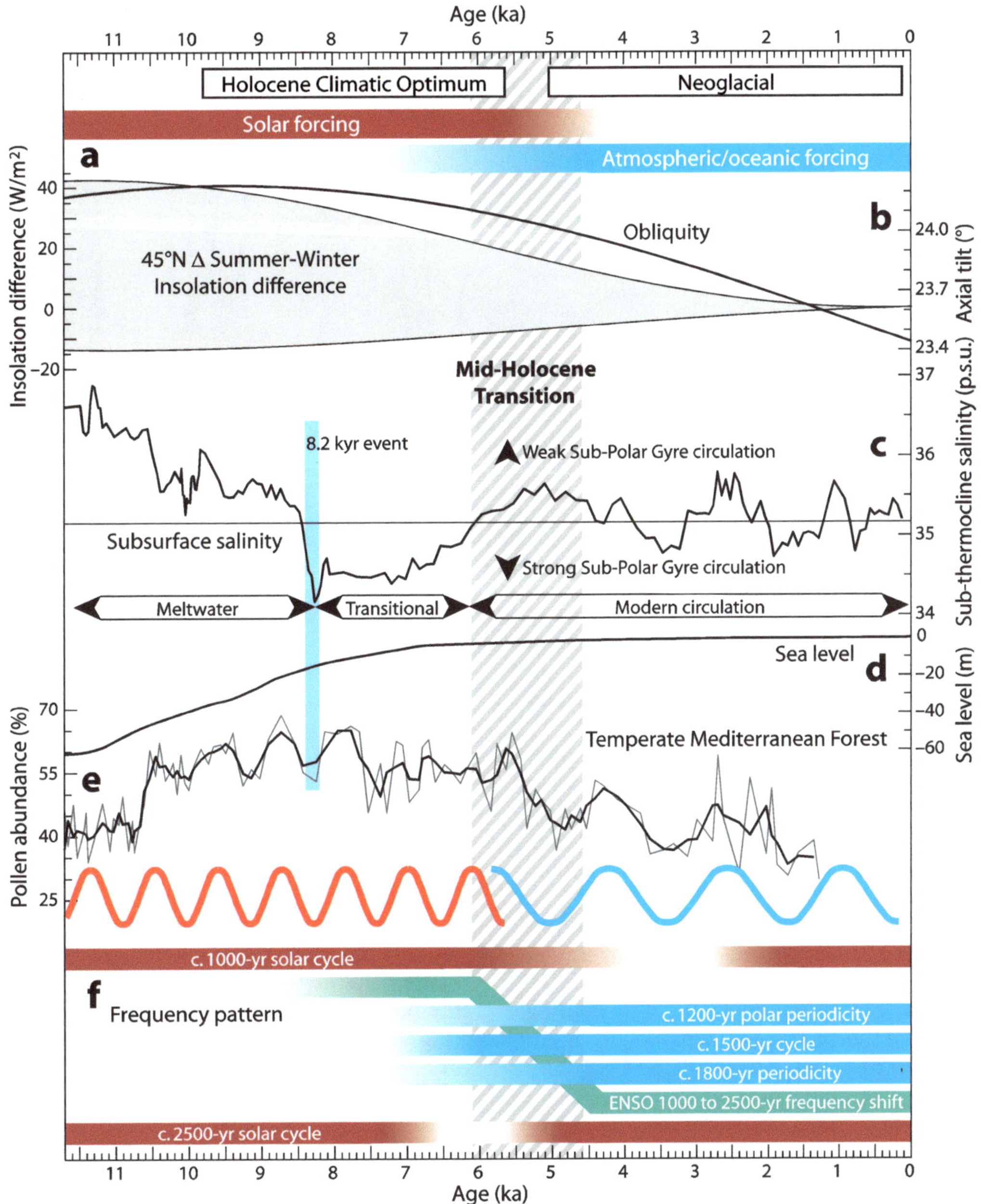

Fig. 4.9 Holocene climate shift at the Mid-Holocene Transition
Holocene climate is dominated by solar-attributed frequencies (red bars) during the Holocene Climate Optimum (HCO), and by atmospheric-oceanic-attributed frequencies (blue bars) during the Neoglacial. **a)** The difference between the summer and winter insolation curves at 45°N reflects the changes in seasonality over time (grey area). Summer-winter seasonality was c. 55 W/m² more marked in the early Holocene that at present, most of that due to higher summer insolation (positive values in left scale). The decrease in summer–winter seasonality was very marked during the Mid-Holocene Transition (MHT). **b)** Obliquity was very high and slow changing during the entire HCO, accelerating its decrease during the MHT. As a result high latitudes received less insolation and tropical latitudes more, and the latitudinal insolation gradient and latitudinal temperature gradient became steeper, leading to more active atmospheric and oceanic circulation and increased heat meridional transport. **c)** Sub-thermocline salinity reconstruction at the RAPiD 12-1K core (south of Iceland) shows that subsurface salinity and the strength of the Sub-Polar Gyre contribution to the Atlantic Meridional Overturning Circulation were dominated by meltwater production until the 8.2 kyr ACE. After a transitional period, modern circulation characterized by oceanic frequency oscillations started at the MHT. After Thornalley et al. (2009). **d)** Sea level greatly reduced its rate of increase at c. 6.8 kyr BP with the melting of the continental ice-sheets, and essentially stopped increasing with the end of the HCO at the MHT. After Lambeck et al. (2014). **e)** Pollen record from western Mediterranean core MD95-2043 as percentage of all temperate and Mediterranean forest taxa, with three-point running mean in bold. The early Holocene displays a c. 1000-yr periodicity (red sinusoid), while the late Holocene displays a c. 1800-yr periodicity (blue sinusoid). After Fletcher et al. (2013). **f)** Changes in frequency patterns displayed by multiple proxies during the Holocene. Solar patterns in red, atmospheric-oceanic patterns in blue, ENSO pattern in aquamarine. The c. 1000 and 2500-yr solar frequencies are not continuous over the Holocene and display gaps at times when the frequency becomes inconspicuous. For an analysis of those gaps see Fig. 8.1. The c. 1200-yr periodicity is found in Antarctic (Crosta et al. 2007) and Greenland (Humlum et al. 2011) Neoglacial records, but not during the HCO. The c. 1500-yr periodicity is very common in multiple atmospheric-oceanic proxies (see Chap. 7). The c. 1800-yr periodicity is found in an Irish speleothem record (McDermott et al. 2001) and in central (Di Rita et al. 2018) and western (Fletcher et al. 2013) Mediterranean forest records. After Debret et al. (2009) and Simonneau et al. (2014).

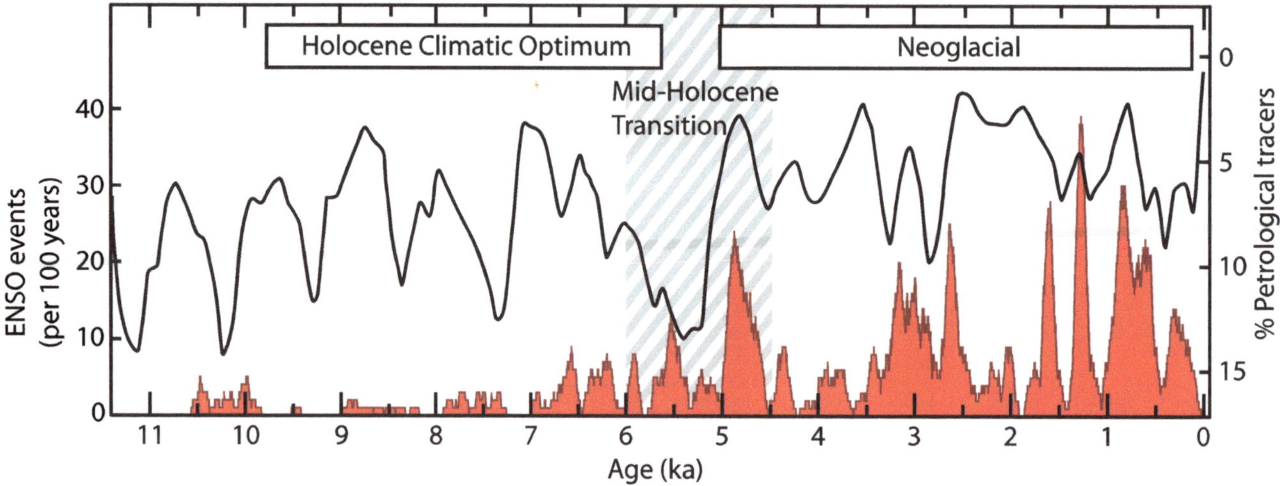

Fig. 4.10 El Niño/Southern Oscillation (ENSO) Holocene activity
Lower red curve, ENSO activity (number of events per century) displays a bimodal distribution, with low ENSO activity during the HCO and high ENSO activity during the Neoglaciation. Upper curve, cold Bond events, marked by increases in ice-rafted debris (inverted), tend to display higher activity preceding and following periods of high ENSO activity, and coincide with periods of low ENSO activity. The effect is as if Bond events would have left an impression in the ENSO frequency curve. After Bond et al. (2001) and Moy et al. (2002).

is linked to the meridional transport of energy, that increases in intensity during cooling periods and decreases during warm periods.

From a thermodynamic point of view high ENSO activity transfers great amounts of heat from the Pacific Ocean sub-surface to the atmosphere, and afterwards a great part of that heat is transported poleward and radiated to space. This constitutes a cooling event from a whole Earth climate system perspective, even if it appears as warming from a lower atmosphere perspective. It is proposed that high ENSO activity is made possible by a high latitudinal (equator-to-pole) temperature gradient (LTG). Under this paradigm, ENSO is the mark of a cooling planet. During the HCO the LTG was kept low by high polar insolation due to high obliquity and ENSO activity was very low. After 7 kyr BP the decrease in polar insolation and the increase in tropical insolation favored a progressive increase in the gradient and in ENSO activity. At periods of planetary cooling immediately preceding a Bond event, the LTG would become steeper, increasing ENSO frequency. But most Bond events are characterized

by very low solar activity (Sect. 4.8, below). As ENSO is under solar control (see Sect. 10.4), long periods of very low solar activity might inhibit ENSO activity, so the periods of highest iceberg activity in the North Atlantic coincide with periods of reduced ENSO activity. Modern Global Warming displays a reduced ENSO activity due to a decrease in the Northern Hemisphere LTG partly due to enhanced Arctic warming.

As we have seen, the MHT was a period of time characterized by profound climatic changes that were the product not only of orbital changes that altered global climatic patterns (Fig. 4.7), but also by a coincident shift in the predominant frequencies of multiple climate forcings (Fig. 4.9), and the emergence of ENSO as a major source of climate variability (Fig. 4.10). At the time the climate of the Holocene was experiencing its biggest transformation and the North Atlantic displayed its most sustained increase in iceberg activity, according to ice-rafted debris proxies (Bond et al. 2001; Fig. 4.11). After the 6.8 and 6.3-kyr ACEs, iceberg activity did not return to lower values, and continued increasing for almost two millennia

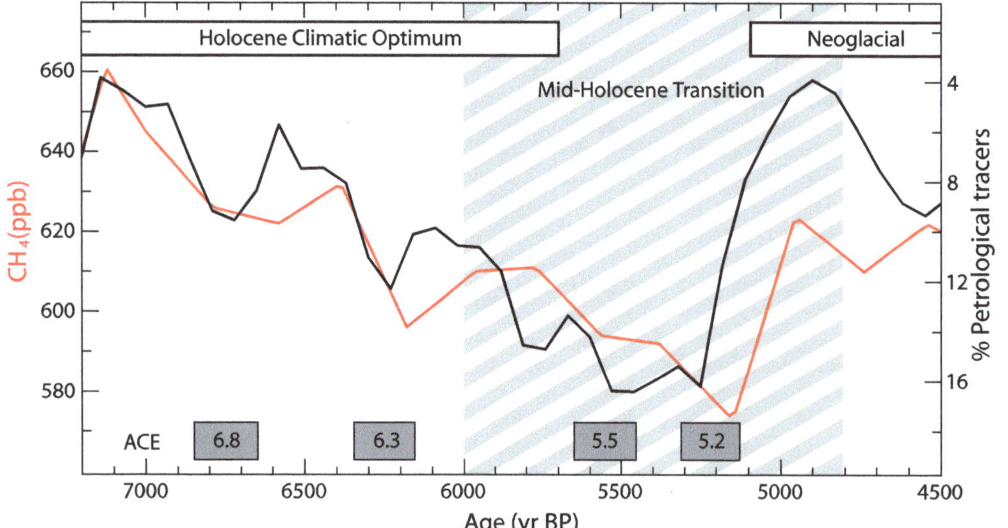

Fig. 4.11 Climate commitment at the Mid-Holocene Transition
After the 6.3-kyr ACE, iceberg proxy records in the North Atlantic indicate a lack of recovery from elevated levels (ice-rafted petrological tracers; black curve, right scale). The increase in iceberg activity, suggestive of cooling, continued until the 5.2-kyr ACE. It was accompanied by a decrease in Holocene methane levels (GRIP ice core parts per billion; red curve, left scale). The recovery from the 5.2-kyr ACE resulted in a different Holocene climate. After Bond et al. (2001) and Blunier et al. (1995).

through the 5.5-kyr ACE, until the 5.2-kyr ACE. It was only after this important event (see below), that iceberg activity decreased to basal levels, according to proxies (Fig. 4.11). This pattern is reproduced by methane levels measured at GRIP ice-core (Blunier et al. 1995; Fig. 4.11), that show a corresponding atmospheric methane decrease at each ACE. After the 5.2-kyr ACE methane levels not only partly recovered, but methane changed its long-term trend in response to the MHT climatic change. In the GRIP record, methane reached its lowest Holocene concentration at 5160 BP and from then on it started to increase. According to Singarayer et al. (2011), and in agreement with the discussed profound climate changes of the MHT, the increase in methane was caused by the described orbital changes that resulted in enhanced precipitation over southern tropical areas.

The MHT was the most important climatic change of the Holocene, and it reached its culmination at the 5.2-kyr ACE. The International Subcommission on Quaternary Stratigraphy decided in 2018 to propose a subdivision of the Holocene according to climatic criteria. It produces stupor that they ignored the MHT and chose instead the 4.2-kyr ACE (see Sect. 7.6). The 4.2-kyr ACE is a regional event that mainly affected the area surrounding the Indian Ocean, with a noticeable effect at the Northern Hemisphere subtropical band, and practically without effect elsewhere. The 4.2-kyr ACE left no global climatic consequences. It didn't change the climate of the planet in any discernible way. The choice of such unimportant climatic event leads to the suspicion that the subcommission just wanted to divide the 12-kyr long Holocene in three 4-kyr

periods to use the early- middle- and late- terms, and just chose the 8.2-kyr and 4.2-kyr ACEs as convenient slice points. Hardly a rigorous scientific decision after 10 years pondering on the question.

4.6 The Neoglacial period

Neoglaciation was the term coined to describe the global glacier advances after the HCO that François Matthes identified in the 1940s. Glacier growth was caused by orbital-driven insolation changes. Although variability in local conditions caused the Neoglacial period to start at different times in different glaciological areas, it is generally agreed that it started c. 5000 years BP in high northern latitudes (Geirsdóttir et al. 2019), and there is evidence also from the tropics and the Southern Hemisphere supporting that date (Thompson et al. 2006). Glaciers fluctuated with major glacier advances followed by shorter glacier retreats, culminating in the LIA when globally glaciers reached their maximum Holocene extent (Fig. 4.12). The Neoglaciation featured global cooling as temperatures responded more to the decrease in solar forcing due to orbital insolation changes than to the increase in GHG forcing. Centurial glacier advances appear to reconstruct the general temperature evolution during the Holocene, and display a clear correspondence to known cooling events (compare Fig. 4.12a and b), lending strong support to the Holocene temperature conundrum. Temperature changes during the Holocene do not appear to follow CO_2 changes. This is strongly supported by implied large perturbations in ocean heat content and Earth's energy budget

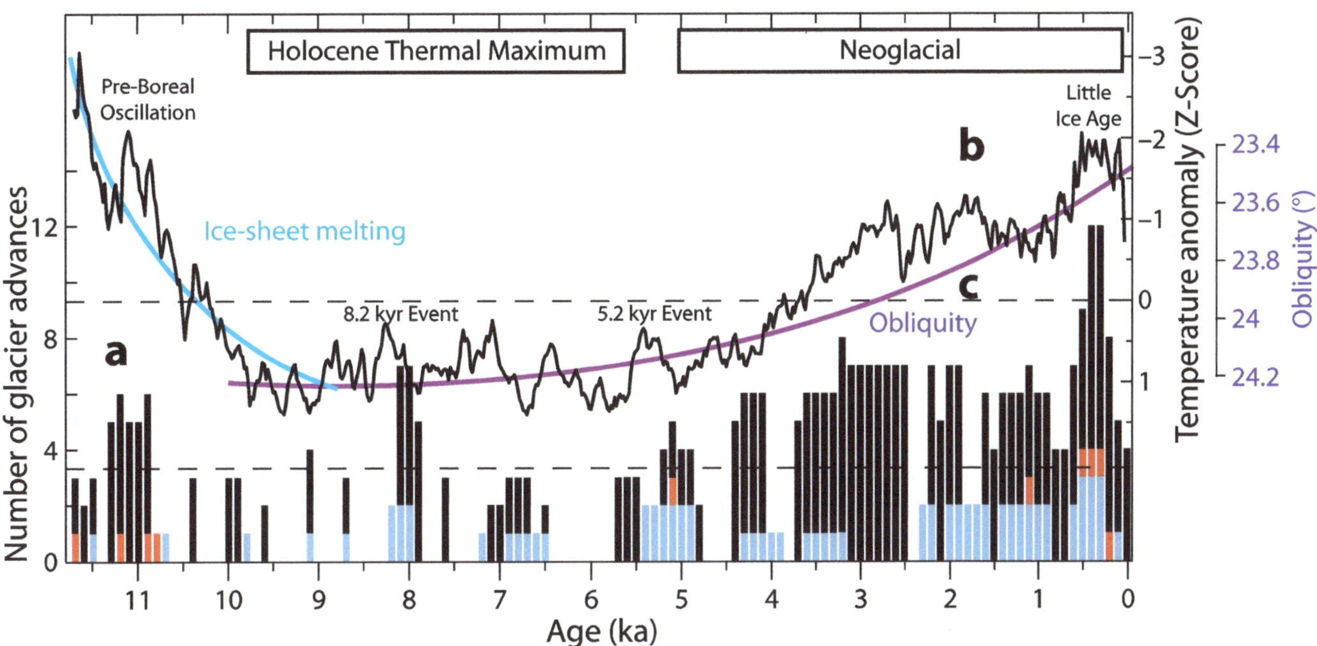

Fig. 4.12 Global glacier advances during the Holocene
Number of areas that display glacier advances for every century during the Holocene versus temperature changes. **a)** World glaciers were distributed by Solomina et al. (2015) between 17 geographical areas. 12 belonging to the Northern Hemisphere are represented in black, 4 from the Southern Hemisphere in light blue, and one for the Low Latitudes in orange. For a geographical representation of the glaciers included in each area see figure 1 in Solomina et al. (2015). Colored bars (left scale) represent number of glacial advances per century for each area. Bottom dashed line, Holocene average centurial glacial advances. After Solomina et al. (2015). **b)** Black curve, Holocene temperature reconstruction (inverted; inner right scale), as in Fig. 4.4. Some well-known cooling periods or events are indicated by their accepted names. The period affected by the melting of the extra-Arctic ice sheets is indicated in aquamarine. The Neoglacial period is characterized by generalized glacier advances that take place coinciding with the decrease in Northern Hemisphere solar forcing. **c)** Inverted changes in obliquity.

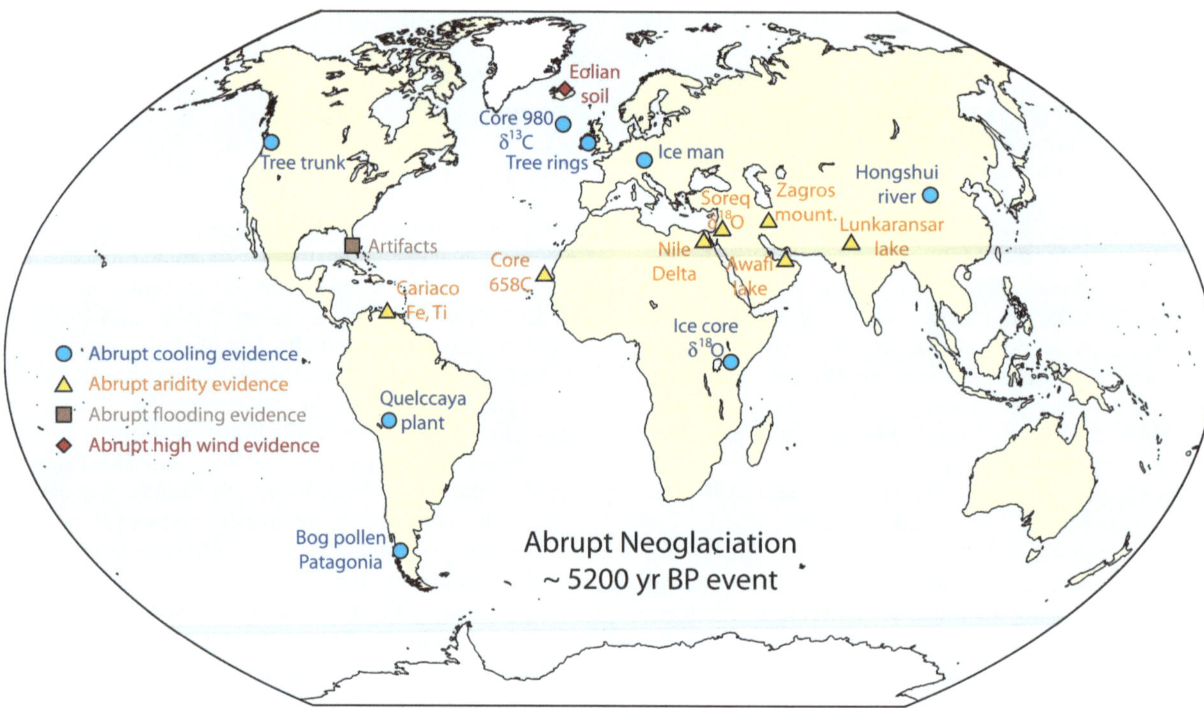

Fig. 4.13 Evidence for an abrupt global cold and arid event at 5.2 kyr BP
High and low latitude locations of proxy evidence for abrupt climate change c. 5,200 yr BP. Evidence for abrupt cooling (circles), aridity (triangles), flooding (square) and high wind (diamond). South-Cascade Glacier rooted tree-trunk (Washington State); remains and artifacts in the Little Salt Spring (Florida); Cariaco Basin metal concentration (Fe, Ti) in ODP site 1002; Quelccaya Glacier ice-buried wetland plant *Distichia muscoides* (*Juncaceae*), dated at 5,138 ± 45 yr BP; bog pollen records of rapid and drastic vegetation changes in Isla Santa Inés (Chile); eolian soil record from Hólmsá (Iceland); North Atlantic benthic core in ODP site 980; dendrochronological records from Irish and Lancashire oaks with some of their narrowest rings during the 3,195 BC decade; Ötzi, the ice-man from South-Tyrol; core S53 palynological record from Burullus Lagoon (Nile Delta); Soreq Cave (Israel) speleothem; Mauritanian coast core 658C; Kilimanjaro ice-core record; Awafi dry lake sediments in SE Arabia; Lake Mirabad sediment in the Zagros Mountains (Iran); Lunkaransar dry lake sediments in NW India; sedimentary section along the Hongshui River, in the southern Tengger Desert, NW China. From multiple sources, some referenced in Thompson et al. (2006).

at odds with the very small GHG radiative forcing anomalies throughout the Holocene (Rosenthal et al. 2017).

Cooling events during the HCO, like the 8.2-kyr ACE, were followed by a complete recovery of temperatures and globally glaciers reached their minimum Holocene extent in most areas between 6–5.5 kyr BP. However there is evidence that the world did not completely recover from the cooling events that took place between 5.6 and 5.1 kyr BP, initiating the Neoglaciation. This Mid-Holocene climate reversal has been recorded globally in multiple proxies both as a decrease in temperature and as hydrological changes (Magny & Haas 2004; Thompson et al. 2006). While the entire sixth millennium BP had a very challenging climate compared to previous millennia, the cooling event that took place at 5.2 kyr BP was particularly abrupt (Fig. 4.13, Thompson et al. 2006). Due to the contemporary change of climate regime and global temperature, some regions became cooler and drier, while others became cooler and wetter, leading to a rapid global glacier advance that buried organic remains, like the Quelccaya Glacier plant (*Distichia muscoides*, Peru), the South-Cascade Glacier rooted tree-trunk (Washington State) and the Ötztal Alps ice-man, that have remained continuously frozen until the present global warming (Thompson et al. 2006).

Coincident with the abrupt cooling and hydrological changes of c. 5,200 yr BP, archaeological studies support a general pattern of abandoned Neolithic human settlements in several areas, including the Andes and the entire Eastern Mediterranean, indicating a widespread climatic crisis that marks the transition from the Chalcolithic to the early Bronze Age (Weninger et al. 2009; see Chap. 6).

4.7 Holocene climate variability

The Last Glacial Maximum and the HCO constitute two extreme metastable states, separated by only 10,000 years, that correspond to essentially the same amount and distribution of incoming energy from the Sun. The main difference between both states is in the redistribution and minimal or maximal exploitation of that energy by the planet. This difference is due to orbital configuration, tectonic disposition, ice and cloud albedo, oceanic-atmosphere response and biological feedback. Since they constitute dramatically different climatic states, the nature of abrupt climatic changes is also different in the two states. Glacial variability comes mainly in the form of warming episodes (Dansgaard–Oeschger events; Fig. 4.14) while interglacial variability comes from cooling episodes (Bond events; Fig. 4.14). There are no global warming abrupt changes in the Holocene once the thermal maximum is reached, just cooling events followed by recovery.

The other major salient characteristic of the Holocene abrupt climatic changes compared to glacial abrupt changes is their much smaller amplitude (Fig. 4.14). It has become a lot more difficult to identify these changes be-

Fig. 4.14 Nature of climatic oscillations during the Ice Age
Oscillations during an interglacial are smaller and of cooling nature (Bond events), and oscillations become larger the colder temperature becomes. During the glacial period oscillations are very large and are of a warming nature (Dansgaard–Oeschger events). Black line represents the obliquity cycle. An asterisk marks the current position.

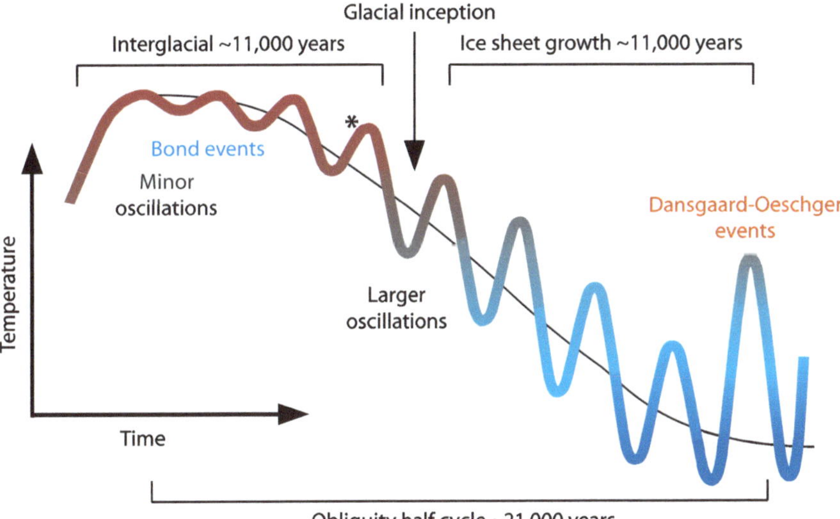

cause their signal is much lower and more difficult to separate from the noise of small high frequency climatic variability. This has created much confusion about the nature and causes of Holocene ACEs and has given many the false impression that the Holocene is characterized by long periods of climate stability. Nothing is further from the truth. The Holocene is a period of almost constant climate change with climatic stability being the exception.

In 1968 climatologist J. Roger Bray recognized several major past cooling episodes and attributed them to a solar cycle. *"A combination of geophysical, biological and glaciological information supports the idea of a 2,600 year solar cycle"* (Bray 1968). Afterwards a spurious c. 2,300-yr solar periodicity was identified in solar cosmo-

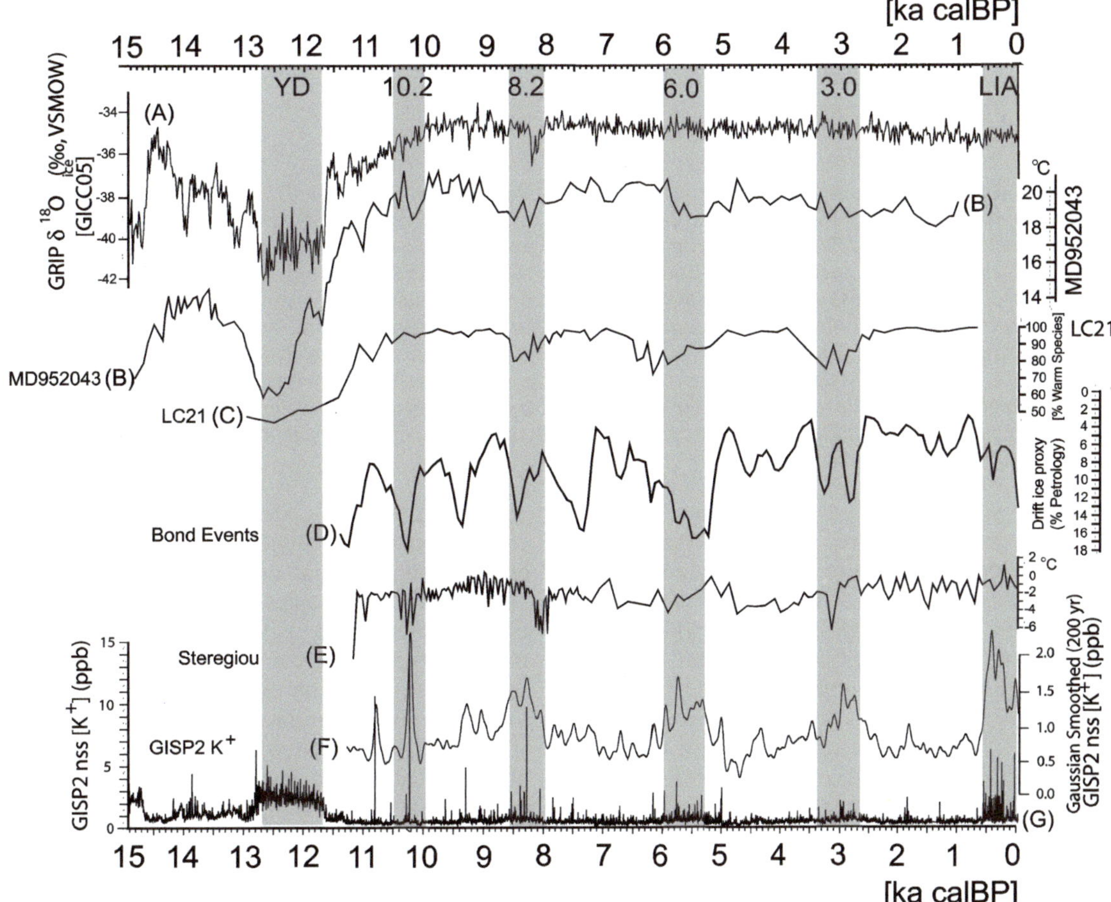

Fig. 4.15 Northern Hemisphere paleoclimate records showing main Holocene abrupt climatic change events
A) Greenland GISP2 ice-core δ18O. B) Western Mediterranean (Iberian Margin) core MD95-2043, sea surface temperature (SST) C37 alkenones. C) Eastern Mediterranean core LC21 (SST) fauna. D) North Atlantic Bond series of drift-ice stacked petrologic tracers. E) Romania (Steregoiu), mean annual temperature of the coldest month. F) Gaussian smoothed (200 yr) GISP2 potassium (non-sea salt) ion proxy for the Siberian High pressure system. G) High resolution GISP2 potassium (non-sea salt). Notice that the main Holocene ACEs are cooling events. Reproduced from Weninger et al. (2009), with the author's permission. © Weninger et al. (2009).

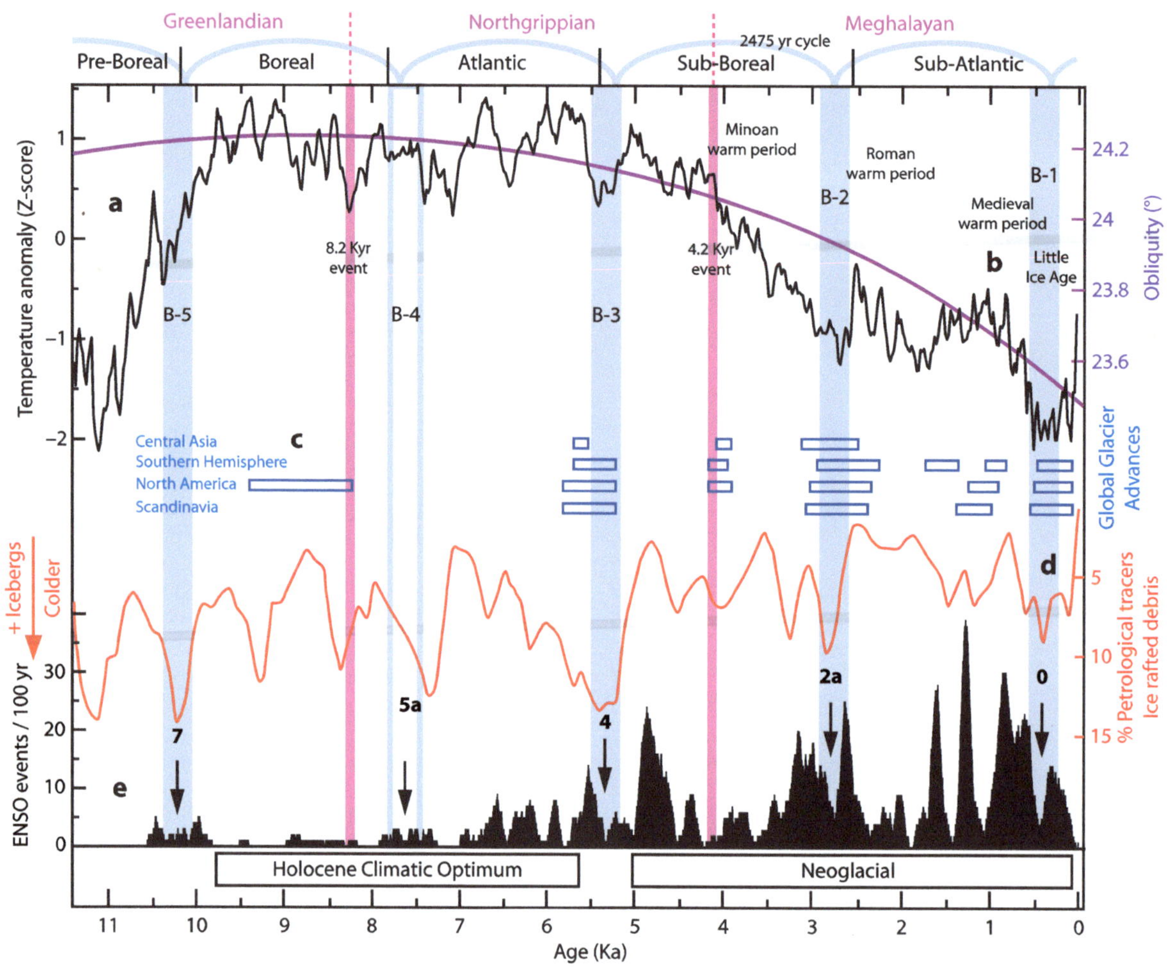

Fig. 4.16 Major periods of the Holocene set by obliquity and the c. 2500-yr Bray cycle
a) Global temperature reconstruction from 73 proxies with temperature anomaly expressed as Z-score (distance to the mean in standard deviations). **b)** Earth's axis obliquity cycle. **c)** Major periods of regional and global glacier advances after Mayewski et al. (2004) and references within. **d)** Ice-rafted debris stack (inverted) from four North Atlantic sediment cores as a proxy for iceberg activity and cold events, after Bond et al. (2001). **e)** ENSO events frequency as determined from a glacial lake core in the Peruvian Andes, after Moy et al. (2002). Blue vertical bars (B-1 to 5), cooling oscillations part of the c. 2500 year Bray cycle. Empty bar (B-4) indicates a low effect oscillation. Narrow pink vertical bars, the 8.2-kyr cooling event coincident with Lake Agassiz outburst and the 4.2-kyr arid-cold event, define the Greenlandian, Northgrippian, and Meghalayan stages (on top). Arches on top constitute a regularly spaced 2475-year periodic marker. The Blytt–Sernander high northern latitude climatic subdivision is shown below the arches. Bold numbers correspond to Bond events that coincide with the Bray cycle cold oscillations. Vertical arrows show the coincidence of Bray cycle lows with suppressed ENSO activity. Some historical climate periods have been indicated.

genic records and wrongly named the Hallstatt cycle, while it should be named the Bray cycle and correctly identified as c. 2500-yr in period. Since Bray's report, other researchers have confirmed the reoccurrence of cooler climates with a periodicity of about 2400–2600 years by different techniques, glacial moraines, temperature-sensitive tree rings widths, and $\delta^{18}O$ isotope and chemical analysis of sea salts and dust in ice cores (O'Brien et al. 1995). Most researchers also ascribe a solar origin to this climatic cycle, since its cooling periods coincide with periods of high $\Delta^{14}C$ formation, which are associated with low solar activity.

By looking at proxy temperature reconstructions, at major global glacier advances, and at other climate proxies, it is easy to recognize the major abrupt cooling changes of the Holocene. Roger Bray identified cooling episodes at 0.4, 2.8, 5.5, 8.2 and 10.2 kyr BP over 45 years ago (Fig. 4.15). These episodes give us an average spacing of c. 2500 years and, at the same time, they define the major climatic states of the Holocene.

The Bray cycle delimits five periods that roughly correspond to the Blytt–Sernander sequence. Vegetation changes suggest that they constitute distinctive climatic states established by insolation conditions from the obliquity and precession cycles (Fig. 4.16). Every abrupt cooling from the Bray cycle would constitute a tipping point in the gradual insolation changes and the world would settle to a different climatic state after recovering. We have just started a sixth period with the proposed name of Anthropocene, that should last around 2,200 years, until about AD 4,200. Every one of the last five periods (since 10.2 kyr ago) started with global warming as a recovery from the depressed temperatures of the cooling oscillations that separate the periods.

4.8 Bond events and other Abrupt Climatic Events

In addition to the major cooling events of the Bray cycle, other cooling ACEs have taken place during the Holocene, and they have been detected in numerous proxies, but particularly in the Bond series of events. The amount of detrital petrological tracers transported by icebergs and deposited in the ice-rafted debris belt (an Atlantic region between 40–50°N) greatly increases during episodes of southward and eastward advection of cold surface waters and drift ice from the Nordic and Labrador seas (Bond et al. 2001; Fig. 4.17a). This sensitive proxy has registered most cold episodes of the Holocene, with a resolution of 50 years.

Gerard Bond attempted to fit the periods of increased drift-ice that he identified during the Holocene into a single cycle related to the Dansgaard–Oeschger cycle, by making two unwarranted assumptions: That every period of cooling responded to the same cause, and that some well-resolved peaks separated by several centuries to a millennium could correspond to a single cold event. The evidence, however, shows that the HCO displays a millennial periodicity in Bond events, with single isolated peaks separated by c. 1000 years, while the Neoglacial period shows a more complex picture with multiple peaks not so well resolved and a more irregular spacing (Debret et al. 2007; Fig. 4.17). It is clear that the Bond record mixed periodicity reflects the climatic shift that took place at the MHT from mainly solar forcing to a mixed solar and atmospheric-oceanic forcing (Fig. 4.9), and therefore it can be concluded that the first assumption of Gerard Bond is incorrect: different peaks represent cooling from different causes, and thus a Bond cycle does not exist in the Holocene. We must reject also his second assumption and treat every peak as a different cooling event and try to identify the cause that originated it. We must move from a Bond series of 8 events (plus number zero) in 12 kyr (one event every 1500 years), to a series of at least 15 cold events with a mixture of periodicities during the Holocene.

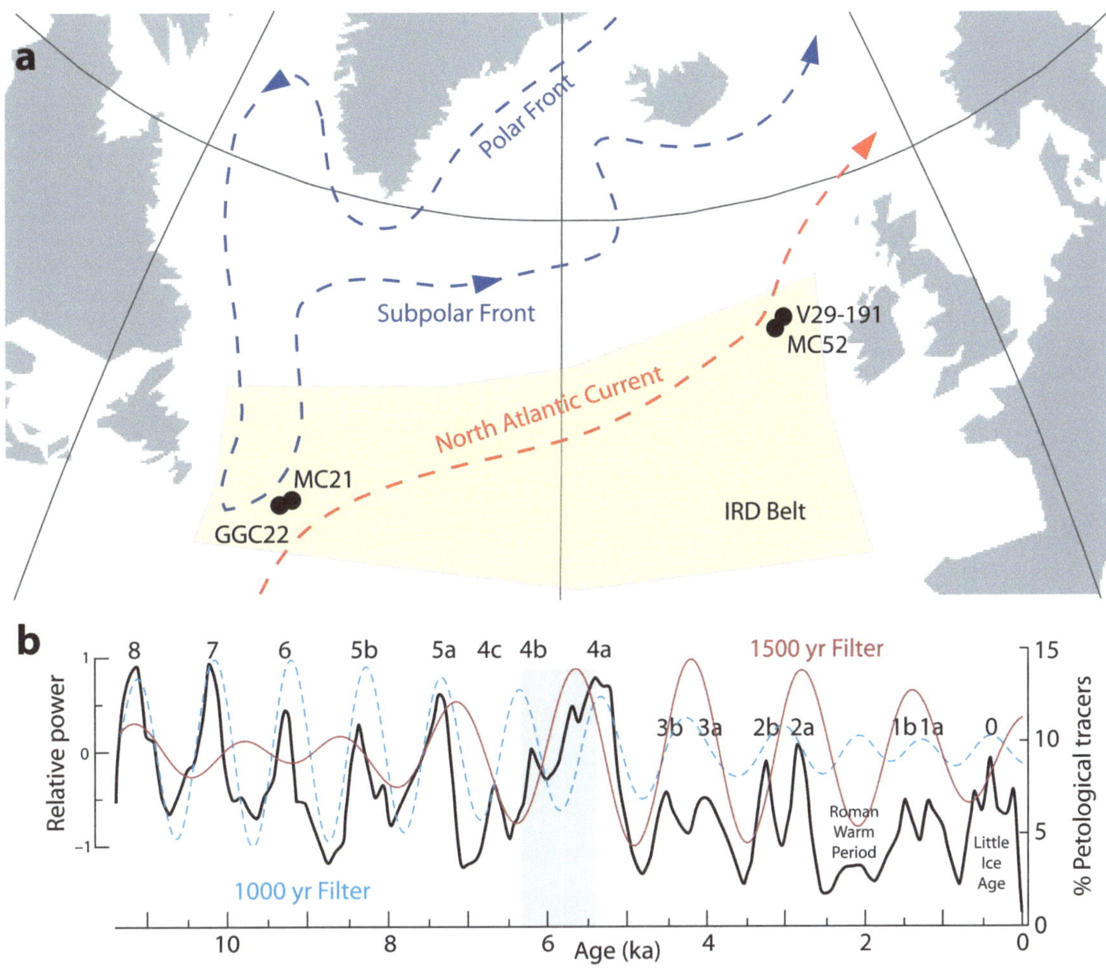

Fig. 4.17 Bond events constitute a record of cold events during the Holocene
a) Map of North Atlantic coring sites. Bond events represent periods of increased deposition of petrological tracers by drift ice at the core locations (black dots) within the ice-rafted debris belt (IRD, yellow/light grey box). They are interpreted as periods of cooler, ice-bearing surface waters displaced eastward from the Labrador Sea and southward from the Nordic Seas. **b)** The Holocene record of iceberg activity (black curve) is a stack of the four cores showing the combined detrended record of hematite-stained grains, detrital carbonate, and Icelandic volcanic glass. The last drift-ice period corresponds to the Little Ice Age, and other known climatic periods of the past can be correlated to this record. The early Holocene period clearly displays a 1000-yr periodicity as shown by a band-pass filter applied on the series (blue dashed curve). A 1500-yr periodicity is only present from 6 kyr BP (red continuous curve). The 1500-yr fit is problematic as it requires considering pairs of peaks as single events and some peaks appear to still follow the 1000-yr periodicity. After Bond et al. (2001); Debret et al. (2007).

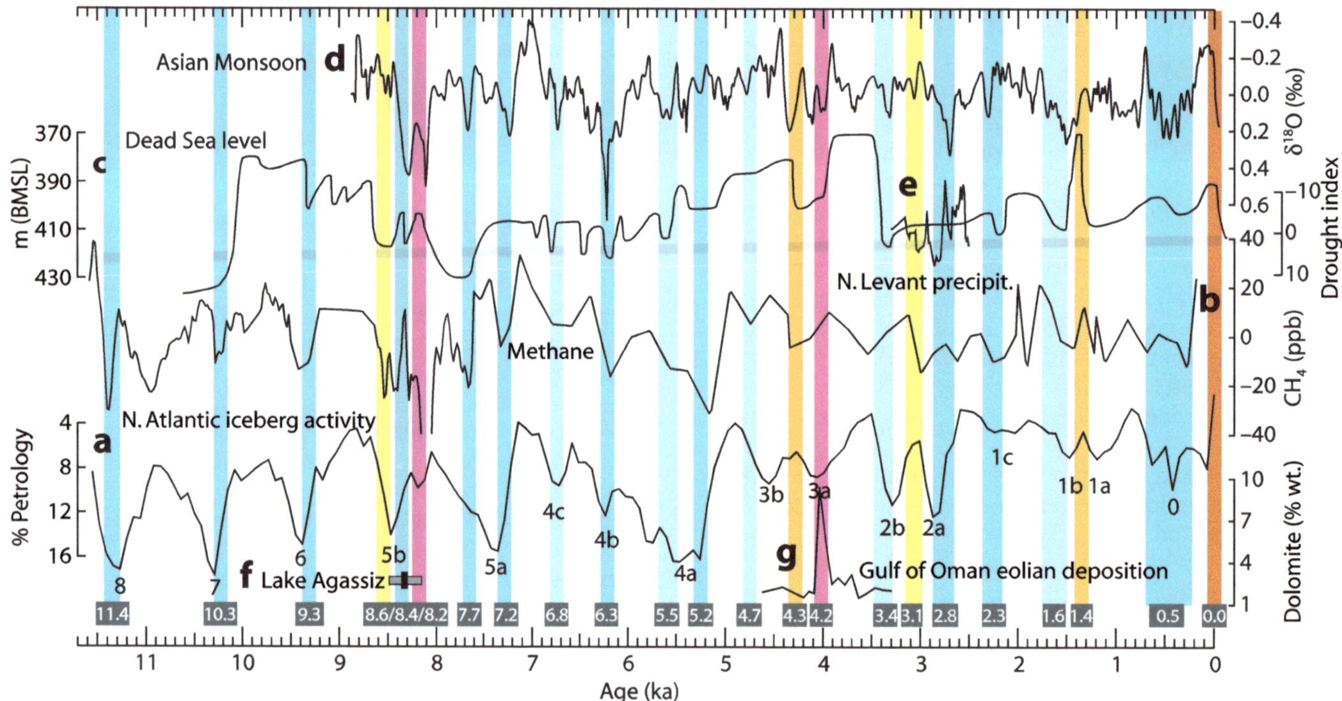

Fig. 4.18 Abrupt Climatic Events during the Holocene
a) Detrended stack of petrological tracers abundance at four different North Atlantic benthic cores as a proxy for ice-rafted deposition and iceberg activity. It is a sensitive proxy for cold events that affected the North Atlantic region. After Bond et al. (2001). **b)** Detrended Greenland methane record built by combining a 180–7620 BP and 8680–9500 BP record from GRIP after Blunier et al. (1995), a 7640–8660 BP record from GISP2 after Kobashi et al. (2007), and a 9520–11580 BP record from NEEM after Chappellaz et al. (2013). The record is truncated at –40 ppb, having reached values of –76 ppb at 8160 BP. It is a sensitive proxy for changes in global precipitation patterns. **c)** Dead Sea level in meters below mean sea-level. After Migowski et al. (2006). It is a sensitive proxy for the North Atlantic storm-track and precipitation levels in the Middle East. **d)** Detrended $\delta^{18}O$ isotope relative abundance in Dongge Cave (China) stalagmite DA speleothem, as a proxy record for the Asian Monsoon. After Wang et al. (2005). It is a sensitive proxy for changes in the Asian Monsoon reflecting changes in land and ocean temperature-difference patterns. **e)** A North Levant precipitation record identifies a 300-yr drought that was a causal factor for the Late Bronze Collapse that affected the Eastern Mediterranean and Black Sea areas and is also reflected in global or hemispheric proxies. After Kaniewski et al. (2015). **f)** Lake Agassiz–Ojibway outburst dating. After Lewis et al. (2012). **g)** Dolomite abundance in Gulf of Oman eolian deposition record, as a proxy for Mesopotamian aridity. After Cullen et al. (2000). Vertical bars identify periods of significant changes in several of the proxies, most of them corresponding to previously identified ACEs, color coded by inferred causes as indicated in table 4.1. Dark grey boxes at bottom give their approximate dates in ka.

The lows of the c. 2500-yr Bray cycle, the main climatic cycle during the Holocene, correspond to Bond events 7, 5a, 4a, 2a, and 0. These events not only show a corresponding age and correct periodicity, but they also constitute the highest petrological tracer peaks for each 2500 year period, suggesting that they were the strongest cooling periods at each time, as glaciological, biological and geophysical evidence also supports.

Paleoclimatic evidence supports the existence of periods of intense climate change (ACEs), that have left their mark in multiple climate proxies of very different nature (Fig. 4.15). To study them it is best to correlate long proxies of different climatic nature (Fig. 4.18). The Bond series of stacked ice-rafted petrological tracers is a sensitive record of cooling ACEs that affected iceberg activity in the North Atlantic (Fig. 4.18a). Methane is naturally produced by northern boreal wetlands and subtropical wetlands in both hemispheres that are sensitive to changing patterns in precipitation (Fig. 4.18b). The Dead Sea drainage area, one of the largest hydrological systems in the Middle East, receives precipitation from the North Atlantic storm-track and the Eastern Mediterranean and is a rain gauge for the Middle East (Fig. 4.18c). The Asian Monsoon is driven by seasonal changes in the land-ocean temperature difference, and is therefore sensitive to changes in the latitudinal tem-

perature gradient that responds primarily (but not exclusively) to changes in the latitudinal insolation gradient (Fig. 4.18d). The combination of these four climate proxies allows the identification of over 20 ACEs during the Holocene (table 4.1). On average a minimum of two ACEs have taken place per millennium. Since ACEs have an average duration of a couple of centuries, climate was either undergoing an ACE or recovering from it for one third to half of the Holocene. The climate of the Holocene has not been stable.

Table 4.1 is a list of the 23 identified Holocene ACEs from the selected proxies shown in figure 4.18. It is clear that this is only a partial list as there is a notable absence of warming events, and early Holocene events are under-represented. A more detailed study could help determine the climate significance of the periods that show a significantly strengthened Asian Monsoon that often correspond to peaks in methane production and troughs in North Atlantic iceberg activity. Do they represent ACEs of a warming nature, or do they just indicate benign stable conditions that didn't last very long? This point has not been clarified in the literature, where there is a notable absence of studies on Holocene warming events. A possibility already mentioned is that interglacial and glacial climates constitute meta-stable conditions that are punctuated by

Date (ka BP)	Name	Temperature effect (1)	Middle East precipitation (2)	Indian/Asian monsoon (3)	Methane (4)	Bond Event (5)	Proposed cause (6)	Periodicity (7)	Reference (8)
11.4	Pre-Boreal Oscillation	Cooling			Decrease	8	Low solar activity	1000	Björck et al. 1997
10.3	Boreal Oscillation 1	Cooling		Weaker	Decrease	7	Low solar activity	2500/1000	Björck et al. 2001
9.3	Boreal Oscillation 2	Cooling	Reduced	Weaker	Decrease	6	Low solar activity	1000	Zhang et al. 2018
8.6		Cooling	Reduced	Unaffected	Decrease	5b			Gavin et al. 2011
8.4		Cooling	Increased	Weaker	Increase	5b	Low solar activity	1000	Rohling & Pälike 2005
8.2	8.2 kyr ACE	Cooling	Increased	Weaker	Decrease	5b	Meltwater pulse		Lewis et al. 2012
7.7		Cooling	Increased	Weaker	Decrease	5a	Low solar activity	2500	Berger et al. 2016
7.2		Cooling	Unaffected	Weaker	Decrease	5a	Low solar activity	1000	Berger et al. 2016
6.8		Cooling	Reduced	Weaker	Decrease	4c			
6.3		Cooling	Reduced	Weaker	Decrease	4b	Low solar activity	1000	Fleitmann et al. 2007
5.5		Cooling	Reduced	Weaker	Decrease	4a			
5.2	5.2 kyr ACE	Cooling	Reduced	Unaffected	Decrease	4a	Low solar activity	2500/1000	Thompson et al. 2006
4.7		Cooling	Unaffected	Unaffected	Decrease	3b			
4.3		Warming?	Reduced	Weaker	Decrease				
4.2	4.2 kyr ACE	Cooling	Reduced	Weaker	Increase	3a	Impact? (9)		Cullen et al. 2000
3.4		Cooling	Reduced	Unaffected	Increase	2b			
3.1	Late Bronze ACE	Warming?	Reduced	Unaffected	Decrease				Kaniewski et al. 2015
2.8	2.8 kyr ACE	Cooling	Increased	Weaker	Increase	2a	Low solar activity	2500	Chambers et al. 2007
2.3		Cooling	Reduced	Weaker	Decrease	1c	Low solar activity	1000	
1.6	Dark Ages ACE	Cooling	Reduced	Weaker	Decrease	1b			Helama et al. 2017
1.4		Warming?	Increased	Unaffected	Increase				Helama et al. 2017
0.5	Little Ice Age	Cooling	Reduced	Weaker	Decrease	0	Low solar activity	2500/1000	
0.0	Global Warming	Warming	Reduced	Weaker	Increase		Greenhouse gases		

Table 4.1 List of Holocene Abrupt Climatic Events. ACEs identified in Fig. 4.18
(1) The temperature effect is deduced from synchronous changes in North Atlantic iceberg activity, after Bond et al. (2001). **(2)** Middle East precipitation effect is deduced from synchronous changes in Dead Sea levels, after Migowski et al. (2006), except for the 3.1-kyr ACE, after Kaniewski et al. (2015). **(3)** Asian Monsoon effect is deduced from synchronous changes in the Asian Monsoon intensity after Wang et al. (2005). For the 10.3 and 9.3-kyr ACEs the effect on the Indian Monsoon is deduced from changes in the Indian Monsoon intensity after Fleitmann et al. (2007). **(4)** The effect on methane is deduced from synchronous changes in the reconstructed Greenland methane record as described in Fig. 4.18 caption. **(5)** The correspondence to Bond events is established from synchrony with peaks in the Bond record, after Bond et al. (2001). As Gerard Bond only numbered eight peaks (plus zero) out of about 20, letters have been added to the numbers to identify the corresponding peaks, as shown in Fig. 4.18. **(6)** One of several causes proposed in the literature that the author considers as most likely. **(7)** A periodicity is indicated when the proposed cause is low solar activity, and the date of the ACE agrees with the known 1000 and/or 2500-yr solar periodicities (see Chaps. 5 & 8). **(8)** A reference is provided that discusses evidence specific for the ACE. Not available for every identified ACE. It is unnecessary for the LIA and Modern Global Warming that have been thoroughly studied. **(9)** The 4.2-kyr ACE constitutes a unique event in several proxies that span most of the Holocene (Cullen et al. 2000; Thompson et al. 2002). Marie–Agnès Courty et al. have been defending an impact-linked dust event (Courty et al. 2008; see Sect. 7.6).

climate variability of the opposite sign.

4.9 Holocene millennial cycles

Table 4.1 lists some ACEs that are proposed to belong to millennial-scale climate periodicities by multiple authors. We have already shown that periodicities are also present in the glacial-interglacial cycle (see Chap. 2), and in the Dansgaard–Oechsger cycle (see Chap. 3). It is clear that most low frequency-high amplitude climate change do not take place in a chaotic manner, but mainly through cycles, quasicycles, and oscillations that respond to periodic changes in the forcings that act over the climate system. Figure 4.19 (adapted from Maslin et al. 2001) shows that these climatic periodicities cover the full spectrum of climate variation, and that, in general, the longer periodicities produce larger variations in climate. Thus Holocene climate change is dominated by periodic variability in the millennial band (grey band, Fig. 4.19).

Within the paleoclimatological scientific community there is widespread acceptance of millennial cycles during the Holocene because their effects are observed in most climatic proxies, and there is ample agreement over certain periodicities that come out of frequency analysis and are in phase from multiple proxies at different locations. Instrumental-era climatologists and astrophysicists are however very skeptical of such periodicities because they have not collected evidence about these long cycles in the short time of modern instrument observations, and we lack a proper understanding of the mechanisms that generate the periodicity and produce the climatic effect. Similar objections were made to Alfred Wegener's continental drift theory that despite solid evidence from geography, geol-

ogy, paleontology, and biology, was shunned until the development of plate tectonics theory could explain how plates, not continents, drifted.

A further complication arises because some climate periodicities do not show the behavior of proper cycles and present gaps when the signal cannot be detected in the data. We already observed that problem when reviewing the Dansgaard–Oeschger cycle, where the oscillations depend on a set of conditions in sea-level, temperature, and obliquity, to become perceptible, and appear to be the result of two different periodicities acting simultaneously. Wavelet analysis of millennial climate cycles during the Holocene shows periods when one or more of the currently operable cycles do not show up in the data. As we do not have a proper knowledge of the mechanisms of these cycles, we do not have an explanation for this behavior. And we also have to consider the awkward nature of most climate proxy data (Witt & Schumann 2005), which is affected by random and systemic errors causing uncertainties along the age axis that grow worse as we go back in time. This data is often unevenly sampled and has increasing compression with growing age, causing a reduction in data density in the older portion of the data. It also suffers from different noise intensity for different paleoclimatic periods and is affected by changing sampling rates. Quite often this awkward nature of paleoclimatic proxy data is not properly accounted for when performing standard time series analyses, which were developed for evenly sampled and stationary time series over a well-defined time axis.

Despite these problems, three relatively well established Holocene millennial-scale climatic periodicities can be described based on evidence. They are the already men-

Fig. 4.19 Climate cycles and periodicities dominate climate change at all temporal scales
Spectrum of climate variance showing the better studied climatic cycles and their proposed forcings, although some are not widely accepted. Cycles, quasicycles, and periodic oscillations are found over the entire temporal range, indicating they are a salient property of climatic variability. As a general rule, the lower the frequency, the more intense the climatic variance produced. The 150-Myr Ice Age cycle has produced three ice ages and a cold age in the last 450 million years. It is proposed to be caused by the crossing of the galactic arms by the Solar system. The 32-Myr cycle has produced two cycles during the Cenozoic era, the first ending in the glaciation of Antarctica and the second in the current Quaternary Ice Age. It is proposed to be caused by the vertical displacement of the Solar system with respect to the galactic plane. The orbital or Milankovitch cycles are the best studied, and between them and the Lunar nodal regression cycle of 18.6 years lies the orbital gap, where no astronomical cycle is known to affect climate. Our knowledge of this range is very insufficient, despite millennial climate cycles (grey band) determining most of Holocene climatic variability. Short term climate variability is dominated by the El Niño/Southern Oscillation. After Maslin et al. (2001).

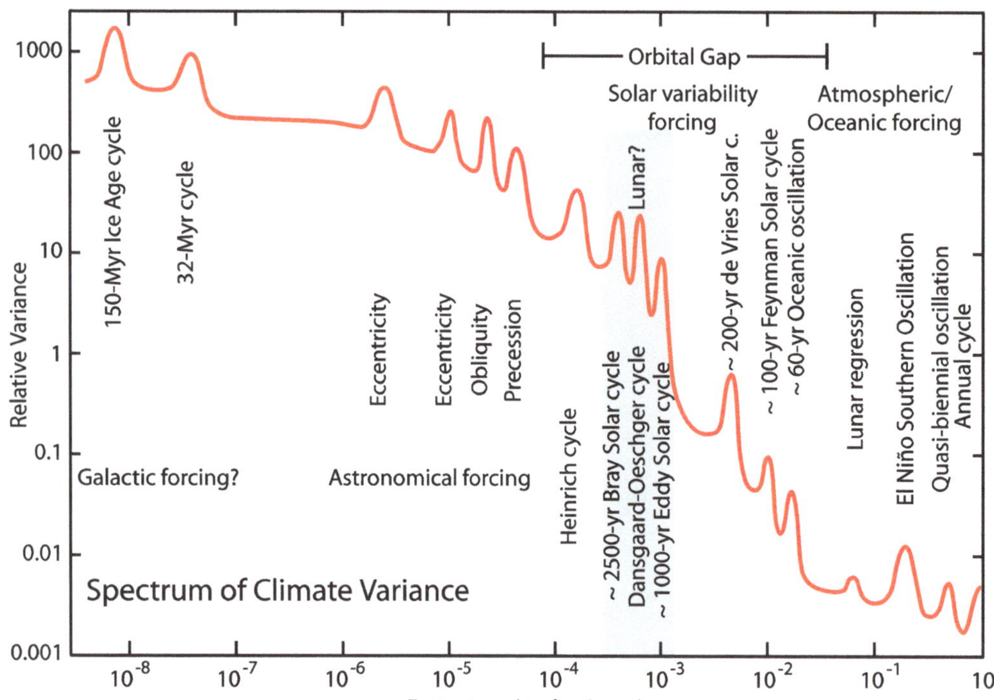

tioned c. 2500-yr Bray climate cycle, a c. 1500-yr atmospheric-oceanic cycle that might be related to the D–O cycle acting during glacial periods, and the c. 1000-yr Eddy climate cycle. As mentioned above, Holocene cycles display abrupt cooling at their lows, creating the conditions for enhanced iceberg activity in the North Atlantic that produces Bond ice-rafting events. As the three cycles have different periodicities, sometimes the lows of two cycles are so close together in time as to make it difficult to resolve them. This is the case in the LIA, when lows for all three cycles took place in close succession, contributing to make this the coldest period in the Holocene, bringing it to the brink of triggering a glacial inception. After each abrupt cooling of the lows of these three cycles comes a warming recovery, that was a complete recovery during the HCO, but only partial during the Neoglacial, as the decline in obliquity prevented a full recovery. The global warming that has taken place during the last 350 years cannot be separated from the previous cooling without losing part of its context. As already indicated in Fig. 4.14, each period of warming during the descent to the next glacial stage should be more intense than the previous ones, as climatic variability increases outside the warm conditions of an interglacial climatic optimum. Underneath the noticeable contribution from anthropogenic greenhouse gases to the 0.0-kyr ACE lies the natural recovery from the extraordinarily strong 0.5-kyr ACE.

4.10 Conclusions

4a. The Holocene is a period of 11,700 years characterized by an intense warming for about 2,000 years and a progressively accelerating cooling for the last 6,000 years, following the changes in obliquity of the Earth's axis.

4b. Fluctuations in greenhouse gases cannot explain Holocene climatic changes and, indeed, their atmospheric changes run opposite to temperature trends for most of the Holocene.

4c. Climate models perform very poorly when trying to reproduce Holocene climate evolution. This is likely due to having too much sensitivity to changes in greenhouse gases and too little sensitivity to insolation and solar variability.

4d. The Holocene Climatic Optimum was a more humid period, 1–2 °C warmer than the Little Ice Age, during which global glaciers reached their minimum extent.

4e. The Mid-Holocene Transition, caused by orbital changes, brought a complete change in climatic mode, with a decrease in solar forcing and an increase in atmospheric-oceanic forcing, displacing the climatic equator and ending the African Humid Period, while increasing ENSO activity.

4f. The Neoglaciation has been a period of progressive cooling, increasing aridity, and advancing glaciers, delimited by the 5.2-kyr event at its beginning and the Little Ice Age at its end.

4g. Holocene climate variability is characterized by periodic cooling events of reduced amplitude compared to glacial climate variability. The main climatic cycle of c. 2500 years delimits five periods of consistent climatic conditions identified over a century ago in the Blytt–Sernander sequence, separated by abrupt climatic changes.

4h. Additional Holocene abrupt climatic variability is reflected in several sensitive climate proxies as being composed of over 20 abrupt events. About half of them appear to respond to solar periodicities in the millennial time frame. Nearly all abrupt Holocene changes identified have been of a cooling nature, followed by a warming recovery. The Little Ice Age is no exception.

4i. Bond events display a mixture of periodicities and constitute cooling events in response to different forcings. A Bond cycle does not exist in the Holocene.

References

Ammann B & Fyfe RM (2014) Blytt–Sernander timescale. In: Matthews JA (ed) Encyclopedia of Environmental Change. SAGE, London, p 107. https://doi.org/10.4135/9781446247501.n457

Berger JF, Delhon C, Magnin F et al (2016) A fluvial record of the mid-Holocene rapid climatic changes in the middle Rhone valley (Espeluche–Lalo, France) and of their impact on Late Mesolithic and Early Neolithic societies. Quaternary Science Reviews 136 66–84

Berke MA, Johnson TC, Werne JP et al (2012) A mid-Holocene thermal maximum at the end of the African Humid Period. Earth and Planetary Science Letters 351 95–104

Björck S, Rundgren M, Ingolfsson O & Funder S (1997) The Preboreal oscillation around the Nordic Seas: terrestrial and lacustrine responses. Journal of Quaternary Science 12 (6) 455–465

Björck S, Muscheler R, Kromer B et al (2001) High-resolution analyses of an early Holocene climate event may imply decreased solar forcing as an important climate trigger. Geology 29 (12) 1107–1110

Blunier T, Chappellaz J, Schwander J et al (1995) Variations in atmospheric methane concentration during the Holocene epoch. Nature 374 (6517) p46–49

Bond G, Kromer B, Beer J et al (2001) Persistent solar influence on North Atlantic climate during the Holocene. Science 294 (5549) 2130–2136

Bray JR (1968) Glaciation and solar activity since the fifth century BC and the solar cycle. Nature 220 (5168) 672–674

Chambers FM, Mauquoy D & Brain SA (2007) Globally synchronous climate change 2800 years ago: proxy data from peat in South America. Earth and Planetary Science Letters 253 (3–4) 439–444

Chappellaz J, Stowasser C, Blunier T et al (2013) High-resolution glacial and deglacial record of atmospheric methane by continuous-flow and laser spectrometer analysis along the NEEM ice core. Climate of the Past 9 (6) 2579–2593

Courty MA, Crisci A, Fedoroff M et al (2008) Regional manifestation of the widespread disruption of soil-landscapes by the 4 kyr BP impact-linked dust event using pedo-sedimentary micro-fabrics. In: Kapur S and Stoops G (eds) New trends in soil micromorphology. Springer, Berlin, p 211–236

Crosta X, Debret M, Denis D et al (2007) Holocene long- and short-term climate changes off Adélie Land, East Antarctica. Geochemistry, Geophysics, Geosystems 8 (11) Q11009

Cullen HM, deMenocal PB, Hemming S et al (2000) Climate change and the collapse of the Akkadian empire: Evidence from the deep sea. Geology 28 (4) 379–382

Debret M, Bout-Roumazeilles V, Grousset F et al (2007) The origin of the 1500-year climate cycles in Holocene North-Atlantic records. Climate of the Past Discussions 3 (2) 679–692

Debret M, Sebag D, Crosta X et al (2009) Evidence from wavelet analysis for a mid-Holocene transition in global climate forcing. Quaternary Science Reviews 28 (25–26) 2675–2688

Denton GH & Karlén W (1973) Holocene Climatic Variations—Their Pattern and Possible Cause. Quaternary Research 3 155–205.

Di Rita F, Fletcher WJ, Aranbarri J et al (2018) Holocene forest dynamics in central and western Mediterranean: periodicity spatio-temporal patterns and climate influence. Scientific reports 8 (1) 8929

Fleitmann D, Burns SJ, Mangini A et al (2007) Holocene ITCZ and Indian monsoon dynamics recorded in stalagmites from Oman and Yemen (Socotra). Quaternary Science Reviews 26 (1–2) 170–188

Fletcher WJ, Debret M & Goñi MFS (2013) Mid-Holocene emergence of a low-frequency millennial oscillation in western Mediterranean climate: Implications for past dynamics of the North Atlantic atmospheric westerlies. The Holocene 23 (2) 153–166

Gagan MK, Ayliffe LK, Hopley D et al (1998) Temperature and surface-ocean water balance of the mid-Holocene tropical western Pacific. Science 279 (5353) 1014–1018

Gavin DG, Henderson AC, Westover KS et al (2011) Abrupt Holocene climate change and potential response to solar forcing in western Canada. Quaternary Science Reviews 30 (9–10) 1243–1255

Geirsdóttir Á, Miller GH, Andrews JT et al (2019) The onset of neoglaciation in Iceland and the 42 ka event. Climate of the Past 15 (1) 25–40

Helama S, Jones PD & Briffa KR (2017) Dark Ages Cold Period: A literature review and directions for future research. The Holocene 27 (10) 1600–1606

Holmgren K, Lee–Thorp JA, Cooper GR et al (2003) Persistent millennial-scale climatic variability over the past 25000 years in Southern Africa. Quaternary Science Reviews 22 (21–22) 2311–2326

Humlum O, Solheim JE & Stordahl K (2011) Identifying natural contributions to late Holocene climate change. Global and Planetary Change 79 (1–2) 145–156

Huybers P (2006) Early Pleistocene glacial cycles and the integrated summer insolation forcing. Science 313 (5786) 508–511

Kaniewski D, Guiot J & Van Campo E (2015) Drought and societal collapse 3200 years ago in the Eastern Mediterranean: a review. Wiley Interdisciplinary Reviews: Climate Change 6 (4) 369–382

Kobashi T, Severinghaus JP, Brook EJ et al (2007) Precise timing and characterization of abrupt climate change 8200 years ago from air trapped in polar ice. Quaternary Science Reviews 26 (9–10) 1212–1222

Kobashi T, Goto–Azuma K, Box JE et al (2013) Causes of Greenland temperature variability over the past 4000 yr: implications for northern hemispheric temperature changes. Climate of the Past 9 (5) 2299–2317

Koch J, Clague JJ & Osborn G (2014) Alpine glaciers and permanent ice and snow patches in western Canada approach their smallest sizes since the mid-Holocene, consistent with global trends. The Holocene 24 (12) 1639–1648

Kullman L (2001) 20th century climate warming and tree-limit rise in the southern Scandes of Sweden. Ambio: A journal of the Human Environment 30 (2) 72–81

Kuper R & Kröpelin S (2006) Climate-controlled Holocene occupation in the Sahara: motor of Africa's evolution. Science 313 (5788) 803–807

Kutzbach JE (1981) Monsoon climate of the early Holocene: Climate experiment with the earth's orbital parameters for 9000 years ago. Science 214 (4516) 59–61

Lamb HH (1977) Climate: Present, past and future. Vol 2. Climatic history and the future. Methuen, London.

Lambeck K, Rouby H, Purcell A et al (2014) Sea level and global ice volumes from the Last Glacial Maximum to the Holocene. Proceedings of the National Academy of Sciences 111 (43) 15296–15303

Leduc G, Schneider R, Kim JH & Lohmann G (2010) Holocene and Eemian sea surface temperature trends as revealed by alkenone and Mg/Ca paleothermometry. Quaternary Science Reviews 29 (7–8) 989–1004

Lewis CFM, Miller AAL, Levac E et al (2012) Lake Agassiz outburst age and routing by Labrador Current and the 82 cal ka cold event. Quaternary International 260 83–97

Liu Z, Zhu J, Rosenthal Y et al (2014) The Holocene temperature conundrum. Proceedings of the National Academy of Sciences 111 (34) E3501–E3505

MacDonald GM, Velichko AA, Kremenetski CV et al (2000) Holocene treeline history and climate change across northern Eurasia. Quaternary Research 53 (3) 302–311

Magny M & Haas JN (2004) A major widespread climatic change around 5300 cal. yr BP at the time of the Alpine Iceman. Journal of Quaternary Science 19 (5) 423–430

Manning K & Timpson A (2014) The demographic response to Holocene climate change in the Sahara. Quaternary Science Reviews 101 28–35

Marcott SA, Shakun JD, Clark PU & Mix AC (2013) A reconstruction of regional and global temperature for the past 11,300 years. Science 339 (6124) 1198–1201

Maslin M, Seidov D & Lowe J (2001) Synthesis of the nature and causes of rapid climate transitions during the Quaternary. In: Seidov D, Haupt BJ and Maslin M (eds) The Oceans and Rapid Climate Change: Past, Present, and Future. American Geophysical Union, Geophysical Monograph Series 126, Washington DC, p. 9–52

Masson V, Vimeux F, Jouzel J et al (2000) Holocene climate variability in Antarctica based on 11 ice-core isotopic records. Quaternary Research 54 (3) 348–358

Mayewski PA, Rohling EE, Stager JC et al (2004) Holocene climate variability. Quaternary research 62 (3) 243–255

McDermott F, Mattey DP & Hawkesworth C (2001) Centennial-scale Holocene climate variability revealed by a high-resolution speleothem $\delta^{18}O$ record from SW Ireland. Science 294 (5545) 1328–1331

Migowski C, Stein M, Prasad S et al (2006) Holocene climate variability and cultural evolution in the Near East from the Dead Sea sedimentary record. Quaternary Research 66 (3) 421–431

Monnin E, Steig EJ, Siegenthaler U et al (2004) EPICA Dome C ice core high resolution Holocene and transition CO_2 data. IGBP PAGES/World Data Center for Paleoclimatology Data Contribution Series 55

Moy CM, Seltzer GO, Rodbell DT & Anderson DM (2002) Variability of El Niño/Southern Oscillation activity at millennial timescales during the Holocene epoch. Nature 420 (6912) 162–165

North Greenland Ice Core Project members (2004) High-resolution record of Northern Hemisphere climate extending into the last interglacial period. Nature 431 (7005) 147–151

O'Brien SR, Mayewski PA, Meeker LD et al (1995) Complexity of Holocene climate as reconstructed from a Greenland ice core. Science 270 (5244) 1962–1964

Pisaric MF, Holt C, Szeicz JM et al (2003) Holocene treeline dynamics in the mountains of northeastern British Columbia Canada inferred from fossil pollen and stomata. The Holocene 13 (2) 161–173

Polissar PJ, Abbott MB, Wolfe AP et al (2013) Synchronous interhemispheric Holocene climate trends in the tropical Andes. Proceedings of the National Academy of Sciences 110 (36) 14551–14556

Porter SC (2000) Onset of neoglaciation in the Southern Hemisphere. Journal of Quaternary Science 15 (4) 395–408

Renssen H, Seä H, Crosta X et al (2012) Global characterization of the Holocene thermal maximum. Quaternary Science Reviews 48 7–19

Rohling EJ & Pälike H (2005) Centennial-scale climate cooling with a sudden cold event around 8,200 years ago. Nature 434 (7036) 975–979

Rosenthal Y, Kalansky J, Morley A & Linsley B (2017) A paleoperspective on ocean heat content: Lessons from the Holocene and Common Era. Quaternary Science Reviews 155 1–12

Rosenthal Y, Linsley BK & Oppo DW (2013) Pacific Ocean heat content during the past 10,000 years. Science 342 (6158) 617–621

Rossignol–Strick M (1985) Mediterranean Quaternary sapropels: An immediate response of the African monsoon to variation of insolation. Palaeogeography, palaeoclimatology, palaeoecology 49 (3–4) 237–263

Schrøder N, Pedersen LH & Bitsch RJ (2004) 10,000 years of climate change and human impact on the environment in the area surrounding Lejre. The Journal of Transdisciplinary Environmental Studies 3 (1) 1–27

Shevenell AE, Ingalls AE, Domack EW & Kelly C (2011) Holocene Southern Ocean surface temperature variability west of the Antarctic Peninsula. Nature 470 (7333) 250–254

Simonneau A, Chapron E, Garçon M et al (2014) Tracking Holocene glacial and high-altitude alpine environments fluctuations from minerogenic and organic markers in proglacial lake sediments (Lake Blanc Huez, Western French Alps). Quaternary Science Reviews 89 27–43

Singarayer JS, Valdes PJ, Friedlingstein P et al (2011) Late Holocene methane rise caused by orbitally controlled increase in tropical sources. Nature 470 (7332) 82–85

Solomina ON, Bradley RS, Hodgson DA et al (2015) Holocene glacier fluctuations. Quaternary Science Reviews 111 9–34

Stott L, Cannariato K, Thunell R et al (2004) Decline of surface temperature and salinity in the western tropical Pacific Ocean in the Holocene epoch. Nature 431 (7004) 56–59

Thompson LG, Mosley–Thompson E, Davis ME et al (2002) Kilimanjaro ice core records: evidence of Holocene climate change in tropical Africa. Science 298 (5593) 589–593

Thompson LG, Mosley–Thompson E, Brecher H et al (2006) Abrupt tropical climate change: Past and present. Proceedings of the National Academy of Sciences 103 (28) 10536–10543

Thornalley DJ, Elderfield H & McCave IN (2009) Holocene oscillations in temperature and salinity of the surface subpolar North Atlantic. Nature 457 (7230) 711–714

Thouret JC, Van der Hammen T, Salomons B and Juvigné E (1996) Paleoenvironmental changes and glacial stades of the last 50,000 years in the Cordillera Central, Colombia. Quaternary Research 46 (1) 1–18

Tinner W, Ammann B & Germann P (1996) Treeline fluctuations recorded for 12,500 years by soil profiles pollen and plant macrofossils in the Central Swiss Alps. Arctic and Alpine Research 28 (2) 131–147

Tzedakis PC, Crucifix M, Mitsui T & Wolff EW (2017) A simple rule to determine which insolation cycles lead to interglacials. Nature 542 (7642) 427–432

Vinther BM, Buchardt SL, Clausen HB et al (2009) Holocene thinning of the Greenland ice sheet. Nature 461 (7262) 385–388

Wang Y, Cheng H, Edwards RL et al (2005) The Holocene Asian monsoon: links to solar changes and North Atlantic climate Science. 308 (5723) 854–857

Wanner H & Brönnimann S (2012) Is there a global Holocene climate mode? PAGES news 20 (1) 44–45

Weninger B, Clare L, Rohling E et al (2009) The impact of rapid climate change on prehistoric societies during the Holocene in the Eastern Mediterranean. Documenta Praehistorica 36 7–59

Werne JP, Hollander DJ, Lyons TW & Peterson LC (2000) Climate-induced variations in productivity and planktonic ecosystem structure from the Younger Dryas to Holocene in the Cariaco Basin, Venezuela. Paleoceanography 15 (1) 19–29

Witt A & Schumann AY (2005) Holocene climate variability on millennial scales recorded in Greenland ice cores. Nonlinear Processes in Geophysics 12 (3) 345–352

Zhang W, Yan H, Dodson J et al (2018) The 9.2 ka event in Asian summer monsoon area: the strongest millennial scale collapse of the monsoon during the Holocene. Climate Dynamics 50 (7–8) 2767–2782

5

THE 2500-YEAR BRAY CYCLE

"From my earlier observations, which I have reported every year in this journal, it appears that there is a certain periodicity in the appearance of sunspots and this theory seems more and more probable from the results of this year. ... If one compares the number of groups with the number of days when no spots are visible, one will find that sunspots have a period of about 10 years, and that for five years of this period they appear so frequently that during that time there are very few or no days when no spots at all are visible."
Heinrich Schwabe (1843), after only 17 years of sunspot observations (less than two periods), while looking for the nonexistent planet Vulcan.

5.1 Introduction

There is strong evidence for a persistent irregular climate cycle of c. 2500 ± 300 years that was first reported in 1968 by Roger Bray. A botanist from Illinois who made a landmark contribution to ecology in the 1950s, Roger Bray moved to New Zealand in the 1960s, where he worked as climatologist at the Department of Scientific and Industrial Research at Palmerston North. Through his research on radioactive isotopes, he followed on Minze Stuiver's 1961 discovery that atmospheric ^{14}C and solar activity were inversely correlated. He reconstructed a 2600-year solar activity index from sunspot and auroral observations that could correlate to glacier re-advances and realized that it constituted a single very long oscillation that had repeated several times through the Holocene, with major glacier advances at multiples of c. 2600 years. As the glaciological cycle was based on biological and geological data and agreed well with climatological periods worked out by Scandinavian palynologists, Bray related the 2500-yr climate cycle to an equal-period solar activity cycle even before there was sufficient evidence to demonstrate the existence of the 2500-yr solar cycle.

The 2500-yr Bray climatic cycle is supported by abundant evidence from vegetation changes, glacier re-advances, atmospheric changes reflected in alterations in wind patterns, oceanic temperature and salinity changes, drift ice abundance, and changes in precipitation and temperature. It is established with proxy records from many parts of the world. This climate cycle correlates in period and phase with a c. 2500-yr cycle in the production of cosmogenic isotopes, that corresponds with clusters of solar grand minima at times of abrupt cooling and climate deterioration. The abrupt cooling events usually coincide with times of human societal stress. The relationship between solar activity and cosmogenic isotope production during the past centuries confirms the c. 2500-yr solar cycle as the origin of the climate cycle.

The existence of the 2500-yr solar cycle that had been proposed by Roger Bray (1968) on the basis of weak data was confirmed by Jan Houtermans (1971) in his Ph.D. thesis using Fourier analysis on bristlecone pine radiocarbon measurements.

5.2 The biological 2500-year climate cycle

The roots of the discovery of the 2500-yr Bray climate cycle can be traced to the early 20th century. Over a century ago Scandinavian botanists started to reconstruct the climate of the Holocene from peat bog stratigraphy. They could distinguish the sediment layers into wet/dry, cold/warm, periods, and developed crude dating methods. Their efforts resulted in an understanding that the Holocene climate could be subdivided into periods of different climatic conditions, like in a diagram by Rutger Sernander from 1912 (Fig. 5.1a).

The development of palynology (pollen studies) by Lenart von Post in the 1930s allowed Knud Jenssen and Johs. Iversen to improve the postglacial period zonation (Fig. 5.1c), and develop a summer vegetation-based temperature scale for the Scandinavian Holocene by the 1940s. This temperature scale allowed Holocene climate reconstructions very similar to our current understanding by 1950 (Fig. 5.1b).

Figure 5.1 summarizes decades of work by botanists to establish vegetation stages in the Northern Hemisphere Holocene. These stages allow us to distinguish a 2500-year vegetation and faunal cycle. Some botanists, like Rutger Sernander, proposed that the transitions between periods were abrupt and not gradual. In particular, he proposed that the last transition between the Sub-Boreal and the Sub-Atlantic, at around 650 BC corresponded to the "Fimbulvintern," or Great Winter of the Sagas, that marks the end of the Nordic Bronze Age (Fig. 5.1a), and made the Nordic countries a colder place.

5.3 The glaciological 2500-year climate cycle

In the early 1950s, researchers noticed a correlation between glacier movements in North America and sunspots for the previous 300 years. In the 1960s Roger Bray constructed a solar index starting in 527 BC by combining telescopic sunspot observations with naked-eye sunspot and auroral observations. He also constructed an index for postglacial major ice re-advances from glaciers all over the world. He compared these two observations and found a high degree of correlation, and good agreement with Icelandic sea-ice, and ^{14}C production variations. He observed in the data a possible 2300–2700-year cycle, that he

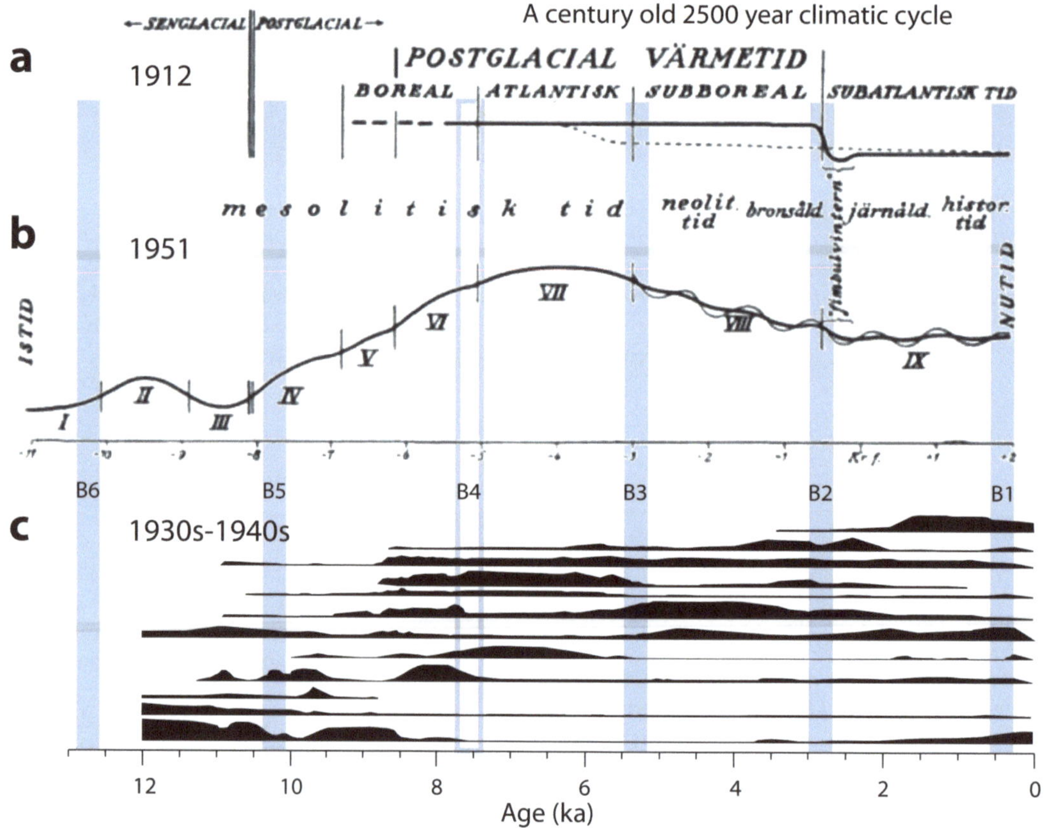

Fig. 5.1 Postglacial vegetation and climate periods as understood during the first half of the 20th century
a) Rutger Sernander's view of postglacial warm climate periods in southern and central Sweden, showing his proposed abrupt climate degradation at the Sub-Boreal/Sub-Atlantic transition, termed "fimbulvintern." The dashed line indicates G. Andersson's opposite view of gradual temperature evolution. After Fries (1956). b) Magnus Fries 1951 diagram of Late Glacial/Postglacial temperature evolution in southern and central Sweden based on biological evidence, showing the temporal disposition of the nine pollen zones in Roman numerals. The thin line represents a near-millennial oscillation in humidity. Dates in CE calendar years. After Fries (1956). c) Example of analytical pollen zones from the 1930s and 1940s for southern and central Sweden confirming Sernander's climatic reconstruction. In this and following figures, blue vertical bars mark the position of the 2500-yr Bray cycle lows identified by the letter B and their order number from the present. B1–B3 coincide with the most recent climate periods identified a century ago. The empty vertical bar marks a less conspicuous B4 in the Bray sequence.

projected into the past from the Little Ice Age, finding that a 2600-year period closely matched both vegetation transitions like the Atlantic/Sub-Boreal, or the Sub-Boreal/Sub-Atlantic transitions, and significant glacier re-advances from the past after the Younger Dryas (Bray 1968). Since he was the first to correctly identify and describe the c. 2500 year climatic and solar cycles, they should be named after him, as tradition indicates.

Bray's glaciological and solar studies were reproduced in 1973 by Denton and Karlén, who did a more detailed study of world glacial advances and came up with essentially the same periodicity of 2500 years (Fig. 5.2a). By then Hans Suess had determined the short-term fluctuations in ^{14}C levels for the past 7000 years from tree rings. They were thought to represent variations in solar activity. As Bray had done previously, Denton & Karlén (1973)

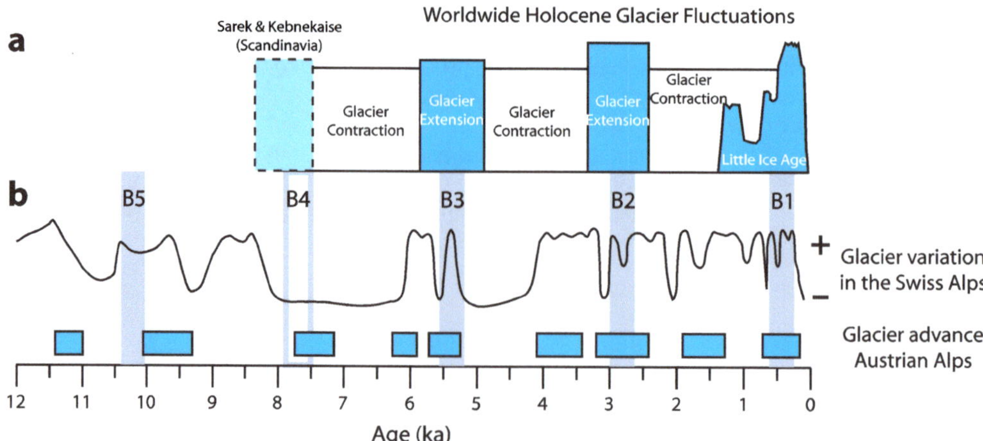

Fig. 5.2 Holocene glacier fluctuations
a) Synthesis of Holocene worldwide glacier fluctuations showing three broad intervals of glacier expansion within the last 6000 years and a fourth one recognized in Scandinavia. After Denton & Karlén (1973). b) Holocene glacier fluctuations in the European Alps showing the complex pattern of advances and retreats that do not always correspond between Austrian and Swiss Alps. After Bressan (2011).

correlated periods of major glacier advances to periods of high ^{14}C production (low solar activity).

Since then glaciologists have reconstructed Holocene glacier movements from hundreds of glaciers all over the planet, and glacier variability has become more complex (Fig. 5.2b). Today we still recognize the major global advances that define the 2500-year cycle (Mayewski et al. 2004; see Fig. 4.16), but there is hardly a century, especially during the Neoglaciation, when glaciers were not advancing somewhere.

By the mid-1970's the scientific community was aware of the existence of a 2500-year climatic cycle that caused glacier advances and recessions, and that separated significantly different vegetation stages. Due to its coincidence with ^{14}C fluctuations, it was inferred that its cause was solar variability. Throughout this book both the climatic and solar 2500-yr cycles are referred to as the Bray cycle, and the lows of the cycle, associated with enhanced ^{14}C production, and climatic changes manifested by cooling, glacier advances, increased drift ice in the North Atlantic, and atmospheric, oceanic, and precipitation changes, are numbered from more recent backwards as B1, B2, ..., with B1 corresponding to the Little Ice Age.

5.4 The atmospheric 2500-year climate cycle

The next great advance in the characterization of the 2500-yr climate cycle came from the study of ice cores. Paul Mayewski, one of George Denton's students, was the sci-

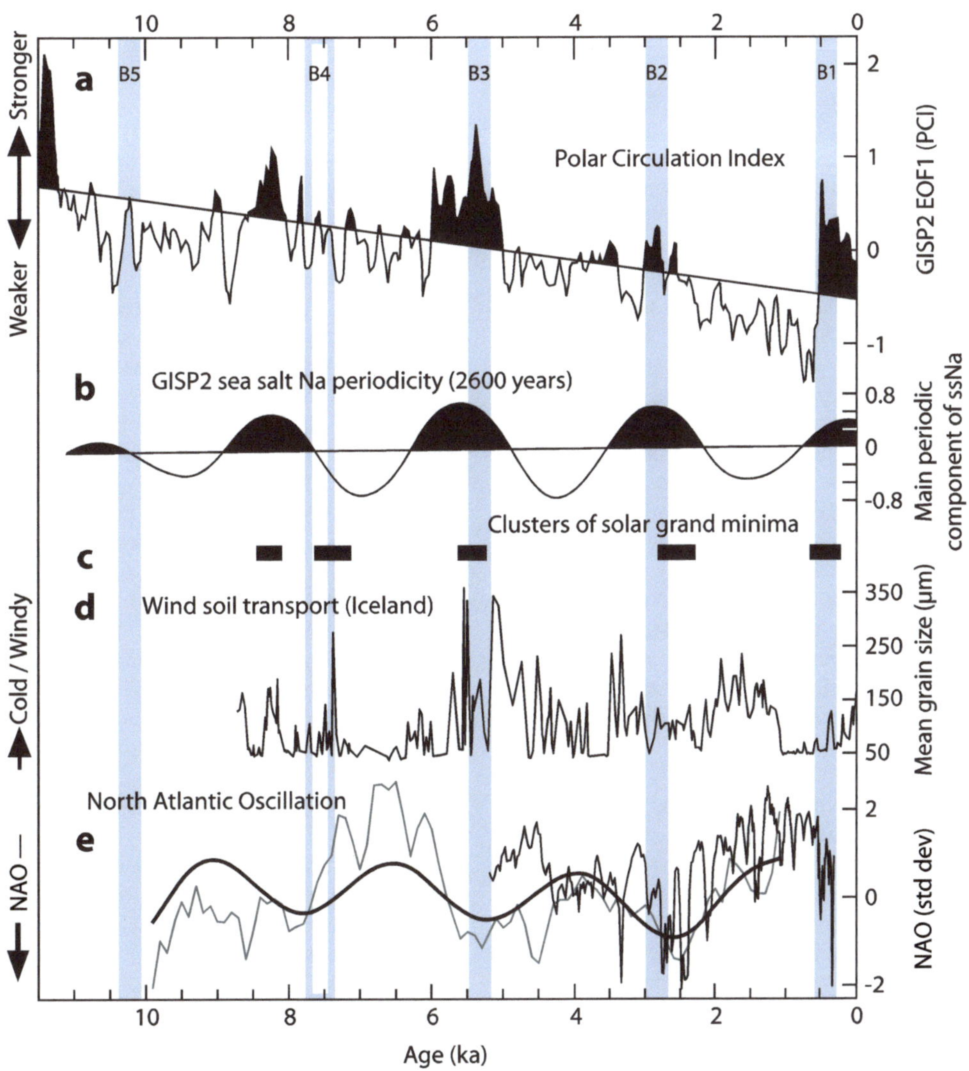

Fig. 5.3 Holocene North Atlantic and Arctic atmospheric changes
a) GISP2 polar circulation index, a time series of the dominant empirical orthogonal function, EOF1, of the major ions in the ice core, that provides a relative measure of the average size and intensity of polar atmospheric circulation. PCI values increase (e.g. more continental dust and marine contributions) during colder portions of the record. After O'Brien et al. (1995). **b)** Main periodic component of the sea salt Na flux in GISP2 ice core with a quasi-2600-year periodicity. **c)** Δ^{14}C intervals that present Maunder- and Spörer-type patterns occurring in clusters. After O'Brien et al. (1995). **d)** Mean grain size of eolian soil deposition at Hólmsá, Iceland, a proxy for wind strength. Windy periods, indicated by the transport and deposition of coarse sediments, are coeval with cool, stormy periods recorded in GISP2 ice and North Atlantic sediment cores. After Jackson et al. (2005). **e)** Thin grey curve, reconstruction of time coefficient by singular spectrum analysis of detrended and normalized alkenone based SST variance, from a NW Africa marine sediment core, as a proxy for AO/NAO oscillation. Thick black sinusoid, main periodic component of the data. After Kim et al. (2007). Thin black curve, inferred NAO circulation pattern from the principal component analysis of redox parameters (Ca/Ti and Mn/Fe ratios) from a Greenland lake sediment record. After Olsen et al. (2012).

entist in charge of coordinating the effort of over 200 scientists in the American Ice Core Program that in 1993 completed the Greenland Ice Sheet Project II (GISP2). He described this effort and its fruits in his 2002 book "The Ice Chronicles: The Quest to Understand Global Climate Change." While other researchers took on studying gases, isotopes, or dust in the GISP2 ice core, Mayewski and colleagues studied the chemical composition of major ions brought to the ice by the wind, using them as tracers for atmospheric circulation. They discovered a strong association between expansions of Northern Hemisphere polar atmospheric circulation systems and the 2500-yr cycle previously described by his former advisor (O'Brien et al. 1995; see Fig. 4.15F & G; Fig. 5.3a & b). An increase in salt deposition is associated with winter atmospheric conditions today. This is when the north polar vortex expands and meridional circulation increases, and thus represents an increase in cold and windy conditions. The periodicity found by Mayewski and colleagues (O'Brien et al. 1995) in GISP2 salts is close to 2600 years (Fig. 5.3b). They noticed a good correlation between the atmospheric maxima and clusters of three solar grand minima (SGM) of the Maunder- and Spörer-type patterns, with the most recent one taking place during the LIA (O'Brien et al. 1995; figure 4.3c). Ram et al. (2009) have demonstrated that dust content in GISP2 ice core is inversely correlated to solar activity and strongly modulated at solar frequencies. As dust arrival to Central Greenland is regulated by atmospheric circulation their result is consistent with a strong solar activity effect on polar circulation.

The atmospheric reorganization that takes place at the lows of the Bray cycle and causes increased polar circulation is partially evident in eolian soil sediments in southern Iceland (Jackson et al. 2005; Fig. 5.3d). Some of the biggest grain sizes transported by the strongest winds are associated with cold periods linked to the passage of the polar front through the proxy location, and coincide with some of the lows of the Bray cycle (B3 & B4, Fig. 5.3d). The authors of the work underscore the wind pattern similarity to the North Atlantic drift-ice Bond record.

The changes in polar circulation and wind strength are suggestive of changes in the Arctic Oscillation/North Atlantic Oscillation (AO/NAO). The AO reflects the state of the atmospheric circulation over the Arctic, through a positive phase, featuring below average geopotential heights, and a negative phase in which the opposite is true. In the negative phase, the polar low-pressure system (also known as the polar vortex) over the Arctic is weaker, which results in weaker upper level winds. Therefore, cold Arctic air and storm tracks move farther south, causing a drop in Northern Hemisphere temperatures and changes in precipitation patterns. The AO often shares phase with the NAO, that reflects differences in the strength of two pressure centers in the North Atlantic: the low pressure near Iceland, and the high pressure over the Azores. Fluctuations in the strength of these pressure centers alter the alignment of the jet stream affecting temperature and precipitation distribution. A NAO negative phase is produced when the weakening of the Iceland low and the Azores high reduces the pressure gradient resulting in weaker more southern westerlies producing colder conditions over much of North America and Northern Europe while moving the storm tracks southward towards the Mediterranean. A NAO negative phase usually features more frequent and

longer blocking conditions when a stationary pressure pattern allows cold Arctic air to spill over mid-latitudes.

The Holocene NAO patterns have been reconstructed from a marine sediment core whose alkenone content has been shown to depend on trade winds intensity-dependent upwelling near the coast of NW Africa (Kim et al. 2007; Fig. 5.3e). NAO intensity for the last millennia has also been reconstructed from lake sediments in Greenland, showing the very low NAO values that characterized the LIA (Olsen et al. 2012; Fig. 5.3e). The evidence indicates a 2500-year periodic variation in SST and upwelling intensity off NW Africa that is associated with a climatic cycle in oceanic circulation that reflects periodic NAO conditions. The lows of this NAO cycle are characterized by NAO negative dominant conditions that produce Northern Hemisphere cooling and precipitation changes. Rimbu et al. (2004), have argued that during the Holocene, the AO/NAO atmospheric circulation was the dominant climate mode at millennial time scales.

5.5 The oceanic 2500-year climate cycle

Given the strong coupling between atmospheric and oceanic variability, it is not surprising that the c. 2500-year climate cycle is prominently displayed by some oceanic current proxies, particularly in the North Atlantic. Oppo et al. (2003) used an established deepwater proxy, the carbon-isotope composition of benthic foraminifera, to evaluate Holocene deepwater variability at a sediment core in the NE subpolar Atlantic. Low ^{13}C values are indicative of a reduction in the $^{13}CO_2$ rich North Atlantic Deep Water (NADW) contribution. Oppo et al. (2003) identify the largest reductions in NADW at 9.3, 8.0, 5.0 and 2.8 kyr ago. The latest three periods correspond with Bray lows 2 to 4 (Fig. 5.4a). Significant reductions in ^{13}C indicative of reduced NADW production have also been reported at 10,300 BP (B5) by Bond et al. (1997), and at the LIA (B1) by Keigwin & Boyle (2000). This means that all the lows in the Bray cycle had been identified as periods of reduced NADW contribution by different authors. Such periods might see a reduction in the northward flux of warm near-surface waters in the North Atlantic to maintain mass balance (that could be the cause of the NADW reduction), and would result also in the cooling of North Atlantic high latitudes.

Temperature and salinity analysis of the Atlantic Meridional Overturning Circulation (AMOC) using a sediment core south of Iceland, where the Faroe and Irmingen currents branch out of the North Atlantic current, shows that episodes of warm saline sub-thermocline conditions are centered at 0.3 (B1), 1.0, 2.7 (B2) and 5.0 (B3) kyr ago, coinciding with known climatic perturbations in the North Atlantic region (Thornalley et al. 2009; Fig. 5.4b). The authors show evidence that the increased salinity, temperature, and water stratification, at times of abrupt climate change, are due to an increase in the Atlantic inflow of warmer saline subtropical gyre waters at the expense of the fresher and colder subpolar gyre waters. They interpret it as a negative feedback from the subpolar gyre, that stabilized the AMOC, at times of freshwater inputs, particularly during the early Holocene when the ice sheets were still melting rapidly, and at the 8.2 kyr event when the outbreak of proglacial Lake Agassiz took place (Thornalley et al. 2009; Fig. 5.4b). They propose solar variability as the forcing behind these oscillations. The increased salinity of the Atlantic inflow observed at the times of reduced NADW formation identified by Oppo et al. (2003; Fig. 5.4a) may have limited the reduction, or helped restart a stronger AMOC.

Andrews (2009) analyzed the distribution of foreign mineral species by drift ice in Icelandic shelf waters. While drift ice has been increasing in the past 6,000 years of Neoglacial conditions off Northern Iceland, the de-trended data supports the existence of a 2500-year climatic periodicity. Periods of high drift ice coincide with the lows of the Bray cycle (Andrews 2009; Fig. 5.4c). As is the case with the Bond series, there is variability in drift ice records, since some cold events do not belong to the Bray sequence.

A more detailed look at millennial-scale oceanic transport changes that took place at the Iceland–Scotland Ridge further confirms the oceanic c. 2500-year cycle.

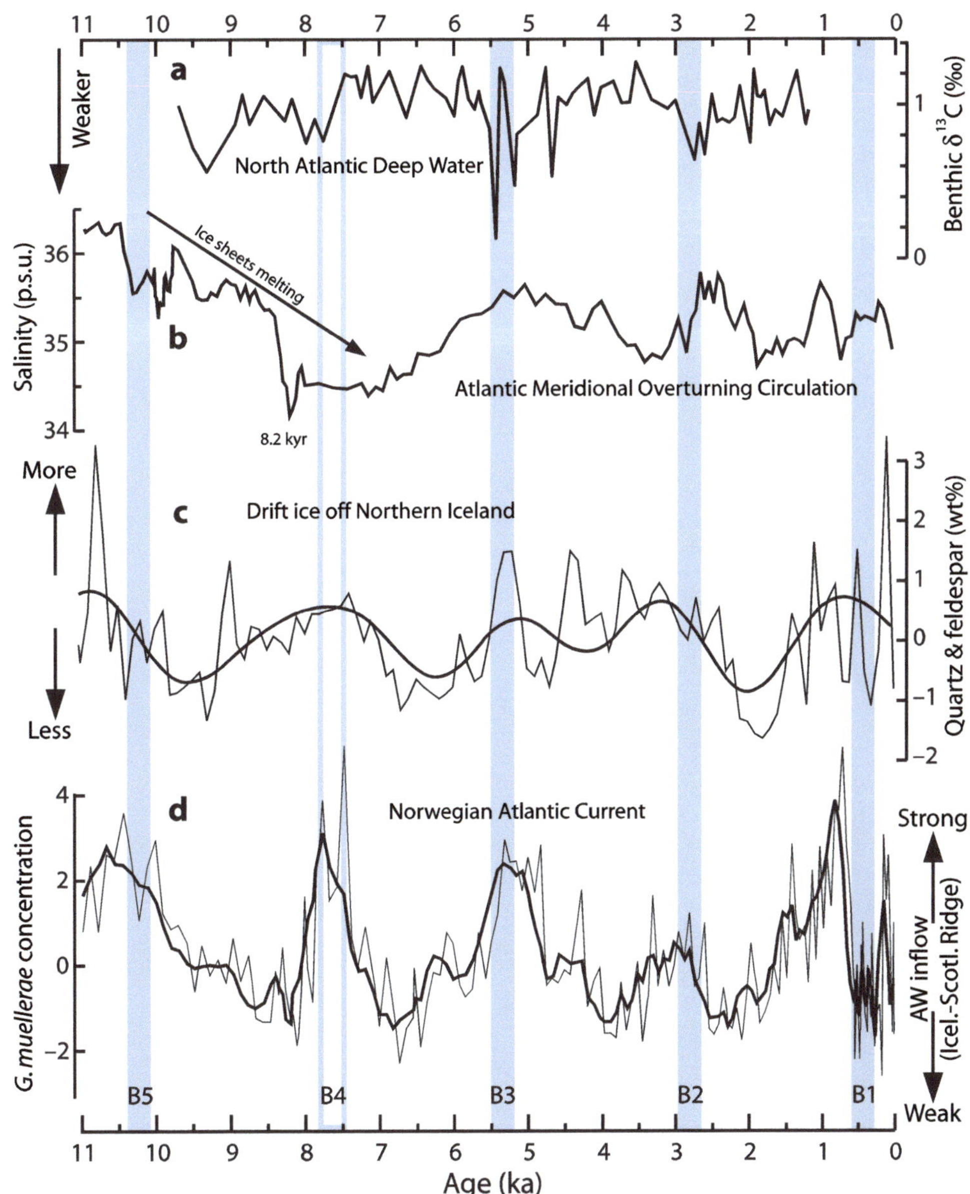

Fig. 5.4 Holocene North Atlantic and Arctic oceanic currents changes
a) Benthic *Cibicidoides wuellerstorfi* δ13C variations, at a marine sediment core in the subpolar NE Atlantic, record total CO_2 δ13C variations in bottom waters, as a proxy for δ13C-rich North Atlantic deepwater (NADW) contribution. The lows in the Bray cycle (vertical bars), correspond to periods of reduced NADW contribution. After Oppo et al. (2003). b) Salinity reconstruction (in practical salinity units) at the base of the thermocline by paired Mg/Ca–δ18O measurements from *Globigerina inflata* from a marine sediment core south of Iceland. During the early Holocene, the sub-thermocline was saltier, but underwent a freshening at a time when the ice sheets were still contributing meltwater. The glacial freshwater discharge event of 8.2 kyr ago can be recognized. Warm saline sub-thermocline conditions took place at 0.3, 1.0, 2.7 and 5.0 kyr ago, coinciding with known climatic perturbations in the North Atlantic region. After Thornalley et al. (2009). c) Quantitative wt% mineralogical (quartz and feldspars) detrended variations from 16 cores from the Iceland shelf (thin line), as a proxy for drift ice from the Arctic Ocean and East Greenland, fitted to a fourth-order polynomial (thick line). Five peaks in residuals from the data are defined by the 2500-year cycle. After Andrews (2009). d) Detrended (grey) and smoothed (black) *Gephyrocapsa muellerae* abundance (n° x 108/g) record as a proxy of warmer Atlantic water flow through the Iceland–Scotland strait of the Nordic Seas from a sediment core off Norway. The fall in abundance during some Bray lows and particularly the LIA (B1) might be due to North Atlantic summer waters being too cold for this warm-loving species. After Giraudeau et al. (2010).

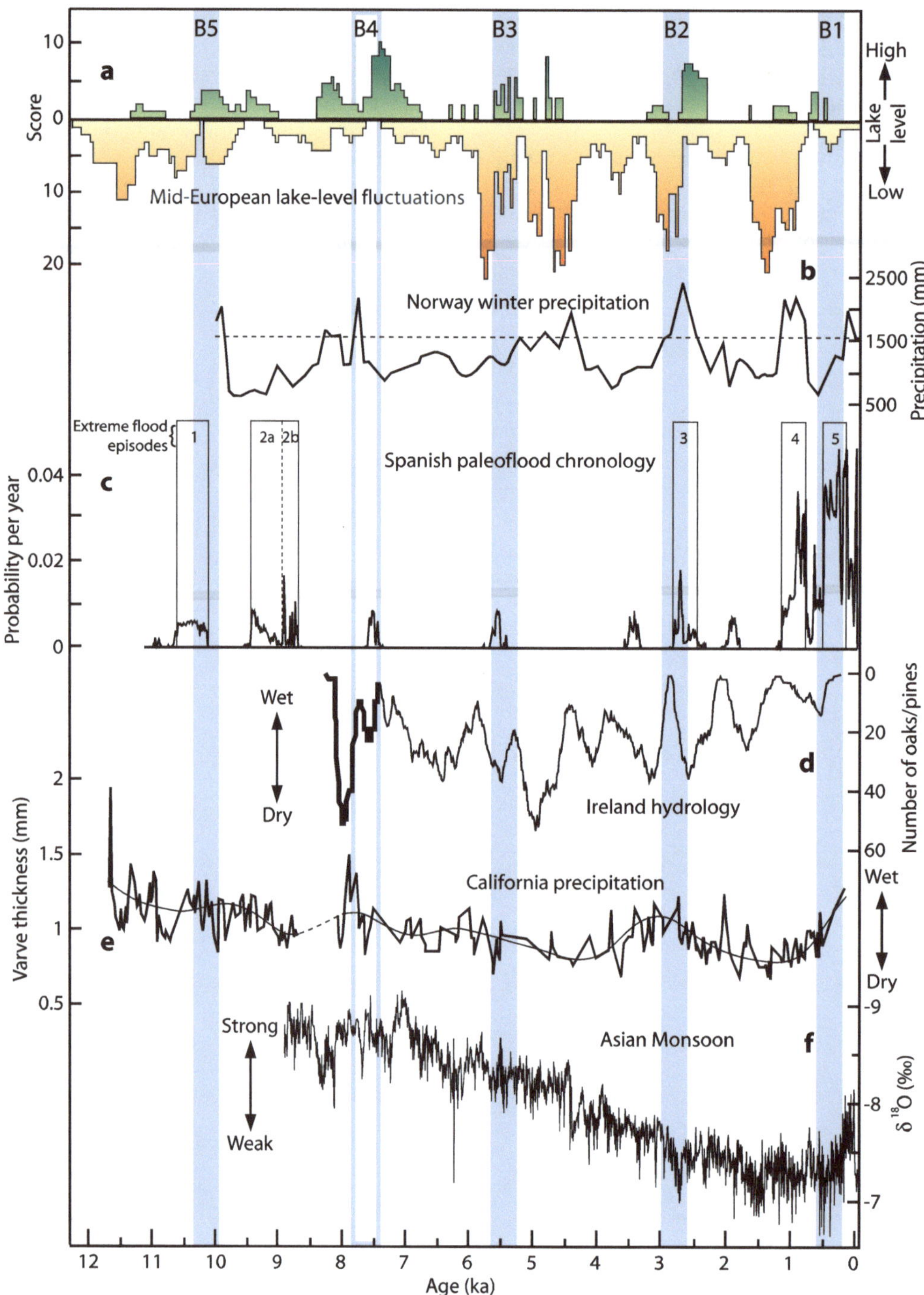

Fig. 5.5 Holocene Northern Hemisphere precipitation changes
a) Holocene mid-European lake-level reconstruction from a data set of 180 radiocarbon, tree-ring and archaeological dates of higher and lower lake-level events based on multiple lines of evidence, obtained from sediment sequences of 26 lakes in the Jura mountains, the northern French Pre-Alps and the Swiss Plateau. The score indicates how well registered the lake-level event is, not its intensity. With a resolution of 50 years, episodes of higher lake-level are defined by a collapse of lower lake-level scores followed by a peak in higher lake-level scores. After Magny (2004). **b)** Winter precipitation reconstruction at Bjørnbreen glacier in Jotunheimen, southern Norway. Precipitation is reconstructed from the known relation between variations in the equilibrium line altitude (ELA, the boundary between the ablation and accumulation areas) and mean July temperature variations reconstructed from palynological data. Winter precipitation is more important than summer temperature for glacier expansion episodes. After Matthews et al. (2005). **c)** Holocene summed probability plot for Spanish fluvial system paleofloods, fine to medium sands deposits on the sides of narrow bedrock canyons that resulted from floods of similar or greater magnitude to those of the largest floods recorded in the instrumental series and are considered evidence of past extreme floods. After Thorndycraft & Benito (2006). **d)** Irish bog-grown oaks (*Quercus spp.*, thin line) and pines (*Pinus sylvestris L.*, thick line) frequency (inverted scale) during the Holocene as evidence of changes in moisture delivery to Ireland. Under humid conditions trees were unable to grow on wetter bogs. After Turney et al. (2005). **e)** 25-year average sedimentary varve thickness record at a marine core in the Santa Barbara Basin as a proxy for annual rainfall in the area. Thin line represents lowpass filter to emphasize millennial scale fluctuations. Data is missing around the 8.2 kyr event when the basin entered a bioturbated non-varved interval similar to glacial stadials. After Nederbragt & Thurow (2005). **f)** 5-year-resolution δ18O isotope record from Dongge Cave (southern China) stalagmite DA as a proxy for the strength of the Asian monsoon over the past 9000 years. After Wang et al. (2005).

Abundance of coccolith species in a marine core off Norway reflects major Holocene changes in Atlantic water transfer to the Nordic Seas with a 2500-year periodicity (Giraudeau et al. 2010; Fig. 5.4d). This proxy depends on the increased supply of warm-loving coccolith species from the North Atlantic with an enhanced advection of warm Atlantic waters, but when the North Atlantic becomes itself too cold the abundance of this species collapses, as can be clearly seen at B1, B4, and B5, and the proxy no longer reports Atlantic water flux correctly (Fig. 5.4d). Millennial-scale Holocene episodes of increased advection of heat by Atlantic waters off Norway are associated with enhanced winter precipitation over Scandinavia, increased sea-salt fluxes over Greenland, and strengthened wind over Iceland. Thereby suggesting common atmospheric forcings, including the location and intensity of the westerlies and the associated changes in mid- to high-latitude pressure gradients. Such atmospheric processes are thought to explain the observed coupling between periods of excess drift ice delivery to Northern Iceland (Andrews 2009; Fig. 5.4c), and intervals of maximum inflow of warm Atlantic water to the Norwegian Sea (Giraudeau et al. 2010; Fig. 5.4d) throughout the last 11,000 years.

5.6 The hydrological 2500-year climate cycle

Precipitation is affected by multiple factors, and in many cases determined by regional or even local climatic and weather patterns. It is clear however that the atmospheric reorganization that accompanies the 2500-yr Bray climate cycle is reflected in precipitation changes in several locations. For decades Michel Magny has been studying Holocene mid-European lake level fluctuations and their impact on prehistoric human settlements (Magny 2004). His research shows very clearly the impact of Holocene climatic change. There is a general trend to increasing dryness during the Neoglacial, after a wetter HCO. Overlapping this general trend attributable to Milankovitch forcing, the 2500-year cycle is characterized by strong transitions from low to high lake levels (Magny 2004; Fig. 5.5a), indicating greatly increased precipitation at the lows of the Bray climatic cycle in this European sub-region.

A winter precipitation reconstruction from Norway's coastal glaciers shows periods of increasing precipitation at the lows of the Bray cycle (Matthews et al. 2005; Fig. 5.5b). Besides feeding glacier advances at these times (Fig. 5.2a), the Norway glacier-derived winter precipitation record matches almost exactly the Norway marine-derived Atlantic warm-water inflow record (Fig. 5.4d), supporting a causal relationship.

Spanish fluvial chronology also supports a 2500-year cycle in precipitation (Thorndycraft & Benito 2006; Fig. 5.5c). Three of the five main flooding periods highlighted by the authors coincide with B1, B2, and B5 lows in the Bray cycle. In addition, B3 and B4 lows are also characterized by significant episodes of slackwater floods or paleofloods, that record periods of increased flood frequency related to Holocene climatic variability (Thorndycraft & Benito 2006; Fig. 5.5c). They are fine-grained sediments produced by large magnitude floods, preserved in valley side rock shelters in bedrock gorge reaches. The last 1300 years register a large increase in the frequency of floods in Spanish rivers. The authors propose an increased preservation potential and/or increased human impact on the landscape as likely cause.

Holocene Ireland hydrology has been reconstructed from oaks and pines collected from bogs. These trees, accurately dated through dendrochronology (oaks) and carbon-dating (pines), provide a record of dry conditions when the decreased water table levels allowed the colonization of these marginal environments by trees (Turney et al. 2005). Although Ireland hydrology shows a complex pattern over an increasingly wet Neoglacial trend, lows in the Bray cycle are associated with periods of increased precipitation (Fig. 5.5d). This is in contrast with a Neoglacial drying trend over much of the rest of Europe and the world

The hydrological changes caused by the 2500-year climatic cycle are not restricted to the North Atlantic region. The same pattern can be found in the Santa Barbara Basin (California), reflected in varve thickness variability, that is known to depend on annual precipitation, and inversely correlates with wind strength (Nederbragt & Thurow 2005). The described c. 2750-year cycle in varve thickness correlates very well with the Bray climate cycle (Fig. 5.5e), with periods of higher varve thickness (increased precipitation) at the Bray lows.

A high-resolution record of the strength of the Asian monsoon was obtained from oxygen isotopic analysis of stalagmite "DA" in Dongge Cave (China; Wang et al. 2005). The record supports episodes of monsoon weakness (dryness) at every one of the Bray lows, most of them highlighted by the authors of the work (Fig. 5.5f). Most of the centennial and millennial variability in the Asian and Indian monsoons has traditionally been linked by multiple authors to solar variability (Wang et al. 2005; Neff et al. 2001).

5.7 The temperature 2500-year cycle

Although temperature variations should not dominate climate change analysis, they are an important indicator of abrupt climatic changes, and therefore one would expect to find traces of the 2500-year climatic cycle in temperature proxy records. In the previous chapter (Sects. 4.2 & 4.4) a global temperature reconstruction based on 73 global proxies selected by Marcott et al. (2013) was presented. In this global temperature reconstruction every low of the Bray cycle coincides with a period when temperatures were experiencing a significant decrease when compared to the previous trend (Fig. 5.6a). Even B5, when the world was still experiencing the fast warming that led to the HCO, shows a significant departure from the warming trend of the previous centuries.

The global temperature reconstruction in figure 5.6a reflects global temperature changes and is not dominated by Northern Hemisphere records. This is confirmed by the Rosenthal et al. (2013) reconstruction of intermediate water temperatures at the equatorial Indo–Pacific Warm Pool, the warmest oceanic region in the world (Fig. 5.6c). Their reconstruction displays a very similar profile to the global reconstruction presented here (compare Fig. 5.6a and c), and shows that every Bray cycle low coincides with a significant downward departure from the general temperature trend. This is confirmed also by the finding in the same

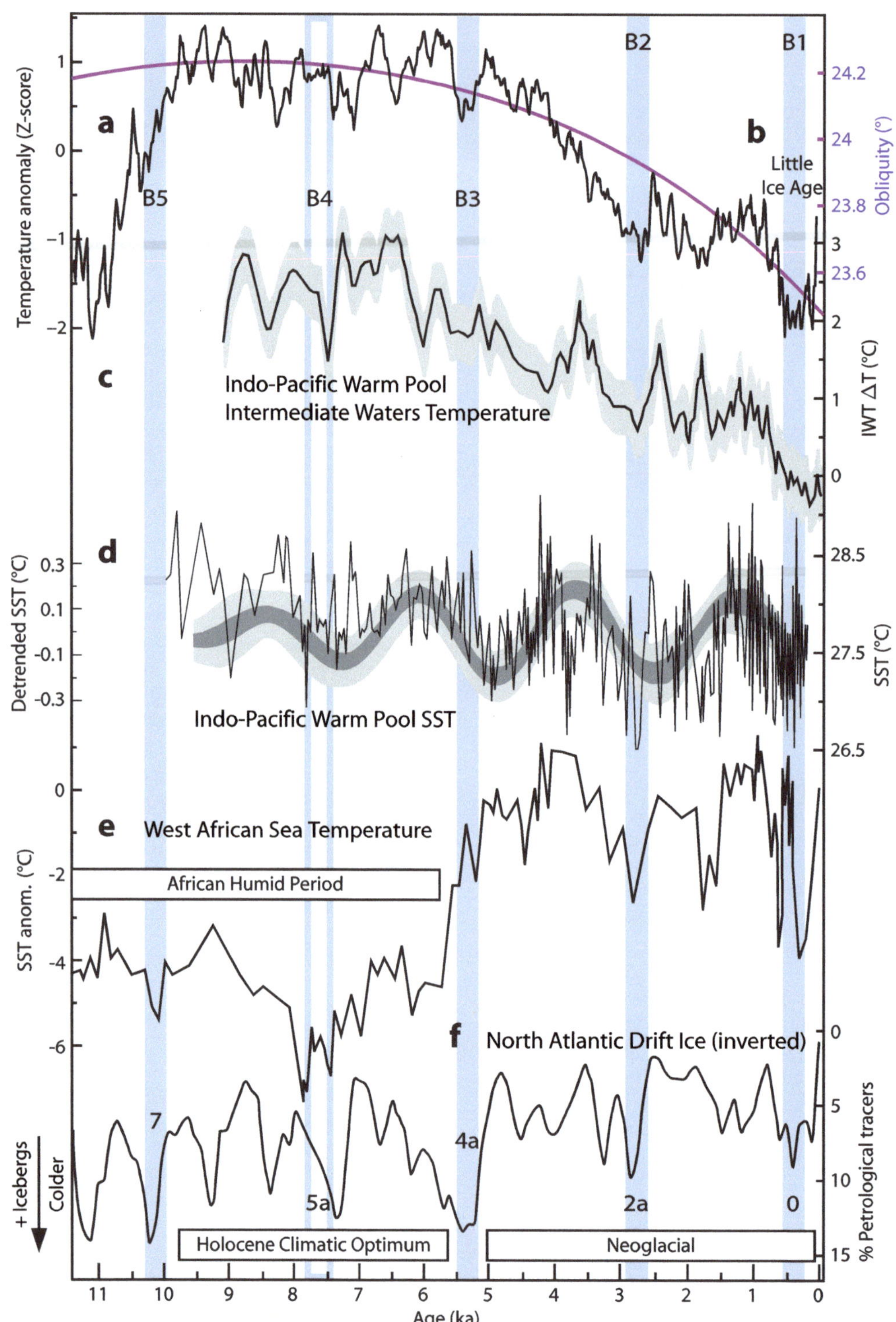

Fig. 5.6 Holocene temperature proxies and reconstruction
a) Global average temperature reconstruction from 73 proxies as in Marcott et al. (2013), using proxy published dates, and differencing average, with temperature expressed as distance to the mean in standard deviations (Z-score), as discussed in Sect. 4.4. **b)** Earth's axis obliquity is shown to display a similar trend to Holocene temperature. **c)** Holocene reconstruction of intermediate-water temperature at 500 m depth from a suite of sediment cores in the Makassar Strait and Flores Sea in Indonesia, at the Indo–Pacific Warm Pool. Temperature expressed as anomaly relative to the temperature at 1850–1880 CE. Shaded bands represent ±1σ. After Rosenthal et al. (2013). **d)** Thin line, right scale, sea surface temperature reconstruction at the Davao Gulf, south of Mindanao, from Mg/Ca levels in the surface foraminifer *Globigerinoides ruber*. Dark grey band (left scale) corresponds to the 2000–3000 years band-pass filter of the detrended data, with the light grey area the 90% confidence level. After Khider et al. (2014). **e)** Holocene variations in subtropical Atlantic SST from marine sediment core 658C. The record documents a well-known shift in African monsoonal climate at 5.5 kyr, when changes in the earth's orbit displaced the African monsoon southward, bringing much drier and warmer conditions to subtropical Africa and ending the African Humid Period. Superimposed on this trend are millennial-scale SST variations coherent with some of the North Atlantic ice-rafting events defined by Bond et al. (2001), including the lows of the Bray cycle (vertical bars). After de Menocal et al. (2000). **f)** Ice-rafted debris stack (inverted) from four North Atlantic sediment cores. The authors proposed that the increase in iceberg activity in the North Atlantic is tied to the increase in cold water advection from the Arctic and Nordic seas. After Bond et al. (2001).

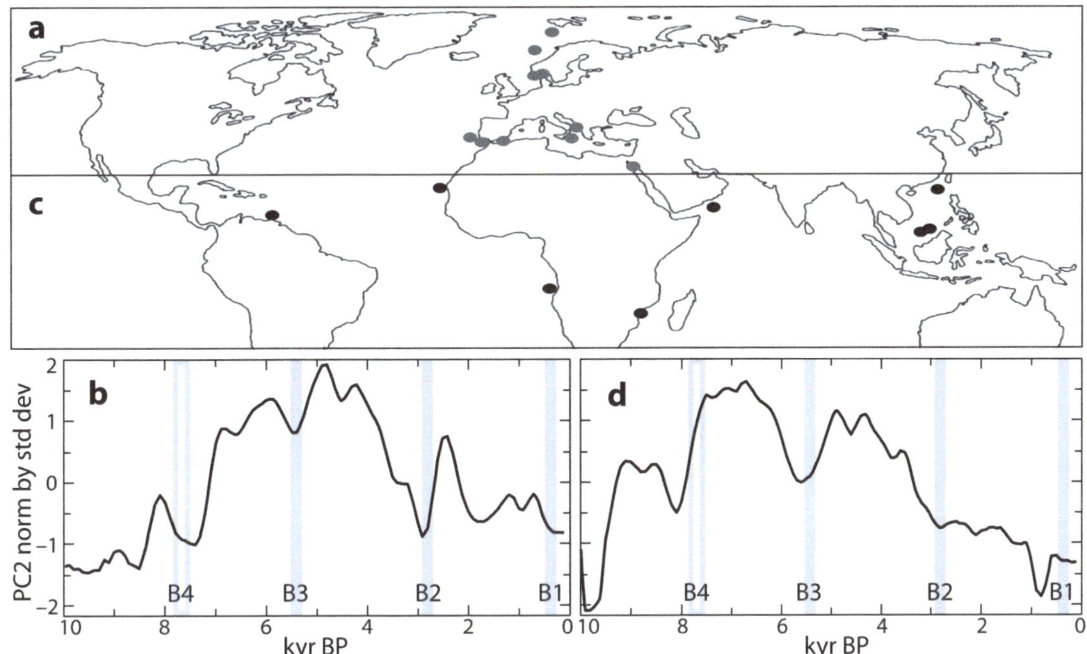

Fig. 5.7 Holocene millennial-scale sea-surface temperature variability
a) and c) Marine sediment core positions for the 10 North Atlantic–Mediterranean–Red Sea cores and the 8 tropical region (25°S to 25°N) cores analyzed, respectively. **b) and d)** Time coefficient (Principal Component Analysis) for the second empirical orthogonal function of the alkenone-based SST variability. The dominant modes of North Atlantic (b) and tropical (d) Holocene SST display a 2.5 kyr cycle linked to the strength of AO/NAO during the Holocene, showing that this cycle has a global character. After Rimbu et al. (2004).

area (south of Magindanao) that Holocene SST display variability in the 1000, 1500, and 2500 periodicities, and the 2500 periodicity coincides very well with the Bray cycle (Khider et al. 2014; Fig. 5.6d). Khider et al. measure the water surface temperature changes associated with the Bray cycle at the Indo–Pacific Warm Pool as 0.3 °C, and calculate a climate sensitivity to millennial solar cycles of 9.3–16.7 °C/Wm^{-2}, an order of magnitude higher than the estimated sensitivity to the 11-year solar cycle.

Temperature proxies at the West African Sea indicate that SST was over 2 °C lower during the African Humid Period (de Menocal et al. 2000; Figs. 4.8 and 5.6e). After it ended, the lack of precipitation due to the southward displacement of the African monsoon produced an abrupt warming of the sea surface before joining the global cooling trend of the Neoglacial. Within this complex general pattern, the lows of the Bray cycle are once more associated with a significant temporal reduction in SST (Fig. 5.6e).

A more complete analysis of SST in the tropical oceans and the North Atlantic region, the Mediterranean, and Red Sea, was performed by Rimbu et al. (2004), using 18 alkenone records. The principal mode of variability reflects Milankovitch forcing, delayed in the case of the North Atlantic by the melting of the ice sheets. The secondary mode of variability (principal temporal component from the second empirical orthogonal function) shows in both regions as a c. 2300-year cycle that agrees well with the Bray cycle (Rimbu et al. 2004; Fig. 5.7). The main disagreement is with B4 due to the 8.2 kyr event, that affected SST in the North Atlantic as early as 8.4 kyr BP, but seems to have had a delayed effect in the tropics around 8.1 kyr BP. As discussed below (Sect. 5.9), B4 constitutes a gap or weakening in the Bray sequence that is reflected in both solar and climate records. By analogy with the

instrumental period records and the analysis of a long-term integration of a coupled ocean-atmosphere general circulation model, Rimbu et al. (2004) suggest that the AO/NAO is one dominant mode of climate variability at millennial time scales. This conclusion agrees well with the other evidence shown here for the Bray climate cycle.

The Bond record of drift-ice petrological deposition in the North Atlantic is also generally considered to correlate to colder conditions in the North Atlantic region that favor more frequent southward moving icebergs (Bond et al. 2001). Most, if not all, Bond events have been linked to cooling and abrupt climate change outside the North Atlantic area. The Bond record also reflects the 2500-yr Bray cycle as the lows in the Bray cycle coincide with Bond events 0 (LIA), 2a, 4a, 5a, and 7 (Fig. 5.6f).

We can conclude that the 2500-year Bray climate cycle is very well established in the proxy record of past climate changes in the North Atlantic region, but affects the entire planet. It is the most important climatic cycle of the Holocene. Although the Bray climate cycle is present in the chemical record of Greenland ice cores, it is not easily seen or, maybe, absent in the Greenland and Antarctic ice core temperature records. This is one of the reasons why it has been ignored for so long despite being present in multiple proxies and recognizable since 1912. Paleoclimatology has come to depend too much on the very reliable and precisely dated polar ice cores at the expense of the often contradictory, unreliable, and imprecisely dated climate proxies. This has had the result that whatever is not prominently displayed in polar ice cores is disregarded. Another complicating factor is that the Bray cycle is not the only cause of climate change during the Holocene and thus proxies are full of signals whose origin is often difficult to ascertain, creating much confusion among researchers that results in contradictory reports.

5.8 The solar variability 2500-year cycle

Radiocarbon dating was developed by Willard Libby in 1952 based on the idea that biological carbon samples that reflected atmospheric $^{14}C/^{12}C$ proportion at the time they were alive would progressively become ^{14}C depleted due to the isotope's radioactive decay, and thus would provide a clock to measure elapsed time. But Libby warned that there was no guarantee that the $^{14}C/^{12}C$ ratio had been constant in time. Therefore, a considerable effort has been ongoing since the 1960s to determine the proportion of ^{14}C in the atmosphere over past millennia. The resulting calibration curve (Fig. 5.8) is used to convert radiocarbon dates into real time dates. But the radiocarbon clock does not run at a constant speed as the real-time clock does. There are times when the radiocarbon clock runs faster and times when it runs slower, creating bumps in the calibration curve (Fig. 5.8, ovals and arrowheads). That the radiocarbon clock runs faster (vertical-axis values decrease faster in Fig. 5.8), implies that the $^{14}C/^{12}C$ ratio is deviating upwards, as samples with more ^{14}C are more recent. This means that either ^{14}C is being produced at a higher rate, or total CO_2 is decreasing while ^{14}C is not. Most scientists believe the first explanation contributes more to the observed changes because the proportion of

^{14}C is so small in the atmosphere (c. 10–12) as to require very large changes in total CO_2 to produce the alterations that can be explained by small increases in ^{14}C. And we know from ice core records that CO_2 changes have been relatively small during the Holocene. A carbon cycle model has been used since the late 1970s to account for the effect of CO_2 variations on radiocarbon dating. Thus, the best explanation for the acceleration observed in the radiocarbon clock is that the ^{14}C production rate from cosmic rays in the atmosphere increased due to a decrease in the solar magnetic flux that takes place when the Sun is in a prolonged period of low activity known as a solar grand minimum (SGM). This conclusion is supported by the same variability shown by a different cosmogenic isotope, ^{10}Be, whose deposition does not depend on the carbon cycle.

From the early ^{14}C production data available in the late 1960s Roger Bray noticed a correspondence between climate change and radiocarbon production (Bray 1968), thus defining both a climate cycle and a solar variability cycle. This caused him to propose that changes in solar activity were responsible for the climatic changes. The solar cycle can be clearly seen in the radiocarbon data from the c. 2500 year spacing of higher ^{14}C production at 12,650, 10,200, 5,250, 2,700, and 500 BP, corresponding to all the Bray lows in the Holocene and Younger Dryas

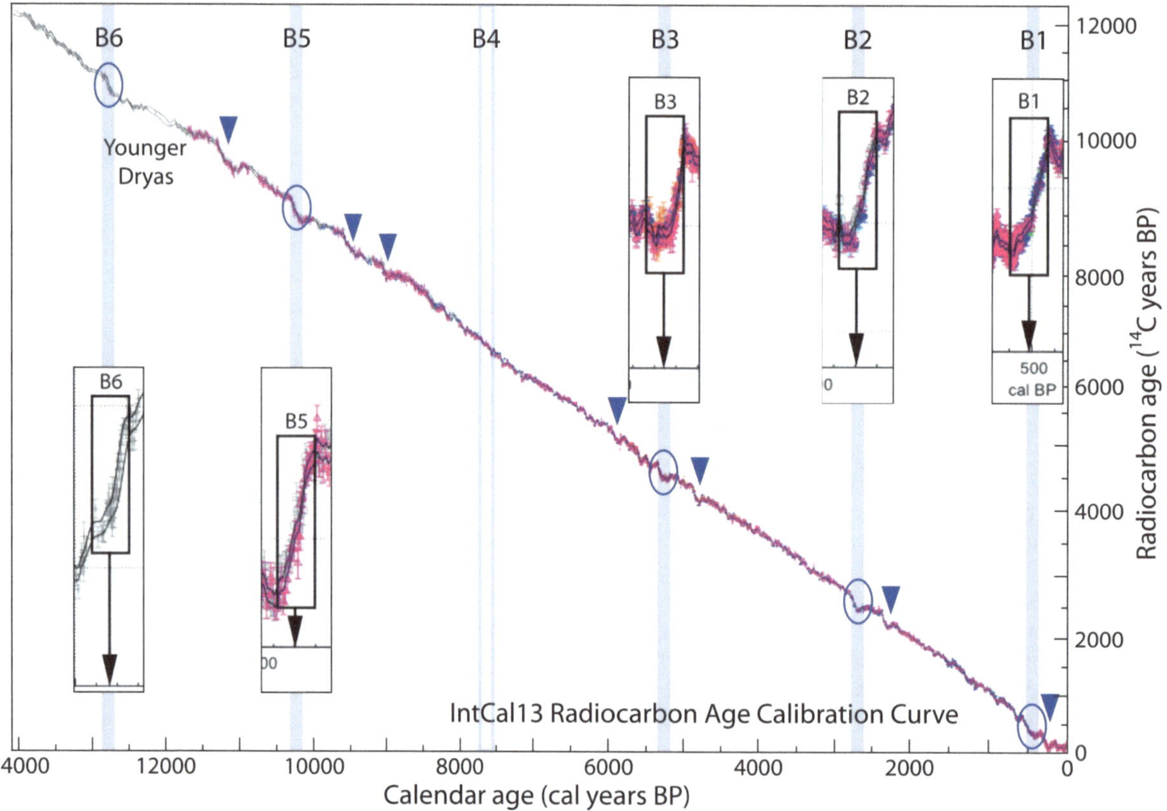

Fig. 5.8 2500-year periodicity in the radiocarbon calibration curve
The radiocarbon calibration curve (IntCal13) is unrelated to climate change and obtained through the efforts of hundreds of researchers over decades to provide an accurate way of measuring the time elapsed since a biological sample stopped living. The calibration curve presents periods of time in the past, when there was a noticeable deviation from linearity (ovals and arrowheads). Five of those periods (ovals) are separated by multiples of c. 2500 years delimitating a ^{14}C cycle. Solar activity reconstruction from cosmogenic ^{10}Be and ^{14}C isotopes shows that those periods correspond to periods of unusually high isotopic production interpreted as solar grand minima, like the Spörer and Maunder minima. Those periods correspond precisely to the lows of the Bray climate cycle (blue bars). Insets show the radiocarbon response to the lows of the Bray cycle from IntCal13, and display the published orientation, with more recent to the left, opposite to the figure. After Reimer et al. (2013).

except B4, that lacks a similarly noticeable ^{14}C production signature (Fig. 5.8 ovals).

The Bray solar cycle was again identified by Jan Houtermans in his PhD thesis of 1971, and has since been confirmed multiple times independently. The missing B4, that should have occurred around 7.8 kyr BP, the very low solar activity at around 8.3 and 7.3 kyr BP, the presence of other periods of very high ^{14}C production between 11.5 and 9 kyr BP (Fig. 5.8 arrowheads), together with a shorter period for the 2700–500 BP oscillation are factors that have contributed to different studies differing in the length of the Bray solar cycle between 2200 and 2600 years depending on the methodology used. Table 5.1 gives the correct central dates for the lows of the Bray cycle, and its periods length. The shortness of the last period, with only 2,200 years, is exceptional, as most periods appear to be between 2400–2600 years. This type of variability, and the absence of some oscillations is not unusual in solar variability, and the 11-year solar cycle presents a proportionally bigger variability (9–14 years), and missed several oscillations between 1650 and 1700.

In the late 1980s Sonett and Damon, despite being aware of Bray's and Houtermans' studies, decided against established custom not to name the cycle after its discoverer, but as Hallstattzei (later Hallstatt) for a late Bronze–early Iron cultural transition in an Austrian archeological site during the cycle's B2 minimum, 2800 years ago. The inappropriateness of a human cultural name from a particular period for a solar cycle that has been acting for tens of thousands, and probably millions of years (Kern et al. 2012), plus the injustice of ignoring its discoverer, demand that the cycle be properly renamed as the Bray cycle.

Bray Low	Date (BP)	Period (yr)
B9	c. 20,500	–
B8	c. 17,500	3,000
B7	c. 15,000	2,500
B6	12,650	2,350
B5	10,200	2,450
B4	–	(2,475)
B3	5,250	(2,475)
B2	2,700	2,550
B1	500	2,200
Average		**2,500**

Table 5.1 Dates and periods for the 2500-year Bray cycle from cosmogenic isotope production rates
Central dates for the Spörer-type solar grand minimum that marks Bray cycle lows are given in the central column rounded to the nearest half century. Period length is calculated from cycle lows, and length between B5 and B3 is divided in two and shown in parentheses.

One peculiarity of the 2500-yr solar cycle is that it modulates the amplitude and phase of the 210-yr de Vries solar cycle (Sonett 1984; Hood & Jirikowic 1990). The amplitude of the de Vries cycle is maximal at the lows of the Bray cycle (Fig. 5.9), and minimal at mid-time between lows, to the point of becoming imperceptible. This modulation has been known since 1984, but its cause is unknown. It demonstrates however that the 2500-yr and 210-yr solar cycles are not independent, and since the 210-

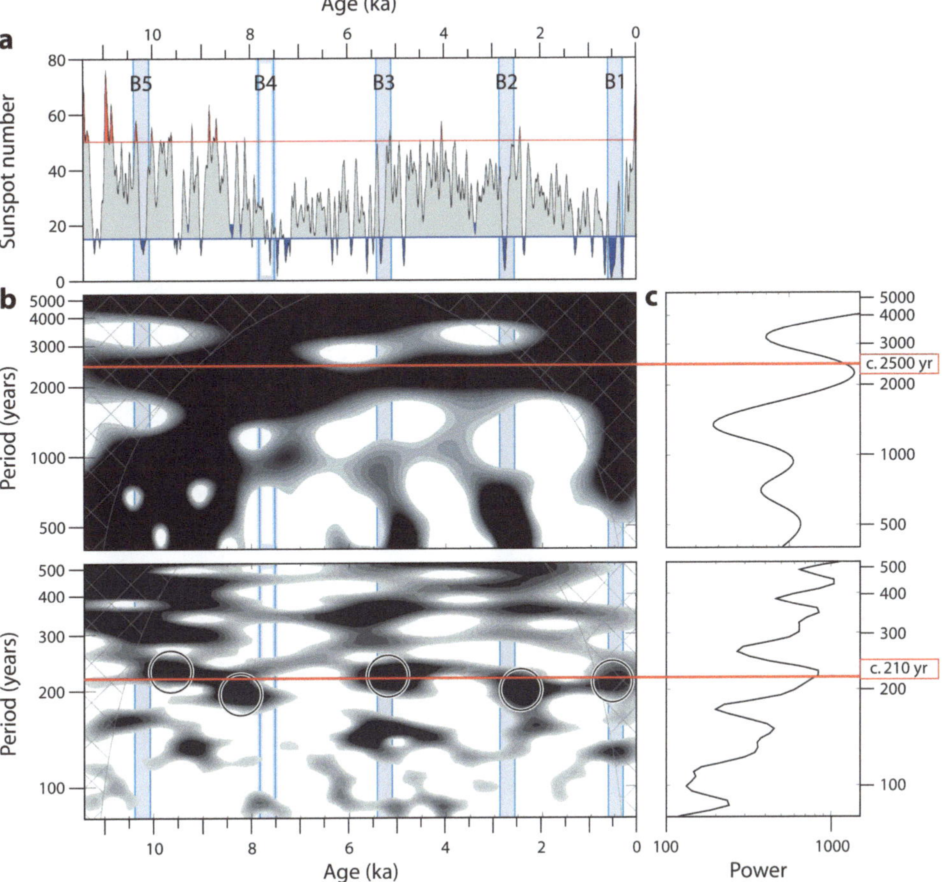

Fig. 5.9 Modulation of the de Vries cycle by the Bray cycle
a) Sunspot activity reconstruction from ^{14}C data. Vertical blue bars show the position of Bray cycle lows. **b)** Wavelet spectrum of data in (a). Upper and lower panels correspond to period ranges of 500–5000 years and 80–500 years. Dark/light shading denotes high/low power. Circles show accumulation of the 210-yr signal close to the lows of the Bray cycle, indicating the 2500-yr cycle modulates the 210-yr cycle. **c)** 2D frequency spectra corresponding to (b). Red horizontal lines mark the position of the 2500 and 210 signals in (b) and (c). Reproduced from Usoskin (2013) with permission. © Usoskin (2013) CC BY-NC 3.0.

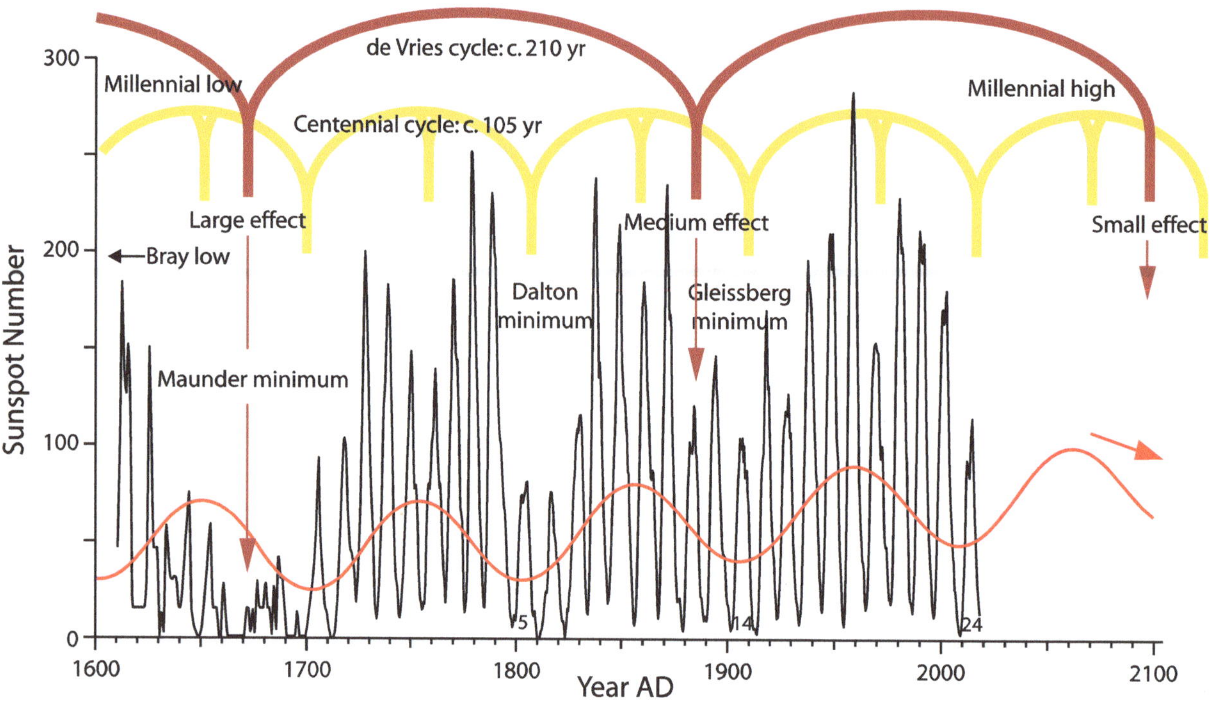

Fig. 5.10 Modulation of the short solar cycles during the telescope era
Sunspot number (black curve, from 1700) reconstructed back to AD 1610 using sunspot groups. Sinusoidal curve, empirical fitted function for the centennial solar cycle with a period of 103 years described by Tan (2011). The centennial cycle (thick orange curve) presents minima of decreasing intensity at 1700, 1805 (SC5), 1910 (SC14), and 2015 (SC24), going from the millennial low at c. AD 1600 to the millennial high at c. AD 2100. The pentadecadal cycle is also shown as thick shorter bars between the centennial lows. The modulation of the de Vries cycle (thick dark red curve) can also be seen as the low of c. AD 1675 is much lower than the low of c. AD 1885. It can be expected that the low of c. AD 2095 should have even less effect. Thus, moving forward from the Bray low at c. AD 1500, we see more solar activity at each successive centennial and bicentennial lows.

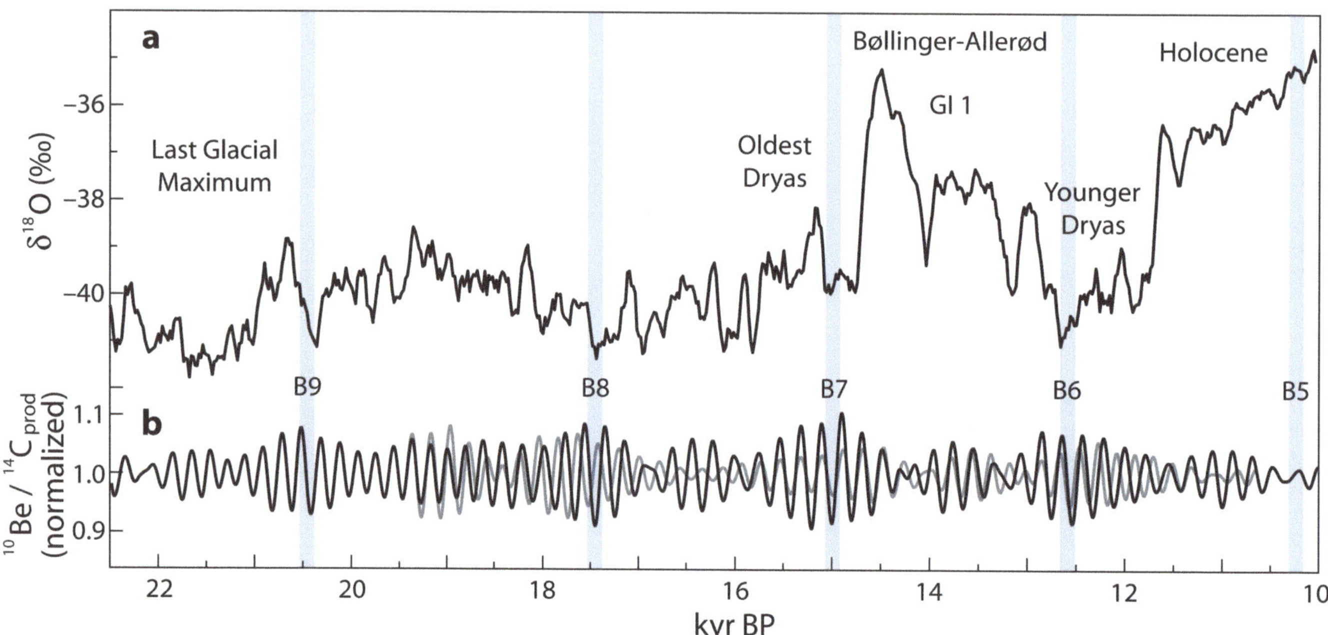

Fig. 5.11 The Bray cycle during the last glacial maximum
a) GISP2 δ^{18}O Greenland temperature proxy on GICC05 timescale (after Seierstad et al. 2014), 80-yr averaged, for the period 10–22.5 ka. **b)** ^{14}C production rate (H82 speleothem, grey line) and ^{10}Be flux (black line), normalized to display only the variability in the 180–230 yr band to capture the solar de Vries cycle (210 yr). After Adolphi et al. (2014). Due to the modulation of the de Vries cycle by the Bray cycle, periods of maximum de Vries variability correspond to the lows of the Bray cycle, and are spaced by c. 2500 years, generally coinciding with cooler periods in Greenland. Those positions also correspond to peaks in ^{14}C production rate produced by solar grand minima. GI, Greenland interstadial.

yr de Vries solar cycle is better known for being present in the sunspot record, it constitutes strong evidence for the 2500-yr solar cycle.

This property of some of the short solar cycles, of being modulated by the long cycles can be observed in the sunspot record of the past 350 years, where we observe that both the de Vries and centennial cycle lows display progressively more activity as we get farther away from the Bray and millennial (Eddy) lows, becoming less conspicuous with time (Fig. 5.10). This is how solar activity has been increasing for the past 400 years, by reducing the periods of below average activity, due to this modulation.

The de Vries cycle modulation by the Bray cycle allows the identification of its lows during the last glacial period, when drastic climatic changes obscured the c 2500-year climatic cycle, and made the cosmogenic record less reliable. Adolphi et al. (2014), isolated the 180–230-year signal containing the de Vries cycle in Δ^{14}C production and ^{10}Be flux data between 22.5 and 10 kyr BP. This signal displays the 2500-year Bray cycle modulation, allowing the identification, albeit less precisely, of the position of Bray lows B7–B9 (Fig. 5.11; table 5.1) at c. 15, 17.5, and 20.5 kyr BP. If correct, these dates support a periodicity for the Bray solar cycle between 2450–2500 years, further substantiating its close association with the climatic cycle that also appears closer to 2500 than 2400 years. The authors also propose that, during the Last Glacial Maximum, solar minima correlate with more negative δ^{18}O values in ice (lower temperatures) and are accompanied by increased snow accumulation and sea-salt input over Central Greenland (Adolphi et al. 2014; Fig. 5.11). This supports the idea that the Bray climate cycle also acts during glacial periods.

SGM do not occur randomly distributed along the cosmogenic record, but tend to happen in clusters. The existence of these clusters of SGM separated by multiples of 200 years has long been known, and the Oort–Wolf–Spörer–Maunder cluster of the past millennium is the last example. Other clusters have taken place at different times during the Holocene (Versteegh 2005). Usoskin et al. (2016), after reconstructing solar activity for the past 9000 years, showed that SGM tend to be closely related with the Bray cycle, being more numerous during its lows (Fig. 5.12b). They propose that the solar dynamo switches between a normal mode and a grand minimum mode, and that the probability of this switch is modulated by the Bray cycle. Given that SGM in clusters tend to be spaced at multiples of 200 years we could add that the combination of a low in the 2500-yr Bray solar cycle and a low in the 210-yr de Vries solar cycle must greatly enhance the probability of the sun entering a grand minimum. As reviewed in chapter 8, the lows of the 1000-yr Eddy solar cycle also tend to be associated to SGM.

5.9 2300-year Hallstatt versus 2500-year Bray

Proper characterization of the 2500-yr solar Bray cycle faces a difficult problem from the unlikely coincidence of an unfortunate set of circumstances. Most authors have described this cycle as a 2300-yr cycle, and that periodicity is incompatible with the 2500-yr cycle described here. Some authors report it as a 2400-yr cycle due to an averaging of both periodicities. The dates given in table 5.1 for the lows of the Bray cycle are consistent with multiple lines of evidence: their increase in ^{14}C production rate

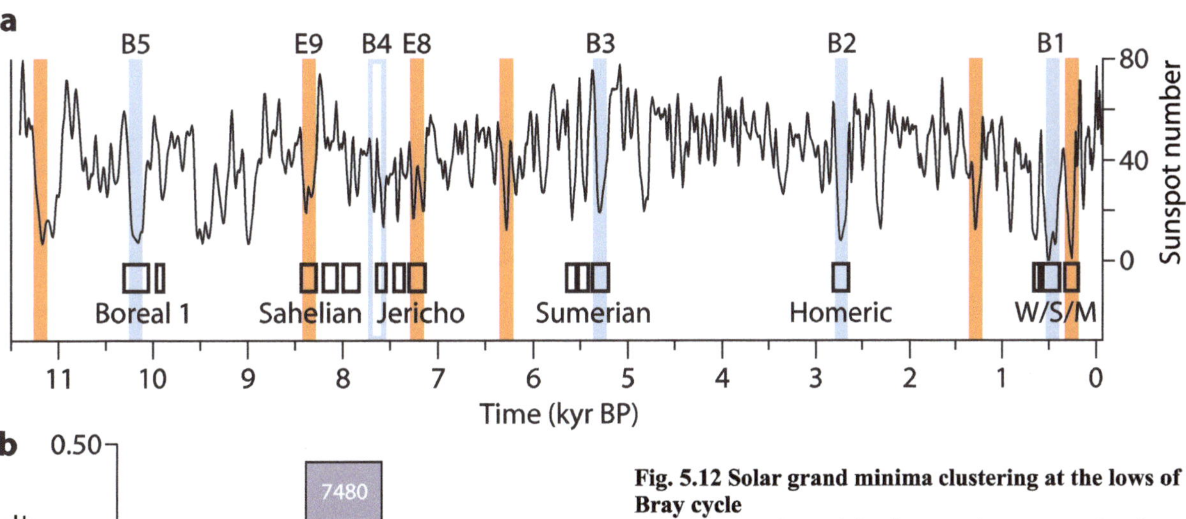

Fig. 5.12 Solar grand minima clustering at the lows of the Bray cycle
a) Holocene solar activity (sunspots) reconstruction from ^{14}C data after Solanki et al. (2004) between 11.4–8.7 kyr BP, and after Wu et al. (2018), from 8.7 kyr BP. Blue bars indicate the lows of the Bray cycle. Black boxes correspond to solar grand minima close to the lows of the Bray cycle, with their names or initials after Davis (1994). Orange bars correspond to some of the lows of the c. 1000-year Eddy solar cycle, with only the lows at 8.3 (E9) and 7.3 (E8) kyr BP numbered. This figure illustrates the difficulty of correctly identifying B4, a cause for the variable length assigned to the cycle by different numerical analyses. **b)** Probability density function (PDF) of the time of occurrence of grand minima relative to the time of occurrence of the nearest low of the Bray cycle, by superposed epoch analysis, after Usoskin et al. (2016). The times of occurrence of the lows of the Bray cycle were defined by considering the average of two second singular spectrum analysis components of the sunspot number reconstruction from ^{14}C and ^{10}Be, and are indicated by the numbers in the figure. They are very close to the dates given in table 5.1.

(Fig. 5.8); their strong climatic effect that results in the Holocene's most prominent climate cycle (Figs. 5.2 to 5.7, & Chap. 6); and for their modulation of the 210-yr de Vries solar cycle (Figs. 5.9 & 5.11). However instead of producing this periodicity, every frequency analysis of cosmogenic isotopes datasets, like IntCal 13, or solar activity reconstructions produces an alternative periodicity of 2300 years. The reasons for this are the following:

- The missing B4 low. The absence or very weak nature of this low is really a problem for the correct identification of the Bray cycle, given its length (only four possible periods in the Holocene). There are two periods of low solar activity close in time from where B4 should be, one at 8.3 kyr BP and one at 7.2 kyr BP, both belonging to the millennial frequency. Climate proxies tend to chose the low at 8.3 for their cycle low, because of the strong climatic effect of the 8.2-kyr event. Solar proxies chose the low at 7.2 as it is associated to a stronger set of solar grand minima. This creates further confusion as the climate cycle might appear 2600-year long (Fig. 5.3a & b), while the solar cycle appears 2300-year long.

- The shortest period in the irregular Bray cycle was the last one, with only 2.2 kyr between B2 at 2700 BP and B1 at 500 BP. The combination of a missing minimum and a very short period prevents frequency analysis from giving a strong signal in the 2500-yr frequency.
- The existence of an alternative periodicity that starting from the short B1–B2 period builds a cycle with a shorter periodicity that continues through the gap at B4. This is the periodicity that is being found at every time series analysis of solar activity (Fig. 5.13). Some evidence however indicates that it is a spurious periodicity, an unhappy coincidence that might have confused solar researchers for decades.

So how do we know which one is the correct periodicity? Both periodicities depend on the Homer grand minimum at 2700 BP for lack of alternatives, but earlier on they must depend on different solar grand minima (SGM) to produce the different periodicities (Fig. 5.13). Several lines of evidence support the 2500-yr periodicity, despite being more irregular during the last 10,000 years.

The 2500-yr periodicity is based on homogeneous SGM. B6, B5, B3, B2, and B1 are all Spörer-type SGM,

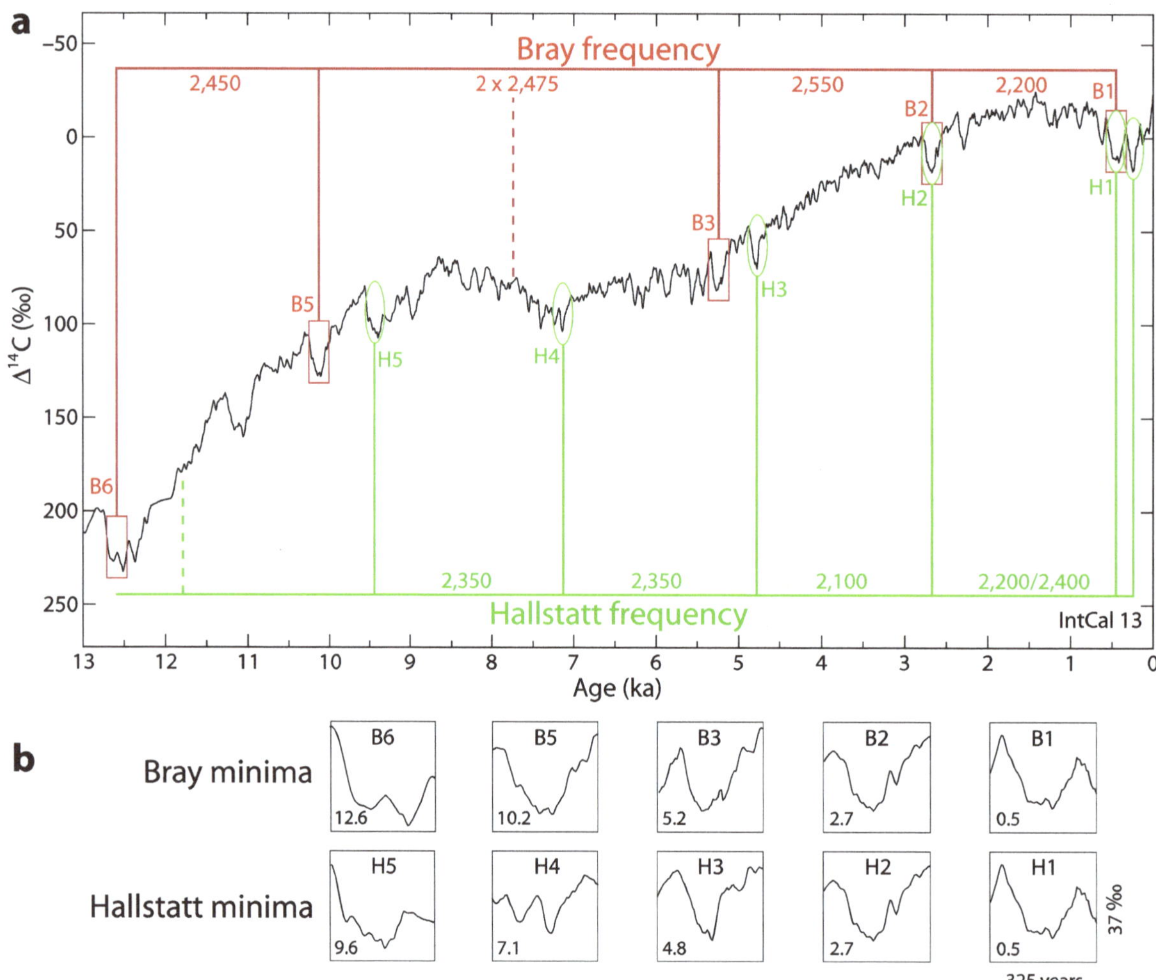

Fig. 5.13 Solar cycle periodicity: 2500 versus 2300 years
a) IntCal 13 atmospheric ^{14}C isotope variability curve for the period 0 to 13,000 yr BP on an inverted Δ^{14}C scale. Identified lows in the 2500-year Bray periodicity (B1 to B6 highs in ^{14}C production) are shown by their solar grand minima (SGM) within rectangles with their separation in years indicated on top. The set of lows producing the 2300-year Hallstatt periodicity (H1 to H5) are shown by their SGM within ovals with their separation in years indicated on bottom. Vertical dashed lines indicate missing SGM at some lows on both periodicities. **b)** Comparison of the SGM at the lows of both periodicities highlighting the homogeneity of the Bray periodicity SGM, composed entirely by Spörer-type SGM, versus the heterogeneity of the Hallstatt periodicity SGM, with a Maunder-type SGM (H3), and an atypical SGM (H4).

characterized for lasting c. 200 years (longer than Maunder-type SGM), and displaying a 20‰ increase in ^{14}C production (Fig. 5.13b). Therefore the 2500-yr Bray solar cycle is characterized by the presence of Spörer-type SGM at its lows. By contrast 2300-yr Hallstatt periodicity SGM are heterogeneous, including Maunder-type SGM or even shorter (H3, H4; Fig. 5.13b), with variable changes in ^{14}C production rates.

All the minima in the 2500-yr Bray cycle for the past 11,000 years (not counting the missing B4), correspond to some of the more intense periods of abrupt climate change in the Holocene known to us (reviewed in Chap. 6). By contrast, the 4.7 kyr ACE that corresponds to H3, although recognizable in climate proxy records (see Fig. 4.18 & table 4.1) is so unremarkable that it is mostly unknown and seldom mentioned in the scientific literature. The 9.6 kyr increase in ^{14}C production corresponding to H5 in the Hallstatt periodicity poses an even bigger challenge. It shows a 28‰ increase in ^{14}C production, by far the biggest within the Holocene. Yet most detectable climatic changes occur later and appear to be associated to the SGM taking place at 9.3 kyr BP (see Fig. 4.18 & table 4.1), that is part of the millennial periodicity. The contrast between the huge increase in ^{14}C production and the lack of synchronous detectable climate effect is so remarkable that it is possible that the great increase in ^{14}C is not due to a corresponding decrease in solar activity but to an increase in galactic cosmic rays. A know source of galactic cosmic rays took place in the early Holocene, although its dating is highly uncertain. It was the formation of the Vela supernova, that according to Sushch & Hnatyk (2014), has a hydrodynamical age of 7–12,000 yr. It is tempting to think that the biggest cosmic ray increase in the Holocene that has no coeval significant climate change was not due to decreased solar activity, but to a nearby supernova.

The Hallstatt 2300-yr periodicity ends at 9.5 kyr BP. Most studies of this periodicity use a too short record to detect its ending. Weiss & Tobias (2016) use 9,400 years; Vasiliev & Dergachev (2002) use only 8,000 years; Beer et al. (2017) go up to 10,000 years. Figure 5.13 shows that there is no suitable SGM for the 2300-yr periodicity after 9.5 kyr. This is further confirmed by wavelet analysis of the last 13,000 years of IntCal13 that shows the 2300-yr frequency absent prior to 12 kyr BP (not shown). By contrast it has been shown here that the 2500-yr periodicity is detectable in cosmogenic records from at least 20.5 kyr BP with that periodicity (Figs. 5.8 and 5.11).

The last line of evidence comes from the ability of the 2500-yr cycle to modulate the 210-yr cycle (Fig. 5.9; see also Fig. 8.11g). The low of the 2300-yr periodicity at 7.2 kyr BP (H4 in Fig. 5.13) fails to show the high amplitude 210-yr oscillations that characterize the lows of the long solar cycle (Beer et al. 2017; Fig. 5.9). This indicates that the H4 low does not belong to the Bray solar cycle. Actually the 7.2 kyr minimum in solar activity belongs to the millennial Eddy solar cycle, that does not display a modulation of the 210-yr solar cycle and that will be reviewed in chapter 8.

The three reasons listed at the beginning of this section, and the four lines of evidence above support that the correct periodicity for the irregular Bray solar cycle is 2500 years. This makes the Bray climate cycle and the Bray solar cycle coincident, both with a weak low at the 8th millennium BP. Time series analysis is notoriously prone to errors, but it might be difficult to accept by many that the clear 2300-yr periodicity in solar proxies is not the real periodicity but the result of an unfortunate set of coincidences. Looking beyond the numerical result into the nature of the data and extending the analysis to data before the Holocene should help accept the reported result. For as long as the wrong 2300-yr periodicity is retained, it will be very difficult to find the cause of this solar cycle and to elucidate its climatic effects as we will be looking at the wrong times.

5.10 The solar–climate relationship

Given the strength of the correlation between past cycles of climate change, and cycles in the production and deposition of cosmogenic isotopes, like the Bray cycle, the solar–climate relationship is accepted in paleoclimatology as non-controversial. Sixteen of twenty-eight (57%) articles whose climatic evidence has been reviewed in this chapter's figures explicitly state that changes in solar forcing are likely to be the cause of the observed climatic changes, and only one explicitly rules them out. Then, why is the solar–climate relationship so controversial outside of the paleoclimatology field?

"The reality of the Maunder Minimum and its implications of basic solar change may be but one more defeat in our long and losing battle to keep the sun perfect, or, if not perfect, constant, and if inconstant, regular. Why we think the sun should be any of these when other stars are not is more a question for social than for physical science" (Eddy 1976).

There are three main objections that opponents of the solar–climate theory raise, and two of them will be reviewed here, as they are pertinent to the Bray cycle. Since the close relationship between climate changes of the past and changes in the cosmogenic isotope record is undeniable, the first objection is to state that the cosmogenic record is likely to be contaminated by climate and therefore is more of a climatic record than a solar activity record. The second objection is that the sun is luckily extraordinarily constant, and therefore the small changes measured in total solar irradiance (TSI) between an 11-year maximum and minimum are of about 0.1% and produce a very small, almost undetectable, effect on climate. Since there is no evidence that the changes were much bigger during the last solar grand minimum, the Maunder Minimum, we know no mechanism to produce the observed climatic changes. The third objection is that for the past four decades solar activity and global temperatures have been going in opposite directions. I will deal with this objection more in detail in chapter 11, but for the time being suffice it to say that solar activity did not decrease between 1970 and 1995, and it is just one of the several forcings that act on climate, and therefore one should not expect temperatures to always follow solar activity, even if the hypothesis that solar variability has a big effect on climate is correct.

That the cosmogenic isotope record is affected by climate changes has been known from the beginning. The Δ^{14}C record is affected by changes in the carbon cycle. When the oceans cool they absorb more CO_2, and for a constant rate of production the $^{14}C/^{12}C$ ratio increases. Changes in vegetation go in the opposite way as plants release CO_2 during periods of cooling. On a scale of years to about one decade the faster plant response dominates, while for periods of decades to millennia the slower ocean response dominates. Solar activity reconstruction from $\Delta^{14}C$ includes a carbon cycle model, usually a box-model, but the sea level changes associated with ice-sheets melting during deglaciation are usually considered too large to be properly modeled and thus solar activity reconstructions from $\Delta^{14}C$ usually span only the Holocene. ^{10}Be deposition at the poles is affected by stratospheric volcanic eruptions and precipitation rates. Volcanic SO_2 and precipitation rates measured from ice cores are taken into account when reconstructing solar activity from ^{10}Be. The generally very good level of agreement between solar activity reconstructions from $\Delta^{14}C$ and ^{10}Be for the Holocene indicates that any remaining contamination must act similarly over the different deposition pathways of both isotopes. This is possible as a significant cooling would increase $\Delta^{14}C$ from enhanced CO_2 uptake by the oceans, while it might increase ^{10}Be by reducing precipitation rates. But as every climate proxy requires careful evaluation of the many factors affecting it, like sedimentation rates, or upwelling strength, to provide accurate information, the question is not if there is climate contamination in the cosmogenic record, but if the reconstructed record provides a good enough proxy for solar activity.

One test available to answer this question is to examine the reconstruction from cosmogenic isotopes over the period where we have information on solar activity from other sources that cannot be affected by climate. Comparison of the cosmogenic records over the past 400 years with the sunspot record shows a very good level of agreement (Fig. 5.14) despite this period undergoing intense climate change, from the depths of the LIA to the present global warming. Aurorae are more frequent the higher the solar activity, and using auroral historical records that extend back 1000 years, we observe that the correlation remains positive for the entire period, and that similar

maxima and minima can be clearly recognized, including a period of high solar activity and frequent aurorae around 1100 AD at the time of the well-known Medieval Warm Period (MWP; Hood & Jirikowic 1990; Fig. 5.14b). The conclusion is that within reasonable expectations the cosmogenic record reflects solar activity and thus is a useful proxy for it.

The cosmogenic record has faithfully registered the solar centennial variability for the past thousand years as determined from auroral records, and for the past 400 years as determined from sunspots numbers. Hood & Jirikowic (1990) provide another argument for the solar origin of the c. 2500-yr Bray solar cycle. If the Bray cycle was terrestrial in origin, the modulation that it produces on the de Vries cycle (Sonett 1984; Fig. 5.9) should not be observable on solar activity records, and the c. 210-yr solar cycle should appear unmodulated in solar activity phenomena, like sunspots or aurorae. However, as figure 5.14 shows, the modulation is clearly observable, as the lows of the de Vries cycle (dV1 to dV5, Fig. 5.14) increase their activity as they become more distant from the Spörer minimum. Again, the only possible conclusion is that the modulation caused by the 2500-yr cycle, and the cycle itself, are also of solar origin.

Further support for the implausibility of a climatic contamination of the cosmogenic record of such magnitude that would render it inadequate to determine past solar activity comes from the study of another climate cycle. A 1500-yr cycle has been identified by several researchers and does not show up in cosmogenic records during the Holocene. Kern et al. (2012) identified this cycle, as well as the Bray and millennial cycles in a Miocene lake sediment 10.5 Myr old. That these cycles are so old speaks of the stability of their causes over time, despite the many changes suffered by the Earth. Within the Holocene, the 1500-yr cycle has been identified in multiple proxy re-

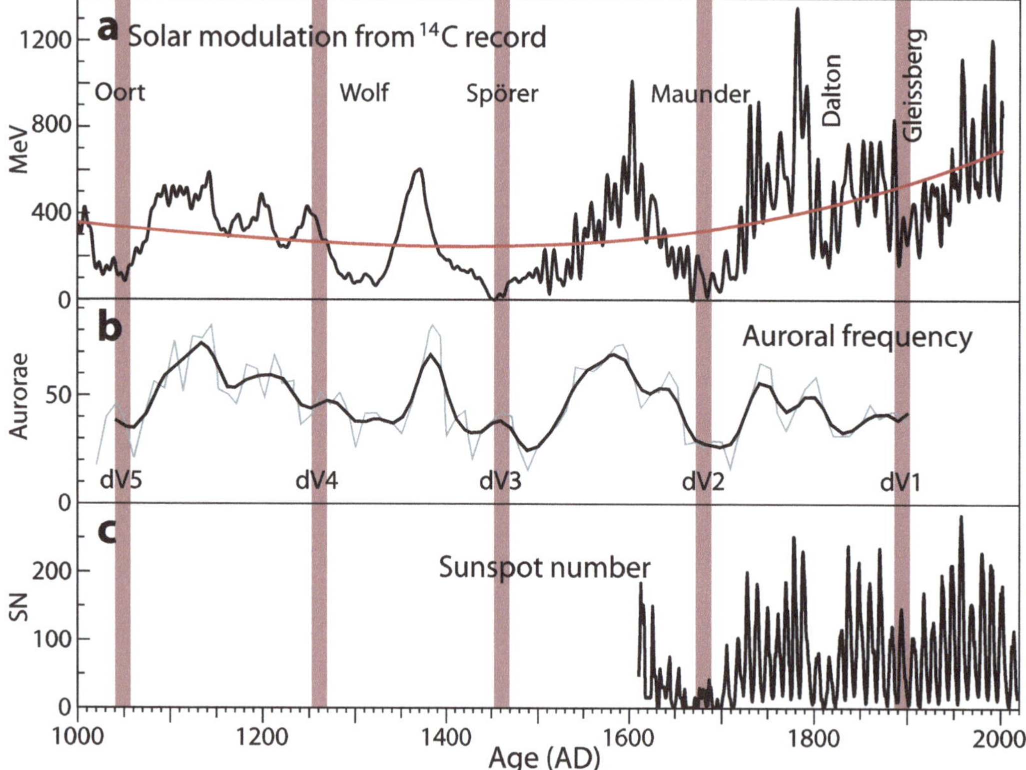

Fig. 5.14 Correlation between cosmogenic isotope production and solar activity
a) Black curve, solar modulation function based on ^{14}C production rate, after low-pass filtering at a cut-off frequency of $1/20$ yr^{-1}. Red line is the quadratic regression, highlighting the great increase in solar modulation (solar activity) since the lows of the Little Ice Age. After Muscheler et al. (2007). **b)** Thin line, auroral frequency record from historic sources. Thick line, three point average. After Hood & Jirikowic (1990). **c)** Sunspot number (WDC–SILSO), and group number prior to 1700. Vertical bars, position of the de Vries lows, spaced c. 210 years. Solar activity agrees well with ^{14}C production, indicating that cosmogenic records are a valid proxy for solar activity.

cords reviewed in chapter 7. The 1500-yr climate cycle has left no trace in the cosmogenic record. It is difficult to argue that some climate cycles are greatly contaminating the cosmogenic record while others do not.

5.11 Solar variability effect on climate

It has been shown that solar activity reconstructions from cosmogenic records are an adequate proxy for solar activity. The second objection that opponents of the solar–climate theory raise is that there is no known mechanism by which small changes in TSI could cause an important effect on climate. This is actually a *non-sequitur* fallacy, because it assumes that the climatic effect must be due to changes in TSI when there is no evidence of it. The effects of the Bray climatic cycle shown by the reviewed proxies provide ample evidence of the mechanism involved, that is confirmed by instrumental measurements, reanalysis data, and climate modeling (reviewed by Gray et al. 2010).

Solar variability is higher at the short-wave part of the spectrum, as UV can change during the 11-year solar cycle by as much as 5–10%. Even though it constitutes a small part of TSI, UV radiation has specific effects in the stratosphere and the oceans. UV radiation of different wavelengths at different heights both creates and destroys ozone in the stratosphere, at the same time warming it. The changes in ozone are difficult to track, because they are affected by ozone transport within the stratosphere and to the troposphere, and by chemical processes that destroy ozone from volcanic eruptions and anthropogenic emissions, but the measured changes in total ozone during the solar cycle are on the order of 3% (Fig. 5.15a). This is 30 times more variation than for TSI during the 11-year solar cycle. As the stratosphere has a very low density, the changes in ozone are accompanied by significant changes in its temperature profile that can be of 0.5–1 °K in the tropical stratosphere for the solar cycle, and by changes in pressure that alter the geopotential height of the tropopause (Fig. 5.15b). As ozone is unequally distributed latitudinally these changes alter both the temperature and pressure gradients between the equatorial and polar stratosphere. The pressure changes are transmitted all the way down to the surface altering the tropospheric pressure distribution and strength. The process is affected both by seasonality and stratospheric oscillations, like the quasi-biennial oscillation. A higher probability of winter blocking days over the North Atlantic during periods of low solar activity has been demonstrated by several authors. This has the effect of increasing the probability of very cold events, like the 2010 Northern Europe snow storm at the solar cycle 23–24 minimum, or the 2021 record snow storm Filomena over Spain at the 24–25 minimum.

The solar-induced changes in the latitudinal pressure gradient cause a global atmospheric reorganization in zonal winds that alters the angular momentum of the atmosphere affecting the speed of rotation of the Earth (Le Mouël et al. 2010; see Sect. 10.6), but since after a few years the direction of the change reverts with the solar cycle, the effect of the 11-year solar cycle on weather is small. However, when the level of average solar activity changes over a period of several decades, the atmospheric reorganization advances, becoming noticeable first and causing important changes in the climate later. One of the effects of the changes induced by solar variability is to

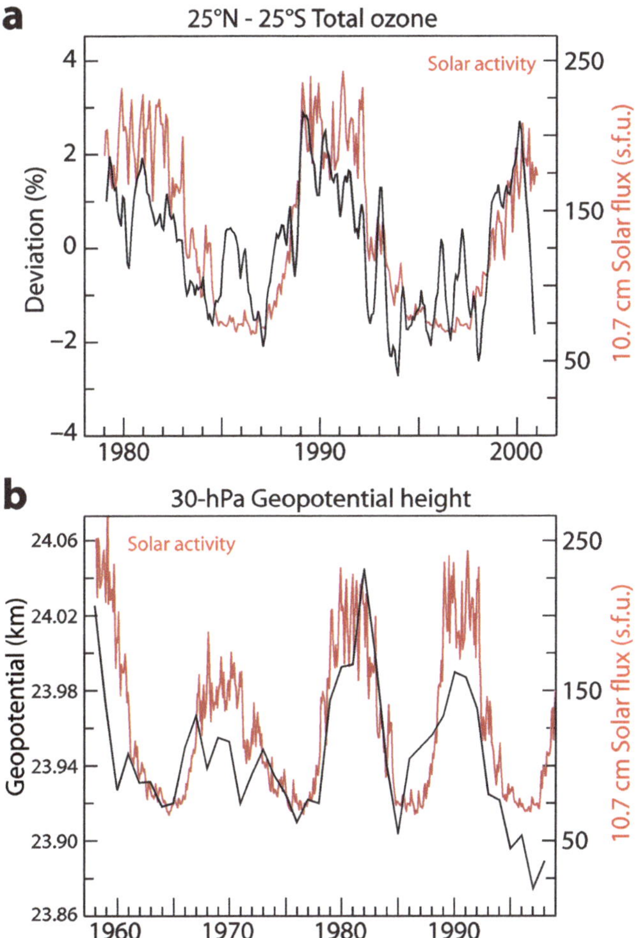

Fig. 5.15 Stratospheric effects of solar activity changes a) Deseasonalized, area-weighted total ozone deviations from merged satellite data for the latitude band 25°S–25°N. The solar flux at 10.7 cm is shown in red as a proxy for solar variability. After Fioletov et al. (2002). **b)** Time series of the 10.7 cm solar flux (red line) and of the annual mean 30-hPa heights (black line) in geopotential km for the gridpoint 30N/150W. After Labitzke (2001).

cause changes in the latitudinal temperature gradient (LTG, also Equator-to-Pole temperature gradient; Soon & Legates 2013). The importance of this gradient cannot be overstated, as it acts as the thermodynamic engine of the planet's climate, and its periodic changes with the Milankovitch obliquity cycle correlate with the glacial cycle and had been proposed as its causative agent (see Fig. 2.17; Raymo & Nisancioglu 2003). Soon & Legates (2013) have shown that the LTG has been decreasing during periods of surface warming and increasing during periods of surface cooling (Fig. 5.16), and convincingly link them to changes in average solar activity. These effects indicate that poleward transport of energy, a fundamental climate stabilizing factor, is affected by changes in solar activity (see Chap. 10 & 11).

It is known that at least since 1979 the Hadley circulation has been expanding poleward at a rate of 0.5–1 °Lat/decade in both hemispheres (Hu & Fu 2007). The cause is uncertain and both ozone changes and GHGs have been proposed. An alternative explanation is proposed in chapter 11. Models indicate that the expansion could have been taking place for most of the 20th century. Using modeling, Sadourny (1994) ascribed to the decrease in solar activity

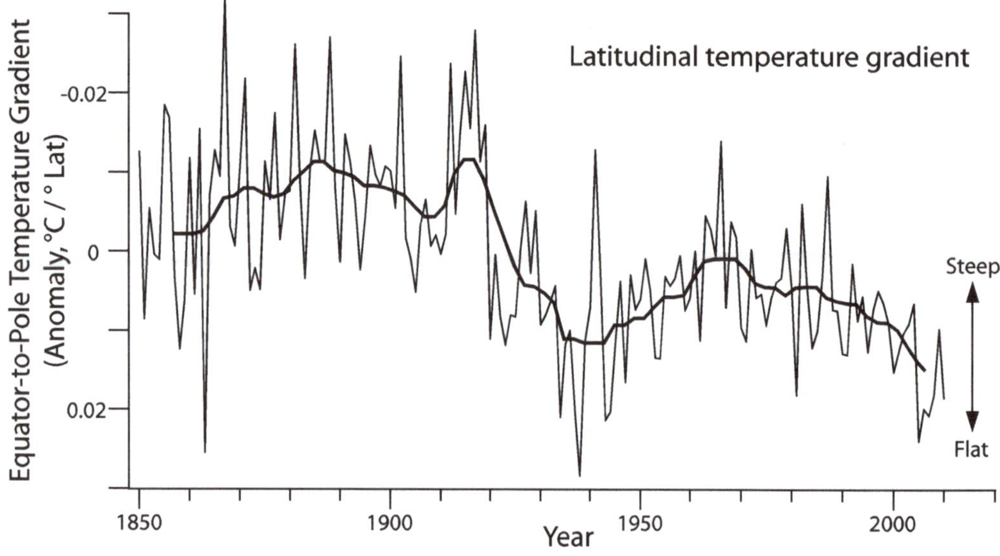

Fig. 5.16 The latitudinal temperature gradient
Annual-mean LTG over the entire Northern Hemisphere (°C/°Latitude; thin line) and smoothed 10-year running mean (thick line) from 1850 to 2010. The values are expressed as anomalies from the average for the 1961–90 period. The scale is inverted. Since the average LTG is strongly negative, positive anomaly values reduce the gradient (warmer Pole and/or colder Equator) while negative anomaly values enhance it (warmer Equator and/or colder Pole). Periods of global warming result in a decrease in LTG. After Soon & Legates (2013).

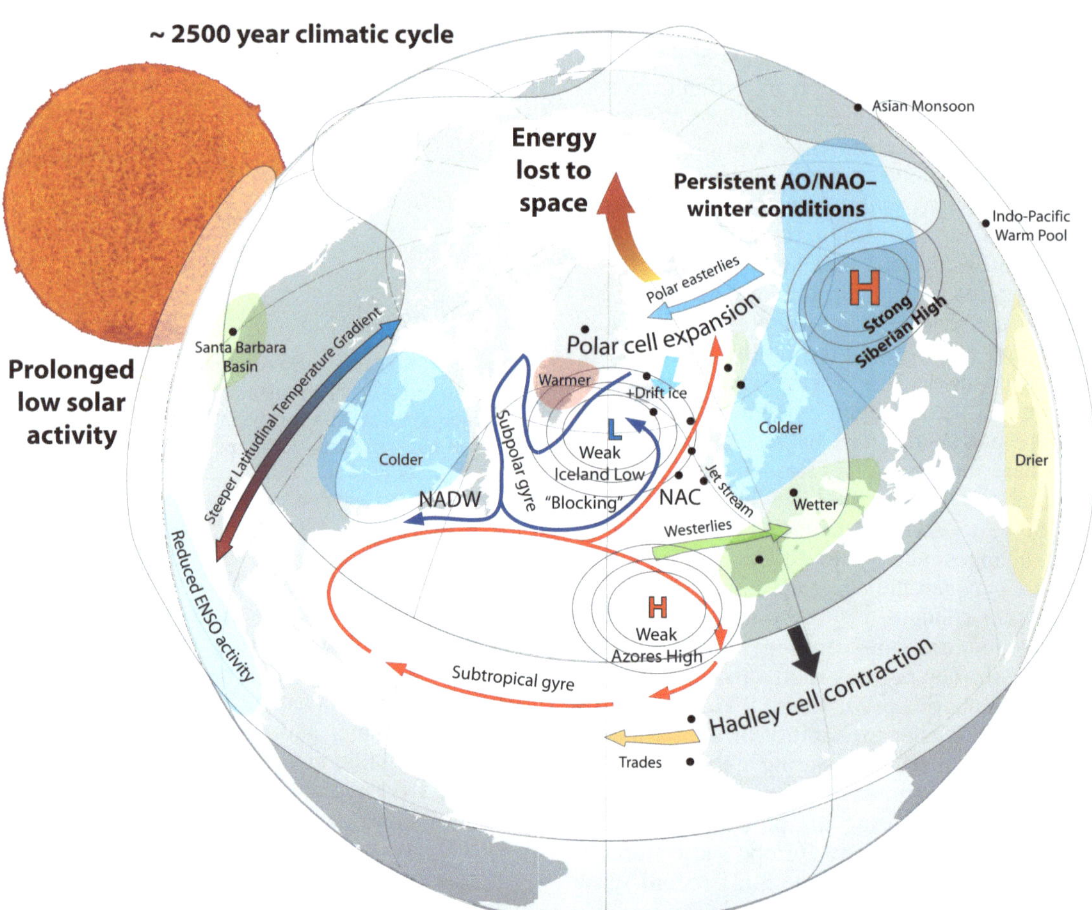

Fig. 5.17 Summary of the climatic effects associated to the lows of the Bray cycle
Global effects are mediated by the steepening of the latitudinal temperature gradient and a more intense poleward transport of energy causing a contraction of the Hadley cells and expansion of polar cells and resulting in global cooling. The Hadley cells contraction also restricts monsoon patterns, causing drier conditions in sub-tropical latitudes. El Niño conditions become infrequent, altering precipitation patterns. The North Atlantic realm is pushed into persistent AO/NAO negative conditions that are characterized by weak Iceland Low and Azores High pressure centers. This reduces the strength of the westerlies that take a more southern path changing precipitation patterns over Europe, and causing blocking conditions over the Atlantic that allow Arctic cold air to drift southward. The contribution of cold fresher subpolar gyre waters to the North Atlantic Current (NAC) decreases, becoming warmer and saltier, and increasing winter precipitation over northern Europe causing glacier advances. A strong Siberian High brings colder conditions over northern Eurasia and increases polar circulation over the Arctic and Greenland, increasing the amount of southward drift ice. Greenland experiences an inversion as masses of cold air are displaced towards northern Europe and North America. NADW labels North Atlantic deep-water currents. Black dots are the location of some of the proxies discussed in this chapter that display a clear c. 2500-year periodicity.

a contraction of the Hadley cell and associated monsoon systems as causing agents for the climatic changes that took place during the Maunder Minimum. The expansion of the Hadley circulation currently observed coincides with the decrease in the LTG (Fig. 5.16) and is a logical explanation as it supposes an expansion of the tropics.

At the ocean surface UV radiation decreases to about 3–5% of TSI, but it can penetrate water as readily as the visible range and is more energetic, so a few meters into the oceans, UV radiation might be responsible for about 7–10% of the ocean warming produced by solar radiation and it can change a few percentage points during the solar cycle. A close correlation between subsurface water temperatures and TSI has been reported south of Iceland between AD 818–1780 (Moffa–Sánchez et al. 2014).

With all this information from the instrumental era, and the information reviewed from proxy records covering past lows of the Bray cycle, an attempt can be made at explaining the effect of prolonged low solar activity on climate change. When the time for a new low in the solar Bray cycle approaches, solar activity starts to decrease, but it does so mainly at the lows of the 210-yr de Vries cycle that become more pronounced due to its modulation by the Bray cycle. The probability of a SGM increases and when it finally takes place it can be of the Spörer (c. 150 years) or Maunder (c. 80 years) types, with a higher tendency to produce a cluster of SGM spaced about 200 years, according to the de Vries cycle. Solar activity goes to minimum values at the SGM and the changes in the stratospheric ozone, temperature and geopotential height induce an atmospheric reorganization characterized by the weakening of the winter stratospheric polar vortex, the progressive expansion of the polar cells and the contraction of the Hadley cells, and as a result a steepening of the LTG that increases the amount of heat lost by the planet. This re-

versible process of atmospheric reorganization is cumulative and proceeds very slowly. This explains why the 18th century, with a solar activity level similar to the 20th century had a different climate. The 20th century expansion of the Hadley cells and reduction of the LTG have been built upon the levels reached over the previous two centuries. The contraction of the Hadley cells at the SGM explains the southward displacement (weakening) of the monsoons associated with the Hadley circulation (Fig. 5.5f), and the decrease in wind strength at the Santa Barbara basin that increases precipitation (Fig. 5.5e). The expansion of the polar cells explains the increase in wind strength over Iceland (Fig. 5.3d), that appears to depend on Milankovitch forcing. With the expansion of the polar cells there is an increase in polar circulation driven by the strengthening of the Siberian High, that produces an increase in salt deposition over Greenland (Fig. 5.3a & b).

In the North Atlantic the decrease in pressure differential causes the atmosphere to enter persistent NAO negative conditions (Fig. 5.3e) as shown in figure 5.17. This causes the Icelandic Low and the Azores High to be in a weak state more often, reducing the strength of both the Westerlies and storm tracks and causing them to move southward. Precipitation levels increased in Central (Fig. 5.5a), and Southwestern Europe (Fig. 5.5c). The weakening of the Westerlies reduces the contribution of fresh cold subpolar gyre waters to the NAC that becomes warmer and saltier (Fig. 5.4b). The Jet stream pushes southward, cooling Northern Europe and Northeast North America, and warming Greenland unless very cold Arctic conditions dominate. The warming of the NAC (Fig. 5.4d) increases precipitation over Ireland (Fig. 5.5d) and Norway (Fig. 5.5b), while colder winter conditions induce glacier expansion at a global scale (Fig. 5.2). The warmer saltier NAC prevents the shutdown of the NADW that experi-

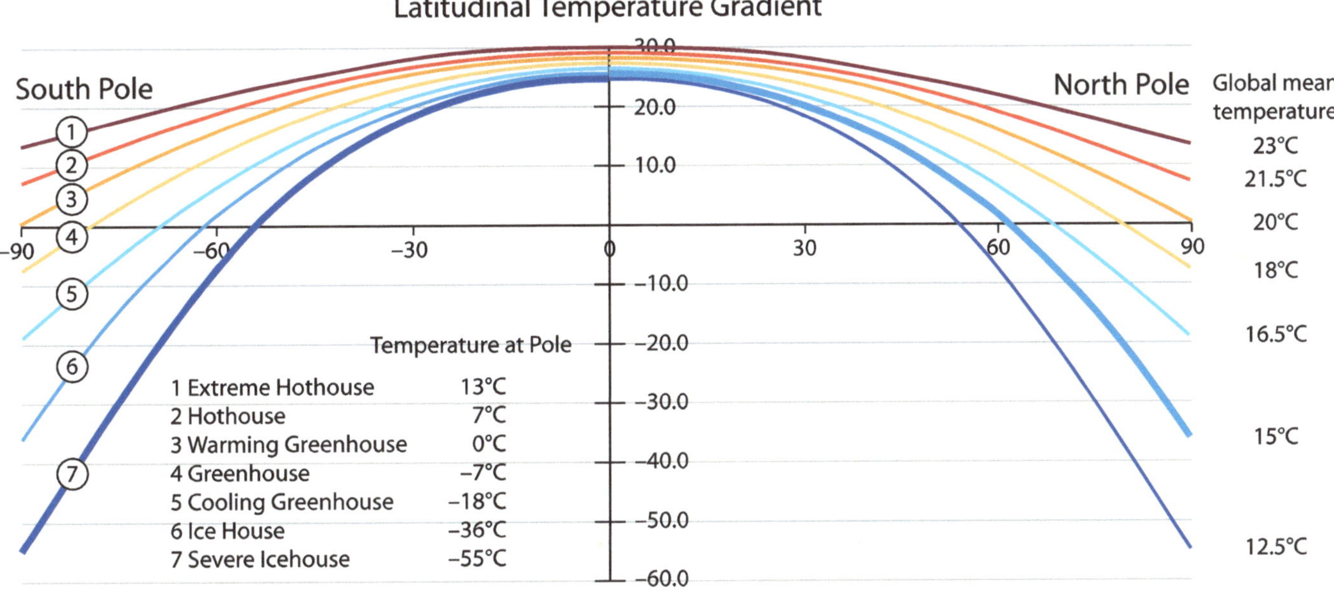

Fig. 5.18 Pole-to-pole temperature gradients for the planet
Latitude versus temperature (°C) Pole-to-pole curves representative of climatic conditions ranging from Extreme Hothouse to Severe Icehouse. The numbers along the right side indicate the corresponding global Mean Annual Temperature for each curve (modern MAT is 14.3°C). Polar temperatures for each of the seven pole-to-pole temperature curves are also listed (modern Antarctica is −55 °C). The average temperature at the Equator has also changed through time, but a lot less than the rest of the planet. The curves for each hemisphere are independent. Current climate is described by curve 7 for the Southern Hemisphere and curve 6 for the Northern Hemisphere (thick curves). We are now in icehouse conditions. After Scotese (2016).

ences a decrease (Fig. 5.4a), due to the AMOC reduction in response to weaker Westerlies. Glacier growth, colder conditions, and the advection of warmer waters to the Arctic, favor an increase in drift ice both North of Iceland (Fig. 5.4c) and in the North Atlantic (Fig. 5.6f).

The global effects of the atmospheric reorganization induced by prolonged low solar activity are thus multiplied because this condition pushes the North Atlantic atmosphere-oceanic system into a persistent NAO negative condition. The hydrological effects of the Hadley cells contraction are mainly zonal in both hemispheres. However, the cooling effect of the increased LTG is global and propagates over nearly all the oceans (Figs. 5.6c, d, e, & 5.7), resulting in global cooling (Fig. 5.6a). The climatic effects of low solar activity over the North Atlantic realm are particularly intense (Fig. 5.17). This is the reason why scientists argue over the regional versus global extent of the MWP and the LIA. The North Atlantic is a hotspot for planetary climate variability during both the glacial (D–O cycle) and interglacial periods (see Sect. 11.7).

The solar–climate debate has been long but not fruitful so far. Astrophysicists and instrumental-era climatologists are entrenched in the low energy changes in TSI as an argument against the connection. But the stratosphere is very rarefied and little energy is needed to alter it significantly. Afterwards the climate system provides the rest of the energy by oscillating to a different state through internal variability. The LTG is a crucial element that determines how much work the energy does on its way out of the planet. For a very similar solar output, as far as we know, the LTG has determined if the planet is in an icehouse, as currently, or in a hothouse as during the Eocene (Fig. 5.18). Finally, the North Atlantic Oscillation amplifies the climatic response, turning a small change in solar output into the cold winters of the Maunder Minimum. Paleoclimatologists are correct in sticking to the evidence that solar variability, when prolonged, has a disproportionate effect on climate change, and in the end, they will win the debate. Astrophysicists are looking to the wrong solar system body to learn about the history of the sun. It is best recorded here on the Earth. The sun might not have a memory, but it is subject to multi-millennial cycles whose cause will have to be elucidated. Recognizing the existence of those cycles is an important first step.

5.12 Conclusions

5a. A 2600-year climate cycle was first proposed in the late 1960s by Roger Bray based on vegetation transitions and major glacier re-advances, and linked to solar activity.

5b. This climate cycle is clearly evident in numerous proxies from the North Atlantic region and other places in the world that reflect c. 2500-year periodic changes in wind patterns, oceanic currents strength and salinity, drift ice, precipitation, and temperature.

5c. This climatic cycle corresponds in period and phase to a cycle in cosmogenic isotopes highlighting the coincidence of abrupt cooling climate change events with clusters of solar grand minima and prolonged periods of low solar activity.

5d. The correct periodicity for the Bray solar cycle is 2500 years, while the 2300-years Hallstatt periodicity does not correspond to a solar cycle.

5e. The solar activity Bray cycle appears to act on climate through changes in the stratospheric temperature and pressure gradients that are transmitted downwards to the troposphere causing an atmospheric reorganization, and through changes in the amount of energy warming the oceans.

5f. Proxy evidence, instrumental era measurements, and reanalysis support the idea that lows in the Bray cycle and prolonged below average solar activity cause a contraction of the Hadley cells, and an expansion of the polar cells, steepening the latitudinal temperature gradient, decreasing global temperatures and changing wind and precipitation patterns.

5g. In the North Atlantic region, in addition, the Arctic and North Atlantic oscillations enter a persistent winter negative phase during the lows of the Bray cycle, causing an intensification of winter climatic effects and making this region particularly sensitive to low solar activity. This helps explain why the Little Ice Age, while global, was particularly strong over Europe and North America.

References

Adolphi F, Muscheler R, Svensson A et al (2014) Persistent link between solar activity and Greenland climate during the Last Glacial Maximum. Nature Geoscience 7 (9) 662–666

Andrews JT (2009) Seeking a Holocene drift ice proxy: non-clay mineral variations from the SW to N-central Iceland shelf: trends regime shifts and periodicities. Journal of Quaternary Science 24 (7) 664–676

Beer J, Tobias SM & Weiss NO (2017) On long-term modulation of the Sun's magnetic cycle. Monthly Notices of the Royal Astronomical Society 473 (2) 1596–1602

Bond G, Showers W, Cheseby M et al (1997) A pervasive millennial-scale cycle in North Atlantic Holocene and glacial climates. Science 278 (5341) 1257–1266

Bond G, Kromer B, Beer J et al (2001) Persistent solar influence on North Atlantic climate during the Holocene. Science 294 (5549) 2130–2136

Bray JR (1968) Glaciation and solar activity since the fifth century BC and the solar cycle. Nature 220 (5168) 672–674

Bressan D (2011) September 19, 1991: The Iceman Natural History. Scientific American. https://blogs.scientificamerican.com/history-of-geology/september-19-1991-the-iceman-natural-history/ Accessed 07 Aug 2017

Davis OK (1994) The correlation of summer precipitation in the southwestern USA with isotopic records of solar activity during the Medieval Warm Period. Climatic Change 26 (2–3) 271–287

Denton GH & Karlén W (1973) Holocene climatic variations.Their pattern and possible cause. Quaternary Research 3 (2) 155–205

de Menocal P, Ortiz J, Guilderson T & Sarnthein M (2000) Coherent high-and low-latitude climate variability during the Holocene warm period. Science 288 (5474) 2198–2202

Eddy JA (1976) The Maunder Minimum. Science 192 (4245) 1189–1202

Fioletov VE, Bodeker GE, Miller AJ et al (2002) Global and zonal total ozone variations estimated from ground-based and satellite measurements: 1964–2000. Journal of Geophysical Research: Atmospheres 107 (D22) ACH-21

Fries M (1956) "Fimbulvintern" ur vegetations-historisk synpunkt. Fornvännen 51 5–10

Giraudeau J, Grelaud M, Solignac S et al (2010) Millennial-scale variability in Atlantic water advection to the Nordic Seas derived from Holocene coccolith concentration records. Quaternary Science Reviews 29 (9–10) 1276–1287

Gray LJ, Beer J, Geller M et al (2010) Solar influences on climate. Reviews of Geophysics 48 (4) RG4001

Hood LL & Jirikowic JL (1990) A probable approx. 2400 year solar quasi-cycle in atmospheric delta C-14. In: Schatten KH and Arking A (eds) Climate Impact of Solar Variability. NASA Conference Publication 3086, p 98–105

Houtermans J (1971) Geophysical interpretations of bristlecone pine radiocarbon measurements using a method of Fourier analysis of unequally spaced data. Dissertation, University of Bern

Hu Y & Fu Q (2007) Observed poleward expansion of the Hadley circulation since 1979. Atmospheric Chemistry and Physics 7 (19) 5229–5236

Jackson MG, Oskarsson N, Trønnes RG et al (2005) Holocene loess deposition in Iceland: Evidence for millennial-scale atmosphere-ocean coupling in the North Atlantic. Geology 33 (6) 509–512

Keigwin LD & Boyle EA (2000) Detecting Holocene changes in thermohaline circulation. Proceedings of the National Academy of Sciences 97 (4) 1343–1346

Kern AK, Harzhauser M, Piller WE et al (2012) Strong evidence for the influence of solar cycles on a Late Miocene lake system revealed by biotic and abiotic proxies. Palaeogeography, Palaeoclimatology, Palaeoecology 329 124–136

Khider D, Jackson CS & Stott LD (2014) Assessing millennial-scale variability during the Holocene: A perspective from the western tropical Pacific. Paleoceanography 29 (3) 143–159

Kim JH, Meggers H, Rimbu N et al (2007) Impacts of the North Atlantic gyre circulation on Holocene climate off northwest Africa. Geology 35 (5) 387–390

Labitzke K (2001) The global signal of the 11-year sunspot cycle in the stratosphere: Differences between solar maxima and minima. Meteorologische Zeitschrift 10 (2) 83–90

Le Mouël JL, Blanter E, Shnirman M & Courtillot V (2010) Solar forcing of the semi-annual variation of length-of-day. Geophysical Research Letters 37 (15) L15307

Magny M (2004) Holocene climate variability as reflected by mid-European lake-level fluctuations and its probable impact on prehistoric human settlements. Quaternary international 113 (1) 65–79

Marcott SA, Shakun JD, Clark PU & Mix AC (2013) A reconstruction of regional and global temperature for the past 11,300 years. Science 339 (6124) 1198–1201

Matthews JA, Berrisford MS, Dresser PQ et al (2005) Holocene glacier history of Bjørnbreen and climatic reconstruction in central Jotunheimen Norway based on proximal glaciofluvial stream-bank mires. Quaternary Science Reviews 24 (1–2) 67–90

Mayewski PA, Rohling EE, Stager JC et al (2004) Holocene climate variability. Quaternary research 62 (3) 243–255

Moffa–Sánchez P, Born A, Hall IR et al (2014) Solar forcing of North Atlantic surface temperature and salinity over the past millennium. Nature geoscience 7 (4) 275–278

Muscheler R, Joos F, Beer J et al (2007) Solar activity during the last 1000 yr inferred from radionuclide records. Quaternary Science Reviews 26 (1–2) 82–97

Nederbragt AJ & Thurow J (2005) Geographic coherence of millennial-scale climate cycles during the Holocene. Palaeogeography, Palaeoclimatology, Palaeoecology 221 (3–4) 313–324

Neff U, Burns SJ, Mangini A et al (2001) Strong coherence between solar variability and the monsoon in Oman between 9 and 6 kyr ago. Nature 411 (6835) 290–293

O'Brien SR, Mayewski PA, Meeker LD et al (1995) Complexity of Holocene climate as reconstructed from a Greenland ice core. Science 270 (5244) 1962–1964

Olsen J, Anderson NJ & Knudsen MF (2012) Variability of the North Atlantic Oscillation over the past 5200 years. Nature Geoscience 5 (11) 808–812

Oppo DW, McManus JF & Cullen JL (2003) Palaeooceanography: Deepwater variability in the Holocene epoch. Nature 422 (6929) 277–278

Ram M, Stolz MR & Tinsley BA (2009) The terrestrial cosmic ray flux: Its importance for climate. Eos, Transactions American Geophysical Union 90 (44) 397–398

Raymo ME & Nisancioglu KH (2003) The 41 kyr world: Milankovitch's other unsolved mystery. Paleoceanography 18 (1) 1–6

Reimer PJ, Bard E, Bayliss A et al (2013) IntCal13 and Marine13 radiocarbon age calibration curves 0–50,000 years cal BP. Radiocarbon 55 (4) 1869–1887

Rimbu N, Lohmann G, Lorenz SJ et al (2004) Holocene climate variability as derived from alkenone sea surface temperature and coupled ocean-atmosphere model experiments. Climate Dynamics 23 (2) 215–227

Rosenthal Y, Linsley BK & Oppo DW (2013) Pacific Ocean heat content during the past 10,000 years. Science 342 (6158) 617–621

Sadourny R (1994) Maunder minimum and the Little Ice Age: impact of a long-term variation of the solar flux on the energy and water cycle. In: Duplessy J–C and Spyridakis M–T (eds) Long-Term Climatic Variations. Springer, Berlin, p 533–550

Schwabe SH (1843) Sonnen-beobachtungen im Jahre 1843 (Solar Observations during 1843). Astronomische Nachrichten 21 495 233

Scotese CR (2016) Some thoughts on global climate change: The transition from icehouse to hothouse. In: Scotese CR (author) The Earth History: The evolution of the Earth System. PALEOMAP Project, Evanston, IL, https://www.researchgate.net/publication/275277369_Some_Thoughts_on_Global_Climate_Change_The_Transition_for_Icehouse_to_Hothouse_Conditions Accessed 18 Oct 2018

Seierstad IK, Abbott PM, Bigler M et al (2014) Consistently dated records from the Greenland GRIP, GISP2 and NGRIP ice cores for the past 104 ka reveal regional millennial-scale $\delta^{18}O$ gradients with possible Heinrich event imprint. Quaternary Science Reviews 106 29–46

Solanki SK, Usoskin IG, Kromer B et al (2004) Unusual activity of the Sun during recent decades compared to the previous 11,000 years. Nature 431 (7012) 1084–1087

Sonett CP (1984) Very long solar periods and the radiocarbon record. Reviews of Geophysics 22 (3) 239–254

Soon W & Legates DR (2013) Solar irradiance modulation of Equator-to-Pole (Arctic) temperature gradients: Empirical evidence for climate variation on multi-decadal timescales. Journal of Atmospheric and Solar-Terrestrial Physics 93 45–56

Sushch I & Hnatyk B (2014) Modelling of the radio emission from the Vela supernova remnant. Astronomy & Astrophysics 561 pA139

Tan B (2011) Multi-timescale solar cycles and the possible implications. Astrophysics and Space Science 332 (1) 65–72

Thornalley DJ, Elderfield H & McCave IN (2009) Holocene oscillations in temperature and salinity of the surface subpolar North Atlantic. Nature 457 (7230) 711–714

Thorndycraft VR & Benito G (2006) The Holocene fluvial chronology of Spain: evidence from a newly compiled radiocarbon database. Quaternary Science Reviews 25 (3–4) 223–234

Turney C, Baillie M, Clemens S et al (2005) Testing solar forcing of pervasive Holocene climate cycles. Journal of Quaternary Science 20 (6) 511–518

Usoskin IG (2013) A history of solar activity over millennia. Living Reviews in Solar Physics 10 (1) 1–94

Usoskin IG, Gallet Y, Lopes F et al (2016) Solar activity during the Holocene: the Hallstatt cycle and its consequence for grand minima and maxima. Astronomy & Astrophysics 587 A150

Vasiliev SS & Dergachev VA (2002) The ~ 2400-year cycle in atmospheric radiocarbon concentration: bispectrum of ^{14}C data over the last 8000 years. Annales Geophysicae 20 (1) 115–120

Versteegh GJ (2005) Solar forcing of climate 2: Evidence from the past. Space Science Reviews 120 (3–4) 243–286

Wang Y, Cheng H, Edwards RL et al (2005) The Holocene Asian monsoon: links to solar changes and North Atlantic climate Science. 308 (5723) 854–857

Weiss NO & Tobias SM (2015) Supermodulation of the Sun's magnetic activity: the effects of symmetry changes. Monthly Notices of the Royal Astronomical Society 456 (3) 2654–2661

Wu CJ, Usoskin IG, Krivova N et al (2018) Solar activity over nine millennia: A consistent multi-proxy reconstruction. Astronomy & Astrophysics 615 A93

THE EFFECT OF ABRUPT CLIMATE CHANGE
ON HUMAN SOCIETIES OF THE PAST

"In view of these findings, we call for an in-depth multi-disciplinary assessment of the potential for solar modulation of climate on centennial scales."
E. Rohling, P. Mayewski, R. Abu–Zied, J. Casford, & A. Hayes (2002)

6.1 Introduction

The role of solar variability on climate change, despite having a very long scientific tradition, is currently downplayed as a climatic factor within the most accepted hypotheses for climate change. At the root of this neglect lie two fundamental problems. Solar variability is quite small (about 0.1% of total irradiance), and there is no generally accepted mechanism by which the solar variability signal could be amplified by the climate system.

The progress being made to solve these problems will be reviewed in chapters 10 and 11, where evidence for the role of solar variability modulating meridional energy transport is presented. Meanwhile there is a growing number of scientific paleoclimatology articles published every year that defend a significant role for solar variability in paleoclimate change. In dynamic-systems identification experiments a much higher contribution from solar activity is required to explain the Medieval Warm Period (MWP) and the Little Ice Age (LIA; de Larminat 2016). There is a clear contradiction between the paleontological proxy evidence for an important solar role in many of the abrupt climatic events (ACEs) of the Holocene (see Fig. 4.18 & table 4.1) and our knowledge of climate change mechanisms. This contradiction should be solved by giving precedence to evidence over theory. There is very solid evidence that periods of low solar activity in the past, identified by a higher rate of cosmogenic isotopes production (^{14}C and ^{10}Be), have a high degree of correlation with periods of climate deterioration manifested mainly as temperature decrease and precipitation changes.

Frequency analysis of solar variability during the Holocene identifies several cycles (McCracken et al. 2013), like the 11.4-yr Schwabe cycle, the centennial Feynman cycle, the 210-yr de Vries cycle, the c. 1000-yr Eddy cycle, and the c. 2500-yr Bray cycle (see Chaps. 5 & 8). Comparison of climate and solar variability records leads to the important observation that the period of the cycle correlates with the amplitude of the climate effect observed and in general the longer the cycle period the more profound effect it appears to have on climate.

In this chapter we will review the evidence for the effect of the c. 2500-yr Bray cycle on climate and people during the Holocene. It is important to highlight two things. First, that solar variability, even if an important factor affecting climate change, is neither the main one,

nor the only one. At the multi-millennial timescale, Earth's temperature appears to depend mainly on orbital changes, firstly obliquity, but also precession and eccentricity. Other factors like oceanic cycles, and volcanic activity also play an important role at times. Therefore, solar variability only partially explains climatic changes. Also, solar cycles are irregular in nature. The Schwabe cycle is a good example. Described as an 11-yr cycle in sunspots, it can be anywhere from 9 to 14 years. Its amplitude is very variable too, and during the Maunder minimum between AD 1620 and 1700 sunspots essentially disappeared. Other solar cycles display a similar irregularity in period and amplitude (see Chap. 8).

6.2 The solar minima of the 2500-yr Bray cycle

The observed effect of the 2500-yr Bray solar cycle is to result in long (Spörer-type) solar grand minima (SGM) or clusters of SGM at its lows. According to Usoskin et al. (2007): *"the occurrence of grand minima depicts a weak (marginally significant) quasi-periodicity of 2000–2400 years, which is a well-known period in ^{14}C data... no clear periodicities are observed in the occurrence of grand maxima."*

In a later work, Usoskin et al. (2016), show that SGM tend to cluster at the lows of the 2500-year cycle. According to this analysis, a low in the Bray cycle would increase the probability of a long SGM or a cluster of SGM that would reduce temperature and cause changes in precipitation patterns bringing about a general worsening of the climate for a few centuries. The end of the low would bring about a return to normal solar activity with a natural increase in temperature that can also take a few centuries.

The climatic effect of the SGM caused by the lows in the Bray cycle, is reviewed in Chap. 5 (see Sect. 5.11), and appears to register mainly as a significant reduction in winter temperatures with a smaller effect on summer temperatures, and profound changes in precipitation patterns, usually registered in the best studied North Atlantic region as a very significant increase in precipitation in Scandinavia, Southern and Central Europe and the Middle East, and by a weakening of the Indian and Asian Monsoons. This combination produces an increase in cold winters, more snow, glacier advances, and spring flooding. The effect on human societies can be postulated as higher frequency of food crises, population decrease, increased migration, increased violence, and higher chance of civilization regression or failure. It is so common that new civilizations emerge after climatic crises that some archaeologists have developed the theory that climate caused environmental stress is an engine to societal change (Weninger et al. 2009; Roberts et al. 2011).

The Holocene can be climatically subdivided in different ways (Fig. 6.1). The biological subdivision (Blytt–Sernander sequence), and the temperature subdivision, initially proposed by Ernst Antevs in 1948, display transitions related to the lows of the Bray cycle. Let's now review what has happened to the planet and people at those times during the Holocene. These have taken place around the following dates:

- B1. 0.5 kyr BP. Little Ice Age
- B2. 2.7 kyr BP. Sub-Boreal/Sub-Atlantic Minimum
- B3. 5.2 kyr BP. Mid-Holocene Transition. Ötzi buried in ice. Start of Neoglacial period
- B4. 7.7 kyr BP. Boreal/Atlantic transition and precipitation regime change
- B5. 10.3 kyr BP. Early Holocene Boreal Oscillation
- B6. 12.7 kyr BP. Younger Dryas cooling onset

The Bray cycle has played an important role in the evolution of human societies. When analyzing Holocene fluctuations in Britain and Ireland human population, Bevan et al. (2017) find pan-regional demographic decline in three episodes that coincide with the lows of the Bray cycle (Fig. 6.1b). At those times societies responded evolving and adopting new food-procurement strategies demonstrating that climate-related disruptions have been quasi-periodic drivers of societal change.

6.3 The 10.3 kyr event. The Boreal Oscillation

The fifth low in the Bray cycle at about 10.3 kyr BP coincides with the SGM known as Boreal 1 and the climate worsening named Boreal Oscillation. The planet was at that time still warming towards the Holocene Climatic Optimum (HCO), due to increasing obliquity and increasing northern latitudinal summer insolation, and the ACE does not appear to have constituted a major trend inversion, but a relatively brief cold and wet period for the areas we have the best evidence.

The 10.3-kyr ACE and its associated Boreal 1 SGM coincide with Bond event 7 of increased iceberg discharge in the North Atlantic (Bond et al. 2001; Fig. 6.2h), and with the highest concentration of non-sea-salt potassium deposition in the Greenland GISP2 ice core for the Holocene (Fig. 6.2c). The presence of non-sea-salt potassium in Greenland is associated to the expansion of the Siberian High pressure system that brings polar temperatures over a wide area in the Northern Hemisphere (Mayewski et al. 2004).

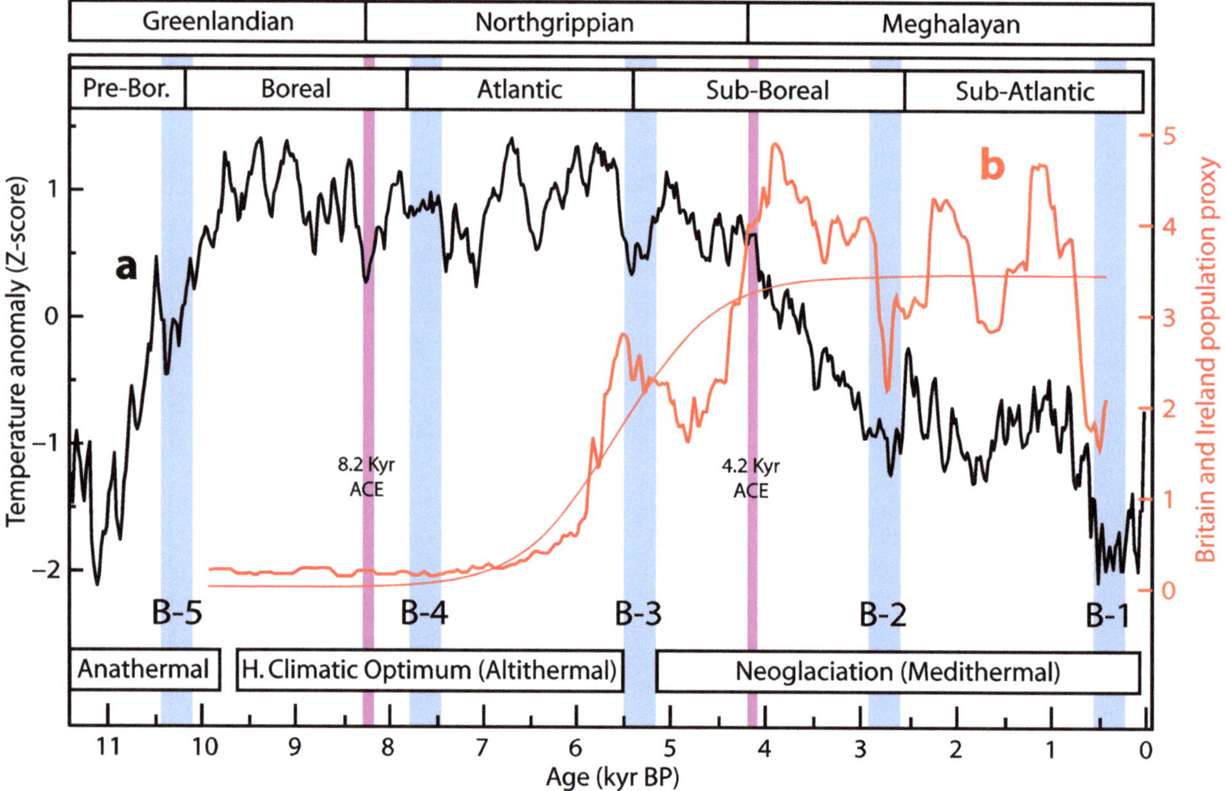

Fig. 6.1 Holocene climate subdivisions
Major Holocene subdivisions: Stratigraphical subdivision on top. Biological subdivision immediately below, displays a c. 2500-yr spacing. After Ammann & Fyfe (2014). Classical subdivision based on temperature at bottom. **a)** Black thick curve, global temperature reconstruction from 73 proxies (See Chap. 4), expressed as distance to the average in standard deviations (Z-score). **b)** Red thick curve, summed probability distribution of anthropogenic radiocarbon dates from Britain and Ireland as a proxy for human population. Red thin curve, fitted logistic model of population growth and plateau. After Bevan et al. 2017. Significant downside population deviations generally match the lows of the 2500-year Bray cycle of solar activity (wide blue bars labeled B-1 to B-5).

The 10.3-kyr ACE has been studied by Björk et al. (2001) and attributed to decreased solar forcing. The event is recorded at proxies from multiple sites in the Northern Hemisphere, like Norwegian sea surface temperature (SST) that shows a drop of >2 °C both in winter and summer temperature that lasted less than 200 years. This Norwegian sea cooling coincides with harsher conditions in the Faroe Islands lacustrine records that show a decrease

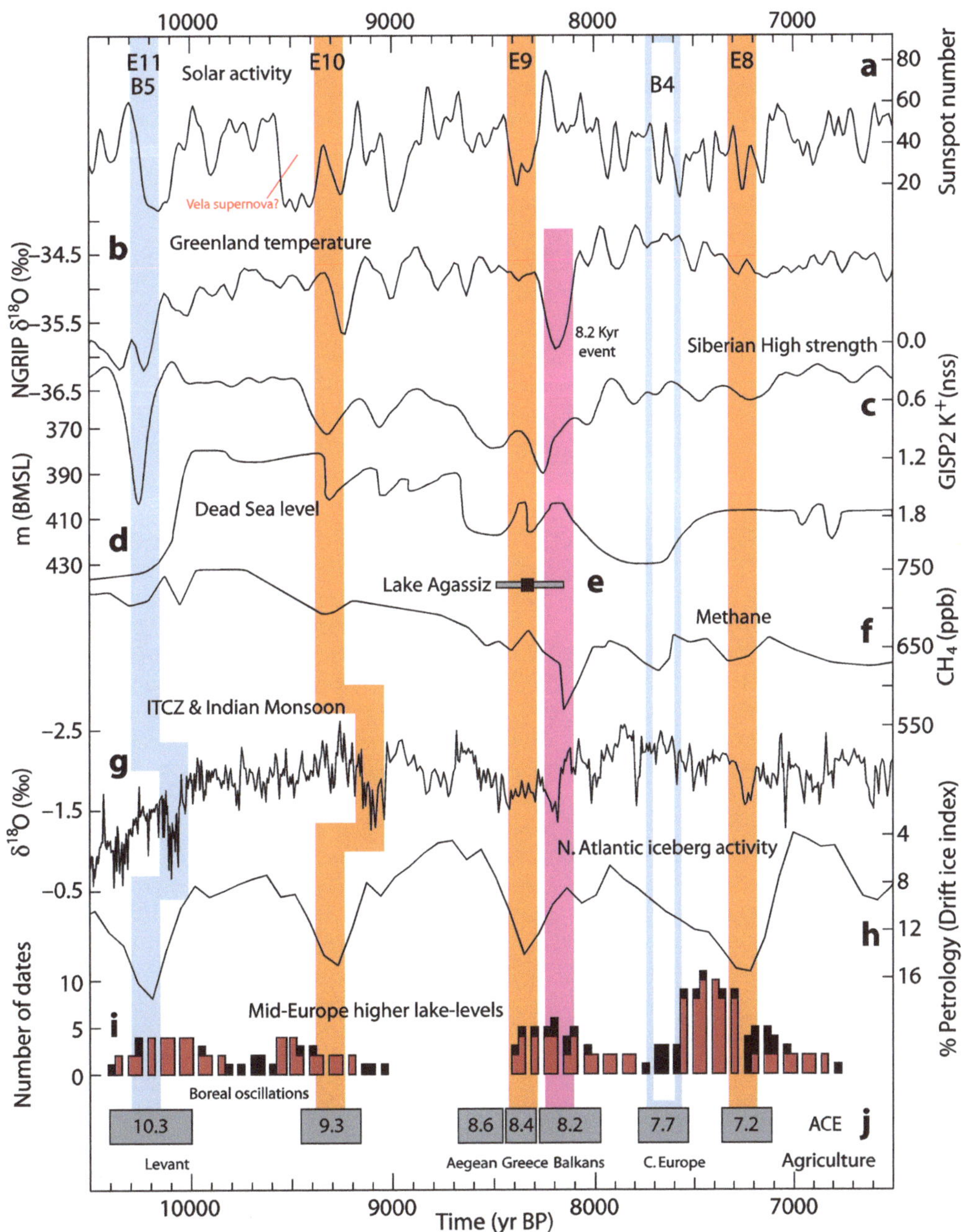

Fig. 6.2 Climate change in the Early Holocene
a) Solar activity reconstruction (sunspot number), after Solanki et al. (2004) until 8700 BP and after Wu et al. (2018) from 8700 BP. The 9500 BP huge increase in cosmic rays without corresponding ACE is proposed to be due to a galactic source, perhaps the Vela supernova. **b)** NGRIP $\delta^{18}O$ temperature proxy (NGRIP members 2004) on GICC05 timescale, smoothed with a 5-point triangular filter. **c)** GISP2 non-sea-salt [K+] (ppb, inverted), a proxy for Siberian High polar conditions. After Mayewski et al. (2004), smoothed with a Gaussian filter. **d)** Dead Sea level in meters below mean sea level, as a proxy for Middle East precipitation. After Migowski et al. (2006). **e)** Dating of the lakes Agassiz and Ojibway outburst after Lewis et al. (2012). Grey bar represents the error range. **f)** Methane levels reconstruction (ppb) from Greenland ice-core data after Kobashi et al. (2007). **g)** $\delta^{18}O$ (‰, inverted) from Qunf Cave speleothem (Southern Oman), a proxy for the strength of the Indian monsoon (as represented, weaker towards bottom). After Fleitmann et al. (2007). **h)** Ice-rafted debris stack (inverted) from four North Atlantic sediment cores as a proxy for iceberg activity and cold events, after Bond et al. (2001). **i)** Mid-European higher-lake-level episodes score of radiocarbon, tree-ring, and archeological dates from 26 lakes in the Jura–Alps region grouped in half-century bins. It is a proxy for Mid-European precipitation. The score indicates confidence, not magnitude. After Magny (2004). **j)** Abrupt climatic events (ACE) identified from multiple proxies with their approximate date in kyr BP. Some of these ACEs have been associated by several authors to the spread of agriculture in the Eurasian region (bottom). See text for details. Vertical blue bar marks the fifth low in the 2500-yr Bray solar cycle, and the eleventh low in the 1000-yr Eddy solar cycle. Empty blue bar marks the inconspicuous fourth low in the 2500-yr Bray solar cycle. Orange bars mark the lows of the 1000-yr Eddy solar cycle. Violet bar marks the 8.2 kyr ACE.

in birch pollen and increase in grass and herb pollen. German pines show at the time a tree-ring width minimum and the Santa Barbara basin shows a cold-related peak of oxygenation. Greenland and Tibetan ice cores display a $\delta^{18}O$ isotope minimum (Björk et al. 2001, and references within). The increase in precipitation in the North Atlantic region is supported by the increase in lake levels in west-central Europe and central Italy (Magny 2004; Fig. 6.2i), and the increased iceberg discharge (Bond et al. 2001). Glacier re-advances took place in Norway (the Erdalen event; Dahl et al. 2002) and Tibet (Seong et al. 2009).

In the Aegean sea, the increase in abundance of the cold-water dinocyst species *Spiniferites elongatus* indicates a strong biological response to the climatic deterioration and lower SST in the Eastern Mediterranean during the 10.3-kyr ACE (Marino et al. 2009). Within dating uncertainties a coincident dryer period for the Indian monsoon can be postulated based on an increase in $\delta^{18}O$ speleothem in the Qunf cave of Oman (Marino et al. 2009; Fig. 6.2g).

The climatic deterioration described must have had an impact on the prehistoric societies, but then most of the world was populated by hunter-gatherer cultures, and there is evidence that hunter-gatherer populations displayed resilience to a previous 11.3-kyr ACE in Northeast England (Blockley et al. 2018). In the Fertile Crescent, humans were in a pre- or proto-agricultural state, with increased population densities. At the time of the 10.3-kyr ACE Göbekli Tepe was being constructed, and Jericho, one of the oldest cities in the world, was the first city known to have built a wall dated precisely at 10.3 kyr BP. The proposed roles for the first city wall fit what we would expect from a climatic deterioration: defensive role against raiders in search of stored food, or protective role against flooding, as mud deposits indicate it had become more common (Bar–Yosef 1986).

Whatever the reason, the walls of Jericho do not appear to have spared the city from the societal changes usually associated to bad climatic conditions. For over 700 years Jericho inhabitants were part of the Pre-Pottery Neolithic A (PPNA) culture, a strange culture characterized for living with their dead (home burials) and construction of the first granaries. The 10.3-kyr ACE marks the end of PPNA in Jericho (Weninger et al. 2009; Fig. 6.3). There is a gap of about 200 years in radiocarbon dates that suggests a hiatus or strong reduction in construction and habitation and afterwards Jericho's inhabitants belong to the PPNB culture that has Anatolian influence, suggesting northern immigration, and is characterized by domesticated sheep, and a different more advanced flint toolkit. The PPNB culture disappears at the 8.2-kyr event, a climate pessimum that does not belong to the 2500-yr Bray cycle.

6.4 The 8.2 kyr climate complex

The second half of the 9th millennium BP (7th millennium BC) witnessed a major climate upheaval, perhaps the biggest in the Holocene. Our ignorance about the causes of abrupt climate change, and a natural tendency to attribute any observed change to a single event is delaying our understanding of this complex climate period. Only recently have some authors started to recognize the heterogeneous nature of the so-called 8.2-kyr event (Rohling & Pälike 2005; Gavin et al. 2011), that appears to comprise three different ACEs in rapid succession. The first ACE appears to have started c. 8600 BP (6650 BC). The climatic signature of this 8.6-kyr ACE is very similar to that of the 3.1-kyr ACE that has been involved in the Late Bronze collapse (Kaniewski et al. 2013). The 8.6-kyr ACE is characterized by a strong reduction in Middle East precipitation, as attested by Dead Sea levels (Fig. 6.2d). At the same time there was an increase in non-sea-salt potassium deposition in Greenland (Fig. 6.2c), indicative of a strengthened Siberian High, and a decrease in methane (Fig. 6.2f). There was an increase in iceberg activity in the North Atlantic (Fig. 6.2h), but the Indian and Asian Monsoons were unaffected (Fig. 6.2h; see also 4.18d). Rohling et al. (2002) have studied the 8.6-kyr ACE in core LC21 from the Aegean Sea where they detect an abrupt drop in warm foraminifer species that represents a 2–4 °C decrease in winter SST. They attribute it to a multidecadal period of predominant cold dry polar air arriving to the Aegean during winters from the Northeast. These conditions would also interrupt the winter North Atlantic storm-track from reaching the Middle East, explaining the drought.

A few years into this centennial Eastern Mediterranean drought, Central Anatolian farmers started a migration to the Asian coastal areas of the Aegean and Marmara seas likely induced by climate-caused famine. At the time the Cappadocian obsidian trade route to Southern Levant was disrupted (Weninger et al. 2014). It was the first step towards the spread of agriculture to Europe. In Southern Levant the Pre-Pottery Neolithic B culture is replaced by the pottery-bearing pastoralist Yarmoukian culture. The similar 3.1-kyr ACE that also caused a profound drought

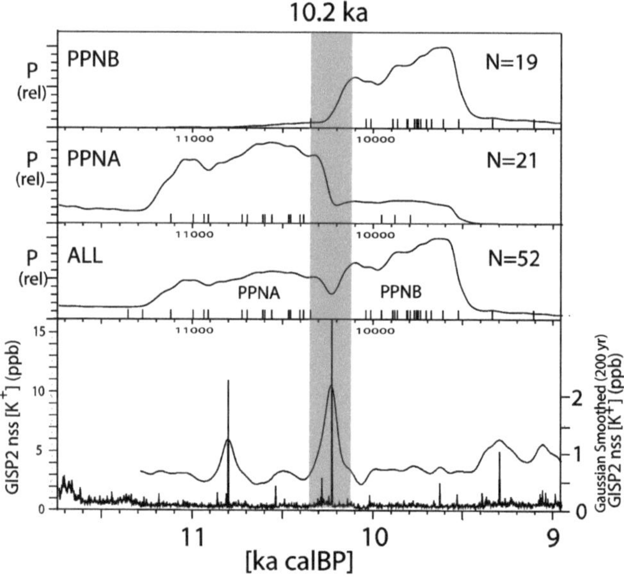

Fig. 6.3 Cultural shift at Jericho coinciding with the 10.3 kyr event
Radiocarbon data from Jericho arranged according to cultural period (Top: Pre-Pottery Neolithic B, PPNB; Middle: PPNA; Lower: Combined PPNA and PPNB), in comparison to (lower graph): Gaussian smoothed (200 yr) and high-resolution GISP2 potassium (non-sea salt [K+]; ppb) ion proxy for the Siberian High. The calibrated ^{14}C-age distribution (radiocarbon periodization) gives reason to assume a hiatus between PPNA and PPNB. Reproduced from Weninger et al. (2009) with permission. © Weninger et al. (2009).

in the Eastern Mediterranean/Black Sea area appears to have contributed to another climate-caused famine-induced migration in the opposite direction by the Sea Peoples during the Late Bronze collapse. The cause of the 3.1 and 8.6-kyr ACEs is unknown. Rohling et al. (2002) suggest solar forcing on the basis of correlations between solar proxy records and polar atmospheric reorganization proxy records that match the Siberian High conditions to produce winter Aegean SST cooling. Both 3.1 and 8.6-kyr ACEs coincide with increases in cosmogenic isotopes production that precede by 200 and 400 years the 8.4-kyr Sahelian 1 SGM and the 2.9-kyr solar low and 2.7-kyr Homeric SGM (grouped in this book as the 2.8-kyr ACE). This spacing indicates that the 3.1 and 8.6-kyr ACEs might coincide with strong lows of the 210-yr de Vries solar cycle, reinforcing the solar origin hypothesis.

200 years after the 8.6-kyr ACE, and coinciding with the first Sahelian SGM, the 8.4-kyr ACE took place. It reverted many of the effects of the 8.6-kyr ACE by increasing precipitation in the Middle East (Fig. 6.2d) and Mid-Europe (Fig. 6.2i). With the increase in precipitation came an increase in methane (Fig. 6.2f) likely due to wetlands recovery. North Atlantic iceberg activity reached a peak (Fig. 6.2h), indicating both cooling and an increase in warm water flow through the North Atlantic Current. Both the Indian Monsoon (Fig. 6.2g) and particularly the Asian Monsoon (see Fig. 4.18d) suffered a strong weakening. Greenland temperature was not very much affected (Fig. 6.2b), as low solar activity cooling of northern high latitudes is mediated by winter polar vortex disorganization that allows cold polar air masses to abandon the Arctic, that as a result becomes warmer (see Fig. 5.17 & Chap. 11). The 8.4-kyr ACE displays many of the characteristics associated with low solar activity ACE (see table 4.1), although the methane response is opposite, perhaps because it took place after a very dry ACE.

Rohling & Pälike (2005), have studied the 8.2-kyr climate complex, and recognized its heterogeneous nature. They present many proxy records, and show that the anomalies started 8,600 yr BP. They show that in many well-dated proxies there was an underlying climatic deterioration between about 8.5 and 8.0 kyr BP that was punctuated by the sharp 8.3 kyr BP proglacial lake outbreak. Rohling & Pälike (2005) attribute the broad deterioration to reduced solar activity due to the temporal coincidence with the three Sahelian SGM.

While the planet was still undergoing the 8.4-kyr ACE, the outburst of the lakes Agassiz and Ojibway took place. It has been dated with some precision at 8.33 kyr BP (error range 8.04–8.49; Lewis et al. 2012), and the cold fresh water outpouring through the Hudson Bay started to affect nearby Greenland at 8.3 kyr BP. Greenland temperature started to drop in 8,320–8,300 BP, and by 8,260 BP had fallen its biggest amount in the entire Holocene. The 8.2-kyr ACE displayed effects clearly attributable to the meltwater pulse beyond Greenland cooling. It also caused the biggest drop in methane levels of the entire Holocene, 100 ppb (Fig. 6.2f), and it reduced iceberg activity in the North Atlantic (Fig. 6.2h), probably because the fresh cold water and reduced SST negatively affected iceberg release. The effects over the Atlantic Meridional Overturning Circulation and North Atlantic Deep Water (NADW) formation were profound, but other effects are more difficult to assign to the meltwater pulse because at the same time the

second Sahelian SGM was taking place, and some of the features of the 8.2-kyr ACE are common with low-solar-activity caused ACEs, like increased Middle East and Mid-Europe precipitation, and monsoon weakening. Rohling and Pälike recommend caution when assigning global climatic effects to the periglacial lakes outburst and the effect of the meltwater on NADW formation, due to this coincidence.

The onset of the third ACE with its profound climatic effect caused another crisis in the Middle East. The centuries between 8250–7950 BP witnessed unprecedented disturbance in Anatolia and the Levant with generalized abandonment of highland settlements and movement to less exposed coastal and lowland settlements. In South Levant this is known as the Late Yarmoukian crisis. A second wave of migration from Northern Levant and Asian Aegean coastal areas expanded into Southeast Europe (Weninger et al. 2014). The 8.2-kyr ACE witnessed the arrival of the Neolithic to Europe. Farmers originating from Anatolia and Levant would soon expand from Greece into the Balkans.

Between 8600 and 8100 BP the planet suffered three different ACEs in rapid succession, each one affecting climate in a different way. Their accumulated effect created the most recognizable climatic event in a multitude of proxy records. Through social distress it prompted human migration that resulted in the spread of agriculture and cultural advancement, with the arrival of the Neolithic to Europe. The last of these ACEs has been chosen as a convenient point to divide the Holocene by the International Union of Geological Sciences, separating the Greenlandian and Northgrippian stages at 8236 b2k (8286 BP; Walker et al. 2018).

6.5 The 7.7 kyr event. The Boreal/Atlantic transition

The low in the Bray cycle at about 7.7 kyr BP coincides with a cluster of four SGM known as Jericho 0 to 3 and a long period of climate change between 7.8 and 7.0 kyr BP. The planet had reached maximum North polar insolation two millennia before but the melting of the ice sheets was completed at about this time so general conditions were still within the Holocene Hypsithermal. The 7.7-kyr ACE marks a climatic change in Holocene conditions in northern Europe from the warm relatively dry Boreal period to the warm more humid Atlantic period, reflected in a significant arboreal species expansion in high northern latitudes.

The 7.7-kyr ACE coincides with an increase in iceberg discharge at the North Atlantic that forms part of Bond event 5a (Bond et al. 2001; Fig. 6.2). It is not however a very strong minimum in the c. 2500-yr Bray cycle and it is completely overshadowed by the strength of the earlier 8.2-kyr climate complex, perhaps the most profound cooling event in the Holocene. Due to that, although detected in most proxy series for climate change, it is seldom studied.

As any Bray low, the 7.7-kyr ACE is characterized both by winter cooling reflected in the Aegean Sea (Marino et al. 2009), and an increase in precipitation in the Mid-European region (Fig. 6.2i). The Scandinavian region was not the only one to experience a multi-millennial dry-

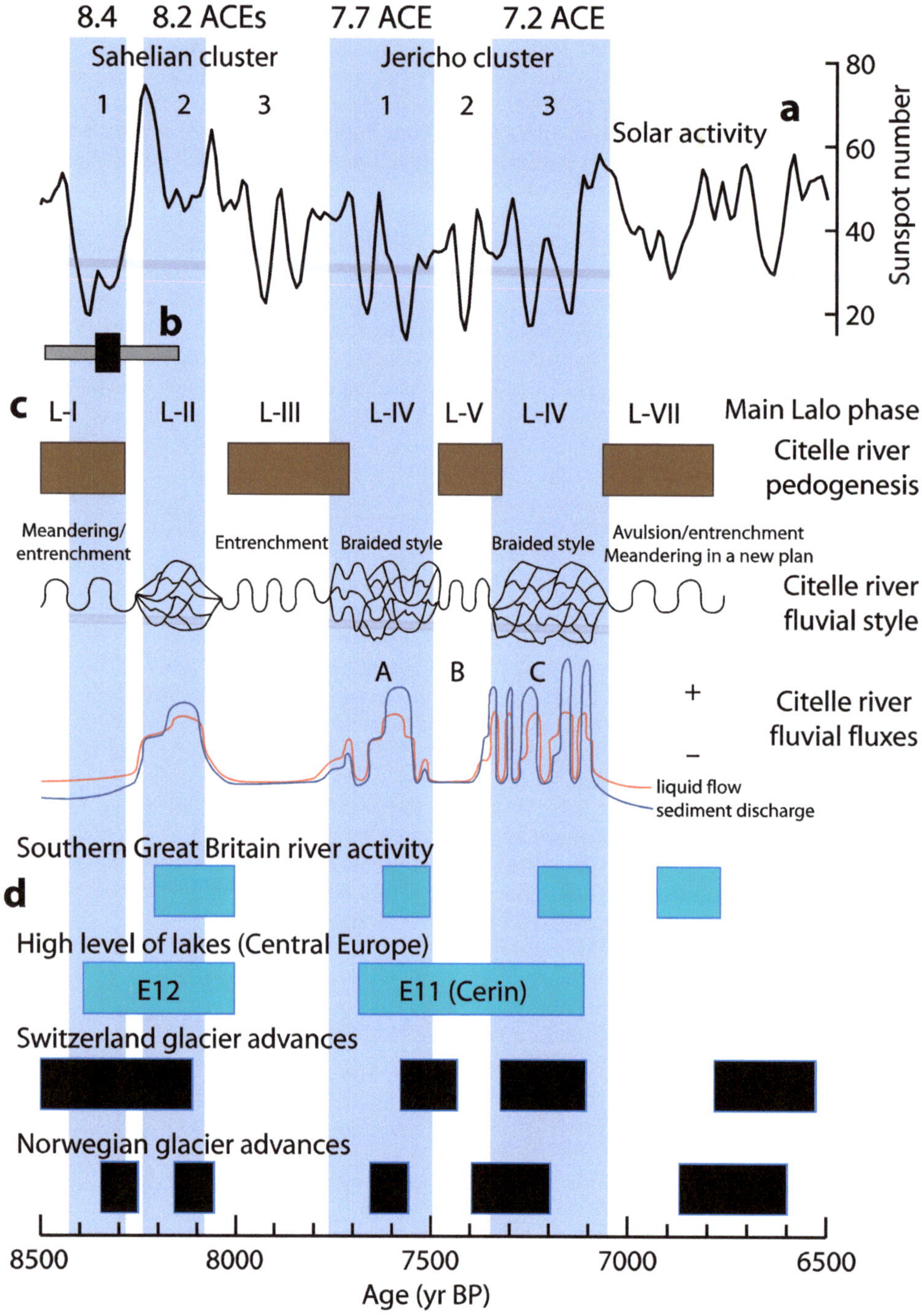

Fig. 6.4 Hydrological and climate indicators during 8.5–6.5 kyr BP
a) Solar activity reconstruction shows the clusters of Sahelian and Jericho solar grand minima. After Wu et al. (2018). **b)** Dating of the lakes Agassiz and Ojibway outburst after Lewis et al. (2012). Grey bar represents the error range. **c)** Hydrological analysis defines seven phases at Lalo site (Rhône Valley, France). Four of them correspond to periods of soil formation (pedogenesis), meandering entrenched Citelle River, and normal sediment discharge. Three periods at 8.2, 7.7, and 7.2 kyr BP show braided Citelle river flow, and enhanced flux and sediment discharges. A, B and C letters indicate the tripartite climate division of the 7.7–7.1-kyr period. **d)** The three periods coincide with periods of reduced solar activity, increased hydrology and glacier advances in Europe. The blue bands correspond to colder periods in the Greenland ice sheet and alpine areas and to moister signals in western/central hydrosystems, defining the known 8.2, 7.7, and 7.2-kyr ACEs. c & d after Berger et al. 2016.

wet transition (Boreal–Atlantic) at the 7.7-kyr ACE. It also coincided with a change in climate mode at the Middle East. This region had been experiencing a decrease in precipitation for 2,200 years (Fig. 6.2d), attributable to the slow decrease in northern insolation due to precession changes. The aridification of the Fertile Crescent pushed farmers and pastoralists in all directions helping the spread of agriculture and pastoralism. The Dead Sea reached its lowest level for the last 10,000 years 7.7 kyr ago, at the time the Asian Monsoon shows a weakening coincident with the 7.7-kyr ACE (see Fig. 4.18c & d). The increase in precipitation that took place after the 7.7-kyr ACE marked a hydrological inflection point and for the next 4000 years Dead Sea levels would increase intermittently (see Fig.

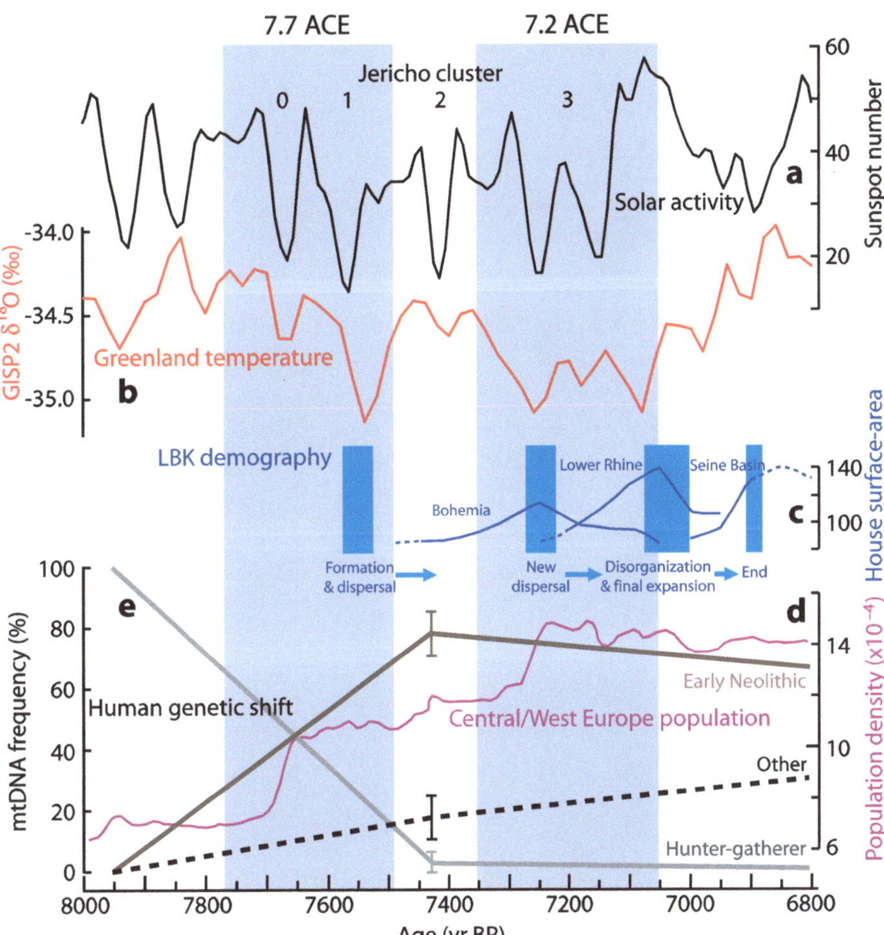

Fig. 6.5 The effect of 8th millennium BP climate changes on human societies of Central Europe
a) Solar activity reconstruction after Wu et al. (2018) shows the cluster of Jericho 0–3 solar grand minima. **b)** Bi-decadal Greenland GISP2 temperature (2-period averaged) highlights a general correspondence between low solar activity and climate cooling during the two periods marked by vertical bands, that correspond also to the same periods in Fig. 6.4. **c)** Linear Pottery LBK culture demographic analysis (lines) shows demographic peaks coincident with temperature valleys just at the time of the major dispersal periods of LBK (boxes). After Dubouloz (2008). **d)** Analysis of Central/West Europe population shows that the population trough at the 8.2-kyr climate complex did not recover until the arrival of the LBK agro-pastoral culture, and that this population increase took place during the cold 7.7 and 7.2-kyr ACE, when LBK was expanding. After Shennan et al. (2013), summed calibrated date probability distributions for radiocarbon dates in western Europe as a demographic proxy. **e)** The agro-pastoral expansion was at the expense of the local hunter-gatherer population that did not get diluted but was outcompeted, according to genetic mitochondrial DNA studies. After Brandt et al. (2013).

4.18c). The end of the aridification that had interrupted urban development, would foster a late Chalcolithic flourishing that will produce the first urban civilizations (Migowski et al. 2006).

A detailed study of the hydrology of the Rhône Valley of France during this time by Berger et al. (2016) shows that the 7.7–7.1 kyr BP period can be climatically subdivided into three subperiods (A, B, and C in Fig. 6.4c): two cold and wet sub-periods separated by a warm and drier interval. During the cold-humid periods the Citelle River changed to a braided fluvial style, greatly increasing the liquid flow and sediment discharges. This fluvial changes coincide with increased hydrological activity elsewhere in Europe, and glacier advances in the Alps (Berger et al. 2016; Fig. 6.4d).

Solar activity cycle timing indicates that the 7.7-kyr ACE belongs to the 2500-yr Bray cycle, while the 7.2-kyr ACE belongs to the 1000-yr Eddy cycle. The Eddy cycle appears to be modulated by a longer cycle resulting in very strong lows (both in solar activity and climate effect) during the early Holocene at 11.2, 10.2, 9.2, 8.2, 7.2, 6.2 and 5.2 kyr BP, followed by subdued lows at 4.3, 3.3, and 2.3 kyr BP, and again increasingly stronger lows at 1.3 and 0.3 kyr BP. Chapter 8 presents a more detailed look at the Eddy solar cycle and its climate effects.

7.7 kyr ago the agro-pastoral system was being introduced in Central and Southern Europe by two routes: One in Central Europe following the river valleys originating in the Hungarian region of the Danube, by a culture known as the Linear Pottery Culture or LBK, and the other in Southern Europe following a maritime route from the coasts of Greece all the way to the Iberian peninsula by the Cardium Pottery Culture. Both group of farmers descended directly from Neolithic people that moved to the Aegean coasts from the Fertile Crescent during the 8.2-kyr climate complex.

The effect on human societies in Central Europe of the cold and wet periods that coincide with periods of reduced solar activity is to reduce the length of the growing season for plants, reducing the output of both natural ecosystems and agro-pastoral systems, resulting in food shortages and food crises. Suddenly the population cannot be sustained. Malnourishment often comes accompanied by plagues. Before increased mortality can reduce the population naturally, some societies resort to emigration to places that are also experiencing food shortages, and are usually subjected to increased violence in the ensuing fight for resources. It is the Four Horsemen scenario. Archeologists are increasingly aware that the pattern of advances of farming in Europe follows stages that coincide, within dating uncertainties, with periods of climate deterioration that are often associated with periods of reduced solar activity. Agro-pastoral societies appear to have spread faster during periods of climate worsening, pushed by human overshooting conditions caused by climate crises.

During the 8.2-kyr climate complex, the Holocene's worst climate change period according to most proxies, farmers from the Levant and Central Anatolia moved to the shores of the Aegean Sea and expanded into the Balkans. The 8.2-kyr climate complex marked a significant population decline in the hunter-gatherer societies of Europe (Shennan et al. 2013), probably facilitating the

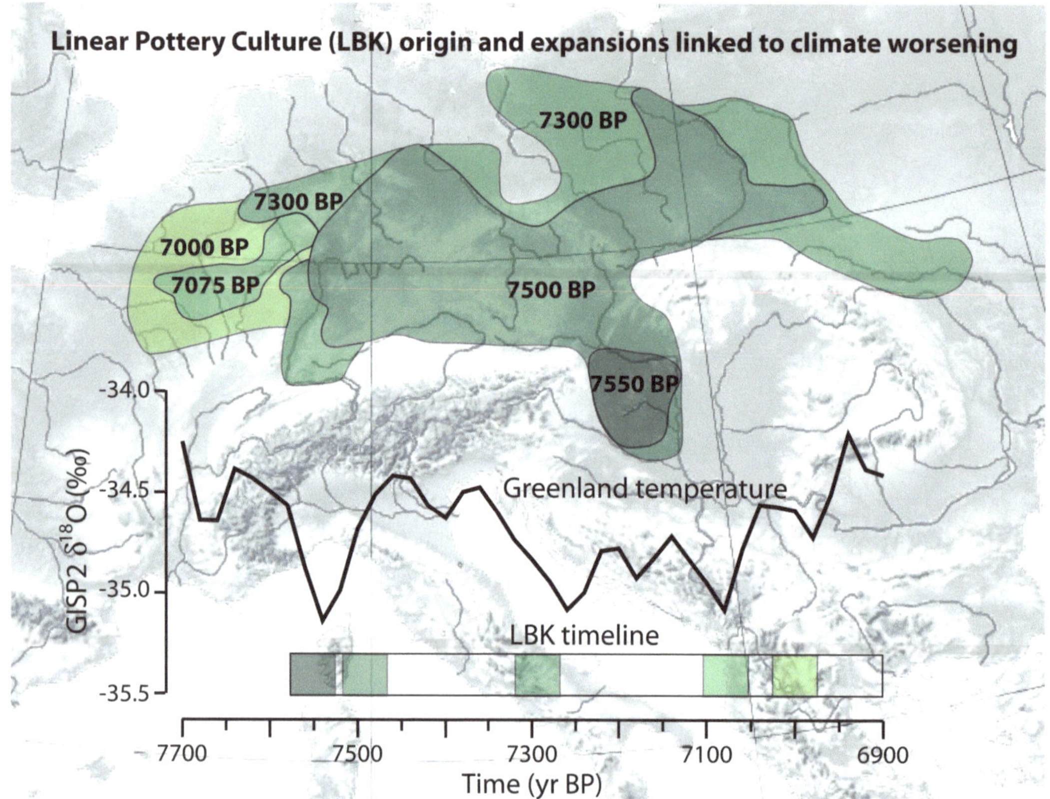

Fig. 6.6 Geographical and temporal expansion phases of Linear Pottery Culture (LBK) Expansion phases coincide with periods of cooling in Greenland, and in fact the entire LBK culture appears to encompass the long period of climate worsening between 7700 and 6900 yr BP, known as the Cerin phase in the Alpine region and defined by some authors as a Little Ice Age within the Holocene Climatic Optimum. After Gronenborn et al. (2013).

invasion. At around 7.7 kyr BP, when the climate deteriorates again, the LBK culture appears and flourishes, expanding into hunter-gatherer areas during the 7.7–7.0-kyr period, replacing the human populations that lived in Central Europe (Fig. 6.5c, d, and e). According to Dubouloz (2008), the LBK culture was well adapted to cold, wet periods through construction of robust buildings, placement of villages in tertiary drainage networks, well away from flood risk areas, emphasizing cattle-herding, reducing the early Balkan Neolithic range of cultivated plants, and the practice of autumn sowing in intensively cultivated plots.

Dubouloz (2008), and Gronenborn et al. (2013), show that LBK expansion follows a climatic rhythm (Fig. 6.5c & 6.6). LBK forms during the increasingly colder 7.7-kyr ACE and initiates its dispersal around 7.5 kyr BP at the peak of cold conditions. During the period of warmer drier climate that followed the 7.7-kyr ACE, LBK consolidates a wide territory. The next period of dispersal initiates again at the next cold period around 7.3 kyr BP when LBK crosses the Rhine into Alsace and the present Dutch area. It is 200 years later during another cold period around 7.1 Kyr BP when LBK experiences its last dispersal into the Seine basin. Demographic analysis of LBK habitation (Dubouloz 2008) indicates that periods of dispersal coincide not only with cold, wet, periods but also with periods of maximal population (Fig. 6.5c dark blue lines), suggesting that the difficult conditions that gave the LBK its edge over other human groups, also caused hardship and population decline that usually instigates climate migration. The arrival of better climate conditions after 7.0 kyr BP probably rendered the harsh climate adaptations of LBK disadvantageous and the culture quickly disorganized, losing its vast circulation networks of raw materials, and disappeared.

The expansion of the agro-pastoral system in Europe marked the end of the hunter-gatherers. Analysis of mitochondrial DNA (mtDNA) frequencies in Central Europe human remains shows that hunter-gatherers mtDNA alleles essentially vanished during the 7.7-kyr ACE with the arrival of the LBK early Neolithic mtDNA alleles (Brandt et al. 2013; Fig. 6.5e). Hunter-gatherer populations probably became residual and restricted to the least productive areas. This mtDNA genetic shift and population substitution takes place even as the general population of Western/Central Europe experiments a great population boom due to the arrival of the agro-pastoral societies (Shennan et al. 2013; Fig. 6.5d). That the two biggest increases in population in Western/Central Europe take place during the 7.7 and 7.2-kyr ACE further confirms the expansion of the agro-pastoral system in Central Europe during periods of climate worsening, and lends support to similar expansions elsewhere, like the Aegean expansion during the 8.2-kyr climate complex.

6.6 The 5.2 kyr event. The Mid-Holocene Transition and the start of the Neoglacial period

During the fifth millennium BC, between 7100 and 5700 BP, the climate of the Atlantic period appears to have been in general warm, humid, and stable, constituting ideal conditions for Neolithic farming. ACEs of that period seem to have had a smaller effect and are seldom studied. In the second half of the fourth millennium BC, however, the entire climate system of the planet changed, driven by orbital changes in precession and producing a reduction in solar forcing, while the oceanic/atmospheric forcing increased in importance. Since 10 kyr BP, northern summer insolation has been decreasing and southern winter insolation increasing. The balance between northern and southern insolation determines the position of the Intertropical Convergence Zone (ITCZ), a low pressure belt around the planet that organizes the wind patterns, separating the hemispheres and determining the location of the monsoons. The southward displacement of the ITCZ and the changes in insolation during the fourth millennium BC completely altered the planet's climate, putting an end to the HCO in what is called the Mid-Holocene Transition, setting the path for the colder, dryer world of the Neoglacial Period (see Sect. 4.5). Among its most notorious effects, the southward displacement of the African monsoon brought an end to the African Humid Period causing the desertification of the Sahara.

Within this context of climate instability and increasingly difficult conditions takes place the next low in the Bray cycle between 5.6 and 5.2 kyr BP, coinciding with Bond event 4a of increased iceberg discharge in the North Atlantic (Bond et al. 2001; B3, see Fig. 5.6). A cluster of three SGM known as Sumerian 1 to 3 are responsible for the reduced solar activity of this low (Fig. 6.7a).

The global temperature reconstruction presented in chapter 4 (Fig. 6.1) displays intense cooling centered at around 5400 BP (Fig. 6.7b), comparable in magnitude to the LIA (c. 0.4 °C), but at warmer temperatures. Other climate proxies also show exceptional climate during this period. Sea and non-sea salts in GISP2 have been used to reconstruct the polar atmospheric circulation (O'Brien et al. 1995; inverted in Fig. 6.7d). The empirical orthogonal function (EOF) that represents the flux of these salts displays at this time its highest values in the entire Holocene respect its baseline (Fig. 6.7d), indicating very strong high latitude winds and a great polar circulation expansion that could have brought a very cold period over the Northern Hemisphere while reducing the cold near the pole. Further confirmation for an extraordinary atmospheric circulation at the time of the Bray 3 cycle low comes from Iceland, where Jackson et al. (2005) detected the strongest winds and associated cooling of the entire Holocene in eolian loess deposits between 5600–5100 BP (Fig. 6.7e). Both polar circulation changes and Icelandic wind intensity proxies also match very well other known periods of strong climate deterioration, including all of the Bray cy-

Fig. 6.7 Climate indicators of the 5.2 kyr event

a) Solar activity reconstruction shows the cluster of Sumerian 1–3 solar grand minima. After Wu et al. (2018). **b)** Temperature reconstruction (Z-score) from 73 proxies displays a significant cooling centered at 5400 BP. **c)** Orange box, mammoth extinction at 5550 ± 100 BP at St. Paul Island (Alaska). After Graham et al. (2016). **d)** Polar Circulation Index determined by sea and non-sea salt fluxes from GISP2 ice core by O'Brien et al. (1995; inverted) manifests at this period one of its biggest departures from baseline (straight line) of the entire Holocene. **e)** Iceland wind strength determined by eolian loess deposit size (inverted) displays the highest values of the entire 8000-years series at this time. After Jackson et al. (2005). **f)** Arboreal/non-arboreal pollen ratio in the Austrian Alps highlights periods of forest retraction (below the baseline) due to colder and wetter climate. After Magny et al. (2006). **g)** Black boxes represent periods of Mid-European higher lake levels. After Magny et al. (2006). **h)** The Alpine glacier advance period known as Rotmoos II in Austria and Piora oscillation in Switzerland. After Magny et al. 2006. Vertical blue bands highlight periods of climate deterioration or abrupt climate events (ACEs).

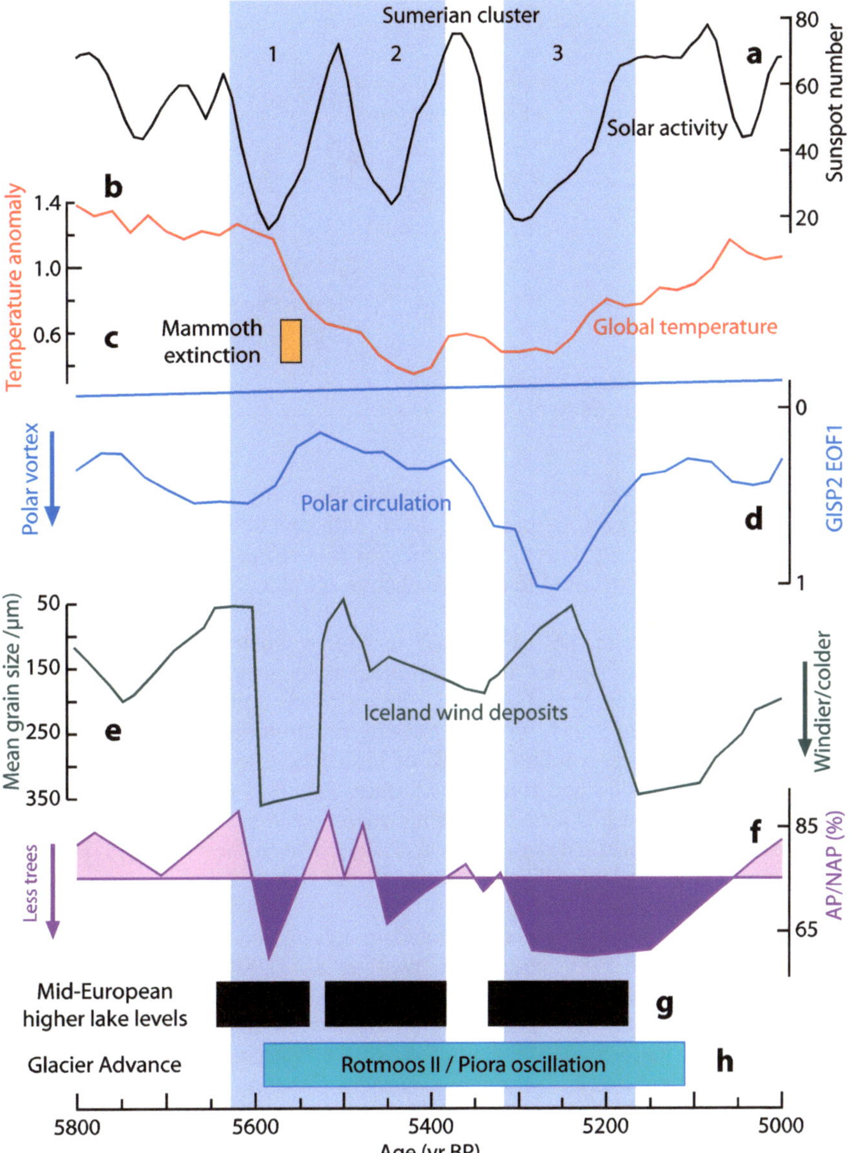

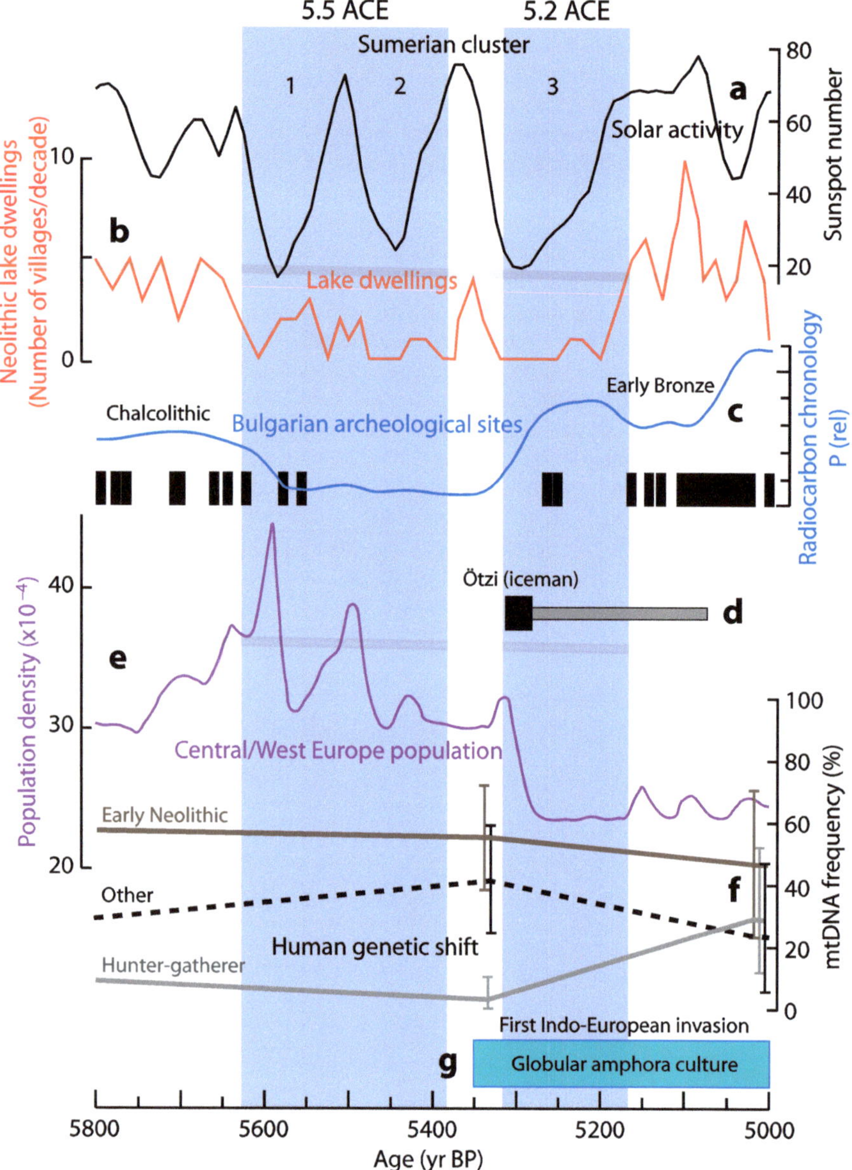

Fig. 6.8 The effect of 4th millennium BC climate changes on human societies of Central Europe

a) Solar activity reconstruction shows the cluster of Sumerian 1–3 solar grand minima. After Wu et al. (2018). **b)** Number of Neolithic lake villages in an area comprising East France and West Switzerland. After Arbogast et al. (2006). **c)** Cultural shift in Northern Greek area (Bulgaria) from Chalcolithic (Copper Age) to Early Bronze based on radiocarbon data (black boxes). After Weninger et al. (2009). The calibrated ^{14}C-age distribution (radiocarbon periodization, curve) supports a hiatus during the 5.5–5.2 kyr ACEs. **d)** The burial dating of Ötzi, the Tyrolean iceman (black box with grey error range). **e)** Analysis of Central/West Europe population reveals a catastrophic decline coincident with the climatic deterioration, with no recovery until the following millennium. After Shennan et al. (2013). **f)** The population decline was accompanied by a shift in mtDNA frequencies that supports a recovery of the descendents of the Paleolithic hunter-gatherer population. After Brandt et al. (2013). **g)** The Globular Amphora Culture period, the first Indo-European culture in Central Europe. Vertical bars correspond to the same periods of climatic deterioration as in Fig. 6.7.

cle lows. The final extinction of the mammoth at its St. Paul Island (Alaska) refuge has been dated with precision at 5550 BP, coinciding with the beginning of this climatic deterioration, and attributed to climate change (Graham et al. 2016; Fig. 6.7c).

In Central Europe, pollen analysis in Alpine Austria shows a retraction of warmer loving arboreal species versus cold-resistant grasses and shrubs in three periods coincident with the three SGM, while increasing precipitation reflects in three corresponding periods of higher lake levels (Magny et al. 2006 and references within; Fig. 6.7f & g). At the same time glacier advances are recorded in several parts of the world (Mayewski et al. 2004), and in the Alpine region this glacier advance receives the names of Rotmoos II, or Piora oscillation (Fig. 6.7h).

The 5.2-kyr ACE was demonstrated to have been a global phenomenon by Lonnie Thompson et al. (2006) through a variety of records that show that at 5200 BP a strong and sudden cooling took place all over the world. Those records include simultaneous freezing of organic remains at glaciers in Tyrol, Peru, and Western Canada, at the same time the Kilimanjaro ice cores display a sudden and profound cooling. The transition from wet to dry con-

ditions is recorded by changes in the water balance in many African lakes and the driest excursion recorded at the Soreq Cave speleothem. Concurrently, global atmospheric CH_4 concentrations recorded in both Greenland and Antarctica stopped decreasing and started to increase for the rest of the Holocene. Dendrochronological records from Irish and Lancashire oaks extending back to 7000 BP exhibit some of their most narrow rings during the decade-long 5145 BP (3195 BC) event (Thompson et al. 2006, and references within).

In Mesopotamia the Uruk culture started to develop around 6000–5800 BP as a response to an increasingly arid period as attested by several Middle East proxies (Soreq Cave, Lake Van, and Gulf of Oman), and flourished based on a system of high yield cereal irrigation, efficient canal transportation, and slavery and forced labor. Uruk culture first developed mass production and writing. By 5500 BP Uruk had become the first city-state and started to expand through "colony" settlements founded across the dry-farming portions of the Middle East. A short but very arid period at around 5200 BP coincides with the abrupt collapse of the Uruk colony system, as the colonies in the north and smaller settlements in the south are abandoned, and formerly cultivated areas are turned into pastures with the changes of river courses (Brooks 2012). The 5200 BP arid event in the Middle East is also reflected in the abrupt cessation of precipitation over the Nile river delta. At that time an Early Bronze culture was flourishing in Egypt based on a very high population density at the Nile Valley boosted by climate refugees from the Sahara aridification over the previous centuries. The sudden 5200 BP dry period must have increased competition over resources resulting in widespread violence that ended with the subjugation of the Lower (northern) Nile by the Upper Nile and the unification of Egypt under the first pharaoh at that date (Brooks 2012).

The difficult climatic conditions through the 5.2-kyr ACE constituted an authentic disaster for Neolithic farmers in Central Europe. There is a widespread record of settlement abandonment at the Late Neolithic/Chalcolithic–Early Bronze transition, as attested by lake dwellings at France and Switzerland (Arbogast et al. 2006; Fig. 6.8b), and the almost complete absence of radiocarbon dates for a period of four centuries in Bulgaria (Weninger et al. 2009; Fig. 6.8c). At the same time, and with a remarkable similarity to the solar activity record, the population of West/Central Europe crashed, revealing the true extent of one of the most difficult periods for humankind (Shennan et al. 2013; Fig. 6.8e). The population fell so hard that it is believed that diseases must have played an important role in bringing down the debilitated Neolithic farmers. About this time, at 5800 BP the pneumonic plague (*Yersinia pestis*) is believed to have emerged for the first time among the Kurgan nomadic herders of the Pontic steppe and could have spread to other populations.

The decline of the Neolithic farmers of Central Europe allowed a return of the Western European hunter-gatherers as attested by the re-appearance of their genetic signature in areas where they had previously disappeared (Brandt et al. 2013; Fig. 6.8f). It was the prelude to the second major invasion and last big population turnover of the Holocene in Europe. Starting around 5350 BP the first nomadic herders from the steppes invaded Central Europe establishing the Globular Amphora Culture (Fig. 6.8g), probably pushed by the climate pessimum conditions and taking advantage of the weak state of Neolithic farmers. A few centuries later came the great invasion by the Battle Axe people (Corded Ware Culture). The Indo-European nomads had domesticated the horse, developed the war chariot, acquired the bronze culture and had a patriarchal war-like culture. The Late-Neolithic farmers did not stand a chance, and according to genetics the third known major genetic shift in Europe took place (Haak et al. 2015), the first was the Neanderthals substitution by Paleolithic hunter-gatherers and the second the replacement of the hunter-gatherers by Neolithic farmers.

Ötzi, the iceman from Tyrol, was a Neolithic farmer, closely related genetically to the LBK people, who met a violent end (Figs. 6.8d & 12.7a). Whether he was killed by other waning Neolithic farmers, by resurging hunter-gatherers now pastoralists, or by invading Indo-European nomadic herders is not possible to know. He fled his enemies only to be buried in ice for over 5000 years, he was a testimonial to both the changing climate of the Mid-Holocene Transition and its devastating effect on human societies.

6.7 The 2.8 kyr event. The Sub-Boreal/Sub-Atlantic Minimum

The second low in the Bray cycle at about 2.8 kyr BP coincides with the Homeric grand solar minimum that took place right after one of the worst climate-induced human catastrophes known to history. This crisis that took place around 3160 BP (1210 BC) has been convincingly linked to a severe drought that affected the Black Sea area and the Eastern Mediterranean in what constituted the 3.1-kyr ACE. The drought is likely to have triggered a massive migration by land and sea of the people that lived North and West of the Black Sea and the Balkans. They are collectively known by historians as the Sea Peoples. They destroyed everything in their way bringing about the Late Bronze Age collapse. In a period of less than 50 years the Hittite Empire and the Mycenaean Kingdoms of Greece were destroyed while the New Kingdom of Egypt was brought to its knees by the combined effect of the climatic crisis and the invaders. Every city between Pylos in the Peloponnese and Ashkelon in Gaza was destroyed, including the famous Troy at the Dardanelles, most of them never to be rebuilt. The Late Bronze Age collapse was so destructive that nothing similar has taken place later. The Fall of the Roman Empire pales in comparison. The palace cultures were replaced by small villages and writing was lost in most areas. Greece entered the Greek Dark Ages and Egypt the Third Intermediate Period.

The 3.1-kyr ACE with its two centuries long megadrought in the Eastern Mediterranean is similar to the 8.6-kyr ACE, also associated with a long drought in the same area, in that none of them displays a strong increase in cosmogenic isotopes production, but both immediately precede what has been identified as a SGM. The 3.1-kyr ACE and its Late Bronze Age collapse were followed by the 2.8-kyr ACE, associated with the Homeric SGM that coincides with Bond event 2a of increased iceberg discharge in the North Atlantic (Bond et al. 2001). At the 2.8-kyr ACE an abrupt climatic change took place in Northwest Europe, from a relatively warm continental climate to cooler and wetter conditions, marking the transition from the Sub-Boreal to the Sub-Atlantic period. This transition is reflected by an abrupt change in moss species in raised bogs at this time (van Geel et al. 1998).

The 2.8-kyr ACE is actually composed of three phases that correlate well with inferred solar activity. The event started with a 50 year reduction in solar activity and temperatures around 2950 yr BP (Fig. 6.9a & b), that brought much needed precipitation to the Eastern Mediterranean, briefly interrupting the drought and allowing a temporary recovery of agriculture in the region (Kaniewski et al. 2013, Fig. 6.9f & g). Since this phase takes place 200 years before the Homeric SGM, it is probable that it corresponds to a low in the 210-yr de Vries solar cycle. After a return to dry warmer conditions for another century, the Homeric minimum started at 2800 yr BP coinciding with an abrupt descent in land and sea-surface temperatures (Fig. 6.9a, b & c). In Europe the event is very well described as a change to colder, wetter conditions from Northwestern Europe to the Eastern Mediterranean (van Geel et al. 1998; Fig. 6.9f). In South America, glacier advances, peat changes, vegetation changes, and sediments, indicate also an abrupt transition to colder, wetter condi-

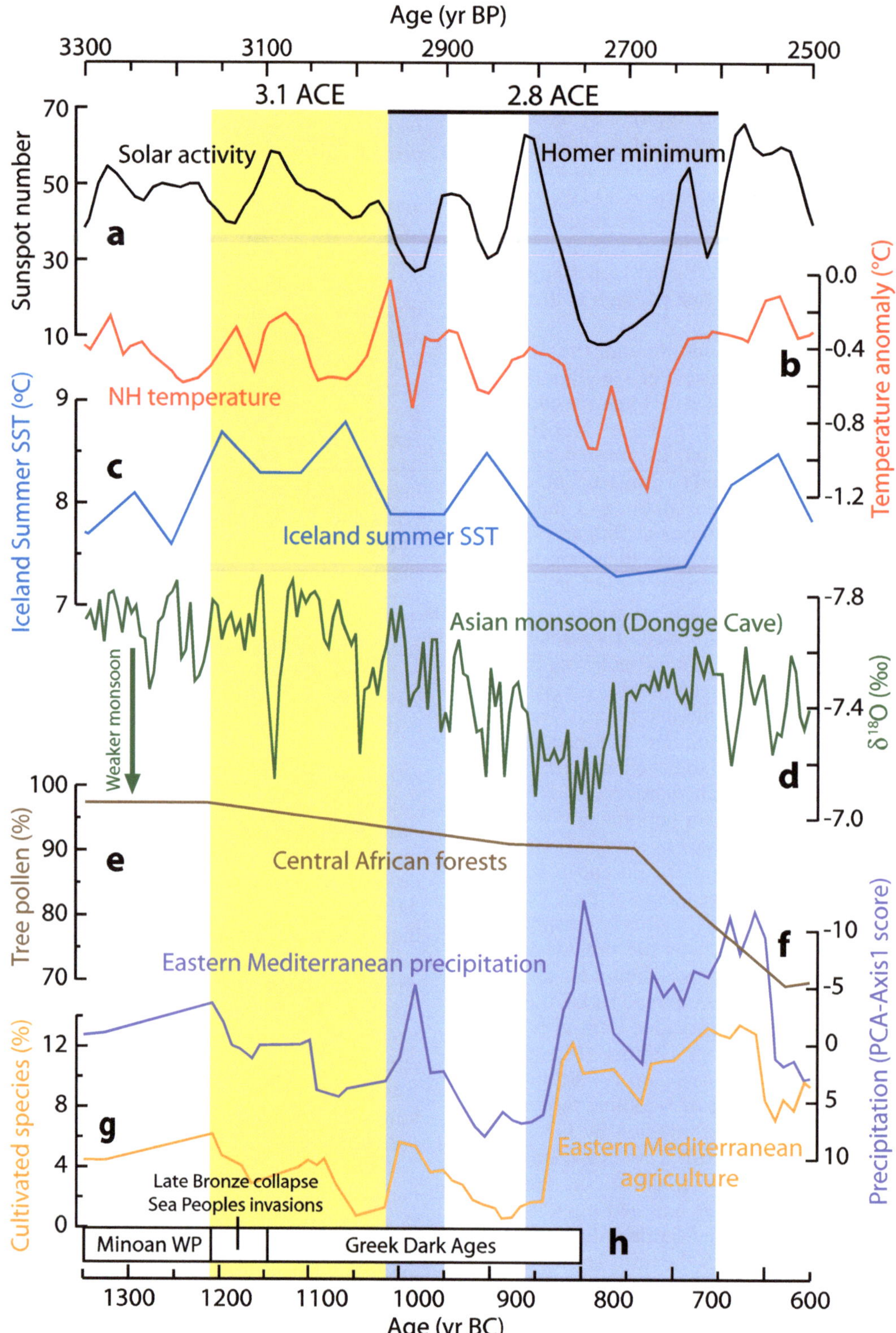

Fig. 6.9 Climate indicators of the 2.8-kyr event

a) Solar activity reconstruction shows the Homeric solar grand minimum. After Wu et al. (2018). **b)** Northern Hemisphere temperature reconstruction displays a significant cooling centered at 2700 BP. After Kobashi et al. (2013). **c)** Iceland summer sea surface temperature diatom-based reconstruction displays a similar simultaneous cooling. After Jiang et al. (2015). **d)** Asian summer monsoon proxy from Dongge Cave stalagmite DA oxygen isotope ratio indicates a weakening of the monsoon starting about 2950 BP and reaching its lowest values at 2750 BP. After Wang et al. (2005). **e)** Tree pollen percentage in a Cameroon lake indicates the biggest Central African forest retraction in the entire Holocene starting at about 2800 BP. After Maley & Brenac (1998). **f)** Pollen-derived proxy of moisture availability at Gibala–Tell Tweini, a city in the ancient Ugarit Kingdom, Northwest Syria, plotted as Principal Components Analysis scores. **g)** Pollen-derived proxy of agriculture showing the percentage of pollen coming from cultivated species at the same location indicating that humid periods at this time in the Eastern Mediterranean coincided with periods of reduced solar activity. After Kaniewski et al. (2013). **h)** White boxes show historic periods in the Eastern Mediterranean. Vertical blue bars correspond to the 2.8-kyr ACE. Light orange bar indicates the 3.1-kyr ACE.

tions dated at 2750 BP. North America was also affected by a general climate change around 3000–2600 BP towards cooler temperatures and increased precipitation. Increased flooding in the Mississippi basin and an hyperactive storm period in the Gulf coast are dated at that time. Pollen and sediment organic contents in Central Asia support also a coincidental increase in precipitation. Lake Pupuke (New Zealand) isotopic levels indicate a 400 year marked increase in precipitation starting at 2800 BP.

While in mid-latitude areas of North America, South America, Europe, Central Asia and Australasia there was an important increase in precipitation, analysis of the Dongge Cave stalagmite DA in China shows one of the biggest reductions in oxygen isotopes ratio of the entire series (Wang et al. 2005; Fig. 6.9d), indicative of an important weakening of the summer monsoon and an increase in aridity in South Asia. This prominent weakening of the Asian summer monsoon coincided with what has been described as the "dramatic forest decline" in Central Africa (Maley & Brenac 1998; Fig. 6.9e), the biggest forest reduction in the area for the entire Holocene, of which the forests of Central Africa are still recovering tree thousand years later. A possible weakening of the West African monsoon is the likely cause, and although drier conditions started at around 3150 BP, it was around 2750 BP that the forests initiated a marked decline accompanied by expansion of grasses and very dry conditions as attested by the complete drying of several lakes such as lake Sinnda in the Niary valley (Congo). The forest cover opened up and fragmented, and enclosed savannas appeared.

In Europe the 2.8-kyr ACE separates the Late Bronze Age from the Early Iron Age. The impact of this climatic crisis is somewhat diluted by the previous dramatic crisis of the 3.1-kyr dry ACE from which there had been no recovery. In fact the increase in precipitation, despite the cooling, was very beneficial for agriculture in drier areas (Fig. 6.9f & g) and probably was a significant factor contributing to the end of the Greek Dark Ages (Fig. 6.9h). In wet marginal areas however the change had a negative impact. In West Friesland (Netherlands), the Late Bronze settlement phase came to an end coincident with rising water tables as houses started to be built on artificial mounds. The rising water and bog expansion caused a loss

of agricultural land forcing the migration of the population to coastal salt marshes, richer in food resources. These marshes had also started to appear around 2700 BP (van Geel et al. 1998). The end of habitation in West Friesland is also coincident with the end of Late Bronze lakeside village construction in Central Europe for half a millennium from 2.8 kyr BP. In Central and Western Europe the Late Bronze Urnfield culture gave way to the Early Iron Hallstatt culture that expanded during the 8th century BC (2750–2650 BP), amid the dramatic changes in flora and fauna that accompanied the 2.8-kyr ACE.

In North America the ACE at 2.8 kyr BP also separates two cultural periods in the Mississippi basin, supporting the theory that abrupt climate change is a motor for cultural change. The Archaic hunter-gatherer period reached an end in the 3000–2600 BP, marked by a hiatus of several hundred years in riverside settlements, suggesting an abandonment of frequently flooded areas, after which the new settlements belong to the Woodland period, characterized by widespread use of pottery and domesticated plants (Kidder 2006). Meanwhile in Central Africa the opening of the forest allowed the migration of Bantu speaking, metal working people into areas that are now completely forested.

It has been recently proposed that the Middle East increase in precipitations at the 2.8-kyr ACE followed by a megadrought 200 years later, were instrumental in the collapse of the Neo-Assyrian Empire (Sinha et al. 2019).

In the Central Asian steppes the increase in precipitation at 2800 BP brought the Scythians into preeminence. They were semi-nomadic herders of the Eastern Iranian language group from the Tuva region at the intersection of Russia, Mongolia, China, and Kazakhstan. With the wetter climate the steppes expanded and could support huge herds of horses, sheep, and goats. The Scythians abandoned any trace of settlement and became nomadic riders. They are credited with developing mounted warfare using composite bows. By 2700 BP they invaded the Northern Black Sea and the Caucasus pushing the Cimmerians southward into conflict with the Assyrians. The Scythians would continue expanding their territory up to Thrace and Eastern Europe, and played a leading role in the destruction of the Neo-Assyrian Empire. To the Greeks they were

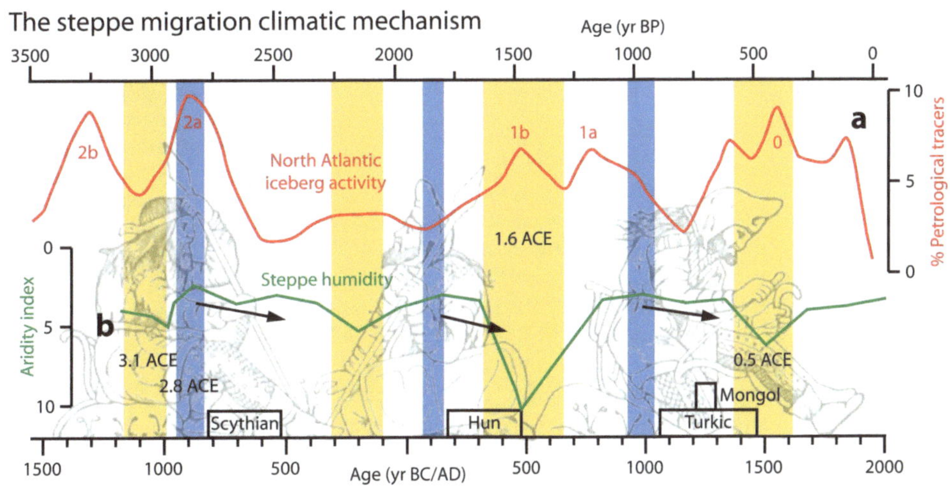

Fig. 6.10 The steppe migration climatic hypothesis

a) North Atlantic stacked percentage of ice-rafted debris, indicative of iceberg activity after Bond et al. (2001). The main peaks have been labeled with their accepted numbers. **b)** Pollen-derived aridity index (inverted) from a Central Mongolian lake after Fowell et al. (2003). Light orange bars indicate drought periods. Blue bars mark millennial humidity maxima. Arrows indicate downward trends in humidity from millennial maxima. Main historic migration events are indicated by boxes, and they took place after humidity maxima within an increasingly arid context. Notice that the Bond events pattern does not correspond to the humidity pattern at the Central Asian steppes, however big changes in humidity at the indicated ACEs tend to coincide with Bond events. Background picture: Scythian king and warriors, drawn after figures on an electrum cup from the Kul'Oba kurgan burial.

the prototype of savage barbarians, as they were very war-like and practiced human sacrifices. They were also their main providers of slaves from selling their captives. They inspired two well known Greek myths, the centaurs from their riding combat skills, and the amazons because their women also fought, as one in three women was buried with weapons and many sustained war wounds.

In the perpetual conflict between nomads and settlers, climate change appears to have played an essential role in setting the stage for numerous conflicts. From the invasion of Central Europe during the 7.7-kyr ACE by LBK agro-pastoralists, to the Sea Peoples invasion at the 3.1-kyr ACE, and the periodical invasions of Eurasia from steppe nomads, we find evidence of climate change creating conditions that resulted in migration as a response, and conflict as a consequence. The productivity of the steppes is very dependent on precipitation and the nomads and their herds cannot rely on stored food during bad years. When analyzing precipitation in Central Asia a common pattern is found for nomadic invasions. They don't take place during arid periods, but following a maximum in humidity (Fig. 6.10b), suggesting that the increase in precipitation, like in the 2.8-kyr ACE, brings the nomad population and their herds to a maximum, and from that point, any decrease in precipitation, even if not pronounced (Fig. 6.10b arrows), places the population in overshooting. The result is a high number of steppe nomads migrating to adjacent areas where easy conquests stimulate further advances, pushing other groups into migration. This pattern is detected not only in the case of the Scythians, but also with the Huns in the 2nd century AD, the Turkic peoples in the 11th century and the Mongols in the 13th century (Fig. 6.10b). This pattern underscores a persistent steppe migration climatic mechanism. A similar pattern is observed with desert locusts, that don't mass migrate during bad years that keep the population in check, but after good years that push the population up creating overshooting conditions.

6.8 The 0.5 kyr event. The Little Ice Age

The first low in the Bray cycle at about 0.5 kyr BP coincides with the Wolf/Spörer/Maunder cluster of SGM that took place during the coldest period of the Holocene that is generally known as the LIA. Due to having taken place during modern historic times, it is also the best studied and known cold event.

The LIA coincides with Bond event 0 of increased iceberg discharge in the North Atlantic (Bond et al. 2001; Fig. 6.10a). Different authors choose a different start for the LIA, since the climate started to deteriorate progressively from its previous warm period at about AD 1150, but did not become significantly colder than the previous four hundred years until after AD 1250. It become a serious problem for human societies of the time after AD 1300. Other authors however wait until after AD 1500, when a relatively warm interlude during the 15th century ended. In this work the start of the LIA is placed at AD 1258, a year after the Rinjani eruption (Lombok, Indonesia), the strongest since writing was invented, and a convenient time marker. As with other lows in the Bray cycle, the LIA displays a pattern of colder phases recognizable in

many climate proxies that in general match the pattern of solar activity (Fig. 6.11a). Some temperature reconstructions (Christiansen & Ljunqvist 2012; Fig. 6.11b) show good agreement with solar activity except for starting the initial cooling before solar activity declined with the Wolf minimum and showing a very cold period before the Maunder minimum. This general pattern of four cold phases for the LIA can be defended on the basis of decreasing Mediterranean SST (Versteegh et al. 2007; Fig. 6.11c), increased Iceland sea ice (Massé et al. 2008; Fig. 6.11d), glacier advances in the Alps (Holzhauser et al. 2005; Fig. 6.11e) and Venezuela (Polissar et al. 2006; Fig. 6.11f), and increased North Atlantic deposition of ice-rafted debris (Bond et al. 2001; Fig. 6.10a). Alpine glaciers do not show an advance during the Spörer minimum, and this requires some explanation. Unlike during most SGM, the LIA SGM do not show a pattern of increased Central European precipitation, and during the Spörer minimum Central Europe experienced a very dry period (Büntgen et al. 2010; Fig. 6.10g). It has been reported that in England, due to the fields not being covered in snow during severe winters around AD 1458, the seeds in the field were killed by the cold resulting in several years of poor crops and famine (Fig. 6.13e, g, & h). The reduction in precipitation would have prevented glacier advances in the region (but not in Venezuela) and might have reduced the growth of Iceland sea-ice that was less during the Spörer minimum than in other minima during the LIA (Fig. 6.11d).

Why would there have been dry conditions in West/Central Europe during the Spörer minimum, when we have seen a general pattern of increased precipitation in this region during previous SGM? In general, Central Europe precipitation correlates with solar activity during the 0.5-kyr ACE (Fig. 6.11a & g), and shows higher level of precipitation when solar activity is not low. Alpine glacier advances took place when there was enough winter precipitation to sustain them. A possible explanation for the contrasting role of solar variability on Central Europe precipitation between the 0.5-kyr ACE and previous ACEs is that as the climate of this interglacial evolves towards its ending and the tropical bands contract and subarctic regions expand, the atmospheric reorganization effects of reduced solar activity must result in differences in precipitation patterns, so regions that previously received increased precipitation during SGM might suffer a precipitation reduction under identical forcing due to the evolving climate conditions of the planet.

The Dalton minimum, is also unusual in some climatic aspects, including precipitation in Central Europe and glacier advances in Venezuela (Fig. 6.11f, g). However the Dalton minimum is not generally considered a SGM as it was brief and showed a lower reduction in solar activity. It is also considered that part of the climatic effects during the Dalton minimum were due to a particularly high volcanic activity at the time (Fig. 6.12a).

Why was the LIA so cold? There was a confluence of causes that made the LIA the coldest period in the Holocene. To start, the Holocene had been cooling since the HCO due to an accelerating reduction in obliquity (Fig. 6.1), so the LIA started from a lower temperature than previous Bray lows. In addition the LIA was a very long cold event (c. 600 years), longer than similar periods during the Holocene (see next section), and its cooling phase was also longer and therefore more profound. A possible reason for the LIA being so long is the coincidence of the lows from three long climate cycles, the 2500-yr Bray solar cycle at around 500 BP, the 1500-yr oceanic cycle also at a mid-cycle low around 500 BP, and the 1000-yr Eddy solar cycle at around 300 BP. The third reason that the LIA was so cold was a very significant contribution from high volcanic activity both at its beginning and at its end. The 1250-1350 and 1750-1850 AD periods are among the 25% most volcanically active centuries in the Holocene according to Greenland volcanic sulfate records

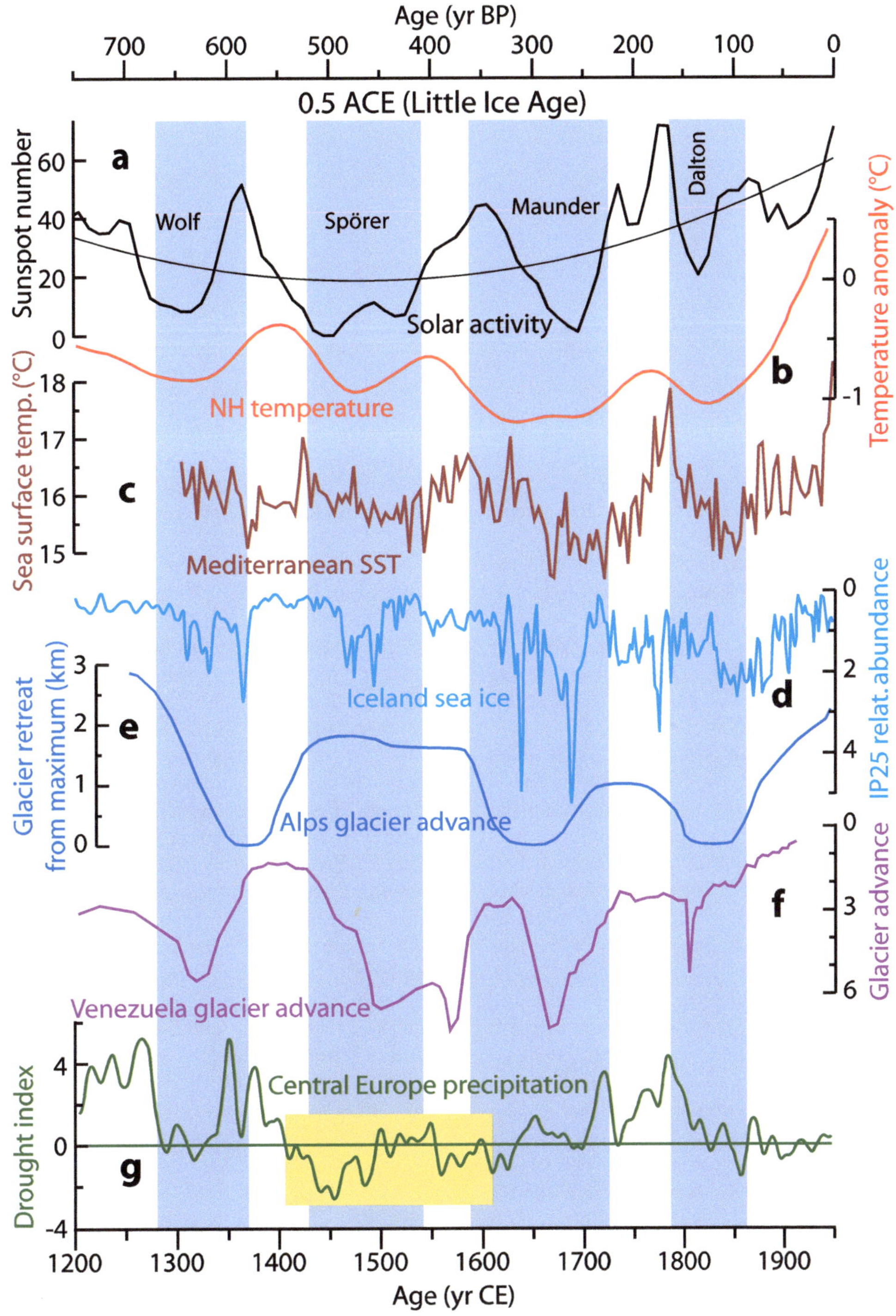

Fig. 6.11 Climate indicators of the 0.5-kyr event
a) Solar activity reconstruction shows the Wolf, Spörer, Maunder, and Dalton solar minima. After Wu et al. (2018). The quadratic regression (thin line) follows the long-term change in solar activity. **b)** Northern Hemisphere temperature reconstruction displaying a pattern that generally matches solar activity. After Christiansen & Ljungqvist (2012). **c)** Mediterranean sea surface temperature proxy record also displays four cooling periods. After Versteegh et al. (2007). **d)** A biomarker sea ice proxy from Iceland agrees well with the sea surface temperature. After Massé et al. (2008). **e)** Glacier retreat in km from maximum extent in the Alps does not show glacier advances during the Spörer minimum (after Holzhauser et al. 2005), while **f)** Venezuelan glaciers show glacier advances at every minima. After Polissar et al. (2006). **g)** Precipitation in Central Europe measured from German oak rings showing a period of very low precipitation during the Spörer minimum (box). After Büntgen et al. (2010). Blue vertical bars highlight the four periods of climatic deterioration within the LIA as determined by climate proxies.

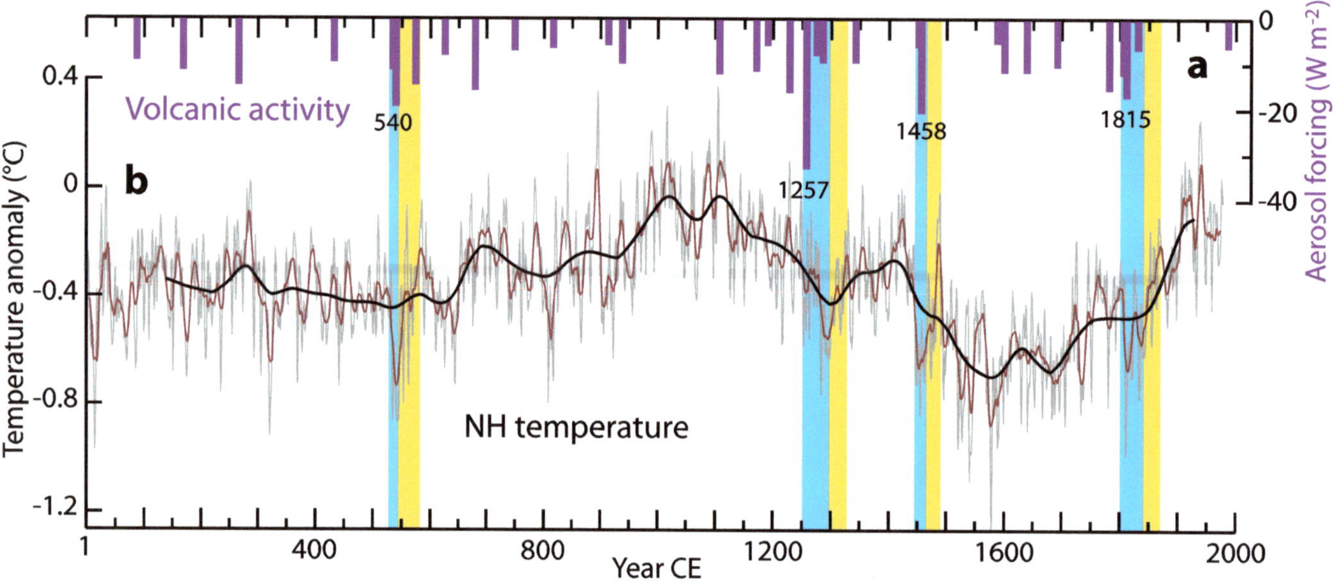

Fig. 6.12 The effect of volcanic forcing on temperature during the past 2000 years
a) Reconstruction of the time and aerosol forcing of major volcanic eruptions from sulfate levels in Greenland and Antarctic ice cores. After Sigl et al. (2015). **b)** Multi-proxy temperature reconstruction between AD 1–1979 (light grey line) with a 10 year moving average (red line) and its >80-yr slow component (black thick line). After Moberg et al. (2005). Light blue bars indicate the temperature reduction after the four biggest volcanic eruptions. Orange bars indicate the temperature recovery after the temperature effect of volcanic eruptions ended.

(Zielinski et al. 1996). In contrast, the 1350-1750 AD period, that includes most of the LIA, displays very low volcanic activity, well below the Holocene average (see Fig. 11.3), and thus cannot explain its coldest period. Volcanic activity might have contributed to the LIA coldness during the 1250-1350 and 1750-1850 AD periods, when solar activity was not particularly low (Fig. 6.11a, b).

While volcanic activity during the past 2000 years does not correlate with solar activity, the concentration of strong eruptions during parts of the LIA is higher than at other periods over the past 2000 years, and it has been proposed that the LIA was mainly due to a volcanic effect on climate. Available evidence, however, indicates that this is not the case. Moberg et al. (2005) Northern Hemisphere temperature reconstruction has very good resolution and allows to investigate this issue (Fig. 6.12a & b). Very strong volcanic eruptions like the AD 536 and 540 eruptions or the AD 1453 and 1458 eruptions (one of them the Kuwae eruption in Vanuatu) had a very clear effect on temperatures that lasted 1 to 2 decades at most (Fig. 6.12a blue bars), while clusters of eruptions, like around 1257 and 1815, can reduce temperatures for about 4–5 decades (see Sect. 11.2). But in every case, after the effect of the volcanic eruption on meridional transport ends, temperatures recover (Fig. 6.12a orange bars), and the general temperature trends continue, whether they were going up or down. It is clear therefore that even though volcanism contributed to the cold and misery of the LIA, and can explain why the cooling at the Wolf and Maunder minima started before there was a significant reduction in solar activity (Fig. 6.13a, b, c), it cannot have been the driving factor behind the LIA.

It has also been shown that much higher levels of volcanic activity have taken place at deglaciation and at the beginning of interglacials, under warming conditions (Huybers & Langmuir 2009; Kutterolf et al. 2013; see Sect. 11.2). The HCO showed a level of volcanic activity

2–4 times higher than the LIA. It is possible that climate change determines volcanic activity more than the opposite, as changes in crust load due to ice melting or build up might be responsible for increases in volcanic activity, not only at deglaciations (Huybers & Langmuir 2009), but perhaps also at the LIA.

The effect of climate deterioration brought about by the LIA on human societies is better known, and the collapse of the Viking colony in Greenland is often mentioned. However it was almost the entire population of the planet who suffered during the LIA. Data for grain production in England shows that yield per acre decreased following a similar pattern to Northern Hemisphere temperature (Fig. 6.13c, e), probably reflecting the shortening of the season. Back to back and even three in a row (highly unusual) years of bad crops took place on this period (Fig. 6.13, vertical grey lines between e and h) causing a marked increase in wheat prices (Fig. 6.13d, inverted) and major famines. The first one at 1315–17 was the worst, affecting most of Northern and Central Europe and initiating the Crisis of the 14th Century. Climatic factors also determined the increase in contact between rodents and humans in Central Asia, giving rise to the bubonic plague, a new manifestation of the plague that had a near 100% mortality. The plague reached Europe in 1347 and in six years killed over one third of the population in the Black Death pandemic. The population decline was so large (Fig. 6.13f), that subsequent crop failures had less effect on people's famine and wheat prices even one century later, like the bad crops of 1459–61. The plague became recurrent in Europe being always present somewhere in the continent until 1750, and causing major epidemics periodically. The spread to other countries of the Hundred Year War between England, France and Burgundy through the Free Companies of mercenary bandits, the start of the Peasants Revolts, and the Western Schism in the Christian Church completed the Crisis of the 14th Century, that manifested as a complete failure of the institutions to cope with climate-related natural disasters that were seeing then as acts of God.

The 15th century was a period of recovery and Renaissance in Europe, despite the severe impact of what is believed to have been the coldest decade of the millennium according to both climate and historic reconstructions, the 1430s during the Spörer minimum (Camenisch et al. 2016). However one can never underestimate the capacity of humans to make a difficult situation worse, and so while Japan was developing successful strategies to cope with the challenges that the LIA posed on food production,

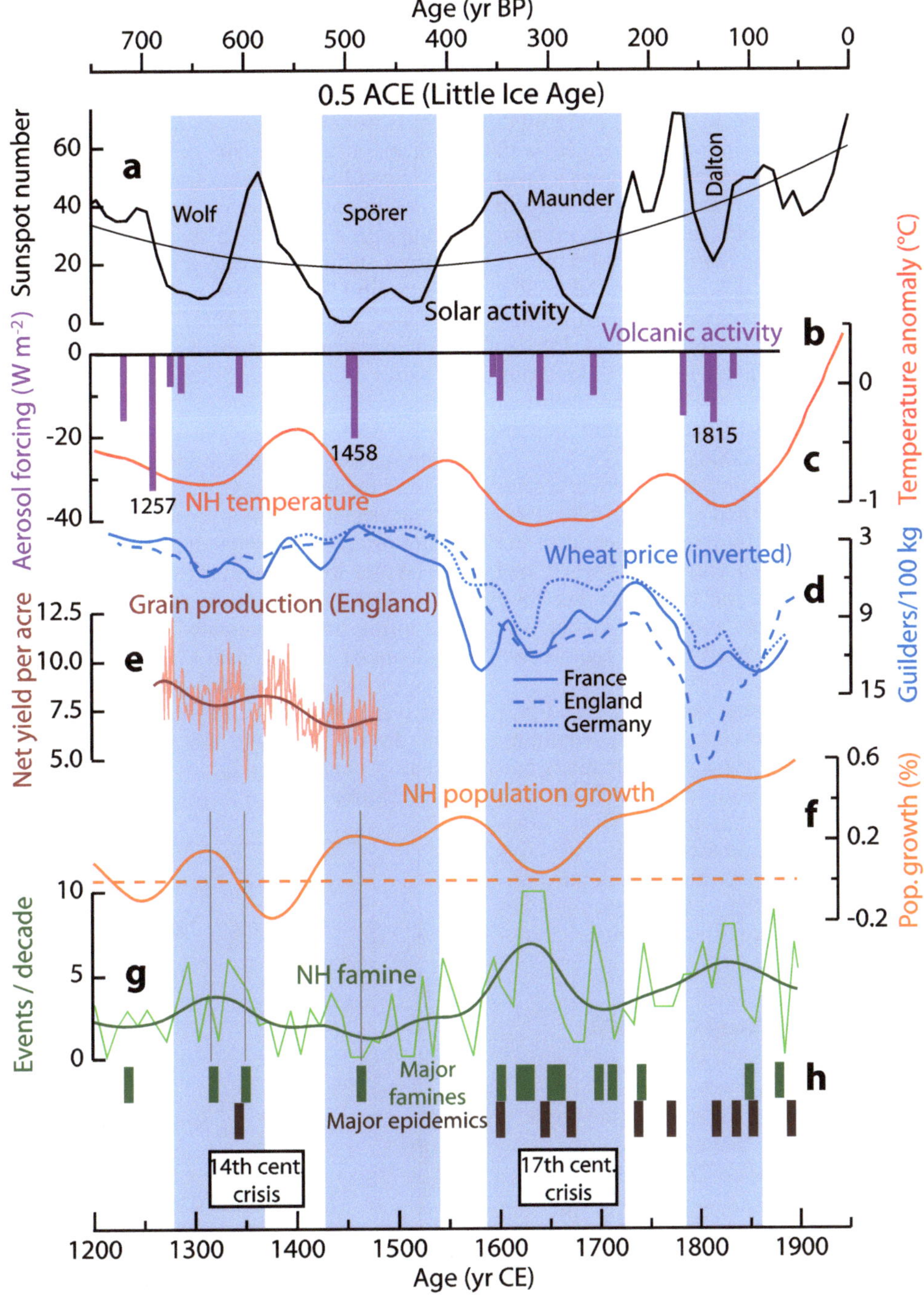

Fig. 6.13 The effect of LIA climate changes on human societies of Europe
a) Solar activity reconstruction shows the Wolf, Spörer, Maunder, and Dalton solar minima. After Wu et al. (2018). The quadratic regression (thin line) follows the long-term change in solar activity. **b)** Volcanic activity reconstruction with dates for the three major eruptions. After Sigl et al. (2015). **c)** Northern Hemisphere temperature reconstruction. After Christiansen & Ljungqvist (2012). **d)** Wheat price in Dutch guilders per 100 kg (inverted), for France (continuous line), England (dashed line) and Germany (dotted line). After Lamb (1995). **e)** Three main crops of grain net yield per acre in England, with annual data (thin line), and long-term trend (thick line). After Campbell & Ó Gráda (2011). **f)** Northern Hemisphere population growth in %. After Zhang et al. (2011). **g)** Northern Hemisphere famine index in events per decade. Thick line 100-yr low-pass filter. After Zhang et al. (2011). **h)** Major famine events (upper boxes) and major epidemic and pandemic events (lower boxes) from multiple sources. Main historical periods of crisis are shown in white boxes at bottom. Grey vertical lines link multiyear crop failures in (e) with major famines in (h). Blue vertical bars are periods of climate deterioration defined in Fig. 6.11.

most of Europe and a great part of the world were again engulfed by man-made crises at the time of the Maunder cold period. The General Crisis of the 17th Century was again a period when the Four Horsemen of Apocalypse rode unchallenged, as the world saw the biggest number, and duration of wars, and war casualties in recorded history to that time. Climatic factors contributed to the general worsening, and glaciers advanced destroying farms, houses, and villages. Climate worsening, together with peasants revolts and war destruction, produced a record number of major famines and accompanying epidemics, to the point of producing a collapse in population growth (Fig. 6.13f, g, & h). It was a period that coincided with major political upheaval, including frequent government replacements and even state failures. One of the biggest countries in Europe, the Polish–Lithuanian Confederation, completely disappeared, together with one third of its population. The Seven Ill Years of Scotland in the 1690s, were caused by a major famine event in Northern Europe that killed half of the population in Finland and 15% in Scotland, and was an important factor leading to its union with England. Geoffrey Parker has analyzed the historical aspects of the global crisis of the 17th century and its connection to climatic events (Parker 2008).

The 18th century was again a period of recovery, after which the climatic deterioration and social problems returned. The inability of the Old Regime to respond to the frequent crises caused the French Revolution when bad crops due to a drought in 1788, and resulting high food prices in 1789 affected the population of France. The French Revolutionary Wars, followed by the Napoleonic Wars engulfed Europe once more.

However the European farmers from the second half of the 18th century had learned to cope with the challenging climatic conditions through a series of adaptations that constituted the Agricultural Revolution, which in turn helped drive the Industrial Revolution. The final disappearance of the plague from Europe around 1750 was followed by the appearance of the recurrent cholera pandemics of the 19th century. By 1850 the LIA had been left behind and much better climatic conditions have accompanied human societies since then.

6.9 Climatic effects of solar grand minima

The considerable amount of information about the climatic effects of SGM essentially points to an atmospheric effect. The observed phenomenology is usually:

- Region-specific increased precipitation in mid and high-latitudes
- Region-specific decreased precipitation in tropical and subtropical areas
- Weakening of tropical monsoons.
- Increase in mid and high-latitude wind strength
- Increase in polar circulation
- Winter cooling
- Sea surface cooling
- Increase in the temperature difference between the Equator and Poles
- Glacier advances
- Increased iceberg activity

These effects are consistent with an expansion of the polar cells, a southward displacement of the polar jet stream, an equatorial shift of the Ferrel cell and subtropical jet, and a similar displacement of the descending parts of the contracting Hadley Cells. The resulting change in wind patterns would be responsible for the alterations in precipitation and temperatures. These atmospheric changes were described by Joanna Haigh (1996) in her landmark article "The impact of solar variability on climate," where she described the changes found in a general circulation model when simulating changes in solar irradiance and stratospheric ozone. Since then Haigh's hypothesis has received support not only from paleoclimatology, as reviewed here, but also from meteorological data reanalysis. The hypothesis states that solar variability affects climate through a bottom-up mechanism from surface changes in irradiance, coupled to a top-down mechanism from stratospheric UV and ozone changes. Chapter 11 includes evidence for a hypothesis explaining how the atmospheric changes might produce intense global climate change capable of explaining the paleoclimatological evidence reviewed here.

Although changes in oceanic circulation have been implicated by some authors in the climatic changes of the Bray cycle cold events, the global nature of these suggests that oceanic changes, although potentially very important, are probably of secondary nature, induced by atmospheric changes in wind patterns.

Analysis of the Holocene climate shows a long term cooling trend punctuated by cold events (Fig. 6.1; see also Chap. 4). For the past seven thousand years, every millennium has been colder on average than the previous one, driven by orbital changes. Obliquity is now decreasing at its fastest rate in 40,000 years and Northern summer insolation is at its minimum value in 20,000 years, an orbital configuration that supports a continuation of the multi-

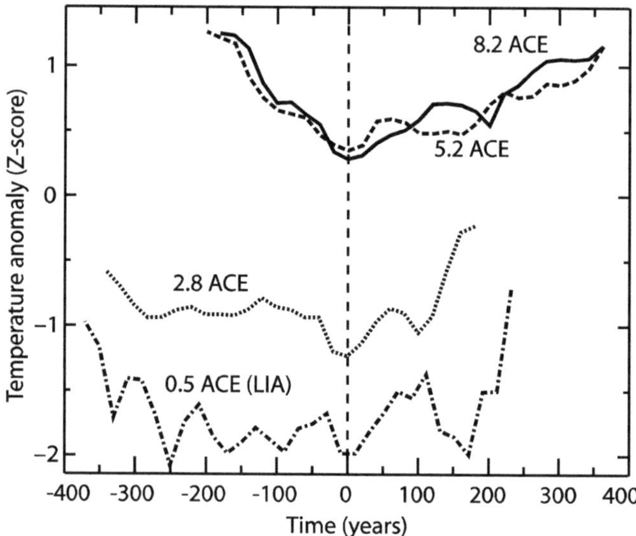

Fig. 6.14 Global temperature change during major Holocene cooling events
Major cooling events from a global temperature reconstruction from 73 proxies. The reconstructed temperature anomaly is expressed as Z-score (see Sect. 4.4). The LIA event curve displayed does not include most of the Modern Global Warming, as only 22% of the proxies reached 1950, leading to statistically weak inferences. Instrumental records show an additional +0.6 °C from the time the reconstruction ends.

millennial cooling trend. Within this background, the Holocene, since reaching the HCO, does not display warming events, as any significant multi-centennial warming period is preceded by a similarly significant multi-centennial cooling period, and Modern Global Warming is no exception, as it is preceded by the LIA.

Comparing the Holocene major cold events that include the lows in the Bray cycle at 0.5, 2.8, and 5.2 kyr, plus the major 8.2-kyr event (Fig 6.14) shows that both cooling and posterior warming can last from 2 to 4 centuries. Given that solar activity usually returns quite quickly to normal levels after a SGM, the slow recovery in temperatures suggests that the atmospheric reorganization induced by the changes in solar activity is a slow process that takes place unforced, and since the entire event usually takes over half a millennia, the climate seems to settle in a different configuration, as the orbital conditions have changed during the time. This could be the reason why the lows in the Bray cycle broadly mark the separation between the different climatic periods of the palynological Blytt–Sernander series (Boreal, Atlantic, Sub-Boreal, Sub-Atlantic periods).

There is paleoclimatological evidence that a poleward atmospheric expansion of the Hadley and Ferrel cells, and associated wind regimes, including the southern westerly winds strengthening and southern displacement associated with persistently positive phases of the Southern Annular Mode (Antarctic Oscillation) has been taking place during the 20th century, as assessed in the Patagonia (Chile; Moreno et al. 2014). The expansion of the Hadley cells, that is usually attributed to ozone depletion, has been measured since 1979 at about 1–2° in latitude. The continuation of the Hadley cells expansion may suggest that natural recovery from the LIA has not ended, since it is believed that greenhouse gases contribute little to this phenomenon (Allen et al. 2012) and it seems to have been taking place for over 100 years. Alternatively, it could be the result of the Modern Solar Maximum that has taken place during the 20th century (1935–2005; see Sect. 11.6).

To the natural warming caused by the post-LIA recovery we have added anthropogenic warming (see Chap. 9). Nowhere is the anthropogenic effect more noticeable than in the status of the cryosphere. Globally glaciers have retreated to a point last seen around 5000 years ago, during the Mid-Holocene Transition at the start of the Neoglacial period. That is the reason why organic remains like Ötzi, the iceman from Tyrol, from 5200 BP are being uncovered all over the world. It is possible that the cryosphere is particularly sensitive to the combination of greenhouse gases increase and light absorbing particles increase from anthropogenic soot. Nevertheless we cannot rule out that the global average temperature is approaching values that took place over 5000 years ago, as Fig. 6.14 suggests.

6.10 Conclusions

6a. Periods of low solar activity that characterize the 2500-year Bray solar cycle coincide in all cases with periods of climate deterioration characterized by global cooling in land and sea, increased iceberg activity, glacier advances, atmospheric changes consistent with equatorward expansion of polar circulation, Hadley cells contraction, and changes in wind and precipitation patterns usually increasing at mid and high latitudes and decreasing at low latitudes with a weakening of the equatorial monsoons.

6b. The periods of abrupt climate change associated to the lows in the Bray solar cycle coincide with periods of crisis for human societies while providing also opportunities for adaptation and advancement, and often coincide with important cultural transitions lending support to the hypothesis that climate change acts as an engine for societal progress.

6c. Despite a clear and intense paleoclimatic effect, changes in solar activity are not properly accounted for in our current understanding of climate forcings due to our ignorance of the underlying physical mechanisms. This underestimation of solar forcing has the inevitable consequence of an overestimation of anthropogenic forcing.

References

Allen RJ, Sherwood SC, Norris JR & Zender CS (2012) Recent Northern Hemisphere tropical expansion primarily driven by black carbon and tropospheric ozone. Nature 485 (7398) 350–354

Ammann B & Fyfe RM (2014) Blytt–Sernander timescale. In: Matthews JA (ed) Encyclopedia of Environmental Change. SAGE, London, p 107. https://doi.org/10.4135/978144624750 1.n457

Arbogast RM, Jacomet S, Magny M & Schibler J (2006) The significance of climate fluctuations for lake level changes and shifts in subsistence economy during the late Neolithic (4300–2400 BC) in central Europe. Vegetation History and Archaeobotany 15 (4) 403–418

Bar–Yosef O (1986) The walls of Jericho: an alternative interpretation. Current Anthropology 27 (2) 157–162

Berger JF, Delhon C, Magnin F et al (2016) A fluvial record of the mid-Holocene rapid climatic changes in the middle Rhone valley (Espeluche–Lalo, France) and of their impact on Late Mesolithic and Early Neolithic societies. Quaternary Science Reviews 136 66–84

Bevan A, Colledge S, Fuller D et al (2017) Holocene fluctuations in human population demonstrate repeated links to food production and climate. Proceedings of the National Academy of Sciences 114 (49) E10524–E10531

Björck S, Muscheler R, Kromer B et al (2001) High-resolution analyses of an early Holocene climate event may imply decreased solar forcing as an important climate trigger. Geology 29 (12) 1107–1110

Blockley S, Candy I, Matthews I et al (2018) The resilience of postglacial hunter-gatherers to abrupt climate change. Nature ecology & evolution 2 (5) 810

Bond G, Kromer B, Beer J et al (2001) Persistent solar influence on North Atlantic climate during the Holocene. Science 294 (5549) 2130–2136

Brandt G, Haak W, Adler CJ et al (2013) Ancient DNA reveals key stages in the formation of central European mitochondrial genetic diversity. Science 342 (6155) 257–261

Brooks N (2012) Beyond collapse: climate change and causality during the Middle Holocene Climatic Transition 6400–5000 years before present. Geografisk Tidsskrift-Danish Journal of Geography 112 (2) 93–104

Büntgen U, Trouet V, Frank D et al (2010) Tree-ring indicators of German summer drought over the last millennium. Quaternary Science Reviews 29 (7–8) 1005–1016

Camenisch C, Keller K, Salvisberg M et al (2016) The Early Spörer Minimum–A Period of Extraordinary Climate and Socio-economic Changes in Western and Central Europe. Climate of the Past 12 2107–2126

Campbell BM & Ó Gráda C (2011) Harvest shortfalls grain prices and famines in preindustrial England. The Journal of Economic History 71 (4) 859–886

Christiansen B & Ljungqvist FC (2012) The extra-tropical Northern Hemisphere temperature in the last two millennia: reconstructions of low-frequency variability. Climate of the Past 8 (2) 765–786

Dahl SO, Nesje A, Lie Ø et al (2002) Timing equilibrium-line altitudes and climatic implications of two early-Holocene glacier readvances during the Erdalen Event at Jostedalsbreen, western Norway. The Holocene 12 (1) 17–25

de Larminat P (2016) Earth climate identification vs. anthropic global warming attribution. Annual Reviews in Control 42 114–125

Dubouloz J (2008) Impacts of the Neolithic demographic transition on Linear Pottery Culture settlement. In: Bocquet–Appel JP and Bar–Yosef O (eds) The Neolithic demographic transition and its consequences. Springer, Dordrecht, p 207–235

Fleitmann D, Burns SJ, Mangini A et al (2007) Holocene ITCZ and Indian monsoon dynamics recorded in stalagmites from Oman and Yemen (Socotra). Quaternary Science Reviews 26 (1–2) 170–188

Fowell SJ, Hansen BC, Peck JA et al (2003) Mid to late Holocene climate evolution of the Lake Telmen Basin, North Central Mongolia, based on palynological data. Quaternary Research 59 (3) 353–363

Gavin DG, Henderson AC, Westover KS et al (2011) Abrupt Holocene climate change and potential response to solar forcing in western Canada. Quaternary Science Reviews 30 (9–10) 1243–1255

Graham RW, Belmecheri S, Choy K et al (2016) Timing and causes of mid-Holocene mammoth extinction on St Paul Island, Alaska. Proceedings of the National Academy of Sciences 113 (33) 9310–9314

Gronenborn D, Strien HC, Dietrich S & Sirocko F (2014) 'Adaptive cycles' and climate fluctuations: a case study from Linear Pottery Culture in western Central Europe. Journal of Archaeological Science 51 73–83

Haak W, Lazaridis I, Patterson N et al (2015) Massive migration from the steppe was a source for Indo-European languages in Europe. Nature 522 (7555) 207–211

Haigh JD (1996) The impact of solar variability on climate. Science 272 (5264) 981–984

Holzhauser H, Magny M & Zumbuühl HJ (2005) Glacier and lake-level variations in west-central Europe over the last 3500 years. The Holocene 15 (6) 789–801

Huybers P & Langmuir C (2009) Feedback between deglaciation, volcanism, and atmospheric CO_2. Earth and Planetary Science Letters 286 (3–4) 479–491

Jackson MG, Oskarsson N, Trønnes RG et al (2005) Holocene loess deposition in Iceland: Evidence for millennial-scale atmosphere-ocean coupling in the North Atlantic. Geology 33 (6) 509–512

Jiang H, Muscheler R, Björck S et al (2015) Solar forcing of Holocene summer sea-surface temperatures in the northern North Atlantic. Geology 43 (3) 203–206

Kaniewski D, Van Campo E, Guiot J et al (2013) Environmental roots of the Late Bronze Age crisis. PLoS One 8 (8) e71004

Kidder TR (2006) Climate change and the Archaic to Woodland transition (3000–2500 cal BP) in the Mississippi River Basin. American Antiquity 71 (2) 195–231

Kobashi T, Severinghaus JP, Brook EJ et al (2007) Precise timing and characterization of abrupt climate change 8200 years ago from air trapped in polar ice. Quaternary Science Reviews 26 (9–10) 1212–1222

Kobashi T, Goto–Azuma K, Box JE et al (2013) Causes of Greenland temperature variability over the past 4000 yr: implications for northern hemispheric temperature changes. Climate of the Past 9 (5) 2299–2317

Kutterolf S, Jegen M, Mitrovica JX et al (2013) A detection of Milankovitch frequencies in global volcanic activity. Geology 41 (2) 227–230

Lamb HH (1995) Climate, history and the modern world. 2nd edn. Routledge, London

Lewis CFM, Miller AAL, Levac E et al (2012) Lake Agassiz outburst age and routing by Labrador Current and the 82 cal ka cold event. Quaternary International 260 83–97

Magny M (2004) Holocene climate variability as reflected by mid-European lake-level fluctuations and its probable impact on prehistoric human settlements. Quaternary international 113 (1) 65–79

Magny M, Leuzinger URS, Bortenschlager S & Haas JN (2006) Tripartite climate reversal in Central Europe 5600–5300 years ago. Quaternary research 65 (1) 3–19

Maley J & Brenac P (1998) Vegetation dynamics, palaeoenvironments and climatic changes in the forests of western Cameroon during the last 28000 years BP. Review of Palaeobotany and Palynology 99 (2) 157–187

Marino G, Rohling EJ, Sangiorgi F et al (2009) Early and middle Holocene in the Aegean Sea: interplay between high and low latitude climate variability. Quaternary Science Reviews 28 (27–28) 3246–3262

Massé G, Rowland SJ, Sicre MA et al (2008) Abrupt climate changes for Iceland during the last millennium: evidence from high resolution sea ice reconstructions. Earth and Planetary Science Letters 269 (3–4) 565–569

Mayewski PA, Rohling EE, Stager JC et al (2004) Holocene climate variability. Quaternary research 62 (3) 243–255

McCracken KG, Beer J, Steinhilber F & Abreu J (2013) A phenomenological study of the cosmic ray variations over the past 9400 years, and their implications regarding solar activity and the solar dynamo. Solar Physics 286 (2) 609–627

Migowski C, Stein M, Prasad S et al (2006) Holocene climate variability and cultural evolution in the Near East from the Dead Sea sedimentary record. Quaternary Research 66 (3) 421–431

Moberg A, Sonechkin DM, Holmgren K et al (2005) Highly variable Northern Hemisphere temperatures reconstructed from low-and high-resolution proxy data. Nature 433 (7026) 613–617

Moreno PI, Vilanova I, Villa–Martínez R et al (2014) Southern Annular Mode-like changes in southwestern Patagonia at centennial timescales over the last three millennia. Nature communications 5 4375

North Greenland Ice Core Project members (2004) High-resolution record of Northern Hemisphere climate extending into the last interglacial period. Nature 431 (7005) 147–151

O'Brien SR, Mayewski PA, Meeker LD et al (1995) Complexity of Holocene climate as reconstructed from a Greenland ice core. Science 270 (5244) 1962–1964

Parker G (2008) Crisis and catastrophe: The global crisis of the seventeenth century reconsidered. The American Historical Review 113 (4) 1053–1079

Polissar PJ, Abbott MB, Wolfe AP et al (2006) Solar modulation of Little Ice Age climate in the tropical Andes. Proceedings of the National Academy of Sciences 103 (24) 8937–8942

Roberts N, Eastwood WJ, Kuzucuoğlu C et al (2011) Climatic vegetation and cultural change in the eastern Mediterranean during the mid-Holocene environmental transition. The Holocene 21 (1) 147–162

Rohling E, Mayewski P, Abu–Zied R et al (2002) Holocene atmosphere-ocean interactions: records from Greenland and the Aegean Sea. Climate Dynamics 18 (7) 587–593

Rohling EJ & Pälike H (2005) Centennial-scale climate cooling with a sudden cold event around 8,200 years ago. Nature 434 (7036) 975–979

Seong YB, Owen LA, Yi C & Finkel RC (2009) Quaternary glaciation of Muztag Ata and Kongur Shan: Evidence for glacier response to rapid climate changes throughout the Late Glacial and Holocene in westernmost Tibet. Geological Society of America Bulletin 121 (3–4) 348–365

Shennan S, Downey SS, Timpson A et al (2013) Regional population collapse followed initial agriculture booms in mid-Holocene Europe. Nature communications 4 2486

Sigl M, Winstrup M, McConnell JR et al (2015) Timing and climate forcing of volcanic eruptions for the past 2500 years. Nature 523 (7562) 543–549

Sinha A, Kathayat G, Weiss H et al (2019) Role of climate in the rise and fall of the Neo-Assyrian Empire. Science advances 5 (11) eaax6656

Solanki SK, Usoskin IG, Kromer B et al (2004) Unusual activity of the Sun during recent decades compared to the previous 11,000 years. Nature 431 (7012) 1084–1087

Thompson LG, Mosley–Thompson E, Brecher H et al (2006) Abrupt tropical climate change: Past and present. Proceedings of the National Academy of Sciences 103 (28) 10536–10543

Usoskin IG, Solanki SK & Kovaltsov GA (2007) Grand minima and maxima of solar activity: new observational constraints. Astronomy & Astrophysics 471 (1) 301–309

Usoskin IG, Gallet Y, Lopes F et al (2016) Solar activity during the Holocene: the Hallstatt cycle and its consequence for grand minima and maxima. Astronomy & Astrophysics 587 A150

van Geel B, Raspopov OM, van der Plicht J & Renssen H (1998) Solar forcing of abrupt climate change around 850 calendar years BC. In: Peiser BJ, Palmer T and Bailey ME (eds) Natural Catastrophes During Bronze Age Civilisations: Archaeological Geological Astronomical and Cultural Perspectives. British Archaeological Reports, Oxford, p 162–168

Versteegh GJM, De Leeuw JW, Taricco C & Romero A (2007) Temperature and productivity influences on U37K′ and their possible relation to solar forcing of the Mediterranean winter. Geochemistry, Geophysics, Geosystems 8 (9) Q09005

Walker M, Head MH, Berklehammer M et al (2018) Formal ratification of the subdivision of the Holocene Series/Epoch (Quaternary System/Period): two new Global Boundary Stratotype Sections and Points (GSSPs) and three new stages/subseries. Episodes 41 (4) 213–224

Wang Y, Cheng H, Edwards RL et al (2005) The Holocene Asian monsoon: links to solar changes and North Atlantic climate Science. 308 (5723) 854–857

Weninger B, Clare L, Rohling E et al (2009) The impact of rapid climate change on prehistoric societies during the Holocene in the Eastern Mediterranean. Documenta Praehistorica 36 7–59

Weninger B, Clare L, Gerritsen F et al (2014) Neolithisation of the Aegean and Southeast Europe during the 6600–6000 calBC period of Rapid Climate Change. Documenta Praehistorica 41 1–31

Wu CJ, Usoskin IG, Krivova N et al (2018) Solar activity over nine millennia: A consistent multi-proxy reconstruction. Astronomy & Astrophysics 615 A93

Zhang DD, Lee HF, Wang C et al (2011) Climate change and large-scale human population collapses in the pre-industrial era. Global Ecology and Biogeography 20 (4) 520–531

Zielinski GA, Mayewski PA, Meeker LD et al (1996) A 110,000-yr record of explosive volcanism from the GISP2 (Greenland) ice core. Quaternary Research 45 (2) 109–118

THE ELUSIVE 1500-YEAR HOLOCENE CYCLE

*"Of course there's climate change. Climate change is part of the normal order of things,
and we know it was happening before humans came."*
Freeman Dyson (2007)

7.1 Introduction

The discovery of the c. 1500-yr glacial Dansgaard-Oechsger (D–O) cycle in Greenland ice cores (see Chap. 3), started a quest to identify this cycle in Holocene records, where its absence was puzzling. In 1995 Gerard Bond and Rusty Lotti demonstrated in deep sea cores that glacial-time D–O events matched periods of increased iceberg activity and that they were linked to Heinrich events (periods of very high iceberg activity in the North Atlantic). The records showed 3–4 D–O events taking place between Heinrich events. This oceanic sedimentary pattern from the glacial period became known as the "Bond cycle." Since 1996 researchers started to report a 1500-yr periodicity in Holocene proxies, and it became Bond's goal to extend his cycle to the Holocene. He claimed to have achieved that in 2001 when he reported his famous drift-ice petrological record that revealed a series of cold events in the Holocene (Bond et al. 2001). Bond's report sparked a paradigm shift because at the time the Holocene was generally viewed as climatically stable, despite contrary evidence from glacier studies. We have seen that Bond's identification of a Holocene cycle was wrong (see Fig. 4.17), and his misleading numbering of the cold events to try to fit them into a 1500-yr series has caused unnecessary confusion in the field. Researchers all over the world have tried to match negative temperature anomalies and even precipitation fluctuations in their proxy records to Bond events, perpetuating the myth. As the evidence is contradictory among reports, the existence of a 1500-yr cycle in the Holocene has become more contentious with time. In the words of Raimund Muscheler (2012) it is *"the enigmatic 1,500-year cycle,"* and according to Maxime Debret et al. (2007) *"one of the outstanding puzzles of climate variability."* After an extensive research Heinz Wanner et al. (2011) concluded that *"multi-century cold events were not strictly regular or cyclic, and one single process cannot explain their complex spatio-temporal pattern."*

Could Bond be wrong, Wanner be right, and still there be a manifestation of the D–O cycle in the Holocene? That is what the evidence supports, and the nature of the evidence points once more to the process that is the likely cause of this climatic cycle.

7.2 What must we expect of a Holocene 1500-year cycle?

Over the last decade we have greatly increased our knowledge of the D–O cycle (see Chap. 3). We know that D–O events are abrupt warming events, that take place at the North Atlantic–Nordic seas area. They appear to take place when the water stratification is abruptly disturbed, allowing a surge of warm subsurface waters that have accumulated for long periods of time, to melt surface ice and release heat to the atmosphere (see Fig. 3.14). They require high ice conditions and low sea level, between 35 and 100 m below the present level (see Fig. 3.10). Their trigger presents a double periodicity of 4.8 and 3.0 kyr (see Fig. 3.9). The 3.0-kyr periodicity is sustained by the sequence spanning Greenland Interstadials (GI) 5.2-7-8-10, of c. 9,000 years in a set of three oscillations, and also by the c. 3.0-kyr distance between GI 19.219.1, GI 17.1–15.2, and GI 1–0 (see Sect. 3.4). If there exists a 1500-yr cycle in the Holocene related to the D–O cycle, it could be due to a sub-harmonic periodicity related to the found 3-kyr periodicity, or both could be harmonics of a shorter elemental frequency. Wolfgang Berger and Ulrich von Rad (2002) proposed that the 1500-yr cycle is a harmonic of the beat between the moon's nodal and apsidal precessions, a hypothesis that fits not only the observed period, but also the required mechanism for vertical water mixing through tidal forcing.

As described, D–O events cannot take place during the Holocene. It is too warm, there is too little ice, and sea level is too high. New D–O events will have to wait until the next glacial period. However, if the trigger is astronomical, its clock is ticking all the time, and it is fair to ask if it might have other effects that could cause a 1500-yr climate cycle during the Holocene. We already know some characteristics that the 1500-yr cycle should present: It should take place when the trigger indicates, and thus in phase with the D–O cycle; its manifestation should be compatible with a mechanism that promotes vertical water mixing (mainly wind and tidal forces); and it might not be due to internal variability of the climate. Given that the warming nature of the D–O events relies on the presence of abundant warm subsurface waters in a specific region, the 1500-yr Holocene cycle doesn't have to be a warming cycle. In fact, if the trigger promotes vertical water mixing, in most places this would mean sea surface cooling. To this author's knowledge no researcher has set any re-

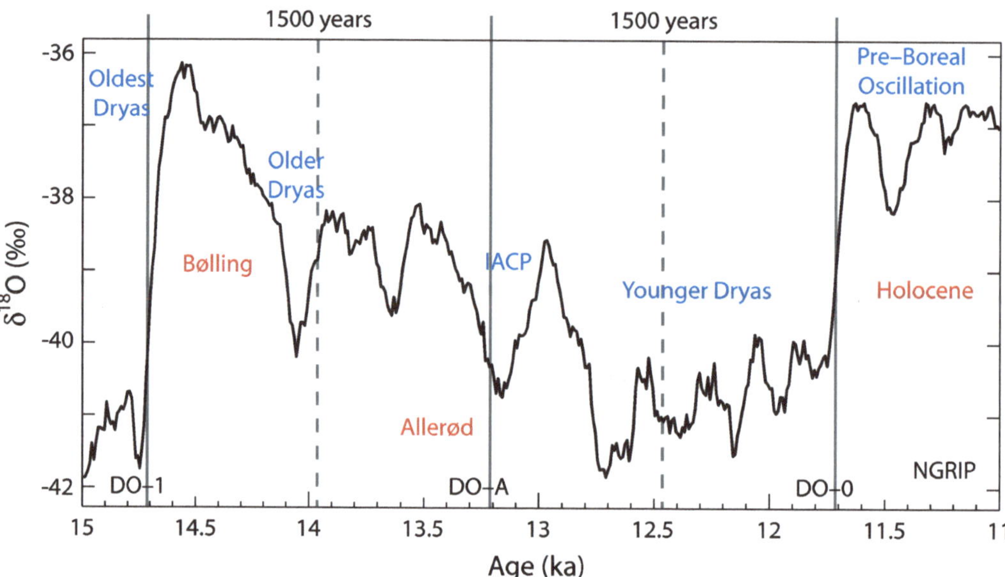

Fig. 7.1 The Dansgaard–Oechsger cycle at the end of the last glacial period
80-yr averaged NGRIP $\delta^{18}O$ proxy for Greenland temperature on the GICC05 timescale, after Seierstad et al. (2014). Proposed D–O events are indicated with continuous grey lines. Dashed grey lines represents the 750-yr harmonic of the D–O periodicity. Cold periods are indicated above the curve, and warm periods below. IACP: Intra-Allerød cold period.

quirements on the 1500-yr cycle, except the periodicity, and since almost every proxy presents some periodicity, often for no obvious reason, several of the reported 1500-yr proxy periodicities might not correspond to the cycle that caused D–O events during glacial periods.

The next question is where to look for the 1500-yr cycle. As Bond identified his glacial cycle in North Atlantic sediments, he set out to identify the 1500-yr Holocene cycle there too. However, as we saw with the 2500-yr Bray cycle (see Chap. 5), the North Atlantic is especially sensitive to low solar activity (see Fig. 5.17), and therefore, most Bond events represent solar-induced cooling events, as Bond et al. (2001) highlighted. Some of the peaks in the Bond series cannot be assigned to solar activity and could represent cooling from the 1500-yr cycle. In fact, it has been shown that during the Neoglacial the Bond series can be better fitted to a 1500-yr periodicity (Debret et al. 2007; see Fig. 4.17). However, given the strength of the solar variability signal in the North Atlantic, proxy records from that area usually present mixed periodicities more difficult to interpret. This has created another problem, as most authors have looked for the cycle in the North Atlantic area, following Bond's steps. The result has been more confusion in the scientific literature and some claims that the 1500-yr cycle was perhaps of solar origin.

Before reviewing the evidence for the 1500-yr cycle let's see where the D–O cycle stands at the start of the Holocene. According to Rahmstorf (2003) the abrupt warming changes that started the Bølling oscillation and the Holocene correspond to D–O events 1 and 0 respectively, and they are separated by 3000 years (Fig. 7.1). He also places another event in the middle (DO–A) as possibly responsible for the warming after the Intra-Allerød cold period. To check the time relationship between proxy records and the D–O clock, the D–O periodicity observed in the 15,000–11,000 year BP interval has been projected to the rest of the Holocene.

7.3 The 1500-year periodicity during the Holocene

For the last 20 years researchers have been reporting a c. 1500-yr periodicity in Holocene climate proxies. In 2007 Maxime Debret et al. carried out a wavelet analysis of some of these proxy records. The wavelet technique allows to determine two-dimensionally not only the periodicities present in the record, but also the times at which those periodicities are found. Debret et al. (2007) conclude from their wavelet analysis that the Holocene millennial variability is composed of three main periodicities of c. 1000, 1500, and 2500 years. Based on the coincidence of the 1000 and 2500-yr cycles with the wavelet analysis profile of ^{14}C and ^{10}Be production rates they defend their solar origin. They assign an oceanic origin for the 1500-yr periodicity because it is absent from solar proxies and present in oceanic proxies. The wavelet analysis also shows that in some proxies the c. 1500-year periodicity is not continuous through the Holocene, being absent or very attenuated during the early Holocene, while in other records this signal appears a few thousand years earlier. However by examining the frequency displayed by these proxies it is clear that they cannot correspond to the same periodicity, as some are c. 1400 years and others c. 1600 years. Over the Holocene a 1500-yr periodicity would present seven oscillations, and a difference of only 100 years would cause a phase shift of 700 years, incompatible with being different manifestations from the same cycle.

The lack of a 1500-yr solar periodicity has not deterred some authors from attributing the 1500-yr periodicity found in oceanic proxies to a more indirect solar cause. Dima & Lohman (2008) proposed that the 1500-yr periodicity arises due to a threshold response of the Atlantic meridional overturning circulation to a low-frequency solar forcing. Ruzmaikin & Feynman (2011), after analyzing several Atlantic proxies, found the periodicity highly nonlinear and unstable but not purely random, and proposed a similar mechanism based on the centennial solar cycle. The evidence that certain climate proxies present a c. 1500-yr periodicity is clear, but many authors doubt that they reflect a true stationary periodical oscillation caused by internal variability or external forcing (Obrochta et al.

2012). Hence, the identification of a persistent 1500-yr periodicity remains controversial both for the last glacial episode and the Holocene.

The approach taken here is to avoid North Atlantic proxy records, heavily influenced by solar forcing, due to the climatic variability hotspot character of the region. It is also assumed that the cycle is coherent from the glacial period to the Holocene, maintaining the period and phase. Proxies presenting a c. 1500-yr periodicity are required to match the 1500-yr D–O pattern shown in Fig. 7.1 extended into the Holocene. Vertical lines in figures will mark exact 1500 years periods since 11,700 BP, with dashed vertical lines indicating the midpoints. This strict requirement ensures that the proxies identified are all oscillating with the same phase and matching the D–O phase through the Holocene.

7.4 The oceanic 1500-year cycle

Among the possible explanations for the 1500-yr cycle proposed by different authors, most appear to attribute it to an oceanic oscillation, whether externally forced or due to internal variability. However most oceanic proxies that contain a 1500-yr frequency signal display a very complex pattern, indicating that the proxy is affected by other climate cycles and changes, and precluding a clear identification of the 1500-yr cycle.

Sea-surface temperature (SST), is affected by changes in wind patterns, wind strength, insolation, cloud cover, pressure, and precipitation patterns, among other factors. Thus, most SST proxy records, especially those from the North Atlantic, do not display a recognizable 1500-yr cy-

cle. One exception is the northwestern Pacific SST alkenone-based proxy record from the coast off Japan reported by Isono et al. (2009; Fig 7.2a). Although the temperature reconstruction does present a 1500-yr periodicity that matches the D–O pattern, the match presents a clear 180° phase shift at 7 kyr BP (Fig. 7.2a, black bar), when higher temperatures went from taking place at mid-cycle before the shift, to lower temperatures taking place at mid-cycle afterwards.

Water temperature right above the thermocline (usually 50–100 m deep) is less affected by precipitation and insolation changes, and can be determined by the oxygen isotopic composition, or the Mg/Ca ratio, of sedimented shells from foraminifera that inhabit that zone. Wang et al. (2016) determined the ^{18}O variations from the foram *Pulleniatina obliquiloculata*, a thermocline dweller, in a marine sediment core from the Okinawa Trough, where the Kuroshio Current transports warm waters from the tropics to higher latitudes. The sea subsurface temperature proxy displays a very clear 1500-yr cycle in phase with the D–O cycle, where for the past 7000 years higher temperatures at the thermocline took place at the times determined by the D–O periodicity (Fig. 7.2b). Curiously this proxy also seems to present a 180° phase shift around 8000 years ago, as prior to that date it was lower temperatures and not higher ones that coincided with the D–O periodicity (Fig. 7.2b, black bar). And it is not only the temperature, but also the strength of the Kuroshio Current that appears to be affected by the 1500-yr cycle, as this periodicity is found also in the 5–18 μm particle fraction that originates from Taiwan rivers and is transported north by the current into the Okinawa Trough (Zheng et al. 2016; Fig. 7.2c). At

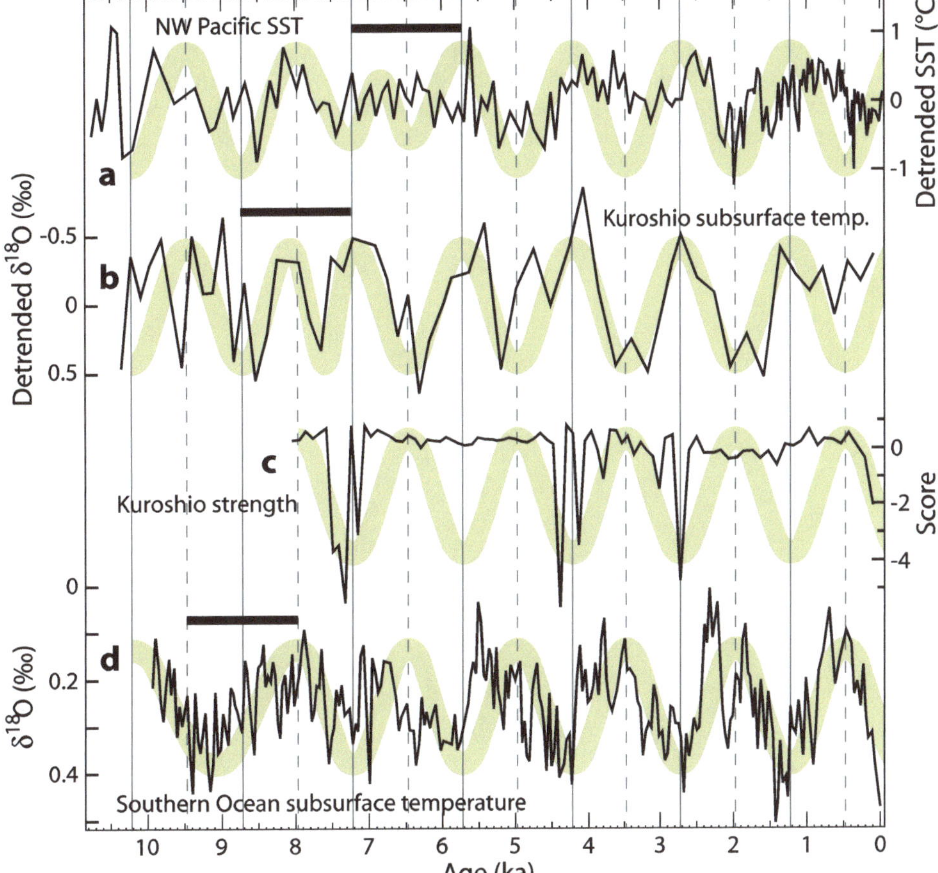

Fig. 7.2 Oceanic proxy records displaying the 1500-year cycle a) Variations in detrended alkenone-derived SST from a core off the coast of central Japan in the northwestern Pacific. After Isono et al. (2009). **b)** Variations of oxygen isotope signature from *Pulleniatina obliquiloculata* calcite in a sediment core at the Okinawa Trough at the southeastern part of the East China Sea, as a proxy for top of the thermocline temperature. After Wang et al. (2016). **c)** Marine sediment core 5–18 μm siliciclastic fraction originated from Taiwan rivers and transported by the Kuroshio Current to the Okinawa Trough as a proxy for the strength of the Kuroshio Current. After Zheng et al. (2016). **d)** δ¹⁸O measurements from *Globigerina bulloides* from a Murray Canyon (Great Australian Bight) marine sediment core as a proxy for intermediate water temperature. After Moros et al. (2009). D–O periodicity (see Fig. 7.1) is indicated with vertical continuous grey lines. Dashed grey lines represent the 750-yr harmonic of the D–O periodicity. Black horizontal bars: Periods with an apparent 180° phase shift in data periodicity. Thick sinusoidal curve: 1500-yr periodicity accounting for the shift.

some of the times that coincide with the D–O periodicity the presence of this fraction collapses, indicating a sudden drop in the current strength, probably due to the Kuroshio Current not entering the Okinawa Trough.

The effect of the D–O cycle on halocline water temperature was already reviewed in Chap. 3 (see Fig. 3.14; Dokken et al. 2013), and was one of the arguments used to support its possible tidal nature. It is therefore very interesting that the Holocene 1500-yr cycle displays a similar manifestation, not only in the Northwest Pacific, but also in the Southern Ocean. Moros et al. (2009) determined thermocline water temperature changes during the Holocene at Murray Canyon (Great Australian Bight) using the oxygen isotopic signature of another thermocline foram, *Globigerina bulloides*. Interestingly the record not only displays a clear 1500-yr periodicity in phase with the D–O cycle (Fig. 7.2d), but it also has a characteristic saw-tooth aspect quite similar to the reported proxy record from the last glacial period in the Norwegian Sea (see Fig. 3.14). And again, we find a period between 9–8,000 years BP when the record displays a 180° phase shift (Fig. 7.2d, black bar).

Although these oceanic proxy records are all very consistent in displaying an in-phase 1500-yr periodicity, they don't give a consistent temperature signal. The NW Pacific SST proxy and the Kuroshio Current subsurface temperature proxy display a warming signal at the specified D–O periodicity, while the Southern Ocean proxy displays the opposite signal. The warming nature of the glacial D–O cycle appears to respond to the specific conditions of warm water accumulation below a fresh cold thermocline layer in the Nordic Seas and is not intrinsic to the cycle nature. At the position where the SST were determined by Isono et al. (2009; Fig. 7.2a) the warm Kuroshio Current and the cold Oyashio Current mix, while the Great Australian Bight record site is close to the present northern limit of sub-Antarctic water. Therefore, the temperature response to the 1500-yr pacing might be determined by regional or hemispheric climatic conditions and

thus the cycle does not appear to convey a temperature signal by itself.

7.5 The atmospheric 1500-year cycle

In 1997 Paul Mayewski and colleagues reported that the chemical ions found in GISP2 ice core presented a 1450-yr periodicity during the last glacial period, that extended into the Holocene. The chemical species that precipitate on polar snow are introduced into the atmosphere by sea salt aerosols and continental dust, and their abundance depends on atmospheric conditions. Using the relative abundance of these chemical tracers, Mayewski et al. (1997) reconstructed a polar circulation index (PCI) that provides a relative measure of the average size and intensity of polar atmospheric circulation. In general terms, PCI values increase (e.g., more continental dusts and marine contributions) during colder portions of the record (stadials) and decrease during warmer periods (interstadials and interglacials). Although the amplitude of the PCI changes decreases markedly during the Holocene, due to its much tamer climate variability, the 1450-yr cycle persists in the record. This indicates that the nature of the periodicity is the same, and that the increase in PCI that accompanies the cycle is associated with more active atmospheric circulation, due to colder winter conditions. A periodogram for the last 11,500 years shows that not only the 1450-yr peak is significant at the >99% level, but also the 725 and 2900-year harmonics, and that neither of these peaks can be assigned to ^{14}C production variability indicative of a solar origin (Mayewski et al. 1997; Fig. 7.3).

Another atmospheric-linked proxy record that displays a very clear, in phase, 1500-yr periodicity comes from the Arabian Sea, where dust from the Arabian desert containing rare-earth elements (REE) is deposited after being transported by northwesterlies (Sirocko et al. 1996). Determination of the variability of a group of these REE in a marine core efficiently balances analytical errors for each of them, and the resulting REE-score shows a very significant 1500-yr periodicity that is in phase with the D–

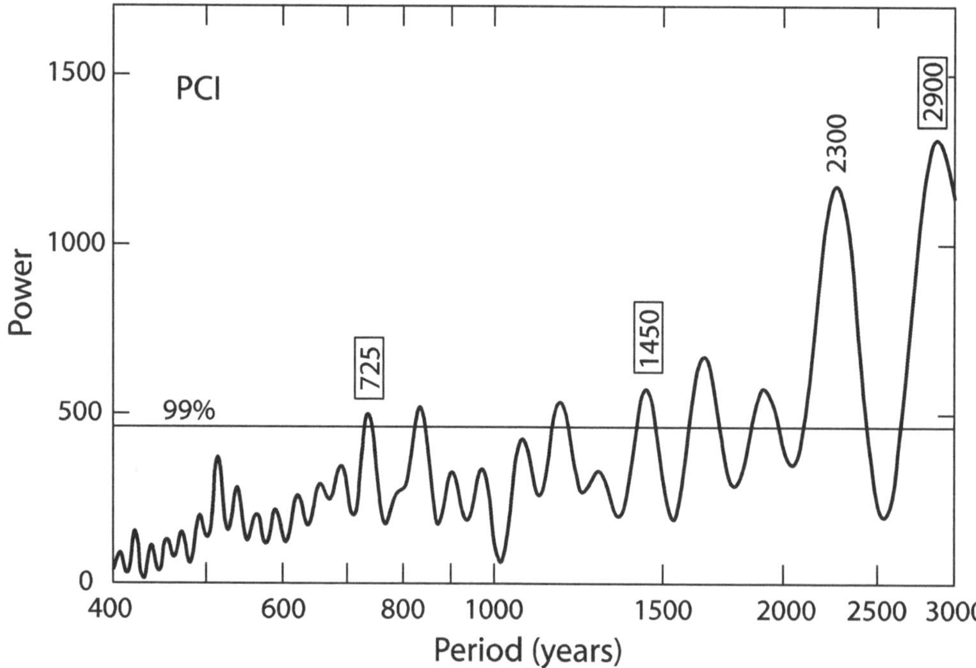

Fig. 7.3 Power spectra for the Polar Circulation Index during the Holocene
PCI series covering the last 11,500 years. Spectral peaks above the horizontal line are significant at the >99% significance level. After Mayewski et al. (1997). The 725 and 1450-yr peaks are sub-harmonics of the much larger 2900-yr peak that likely corresponds to the 3.0 kyr frequency found in the D–O cycle (see Chap. 3). The 2300-yr peak likely corresponds to the 2500-yr solar cycle, affected by cycle irregularities (see Chap. 5).

Fig. 7.4 Atmospheric proxies for the 1500-year cycle
a) Changes in Arabian dust rare-earth elements abundance in an Arabian Sea sediment core. Black bar: Period with an apparent 180° phase shift in data periodicity. After Sirocko et al. (1996). **b)** Isothermal Remanent Magnetization as a moisture proxy in two lake cores (WL 03-1 continuous, WL 03-2 dotted) from White Lake, New Jersey, USA. After Li et al. (2007). Pink vertical bar: Position of the 4.2 kyr arid event. D–O periodicity (see Fig. 7.1) is indicated with vertical continuous grey lines. Dashed grey lines represent the 750-yr harmonic of the D–O periodicity. Black horizontal bar: Period with an apparent 180° phase shift in data periodicity. Thick sinusoidal curve: 1500-yr periodicity.

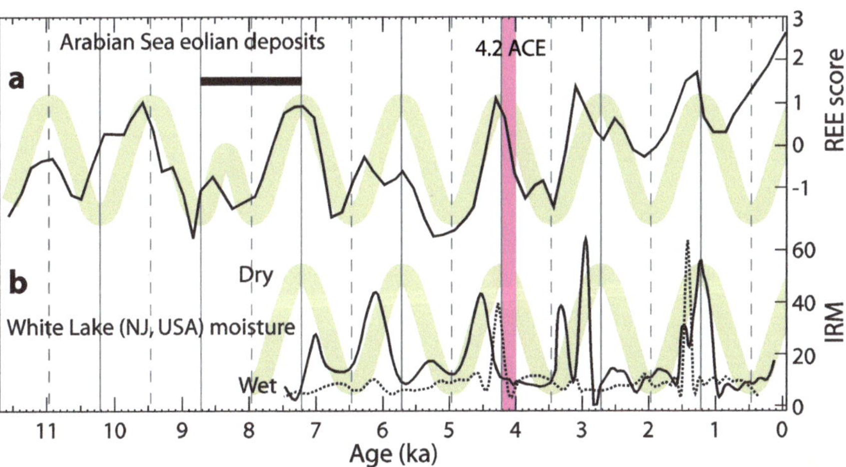

O cycle (Fig. 7.4a). In addition, this proxy record displays a 1500-yr cycle continuously for the past 19 kyr (11.7 kyr shown in Fig. 7.4a) supporting that the pacing mechanism is always ticking and not affected by the drastic climatic changes that took place between 19–11 kyr ago. Furthermore, for the first 10,000 years (19–9 kyr BP) the increase in dust took place at mid-cycle, while for the last 8,000 years the mid-cycle showed a decrease in dust. Between 9 and 8 kyr BP the cycle displayed a 180° phase shift (Fig. 7.4a, black bar) as in previous cases. This phase shift is probably affecting the cycle periodicity determination, reported as 1450 years (Sirocko et al. 1996). When plotted, the oscillations are clearly spaced at 1500 years on both sides of the shift, but the introduction of a half length oscillation, to produce the 180° phase shift, changes the average to c. 1450 years.

The cyclic increase in Arabian REE-containing dust probably reflects an increase in northwesterlies strength, although a cyclic increase in aridity reflected in a higher dust production cannot be ruled out. In fact, evidence of the association between the 1500-yr cycle and precipitation has been found in the Mid-Atlantic region of the US, where two lake sediment cores show that periods of low lake levels took place at a 1500-yr periodicity (Li et al. 2007; Fig. 7.4b). Exposure of marls due to low lake levels led to their oxidization and magnetic intensity increase, followed by their transport and re-sedimentation. Despite the age-depth model imprecision of this proxy and both cores showing different peaks, the coincidence of the peaks with the D–O periodicity appears clearly within dating error. The question of the possible 1500-yr cycle association with a precipitation cycle in certain regions is an interesting one that deserves more research, as we approach a new cycle peak (c. AD 2180) and precipitation is so critical to our society.

7.6 The 4.2 kyr event

At about 4,200 yr BP an abrupt climatic event (ACE) took place that had a strong aridity effect at middle and low latitudes in Africa, the Middle East and southern Asia. The intense drought reduced precipitation by about 30% for about 100–200 years likely causing the end of the Egyptian Old Kingdom, the collapse of the Akkadian Empire in Mesopotamia, and initiated the dispersion of the urban Harappan civilization in the Indus Valley. The 4.2-kyr ACE is also seen throughout the Northern Hemisphere but in a more complex and irregular manner, unlike most Holocene cold events. Although intense cooling is detected in Iceland lake sediments at 4.2 kyr BP (Geirsdóttir et al. 2013), it is brief and completely reversed in about 100 years. Glacier advances are also recorded at the time in Central Asia, the Southern Hemisphere and North America (Mayewski et al. 2004). Interestingly, the 4.2-kyr ACE is also seen in the GISP2 Greenland ice core. It shows as a significant drop in chlorides (sea salt) concentration (a sea-ice proxy; Mayewski & White 2012), unlike most cold events of the Holocene, suggesting that the cold might have been accompanied by reduced precipitation.

Proxies indicate that the 4.2-kyr ACE is centered in the Arabian sea region, affecting the East African and Asian monsoons, the Mediterranean and Southern Europe, with a smaller effect on the North Atlantic region and South America, while the cooling appears global. A Kilimanjaro (East Africa) ice core presents a 200-fold increase in dust particles at the time (Thompson et al. 2002; Fig. 7.5a), while a marine sediment core in the Gulf of Oman presents a 10-fold increase in wind transported dolomite from the Mesopotamian region (Cullen et al. 2000; Fig. 7.5b).

Tierney at al. (2011) analyzed in detail the hydrology of Lake Challa, close to Kilimanjaro. One of the proxies they used was the proportion of deuterium in lake sediment plant leaf waxes, interpreted as a proxy for the strength of the East African monsoon. While other proxies indicate Lake Challa did not have low lake levels at the time, δDwax indicates the monsoon decoupled at the time from the total rainfall amount in the local basin. The East African monsoon showed at the time its weakest values in the entire Holocene (Tierney et al. 2011; Fig. 7.5c). The 4.2-kyr ACE coincides also with a period of weakness of the Asian monsoon (see Fig. 4.18). The general monsoonal weakness during the 4.2-kyr ACE must have contributed to its unusual aridity.

We must conclude that the 4.2-kyr event is a uniquely abrupt regional arid event that also caused global cooling. Regional proxies that show it best do not display a clear periodicity, indicating that Holocene climate cycles were not the cause. Furthermore, regional proxies support the unique nature of the 4.2-kyr ACE within the Holocene in terms of dust and mineral production (Fig. 7.5a, b). The strong monsoon weakening and severe regional aridification (Fig. 7.5c) are different to the rest of Holocene cooling events and underscore a primary atmospheric manifes-

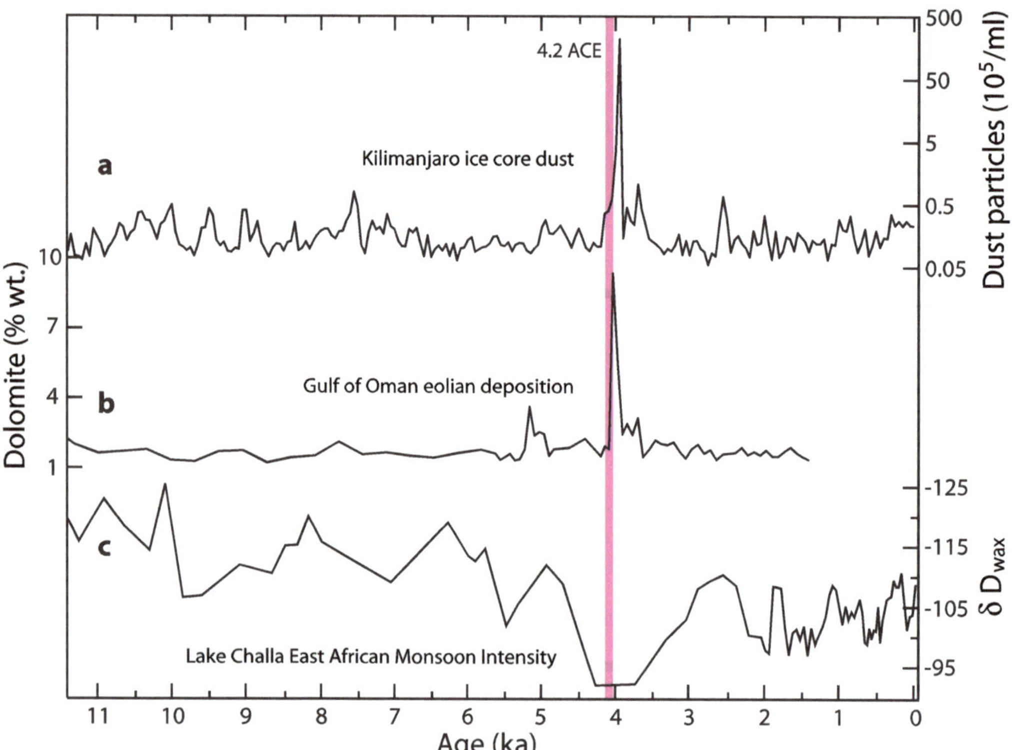

Fig. 7.5 The 4.2 kyr abrupt climatic event a) 50-year average of the Holocene dust history from Kilimanjaro ice core NIF3. Dust concentration measured as 0.63–16.0 μm diameter particles per ml sample. After Thompson et al. (2002). **b)** Gulf of Oman core M5-422 changes in dolomite which reflect eolian mineral supply from Mesopotamian sources. After Cullen et al. (2000). **c)** Deuterium changes in sedimentary plant leaf wax δDwax measured as ‰ vs. Vienna standard, as a proxy for East African monsoon strength at Lake Challa (Kenya). After Tierney et al. (2011). Pink vertical bar: Position of the 4.2 kyr arid event.

tation. Its cause is a complete mystery. Most authors talk about shifts and thresholds in oceanic/atmospheric systems. No big volcanic eruption or asteroid impact capable of such global effect has been convincingly linked to the event, although the abruptness, nature and development of the arid-cold event is compatible with a big tropical volcanic eruption or asteroid impact. Since 1998 soil scientist Marie–Agnès Courty has been defending that soil microfabrics bear the signature of a cosmic impact at the time (Courty et al. 2008). However, the lack of more substantive evidence, like iridium, nickel or platinum spikes, or a well-dated crater, has made her research largely ignored.

Whatever its cause, the 4.2-kyr event had a brutal impact on human societies, crippling the most advanced civilizations at the time and changing the course of history. The world is now 100 times more populated and, despite civilization advances, no less vulnerable to the effects of the changes described. The success of the Akkadian empire was partly due to the sophisticated measures (at the time) they implemented to cope with recurrent droughts in the region. They were just unprepared for the unimaginable scale of what came their way. A Sumerian literary text, "The cursing of Agade" (Akkad; Black et al. 1998), perhaps constitutes the first written source describing a widespread climate catastrophe. It narrates the confrontation of king Naram–Sin of Akkad (2254–2218 BC) with Enlil, god of wind and storms.

"As if it had been before the time when cities were built and founded, the large arable tracts yielded no grain, the inundated tracts yielded no fish, the irrigated orchards yielded no syrup or wine, the thick clouds did not rain, the macgurum plant did not grow. In those days, oil for one shekel was only half a liter, grain for one shekel was only half a liter, wool for one shekel was only one mina, fish for one shekel filled only one ban measure -- these sold at such prices in the markets of the cities! Those who lay down on the roof, died on the roof; those who lay down in the house were not buried. People were flailing at themselves from hunger".

The 4.2-kyr ACE has been chosen as a convenient point to divide the Holocene by the International Union of Geological Sciences, separating the Northgrippian and Meghalayan stages at 4250 b2k (4200 BP; Walker et al. 2018). It is a questionable choice (Voosen 2018), as the event was mainly regional, with hemispheric repercussions, and it does not constitute a geological or climatic divide. The 5.2-kyr ACE (Thompson et al. 2006) can be considered the start of the Neoglaciation, and had global manifestations. It can be argued that constitutes a better choice, but it does not neatly divide the post-Greenlandian period into two equal parts.

7.7 Storminess, drift ice and tidal effects

The proposed lunisolar tidal basis for the 1500-yr cycle (see Sect. 3.9) has an outstanding prediction. One of the most salient effects of high tides is that they multiply the water rise due to storms (storm surge). Hurricane Sandy in New York City, 2012, had a storm surge of 4.2 m (14 ft) due to high tides at the time. By analyzing past storminess records we should be able to detect the effect of the 1500-yr cycle if indeed it is a tidal cycle. The difficulty is that storms have a random nature and it is necessary to combine multiple records from different locations. Sorrel et al. (2012) analyzed high-energy estuarine and coastal sedimentary records from the macrotidal Seine Estuary and Mont-Saint–Michel Bay in the southern coast of the English Channel and defined five Holocene storm periods that also reflected periods of high storm activity at other northern European locations (Sorrel et al. 2012; Fig. 7.6a). According to the authors, these periods of high storm activity occurred periodically with a frequency of about 1,500

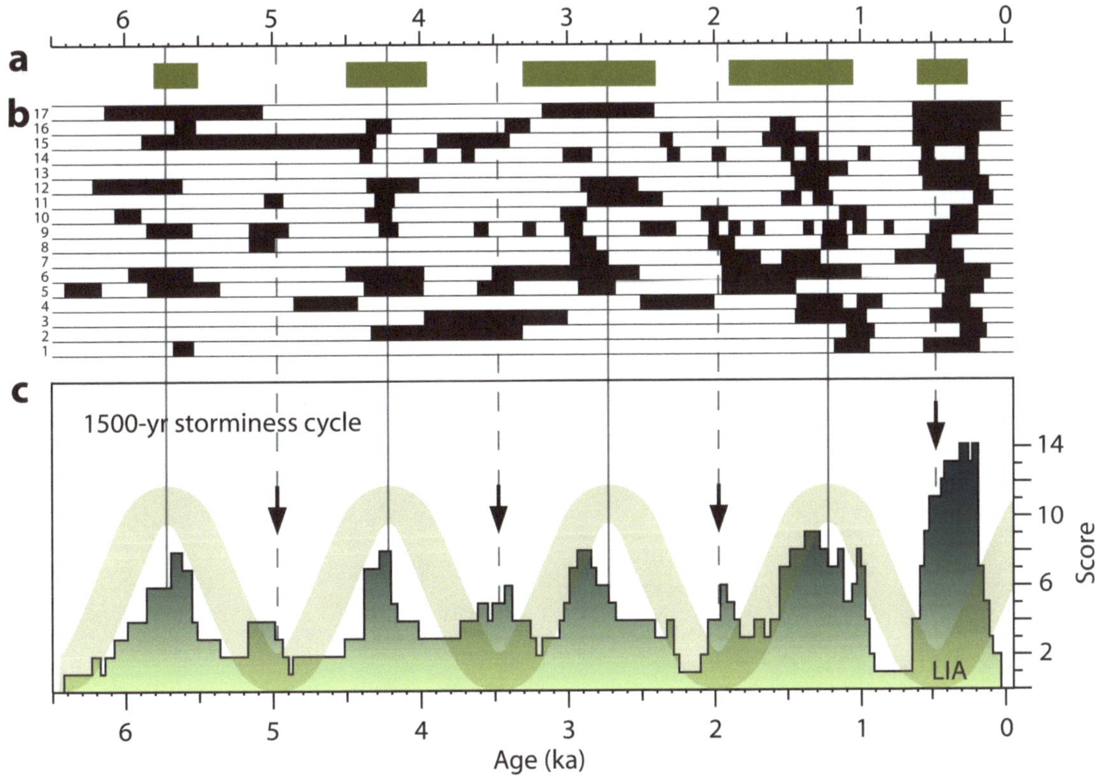

Fig. 7.6 The 1500-year storminess cycle
a) Five Holocene widespread storm intervals (boxes) defined on the basis of nine independently dated records of northern coastal Europe storm activity. After Sorrel et al. (2012). **b)** Periods of high storm activity (black bars) from seventeen independent storminess studies for the North Atlantic and Western Mediterranean compiled from the literature by Costas et al. (2016). The numbers correspond to the references given in Costas et al. (2016) their figure 10. **c)** Score of the number of studies in (b) that identify any point in time as belonging to a high storm activity period. This storminess meta-figure reveals the 1500-yr cycle. D–O periodicity (see Fig. 7.1), indicated with vertical continuous grey lines. Dashed grey lines represent the 750-yr harmonic of the D–O periodicity. Thick sinusoidal curve: 1500-yr periodicity. Arrows indicate storminess power at the 750-yr harmonic.

years, closely related to cold and windy periods identified by Bond et al. (2001), and Wanner et al. (2011).

The temporal resolution of the Sorrel et al. (2012) findings can be improved if we use more records. Costas et al. (2016), in their SW Europe Holocene windiness study, cite 17 storminess studies from Iceland and Scandinavia to the western Mediterranean (Costas et al. 2016, their figure 10; Fig. 7.6b). The result of scoring these 17 studies is displayed in Fig. 7.6c, showing that the storminess cycle has not only a 1500-yr periodicity, but the periodicity is coherent with the D–O cycle. This semiquantitative reconstruction shows an increase in storminess levels with time. This increase could simply be due to a better preservation of more recent records or alternatively reflect the effect of the increasingly colder Neoglacial period on cyclone frequency. This interpretation is supported by the high frequency of storms during the Little Ice Age, the coldest period of the Holocene.

The tidal hypothesis for the 1500-yr cycle is strengthened by the storminess evidence. As expected, the cycle displays a clear association with storminess intensity. In the tidal hypothesis, as proposed by Berger & von Rad (2002), the 1500-yr periodicity is a harmonic of the beat between the moon's nodal and apsidal precession. This configuration allows the manifestation of power at the other harmonic periodicities, and indeed that is the case, as storminess also displays an increase at the 750-year periodicity (Fig. 7.6c arrows), and the LIA storminess maximum coincides with this half cycle periodicity.

Wolfgang Berger proposed the tidal hypothesis of the 1500-yr cycle after studying the varved sediments in an oxygen-minimum zone of the continental slope off the coast of Pakistan in the Arabian Sea. Varve thickness and the presence of turbidites (sedimentary storm deposits) display a large proportion of multiples of the basic tidal cycles of the lunar perigee and the lunar half-nodal. Three of the four longest cycles detected in the 5000-year sedimentary record are 366, 490 and 750 years which are one fourth, one third, and half the 1500-yr cycle (Berger & von Rad 2002). The authors link the tidal cycle to Bond events of increased drift ice through a periodical removal of marine-based glacial ice from the shelves by unusually high tidal waves.

This conjecture might have support from drift ice data off the coast of Alaska obtained by Darby et al. (2012), although the authors appear unaware of Berger's work. Darby and colleagues assessed patterns of sea-ice drift in the Arctic Ocean over the past 8,000 years by geochemically determining the source of ice-rafted iron grains in a sediment core off the coast of Alaska. They identified pulses of sediment carried by sea ice from the Kara Sea, that display a 1500-yr periodicity (Darby et al. 2012; Fig. 7.7). The periodicity is coherent and in phase with the D–O periodicity (Fig. 7.7a) supporting the same causality. Furthermore, spectral analysis of the data shows not only the 1500-yr peak, but also the two lower harmonics (Fig. 7.7b). This composite nature of the 1500-yr cycle also agrees well with the tidal hypothesis.

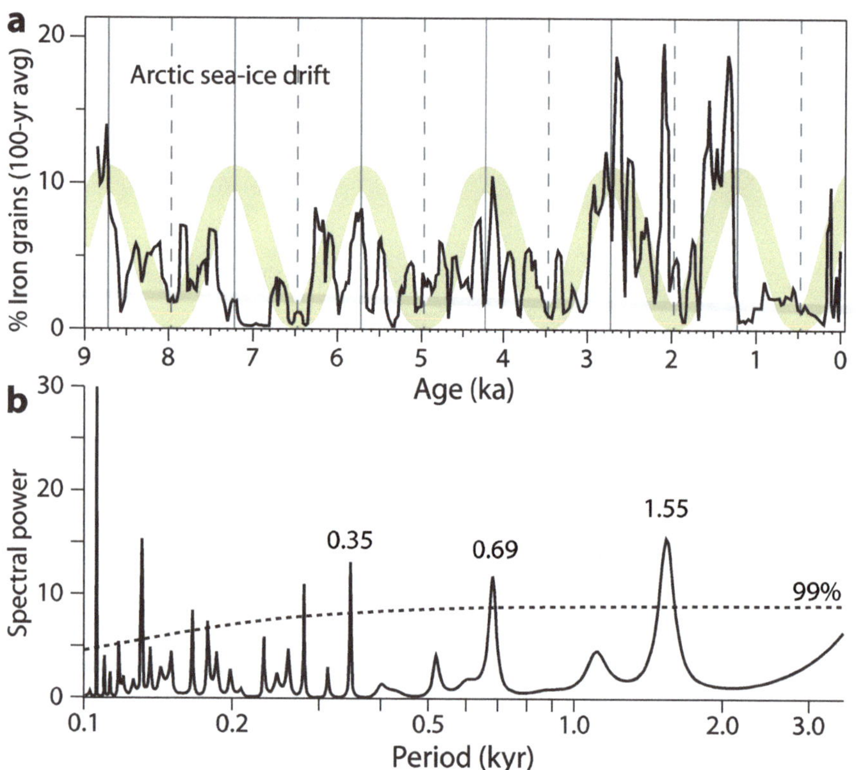

Fig. 7.7 The 1500-year cycle in Holocene Arctic sea ice drift
a) 100-year average of the Kara Sea iron grain-weighted percentage in core JPC16. The iron grains originated in the Kara Sea and were deposited on the coast off Alaska by sea-ice drift. D–O periodicity is indicated with continuous grey lines. Dashed grey lines represent the 750-yr harmonic of the D–O periodicity. After Darby et al. (2012). Thick sinusoidal curve: 1500-yr periodicity. **b)** Time series analysis of the iron grains, by the maximum entropy method. The dashed curve is the 0.99 confidence limit for the record. A prominent 1.5 kyr cycle and its harmonics are present in the Kara dataset.

7.8 Ending the confusion about the 1500-year cycle

For historical reasons the Holocene 1500-yr cycle has been mired in confusion since it was first proposed. The evidence to clarify its timing, cause and effects has already been made available by the many researchers studying it. But first it is necessary to separate this cycle manifestations from those coming from other oscillations within the abundant noise resulting from climate variability. This chapter has dealt only with the 1500-yr cycle that relates to the 3000-yr tidal periodicity present in the D–O cycle (see Chap. 3). But the D–O cycle also presents a 4800-yr tidal periodicity, and it was shown in Chap. 3 that both periodicities relate through the 1800-yr lunisolar tidal cycle. This other cycle, studied by Keeling & Whorf (2000), is likely to be present also during the Holocene and proxies displaying that periodicity have been reported (Fletcher et al. 2013; Di Rita et al. 2018).

As the glacial D–O cycle (Fig. 7.1) and the Holocene 1500-yr cycle display phase coherence and similar manifestations (subsurface water-stratification disruption at the halocline, polar atmospheric circulation intensification, and enhanced Arctic drift ice), it can be assumed that they represent different states of the same cycle, due to the very different climatic conditions at those two periods. Over the last decades the D–O cycle has become increasingly questioned (Ditlevsen et al. 2007; Obrochta et al. 2012). The initially described D–O 1470-yr periodicity based on the original GISP2 timescale disappears when using the more recent GICC05 age model (see Fig. 3.8b). However the 3000-yr periodicity is strengthened, of which the 1500-yr periodicity appears to be a harmonic. A 1470-yr periodicity does not fit the evidence presented here, as over the course of the Holocene the accumulated drift would be of 240 years which is larger than what is observed in the

proxy data presented. When discussing the cause of the 1500-yr cycle we must take into account the nature and distribution of the proxies that display it. These proxies are spread all over the world (Fig. 7.8), making it very difficult to argue that the cycle might be caused by internal variability or specific oceanic currents.

Any hypothesis for this cycle must explain the exact timing and global synchroneity of these events. Authors of the studies on the Northwest Pacific proxies have proposed teleconnections linking the North Pacific Gyre with the North Atlantic through the westerlies, but this explanation falls short since this also requires teleconnections with the Southern Ocean. As there are no inter-hemisphere tropospheric winds, and currents would have a delay of decades to centuries, this seems implausible. The only consistent explanation is an external pacer. This pacer cannot be the sun for multiple reasons. There is no 1500-yr solar cycle; the effect on subsurface waters takes place at the halocline, at 50–100 m depth, and even below sea ice during the glacial period. These characteristics are difficult to explain with a solar cause. Further, the cycle can cause either cooling or warming depending on location and conditions.

The lunisolar tidal hypothesis, however, can explain a global synchronous effect and the variety of manifestations, as it enhances both atmospheric and oceanic tides. Unlike a solar cycle, a tidal cycle does not carry a temperature signal, and temperature changes are determined within the climate system depending on conditions. Furthermore, the tidal hypothesis is built over shorter harmonics, and some of those harmonics described by Berger and von Rad in 2002 have been observed (Figs. 7.3, 7.6 & 7.7; Mayewski et al. 1997; Darby et al. 2012). One of the problems of the tidal hypothesis is that the cycle is composed of a basic unit of c. 375 years, while most of the power is displayed at the 1500-yr harmonic. Berger & von Rad (2002), advanced a possible explanation:

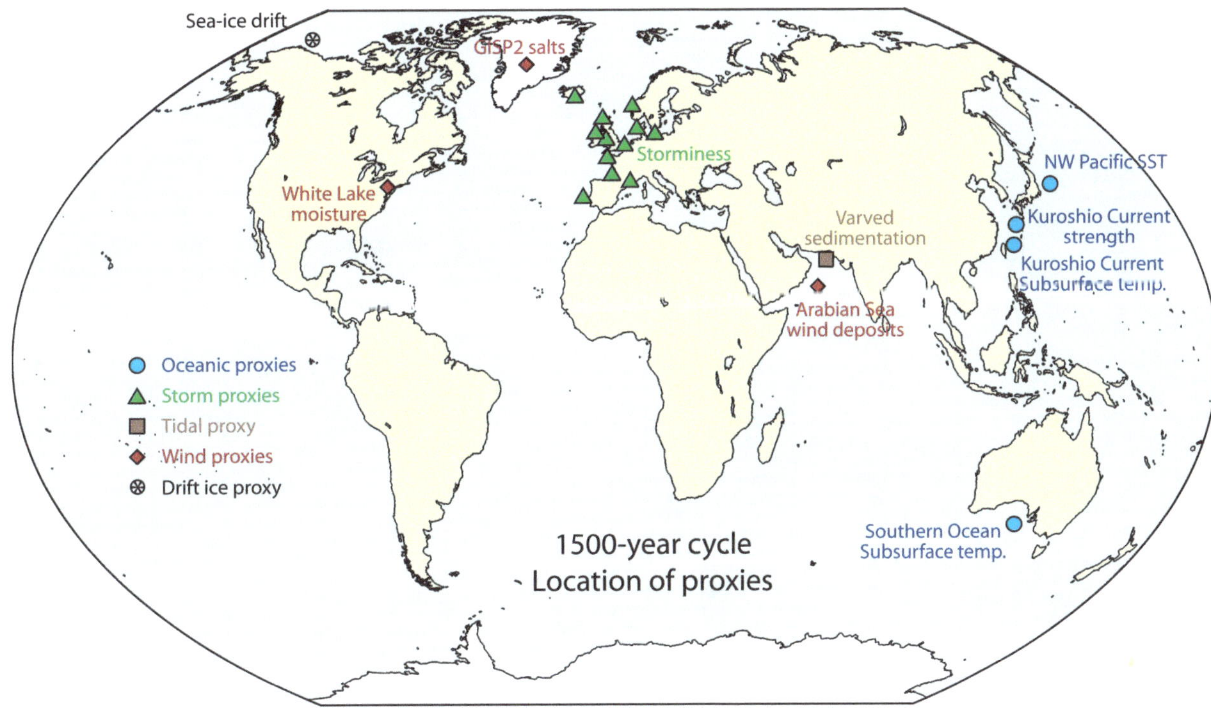

Fig. 7.8 Global distribution of proxies displaying the 1500-year cycle
Evidence from proxies reflecting oceanic (circles), wind (diamonds), storm (triangles), tidal (square) and drift ice (circled asterisk) manifestations. References for these proxies are given in figures 7.2 to 7.7.

"...these tidal maxima [have to] occur in the correct seasonal window. The [1500-year cycle] would thus have a plausible mechanism, that is, lunar forcing tied to season."

During the glacial cycle that season would be the summer, when sea ice is at a minimum, while during the Holocene the season could be the winter, when the cooling effect would be maximal. If this explanation is correct it should have two important predictions. The first one is that in the transition from the glacial period to the Holocene, the change to the opposite season should cause a 180° phase shift in the cycle. This is exactly what we find in several proxies (Fig. 7.2a, b & d; Fig. 7.4a; black bars). This 180° phase shift in most 1500-yr cycle proxies is unlikely to be a coincidence. In the four cases shown it happens within a period of 3750 years (between 9500 and 5750 BP) that coincides with the Holocene Climatic Optimum. The shift in phase manifested by proxies supports a seasonally-tied forcing that bears no temperature signal, allowing the cycle to adapt to the different climate conditions between glacial periods and interglacials. Another consequence of the shift in phase is that it affects the periodicity of the cycle in frequency analyses. The periodicity has to be determined outside the shift. This explains why the REE abundance in Arabian Sea core 74KL was reported at 1450 years and when plotted it shows a 1500-yr periodicity compatible with the D–O periodicity (Fig. 7.4a). The other prediction is that if the effect of the 1500-yr cycle is tied to the season, it should be opposite in each hemisphere, as the seasons are inverted. Again, that is what is found, as the cycle tied to subsurface water temperatures is inverted between the Southern Ocean and the Northwest Pacific (compare Fig. 7.2b & d).

Regardless of its cause, by knowing its timing, we can analyze the effects of the 1500-yr cycle during the Holo-cene. The polar circulation index reconstruction (Mayewski et al. 1997) suggests the cycle is associated in the Arctic region with increased cooling and more winter-like conditions. North Atlantic proxies, however, support that the cooling effect was muted during the Holocene Climatic Optimum (Debret et al. 2007), a period when we see some of the proxies experiment a 180° phase shift. Comparison of the 1500-yr cycle with the Bond series of ice-drift activity in the North Atlantic further confirms the lack of effect during this period (Fig. 7.9, downward arrow). It is only during the Neoglacial cooling of the past 6000 years when the cycle is consistent with an increase in drift ice (Fig. 7.9, upward arrows) responsible for giving the record its 1500-yr apparent periodicity, but indicative that the 1500-yr cycle is only contributing to the drift ice record, and not its primary driver.

Unlike in the case of the 2500-yr cycle, a Holocene temperature reconstruction (see Fig. 4.4) does not show a clear effect of the 1500-yr cycle on global temperature. If Berger's tidal, season-linked, hypothesis is correct, global temperatures would see less effect from the cycle than hemispheric temperatures.

The 1500-yr climate cycle is the millennial cycle whose peak effect is scheduled to take place next, at c. AD 2180. What should we reasonably expect from such cyclic occurrence? If our knowledge of the cycle is correct we should see bigger tides and an increase in storm flooding events. There should be an increase in Arctic sea ice and iceberg activity. We should also see an increase in zonal wind circulation and associated precipitation changes. Most of the effects could be smaller than in previous instances of the cycle if global temperature continues at the current level or increases over the next 150 years. Any decrease in Northern Hemisphere surface temperature caused by the cycle should be limited, and global average

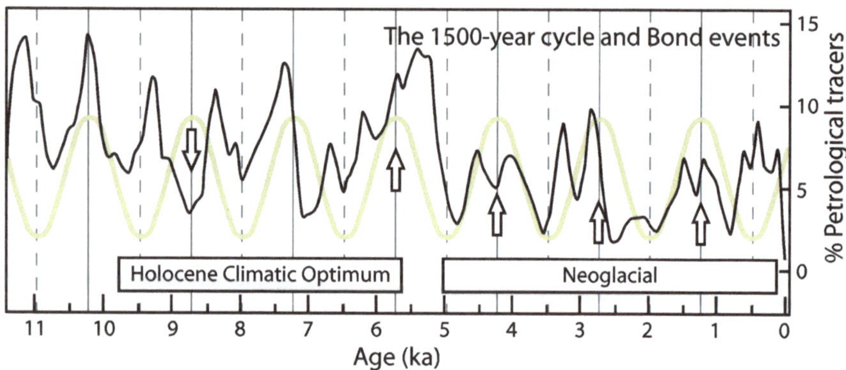

Fig. 7.9 The 1500-year cycle and Bond events
Holocene record of North Atlantic iceberg activity (black curve) determined by the presence of drift-ice petrological tracers. After Bond et al. (2001). The 1500-yr cycle is represented by vertical grey lines indicating the D–O periodicity (continuous lines) or its harmonic (dashed line), and by the fitted 1500-yr cyclic periodicity (sinusoidal curve). A downward arrow marks the lack of correlation during the first half of the Holocene, while upward arrows suggest some effect of the 1500-yr cycle on iceberg activity during the last 6000 years.

temperature should not be reduced by much, probably not more than 0.2 °C, as suggested by the standard deviation in a global reconstruction where its effect cannot be detected (Marcott et al. 2013). It will be however an outstanding opportunity to study the 1500-yr cycle and establish the reality of climate cycles at the millennial level, if any observed effect is correctly attributed.

7.9 Conclusions

7a. The 1500-year cycle displayed in Northern Hemisphere records from the last glaciation is also observed in Holocene records from all over the world.

7b. The cycle is most prominently displayed in oceanic subsurface water temperature, Arctic atmospheric circulation, wind deposits, Arctic drift ice, and storminess records.

7c. Proxies indicate the cycle also displays power at the 750-year harmonic and might have undergone a phase shift during the Holocene Climatic Optimum.

7d. A lunisolar tidal hypothesis currently best explains the cycle's timing, features, and effects. The hypothesis proposes that the cycle is season linked and thus has opposite manifestations in each hemisphere.

7e. The tidal hypothesis also provides an explanation for the cycle's lack of a temperature signal, a phase shift observed during the Holocene Climatic Optimum, as well as the apparent opposite response from a Southern Hemisphere proxy.

7f. Although the 1500-year cycle is associated with cooling indicators in the Northern Hemisphere, its effect on global temperatures remains to be determined.

7g. The next peak effect of the 1500-year cycle is expected in about 160 years, and will provide a rare opportunity to clarify its causes and effects.

References

Berger WH & Von Rad U (2002) Decadal to millennial cyclicity in varves and turbidites from the Arabian Sea: hypothesis of tidal origin. Global and Planetary Change 34 (3–4) 313–325

Black JA, Cunningham G, Fluckiger–Hawker E et al (1998) The Electronic Text Corpus of Sumerian Literature (https://etcsl.orinst.ox.ac.uk/). Oxford

Bond GC & Lotti R (1995) Iceberg discharges into the North Atlantic on millennial time scales during the last glaciation. Science 267 (5200) 1005–1010

Bond G, Kromer B, Beer J et al (2001) Persistent solar influence on North Atlantic climate during the Holocene. Science 294 (5549) 2130–2136

Costas S, Naughton F, Goble R & Renssen H (2016) Windiness spells in SW Europe since the last glacial maximum. Earth and Planetary Science Letters 436 82–92

Courty MA, Crisci A, Fedoroff M et al (2008) Regional manifestation of the widespread disruption of soil-landscapes by the 4 kyr BP impact-linked dust event using pedo-sedimentary micro-fabrics. In Kapur S and Stoops G (eds) New trends in soil micromorphology. Springer, Berlin, p 211–236

Cullen HM, deMenocal PB, Hemming S et al (2000) Climate change and the collapse of the Akkadian empire: Evidence from the deep sea. Geology 28 (4) 379–382

Darby DA, Ortiz JD, Grosch CE & Lund SP (2012) 1500-year cycle in the Arctic Oscillation identified in Holocene Arctic sea-ice drift. Nature Geoscience 5 (12) 897–900

Debret M, Bout–Roumazeilles V, Grousset F et al (2007) The origin of the 1500-year climate cycles in Holocene North-Atlantic records. Climate of the Past Discussions 3 (2) 679–692

Di Rita F, Fletcher WJ, Aranbarri J et al (2018) Holocene forest dynamics in central and western Mediterranean: periodicity spatio-temporal patterns and climate influence. Scientific reports 8 (1) 8929

Dima M & Lohmann G (2009) Conceptual model for millennial climate variability: a possible combined solar-thermohaline circulation origin for the ~1,500-year cycle. Climate Dynamics 32 (2) 301–311

Ditlevsen PD, Andersen KK & Svensson A (2007) The DO-climate events are probably noise induced: statistical investigation of the claimed 1470 years cycle. Climate of the Past 3 (1) 129–134

Dokken TM, Nisancioglu KH, Li C et al (2013) Dansgaard-Oeschger cycles: Interactions between ocean and sea ice intrinsic to the Nordic seas. Paleoceanography and Paleoclimatology 28 (3) 491–502

Dyson F (2007) Our rosy future, according to Freeman Dyson. Interview published in Salon, 29 Sep 2007 https://www.salon.com/2007/09/29/freeman_dyson/ Accessed 04 Jun 2022

Fletcher WJ, Debret M & Goñi MFS (2013) Mid-Holocene emergence of a low-frequency millennial oscillation in western Mediterranean climate: Implications for past dynamics of the North Atlantic atmospheric westerlies. The Holocene 23 (2) 153–166

Geirsdóttir Á, Miller GH, Larsen DJ & Ólafsdóttir S (2013) Abrupt Holocene climate transitions in the northern North Atlantic region recorded by synchronized lacustrine records in Iceland. Quaternary Science Reviews 70 48–62

Isono D, Yamamoto M, Irino T et al (2009) The 1500-year climate oscillation in the midlatitude North Pacific during the Holocene. Geology 37 (7) 591–594

Keeling CD & Whorf TP (2000) The 1800-year oceanic tidal cycle: A possible cause of rapid climate change. Proceedings of the National Academy of Sciences 97 (8) 3814–3819

Li YX, Yu Z & Kodama KP (2007) Sensitive moisture response to Holocene millennial-scale climate variations in the Mid-Atlantic region, USA. The Holocene 17 (1) 3–8

Marcott SA, Shakun JD, Clark PU & Mix AC (2013) A reconstruction of regional and global temperature for the past 11,300 years. Science 339 (6124) 1198–1201

Mayewski PA & White F (2012) The ice chronicles: The quest to understand global climate change. University Press of New England, Hanover

Mayewski PA, Meeker LD, Twickler MS et al (1997) Major features and forcing of high-latitude northern hemisphere atmospheric circulation using a 110,000-year-long glaciochemical series. Journal of Geophysical Research: Oceans 102 (C12) 26345–26366

Mayewski PA, Rohling EE, Stager JC et al (2004) Holocene climate variability. Quaternary research 62 (3) 243–255

Moros M, De Deckker P, Jansen E et al (2009) Holocene climate variability in the Southern Ocean recorded in a deep-sea sediment core off South Australia. Quaternary Science Reviews 28 (19–20) 1932–1940

Muscheler R (2012) Palaeoclimate: The enigmatic 1500-year cycle. Nature Geoscience 5 (12) 850

O'Brien SR, Mayewski PA, Meeker LD et al (1995) Complexity of Holocene climate as reconstructed from a Greenland ice core. Science 270 (5244) 1962–1964

Obrochta SP, Miyahara H, Yokoyama Y & Crowley TJ (2012) A re-examination of evidence for the North Atlantic "1500-year cycle" at Site 609. Quaternary Science Reviews 55 23–33

Rahmstorf S (2003) Timing of abrupt climate change: A precise clock. Geophysical Research Letters 30 (10) 1510–1514

Ruzmaikin A & Feynman J (2011) The 1500 year quasiperiodicity during the Holocene. In: Rashid H, Polyak L & Mosley–Thompson E (eds) Abrupt Climate Change: Mechanisms, Patterns, and Impacts. Geophysical Monograph Series 193 161–171

Seierstad IK, Abbott PM, Bigler M et al (2014) Consistently dated records from the Greenland GRIP, GISP2 and NGRIP ice cores for the past 104 ka reveal regional millennial-scale $\delta^{18}O$ gradients with possible Heinrich event imprint. Quaternary Science Reviews 106 29–46

Sirocko F, Garbe–Schönberg D, McIntyre A & Molfino B (1996) Teleconnections between the subtropical monsoons and high-latitude climates during the last deglaciation. Science 272 (5261) 526–529

Sorrel P, Debret M, Billeaud I et al (2012) Persistent non-solar forcing of Holocene storm dynamics in coastal sedimentary archives. Nature Geoscience 5 (12) 892–896

Thompson LG, Mosley–Thompson E, Davis ME et al (2002) Kilimanjaro ice core records: evidence of Holocene climate change in tropical Africa. Science 298 (5593) 589–593

Thompson LG, Mosley–Thompson E, Brecher H et al (2006) Abrupt tropical climate change: Past and present. Proceedings of the National Academy of Sciences 103 (28) 10536–10543

Tierney JE, Russell JM, Damsté JSS et al (2011) Late Quaternary behavior of the East African monsoon and the importance of the Congo Air Boundary. Quaternary Science Reviews 30 (7–8) 798–807

Voosen P (2018) Massive drought or myth? Scientists spar over an ancient climate event behind our new geological age. Science doi: 10.1126/science.aav0432

Walker M, Head MH, Berklehammer M et al (2018) Formal ratification of the subdivision of the Holocene Series/Epoch (Quaternary System/Period): two new Global Boundary Stratotype Sections and Points (GSSPs) and three new stages/subseries. Episodes 41 (4) 213–224

Wang L, Li J, Zhao J et al (2016) Solar- monsoon- and Kuroshio-influenced thermocline depth and sea surface salinity in the southern Okinawa Trough during the past 17,300 years. Geo-Marine Letters 36 (4) 281–291

Wanner H, Solomina O, Grosjean M et al (2011) Structure and origin of Holocene cold events. Quaternary Science Reviews 30 (21–22) 3109–3123

Zheng X, Li A, Kao S et al (2016) Synchronicity of Kuroshio Current and climate system variability since the Last Glacial Maximum. Earth and Planetary Science Letters 452 247–257

8

CENTENNIAL TO MILLENNIAL SOLAR CYCLES

"If the prolonged maximum of the 12th and 13th centuries and the prolonged minima of the 16th and 17th century are extrema of a cycle of solar change, the cycle has a full period of roughly 1000 years. ... These coincidences suggest a possible relationship between the overall envelope of the curve of solar activity and terrestrial climate in which the 11-year solar cycle may be effectively filtered out or simply unrelated to the problem."
John A. Eddy (1976)

8.1 Introduction

The study of solar cycles and their climatic effect is hampered by a very short observational record (c. 400 years), an inadequate understanding of the physical causes that might produce centennial to millennial changes in solar activity, and an inadequate knowledge of how such changes produce their climatic effect. Despite this lack of a solid theoretical framework, paleoclimatologists keep publishing article after article where they report correlations between solar proxy periodicities and climate proxy periodicities, and the observational evidence is now so abundant as to obviate the lack of a theory or well defined mechanism.

Study of the atmospheric effects of the 11-year solar cycle reveals a tropospheric reorganization characterized by wintertime changes in zonal wind circulation, the latitudinal temperature gradient, and poleward heat transport (Brugnara et al. 2013; Zhou & Tung 2013). Since the effects are season-specific, affected by other factors like El Niño/Southern Oscillation and stratospheric circulation, and reverted from low to high solar activity over the course of a few years, the climatic signal of the 11-year cycle is faint, amounting to 0.1–0.2 °C surface temperature change over the cycle (Tung & Camp 2008). This small effect of the 11-year cycle contrasts with the very large climatic effect of long solar cycles revealed by proxy evidence (see Chap. 5). The obvious explanation is that solar-induced climatic changes can be cumulative when solar activity is affected for a long time. Consistent with this possibility, a multidecadal lagged cumulative effect has been observed in Greenland temperature response to solar forcing (Kobashi et al. 2015).

If the climatic effect of centennial to millennial solar cycles is the result of cumulative effects from persistently affected solar activity, the effect must present a bias towards being more intense during periods of persistently low solar activity. The reason is that high solar activity is always interrupted every 5–6 years by a solar minimum, while low solar activity during solar grand minima (SGM) can be uninterrupted for decades. This reasoning is consistent with the outsized climatic effect found during the lows of the millennial and multi-millennial solar cycles (see Chap. 5), exemplified in the Little Ice Age (LIA). By contrast centennial and multi-centennial solar cycles display a much smaller climatic effect when their lows take place centuries away from millennial lows, like in the present.

Solar cycles are non-stationary, quasi-periodic oscillations in solar activity that display great variability in their period and amplitude. The 11-year solar cycle since 1750 has comprised oscillations from 9 to 13.6 years in length, and from 80 to 280 monthly smoothed maximum sunspots in amplitude (WDC–SILSO international sunspot number). This cycle even disappeared completely during the Maunder Minimum, when over a decade could pass between observations of a single sunspot (Eddy 1976). Similar properties can be observed in the long solar cycles, that might present irregular oscillations both in length and amplitude, to the point of having missing oscillations. The cycle might disappear for a time before returning with the same quasi-periodicity as the 11-yr cycle did during the Maunder Minimum. Such properties cannot be presently explained, as we ignore the origin of the sun cyclical behavior, but as they are accepted as properties of the 11-year cycle, they should also be accepted as properties of long solar cycles, instead of being used to refute their existence.

8.2 The millennial Eddy solar cycle

Frequency analysis of Holocene solar activity reconstructions shows a peak at c. 1000 years that displays modulation by the 2500-year cycle (Usoskin 2013; see Fig. 5.9b & c). Wavelet analysis shows the c. 1000-year periodicity having a strong signal between 11,500 and 4,500 yr BP, and between 2,500 and present time, but a much lower signal between 4,500 and 2,500 yr BP (Ma 2007; Fig. 8.1 grey area). This period of two millennia displays an interruption of the SGM that can be identified as belonging to the Eddy cycle (Fig. 8.1a). The average period of the c. 1000-year cycle can be calculated from the SGM at 11,115 yr BP to the one at 1,265 yr BP (dates from Usoskin et al. 2016) for ten periods at 985 years, a span in very good agreement with the calculated 970 years from frequency analysis, and the calculated 983.4 years from astronomical cycles (Scafetta 2012).

The 1000-yr solar cycle, despite its shorter period and variable amplitude compared to the Bray solar cycle, seems to have dominated Holocene climate variability between 11,500 and 4,500 yr BP. Several authors have noticed this solar forcing dominance during the early Holocene (see Fig. 4.9; Debret et al. 2007; Simonneau et al. 2014). The Bond series of North Atlantic drift-ice record reflects a clear c. 1000-year periodicity during the first 6,500 years of the Holocene that correlates with the

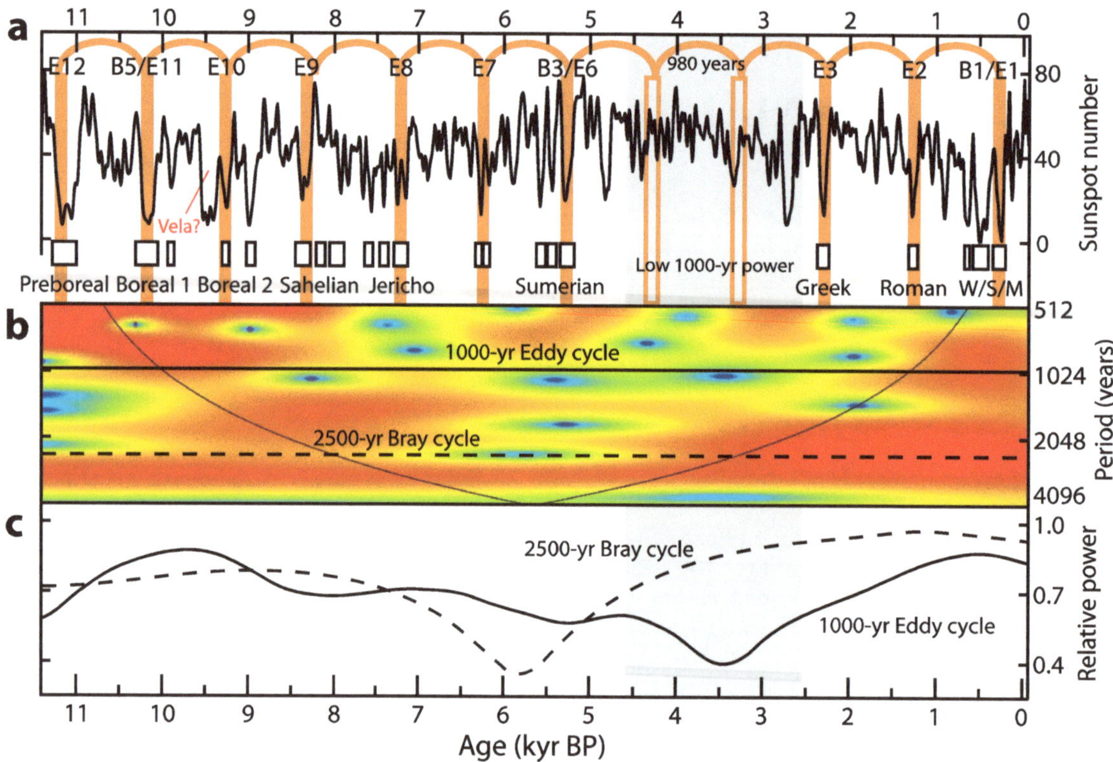

Fig. 8.1 The 1000-year Eddy cycle in solar activity reconstructions
a) Sunspot number reconstruction, after Solanki et al. (2004) until 8700 BP and after Wu et al. (2018) from 8700 BP. The 9500 BP huge increase in cosmic rays without corresponding climate effect is proposed to be due to a galactic source, perhaps the Vela supernova. A regularly spaced 980-year periodicity is shown as arches above. The Eddy lows that correspond to this periodicity (orange bars) are numbered from most recent. Solar grand minima that correspond to these lows are indicated with boxes with their names. W/S/M correspond to the Wolf, Spörer, and Maunder minima. **b)** Wavelet analysis of the sunspot number reconstruction, with the Eddy periodicity indicated by a continuous line, and the Bray periodicity by a dashed line. **c)** Wavelet power along the 1000 years transect (Eddy periodicity, continuous line), and the 2500 years transect (Bray periodicity, dashed line). The non-stationary nature of solar cycles makes them disappear at times. Empty bars mark missing lows in the 1000-yr Eddy cycle during the 4500–2500 yr BP period of low 1000-yr frequency power indicated by the light grey area.

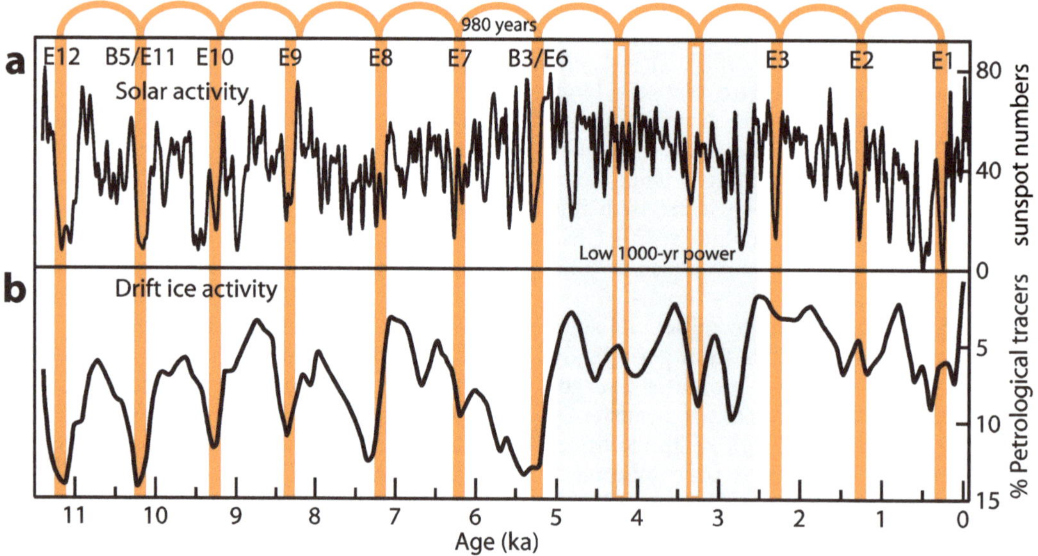

Fig. 8.2 The 1000-year Eddy cycle correspondence to Bond events
a) Solar activity reconstruction (sunspot number), after Solanki et al. (2004) and Wu et al. (2018). A regularly spaced 980-year periodicity is shown as arches above. The Eddy lows that correspond to the 1000-yr solar periodicity (vertical bars) are numbered from most recent. Light grey area represents the period when the Eddy frequency presents low power in frequency analysis of solar activity, coincidental with Eddy lows that do not correspond to low solar activity (empty bars). **b)** Holocene record of North Atlantic iceberg activity determined by the presence of drift-ice petrological tracers. After Bond et al. (2001). The correspondence is high except for the period when the Eddy cycle displays low power.

Fig. 8.3 North Atlantic iceberg activity and the Eddy solar cycle
a) Black curve, inferred iceberg activity in the North Atlantic (inverted) from petrological tracers. After Bond et al. (2001). Red sinusoid curve, a 1000-yr period sinusoid representing solar activity for that periodicity, whose amplitude reflects the relative power of that frequency in a solar activity wavelet analysis. Light grey box marks the period of low 1000-yr frequency power. The last three warm periods (orange bars) and 2 cold periods (blue bars) are indicated. RWP, Roman Warm Period. DACP, Dark Ages Cold Period. MWP, Medieval Warm Period. LIA, Little Ice Age. MGW, Modern Global Warming. **b)** Relative power at the 1000-yr frequency transect of a solar-activity wavelet analysis. Sunspot reconstruction after Usoskin et al. (2004) and Wu et al. (2018). Agreement between the 1000-yr solar cycle and climate oscillations is exclusive of periods when the 1000-yr solar signal displays high power.

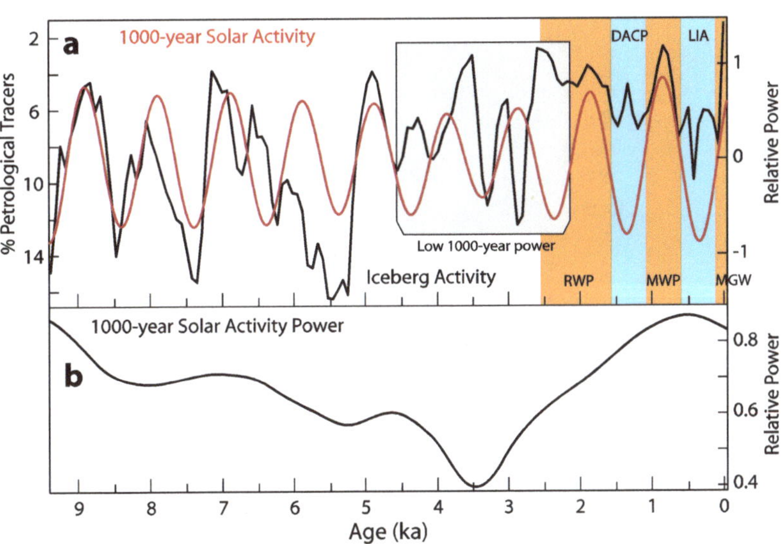

1000-year Eddy solar cycle (Fig. 8.2; Debret et al. 2007).

The 1000-yr periodicity displays low power in solar activity wavelet analysis during two millennia (Figs. 8.1 & 8.3; Ma 2007). When the amplitude of the 1000-yr solar signal over time is adjusted by its wavelet power (Fig. 8.3), a high correlation between North Atlantic iceberg activity and the 1000-yr Eddy solar cycle corresponds to the periods when the 1000-yr solar signal is high, while the correlation is low at periods of weak 1000-yr solar signal. The strong relationship between climatic Bond events and solar activity has been acknowledged by multiple authors, starting with Gerard Bond himself (Bond et al. 2001). The unusually long Roman Warm Period (2500–1600 BP; Wang et al. 2012) coincided with the final part of this interval of low Eddy solar cycle activity, while known warm and cold periods have faithfully followed the since strengthened 1000-yr Eddy solar cycle (Fig. 8.3). When looking from the present, the 1000-yr cycle breaks down at the Roman Warm Period because that is when cycle started manifesting again, and one has to go before 4500 BP to see the cycle continuation preserving its phase.

The 1000-yr solar cycle was named the Eddy cycle by Abreu et al. (2010), and its lows have been numbered here, from more recent, as E1, E2, … (Fig. 8.1). The climatic effect of the Eddy cycle should manifest in the two periods when solar activity was most affected by this millennial periodicity. In the most recent period, we observe a millennial separation between warm periods: Modern Global Warming (present), Medieval Warm Period (c. AD 1100), Roman Warm Period (c. AD 100); and between cold periods: LIA (c. AD 1650; E1), and Dark Ages Cold Period (c. AD 650; E2). During the early Holocene, the lows of the Eddy cycle coincide with prominent climate change epi-

Table 8.1 Solar grand minima of the Holocene
Conservative list with approximate dates (in CE and BP scales) of solar grand minima (SGM) in reconstructed solar activity. The name refers in some cases to a SGM cluster (cl.). The cycle states if the SGM shows a temporal coincidence with a low from the Bray (B), or Eddy (E) cycle. References: SGM listed in 1, Usoskin et al. (2007); 2, Inceoglu et al. (2015); 3, Usoskin et al. (2016). After Usoskin (2017). Question mark indicates the biggest increase in ^{14}C production during the Holocene. In the opinion of the author, it does not correspond to a SGM because it does not have a corresponding climate effect.

sodes defining a clear millennial periodicity (Fig. 8.4; Marchitto et al. 2010). E12 (11,250 BP) coincides with a particularly humid phase in northwestern and central Europe towards the end of the Preboreal oscillation (van der Plicht et al. 2004; Magny et al. 2007). E11 (10,300 BP) coincided with the first cold, humid event, of the Boreal phase (Björck et al. 2001; Magny et al. 2004), while

Date CE	Date BP	Name	Cycle	Ref.
1680	270	Maunder	E1	
1470	480	Spörer	B1	
1310	640	Wolf	B1	
1030	920	Oort		
690	1260	Roman	E2	1–3
−360	2310	Greek	E3	1–3
−750	2700	Homeric	B2	1–3
−1385	3335		E4	1–3
−2450	4400		E5	2, 3
−2855	4805	Noachian		1–3
−3325	5275	Sumerian cl.	B3/E6	1–3
−3495	5445	Sumerian cl.	B3	1–3
−3620	5570	Sumerian cl.	B3	1–3
−4220	6170		E7	1–3
−4315	6265		E7	1–3
−5195	7145	Jericho cl.	E8	2, 3
−5300	7250	Jericho cl.	E8	1–3
−5460	7410	Jericho cl.	B4	1–3
−5610	7560	Jericho cl.	B4	1–3
−6385	8335	Sahelian	E9	1–3
−7035	8985			1
−7305	9255	Boreal 2 cl.	E10	1
−7515	9465	Boreal 2 cl.	?	1
−8215	10165	Boreal 1	B5/E11	1
−9165	11115	Preboreal	E12	1

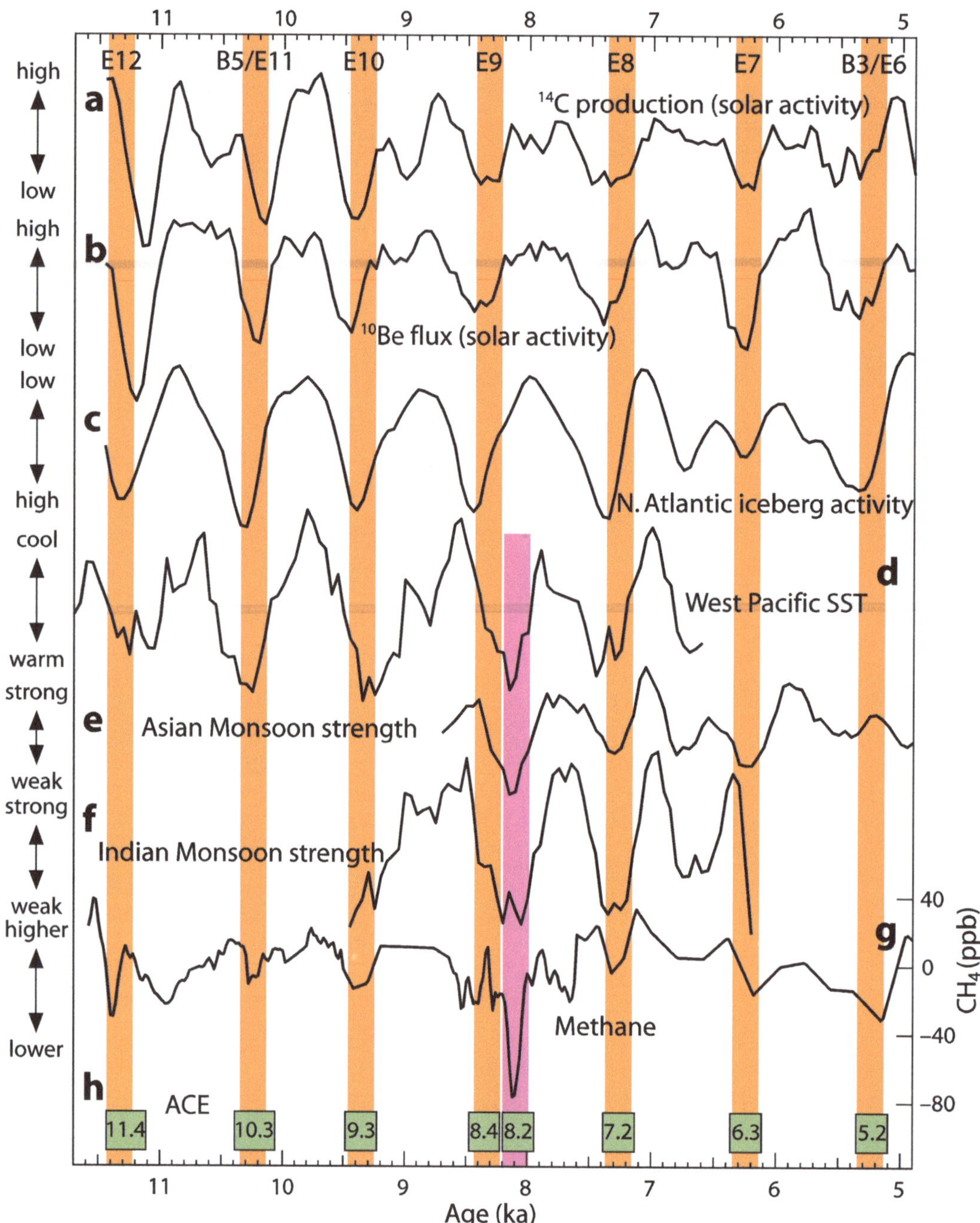

Fig. 8.4 Millennial climate change periodicity
a–f) Climatic and solar proxy records, spanning the early Holocene, 250-year smoothed and 1800-year high-pass filtered. Data from Marchitto et al. (2010). Records are: **a)** Inverted tree-ring–derived ^{14}C production rate (solar activity proxy). **b)** Inverted ice core ^{10}Be flux (solar activity proxy). **c)** Inverted North Atlantic stack of IRD petrologic tracers (North Atlantic iceberg activity proxy). **d)** Soledad Basin *G. bulloides* Mg/Ca (sea-surface temperature proxy). **e)** Inverted Dongge Cave (southern China) stalagmite δ^{18}O (Asian monsoon proxy). **f)** Inverted Hoti Cave (Oman) stalagmite δ^{18}O (Indian monsoon proxy). **g)** Detrended Greenland methane record built by combining a 180–7620 BP and 8680–9500 BP record from GRIP after Blunier et al. (1995), a 7640–8660 BP record from GISP2 after Kobashi et al. (2007), and a 9520–11580 BP record from NEEM after Chappellaz et al. (2013). The record is a sensitive proxy for changes in global precipitation patterns. **h)** Abrupt climatic events (ACE; boxes) belonging to the 1000-yr climate cycle, marked with vertical orange bars, and the 8.2-kyr event with a pink bar.

E10 (9,300 BP) matches the second Boreal event (Rasmussen et al. 2007; Magny et al. 2004). E9 (8,300 BP) coincided with the outbreak of Lake Agassiz, and researchers are trying to differentiate the relative climatic contribution to the 8.2 kyr event from the solar minimum and the proglacial lake outbreak (Rohling & Pälike 2005; see Sect. 6.4). E8 (7,300 BP) coincides with the last cold, humid phase of the sixth millennium BC (Berger et al. 2016). E7 (6,300 BP) is less well established in the literature, although clearly identified as a dry event in Oman caves speleothems (Fleitmann et al. 2007). E6 (5,200 BP) has been well described worldwide as an abrupt cold event (see Fig. 4.13; Thompson et al. 2006).

The identification of the Eddy cycle lows, as well as the Bray cycle lows, allows an examination of SGM distribution according to the two main solar cycles of the

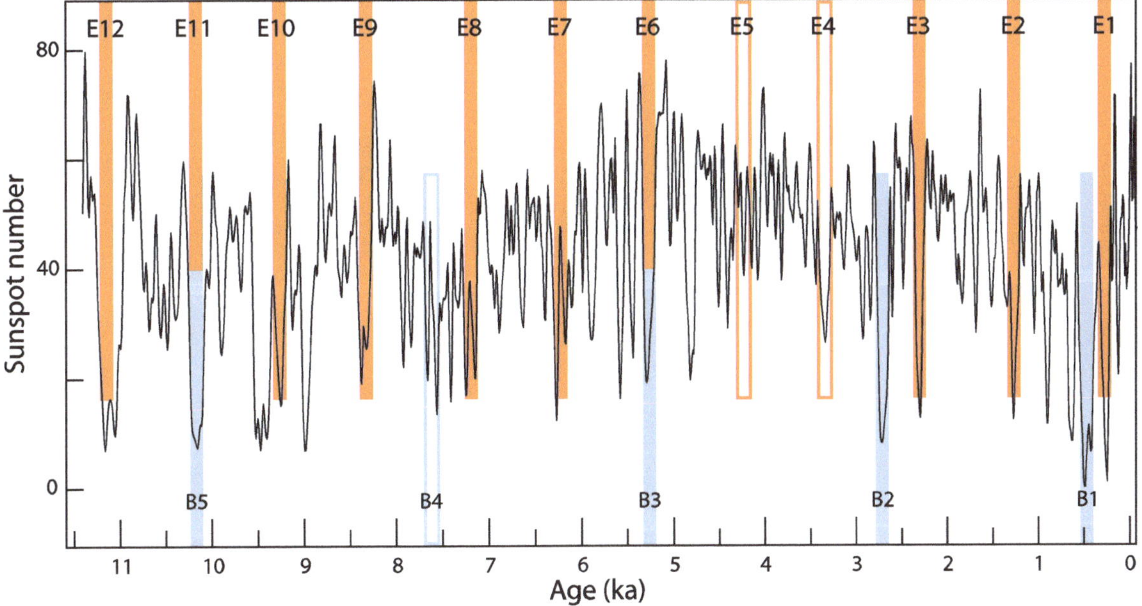

Fig. 8.5 Solar grand minima of the Holocene
Solar activity reconstruction (sunspot number), after Solanki et al. (2004) and Wu et al. (2018). showing the disposition of the SGM associated with the Bray (blue bars) and Eddy (orange bars) cycle lows.

Holocene. Usoskin (2017) gives a conservative list of 25 SGM that were identified in previous studies by different researchers for the past 11,500 years. They are listed in table 8.1. There is a notable coincidence. Since the Eddy cycle is so close to one thousand years, all the lows of the cycle take place at c. X,300 yr BP, with X being every millennia of the Holocene. We can observe in the list of SGM that 15 of them take place at c. X,300 ± 80 yr BP (table 8.1; Usoskin 2017). Those SGM are assigned to the Eddy cycle given the good temporal coincidence (Fig. 8.5). Next, we have 9 SGM that coincide with the lows of the 2,500-yr Bray cycle, and in fact define it (Fig. 8.5). Spörer-type SGM belong to the Bray series. Two of these SGM, at 10,165 and 5,275 years BP, also coincide with the Eddy cycle, as both cycles tend to coincide in phase when two Bray periods (4,950 years), and five Eddy periods

(4,900 years) have passed.

Of the 25 SGM identified by Usoskin (2017) during the Holocene, only three are not located close to the lows of the Eddy or Bray cycles. The Oort (920 BP), Noachian (4805 BP), and an unnamed SGM at 8995 BP, that could be considered part of the Boreal 2 cluster. Since 88% of SGM occur during an Eddy or Bray low, it is unlikely that the next SGM will take place before around AD 2600, when the next Eddy cycle low is expected.

8.3 The 210-year de Vries solar cycle

As previously described (see Sect. 5.8), the de Vries solar cycle is strongly modulated by the Bray solar cycle. For about a millennium centered in each Bray cycle low, the de Vries cycle reduces solar activity every c. 210 years,

Fig. 8.6 Bi-centennial solar influence on Northern Hemisphere summer temperatures from tree-rings
a) Left scale: Reconstructed Northern Hemisphere mean May–June–July–August temperature anomaly time series (black line), smoothed with a 30-year Gaussian filter. After Anchukaitis et al. (2017). Right scale: Total solar irradiance (TSI) reconstruction after Vieira et al. (2011), expressed as anomaly versus the 1961–1990 average. **b)** Wavelet coherence between the Northern Hemisphere mean MJJA temperature anomaly time series and solar forcing variability in (a). After Anchukaitis et al. (2017). Arrows indicate the phase of the relationship where coherence >0.65. In-phase signals point directly to the right of the plot. A continuous in phase coherence between tree-ring-deduced temperatures and solar activity is seen at the de Vries periodicity.

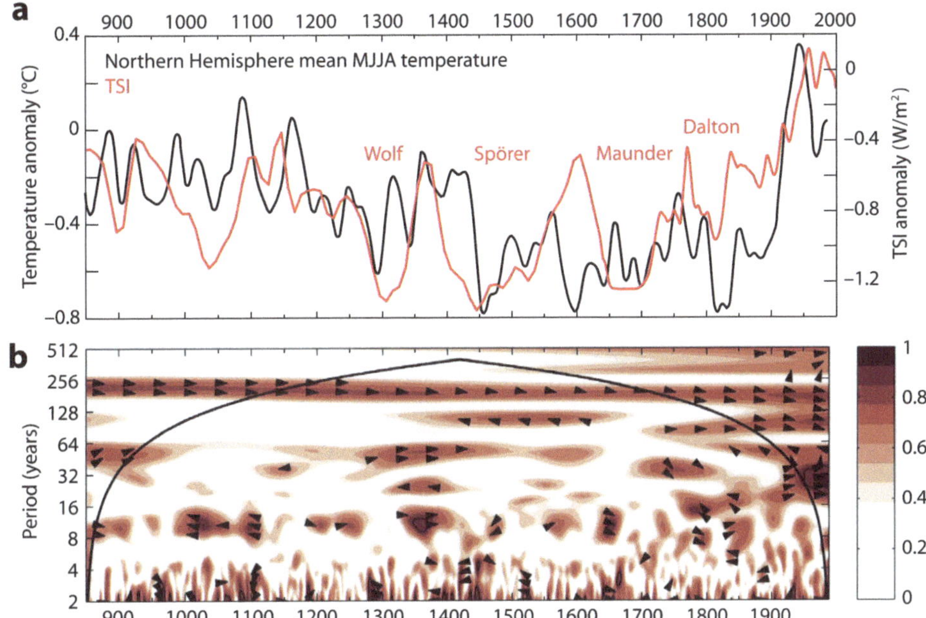

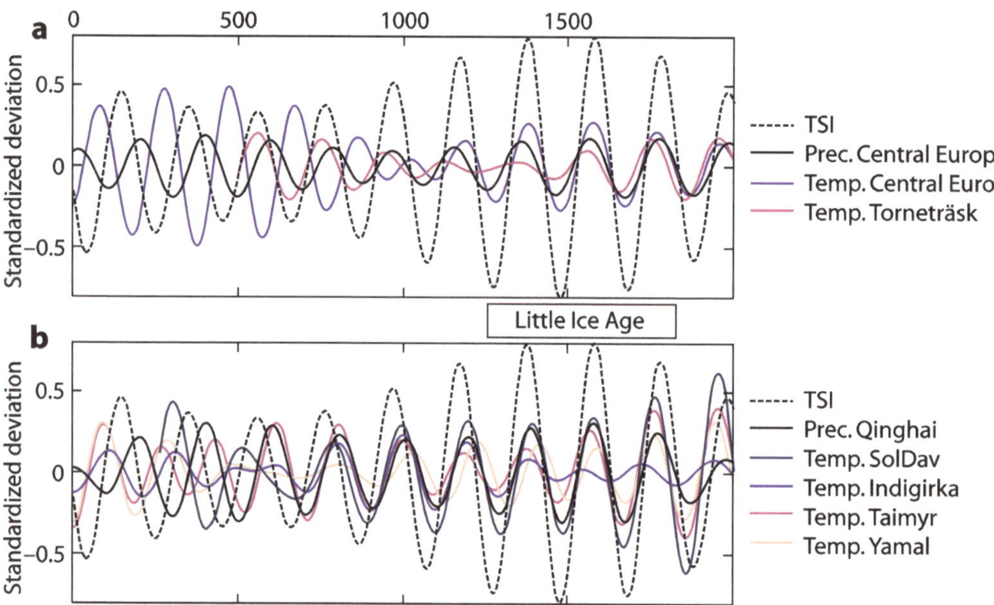

Fig. 8.7 Climate response to the de Vries solar cycle in tree-ring chronologies over the past 2000 years Band-pass filtered total solar irradiance (dotted line) and tree-ring-derived climate data series in the range of periods 180–230 years for **a)** Europe, and **b)** Asia. The synchronization, and in some cases amplitude, of the climatic signal correlates with the strength of the solar signal, indicating that the modulation of the de Vries cycle by the Bray cycle extends to its climatic effect. After Breitenmoser et al. 2012. See source for details on the proxies.

and when a cluster of SGM takes place, it establishes the average spacing between them (see Fig. 5.14). Outside these windows centered in the Bray cycle lows, the de Vries periodicity has very low power in wavelet analysis indicating that by itself it has a smaller effect on solar activity (see Fig. 5.9). The climatic effect of the de Vries cycle matches its solar (cosmogenic isotope) signature.

Charles Sonett and Hans Suess (1984) proposed that the 210-yr cycle seen in solar activity proxies could be related to c. 200-year periodicity changes detected in tree-rings width. This finding has been confirmed for tree-rings, which reflect changes in temperature or precipitation, in several regions of the planet. Anchukaitis et al. (2017) have constructed a tree-ring multi-proxy (54 series), extra-tropical Northern Hemisphere, warm season (MJJA), temperature record spanning 1,200 years (AD 750–1988). The record shows high and stable coherence and consistent phasing with solar irradiance estimates at bi–centennial time scales (194–222-year periods), the c. 210-yr de Vries solar cycle frequency (Fig. 8.6; Anchukaitis et al. 2017).

The modulation of the de Vries cycle by the Bray cycle is also apparent in the climatic data. Breitenmoser et al. (2012) analyzed the c. 200-year periodicity during the past two millennia using seventeen near worldwide distributed tree chronologies, and found significant periodicities in the 210-yr frequency band, corresponding to the de Vries cycle of solar activity, indicating a solar contribution in the temperature and precipitation series. The result continued being significant after the removal of the volcanic-attributed signal, and was most prominent in records from Asia and Europe (Fig. 8.7; Breitenmoser et al. 2012). When the 180–230 years band-pass filtered variability was compared with that of solar variability, highlighting the de Vries cycle, it can be seen that as the de Vries signal increases after about AD 800 due to its modulation by the Bray cycle, the climatic signals start to synchronize with the solar signal and in some cases also increase their amplitude (Fig. 8.7). This synchronization means that after AD 800 the geographical region is responding to solar forcing, changing the climate according to the 210-yr solar periodicity.

Phase relationships between hemispheric and global climate reconstructions from tree-rings and the solar irradiance time series indicate a lag of c. 10 years (range, 5–20 years), with solar changes leading temperature anomalies, consistent with both climate modeling and other climate and solar variability studies (Eichler et al. 2009; Breitenmoser et al. 2012; Anchukaitis et al. 2017).

Other studies link the 210-yr de Vries cycle to climate change, including Central Asian ice-cores (Eichler et al. 2009), Asian (Duan et al. 2014) and South American (Novello et al. 2016) monsoon-record speleothems, Mesoamerican lake-sediment cores as drought proxies (Hodell et al. 2001), and Alpine glaciers (Nussbaumer et al. 2011). The climatic effect of the de Vries solar cycle is thus well established.

8.4 The 88-year Gleissberg solar cycle

Despite the popularity of the Gleissberg solar cycle in the literature the author has not been able to unambiguously identify this cycle as important for solar–climate effects. This is partly due to the Gleissberg cycle being different things for different researchers.

Wolfgang Gleissberg (1944), working at the University of Istanbul observatory, described a long solar cycle that could only be revealed by applying what he called a "secular smoothing" (a trapezoidal 1-2-2-2-1 filter) to a numerical sequence formed by the maximum sunspot values of the known 11-year solar cycles. According to him this numerical procedure revealed *"a long cycle which produces systematic changes of the features of the 11-year cycle and which includes seven 11-year cycles, or 77.7 years."* The cycle thus described is not apparent in the sunspot record, and cannot be produced from it by frequency analysis.

As originally described, the Gleissberg cycle is unacceptable by modern scientific standards (the author would dare to say inexistent), and due to it the term "Gleissberg cycle" means different things to different authors. For some authors it is a frequency peak of c. 88 years that appears in frequency analysis of the cosmogenic record

(McCracken et al. 2013b; Knudsen et al. 2011 supplemental information). Other researchers have found that applying the trapezoidal filter of Gleissberg separately to dates of solar cycle minima and maxima from sunspot records then merging them, one also obtains a c. 80-year time-domain periodicity (Peristykh & Damon 2003). They interpret this result as confirmation of the cycle, that would simultaneously regulate the 11-year cycle amplitude and period. Yet the biggest group of researchers just call any periodicity between 50 and 150 years the Gleissberg cycle, often giving the name simultaneously to two different bands. Joan Feynman, sister of the famous physicist, studied the centennial solar cycle under the Gleissberg flag of convenience (Feynman & Ruzmaikin 2014).

Of interest to the analysis of solar cycles is only the c. 88-year periodicity present in cosmogenic records that we can also call the Gleissberg cycle, if only to avoid further confusion. The problem is that wavelet analysis shows that this periodicity was only apparent between 6,500 and 3,500 BP (Knudsen et al. 2011 supplemental information). This explains why the cycle cannot be detected in the sunspot record. Whether it is a real cycle subjected to a very long modulation, or a temporal pseudo-periodicity that emerged from the unknown interactions that generate long term solar variability, cannot be determined. It is also very unlikely that we will be able to determine if it played a significant role in the climate of the period. As the evidence indicates this periodicity is not currently relevant, we will not consider it further.

8.5 Other solar periodicities

It is clear that solar cycles are pseudo-cycles or quasi-periodicities that display a relatively high level of period and amplitude variability. Some of the cycles, like the 2500-yr Bray and the 1000-yr Eddy cycle, appear to be featured in records several million years old (Kern et al. 2012). The 210-year de Vries cycle has been detected in ice-cores for at least the past 50,000 years (Raspopov et al. 2008; Adolphi et al. 2014; see Fig. 5.11). Other periodicities however, like the 88-yr Gleissberg cycle, have only been found for a few millennia.

Frequency analysis of ^{14}C and ^{10}Be display other clear peaks at 52, 104, 130, 150, 350, 515, and 705 years (McCracken et al. 2013b). Some of them could be harmonics of longer cycles. 6,000 and 9,500-year solar cycles have also been proposed (Xapsos & Burke 2009; Sánchez–Sesma 2015). Is the Sun subject to over a dozen different cycles? Or are some of them simply artifacts and not solar variability cycles? Instead of assuming every peak in a frequency analysis constitutes sufficient evidence for the existence of a cycle, this author only considers those where abundant evidence exists in the scientific literature that solar cycles match the climate evidence precisely. They are the Bray, Eddy, and de Vries cycles. Of interest are also the periodicities recognizable in the sunspot record, the Schwabe (11-year), Pentadecadal, and Centennial (Feynman) cycles. These last two might be simply harmonics of the de Vries cycle, but as they are currently observable, they may be useful to interpret the past, as well as project future solar activity.

It is worth noting, however, that multiple harmonic constituents in complex astronomical phenomena are a reality. Until the advent of computers, tides were predicted by complex "brass brain" machines. The first of these was built by Lord Kelvin in 1873. After identifying the spectral harmonic components from a long tidal data series at a specific port, machines that could handle up to 40 tidal constituents would produce a year of tidal predictions for that port in a few hours (Parker 2011).

8.6 The 100-year Feynman and 50-year Pentadecadal solar cycles

In 1862 Rudolf Wolf, after completing the first continuous record of sunspot numbers, *"concluded from the sunspot observations available at that time that high and low maxima did not follow one another at random: a succession of two or three strong maxima seemed to alternate with a succession of two or three weak maxima"* (as cited by Peristykh & Damon 2003). That observation led to the suggestion of the existence of a long cycle, or secular variation, the length of which was estimated at that time to be equal to 55 years. Thus, the Pentadecadal solar cycle is the oldest discovered secular variation of the sun. Although the Pentadecadal solar cycle displays low power and is statistically non-significant in the sunspot record, it is very prominent in the ^{10}Be record from the one year resolution Dye 3 ice core for the period AD 1420–1992 (McCracken et al. 2013a; Fig. 8.8).

The Pentadecadal cycle should be responsible for the decrease in solar activity at solar cycle 20 (SC20) between 1965 and 1976. This periodicity is interesting in that it could be related to the pentadecadal variability described in sea level pressure and temperatures in the North Pacific (Minobe 2000). Besides having the same length, the pentadecadal solar change that took place at SC20 was shortly followed by the well-known and studied Pacific climate shift that took place in 1976 (Miller et al. 1994). For a hypothesis of their possible relation see Sect. 11.7 (see also Fig. 11.14).

Joan Feynman studied the centennial solar cycle from the early 1980s (Feynman 1982), until her passing in 2020, identifying its properties and correct periodicity. In the long established tradition of naming solar cycles after their discoverer, the centennial solar cycle should bear her

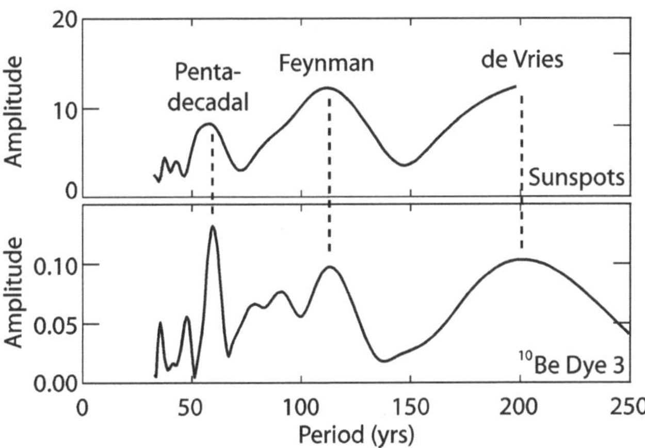

Fig. 8.8 Solar activity spectra during the last centuries
Fourier spectra showing periodicities above 33 years of **a)** the 1610–2010 sunspot number and **b)** the annual ^{10}Be data from Dye 3 ice core for the interval 1420–1992. Main periodicities names indicated. After McCracken et al. (2013a)

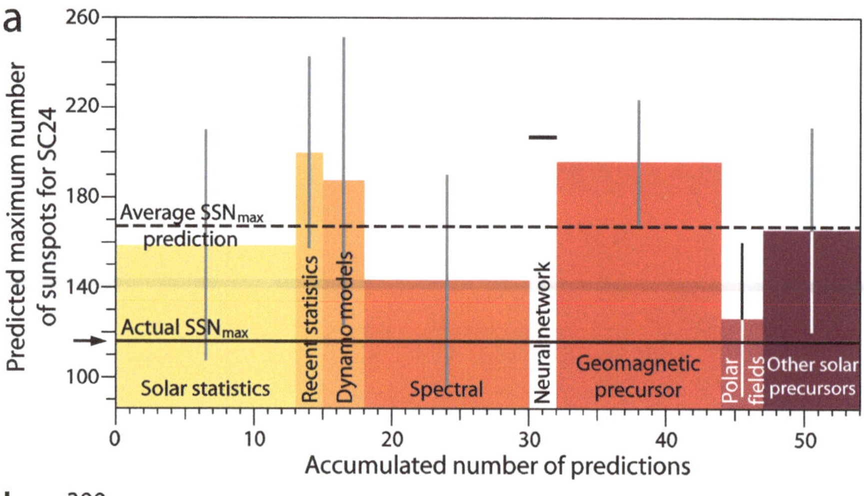

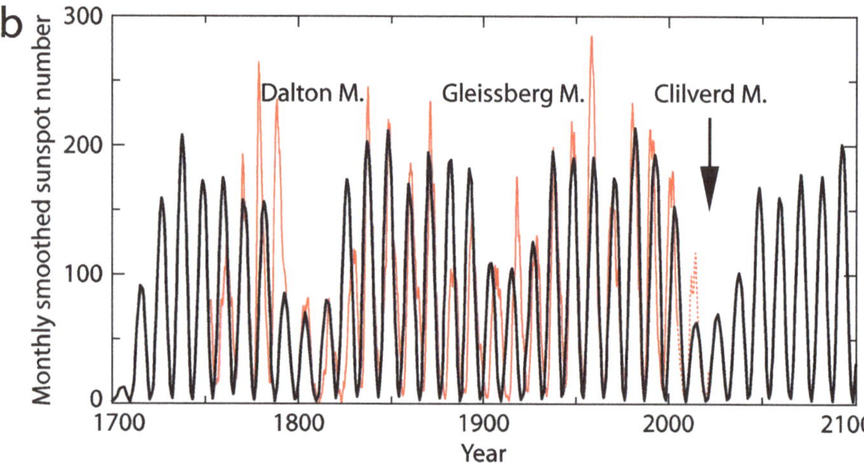

Fig. 8.9 Solar Cycle 24 prediction Solar Cycle 24 corresponds to a low in the Feynman (100-yr) solar cycle. **a)** 54 Cycle-24 predictions grouped by category or subcategory after Pesnell (2008). The average forecast was 45% too high. **b)** Average sunspot number prediction by a low-frequency modulation model (thick black curve) based on frequency analysis from sunspot and cosmogenic isotope records, compared to the average sunspot number since 1750 (thin red curve, dotted since 2006). After Clilverd et al. (2006). Despite a low bias, the model predicted for the first time the current centennial minimum for cycles 24 and 25, that should therefore receive the name "Clilverd minimum." Sunspot numbers in this figure have been converted from the original Boulder sunspot number to the international sunspot number.

name, and it will be referred to here as the Feynman cycle. This solar cycle appears as a peak of c. 104 years in cosmogenic isotopes frequency analysis, and as a decrease in maximum and minimum sunspot numbers at the beginning of each century since there have been telescopic sunspot observations. Despite this precedent, most solar physicists were expecting SC24 to have a slightly lower level of activity than SC23 and were surprised by the depth and duration of the 2008 minimum and the subsequent low activity of SC24. Of the 54 SC24 predictions published or submitted to the SC24 Prediction Panel in six general categories, spectral analysis predictions (Fig. 8.9a; Pesnell 2008) based on Fourier, wavelet, or autoregressive-based forecasts, outperformed all other categories, predicting below average SC24 activity. In this real test, the use of long periodicities found in solar activity records, for which we have no explanation, fared better than methods based on our clearly inadequate understanding of solar physics. A subcategory based on polar fields produced a better prediction, but it can only predict the next cycle when it is close to the minimum, while spectral methods can predict multiple cycles in advance.

Of the spectral predictions, the one published (Clilverd et al. 2006) used a low-frequency modulation model that has some clear inadequacies, like including the Gleissberg 88-yr cycle that is no longer observable, assigning an extremely low amplitude to the 210-yr de Vries cycle, and not including the modulation by the 2500-yr Bray cycle that has been discussed here. However, since it included the 104-yr Feynman periodicity, it predicted very low activity for SC24 (Fig. 8.9b). Indeed, SC24 turned out

to be the least active cycle in 100 years. With its faults corrected the model would have predicted accurately slightly more activity for SC24 than for SC14 (in 1904), instead of less. Importantly, the model also predicted in 2006 that SC25 will again be a below average cycle of similar amplitude to SC24. The polar field method predicted the same 10 years later (Hathaway & Upton 2016), and after the December 2019 solar minimum the increase in solar activity at the time of this writing appears to confirm it. A new prolonged solar minimum, like the Gleissberg minimum of 1879–1914, is being established by the Feynman cycle, and should last at least until around 2032. The earliest prediction of this extended centennial minimum is by Mark Clilverd et al. (2006), and it should be called the Clilverd minimum (Fig. 8.9b).

Feynman and Ruzmaikin (2014) showed that the centennial periodicity is observable on the Sun, in the solar wind, at the Earth, and throughout the Heliosphere. It is supported by the very weak solar wind at the SC23–24 transition, the weakest observed in the space age. Feynman cycle lows (extended minima) are characterized by very low annual sunspot numbers (less than 5, Fig. 8.10a, black arrows & asterisks), and a slight increase in the duration of the 11-year cycle. The Pentadecadal cycle can be seen, non–statistically significant, in Feynman & Ruzmaikin (2014) sunspot analysis (Fig. 8.10 red arrows).

The Feynman cycle is the only long periodicity whose lows have been observed with modern instrumentation. The aa (antipodal amplitude) index that starts in 1868 and measures the disturbance of the Earth's magnetic field by

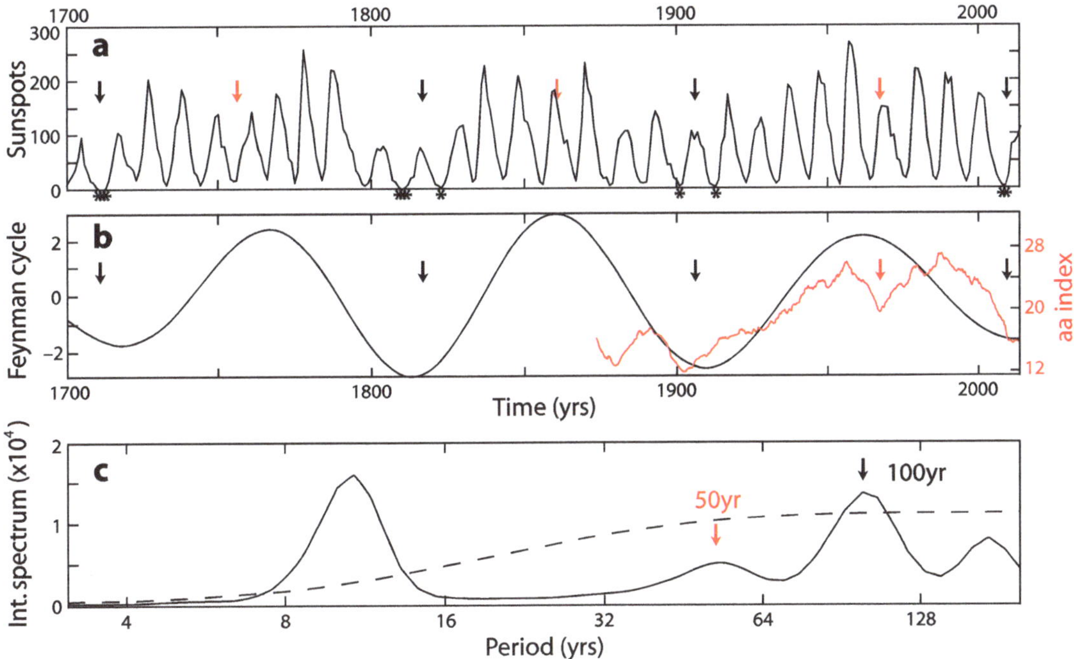

Fig. 8.10 The Feynman solar cycle
a) The sunspot number record. Asterisks mark years when the annual sunspot number was less than 5 (WDC–SILSO data), Black arrows indicate Feynman minima. Red arrows indicate pentadecadal minima. **b)** Black line, left scale, the time series of the 80–110-year band from a wavelet spectrum of 1700–2013 annual sunspots. Red line, 11-year average of the antipodal amplitude (aa) geomagnetic index (data from ISGI). **c)** The integral spectrum obtained by averaging over the time axis the wavelet spectrum of sunspots. The dashed line shows the significance of this spectrum at the 1σ level. After Feynman & Ruzmaikin (2014)

solar wind, clearly displays the last full period of the Feynman cycle, with its lowest values in 1901 and 2009 (Fig. 8.10b). In the Sun, surface differential rotation changes on a centennial timescale coincide with the observed phase change between the toroidal and poloidal magnetic field components and the time dependence of the dipole and quadrupole components of the poloidal magnetic field (Feynman & Ruzmaikin 2014). Solar dynamo models still have to accommodate these centennial variations in the Sun.

Previous Feynman lows are associated with colder periods at the early decades of each of the past three centuries. The ongoing extended minimum is associated with an unexpected hiatus in global warming that has yet to be adequately explained (see Chap. 11).

8.7 Solar cycles interrelation

Spectral analysis of a complex time series, like ^{14}C production rates, usually results in a number of significant frequencies identified on statistical grounds. McCracken et al. (2013a) report 15 significant frequencies between 65 and 2500 years in the Fourier spectra of cosmogenic records. Most of these frequencies are unresearched and unnamed, and it is unknown if they all represent variable solar modulation of galactic cosmic rays or if some could have a different origin.

The approach taken in this work has been to focus on abrupt climate change events (ACEs) identified from selected Holocene climate proxies (see Sect. 4.8), under the assumption that if solar variability has a climatic effect, then climate variability should confirm solar variability. This approach has been very fruitful in recognizing several solar cycles for their coincidence with climatic frequen-

cies, resulting in an alternative interpretation of the cosmogenic record in climate terms. The correct periodicity for the bi–millennial solar cycle is the 2500-yr Bray periodicity, not the 2300-yr Hallstatt periodicity identified in frequency analysis (Fig. 8.11a–c). This frequency analysis artifact is mainly due to a missing minimum in the Bray cycle (B4) that should have taken place around 7.7 kyr BP, and other irregularities in the cycle (see Fig. 5.13). This approach also serves to identify mismatches between the climate record and the solar activity record, like the large maximum in ^{14}C production at 9500 yr BP that lacks a proportional climate effect, suggesting an extra-solar galactic cosmic ray source, proposed here to be the Vela supernova (Sushch & Hnatyk 2014). Also the 1000-yr solar frequency displays a period of two millennia when its effect on climate was very subdued, resulting in the long Roman Warm Period (Fig. 8.3). The very exact match between very low solar activity at the minima of the 2500 and 1000-yr solar cycles and ACEs leaves two possibilities: The existence of climatic cycles of unknown origin that contaminate the solar proxy record, or the existence of an unknown mechanism by which solar variability profoundly affects climate. The first one is highly unlikely as it contradicts all the information available since there are written records on solar activity from aurorae and naked-eye sunspot observations, well over a millennium. This includes the last 400 years since the telescope was invented, that have coincided with the most drastic period of climate change within the Holocene, from the depths of its coldest period, the LIA, to the present warm period, one of the warmest within the Neoglaciation. We know that over this period of intense climate change the cosmogenic record reflects primarily changes in solar activity, not changes in climate, as it agrees very well with sunspots

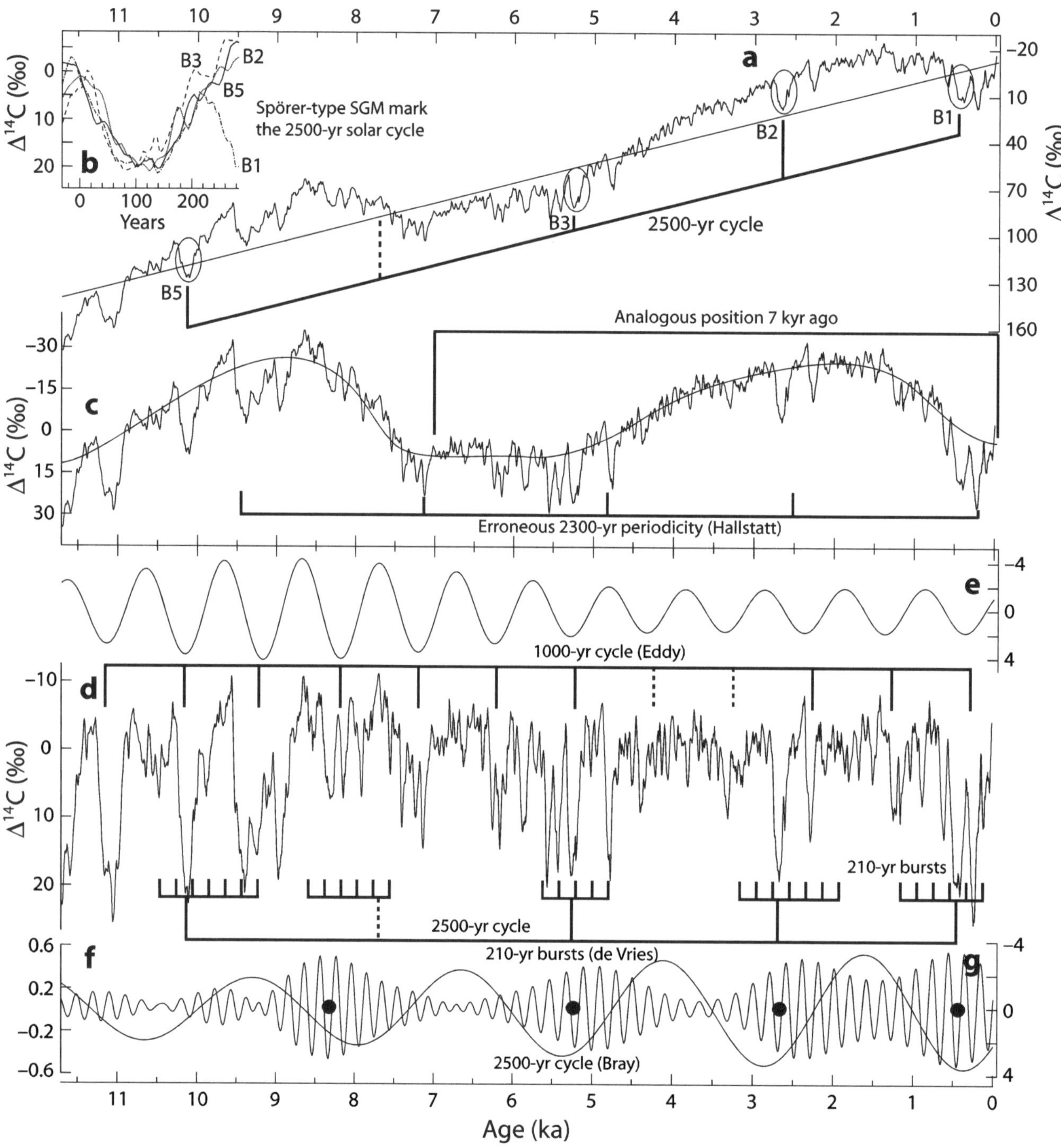

Fig. 8.11 Climate-relevant solar periodicities in the radiocarbon record
a) IntCal13 record of ¹⁴C variation (Δ^{14}C per mil, inverted scale), after Reimer et al. (2013). Thin line is a least-squares regression. Ovals include Spörer-type solar grand minima (SGM) that mark the lows of the 2500-yr Bray solar cycle, labeled B1 to B5. Vertical dashed line marks the center position between B3 and B5 due to the absence of a recognizable B4. b) Comparison between B1 (dash-dotted line), B2 (dotted line), B3 (dashed line), and B5 (continuous line) showing the great similitude in the SGM that take place at the lows of the Bray cycle, spanning c. 200 years and displaying a 20‰ increase in radiocarbon. c) Linearly detrended IntCal13 record showing on top the similitude of the present position in the radiocarbon record with 7 kyr ago. The scale at the bottom shows the reason a strong 2300-yr periodicity is found in frequency analysis. By several criteria (see Sect. 5.9) the Hallstatt periodicity does not correspond to a true solar cycle. d) Detrended IntCal13 record after subtracting from the linearly detrended record the empirical curve shown in (c). The scale above marks the SGM identified as belonging to the 1000-yr Eddy cycle on the basis of their position, with two dashed lines for missing SGM. The scale below marks the SGM identified as belonging to the 2500-yr Bray cycle on the basis of their position, with a dashed line for a missing SGM; and the solar variability that presents clear c. 200-yr spacing constituting the de Vries cycle bursts associated to the modulation of this cycle by the Bray cycle. e) 900–1100-yr band-pass filter of the detrended IntCal13 record. Notice the close match of the lows of the cycle with the SGM identified as belonging to the cycle in (d). f) 2344–2656-yr band-pass filter of the detrended IntCal13 record. There is also a good match between the lows of the cycle and the SGM identified as belonging to the cycle in (a) and (d). g) 180–220-yr band-pass filter of the detrended IntCal13 record. The bursts in signal amplitude match well the lows in the 2500-yr Bray cycle, except in the early Holocene when the presence of long periods of very large radiocarbon increases prevents the identification of the signal. Black dots indicate the position of the larger SGM within the bursts.

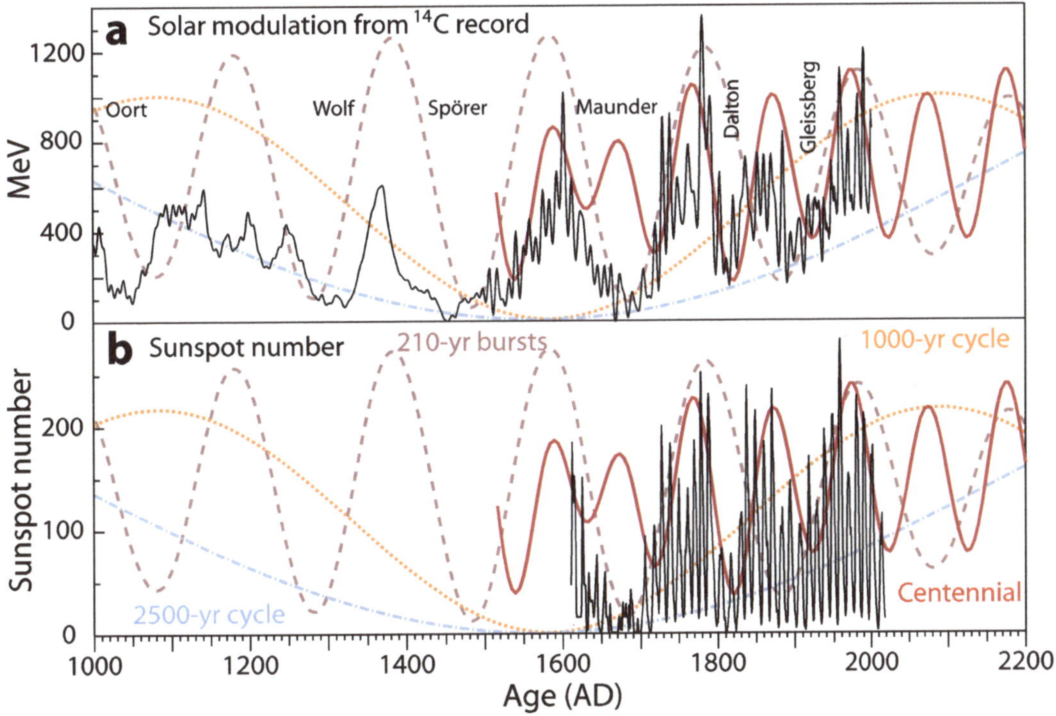

Fig. 8.12 Solar cycle interrelation during the past millennium
a) Solar activity for the past 1000 years reconstructed from the ¹⁴C record as solar modulation potential by Muscheler et al. (2007; thin black line). **b)** International sunspot number since 1700 from the Royal Observatory of Belgium (WDC–SILSO, yearly data to 1749 and 13-month average from 1749; thin black line), with scaled (12.08 factor) group sunspot number between 1610 and 1700 from Svalgaard & Schatten (2016). The 2500-yr (dash-dotted), 1000-yr (dotted), 200-yr (dashed), and 100-yr (continuous) sinusoid curves in (a) and (b) are the band-pass filtered corresponding frequencies from the IntCal13 radiocarbon reconstruction after Reimer et al. (2013), as explained in the text. They have been extrapolated between 1950 and 2200.

observations (Muscheler et al. 2016). The only possible conclusion is that solar variability profoundly affects climate on centennial and millennial timescales.

The matching of climate frequencies identified in proxy records to solar frequencies identified in cosmogenic records leads to a strong confirmation of the very significant climate effect of some of the long solar periodicities. These frequencies are the 2500-yr Bray, 1000-yr Eddy, and 210-yr de Vries cycles. Other frequencies (1770, 1300, 1125, 710, 510, and 350 years; McCracken et al. 2013a) have not been identified so far for their climatic effect, although it cannot be ruled out that they have it. To the well studied Bray, Eddy, and de Vries cycles we can add the periodicities found in 400-years of direct solar activity observations, the 100-yr Feynman, and 50-yr Pentadecadal cycles, that can be related to lesser effects on climate that could escape rigorous identification in climate proxies. The 150, 130, 88, 75, and 65-yr periodicities (McCracken et al. 2013a) are not observed in the sunspot record, and therefore are not active at present. To their existence as solar cycles we cannot attest at this time, as they cannot be independently confirmed by solar activity observations or climate records.

After identifying the relevant solar frequencies that affect climate, from proxies (2500, 1000, and 210 years), or from solar observations (100 years), it is necessary to study their recent evolution to see how they can contribute to explain both recent solar activity and recent climate change. For this purpose it is sufficient to use the ¹⁴C record in its 2013 release, IntCal13 (inverted in Fig. 8.11a; Reimer et al. 2013). The 2500-yr cycle can be clearly identified by four c. 200-yr long Spörer-like SGM that mark its

lows (Fig. 8.11b). A least-squares linear detrending reveals an already described c. 6000-yr quasi-periodicity (Fig. 8.11c; Xapsos & Burke 2009), showing also why the erroneous 2300-yr Hallstatt periodicity gives a better signal in frequency analysis than the correct 2500-yr Bray periodicity. Further detrending of this 6000-yr periodicity produces a record whose main characteristic is the presence of the 2500, 1000, and 210-yr periodicities in the form of SGM that cause an abrupt increase in ¹⁴C (Fig. 8.11d), that, as the Maunder Minimum showed, corresponds to a fall in solar activity. Filtering of this data with a band-pass filter at 900–1100-yr period reveals the Eddy cycle with its lows aligning with the SGM that usually take place 300–200 years before the change of millennium in BP scale (Fig. 8.11e). The two lows of this cycle that lack a prominent SGM are indicated in the scale above Fig. 8.11d with dashed lines. The use of a 2344–2656-yr band-pass filter reveals the Bray cycle with its lows close to the identified Spörer-type SGM (Fig. 8.11f), while a 180–220-yr filter shows the de Vries cycle bursts associated to the lows of the Bray cycle (Fig. 8.11g). The missing B4 low (vertical dashed line in Fig. 8.11a and at the bottom scale of 8.11d) displaces the fourth burst to the SGM at 8.3 kyr BP, while the very large SGM of the early Holocene hides the fifth burst. An 85–115-yr filter was also used to identify the Feynman cycle in the last 500 years of data (Fig. 8.12).

The detailed analysis of solar cycles interrelation for the past 1000 years compared to solar activity records (Fig. 8.12) produces some interesting observations. Data filtering places the last low of the Bray cycle at c. AD 1580, but as we have seen this low should coincide with the Spörer Minimum centered at 1460 AD. This earlier position is confirmed by the de Vries burst, that makes the lows of this cycle coincide with every SGM between AD 1000–1750. By 1880 the de Vries extended minimum (Gleissberg Minimum), is already a weak one, closer to the end of the burst. The center of the burst therefore can be placed at the beginning of the Spörer Minimum. The Eddy cycle shows a maximum c. AD 1100, coinciding with the Medieval Warm Period, and a minimum at c. AD 1600. The near-coincidence of the lows of the Eddy and Bray cycles takes place every c. 5000 years, defining periods of very intense climate change like the LIA. Since 1700 solar activity has been increasing according to the ^{14}C and sunspot records, in agreement with the Bray and Eddy cycle, with a maximum in this last one expected c. 2100.

The lack of synchronization between the de Vries and Feynman cycles that can be seen in sunspot records (see also Fig. 5.10), suggests they are not true harmonics despite de Vries period being double of Feynman period. Their lows appear not to coincide exactly, and comparison of band-pass filtered data at both frequencies confirms this difference (Fig. 8.12). An important conclusion from this comparison is that low solar activity in one cycle cannot be compensated by high activity in other cycles. Thus, the de Vries low during the Maunder Minimum coincides with a high in the Feynman cycle, while the Feynman low during the Dalton Minimum coincides with high activity in the de Vries cycle. In all these cases the effect of the cycle undergoing a low in activity prevails over cycles showing higher activity. The consequence of this predominance of cycles undergoing a minimum, over cycles that are not, is tantamount to solar cycles having only a negative effect on solar activity, and thus acting mainly at their lows. If correct, this would explain why the Eddy lows produce SGM even when the Bray cycle is in a high phase. It would also explain the modulation of the de Vries cycle by the Bray cycle, as only the negative effect on solar activity is cumulative between different cycles and thus the small effect of the shorter de Vries cycle is amplified when added to the larger effect by the Bray cycle, the biggest one that produces the largest SGM. This is another bias displayed by solar activity cycles that indicates that solar activity is maximal in the absence of cycles at lows, and the only effect of solar cycles is to reduce solar activity from its maximum potential. If this concept is correct, solar cycles can be considered inactive when sufficiently far from their lows, and solar maxima do not really exist, but correspond to periods when no cycle has a negative effect on solar activity. The Modern Solar Maximum (Solanki et al. 2004) would be the result of the absence of a negative effect from the Bray, Eddy, and de Vries cycles, that would have been inactive during that time.

The other conclusion from this comparison is that the de Vries cycle is already almost inactive, being already 600-years away from its burst center, and modern solar activity variability is being negatively affected mainly by the weakened but still active Feynman cycle (Fig. 8.12). This has important implications to project future solar activity based on cycles, as it is done in chapter 13 (Figs. 13.6 & 13.7). Since past climate depended on solar activity, future climate should also depend on it.

8.8 Conclusions

8a. The c. 1000-yr Eddy solar cycle seems to have dominated Holocene climate variability between 11,500–4,500 years BP and during the last two millennia, where it defines the Roman, Medieval, and Modern warm periods.

8b. The c. 210-yr de Vries solar cycle displays strong modulation by the 2500-yr Bray solar cycle, both in its cosmogenic isotope signature and in its climatic effects.

8c. The c. 88-yr Gleissberg solar cycle is ill-defined in the literature and hasn't manifested itself for the past 3,500 years. It is non detectable in sunspots records.

8d. The c. 11-yr Schwabe, the c. 100-yr Feynman, and the c. 50-yr Pentadecadal solar cycles are observable in the sunspot record. The Feynman solar cycle is responsible for the present extended solar minimum.

8e. In all cases a prolonged multi-decadal reduction in solar activity below average levels is associated with a decrease in temperature and a change in precipitation patterns. A delay between solar changes and climatic changes is observed in some studies.

8f. Solar variability has an asymmetric effect, with low activity displaying the most obvious climatic effects.

References

Abreu JA, Beer J & Ferriz–Mas A (2010) Past and future solar activity from cosmogenic radionuclides. In: Cranmer SR, Hoeksema JT and Kohl JL (eds) SOHO-23: understanding a peculiar solar minimum. ASP Conference Series 428 p 287

Adolphi F, Muscheler R, Svensson A et al (2014) Persistent link between solar activity and Greenland climate during the Last Glacial Maximum. Nature Geoscience 7 (9) 662–666

Anchukaitis KJ, Wilson R, Briffa KR et al (2017) Last millennium Northern Hemisphere summer temperatures from tree rings: Part II spatially resolved reconstructions. Quaternary Science Reviews 163 1–22

Berger JF, Delhon C, Magnin F et al (2016) A fluvial record of the mid-Holocene rapid climatic changes in the middle Rhone valley (Espeluche–Lalo, France) and of their impact on Late Mesolithic and Early Neolithic societies. Quaternary Science Reviews 136 66–84

Björck S, Muscheler R, Kromer B et al (2001) High-resolution analyses of an early Holocene climate event may imply decreased solar forcing as an important climate trigger. Geology 29 (12) 1107–1110

Blunier T, Chappellaz J, Schwander J et al (1995) Variations in atmospheric methane concentration during the Holocene epoch. Nature 374 (6517) 46–49

Bond G, Kromer B, Beer J et al (2001) Persistent solar influence on North Atlantic climate during the Holocene. Science 294 (5549) 2130–2136

Breitenmoser P, Beer J, Brönnimann S et al (2012) Solar and volcanic fingerprints in tree-ring chronologies over the past 2000 years. Palaeogeography, Palaeoclimatology, Palaeoecology 313 127–139

Brugnara Y, Brönnimann S, Luterbacher J & Rozanov E (2013) Influence of the sunspot cycle on the Northern Hemisphere wintertime circulation from long upper-air data sets. Atmospheric chemistry and physics 13 (13) 6275–6288

Chappellaz J, Stowasser C, Blunier T et al (2013) High-resolution glacial and deglacial record of atmospheric methane by continuous-flow and laser spectrometer analysis along the NEEM ice core. Climate of the Past 9 (6) 2579–2593

Clilverd MA, Clarke E, Ulich T et al (2006) Predicting solar cycle 24 and beyond. Space weather 4 (9) 1–7

Debret M, Bout–Roumazeilles V, Grousset F et al (2007) The origin of the 1500-year climate cycles in Holocene North-Atlantic records. Climate of the Past Discussions 3 (2) 679–692

Duan F, Wang Y, Shen CC et al (2014) Evidence for solar cycles in a late Holocene speleothem record from Dongge Cave, China. Scientific Reports 4 5159

Eddy JA (1976) The Maunder Minimum. Science 192 (4245) 1189–1202

Eichler A, Olivier S, Henderson K et al (2009) Temperature response in the Altai region lags solar forcing. Geophysical Research Letters 36 (1) L01808

Feynman J (1982) Geomagnetic and solar wind cycles 1900–1975. Journal of Geophysical Research: Space Physics 87 (A8) 6153–6162

Feynman J & Ruzmaikin A (2014) The Centennial Gleissberg Cycle and its association with extended minima. Journal of Geophysical Research: Space Physics 119 (8) 6027–6041

Fleitmann D, Burns SJ, Mangini A et al (2007) Holocene ITCZ and Indian monsoon dynamics recorded in stalagmites from Oman and Yemen (Socotra). Quaternary Science Reviews 26 (1–2) 170–188

Gleissberg W (1944) Evidence for a long solar cycle. The Observatory 65 (282) 123–125

Hathaway DH & Upton LA (2016) Predicting the amplitude and hemispheric asymmetry of solar cycle 25 with surface flux transport. Journal of Geophysical Research: Space Physics 121 (11) 1–10

Hodell DA, Brenner M, Curtis JH & Guilderson T (2001) Solar forcing of drought frequency in the Maya lowlands. Science 292 (5520) 1367–1370

Inceoglu F, Simoniello R, Knudsen MF et al (2015) Grand solar minima and maxima deduced from ^{10}Be and ^{14}C: magnetic dynamo configuration and polarity reversal. Astronomy & Astrophysics 577 pA20

Kern AK, Harzhauser M, Piller WE et al (2012) Strong evidence for the influence of solar cycles on a Late Miocene lake system revealed by biotic and abiotic proxies. Palaeogeography, Palaeoclimatology, Palaeoecology 329 124–136

Knudsen MF, Seidenkrantz MS, Jacobsen BH & Kuijpers A (2011) Tracking the Atlantic Multidecadal Oscillation through the last 8,000 years. Nature communications 2 178

Kobashi T, Severinghaus JP, Brook EJ et al (2007) Precise timing and characterization of abrupt climate change 8200 years ago from air trapped in polar ice. Quaternary Science Reviews 26 (9–10) 1212–1222

Kobashi T, Box JE, Vinther BM et al (2015) Modern solar maximum forced late twentieth century Greenland cooling. Geophysical Research Letters 42 (14) 5992–5999

Ma LH (2007) Thousand-year cycle signals in solar activity. Solar Physics 245 (2) 411–414

Magny M (2004) Holocene climate variability as reflected by mid-European lake-level fluctuations and its probable impact on prehistoric human settlements. Quaternary international 113 (1) 65–79

Magny M, Vannière B, De Beaulieu JL et al (2007) Early-Holocene climatic oscillations recorded by lake-level fluctuations in west-central Europe and in central Italy. Quaternary Science Reviews 26 (15–16) 1951–1964

Marchitto TM, Muscheler R, Ortiz JD et al (2010) Dynamical response of the tropical Pacific Ocean to solar forcing during the early Holocene. Science 330 (6009) 1378–1381

McCracken KG, Beer J, Steinhilber F & Abreu J (2013a) A phenomenological study of the cosmic ray variations over the past 9400 years, and their implications regarding solar activity and the solar dynamo. Solar Physics 286 (2) 609–627

McCracken K, Beer J, Steinhilber F & Abreu J (2013b) The heliosphere in time. Space Science Reviews 176 (1–4) 59–71

Miller AJ, Cayan DR, Barnett TP et al (1994) The 1976–77 climate shift of the Pacific Ocean. Oceanography 7 (1) 21–26

Minobe S (2000) Spatio-temporal structure of the pentadecadal variability over the North Pacific. Progress in Oceanography 47 (2–4) 381–408

Muscheler R, Joos F, Beer J et al (2007) Solar activity during the last 1000 yr inferred from radionuclide records. Quaternary Science Reviews 26 (1–2) 82–97

Muscheler R, Adolphi F, Herbst K & Nilsson A (2016) The revised sunspot record in comparison to cosmogenic radionuclide-based solar activity reconstructions. Solar Physics 291 (9–10) 3025–3043

Novello VF, Vuille M, Cruz FW et al (2016) Centennial-scale solar forcing of the South American Monsoon System recorded in stalagmites. Scientific Reports 6 24762

Nussbaumer SU, Steinhilber F, Trachsel M (2011) Alpine climate during the Holocene: a comparison between records of glaciers lake sediments and solar activity. Journal of Quaternary Science 26 (7) 703–713

Parker B (2011) The tide predictions for D-Day. Physics Today 64 (9) 35–40

Peristykh AN & Damon PE (2003) Persistence of the Gleissberg 88-year solar cycle over the last ~ 12,000 years: Evidence from cosmogenic isotopes. Journal of Geophysical Research: Space Physics 108 (A1)

Pesnell WD (2008) Predictions of solar cycle 24. Solar Physics 252 (1) 209–220

Rasmussen SO, Vinther BM, Clausen HB & Andersen KK (2007) Early Holocene climate oscillations recorded in three Greenland ice cores. Quaternary Science Reviews 26 (15–16) 1907–1914

Raspopov OM, Dergachev VA & Guskova EG (2008) On a combined influence of long-term solar activity variations and geomagnetic dipole changes on climate change. In: Proceedings of the 7th International Conference Problems of Geocosmos, St Petersburg, 26–30 May 2008 229–234

Reimer PJ, Bard E, Bayliss A et al (2013) IntCal13 and Marine13 radiocarbon age calibration curves 0–50,000 years cal BP. Radiocarbon 55 (4) 1869–1887

Rohling EJ & Pälike H (2005) Centennial-scale climate cooling with a sudden cold event around 8,200 years ago. Nature 434 (7036) 975–979

Sánchez–Sesma J (2015) Multi-millennial-scale solar activity and its influences on continental tropical climate: empirical evidence of recurrent cosmic and terrestrial patterns. Earth System Dynamics Discussions 6 (2)

Scafetta N (2012) Multi-scale harmonic model for solar and climate cyclical variation throughout the Holocene based on Jupiter–Saturn tidal frequencies plus the 11-year solar dynamo cycle. Journal of Atmospheric and Solar-Terrestrial Physics 80 296–311

Simonneau A, Chapron E, Garçon M et al (2014) Tracking Holocene glacial and high-altitude alpine environments fluctuations from minerogenic and organic markers in proglacial lake sediments (Lake Blanc Huez, Western French Alps). Quaternary Science Reviews 89 27–43

Solanki SK, Usoskin IG, Kromer B et al (2004) Unusual activity of the Sun during recent decades compared to the previous 11,000 years. Nature 431 (7012) 1084–1087

Sonett CP & Suess HE (1984) Correlation of bristlecone pine ring widths with atmospheric ^{14}C variations: a climate–Sun relation. Nature 307 (5947) 141–143

Sushch I & Hnatyk B (2014) Modelling of the radio emission from the Vela supernova remnant. Astronomy & Astrophysics 561 pA139

Svalgaard L & Schatten KH (2016) Reconstruction of the sunspot group number: the backbone method. Solar Physics 291 (9–10) 2653–2684

Thompson LG, Mosley–Thompson E, Brecher H et al (2006) Abrupt tropical climate change: Past and present. Proceedings of the National Academy of Sciences 103 (28) 10536–10543

Usoskin IG (2013) A history of solar activity over millennia. Living Reviews in Solar Physics 10 (1) 1–94

Usoskin IG (2017) A history of solar activity over millennia. Living Reviews in Solar Physics 14 (1) 3

Usoskin IG, Solanki SK & Kovaltsov GA (2007) Grand minima and maxima of solar activity: new observational constraints. Astronomy & Astrophysics 471 (1) 301–309

Usoskin IG, Gallet Y, Lopes F et al (2016) Solar activity during the Holocene: the Hallstatt cycle and its consequence for grand minima and maxima. Astronomy & Astrophysics 587 A150

van der Plicht J, van Geel B, Bohncke SJP et al (2004) The Preboreal climate reversal and a subsequent solar-forced climate shift. Journal of Quaternary Science 19 (3) 263–269

Vieira LEA & Solanki SK (2010) Evolution of the solar magnetic flux on time scales of years to millenia. Astronomy & Astrophysics 509 A100

Vieira LEA, Solanki SK, Krivova NA & Usoskin I (2011) Evolution of the solar irradiance during the Holocene. Astronomy & Astrophysics 531 A6

Wang T, Surge D & Mithen S (2012) Seasonal temperature variability of the Neoglacial (3300–2500 BP) and Roman Warm Period (2500–1600 BP) reconstructed from oxygen isotope ratios of limpet shells (Patella vulgata), Northwest Scotland. Palaeogeography, Palaeoclimatology, Palaeoecology 317 104–113

Wu CJ, Usoskin IG, Krivova N et al (2018) Solar activity over nine millennia: A consistent multi-proxy reconstruction. Astronomy & Astrophysics 615 A93

Xapsos MA & Burke EA (2009) Evidence of 6 000-year periodicity in reconstructed sunspot numbers. Solar Physics 257 (2) 363–369

Zhou J & Tung KK (2013) Observed tropospheric temperature response to 11-yr solar cycle and what it reveals about mechanisms. Journal of the Atmospheric Sciences 70 (1) 9–14

GREENHOUSE GASES AND CLIMATE CHANGE

"Human beings are now carrying out a large scale geophysical experiment of a kind that could not have happened in the past nor be reproduced in the future."
Roger Revelle and Hans E. Suess (1957)

9.1 Introduction

The greenhouse effect (GHE) was established in the 19th century from an initial proposition by Joseph Fourier in 1824. It is based in solid physical principles and provides an explanation for the average surface temperature of the Earth, Venus, and Titan, taking into account the solar irradiance they receive and the greenhouse gases (GHGs) content of their atmospheres. The CO_2 hypothesis attempts to explain Earth's changing climate in terms of changes in atmospheric CO_2. It assumes that water vapor (humidity) only changes as a result of temperature changes, and tries to demonstrate a correlation between long-term changes in temperature and CO_2. It was initially proposed by Svante Arrhenius in 1894 to try to explain Pleistocene glaciations, but it was refuted as their cause in 1976, when Hays, Imbrie and Shackleton demonstrated that Pleistocene glaciations respond to orbital changes, as proposed by Milutin Milanković in 1920. The demonstration by Suess, Revelle and Keeling between 1955 and 1960 that human activities were fundamentally altering CO_2 levels in the atmosphere, started to matter politically only when surface warming was detected in the early 1980s.

Starting in the late 1960s the Western World became worried about the evident environmental degradation, wildlife population decline, and human encroachment on remaining natural spaces. These observations were linked to population growth and human activities. The discovery by Molina and Rowland in 1974 that chlorofluorocarbons destroyed ozone was followed by a US ban of chlorofluorocarbons in aerosol cans in 1978, and their ban by 43 nations in the Montreal Protocol of 1987. This political success story created a good precedent to tackle the next global problem. The world not only needed saving from us, but the process could lead to a strengthening of supranational organizations, reduce the chances for conflict and level global economic differences. The following year global warming entered politics with the June 1988 Toronto Conference on the Changing Atmosphere, that outlined the dangers from global warming and sea level rise and made the first call for a reduction in CO_2 emissions. Five days before the conference started, James Hansen, director of NASA GISS, testified before the US Senate Committee on Energy and Natural Resources, saying that NASA was 99% confident that the warming was caused by the accumulation of GHGs in the atmosphere. Five months later, and before the year was over, the World Meteorological Organization (WMO) and the UN Environment Programme (UNEP) created the Intergovernmental Panel on Climate Change (IPCC) to provide governments with an assessment of the available scientific knowledge on the issue.

In the 30 years since the IPCC was created a strong scientific consensus has developed about the CO_2 hypothesis. This is despite a disappointing lack of advance in quantifying the effect of CO_2 changes on global temperature (Fig. 9.1). There is clear evidence for the observed warming, clear evidence for the human-caused increase in CO_2, and abundant evidence of the CO_2 effects on the biosphere. Most climatic changes detected can be clearly tied, at least in part, to the warming observed, although internal variability is often underestimated as a factor for climate change. What is crucially missing is clear indisputable evidence that the warming observed is the result of the measured increase in CO_2, which the IPCC believes to be unequivocal.

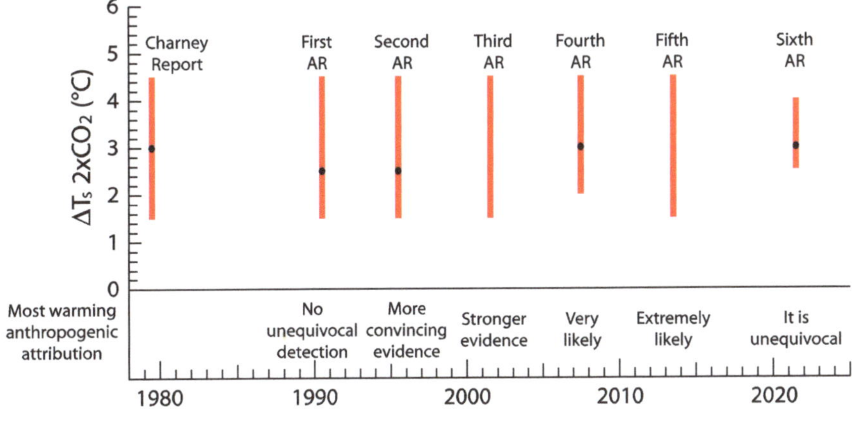

Fig. 9.1 Lack of progress in quantifying the effect of CO_2 increase on global temperature Top: Vertical bars represent estimates of the Equilibrium Climate Sensitivity (ECS), or global average temperature increase for a doubling of CO_2 levels in the reports indicated. Black dots represent best estimate values. AR: Assessment Reports by IPCC. **Bottom:** Confidence statements on the attribution of most of the observed warming to the human-caused CO_2 increase in the IPCC reports. Lack of progress in determining the climatic effect of CO_2 contrasts vividly with increasing certainty that said climatic effect is due to human-emitted CO_2. After Arias et al (2021) and IPCC reports.

9.2 Towards a greenhouse theory of climate

Early 19th century people had no reason to believe the Earth's climate had been significantly different in the distant past. The then dominant Plutonist theory of rock formation defended by James Hutton, suggested that the planet had been cooling from its igneous origin so, if anything, the past must have been warmer. Since the mid–18th century there were reports by naturalists that erratic rock blocks found all over North and Central Europe, were being explained by some perspicacious Alpine people as transported by glaciers that must have been much larger in the past. But the scholars of the time were not prepared to accept such ideas, and the great Biblical Flood was the consensus theory of the time. In the early 19th century the first hypotheses of a glacial past were taking shape but were vigorously rejected by the orthodoxy. A reluctant Jean de Charpentier was convinced by the evidence of the glacial transport idea presented to him by an engineer and a peasant, and built a solid hypothesis of greatly expanded Alps glaciers in the past. At the same time Karl Friedrich Schimper, botanist and poet, also interested in astronomy, developed the hypothesis that the erratic blocks had been transported by glaciers during global winters that alternated with global summers, that he called activation and stagnation periods. He coined the term Ice Age in a poem in 1837. The evidence from moraines indicated that several cold periods must have taken place. However, in modern geological terminology, the scholars of the 19th century were not studying ice ages, but glacial periods within the present Pleistocene Glaciation, part of the Late Cenozoic Ice Age.

De Charpentier and Schimper had in common a friendship with Louis Agassiz and together convinced him of the glacial hypothesis. At the 1837 Swiss Scientific Society meeting at Neuchâtel, Louis Agassiz abandoned his fossil fish studies and presented a synthesis of his friends'

hypotheses. From then on, he would defend the glacial theory as his own, without giving any credit to his ex–friends (Krüger 2013). The glacial theory created a huge scientific controversy and was not generally accepted until 1860–80. But the search for a cause for the glacial periods started as soon as the theory was widely known. The first to advance an explanation, in 1842, was Joseph Adhémar. He had been working since 1830 on an astronomical hypothesis that could explain cyclical climatic changes, whether floods or glaciations, by way of orbital changes.

In mid–19th century any theory of climate had only one thing to explain: the glacial periods recently discovered. At that time paleontologists were describing animals that could only have lived in a warmer world, but it was generally accepted that the progressive cooling of the Earth from a hot origin, was the explanation for a warmer past interrupted by several glaciations. Joseph Fourier had proposed in 1824 that part of the heat produced by solar radiation was being intercepted by the atmosphere and reflected back to the Earth. In this context, John Tyndall had worked on glacier movements in the 1850s and, interested by the cause of glaciations, considered that changes in the atmosphere could be responsible. In 1859 he started experimenting with heat absorption by gases. He quickly found that some complex gases, water vapor, CO_2, and ozone, were much better absorbers than oxygen or nitrogen. He concluded that water vapor was responsible for most of the heat retained by the atmosphere, and thus the most important gas in regulating the temperature of the planet's surface. He inferred that a reduction in water vapor by any cause should be sufficient to trigger a glaciation.

The GHE (Fig. 9.2) is due to the presence in the atmosphere of a significant amount of gases that are good IR absorbers and emitters (H_2O, CO_2, CH_4, O_3). As a result, the effective radiation to space takes place higher in the atmosphere. This makes the upper atmosphere cool and the surface and lower atmosphere warm. The temperature difference between upper and lower atmosphere due to the

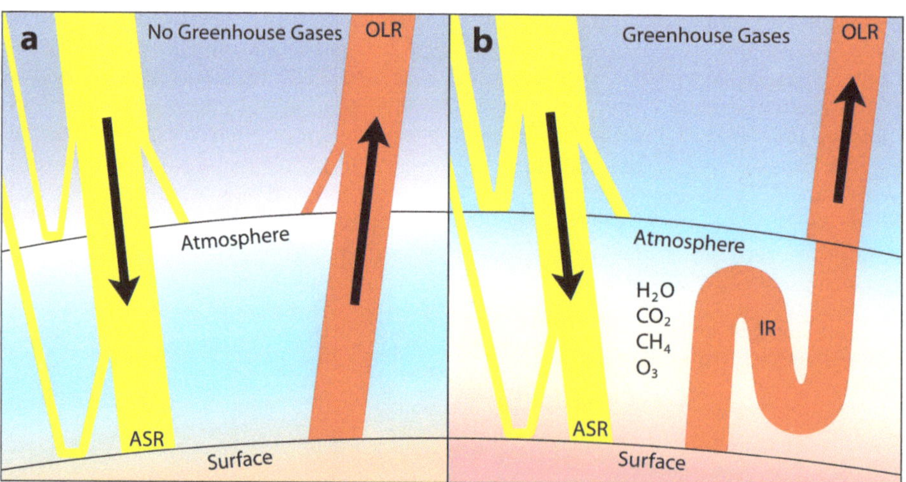

Fig. 9.2 The Greenhouse effect
a) In the absence of GHGs, an atmosphere is largely transparent to both shortwave and longwave radiation. Absorbed shortwave radiation (ASR) warms the surface, and outgoing longwave radiation (OLR) cools it. **b)** When the atmosphere contains GHGs infrared radiation (IR) from the surface is intercepted by the atmosphere, and part of it redirected to the surface. A temperature gradient is established in the atmosphere, and OLR emission takes place mostly from the upper atmosphere. As the temperature of the upper atmosphere where effective emission takes place is lower, the surface and lower atmosphere must warm to increase emission so the balance between ASR and OLR is maintained. The average temperature at the surface is not only higher, but daily, seasonal, and latitudinal temperature differences display a narrower range.

GHE is responsible for most of the convective overturning that manifests itself as weather.

In Sweden, the interest in glacial periods coincided with technical progress in CO_2 measurements and with the first rough calculations of the carbon cycle by Arvid Högbom. The first consideration that changes in atmospheric CO_2 could be responsible for glaciations and deglaciations was born among Stockholm members of the Swedish Physicists Society. Svante Arrhenius proposed in 1894 that, as the amount of water vapor in the atmosphere depends on its temperature, the real trigger of ice ages must be CO_2. He calculated that a 40% reduction in CO_2 was sufficient to reduce the temperature 5 °C and start a glaciation. Nils Ekholm, a close friend of Arrhenius, was the first to introduce the greenhouse analogy as a partial explanation to the effect of CO_2 (Ekholm 1901). The analogy, although inexact, proved popular, and in 1907 John Poynting introduced the "Greenhouse Effect," and soon after Frank Very named "The Greenhouse Theory" in his response to him (Poynting 1907; Very 1908). But the simplifications taken by Arrhenius, the suspicion that CO_2 role could be overestimated due to spectrum masking by water vapor, and experiments showing CO_2 effect saturating at amounts considered relevant, caused the Arrhenius theory to be disregarded in the early 20th century.

Between 1840 and 1970 scientific consensus swung back and forth between the astronomical and greenhouse theories to explain glaciations. From 1938 Guy Callendar revived the CO_2 hypothesis taking advantage of a warming period (1910–1940), that he attributed to an increase in CO_2. Objections to the theory were numerous due to the complexity of the climate physics, but Callendar and others were able to improve the theory by showing that the spectral masking at sea level did not take place in the upper atmosphere where the spectral lines were separated. By the early 1950s it was clear that the increase in CO_2 mattered in the upper atmosphere and should necessarily have a warming effect on surface temperature.

The other main roadblock to the acceptance of the CO_2 effect as an explanation was the argument that the ocean could easily dispose of the added CO_2. Hans Suess was able to detect the reduction in $^{14}CO_2$ produced by the burning of fossil fuel (the Suess effect; Suess 1955). It was direct evidence that fossil fuel CO_2 was producing a change in the atmosphere. Suess was hired by oceanographer Roger Revelle, director of Scripps Institution of Oceanography, who had established from his work on radioactive fallout that only the upper layer of the ocean was available for CO_2 exchange over relevant timescales. Revelle's ocean chemistry knowledge led them to propose that the buffering capacity of the ocean would result in a reduced capacity to absorb atmospheric CO_2 (the Revelle effect; Revelle & Suess 1957). As Revelle noticed, his discovery implied that the world was destined to undergo a rapid and unexpected increase in atmospheric CO_2. Revelle also hired Charles Keeling, who carefully set up a research station at the Mauna Loa volcano in Hawaii to reliably measure CO_2 concentrations for the first time. In just two years Keeling was able to report that CO_2 levels in the atmosphere were increasing, confirming that the ocean was not capable of absorbing all industrial CO_2 emissions (Keeling 1960).

Meanwhile the astronomical theory of glaciations had fallen out of favor. Milankovitch theory had been tied to Albrecht Penck and Eduard Brückner's reconstruction of Alpine glacier history. This reconstruction fit the astronomical calculations particularly well (Köppen & Wegener 1924). This helped Milankovitch theory at first, and during the 1930s and 1940s most European geologists accepted the astronomical explanation. But all that changed in 1951, when radiocarbon dating became available. The Alpine reconstruction with four glacial periods in the past 600 Ka was shown incorrect, and ice fluctuations for the past 80 Ka appeared more frequent than Milankovitch theory allowed. By 1955 Milankovitch theory had lost nearly all its supporters.

In 1960 the principles of a CO_2 hypothesis based on the greenhouse effect capable of explaining glacial periods had been solidly established. It was clear that the increase in CO_2 was a cause for warming by increasing the height of radiation to space. It had been demonstrated that CO_2 was rapidly increasing as the ocean could not absorb the CO_2 quickly enough. Therefore, CO_2 levels in the atmosphere could change significantly over a short period of time, and so could the temperature of the Earth. At the same time few people believed in the competing astronomical theory. But as is often the case, nature refused to cooperate and the world entered a period of cooling. The cooling began in the mid–1940s, but became very noticeable in the mid–1960s. The lack of warming despite increasing CO_2 levels put the CO_2 hypothesis in limbo. The greenhouse effect was accepted, as it is based on solid science, but the changes in CO_2 could not explain the changing climate, so many scientists believed other significant factors were at play. And then incontrovertible evidence was produced in favor of the astronomical theory. Scientists pulled out benthic cores from the Indian Ocean that demonstrated without doubt that the climate of the past 450 Ka responded within 5% of error to the Milankovitch orbital frequencies (Hays et al. 1976; see Chap. 2). The astronomical theory of Milankovitch finally provided the long-sought explanation for glacial periods of the past. A 140-year quest had ended.

The CO_2 hypothesis had been developed to explain glaciations but could not explain why they took place at exactly the frequencies predicted by Milankovitch theory. Undaunted, its supporters soon regrouped. Perhaps the hypothesis could not explain when glaciations took place, but it could still explain their magnitude and effects. An ice core from Antarctica showed in 1985 that temperature, CO_2, and methane, correlated through the glacial cycle. Despite being something that should be expected in any case, it was proposed that the correlation suggested GHGs were required to amplify the effects of orbital shifts, and to maintain inter-hemispheric synchroneity (Lorius et al. 1985). GHGs role in the glacial cycle remains controversial (see Sect. 2.11) and supported mainly with models. The debate about CO_2 controlling the planet's climate was forced out of the Pleistocene by the evidence. Ice cores provided evidence that Holocene climate had not been under CO_2 control either (see Sect. 4.3). Our current understanding is that the 1910–40 warming period was not due to an increase in GHGs, refuting Callendar's belief (Meehl et al. 2004; see Fig. 12.2). The scope of the CO_2 hypothesis has been reduced to the distant past, before the Pleistocene, and the last warming period since 1976.

9.3 Past atmospheric changes and climate evolution

9.3.1 The Faint Sun Paradox

As the Sun transforms hydrogen into helium, the proportion of hydrogen at its core decreases over time. This should reduce the production of energy, but as the energy produced is what keeps the Sun from collapsing from its gravitational pull, the core contracts increasing the density of hydrogen by the amount needed to keep equilibrium. The slow change results in the outer core becoming progressively hotter and brighter, the increased radiation pushes the outer layers making the Sun bigger and brighter over time. It has been calculated that the Sun must have expanded by 10%, warmed 200 K, and increased in brightness by 30% in the 4.57 Gyr since it reached the main sequence after igniting. During this time it has consumed half of the hydrogen in the core, allowing the development of life and climate on Earth. The increase in solar brightness with time is the inevitable consequence of well understood physical principles.

In the early 1970s it was recognized that the faint young Sun created a paradox (Sagan & Mullen 1972). Under present conditions, the reduced brightness of the young Sun implied the Earth should have been frozen until 2.3 Ga, yet there is very solid geological and paleontological evidence of substantial surface liquid water presence between 3–4 Ga. Conditions in the early Earth must have been very different, and the two most likely factors to explain the Faint Sun Paradox are differences in GHGs and in albedo, that currently reflects about 30% of incoming solar radiation. Other changes in the Sun and Earth complicate the interpretation of the paradox (Fig. 9.3). The Sun has also been decreasing its rate of rotation, the process responsible for driving its magnetohydrodynamic dynamo, with the concomitant decrease in solar magnetic activity and solar wind strength over time (Feulner 2012). A much higher rate of solar superflares and coronal mass ejections from the young Sun has also been proposed as a solution to the paradox (Airapetian et al. 2016). The Earth's magnetic field should have been 30–75% weaker in the distant past, before its inner core started solidifying about 2.5–0.5 Ga (Labrosse et al. 2001), indicating the Earth was subjected to a more intense high-energy flux, which affected the ancient atmosphere composition. The Earth has also been decreasing its rate of rotation and this has important implications because the size of atmospheric eddies, and thus the amount of energy they transport, depends on the speed of rotation. Models show a faster speed of rotation results in smaller eddies, decreasing latitudinal heat transport (Kuhn et al. 1989). The continental area has also grown enormously from about 10% of its present value 3.8 Gyr ago, and this growth should have resulted in an increase in albedo over time. Lack of biologically-derived condensation nuclei has been proposed as another cause for an ancient reduced albedo that could help explain the paradox (Rosing et al. 2010).

The Faint Sun Paradox cannot be considered solved (Feulner 2012). There is controversy over the importance of GHGs, albedo and other factors in warming the young Earth. Our knowledge of the conditions so long ago is very limited. Nevertheless, it appears that a combination of factors including higher levels of GHGs and a reduced albedo is the most likely answer to a liquid water world under a faint Sun. In any case, the Faint Sun Paradox cannot provide support for the CO_2 hypothesis, as we don't know how much CO_2 there was in the ancient atmosphere, nor can we assess the contribution from GHGs to the warming of the Earth 4–2 Ga, versus other factors.

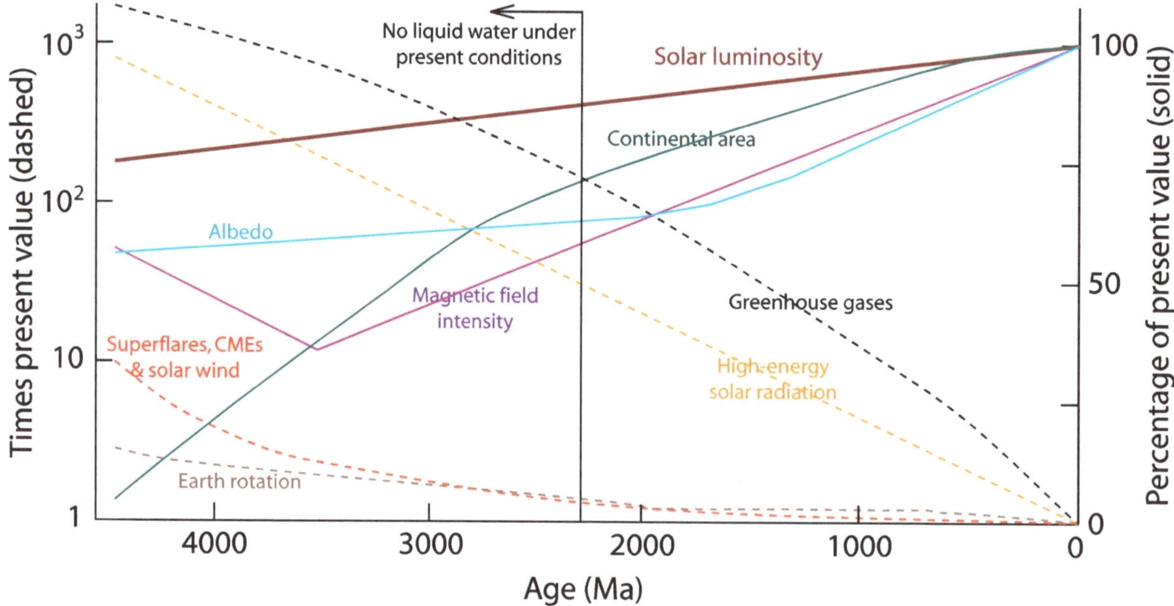

Fig. 9.3 Factors affecting the Faint Sun Paradox
Due to stellar evolution, the Sun has been increasing in luminosity (thick line). Under a fainter Sun the Earth could not have had surface liquid water with present conditions prior to 2.3 Ga, yet there is evidence for liquid water as early as 3.8 Ga. Factors that could affect temperature and are believed to have changed are shown, with those decreasing over time with dashed lines (left scale), and those increasing over time with solid lines (right scale). Values shown are highly speculative. See Feulner (2012), and references therein. CMEs: coronal mass ejections.

9.3.2 Phanerozoic climate

The Phanerozoic eon spans the last 541 Ma and begins with the relatively sudden appearance of complex metazoans in the Cambrian explosion. It has been speculated that solidification of the Earth's inner core, and the resulting strengthening of the magnetic field, caused a decrease in high-energy solar flux allowing the development of a more protective atmosphere and of complex life on land (Doglioni et al. 2016). Several Phanerozoic geochemical proxy records have prompted attempts at reconstructing its temperature and GHG evolution. In his 1979 book "Climates throughout geologic time," Lawrence Frakes presented one of the first temperature reconstructions for this period of the Earth, proposing an alternation of glacial periods and comparatively warmer periods supported on abundant geological data. His conclusion attests to the predominant view at the time:

"Retention of heat energy on the earth has varied through time and has given rise to variable climates. Retention appears to be in part a function of the global climatic state, including global albedo and distribution of land and sea. To a lesser extent, factors such as the magnitude of the greenhouse effect and atmospheric and interstellar interference with solar radiation may have influenced the history of global climate by contributing positively or negatively to the net heat flux" (Frakes 1979).

In their follow-up 1992 book "Climate modes of the Phanerozoic," Frakes et al. translate this climatic alterna-

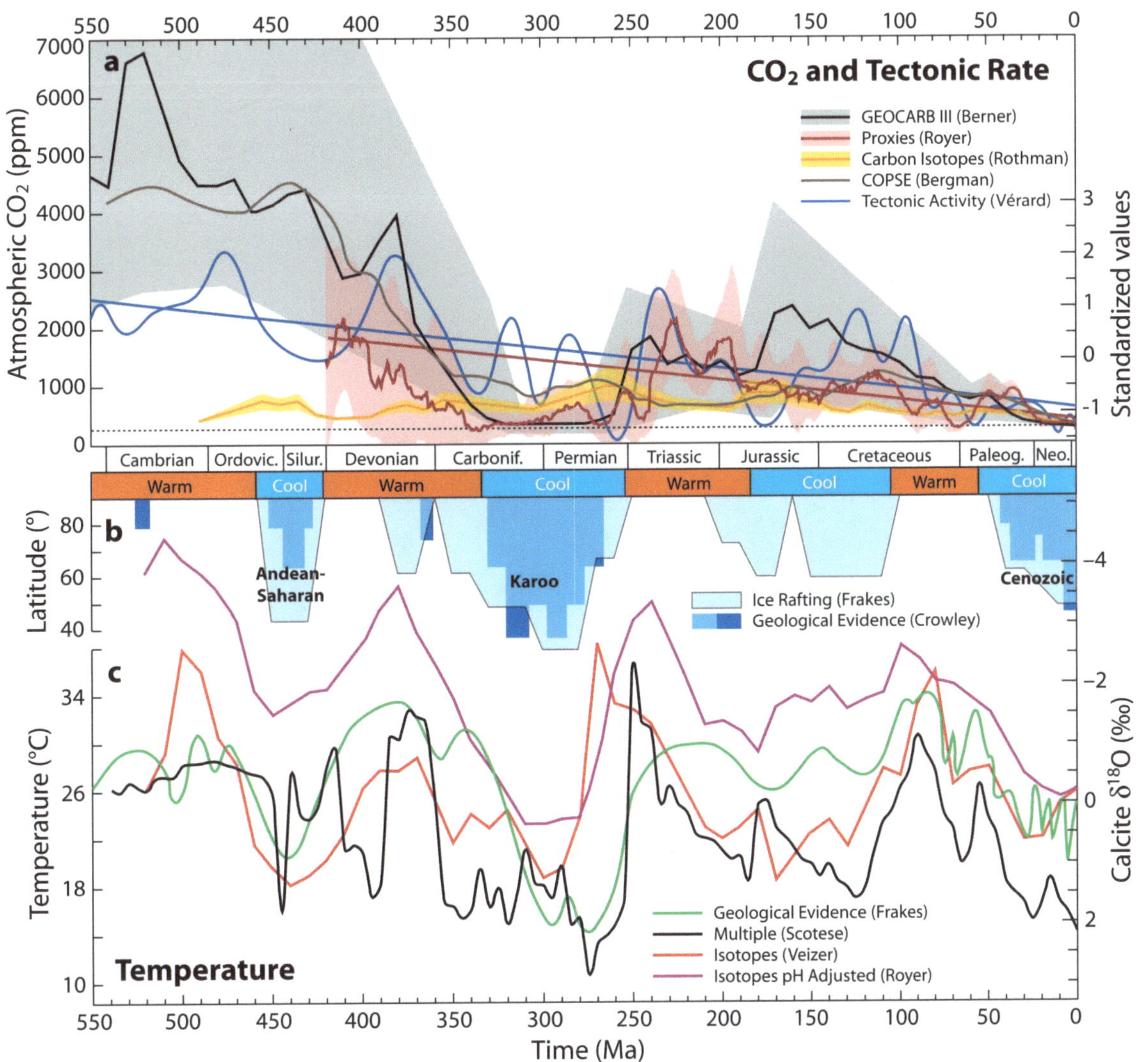

Fig. 9.4 Phanerozoic CO₂ levels, tectonic activity and climate indicators
a) Phanerozoic atmospheric CO₂ levels in ppm (left scale) from GEOCARB III model (black curve with grey estimate of error; after Berner & Kothavala 2001), multi-proxy reconstruction (dark red curve with pink 95% confidence interval, red line is least squares fit; after Foster et al. 2017), CO₂ reconstruction from Δ¹³C isotopic inorganic/organic carbon fractionation (orange curve with gold uncertainty; after Rothman 2002), and COPSE model (brown curve; after Bergman et al. 2004). Horizontal dotted line indicates AD 1750 CO₂ level of 270 ppm. Blue curve represents combined tectonic rate in standardized values, and blue line its linear regression (right scale; after Vérard et al. 2015). **b)** Alternating Warm (orange) and Cool (blue) Modes, after Frakes et al. (1992), and glacial indicators, measured by the latitude reached from one of the poles. Ice rafting evidence (ice blue areas; after Frakes & Francis 1988), and direct geological evidence for glaciation (azure and blue areas; after Crowley 1998). **c)** Phanerozoic temperature reconstructions: Mean global temperature estimate (no scale) based on sedimentology and paleoecology (green curve; after Frakes 1979, and Frakes et al. 1992). Mean global temperature reconstruction based on geological evidence and isotopic studies (black curve, left scale; after Scotese 2018). Tropical marine shallow water calcite ¹⁸O levels detrended (red curve, right scale; after Veizer et al. 2000). Tropical marine shallow water calcite ¹⁸O levels detrended and pH adjusted with GEOCARB III CO₂ data (violet curve; after Royer et al. 2004).

tion into four warm modes alternated with four cool modes (Fig. 9.4, orange and blue boxes), more clearly delimitated by their seminal study on Phanerozoic ice rafting (Frakes & Francis 1988; Fig. 9.4b, light blue boxes).

"Phanerozoic climate history has been divided into four Warm Modes and four Cool Modes. Among Warm Modes possibly the warmest was the late Cretaceous to early Tertiary, with the Silurian–Devonian and early Mesozoic ranked equal second. The position of the Cambrian–Ordovician is uncertain owing to the lack of information from high latitudes. The most extreme cold climates prevailed during the late Palaeozoic and the late

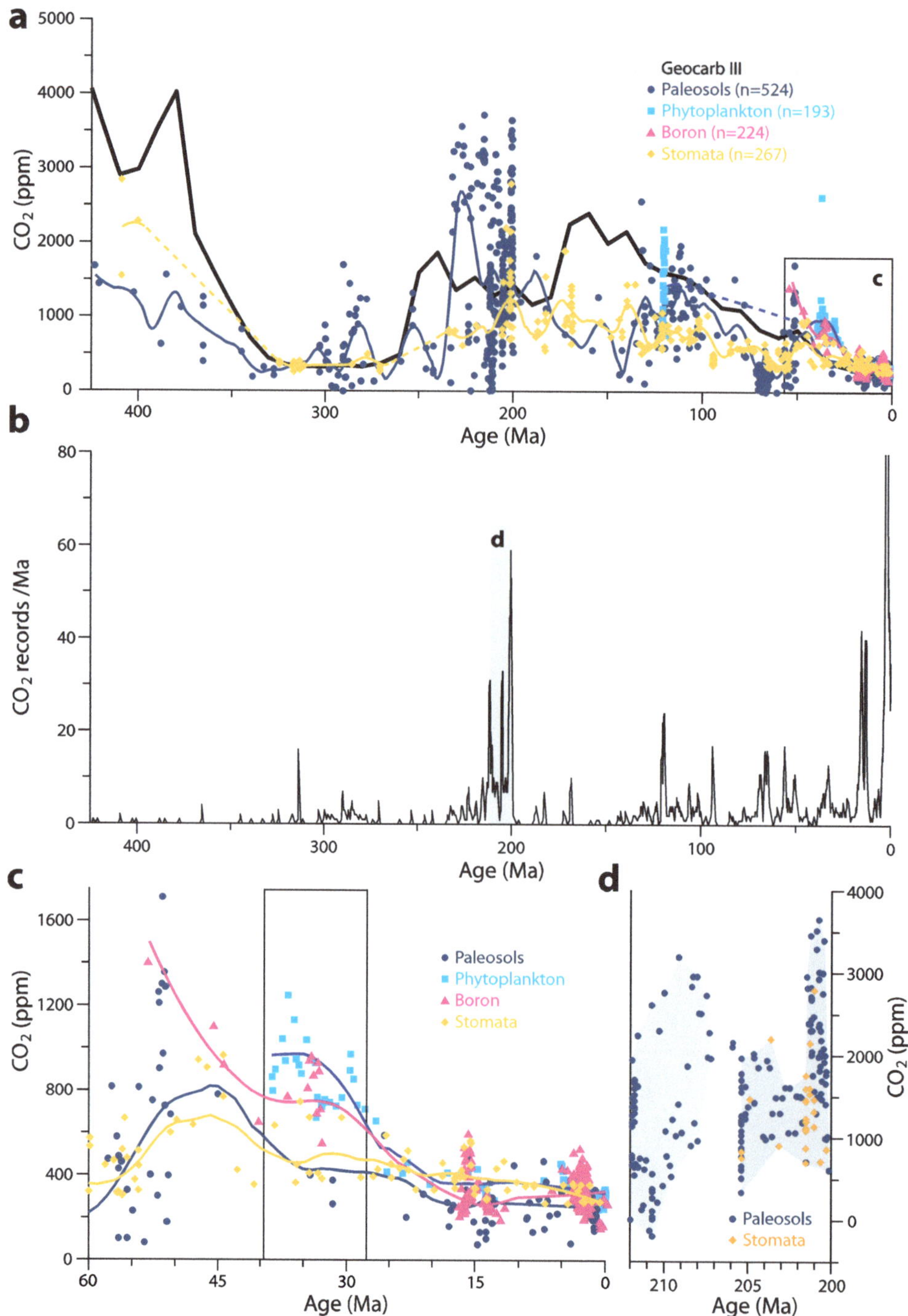

Fig. 9.5 Disagreement and uncertainty in Phanerozoic CO$_2$ proxies
a) Atmospheric CO$_2$ from Individual proxies. Black thick line GEOCARB III model (Berner & Kothavala 2001). Dark blue dots, paleosols (pedogenic carbonate δ^{13}C). Light blue squares, phytoplankton (δ^{13}C of alkenones). Pink triangles, boron isotopes. Yellow diamonds, leaf stomata. Curves represent local least-squares 2nd degree polynomial (Savitzky–Golay) smoothing of individual proxy data after averaging data with the same date, followed by a 1 Myr window averaging and data interpolation at 1 Myr intervals. Dashed segments correspond to long periods without data. **b)** Data coverage in number of CO$_2$ records per Myr is very uneven, with most periods very poorly represented and a few hotspots for data. **c)** Magnification of the squared area in (a). Boxed area shows a period of disagreement between the four proxies that support CO$_2$ levels between 400 and 1000 ppm. **d)** Plot of data from the grey area in (b) showing that the high number of records between 200 and 212 Ma actually results in a bigger uncertainty of the CO$_2$ levels at the time. Medium grey area is the envelop of the central measurements. Light grey area is the envelop of the low/high estimate ranges. Levels of CO$_2$ between a few hundreds to many thousands of ppm are possible. Data from Foster et al. (2017).

Cenozoic, followed by the Ordovician–Silurian, which was more intensely cold than both the middle Mesozoic and the anomalous late Devonian local cooling. An important conclusion to be drawn from this ranking is that the great variability of Phanerozoic climates has not seen a clear trend towards overall warming or cooling in the last 570 m.y, but rather can be characterized as alternating cool and warm intervals of long period" (Frakes et al. 1992).

Phanerozoic temperature reconstructions agree quite well with this general view. Frakes et al. (1992) estimated reconstruction (Fig. 9.4b, green curve) is based on latitudinal distribution of geological indicators. Shaviv & Veizer (2003) low-latitude shallow water reconstruction (Fig. 9.4b, red curve) is based on low-Mg calcite $\delta^{18}O$ isotope data. Scotese (2018) integrates fossil, geological, and isotopic information to reconstruct tropics to pole hemispheric profiles that are transformed into a mean average temperature (Fig. 9.4b, black curve). It is the reconstruction that most faithfully represents the present very cold icehouse condition with an average temperature of only 14.5 °C, lower than 95% of Phanerozoic temperatures. Royer et al. (2004) reconstruction (Fig. 9.4b, purple curve) uses the same data as Shaviv & Veizer (2003), but applies a pH-correction based on CO_2 levels. It really stands out as the different one, making for an improbably warm Andean–Saharan Ice Age, as well as equally unlikely very warm Cambrian and Devonian. Another issue is that a temperature adjustment based on CO_2 data runs into circulus in demonstrando when the result is used to argue the similitude between temperature and CO_2 over time.

By contrast, Phanerozoic CO_2 reconstructions agree mainly on a general decreasing trend and very low values during the Karoo and Cenozoic ice ages. Geocarb III model of Berner & Kothavala (2001. Fig. 9.4a, black curve), long considered the state-of-the-art in CO_2 modeling, shows bigger differences with proxy data in the middle Mesozoic and Devonian with each new release of proxy reconstructions by Dana Royer et al., casting doubts about its general validity for more ancient times. Bergman et al. (2004) COPSE model (Fig. 9.4a, brown curve), looks similar to Geocarb III only until the Carboniferous, and to Royer et al. latest proxy reconstruction (Foster et al. 2017; Fig. 9.4a, dark red curve) only during the middle Mesozoic cool period, showing important differences during the Carboniferous, Permian and Triassic.

The problem with establishing CO_2 levels during the Phanerozoic goes deeper than the lack of agreement between proxy reconstructions and models. As Royer et al. (2004) show, CO_2 data from four different proxies give significantly different reconstruction curves (Fig. 9.5a & c). If we assume that one of the proxies must be better than the others, a reconstruction from all four is a worse approximation to the actual values. And the problem is worse if we consider the variability of the data that the proxies reveal. CO_2 data from proxies is not only very sparse during the Phanerozoic, it is highly uneven also, with the presence of certain hotspots where most of the data accumulates (Fig. 9.5b), leaving the rest very poorly sampled. And it gets even worse, as the accumulation of data at these hotspots does not lead to a reduction in the uncertainty of CO_2 levels. Quite the contrary, the uncertainty at the hotspots is much higher (Fig. 9.5d). In Foster et al. (2017) CO_2 proxy database there are 80 records between 200 and 202 Ma with central values between 610 and

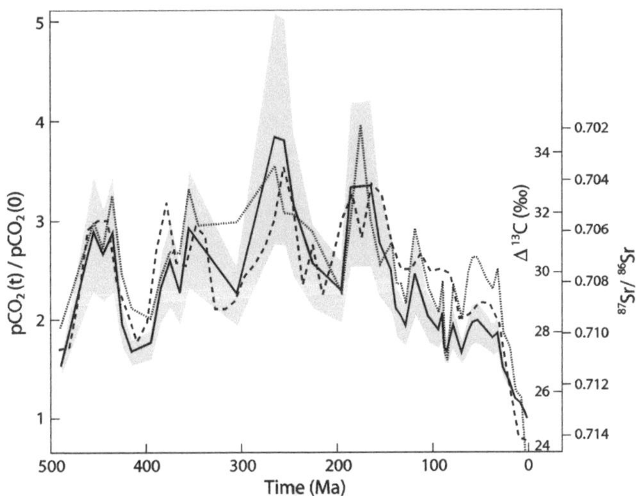

Fig. 9.6 Rothman's CO_2 reconstruction
After eliminating the effect of sedimentary recycling, the $^{87}Sr/^{86}Sr$ data (dashed line) shows a significant correlation to the $\Delta^{13}C$ isotopic inorganic/organic carbon fractionation (dotted line), attributed to both similarly depending on weathering and magmatism. The CO_2 partial pressure, pCO_2, relative to present value (continuous line) is estimated from the other two curves for a reference constant total organic content ratio, with the grey area representing upper and lower reference ratios. This curve and range correspond to the orange curve in Fig. 9.4a. After Rothman (2002).

3650 ppm, extended to 300 to 7000 ppm if the high and low estimates are considered (Fig. 9.5d). This ridiculously high uncertainty cannot be reduced by the use of a different proxy, as it is present in both proxies available at the time, paleosols and stomata. If the uncertainty at data hotspots is representative we cannot have any confidence that a lower number of measurements at other times better constrain our knowledge of past CO_2 levels. It becomes clear that the uncertainty over CO_2 levels during the Phanerozoic precludes any inference on the role of CO_2 on climate beyond a correlation between low CO_2 levels and low temperature during glaciations (Fig. 9.4). The hypothesis that CO_2 was a primary driver of Phanerozoic climate (Royer et al. 2004) has insufficient evidence.

Rothman (2002), took a completely different approach to reconstructing Phanerozoic CO_2 levels. He removed the effects of sediment recycling from the $^{87}Sr/^{86}Sr$ isotopic record showing that it then has a very strong correlation to $\Delta^{13}C$, the isotopic fractionation between inorganic and organic $\delta^{13}C$ (Fig. 9.6). The striking similarity is due to both signals' dependence on magmatic processes and weathering. CO_2 levels are then inferred from the shared oscillations. Rothman's CO_2 reconstruction (Fig. 9.4a, orange curve; Fig. 9.6) shows no correspondence to other CO_2 or temperature reconstructions and has been criticized by Royer et al. (2004) on the grounds of $\Delta^{13}C$ being affected by other factors besides atmospheric CO_2 photosynthesized into organic carbon. While the significance of Rothman's results remains to be settled, it underscores our poor knowledge of carbon and CO_2 changes over the Phanerozoic.

Another source of information comes from the development of a 600 Ma global tectonic model at the University of Lausanne (Vérard et al. 2015). The model shows a decreasing trend over time in global tectonic activity (Fig.

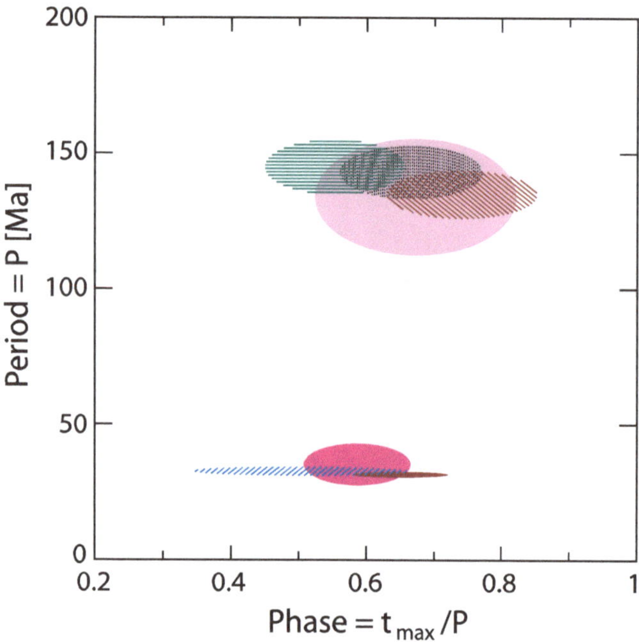

Fig. 9.7 The galactic climate hypothesis
This hypothesis proposes two climatic cycles determined by celestial drivers. The basis of the 150-Ma cycle hypothesis is the coincidence in period and phase between two independent climate records and two celestial datasets. The climate records are the geological alternation between greenhouse and icehouse proposed by Frakes et al. 1992 (horizontal hatching ellipse), and the tropical sea temperature isotopic reconstruction (Veizer et al. 2000; left-slanted hatching ellipse). The celestial datasets are the reconstructed galactic cosmic ray flux from meteorite exposure (dotted ellipse), and the astronomical data on solar system passage through the galaxy spiral arms (light colored ellipse). The 32-Ma cycle hypothesis relies on the crossing of the galactic plane (medium colored ellipse) coinciding with a climate indicator, the tropical sea temperature isotopic reconstruction (Shaviv & Veizer 2003; dark colored ellipse), and ^{3}He in pelagic sediments, an indicator of interplanetary dust influx (right-slanted hatching ellipse; Farley 1995). After Shaviv (2006).

9.4a, blue curve). As magmatism is the accepted long-term source of most atmospheric CO_2 prior to fossil fuel use, the result is compatible with the decreasing trend in CO_2 over the Phanerozoic displayed by models and proxy reconstructions. The authors propose that the decrease in tectonic activity reflects the progressive cooling of the solid Earth, and they extrapolate it to a complete stop in tectonic activity in 500 Ma. Long before that time atmospheric CO_2 will have reached such low levels as to make the existence of complex plants and all the trophic chains depending on them impossible. The authors point at the model producing below-average tectonic rates during cool periods. They attribute the cool periods to a lower rate of CO_2 production. However, an opposite explanation cannot be ruled out. Since a cooler Earth partly covered with huge ice sheets shows reduced tectonic rates that result in lower CO_2 production, the cooling could be the cause of the CO_2 decrease.

The Phanerozoic temperature record oscillates within a narrow range compatible with life, displaying a c. 150-Myr periodicity (Fig. 9.4b). G.E. Williams was the first to report this periodicity in 1975, identifying also the middle Mesozoic cool period from faunal and geological evi-

dence, that was later confirmed by Frakes et al. (1992) as the middle Jurassic to early Cretaceous Cool Mode. Williams postulated that the cause of this climatic periodicity was related to the galactic year (Williams 1975). The relation between a c. 150 Ma climate cycle and the crossing of the galactic spiral arms by the solar system is currently defended by Nir Shaviv and Ján Veizer (2003). Shaviv has showed the coincidence in period and phase between two climatic datasets, one sedimentary and another isotopic, and two celestial datasets, one the reconstructed galactic cosmic rays from meteoritic exposure and another the astronomical data on galactic arms crossing (Fig. 9.7). The hypothesis is controversial and the path, solar system speed, and position of the galactic arms in the distant past, are not well constrained. However, the hypothesis cannot be rejected, as the c. 150 Ma periodicity in the temperature data is defensible and appears absent in CO_2 data and models.

9.3.3 Earth's proposed thermostat

Sagan and Mullen solution to the Faint Sun Paradox based on ammonia's strong greenhouse effect was soon discarded as ammonia's photolysis rate by solar irradiance is extraordinarily high. Hypotheses based on CO_2, CH_4 and CO would prove more stable until oxygen became abundant in the atmosphere. Afterwards, CO_2 would have to do most of the low solar luminosity compensation. Implicit in these hypotheses was the necessity that greenhouse gases have adjusted their abundance to match the changes in solar luminosity over time. Walker et al. (1981) were the first to make explicit this prerequisite. This conclusion leads directly to the anthropic principle, as we would be here only because CO_2 was able to match successfully the forcing required to compensate the change in solar luminosity within the narrow range needed for the evolution of complex life over billions of years. Foster et al. (2017) express it in the following terms: *"the long-term decrease in CO_2 over the last 420 Myrs has largely compensated for the increase in solar output over the interval."*

To avoid leaving this strong atmospheric requirement to chance, Lovelock and Margulis (1974) developed the Gaia hypothesis. This hypothesis states that from its development the biosphere took a leading role in keeping the planet's life-compatible conditions over time. However, the Gaia hypothesis has fallen out of favor among scientists, if it ever had any. The development of oxygenic photosynthesis by cyanobacteria is believed to have caused the Great Oxygen Crisis that probably plunged the planet into the Huronian Glaciation that lasted c. 300 Ma. And the discovery of lignin by vascular land plants in the Devonian caught the decomposing organisms unprepared, leading to the burial of large amounts of carbon into coal deposits. The resulting decrease in atmospheric CO_2 levels probably contributed to the Karoo Ice Age being the coldest and longest (so far) in the Phanerozoic. Clearly, living organisms have not been kind to the planet they inhabit, and have tested its thermal homeostasis to the limit.

Walker et al. (1981) proposed a solution in the form of carbon cycle weathering control by temperature. Over timescales of millions of years CO_2 abundance is controlled only by its magmatic production and by its incorporation into carbonates, via weathering, that are subsequently buried, as all other processes have time to equilibrate. Continental silicate weathering by CO_2 is a process dependent among other things on temperature. If the planet's temperature decreases, silicate weathering rate slows down leading to CO_2 accumulation and global warming, and vice versa for a temperature increase. This proposed thermostat that would stabilize Earth's climate is now part of textbooks despite a lack of evidence for a global silicate weathering control by temperature. In fact, a global analysis of large rivers failed to show temperature as an important parameter controlling silicate weathering (Gaillardet et al. 1999), casting doubts on the validity of the hypothesis. The existence of an effective silicate weathering negative feedback thermostat capable of stabilizing Earth's climate, compensating for long-term changes in solar output, requires that on a timescale of millions of years CO_2 levels and temperature correlate. It also means that on that timescale CO_2 levels are controlled by temperature (via weathering), and not the opposite.

9.3.4 Cenozoic climate

Our knowledge of the past improves as we get closer to the present. If Earth's climate has been controlled by changes in CO_2 this should be more evident over the last 66 million years than in previous times. The Cenozoic has been very climatically varied, from the Eocene hothouse to the Pleistocene icehouse, and therefore constitutes an ideal test for a CO_2 control of these climatic changes. A certain degree of agreement is expected from the long-term decrease in CO_2 levels over time (Fig. 9.4a) and the Cenozoic hothouse to icehouse decrease in temperature. The question is if that agreement is close enough to support a first-order control of temperature by CO_2 levels. And the answer is absolutely not. The end of the Cretaceous was a very warm period characterized by low CO_2 levels, probably below 500 ppm and similar to the present, as indicated by proxies (Fig. 9.4; Foster et al. 2017). Such a drop in CO_2 levels at one of the warmest periods in Earth history might be a consequence of the Chicxulub impact that marks the end of the Cretaceous and might have caused severe global cooling. In any case temperature seems to have recovered faster than CO_2, as the Paleocene showed stable very warm temperatures at a time of increasing CO_2 levels (Fig. 9.8). The early Eocene, between 56 and 48 Ma, showed the warmest temperatures of the Cenozoic with CO_2 levels estimated between 500–1000 ppm. However, starting with the Lutetian, c. 48 Ma, temperature dropped almost continuously for 14 Myr until the Oligocene. This drastic cooling, estimated at –5 °C in global

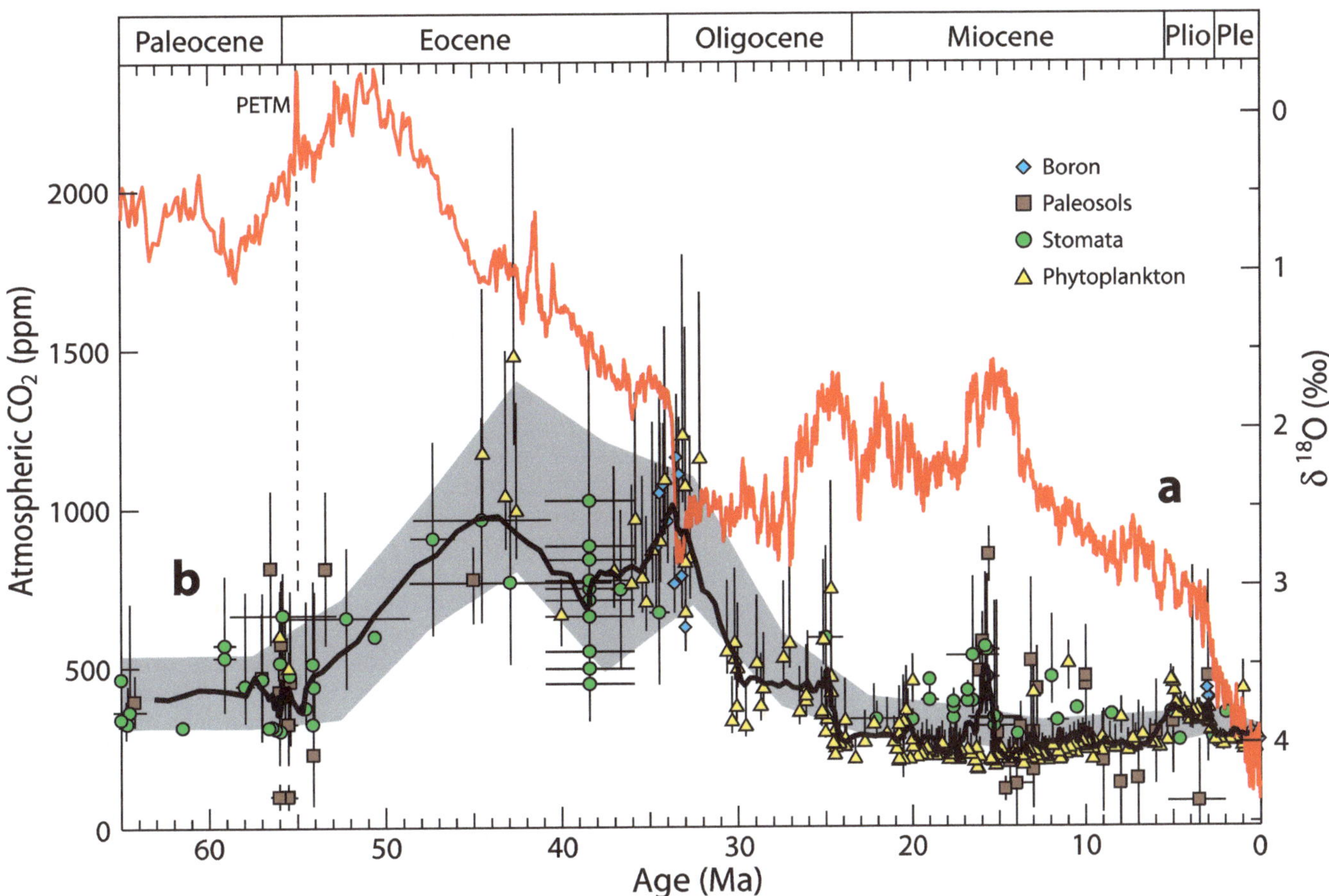

Fig. 9.8. Cenozoic temperature and CO_2 evolution
a) Upper curve (right scale), global deep-sea $\delta^{18}O$ curve from over 40 drilling projects as a temperature and continental ice proxy (Zachos et al. 2001). Data originally smoothed with a five-point running mean was also averaged with an 80,000-year running window to eliminate obliquity-linked variability and compensate for differences in data density. **b)** CO_2 data for the indicated proxies from Beerling & Royer (2011). Thick black curve is the 9-point average, and the grey area represents the 5 Myr binned average of the high and low CO_2 estimate. PETM, Paleocene–Eocene Thermal Maximum, indicated by the vertical dashed line. A change in CO_2 levels at that time must have been too transient to leave an impression in CO_2 records. No correlation is evident between temperature and CO_2 data. Also, the CO_2 response to Antarctica's glaciation at the Eocene–Oligocene boundary is surprising.

average temperature, took place while CO_2 levels increased to above 1000 ppm (Fig. 9.8). At 34 Ma and in less than one million years Antarctica glaciated causing a further drop of another –2.5 °C, while CO_2 proxies support levels above 1000 ppm for two million years after the Antarctic glaciation, before decreasing back to c. 500 ppm between 32–25 Ma. As temperatures recovered during the Oligocene and were 2–4 °C warmer than now from the late Oligocene to the mid-Miocene climate optimum, CO_2 levels overall decreased to a minimum c. 15 Ma, but showing significant increases at c. 17 Ma coincident with the Columbia River Basalt eruptions, and at c. 25 Ma, associated with warmer periods. For the past 15 Myr CO_2 levels appear to have oscillated between 250–400 ppm (similar to Late Holocene changes) most of the time, while temperatures display another important drop estimated at – 5 °C from the mid-Miocene climate optimum into the depths of the late Pleistocene glaciation.

Therefore, there is a complete lack of correlation between temperature changes and CO_2 changes during the Cenozoic's two most important climatic change periods, the 50–34 Ma and 15–1 Ma strong cooling periods. And there is anti-correlation during the 33–17 Ma period of warming that cannot be explained in terms of the decreasing CO_2 levels that the proxies suggest. Considering this evidence the assertion that *"there is a strong CO_2-temperature coupling during the Cenozoic"* (Royer 2014) is most surprising. While the effect of CO_2 on temperature is based on solid physical principles, and it cannot be ruled out that the effect might be dominant at certain times, like during the c. 17 Ma Columbia River Basalt eruptions at the mid-Miocene climatic optimum; Cenozoic climate evolution demonstrates that CO_2 cannot be a primary driver of climate over the long-term as it has been repeatedly claimed (Royer et al. 2004; Royer 2014). Comparative analysis of CO_2 data, marginal radiative forcing at the top of the troposphere from CO_2 changes, and $\delta^{18}O$-derived temperature data over the past 425 million years (Davis 2017), leads to the same conclusion as the examination of Cenozoic changes: *"neither the atmospheric concentration of CO_2 nor $\Delta RFCO_2$ is correlated with temperature over most of the ancient (Phanerozoic) climate"* (Davis 2017).

9.3.5 Phanerozoic climatic cycles

$\delta^{18}O$ benthic isotopic data for the past 67 Myr (Zachos et al. 2001) displays a clear autocorrelation between its left and right halves (Fig. 9.9). Already in 1977 Fischer and Arthur had demonstrated that the biological diversity of pelagic biotas varied with a rhythm of 32 Myr between times of higher diversity associated with warmer temperatures, widespread anaerobism, eustatic sea-level rises, and heavier carbon isotope values in marine calcareous organisms, and times of lower diversity associated to opposite conditions (Fischer & Arthur 1977).

Using a database of 17,797 fossil marine animal genera, Rohde & Muller (2005) showed that the main periodicity in marine metazoan diversity is a 62 ± 3 Myr cycle (Fig. 9.10b, d), with a second periodicity at 140 ± 15 Myr (Fig. 9.10c, d). The 32 Myr periodicity could be a low power harmonic as suggested in the Cenozoic part of the data (Fig. 9.10b, d, arrows).

A 30 Myr periodicity in diverse geological phenomena had been proposed multiple times since the early 20th century, usually on weak statistical grounds. In 1984, Rampino & Stothers (1984a) proposed that a 33 ± 3 Myr periodicity occurred in major geological perturbations (low sea levels, sea-floor spreading discontinuities, tectonic episodes, and geomagnetic reversals), and in dates of terrestrial impact craters during the Mesozoic and Cenozoic. The same year, Raup & Sepkoski (1984) published the existence of a 26 Myr periodicity in marine fossil fauna extinctions, and their article was immediately followed by the proposition that impact craters were responsible, and that the crossing of the galactic plane by the solar system was the ultimate cause (Rampino & Stothers 1984b; Schwartz & James 1984). The controversy over this hypothesis continues to this day. After three decades the periodicity in marine fauna extinctions appears solid and has been refined to the genus level, and using newer geological timescales established at 27 Myr (Melott & Bambach 2014). The terrestrial impact cratering periodicity is now attributed to dark matter at the galactic mid–plane (Randall & Reece 2014; Rampino & Caldeira 2015), but is still disputed on statistical grounds (Erlykin et al. 2017). The cause of mass extinctions is of peripheral interest to the climatic question, as different mass extinctions appear to respond to different causes. The association of many of them to large igneous province eruptions appears clear (Bond & Grasby 2017), and no 27-Myr climatic cycle is apparent in the climatic record.

The correlation between biodiversity and climate is, however, a critical issue. Nutrient productivity and availability depends strongly on climatic factors like temperature, precipitation, erosion, aerial fertilization, ocean upwelling, and CO_2 levels. It is expected that climate changes drive biodiversity changes both in the past and during Modern Global Warming. The evidence shows this correlation. The marine fossil fauna biodiversity data (Rohde & Muller 2005), after detrending and subtraction of the 62-Myr signal, displays primarily its 140-Myr periodical signal (Fig. 9.10c). This 140-Myr cycle in biodiver-

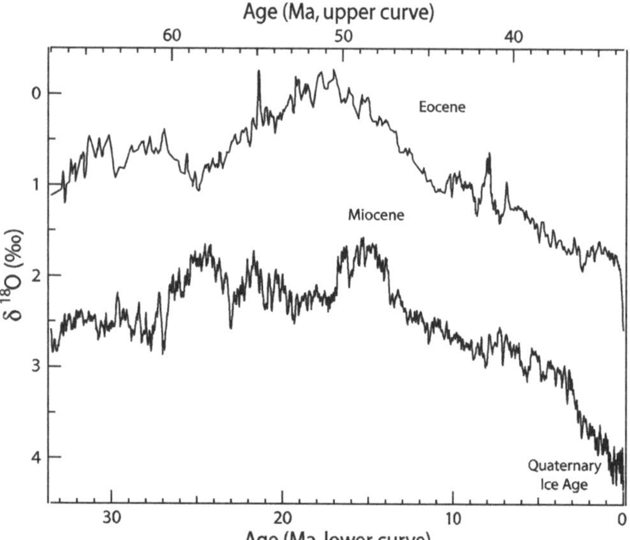

Fig. 9.9 Symmetry in Cenozoic temperature proxy record
$\delta^{18}O$ temperature proxy data for the past 67 Myr from Zachos (2011), as in Fig. 9.8. The 67–33.5 and 33.5–0 Ma periods have been superposed to show the striking similarity.

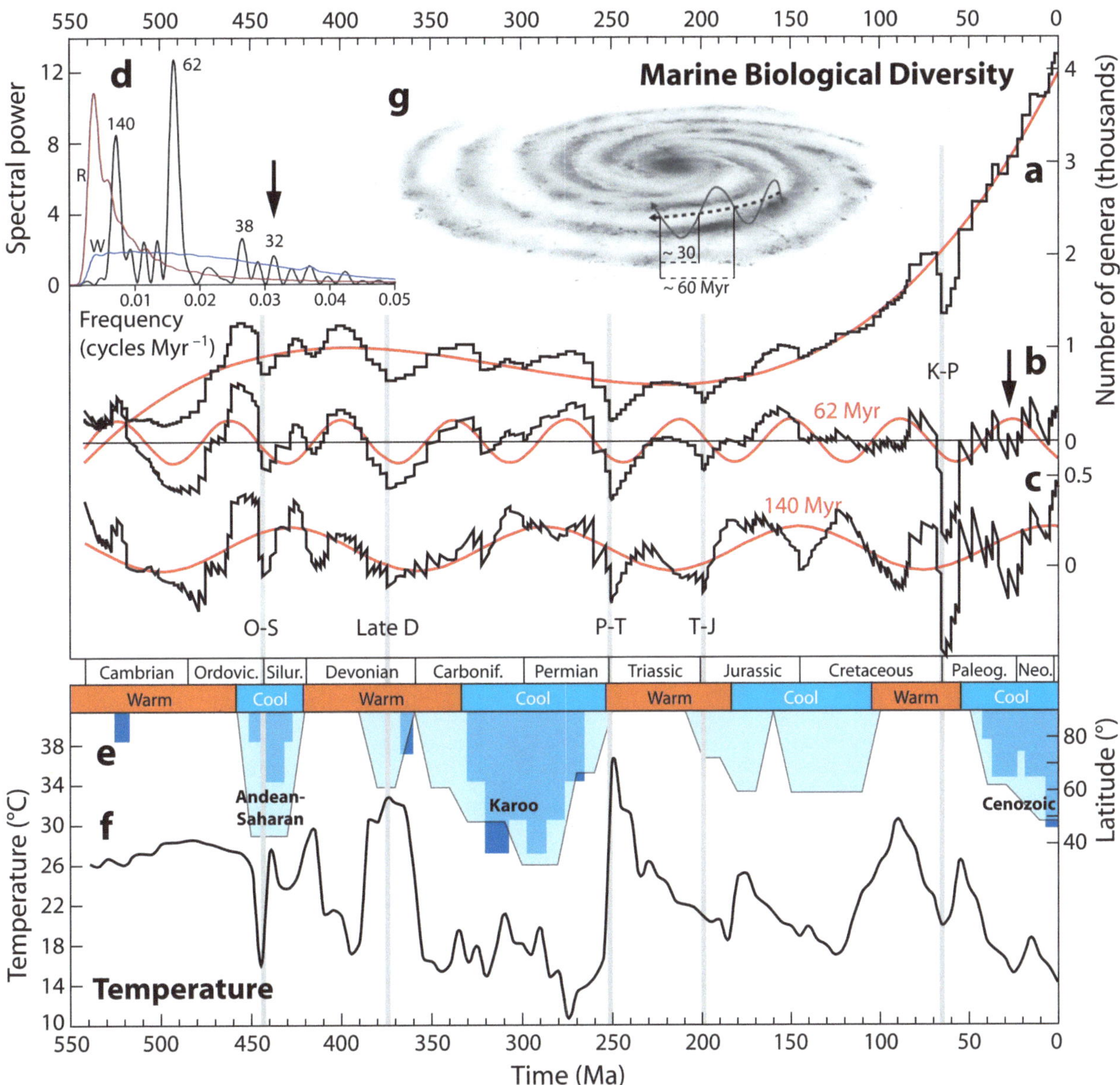

Fig. 9.10 Biological diversity and climate cycles
a) Number of marine animal genera over time in the 17,797-fossil dataset used by Rohde & Muller (2005). Red line is a third-order poly-nomial fit. **b)** Same data after detrending. Red line is a 62-Myr sine wave fitted to the data. **c)** Detrended data after also subtracting the 62-Myr cycle, and with a 140-Myr sine wave fitted to the data. **d)** Fourier spectrum of the detrended data in (b). R (red) and W (blue) curves are estimates of red and white spectral noise. (a) to (d) after Rohde & Muller (2005). Arrows in (b) and (d) hint at the possibility of a 32-Myr sub–harmonic periodicity. Grey vertical bars indicate the five biggest mass extinctions in marine fauna. O–S, Ordovician-Silurian; Late D, late Devonian; P–T, Permian–Triassic; T–J, Triassic–Jurassic; K–P, Cretaceous–Paleogene. The 140-Myr marine bio-logical diversity periodicity in (c) shows a clear correlation to climatic indicators. **e)** Alternation in Warm and Cool Modes (after Frakes et al. 1992), and glacial indicators (right scale in latitude reached; Ice rafting evidence (ice blue areas; after Frakes & Francis 1988), and direct geological evidence for glaciation (azure and blue areas; after Crowley 1998) **f)** Mean global temperature reconstruction based on geological evidence and isotopic studies (black curve, left scale; after Scotese 2018). **g)** Diagram showing Solar System c. 60-Myr oscil-lation with respect to the galactic plane, crossing it every c. 30 Myr, as it circles its core, that has been proposed as a cause for periodici-ties in biological extinctions (Schwartz & James 1984), and climate oscillations (Shaviv 2006).

sity is a mirror image of the temperature reconstruction (Fig. 9.10f; Scotese 2018), and shows a near perfect coin-cidence with the alternation between warm and cold peri-ods (Frakes et al. 1992), and with glacial indicators (Frakes & Francis 1988; Fig. 9.10e). This is an important result because two indicators that should be related ac-cording to theoretical knowledge, are independently show-ing the same pattern. To our best knowledge this pattern is not displayed by CO_2 reconstructions or models (Fig. 9.4). Marine faunal biodiversity anticorrelates to temperature, since it is higher usually during cold periods and lower

during warm periods. This is consistent with our knowl-edge that land and ocean productivity respond differently to global temperature. Cold periods favor water oxygena-tion and increase CO_2 solubility. With cooling, the tem-perature gradient between the poles and the equator in-creases, favoring the development of strong zonal winds and a stronger thermocline resulting in upwelling increase and enhanced nutrient recycling (Vincent & Berger 1985). Also, phosphorus availability in the open ocean increases (Palastanga et al. 2013), and a stronger dust flux produces higher soluble iron levels (Conway et al. 2015), resulting in higher productivity and more efficient nutrient utilization (Martínez–García et al. 2014). As an illustration of this enhanced ocean productivity during cold periods, responsible for the biodiversity increase, baleen whales, the biggest animals to ever inhabit the planet, developed their gigantism during the Late Cenozoic Ice Age due to intensified coastal upwelling (Slater et al. 2017).

9.4 Radiative forcing and anthropogenic effect

The effect that changes in atmospheric greenhouse gases have on climate is defined and quantified through changes in radiative forcing. Conceptually, climate change is assumed to be due either to an external cause (forcing), or to internal variability. Figure 9.11 shows a schematic representation of the climate system with many important subsystems and processes. Anything that is not affected by the climate system is considered external, although the distinction is not absolute. For example, volcanoes are often external to the climate system, however it is known that their frequency responds to changes in sea level and the ice-sheet load during deglaciations (Huybers & Langmuir 2009; see Sect. 11.2). Forcings cause climate change, while feedbacks respond to climate change, affecting the amplitude of the change positively or negatively with respect to the direction of the change caused by the forcing. It becomes confusing because the same factor can be a feedback if produced naturally in response to climate change, and simultaneously a forcing if produced by humans. Several greenhouse gases are in that situation.

The thermodynamic nature of climate establishes that the available energy to the climate system to realize work is the energy that enters the system, and to be in equilibrium, it must match the energy that exits it. The work is realized through the conversion of short-wave radiation (SR) into infra-red radiation (IR), a more degraded form of energy. The match between incoming and outgoing energy is known as the energy balance. In the absence of energy balance alterations, climate change can still be driven by internal variability, but significant differences in the en-

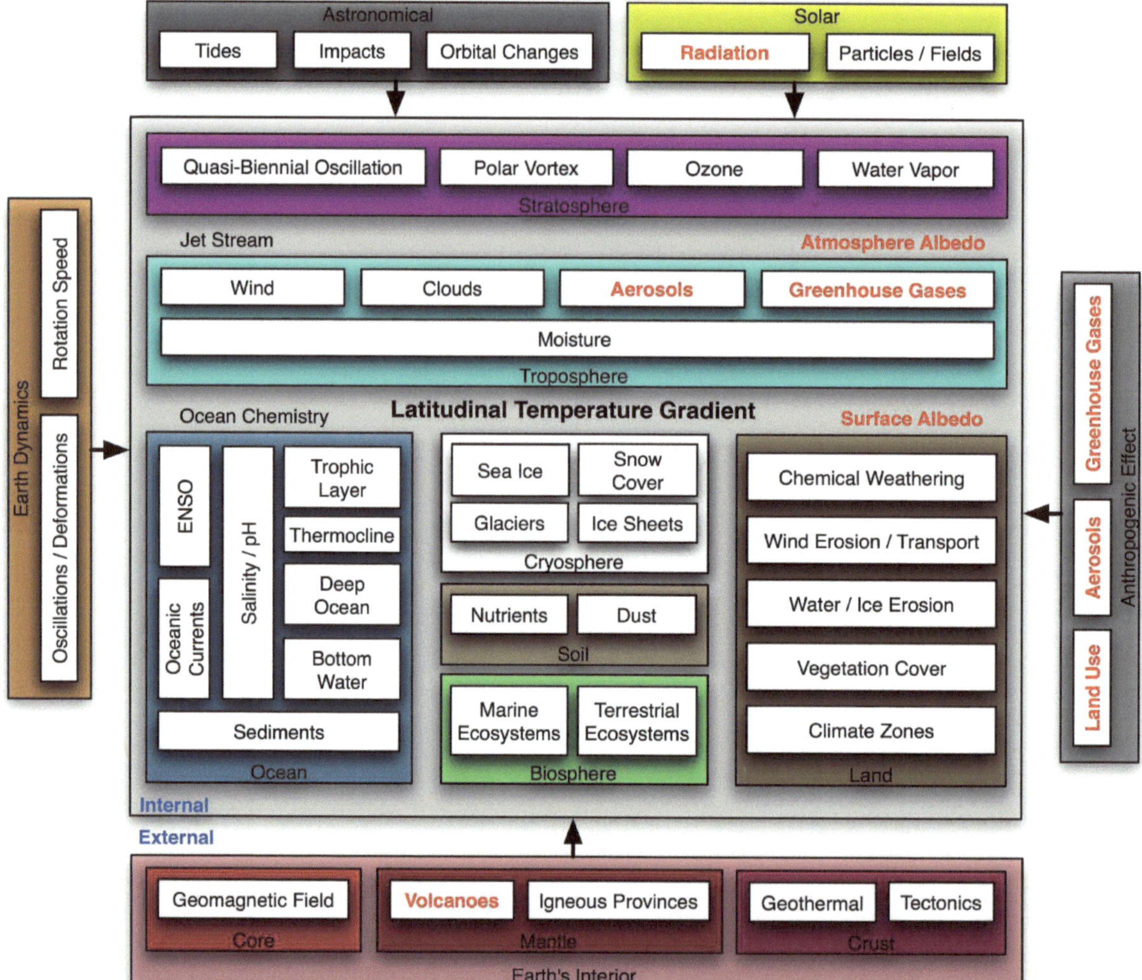

Fig. 9.11 Simplified schematic representation of Earth's climate system
Different subsystems are shown with different background colors. Climatic phenomena and processes affecting climate in white boxes. Subsystems and phenomena within the central pearl-colored box are generally considered internal to the climate system, and everything else, as usually not affected by climate (with some exceptions), external. Some important properties or phenomena at the interface between subsystems are placed outside boxes. The Latitudinal (Equator-to-Pole) Temperature Gradient is a central property of the climate system that changes continuously and defines the thermal state of the planet (see Scotese 2016). For simplification, lines joining related boxes have been omitted. Bold names in red are the main contributors to changes in the radiative budget that according to the IPCC are almost exclusively responsible for Modern Global Warming.

ergy balance should always result in climatic changes. Geothermal and tidal energies are known to be too small or vary by too little to affect the system (Davies & Davies 2010; Egbert & Ray 2000). It is assumed that the only changes that can affect the energy balance are radiative changes. This presupposition is only correct if we have enough knowledge about other known forcings to rule them out, and if there are no other significant forcings unknown. It is debatable that our level of knowledge is sufficient for that presupposition to be correct, and non-linear effects of solar variability (Gray et al. 2010; see Sects. 11.6 & 11.7), and the effect of galactic cosmic rays (Svensmark et al. 2016), are active areas of research that question it. Radiative forcing (RF) is nevertheless defined as the net change in the energy balance of the Earth system due to some imposed perturbation (Myhre et al. 2013):

$$RF = \Delta(E_{in} - E_{out}) = \Delta E_{in} - \Delta E_{out}$$

At this point the entire climate system complexity represented (in a simplified form) in Fig. 9.11 has been reduced to the factors shown in red, the rest of the factors are considered subordinate. However, the advantage of RF is that it allows us to calculate the theoretical effect of every single factor that affects the energy balance. The IPCC presented a surprising result for this theoretical calculation (Myhre et al. 2013; IPCC 2014 SPM; Fig. 9.12). The net radiative forcing for the combined natural factors known to affect the energy budget for the six decades between 1951–2010 was calculated at 0 W/m². In other words, the theoretical framework finds no natural contribution (positive or negative) to the observed net warming. Under this paradigm humans are responsible for 120% of

the observed warming, as well as the 20% cooling from aerosols required to settle at 100%.

The surprising proposition that nearly all radiative forcing change, and hence climate change, since 1750 has been due to human contribution (Fig. 9.12c) should be met with due skepticism. Yet it is the unavoidable conclusion from IPCC's climate paradigm. The complexity of the climate system (Fig. 9.11 is a strong simplification) guarantees chaotic oscillations and random walks that must result in intrinsic oscillations on top of which continuously varying forcings impose greater changes through interference and resonance. The proxy record supports continuous climatic changes at all time scales. *Climate "is" climate change*. The change is intrinsic to a very complex, strongly-regulated dynamic system that has maintained life-compatible conditions for eons. The belief in a stable benign climate suddenly thrown out of equilibrium by human actions is, in all probability, wrong. The theoretical result that natural climate change ceased at the time anthropogenic GHG emissions started to alter atmospheric levels is an indication of a faulty hypothesis. It raises the suspicion that anthropogenic forcing has been seriously overestimated. There are signs of this overestimation. Models have been constrained to reproduce historical temperature anomaly changes, but they are diverging rapidly from observations in the unconstrained 21st century (see Fig. 13.10). In the 2018 IPCC special report "Global Warming of 1.5 °C" (Chap. 2, 2.2.2.2 in the report) the CO_2 budget to reach +1.5 °C is more than doubled from the one published only four years before in the IPCC Fifth Assessment Report. It goes from 400 to 860 Gt CO_2. This increase, produced by a change in methodology, is equiva-

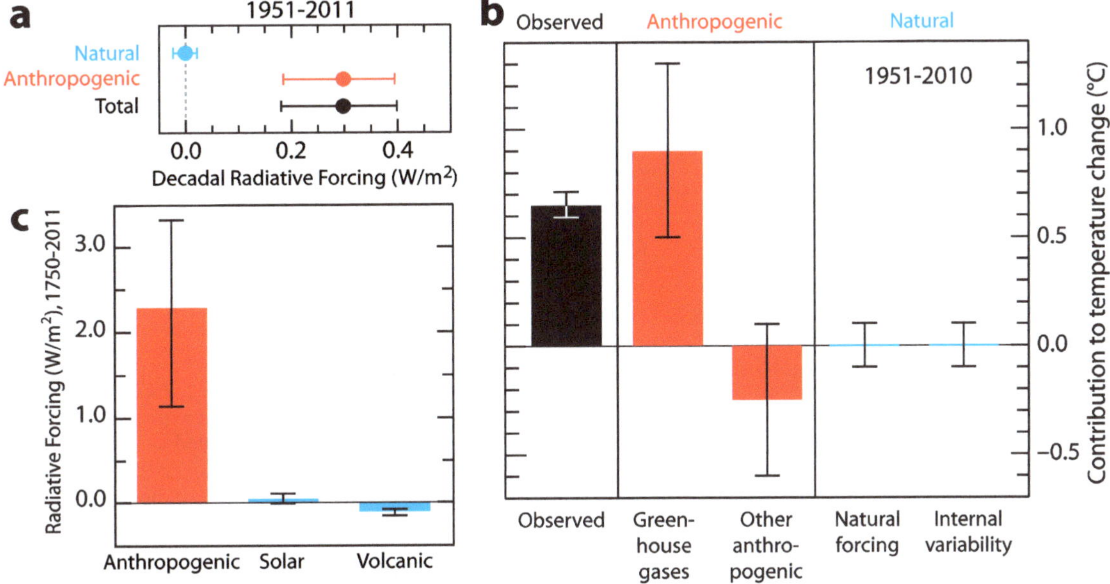

Fig. 9.12 Anthropogenic and natural radiative forcing contribution to climate change
a) Linear trend in anthropogenic, natural and total forcing for the period 1951–2011 (with 5–95% confidence range). After Myhre et al. 2013. **b)** Assessed contributions to observed surface temperature change over the period 1951–2010 (black bar with 5–95% confidence range). The relative contributions (colored bars) are based on observations combined with climate model simulations. Combined anthropogenic forcings account for essentially all warming, according to models. After IPCC's Summary for Policymakers of the Synthesis Report (2014). **c)** Global annual average radiative forcing change from 1750 to 2011 due to human activities, changes in total solar irradiance, and volcanic emissions. Bars indicate 5 to 95% confidence range. Over the time period, solar forcing has oscillated from solar minimum to solar maximum between −0.11 and +0.19 W/m². Radiative forcing from volcanic emissions is small on average because it is short-lived. After Wuebbles et al. (2017).

lent to reducing CO_2 RF by half for the period considered, or to a decade of human emissions free of climatic effect.

9.5 Climate feedbacks

A feedback occurs when part of the output from a system is added or subtracted to the input modifying the result. Amplifying feedbacks are positive and dampening feedbacks negative. Systems dominated by negative feedbacks are inherently stable and systems where positive feedbacks predominate are unstable. Earth climate system is known to be stable as, unlike other planets, it has maintained a narrow range of life-compatible temperature variability for nearly four billion years, during which all sort of events have pushed it towards both temperature extremes. Climate feedbacks cannot be observed, since by definition they are the effect of one variable on another. Feedback relationship between correlating variables can only be inferred probabilistically or estimated with models. In the same sense feedbacks cannot be deduced from observations, they must be deduced from theory or from model observations. In a complex system such as climate there is a great uncertainty that all relevant feedbacks have been correctly identified.

Climate models are mathematical representations of the energy exchange between the climate system and space. The models attempt to determine the effects of radiative changes on a hypothetical average surface temperature. The starting point of every climate model is that changes in enthalpy (heat content) are the difference between absorbed shortwave and outgoing longwave radiation (ASR and OLR; Fig. 9.2). In equilibrium both are equal and there is no net flux at the top of the atmosphere ($F_{TOA} = 0$). When there is a RF change that causes a radiation flux into the system (R), such as an increase in GHGs or in solar irradiance, it causes a change in surface temperature (T_s) proportional to the effective heat capacity of an atmosphere-ocean column (C). This surface warming causes a compensating change in top-of-the-atmosphere emissions (ΔF_{TOA}):

$$C\, \partial T_s/\partial t = ASR - OLR$$

At equilibrium both terms are zero and the temperature doesn't change. When perturbed by a radiative forcing we have:

$$C\, \partial T_s/\partial t = R + \Delta F_{TOA}$$

The key assumption in climate feedback analysis is that changes in radiative flux are proportional to surface temperature changes.

$$\Delta F_{TOA} = \alpha \Delta T_s \ [1]$$

where α is a constant of proportionality between the change in temperature and the resulting change in radiative flux, with units of W m^{-2} K^{-1}. It is the climate feedback parameter and represents the increase in OLR per unit of global warming. It is reciprocal to the climate sensitivity parameter, λ, the steady state global warming per unit of radiative forcing in m^2 K W^{-1}, that for a doubling of CO_2 constitutes the equilibrium climate sensitivity (ECS).

Theory-derived climate feedbacks have been investigated with models, as they cannot be measured. In principle, observations should be able to constrain the range in climate feedback values, but we have not been very successful in doing so. That is the reason that after 40 years we have not been capable of constraining the range of possible ECS values presented in the Charney Report (Charney et al. 1979; Fig. 9.1). However, this part of climate science is critical, as most of the temperature-related effects from anthropogenic emissions are dependent on feedbacks amplifying the CO_2 radiative forcing effect. The assumption that changes in temperature are linearly dependent solely on changes in radiative flux [1], allows the decomposition of α into individual feedback processes. The most important considered are:

- The Planck feedback. The warmer an object the more it radiates. All models agree on the value of this feedback at -3.3 W m^{-2} K^{-1}. The Planck feedback is often not considered with the rest of climate feedbacks (Fig. 9.13).
- Fast positive feedbacks: The water-vapor feedback is the largest of the remaining climate feedbacks, ac-

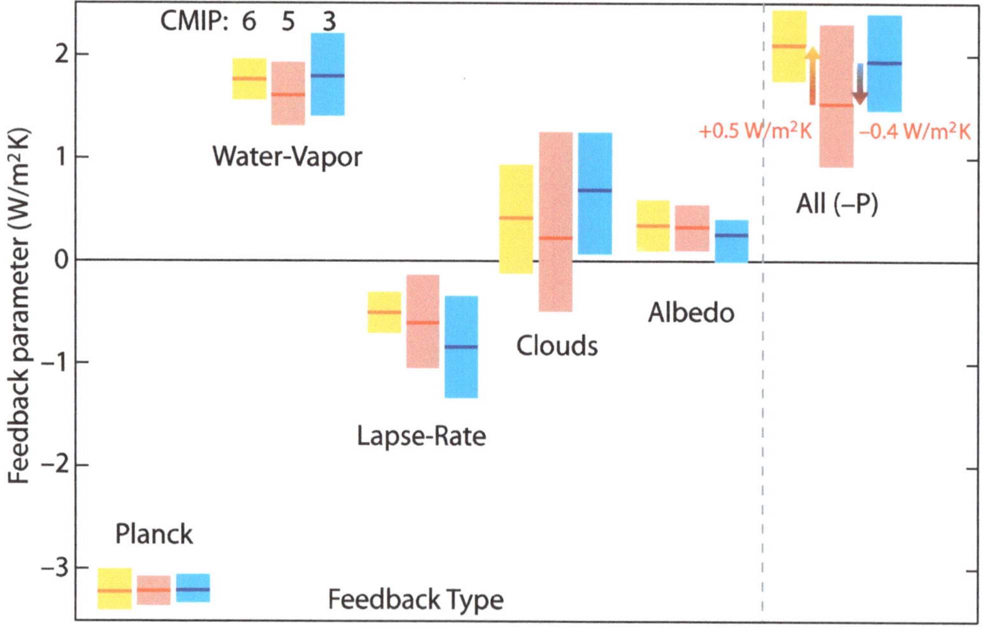

Fig. 9.13 Climate feedback strength
Feedback parameter value of individual feedbacks for CMIP3 (right columns), CMIP5 models (center columns), CMIP6 models (left columns), and model inter-comparison average (central thick lines). All (−P), sum of feedbacks except Planck. Feedbacks are for an abrupt fourfold increase in CO_2 concentration. Arrow marks the 20% decrease in feedback strength from CMIP3 to CMIP5 mean value and the 33% increase from CMIP5 to CMIP6. After Flato et al. (2013) and Forster et al. (2021).

cording to models. Warming of the atmosphere increases its absolute moisture content, and as water vapor is a strong GHG it produces a positive feedback loop. Models consider that this feedback roughly doubles the warming from CO_2. However, the water-vapor response to global warming has not been properly determined so far. The ice-albedo feedback produces another loop where ice melting creates additional surface warming causing more ice to melt. It is supposed to be important for Arctic amplification, but overall is less important because most of the global albedo takes place at the atmosphere.

- The sign of the cloud feedback is unknown, as low clouds create a negative feedback by increasing albedo, while high clouds are supposed to be a positive feedback by reflecting more IR downwards than SR upwards. Also, at night all clouds direct more IR to the ground than clear skies. Some models give a small net negative feedback value while others give a large positive feedback.
- A fast negative feedback is the lapse-rate feedback. The lapse rate is the rate at which air temperature decreases with height. As the greenhouse effect warms the atmosphere more than the surface at the tropics, the lapse rate decreases. As a result, OLR emission increases with respect to an homogenous vertical temperature increase, reducing warming at the surface. At high latitudes, lower water vapor and a colder surface result in higher surface warming and a positive lapse rate feedback, but the tropical effect dominates, and the lapse rate feedback is believed to be negative.

Overall fast feedbacks, not including the Planck feedback, are estimated at c. 1.5–2.0 W m^{-2} K^{-1} (Fig. 9.13), so most of the warming in the near term is expected to come from them. Then we have the slow feedbacks that are highly speculative and difficult to estimate. On relevant time scales most of the ones considered are positive. Carbon-cycle feedback is expected to release large stores of sequestered carbon (permafrost and peat decomposition,) and methane (permafrost and clathrate from sea floors). Changes in vegetation cover are also expected to occur (rainforest decay from drying, forest fires, desertification), with a more mixed result as carbon is expected to be released, but albedo could increase.

Climate feedback studies are clearly an immature scientific subject and the results should be taken with a grain of salt, particularly due to a possible skew towards publication of reports supportive of large positive feedbacks. If climate feedbacks were truly so skewed towards the positive it is reasonable to ask why the Earth has not been repeatedly frozen and heat-sterilized in such a long history. Currently the planet surface average temperature is close to 14.5 °C, which is within the coldest 10% of the Phanerozoic (Fig. 9.4b). The average temperature of the planet for the past 540 million years has been around 19 °C (c. 4 °C warmer than now; Scotese 2018), and several of the large δ^{13}C excursions associated with large volcanic activity have taken place at times that were even warmer. It is possible that given the difficulty of identifying and quantifying feedbacks the overall positive feedback response might have been overestimated. From CMIP3 (2010) to CMIP5 (2014) the overall fast feedback decreased by –0.4 W m^{-2} K^{-1}, a 20% reduction in just four years. From CMIP5 to CMIP6 (2021) the overall fast feedback increased by +0.5 W m^{-2} K^{-1}, a 33% increase (Fig. 9.13). A problem with the latest increase is that models become implausibly hot (Voosen 2021).

9.6 The CO_2 hypothesis of climate change

When we analyze the global energy budget we observed that water is involved in most of the terms:

- Evaporation
- Latent heat of fusion
- Latent heat of vaporization, equal and of opposite sign in the global mean, but vaporization at the ocean surface and condensation at the troposphere
- Cloud albedo and IR reflection
- Tropospheric greenhouse effect
- Stratospheric greenhouse effect
- Atmospheric SR absorption
- Surface Albedo
- Enthalpy transport by the atmosphere and oceans

On top of that, water is extraordinarily abundant in its three states. Its role in climate change has not been properly explained. Oxygen isotope proxies indicate that the ratio between its three states has closely followed climate change for millions of years. Starting with Arrhenius it was assumed that, since water vapor abundance and water change of state respond to temperature, water vapor could not be responsible for climate change forcing and should only be a feedback. However, this assumption rules out, without evidence, the existence of changes in water state and/or water vapor distribution that are non–linear or independent of temperature changes. Our limited knowledge of the climate system, might make us overlook phenomena that, given the importance and abundance of water, might significantly affect climate change. One such example was the discovery in 2010 that stratospheric water vapor decadal changes made a significant contribution to the rate of global warming (Solomon et al. 2010). According to estimates, 20–30% of the warming during the 1990s could be attributed to a factor that was not included in climate models, since until recently general circulation models completely neglected the stratosphere. Stratospheric water vapor appears to be controlled mainly by Indo–Pacific Warm Pool sea-surface temperature, the tropical cold point in tropospheric temperature, and El Niño/Southern Oscillation, and as the authors state: *"It is therefore not clear whether the stratospheric water vapor changes represent a feedback to global average climate change or a source of decadal variability"* (Solomon et al. 2010).

On planet Earth the warming resulting from a change in GHGs is the consequence of the interplay between the greenhouse effect and the hydrological cycle. And while the first is well understood, the second is not. All models that predict strong warming due to the atmospheric CO_2 increase from human emissions do so because they include a strongly positive hydrological feedback (water-vapor, cloud, and albedo feedbacks). If the hydrological feedback is weak the warming would be modest. And we are having trouble measuring the hydrological cycle response to warming. If temperature data is sparse, incomplete and full of inhomogeneities, water vapor data is much worse. It has

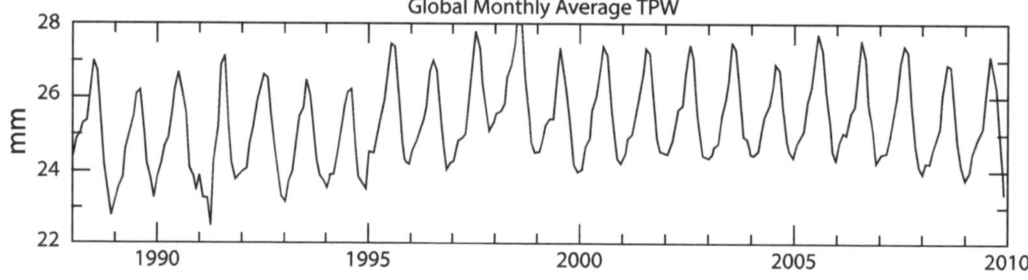

Fig. 9.14 Seasonal and interannual variability in atmospheric water vapor content
1988–2009 global monthly average in total precipitable water, the amount of water vapor in a 1 m^2 atmospheric column measured in mm. Data from multi–method NASA Water Vapor Project NVAP–MEaSUREs. After Vonder Haar et al. (2012).

only been measured globally very recently and different ways of measuring it give statistically different results (Schröder et al. 2016). The NASA Water Vapor Project global water vapor multi–method dataset shows no discernible trend in the 1988–2009 total precipitable water (Vonder Haar et al. 2012; Fig. 9.14). Radiosonde data, passive microwave satellite observations, and ground-based global positioning system measurements, however, show different positive trends (Wang et al. 2016). While data suggests water vapor has increased more consistently over the ocean, the situation over land is more complex, with large regions showing opposite trends (Zhang et al. 2018). It might take decades to obtain sufficiently long, reliable, water-vapor datasets to settle the role of the hydrological cycle in global warming.

It appears that discounting a role for the hydrological cycle in climate change, other than simply responding to temperature changes is a little premature. Yet that is what the CO_2 hypothesis of climate change does. By focusing exclusively on radiative changes due to GHGs and assigning water vapor exclusively a feedback role according to the Clausius–Clapeyron relation, the hypothesis claims that CO_2 is *"the principal control knob governing Earth's temperature"* (Lacis et al. 2010), and *"the primary driver of Phanerozoic climate"* (Royer et al. 2004). It is important to distinguish between the GHE that explains planetary temperatures on the basis of their GHGs composition, and the CO_2 hypothesis that explains climate changes in terms of CO_2 changes. There is plenty of evidence for the GHE, but only circumstantial evidence at best for the CO_2 hypothesis. The important non-condensing role of preventing water from freezing out of the atmosphere almost completely, resulting in a Snowball Earth (Lacis et al. 2010) can probably be accomplished with CO_2 levels as low as 40–80 ppm.

The CO_2 hypothesis provides a simple answer to a complex question. Why does climate change? Mainly because atmospheric CO_2 levels change. Since humans have changed CO_2 levels, we are responsible for all global warming observed since the bottom of the LIA (Fig. 9.12c). Since CO_2 levels correlate with temperature changes in Antarctic ice-core records, the Milankovitch-forced glacial cycle is *"consistent with an important role for CO_2 in driving global climate change over glacial cycles"* (Shakun et al. 2012). That such a simplistic explanation for such a complex issue as climate change has been able to convince so many scientists and become the dominant hypothesis in the field is something likely to be studied by future science historians. Meanwhile, to restore some sanity, let's remember that the CO_2 hypothesis was

beaten out of the Pleistocene by Milankovitch (Hays et al. 1976), not supported by evidence during the Holocene (Liu et al. 2014), contradicted by the CO_2 proxy record during the Cenozoic (Fig. 9.8), and inconsistent with climatic records for most of the Phanerozoic (Fig. 9.4). All that is left is the coincidence of very low CO_2 values during the Karoo and Cenozoic Ice Ages, a very speculative solution to the Faint Sun Paradox, and the coincidence in CO_2 increase and temperature increase in the 1975–2000 period. Very little to show for such grandiose claim. A more realistic CO_2 hypothesis would be to consider it the pilot flame that keeps the thermostat alive during ice ages only to fire up the real thermostat, the hydrological cycle, once they are past. Without doubt, important increases in CO_2 have a significant effect on temperatures. There is a solid theory and enough evidence for that. But the evidence does not support an effect as large as models claim. In the words of the British statistician George Box: *"Remember that all models are wrong; the practical question is how wrong do they have to be to not be useful"* (Box & Draper 1987, pg. 74).

9.7 Climate change attribution

A scientifically relevant aspect of Modern Global Warming that becomes critical for policy implementation trying to deal with its negative consequences is the question of attribution. While science is firmly imbedded in uncertainty, policymaking demands certainty. Curry & Webster (2011a) essay on climate uncertainty claimed that it is too big to be simplified, as the IPCC does, or hidden as in the media portraits of global warming. Their claim was contested by a group of IPCC authors (Hegerl et al. 2011, followed by Curry & Webster 2011b reply). While many scientists fear that uncertainty recognition will lead to inaction and are furiously opposed to anyone raising the issue as a *"merchant of doubt"* (Oreskes & Conway 2010), Curry and Webster propose ideas for dealing with uncertainty, and convincingly argue that trust is more important than certainty for public confidence (Curry & Webster 2011).

The IPCC was established in 1988 by the WMO and UNEP to assess the available scientific, technical, and socio-economic information relevant for the understanding of the risk of "human-induced climate change" (Principles governing IPCC work). Through its invitations to experts, selection of authors, and monumental assessment reports, the IPCC exerts an enormous influence on climate science. Yet it is focused on human-induced climate change, which is the reason for its existence but only a part of the science of climate. Before the IPCC was established some aspects of climate science were well accepted and considered to have low uncertainty. They were:

- The GHE, solidly grounded in basic science and well-understood conceptually and mathematically
- The human-caused increase in CO_2, well constrained from a combination of the Suess effect, the Revelle effect, and Keeling's measurements. There is absolutely no doubt that humans have produced more CO_2 than necessary to explain the atmospheric increase, particularly since c. 1960
- The warming of the planet's surface since the Little Ice Age and, more relevant to a pronounced human effect, since 1976. The amount of warming was uncertain, but the warming itself was clear

The easy path for the IPCC was to define climate change in terms of radiative change, knowing well that humans were changing radiative forcing. Under this paradigm attribution is just a question of linking detection of changes to radiative forcing using models built under the assumption that changes in temperature are linearly proportional to changes in radiative forcing. The predetermined conclusion from the established assumptions is that all climate change is human-induced (Fig. 9.12), and the mission can be considered a success. Natural climate change has been abolished and we have entered a new era of human climate control, the Anthropocene. Confirmation is done through attribution studies, popularly referred to as fingerprinting. In the Fifth Assessment Report the IPCC redefines attribution as *"the process of evaluating the relative contributions of multiple causal factors to a change or event with an assignment of statistical confidence."* Also, *"a component of an observed change is attributed to a specific causal factor if the observations can be shown to be consistent with results from a process-based model that includes the causal factor in question, and inconsistent with an alternate, otherwise identical, model that excludes this factor."* (Bindoff et al. 2013, p 872–873). In other words, non-validated models built on the same set of assumptions are the final arbiters of attribution studies. Since changes in GHGs are known to cause temperature changes (from the GHE), then attributing temperature-dependent climatic changes to human-caused radiative changes becomes an exercise in circularity. But let's examine the causality issue in more detail, starting from the first principles generally accepted:

I. The greenhouse effect states that an increase in GHGs must cause an increase in temperature

II. An increase in temperature causes an increase in water vapor in a positive feedback loop responsible for a large part of the final temperature increase

III. There is an important human-caused increase in CO_2, particularly since 1960. Land use changes since earlier times have contributed to radiative changes

IV. The planet surface has been warming for the past 350 years since the LIA. The rate of warming has been highest between 1976–2000

How do we set out to attribute IV to III, versus hypothetical cause B? An enthalpy increase carries no signature of its origin. We have little uncertainty that part of IV has been caused by III, due to I. But radiative forcing analysis produces the surprising result that nearly all warming since 1750 is due to III (Fig. 9.12c), casting doubts that it properly reflects all participating causes of climate change. The attribution analysis is therefore marred because:

- Evidence of warming or changes in temperature-related phenomena only proves that warming has taken place, not its cause
- Evidence of CO_2 increase and related phenomena only proves that humans have increased CO_2 levels
- Evidence of an increased greenhouse effect only proves that warming has taken place, as in a watery planet warming increases the greenhouse effect. That is where most of future warming is expected to come from

The supposed human fingerprints cannot unequivocally tie the warming and increased greenhouse effect to the human-caused increase in CO_2, and models cannot be declared judges in the matter since they are partial due to the assumptions that went into their coding. The IPCC cannot demonstrate its proposition that *"the best estimate of the human-induced contribution to warming is similar to the observed warming over this period"* (IPCC 2014 SPM). The claim is reduced to the statement that we don't know other causes that could explain the warming in physically plausible terms. A similar answer to Wegener's continental drift hypothesis that, despite being based on evidence, could not provide a physically plausible mechanism until plate tectonics was discovered forty years later.

From AR4 to AR5 the IPCC increased its level of confidence. In 2007, *"Most of the observed increase in global average temperatures since the mid-20th century is very likely [>90%] due to the observed increase in anthropogenic GHG concentrations"* (IPCC 2007 SPM). In 2014, *"The evidence for human influence on the climate system has grown since the IPCC Fourth Assessment Report (AR4). It is extremely likely [>95%] that more than half of the observed increase in global average surface temperature from 1951 to 2010 was caused by the anthropogenic increase in GHG concentrations and other anthropogenic forcings together"* (IPCC 2014 SPM). In the 2007–2014 period:

- Global average temperature did not increase, showing a bigger divergence with model predictions
- Global CO_2 emissions increased on average 2.8% every year from 30 to 36 Gt CO_2/year, a 20% increase
- Model mean fast-feedback forcing was reduced by -0.4 W m^{-2} K^{-1} from CMIP3 to CMIP5, a 20% decrease

The IPCC has not explained how it was able to reduce uncertainty to ridiculously low values when the evidence was changing in the opposite direction to the hypothesis, and major revisions were done to models that reduced the predicted warming.

It is interesting that when the anthropogenic, solar and volcanic contributions over the past one thousand years are allowed free scaling to minimize the difference by mean squares between proxy-derived observation data and model output, the volcanic contribution remains the same, while the anthropogenic contribution is reduced and the solar contribution is increased, in comparison to running the experiment constrained by IPCC-established CO_2 and TSI radiative values (de Larminat 2016). In the free identification experiment, solar activity partly explains the

MWP and the LIA, and the total error is reduced. In the constrained identification, a significant cross-correlation appears between solar activity and output error, an indication of a causality not being correctly accounted for. A great deal of the science discussed in this book indicates that the climatic effect of solar variability has been significantly underestimated, out of ignorance and neglect, leading to an overestimation of the role of CO_2 and to the incorrect CO_2 hypothesis.

9.8 Conclusions

9a. The greenhouse effect explains a planet's surface temperature in terms of solar irradiance and atmospheric greenhouse gas content. The CO_2 hypothesis proposes that CO_2 is Earth's main temperature control knob and changes in CO_2 levels are responsible for most of the past climate history.

9b. The Faint Sun Paradox cannot be considered solved. It probably requires an ancient atmosphere with more GHGs and less albedo, but other important factors cannot be ruled out.

9c. Climate for the past 540 million years (Phanerozoic) has been a succession of warm periods and cool periods. CO_2 levels have followed a generally decreasing trend from above 2000 ppm to below 500 ppm, with very low values only occurring during the two longest ice ages. There is no general agreement between Phanerozoic CO_2 levels and climate.

9d. Temperature evolution for the past 65 million years shows an almost complete lack of correlation to changes in CO_2 levels. CO_2 decreased between 34–20 Ma, while temperature decreased between 50–34 Ma and 15–1 Ma, and increased between 34–15 Ma. Temperature during the Cenozoic cannot have been under CO_2 control.

9e. Marine biological diversity shows a periodical variability that is congruent with climate reconstructions, providing independent support for a cyclical climate over Earth's past 540 million years.

9f. Earth's climate system is extraordinarily complex. The IPCC has assumed that only factors known to affect the radiative balance matter and concluded that the anthropogenic contribution is likely to have been the only relevant climatic factor during Modern Global Warming.

9g. A large uncertainty about climate fast feedbacks prevents determination of the climatic effect from CO_2 changes, and confounds the attribution of observed temperature changes.

References

Airapetian VS, Glocer A, Gronoff G et al (2016) Prebiotic chemistry and atmospheric warming of early Earth by an active young Sun. Nature Geoscience 9 (6) 452–456

Arias PA, Bellouin N, Coppola E et al (2021) Technical Summary. In: Masson–Delmotte V, Zhai P, Pirani A et al (eds.) Climate Change 2021: The Physical Science Basis. Contribution of Working Group I to the Sixth Assessment Report of the Intergovernmental Panel on Climate Change. Cambridge University Press, Cambridge. In Press

Beerling DJ & Royer DL (2011) Convergent Cenozoic CO_2 history. Nature Geoscience 4 (7) 418–420

Bergman NM, Lenton TM & Watson AJ (2004) COPSE: a new model of biogeochemical cycling over Phanerozoic time. American Journal of Science 304 (5) 397–437

Berner RA & Kothavala Z (2001) GEOCARB III: a revised model of atmospheric CO_2 over Phanerozoic time. American Journal of Science 301 (2) 182–204

Bindoff NL, Stott PA, AchutaRao KM et al (2013) Detection and Attribution of Climate Change: from Global to Regional. In: Stocker TF, Qin D, Plattner G–K et al (eds) Climate Change 2013: The Physical Science Basis. Contribution of Working Group I to the Fifth Assessment Report of the Intergovernmental Panel on Climate Change. Cambridge University Press, Cambridge, p 867–952

Bond DP & Grasby SE (2017) On the causes of mass extinctions. Palaeogeography, Palaeoclimatology, Palaeoecology 478 3–29

Box GE & Draper NR (1987) Empirical model-building and response surfaces. John Wiley & Sons, New York

Charney JG, Arakawa A, Baker DJ et al (1979) Carbon dioxide and climate: a scientific assessment. National Academy of Sciences, Washington DC, p 2030–2050

Conway TM, Wolff EW, Röthlisberger R et al (2015) Constraints on soluble aerosol iron flux to the Southern Ocean at the Last Glacial Maximum. Nature communications 6 7850

Crowley TJ (1998) Significance of tectonic boundary conditions for paleoclimate simulations. In: Crowley TJ and Burke K (eds) Tectonic boundary conditions for climate reconstructions. Oxford monographs on geology and geophysics 39. Oxford University Press, New York, p 3–20

Curry JA & Webster PJ (2011a) Climate science and the uncertainty monster. Bulletin of the American Meteorological Society 92 (12) 1667–1682

Curry JA & Webster PJ (2011b) Reply to Comment on "Climate Science and the Uncertainty Monster." Bulletin of the American Meteorological Society 92 (12) 1686–1687. doi:10.1175/BAMS-D-11-00195.1

Davies JH & Davies DR (2010) Earth's surface heat flux. Solid Earth 1 (1) 5–24

Davis WJ (2017) The relationship between atmospheric carbon dioxide concentration and global temperature for the last 425 million years. Climate 5 (4) 76

de Larminat P (2016) Earth climate identification vs. anthropic global warming attribution. Annual Reviews in Control 42 114–125

Doglioni C, Pignatti J & Coleman M (2016) Why did life develop on the surface of the Earth in the Cambrian? Geoscience Frontiers 7 (6) 865–873

Egbert GD & Ray RD (2000) Significant dissipation of tidal energy in the deep ocean inferred from satellite altimeter data. Nature 405 (6788) 775–778

Ekholm N (1901) On the variations of the climate of the geological and historical past and their causes. Quarterly Journal of the Royal Meteorological Society 27 (117) 1–62

Erlykin AD, Harper DA, Sloan T & Wolfendale AW (2017) Mass extinctions over the last 500 myr: an astronomical cause? Palaeontology 60 (2) 159–167

Farley KA (1995) Cenozoic variations in the flux of interplanetary dust recorded by 3He in a deep-sea sediment. Nature 376 (6536) 153–156

Feulner G (2012) The faint young Sun problem. Reviews of Geophysics 50 (2) RG2006

Fischer AG & Arthur MA (1977) Secular variations in the pelagic realm. The Society of Economic Paleontologists and Mineralogists, Special publication 25 19–50

Flato GJ, Marotzke B, Abiodun P et al (2013) Evaluation of Climate Models. In: Stocker TF, Qin D, Plattner G–K et al (eds) Climate Change 2013: The Physical Science Basis. Contribution of Working Group I to the Fifth Assessment Report of the

Intergovernmental Panel on Climate Change. Cambridge University Press, Cambridge, p 741–866

Forster P, Storelvmo T, Armour K et al (2021) The Earth's Energy Budget, Climate Feedbacks, and Climate Sensitivity. In: Masson–Delmotte V, Zhai P, Pirani A et al (eds.) Climate Change 2021: The Physical Science Basis. Contribution of Working Group I to the Sixth Assessment Report of the Intergovernmental Panel on Climate Change. Cambridge University Press, Cambridge. In Press

Foster GL, Royer DL & Lunt DJ (2017) Future climate forcing potentially without precedent in the last 420 million years. Nature Communications 8 p14845

Frakes LA (1979) Climates throughout geologic time. Elsevier, Amsterdam

Frakes LA & Francis JE (1988) A guide to Phanerozoic cold polar climates from high-latitude ice-rafting in the Cretaceous. Nature 333 (6173) 547–549

Frakes LA, Francis JE & Syktus JI (1992) Climate modes of the Phanerozoic. Cambridge University Press, Cambridge

Gaillardet J, Dupré B, Louvat P & Allegre CJ (1999) Global silicate weathering and CO_2 consumption rates deduced from the chemistry of large rivers. Chemical geology 159 (1–4) 3–30

Gray LJ, Beer J, Geller M et al (2010) Solar influences on climate. Reviews of Geophysics 48 (4) RG4001

Hays JD, Imbrie JN & Shackleton J (1976) Variations in the Earth's Orbit: Pacemaker of the Ice Ages. Science 194 1121–1132

Hegerl G, Stott P, Solomon S & Zwiers F (2011) Comment on "Climate Science and the Uncertainty Monster" by JA Curry and PJ Webster. Bulletin of the American Meteorological Society 92 (12) 1683-1685

Huybers P & Langmuir C (2009) Feedback between deglaciation, volcanism, and atmospheric CO_2. Earth and Planetary Science Letters 286 (3–4) 479–491

Intergovernmental Panel on Climate Change (2007) Summary for Policymakers. In: Core Writing Team, Pachauri RK & Reisinger A (eds) Climate Change 2007: Synthesis Report. Contribution of Working Groups I, II and III to the Fourth Assessment Report of the Intergovernmental Panel on Climate Change. IPCC, Geneva, p 11

Intergovernmental Panel on Climate Change (2014) Summary for Policymakers. In: Core Writing Team, Pachauri RK & Meyer LA (eds) Climate Change 2014: Synthesis Report. Contribution of Working Groups I, II and III to the Fifth Assessment Report of the Intergovernmental Panel on Climate Change. IPCC, Geneva, p 4–6

Keeling CD (1960) The concentration and isotopic abundances of carbon dioxide in the atmosphere. Tellus 12 (2) 200–203

Köppen W & Wegener A (1924) Die klimate der geologischen vorzeit. Gebruder Borntraeger, Berlin

Krüger T (2013) Discovering the Ice Ages: International reception and consequences for a historical understanding of climate. Brill, Leiden

Kuhn WR, Walker JCG & Marshall HG (1989) The effect on Earth's surface temperature from variations in rotation rate, continent formation, solar luminosity, and carbon dioxide. Journal of Geophysical Research: Atmospheres 94 (D8) 11129–11136

Labrosse S, Poirier JP & Le Mouël JL (2001) The age of the inner core. Earth and Planetary Science Letters 190 (3–4) 111–123

Lacis AA, Schmidt GA, Rind D & Ruedy RA (2010) Atmospheric CO_2: Principal control knob governing Earth's temperature. Science 330 (6002) 356–359

Liu Z, Zhu J, Rosenthal Y et al (2014) The Holocene temperature conundrum. Proceedings of the National Academy of Sciences 111 (34) E3501–E3505

Lorius C, Jouzel J, Ritz C et al (1985) A 150,000-year climatic record from Antarctic ice. Nature 316 (6029) 591–596

Lovelock JE & Margulis L (1974) Atmospheric homeostasis by and for the biosphere: the Gaia hypothesis. Tellus 26 (1–2) 2–10

Martínez–García A, Sigman DM, Ren H et al (2014) Iron fertilization of the Subantarctic Ocean during the last ice age. Science 343 (6177) 1347–1350

Meehl GA, Washington WM, Ammann CM et al (2004) Combinations of natural and anthropogenic forcings in twentieth-century climate. Journal of Climate 17 (19) 3721–3727

Melott AL & Bambach RK (2014) Analysis of periodicity of extinction using the 2012 geological timescale. Paleobiology 40 (2) 177–196

Myhre GD, Shindell F–M, Bréon W et al (2013) Anthropogenic and Natural Radiative Forcing. In: Stocker TF, Qin D, Plattner G–K et al (eds) Climate Change 2013: The Physical Science Basis. Contribution of Working Group I to the Fifth Assessment Report of the Intergovernmental Panel on Climate Change. Cambridge University Press, Cambridge, p 659–740

Oreskes N & Conway EM (2010) Defeating the merchants of doubt. Nature 465 (7299) 686–688

Palastanga V, Slomp CP & Heinze C (2013) Glacial-interglacial variability in ocean oxygen and phosphorus in a global biogeochemical model. Biogeosciences 10 (2) 945–958

Poynting JH (1907) On Prof Lowell's method for evaluating the surface-temperatures of the planets; with an attempt to represent the effect of day and night on the temperature of the Earth. Philosophical Magazine 14 (84) 749–760

Rampino MR & Caldeira K (2015) Periodic impact cratering and extinction events over the last 260 million years. Monthly Notices of the Royal Astronomical Society 454 (4) 3480–3484

Rampino MR & Stothers RB (1984a) Geological rhythms and cometary impacts. Science 226 (4681) 1427–1431

Rampino MR & Stothers RB (1984b) Terrestrial mass extinctions cometary impacts and the Sun's motion perpendicular to the galactic plane. Nature 308 (5961) 709–712

Randall L & Reece M (2014) Dark matter as a trigger for periodic comet impacts. Physical review letters 112 (16) 161301

Raup DM & Sepkoski JJ (1984) Periodicity of extinctions in the geologic past. Proceedings of the National Academy of Sciences 81 (3) 801–805

Revelle R & Suess HE (1957) Carbon dioxide exchange between atmosphere and ocean and the question of an increase of atmospheric CO_2 during the past decades. Tellus 9 (1) 18–27

Rohde RA and Muller RA (2005) Cycles in fossil diversity. Nature 434 (7030) 208–210

Rosing MT, Bird DK, Sleep NH & Bjerrum CJ (2010) No climate paradox under the faint early Sun. Nature 464 (7289) 744–749

Rothman DH (2002) Atmospheric carbon dioxide levels for the last 500 million years. Proceedings of the National Academy of Sciences 99 (7) 4167–4171

Royer DL (2014) Atmospheric CO_2 and O2 during the Phanerozoic: Tools patterns and impacts. In: Holland HD and Turekian HK (eds) Treatise on Geochemistry, 2nd ed. Elsevier, Vol 6 p 251–267

Royer DL, Berner RA, Montañez IP et al (2004) CO_2 as a primary driver of phanerozoic climate. GSA today 14 (3) 4–10

Sagan C & Mullen G (1972) Earth and Mars: Evolution of atmospheres and surface temperatures. Science 177 (4043) 52–56

Schröder M, Lockhoff M, Forsythe JM et al (2016) The GEWEX water vapor assessment: Results from intercomparison trend and homogeneity analysis of total column water vapor. Journal of Applied Meteorology and Climatology 55 (7) 1633–1649

Schwartz RD & James PB (1984) Periodic mass extinctions and the Sun's oscillation about the galactic plane. Nature 308 (5961) 712–713

Scotese CR (2016) Some thoughts on global climate change: The transition from icehouse to hothouse. In: Scotese CR (author)

The Earth History: The evolution of the Earth System. PALEOMAP Project, Evanston, IL, https://www.researchgate.net/publication/275277369_Some_Thoughts_on_Global_Climate_Change_The_Transition_for_Icehouse_to_Hothouse_Conditions Accessed 18 Oct 2018

Scotese CR (2018) Phanerozoic Temperatures: Tropical Mean Annual Temperature (TMAT), Polar Mean Annual Temperature (PMAT) and Global Mean Annual Temperature (GMAT) for the last 540 million Earth's Temperature. History Research Workshop, Smithsonian National Museum of Natural History, Washington DC 30–31 Mar 2018 https://www.researchgate.net/publication/324017003_Phanerozoic_Temperatures_Tropical_Mean_Annual_Temperature_TMAT_Polar_Mean_Annual_Temperature_PMAT_and_Global_Mean_Annual_Temperature_GMAT_for_the_last_540_million_years Accessed 13 Jun 2022

Shakun JD, Clark PU, He F et al (2012) Global warming preceded by increasing carbon dioxide concentrations during the last deglaciation. Nature 484 (7392) 49–54

Shaviv NJ (2006) Long-term variations in the galactic environment of the Sun. In: Frisch PC (ed) Solar journey: The significance of our galactic environment for the heliosphere and Earth. Springer, Dordrecht, p 99–131

Shaviv NJ & Veizer J (2003) Celestial driver of Phanerozoic climate? GSA today 13 (7) 4–10

Slater GJ, Goldbogen JA & Pyenson ND (2017) Independent evolution of baleen whale gigantism linked to Plio–Pleistocene ocean dynamics. Proceedings of the Royal Society B: Biological Sciences 284 (1855) 20170546

Solomon S, Rosenlof KH, Portmann RW et al (2010) Contributions of stratospheric water vapor to decadal changes in the rate of global warming. Science 327 (5970) 1219–1223

Suess HE (1955) Radiocarbon Concentration in Modern Wood. Science 122 (3166) 415–417

Svensmark J, Enghoff MB, Shaviv NJ & Svensmark H (2016) The response of clouds and aerosols to cosmic ray decreases. Journal of Geophysical Research: Space Physics 121 (9) 8152–8181

Veizer J, Godderis Y & François LM (2000) Evidence for decoupling of atmospheric CO_2 and global climate during the Phanerozoic eon. Nature 408 (6813) 698–701

Vérard C, Hochard C, Baumgartner PO et al (2015) Geodynamic evolution of the Earth over the Phanerozoic: Plate tectonic activity and palaeoclimatic indicators. Journal of Palaeogeography 4 (2) 167–188

Very FW (1908) The Greenhouse Theory and planetary temperatures. Philosophical Magazine 16 (93) 462–480

Vincent E & Berger WH (1985) Carbon dioxide and polar cooling in the Miocene: The Monterey hypothesis. In: Sundquist ET and Broecker WS (eds) The carbon cycle and atmospheric CO_2: Natural variations Archean to present. American Geophysical Union, Washington DC. Geophysical Monograph Series 32 p 455–468

Vonder Haar TH, Bytheway JL & Forsythe JM (2012) Weather and climate analyses using improved global water vapor observations. Geophysical Research Letters 39 (15)

Voosen P (2021) Climate panel confronts implausibly hot models. Science 373 (6554) 474–475

Walker JC, Hays PB & Kasting JF (1981) A negative feedback mechanism for the long-term stabilization of Earth's surface temperature. Journal of Geophysical Research: Oceans 86 (C10) 9776–9782

Wang J, Dai A & Mears C (2016) Global water vapor trend from 1988 to 2011 and its diurnal asymmetry based on GPS radiosonde and microwave satellite measurements. Journal of Climate 29 (14) 5205–5222

Williams GE (1975) Possible relation between periodic glaciation and the flexure of the galaxy. Earth and Planetary Science Letters 26 (3) 361–369

Wuebbles DJ, Fahey DW & Hibbard KA (2017) Executive summary. In: Climate Science Special Report: Fourth National Climate Assessment, US Global Change Research Program, Washington DC, Vol I pp 12–34

Zachos J, Pagani M, Sloan L et al (2001) Trends, rhythms, and aberrations in global climate 65 Ma to present. Science 292 (5517) 686–693

Zhang Y, Xu J, Yang N & Lan P (2018) Variability and Trends in Global Precipitable Water Vapor Retrieved from COSMIC Radio Occultation and Radiosonde Observations. Atmosphere 9 (5) 174

10

MERIDIONAL TRANSPORT, A SOLAR-MODULATED FUNDAMENTAL CLIMATE PROPERTY

"It is not easy to convey, unless one has experienced it, the dramatic feeling of sudden enlightenment that floods the mind when the right idea finally clicks into place. One immediately sees how many previously puzzling facts are neatly explained by the new hypothesis. One could kick oneself for not having the idea earlier, it now seems so obvious. Yet before, everything was in a fog."

Francis Crick (1988)

10.1 Introduction

The energy flux at the top of the atmosphere (ToA) depends upon the irregular distribution of the incoming shortwave solar radiation over a rotating planet with its axis tilted with respect to the ecliptic. The amount of incoming radiation that is reflected by the planet is also irregularly distributed spatially, as planetary albedo is minimal at the tropical bands, increases with latitude, and varies with the seasons, being highest in winter outside the tropics. The result is that the net absorbed solar radiation at the ToA falls rapidly with latitude and the decrease is more marked in winter than in summer. The Earth also emits radiation at the ToA. This outgoing longwave radiation (OLR) depends on surface temperature and atmospheric greenhouse effect and is also irregularly distributed. But it varies less with latitude and seasons than the absorbed solar radiation mainly because, on the absolute scale, surface temperature differences are small. As a result the net radiation flux at the ToA is positive (energy gain) on an annual average from the equator to c. 35°, and negative from c. 35° to the poles. However, in winter the net flux becomes negative c. 15°N and 10°S (Hartmann 2016).

To maintain a net negative flux over time at mid-high latitudes without a continuous fall in surface temperature, a meridional transport (MT) of energy from the equator towards the poles is implied. This MT of energy is carried out by the atmosphere and the ocean along a latitudinal temperature gradient (LTG), and results in a redistribution of mass, moisture, aerosols and chemicals. The total MT is usually calculated from the measured net radiative flux at the ToA. Direct measurement of heat MT by the atmosphere and ocean is hampered by sparse observations and big uncertainties. An indirect estimation method is often used based on the assumption that energy storage in the ocean is near-constant, and thus the energy flux across the sea surface, easier to measure, properly reflects the MT partition between both. With the indirect method the atmospheric and oceanic MT are not independently measured, and the hypothesis that changes in one are fully compensated by changes in the other, known as the Bjerknes compensation, is also assumed.

A postulate of the global energy balance is that horizontal transport integrated over the globe is zero because energy can only be removed from the Earth at the ToA.

The annual global average net radiation is considered very close to zero, and any global mean net radiative imbalance will lead to either heating or cooling of the Earth (Hartmann 2016). While it is clear that the integral of the global horizontal energy transport must be zero, the assumption that horizontal transport, and more properly MT, does not affect the net radiation flux at the ToA is the most fundamental mistake of modern climatology.

The distribution of greenhouse gases (GHGs) over the planet is very irregular and, latitudinally averaged, the tropics are GHG-rich due to high humidity while the poles are GHG-poor for the opposite reason. Since a change in GHG content affects the radiative flux balance at the ToA, transporting more energy from a GHG-rich region to a GHG-poor region increases the amount of energy lost at the ToA, as the effect is similar to reducing the GHG content. The modulation of the energy flux of the planet by the strength of poleward energy transport makes the climate very sensitive to changes in MT. Energy transport by the atmosphere and ocean are not properly measured and significant errors and uncertainty prevent adequate knowledge of its variability (Wunsch 2005). This uncertainty obscures existing evidence of the modulation of MT by solar activity. Willful ignorance of this important contribution prevents us from understanding the climate and its changes.

10.2 Planetary transport of energy by the atmosphere

The depth of the troposphere (c. 10 km) is just 1/600 of the Earth's radius, but this thin film of gas has the crucial role of always maintaining a land surface temperature compatible with complex life, something it has done for at least the past 540 Myr. To fulfill this role the atmosphere must accomplish two tasks. The first is to compensate for the daily changes in insolation due to Earth's rotation by reducing the incoming solar radiation on the bright side through albedo, and by reducing the outgoing longwave radiation on the dark side through the greenhouse gas effect. Without this role, the land surface temperature of the Earth would quickly reach values incompatible with complex life. The second task is to compensate for the latitudinal decrease in solar radiation due to the angle of incidence, and the seasonal changes in insolation due to the axial tilt of the planet. The atmosphere accomplishes this task by transporting energy from the tropics, which receive the highest insolation, towards the poles that receive the lowest insolation. The ocean participates in the MT of energy, but outside the tropics the bulk of the transport is done by the atmosphere.

The dominant CO_2 climate hypothesis remains focused on the first task, giving extreme importance to small changes in GHG forcing and attributing the observed changes almost exclusively to them through not very well constrained feedbacks. However, the second task, also crucial for the planetary energy budget, has received much less attention due to the difficulties of measuring and modeling hemispheric and inter-hemispheric MT. It is generally assumed that moving energy around within the climate system does not affect the long-term radiative budget, but that assumption is not warranted.

MT takes place because there is a LTG resulting from the latitudinal insolation gradient. The axial tilt introduces an asymmetry, leaving the winter pole in constant darkness. As a result, the LTG becomes steeper towards the winter pole and MT is stronger in that direction. More energy is transported in winter than in summer through each hemisphere. The LTG is a defining feature of the planet's climate. For the past 540 Myr the planet has gone through hot periods (hothouse climate), temperate periods (greenhouse climate) and cold periods (icehouse climate). Through all of them the temperature and width of the equatorial wet and the subtropical arid belts have not changed significantly, because the open deep-ocean sea surface temperature cannot go much higher than 30 °C due to deep convection (Sud et al. 1999), and because their width is controlled by the Hadley Cell circulation (Scotese et al. 2021). So, the changing climate of the Earth has been characterized by the changing width of the Warm Temperate, Cool Temperate and Polar Belts (light green, brown,

and light blue areas in Fig. 10.1a). From fossil and geochemical evidence, Scotese et al. (2021) have reconstructed the climatic zones of the planet at 5 Myr intervals. The Equator-to-Pole LTG of the different epochs describes the different climates (Fig. 10.1b). At present (21st century), the Earth is in a severe icehouse climate; it has an average temperature of 14.4 °C and the LTG is very steep. The temperature falls by 0.6–1 °C/°latitude from the equator to the winter pole. Such icehouse conditions have been relatively rare during the past 540 Myr (less than 10% of the time). During the early Eocene the average temperature is estimated at 23.8 °C and the LTG was very shallow, at 0.25–0.45 °C/°latitude, with temperatures at the North Pole above freezing all year round, as attested by the presence of frost-intolerant biota. These hothouse conditions have been even rarer. Over 80% of the Phanerozoic Eon the Earth had an average temperature of 17–20 °C (Scotese et al. 2021).

Past hothouse climates of the Earth cannot be explained with our current climate theory. It has been defined as the *"equable climate problem"* (Huber & Caballero 2011). In order to reproduce the early Eocene warm continental interior temperatures and above freezing winter high latitudes, models have to raise CO_2 levels to 4700 ppm and tropical temperatures to 35 °C. However, the best CO_2 estimates for the early Eocene climatic optimum (Beerling & Royer 2011; Steinthorsdottir et al. 2019) place CO_2 levels at 500–1000 ppm, and it is unclear that a tropical temperature above 30 °C is possible. The survivability wet-bulb temperature limit for mammals is 35 °C, at which point they become unable to lose heat (Sherwood & Huber 2010). The highest wet-bulb temperature on Earth today is 30 °C, and there is no reason to think that it has been higher at any time in the past at places where mammal fossils are found.

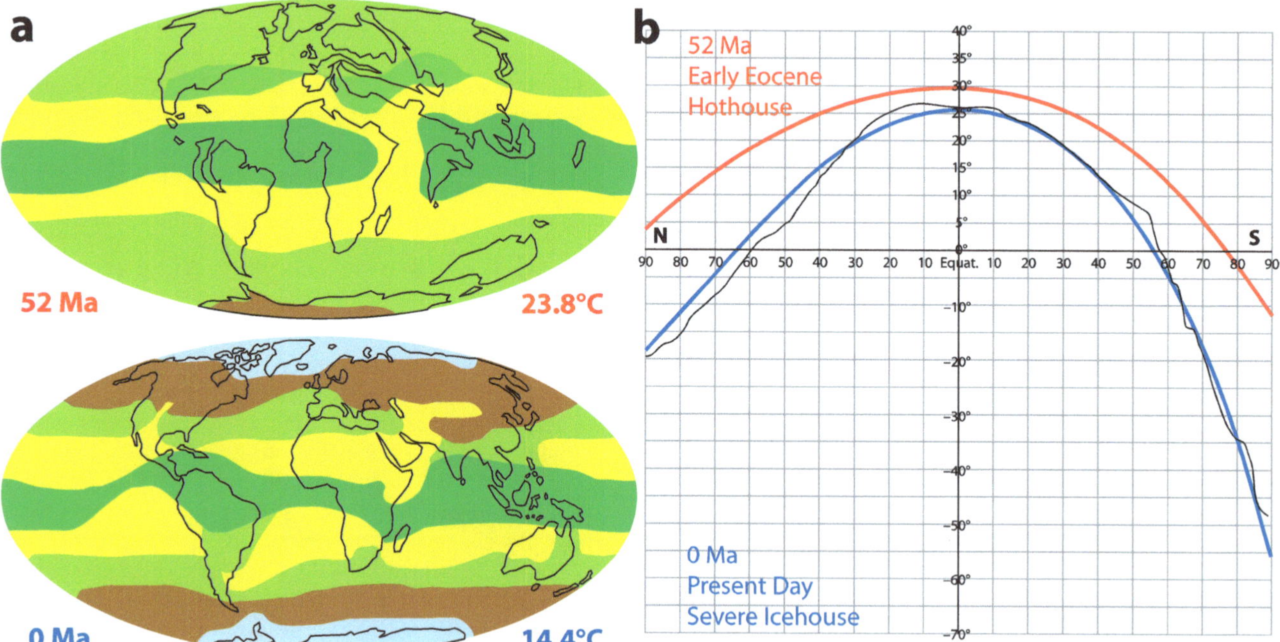

Fig. 10.1 The Earth's climate is defined by its latitudinal temperature gradient
a) Climatic belts of the early Eocene hothouse (top) deduced from fossil and geochemical evidence by Scotese et al. 2021, and the present severe icehouse (bottom). Equatorial wet (dark green), subtropical arid (yellow), warm temperate (light green), cool temperate (brown) and polar (light blue) belts. Temperature is the estimated global mean average. **b)** Latitudinal temperature gradient inferred for the early Eocene (red) and the present (blue) versus measured (black, fine line). After Scotese et al. 2021.

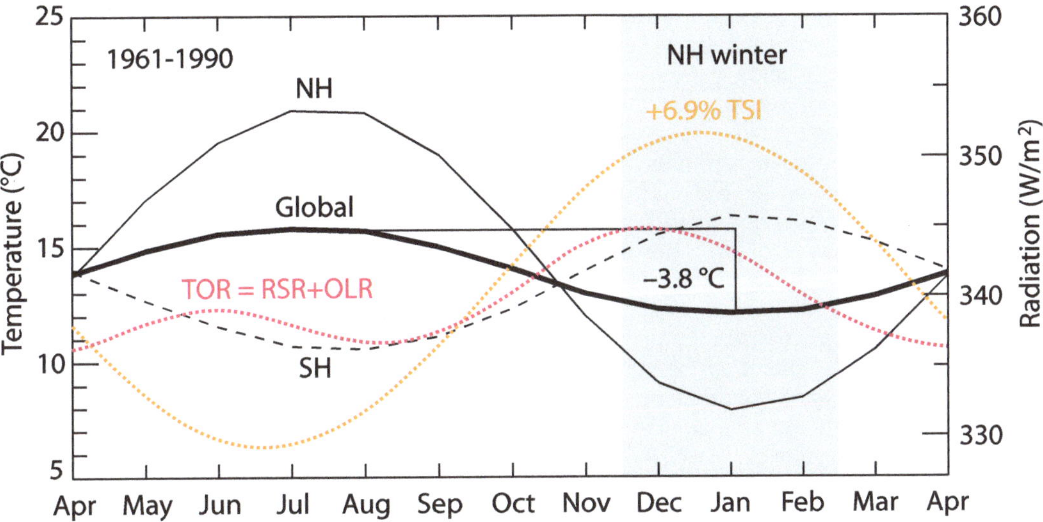

Fig. 10.2 Yearly temperature and radiation change
The global surface average temperature of the planet (thick line) changes by 3.8 °C over the course of a year, mostly because the NH (thin line) varies by 12 °C. The planet is coldest during the month of January, despite receiving 6.9% more total solar irradiance (TSI, dotted yellow line) in early January when the Earth is in perihelion. The planet has two peaks of energy loss (TOR, Total Outgoing Radiation, outgoing longwave and reflected shortwave, dotted red line) when each hemisphere cools, with the highest during the cooling of the NH. Between November and January the planet emits more energy (TOR) than at any other time. SH, dashed line. NH winter, light grey area. 1961–1990 temperature data from Jones et al. 1999. Radiation data from Carlson et al. 2019.

Conceptually, we believe that to have warm poles more heat must be transported there, to compensate for the insolation deficit. Heat MT is a very important part of the planetary energy budget, and it is generally believed that without it the poles would be much colder. But MT depends on the LTG since much of the poleward transport in the present climate is carried by atmospheric eddies resulting from baroclinic instability. Somewhat counterintuitively the warm poles of the early Eocene and their much shallower LTG imply a very reduced MT, in what has been termed the *"low gradient paradox"* (Huber & Caballero 2011). The climate of the early Eocene was warmer and equable (reduced LTG and seasonality) despite (or because of) a lower heat transport to the poles. It is no wonder that climate models have such a problem reproducing it.

In essence, we go back to the two tasks the atmosphere must perform to maintain land surface life-compatible conditions, reduce the day-night contrast (GHGs role) and reduce the winter-summer contrast at high latitudes (MT role). The ruling CO_2 hypothesis has focused mainly on the first one, assuming that the average net radiation flux at the ToA depends critically on non-condensing GHG levels and assuming that MT does not affect the flux significantly. It is clear that we need a better understanding of meridional heat transport.

In terms of temperature and MT the planet is highly asymmetric (Fig. 10.1b, blue curve). According to reanalysis, the NH is 1.2 °C warmer than the SH (Kang et al. 2015), but the seasonal variation is much higher for the NH ($\Delta11.7$ °C) than the SH ($\Delta5.1$ °C). The NH in winter is

2.2 °C colder than the SH in winter, notwithstanding Antarctica being so cold. The boreal winter is not only the coldest period for the NH, but also for the planet (Jones et al. 1999; Fig. 10.2), despite the Earth receiving 6.9% more insolation in January than in July. Planetary albedo is higher during the austral summer due to oceanic clouds and Antarctic albedo, so more shortwave radiation is reflected at perihelion compensating in part for the higher incoming shortwave radiation. The planet surface cools

Fig. 10.3 Asymmetry in meridional profiles
Annual mean, thick line; DJF mean, thin line; JJA mean, dashed line. **a)** The zonal-mean cloud cover (in %). **b)** The east-west standard deviations of the time-mean geopotential height in gpm (1 gpm ≈ 1 m). **c)** The zonal- and vertical-mean northward flux of sensible heat by stationary eddies in units of °C m/s. **d)** The divergence of the vertically integrated atmospheric transport of energy in W/m². After Peixoto & Oort 1992.

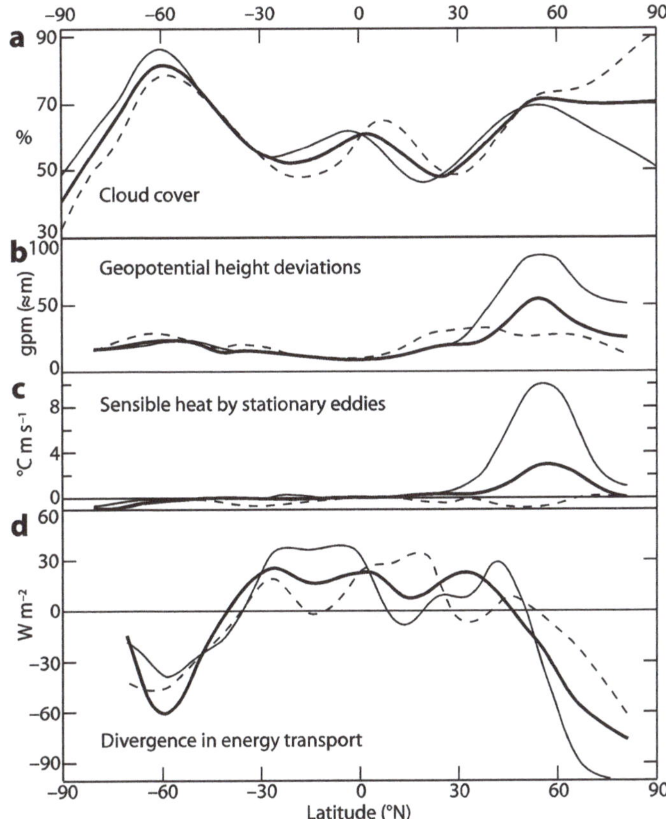

down during the boreal fall by 3.8 °C between July and January every year (Fig. 10.2). During the entire year the planet is in a profound energy imbalance, gaining energy between October and May (its cold period), and losing it between May and October (its warm period). It is believed that the hemispheric difference in temperature is due mainly to the larger land fraction in the NH that warms and cools more than the ocean surface. Some of the consequences are the preferential location of the Intertropical Convergence Zone (the climatic equator) in the NH and a net inter-hemispheric heat transport from the SH to the NH. Hemispheric transport asymmetry results also from the reduction in MT to the South Polar Cap, hindered by the Antarctic Circumpolar Current and the Southern Annular Mode, that climatically isolate Antarctica. The result from these asymmetries is that despite the South Pole being much colder, more energy is transported to the North Pole (Fig. 10.3 & 10.4; Peixoto & Oort 1992).

Peixoto and Oort (1992) show a synthesis of the polar energy flows in a diagram (Fig. 10.4). Considering the 70–90° region, there is very little incoming short-wave radia-

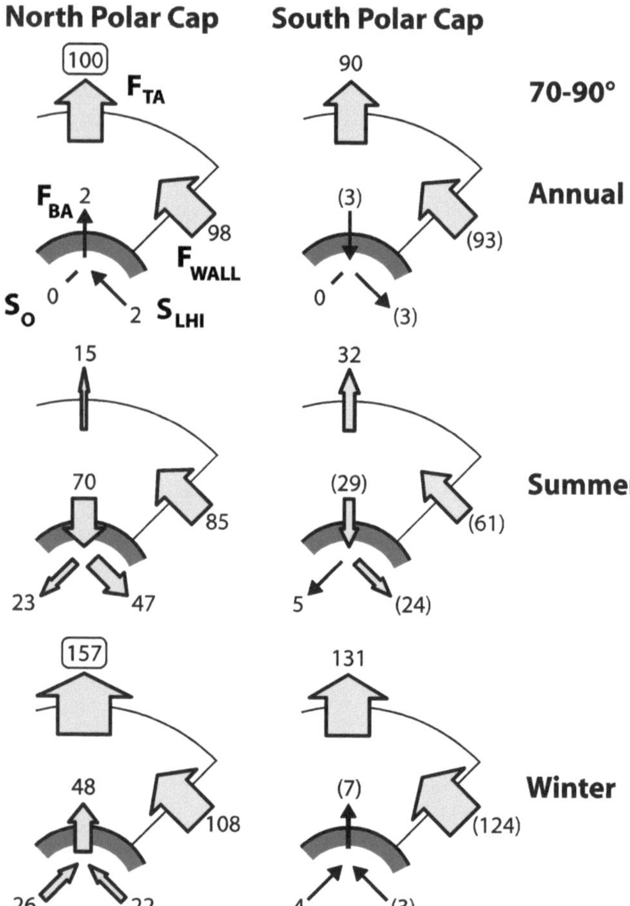

Fig. 10.4. Observed polar heat budgets during typical annual, summer, and winter mean conditions for the two polar caps While the annual-mean budget is similar, in winter the radiative loss at the NPC top of the atmosphere is very large, one third from cooling and freezing the ocean, and two thirds transported by the atmosphere. Values in W/m². The numbers in parentheses are indirect estimates. F_{TA} net energy flux at top of atmosphere. F_{BA} downward flux of energy at bottom of atmosphere. F_{WALL} flux across 70° latitudinal wall. S_O rate of storage of sensible heat in the oceans. S_{LHI} rate of storage of latent heat in the form of snow and ice. After Peixoto & Oort 1992.

tion during the winter, and the net flux at the ToA is dominated by the loss of long-wave radiation that depends on atmospheric temperature. The loss is smaller in the South polar cap because the atmosphere is colder. Over the year, the North polar cap loses about 10 W/m² more heat than the SPC due to its warmer atmosphere.

The atmosphere provides more heat to the North polar region in winter (120 W/m² across 70°N) than in summer (80 W/m² across 70°N). Most of the transport is carried out by transient eddies and the mean meridional circulation, but the winter-summer difference is mostly due to stationary eddies that in winter are responsible for most of the increase (Fig. 10.3c). Stationary eddies are characterized by persistent zonal asymmetries in atmospheric variables that average to zero along a latitude band. They arise from zonal surface asymmetries (land/sea contrasts and orography) that are more pronounced in the NH. Stationary eddies can shape winter storm tracks and control where they terminate (Kaspi & Schneider 2013).

The energy flux between the atmosphere and the surface in the Arctic is characterized by a large upward transport of energy from the surface during the winter (c. 50 W/m²) due to the absence of solar radiation. The equilibrium temperature of sea water in contact with ice is practically constant regardless of the atmospheric temperature and sea ice thickness. Water temperature below the ice can reach –1.9 °C, while the atmosphere can be –30 to –60 °C. Sea ice is a very good insulator (K ≈ 2.2 W/m K), but when the ice is thin there is some loss. For example, for an ice layer 2 m thick and a temperature difference of 30 °C the amount of heat transferred is on the order of 30 W/m² (Peixoto & Oort 1992). For exposed waters the loss is much higher and can amount to 250 W/m² of sensible heat and 60 W/m² of latent heat transferred to the atmosphere.

For the North polar cap the heat arriving in summer to the polar atmosphere from lower latitudes is largely balanced by the heat flowing to the surface (snow/ice melting and ocean warming). In winter the radiative loss at ToA is very large. Two thirds of it carried by the atmosphere from lower latitudes, and one third through the surface, about equally from lowering the ocean temperature and freezing the surface water. For the South polar cap the surface exchange is very small, as it is 78% land, most of it permanently frozen.

10.3 Arctic winter heat transport by the atmosphere

Dry static (sensible + geopotential) heat is brought into the winter Arctic by both the middle (20–100 km height) and lower atmosphere, while latent heat (moisture) is transported almost exclusively by the lower atmosphere. In an undisturbed atmosphere air would rise in the summer hemisphere, cooling adiabatically, and would descend in the winter hemisphere, warming, and in the process transporting dry static heat. This is what is observed in the upper layers of the atmosphere, down to the mesosphere, where this interhemispheric transport takes place. However, the low mass of the upper atmosphere (0.1%) makes this transport irrelevant for the energy flow at the winter pole.

There are three branches of mass circulation at 60°N in the lower atmosphere (0–20 km height): the strato-

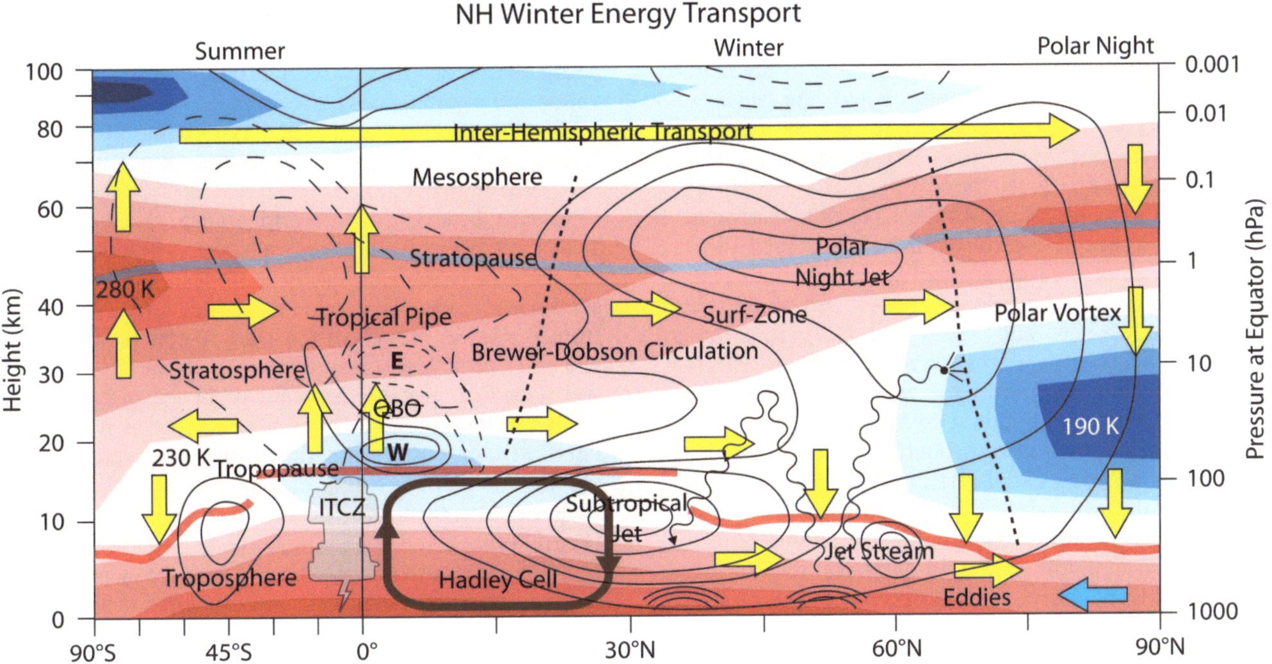

Fig 10.5 Schematic of atmospheric circulation at the December solstice in a two-dimensional lower and middle atmospheric view Background colors indicate relative temperatures at 10 K steps, with red being warmer and dark blue being cooler. Vertical scale is logarithmic, and the SH latitudinal scale is compressed. Westerly winds represented by thin lines; easterly winds by thin dashed lines. The tropopause (thick orange line) separates the troposphere and stratosphere, and the stratopause (thick steel blue line) the stratosphere and the mesosphere. Thick dotted lines separate the tropical pipe (ascent zone), the surf-zone (wave-breaking zone), and the polar vortex. Planetary waves (undulating lines) generate at areas of contrast (concentric lines at surface) and can pass through the stratosphere, be deflected and break at the stratosphere or be refracted back to the troposphere. The quasi-biennial oscillation (QBO) is shown with its easterly and westerly components close to the Equator. The intertropical convergence zone (ITCZ) is shown as a tall stormy cloud. The Hadley circulation is displayed in dark brown (only shown in the NH). Other atmospheric circulation is represented by yellow arrows except the lower tropospheric equatorward circulation in turquoise. The stratospheric circulation is termed Brewer–Dobson circulation. Its deep branch (upper stratospheric) and mesospheric circulation are interhemispheric from the summer to the winter pole. Tropospheric circulation is carried out mainly by eddies, and the rest by the mean residual circulation. At the December solstice, regions North of 72° are in polar night.

spheric poleward warm air branch, the poleward warm air branch in the upper troposphere, and the equatorward cold air branch in the lower troposphere (Fig. 10.5; Yu & Ren 2019). They are driven by both diabatic heating/cooling and eddy-induced forcing. Positive surface pressure anomalies in the Arctic occur when both the cold and warm mass circulation branches strengthen, leading to a transport of warm air mass into the polar region in the upper troposphere, and cold air into the mid-latitudes near the surface.

First let us consider the poleward winter stratospheric transport. The stratosphere contains 15% of the atmospheric mass and the other 85% is in the troposphere. Air enters the stratosphere through the tropical pipe (Fig. 10.5) and goes through a cold region above the tropical tropopause where it loses most of its water vapor. Stratospheric meridional transport has been termed the Brewer–Dobson circulation (BDC). In the lower stratosphere, the shallow branch of the BDC has a poleward direction, although it is stronger towards the winter pole. In the middle and high latitudes the BDC air descends through the tropopause toward the surface. In the upper stratosphere the deep branch of the BDC is inter-hemispheric and moves toward the winter pole (Fig. 10.5). The BDC takes place through a meridional thermal wind balance established by the LTG, and is powered by planetary and synoptic waves that release energy and momentum when they dissipate. Atmos-

pheric waves are periodic changes in atmospheric properties (pressure, temperature, geopotential height or wind speed) that originate from a variety of dynamic disturbances to those properties (convection, temperature contrasts, flows over mountain ranges) in the troposphere and propagate horizontally, vertically or remain stationary. They receive different names depending on their cause, direction of movement, temporal, and spatial scale. At the time they form, atmospheric waves acquire momentum from the troposphere or the ocean-solid Earth, and when they dissipate (break) they deposit their momentum and energy to the medium.

During summer, the permanent insolation over the North Pole creates an anticyclone (high pressure center with clockwise rotating winds). The summer easterly winds do not allow the vertical propagation of planetary waves. In autumn, the faster cooling of the Arctic atmosphere replaces the anticyclone for a pole-centered cyclone (low pressure center) surrounded by westerly winds known as the polar vortex (PV). The Charney–Drazin criterion recognized that wave propagation is strongly controlled by zonal winds, and upward wave propagation is only possible when they are westerly and weaker than a critical wind speed, inversely proportional to zonal wave number (amplitude classification). During winter, westerly winds are so strong that only the longest waves (numbers 1 and 2) can propagate upwards to the stratosphere. Planetary waves carry momentum and energy that can be released in the stratosphere when they dissipate (or break); alternatively, the wave can be reflected back to the troposphere. The area of the stratosphere where the waves release their momentum and energy is known as the "surf-zone" (McIntyre & Palmer 1984). The effect on the zonal mean circulation is a deceleration of westerly winds disrupting the thermal structure. As the LTG cannot be maintained under weaker westerly winds, air is forced down inside the PV, warming adiabatically, and up outside the PV, cooling. The Arctic atmosphere can warm by 30 °C in the lower stratosphere and up to 100 °C in the upper stratosphere. Afterwards, as the Arctic atmosphere is under strong radiative cooling during the winter, the stratosphere cools and the westerlies regain speed.

As the wave-transmitting properties of the atmosphere change, there is strong variability in the wintertime stratosphere (Fig. 10.6). In extreme cases, when wave propagation is particularly strong, the LTG in the stratosphere inverts between the pole and mid–latitudes due to very strong warming of the polar atmosphere, to the point of reversing the winds direction to easterly. These events are known as sudden stratospheric warming (SSW). During them the PV relocates outside of the pole and can even break into two smaller vortices. SSWs not only impact tropospheric weather for up to three months but also influence oceanic variability through wind stress and heat flux anomalies and display a robust link to negative North Atlantic Oscillation (NAO) state.

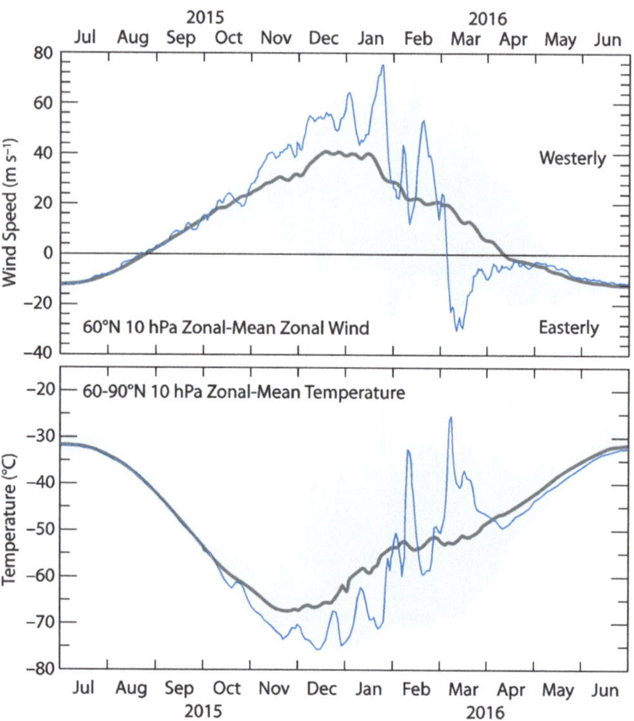

Fig. 10.6 Changes in the polar vortex due to wave propagation during the 2015 El Niño

Top, zonal-mean zonal wind speed at 60°N and 10 hPa (blue fine line). The polar vortex is formed by westerly winds. After strong wave pounding, a final SSW caused the earliest vortex termination on record. **Bottom**, corresponding 60–90°N 10 hPa temperature (blue fine line), showing the sudden warming of 30 °C. Thick line, 1979–2014 mean; grey area, 1979–2014 maximum-minimum. Data from NASA Goddard Space Flight Center. https://ozonewatch.gsfc.nasa.gov/meteorology/figures/merra/

When wave propagation weakens, the opposite happens and temperature at 30 km above the Arctic can become as low as –80 °C. At that temperature, frozen water vapor and nitrous acid can form polar stratospheric clouds.

In the troposphere the Hadley circulation is very inefficient in transporting heat poleward, as its upper branch poleward transport of sensible heat is almost matched by its lower branch equatorward transport of latent heat. That's why most of the heat transported poleward between 30°S–30°N is transported by the ocean. At the equatorial sea surface, heat MT is even more inefficient. Trade winds from the lower branch of the Hadley cell are equatorward and then turn West (easterlies) due to Earth's rotation, so warm surface water is moved westward instead of poleward by the wind-driven circulation. When it reaches the Pacific basin western margin it turns poleward in the Kuroshio and East Australia currents. In the Atlantic Ocean western margin, it turns poleward in the Gulf Stream and Brazil current, and in the Indian Ocean in the Agulhas and Mozambique currents. As the Pacific Ocean is very big, a large amount of warm water accumulates in the Indo–Pacific Warm Pool (IPWP) producing a very slanted thermocline. El Niño/Southern Oscillation (ENSO) acts as a heat pump linked to MT. The accumulated heat prior to the onset of El Niño comes from surface heating that exceeds poleward MT. El Niño increases poleward MT by the ocean and atmosphere while decreasing surface equatorward heat flux (Sun 2000). During La Niña the poleward transport is reduced and heat starts to accumulate again in the equatorial upper ocean.

Northward of c. 30°N, the atmosphere takes over as the main MT agent. Unlike the Hadley and Polar cells that are thermally driven, the Ferrer cell is mechanically driven, connecting the Hadley and Polar cells, with the ascending branch in a colder region than the descending branch, which results in a weak highly variable zone of mixing where westerlies are stronger when the pressure at the pole is lower. During the NH winter, heat is transported to the Arctic mainly by stationary eddies (planetary waves) and transient eddies (cyclones). Cyclones preferentially generate, propagate and dissipate in storm tracks and tend to form where surface temperature gradients are large (Shaw et al. 2016). The jet stream influences their speed and direction of travel. The winter eddy heat flux reveals the preferred storm track areas (Fig. 10.7; Hartmann 2016).

A large part of the heat and moisture transported into the Arctic winter is attributable to a few extreme events per season associated with individual weather systems. Papritz and Dunn–Sigouin (2020) showed that the planetary-scale moisture transport is closely linked to large-scale atmospheric blocking situations that deflect cyclone tracks poleward. Figure 10.8 shows one of these extreme events that took place in the last days of 1999 and first days of 2000, a case studied by Woods and Caballero (2016).

Three main pathways have been identified for the moist static energy transport into the Arctic, the North Atlantic (300–60°E), North Pacific (150–230°E), and Siberian (60–130°E) pathways (Mewes & Jacobi 2019; Woods et al. 2013). For the overall winter MT, the pathways over both ocean basins are more important, and the North Atlantic pathway is the main one (Woods et al. 2013; Alekseev et al. 2019). The entry gateways for MT into the Arc-

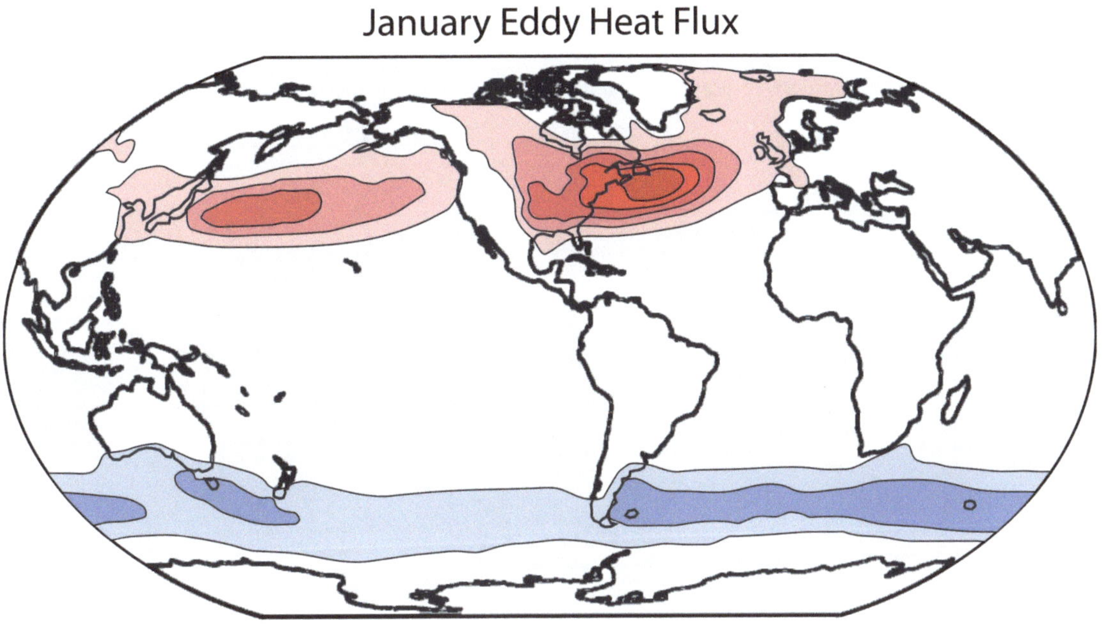

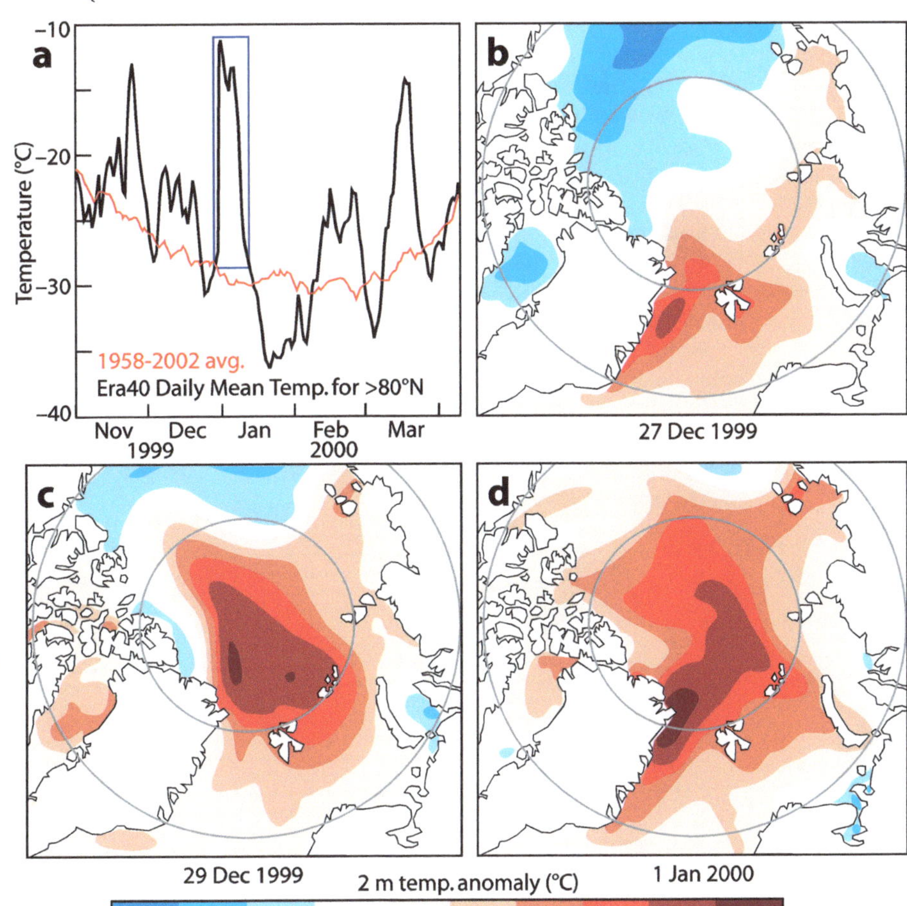

Fig. 10.7 January northward heat flux by eddies During boreal winter the NH subtropical jet has two maxima downstream of the Himalaya and Rocky Mountains over the Pacific and Atlantic oceans, respectively. These wind speed maxima result in vigorous mid–latitude cyclones following storm tracks that define the main gateways into the Arctic. Contour is 5 K m/s. Blue shading in the SH indicates southward flux. After Hartmann 2016.

tic arise because large-scale blocking conditions develop to the east of each basin, deflecting midlatitude cyclones poleward (Woods et al. 2013). According to Nakamura and Huang (2018) blocking develops like a traffic jam when the jet stream capacity for the flux of wave activity (a measure of meandering) is exceeded. The atmospheric modes of variability that favor blocking over the Pacific are different from the Atlantic (Gollan & Greatbatch 2017). Over the Atlantic, winter blocking strongly anti-correlates with the North Atlantic Oscillation (Wazneh et al. 2021).

On annual average, the Arctic region North of 70°N loses radiative energy to space, acting as the main heat sink for the planet (i.e. most negative energy gain/loss ratio at ToA; see Fig. 10.4), but most of the energy loss is concentrated in the cold season (Nov–Apr; Fig. 10.9), when the Arctic energy budget is more negative. In summer the heat transported to the Arctic is mostly stored, by warming the ocean and melting ice and snow, until the arrival of the cold season when it is returned to the atmos-

Fig. 10.8 Intense intrusion event of moist warm air into the Arctic in winter **a)** Daily mean temperature North of 80°N for Nov 1999–Mar 2000 (black line) from ERA40 reanalysis, and the 1958–2002 average (red line). A rectangle marks the event. Data from the Danish Meteorological Institute (2021). **b–d)** Surface air temperature anomaly in the Arctic at different times during the intrusion event. After Woods & Caballero 2016.

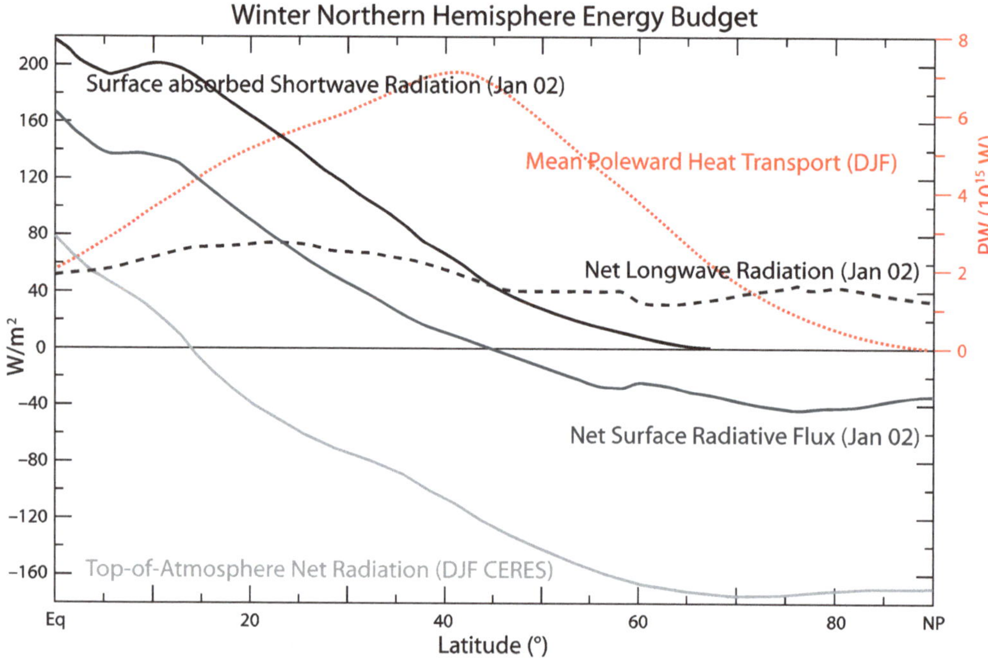

Fig. 10.9 The Arctic in winter is the biggest heat sink of the planet
In January there is no surface shortwave radiation (black line) North of 70°N, but the surface still emits longwave radiation (dashed line). As a result, the net surface radiative flux (dark grey line) is negative. Although the poleward heat transported in winter (dotted line, right scale) warms the surface and atmosphere, it also results in a much bigger loss of energy at the top of the atmosphere (light grey line). Radiation curves after Randall 2015, winter seasonal transport after Liang et al. 2018.

phere, by the reverse processes. The stored summer heat is lost together with the heat transported by the atmosphere from lower latitudes by radiative cooling during the cold season. Arctic precipitable water is c. 1.5 cm in summer, but in winter it drops to c. 0.2 cm (Wang & Key 2005), the lowest value outside Antarctica. As a result, cloud cover becomes lower in winter (Fig. 10.3a; Wang & Key 2005), increasing the energy loss. With a reduced cloud cover, almost no water vapor, and no albedo effect, the Arctic in winter has essentially no positive feedback to the greenhouse effect from CO_2.

Arctic amplification, the enhanced warming of the Arctic compared to lower latitudes, is the consequence of an increase in MT, as the Arctic has a negative annual energy budget. Looking at the seasonal surface temperature change North of 80°N it is clear that Arctic amplification is a seasonal phenomenon, since summer temperatures are not increasing (Fig. 10.10). Any increase in summer net heat MT to the Arctic goes into melting ice and snow, a form of energy storage until the cold season, and increased OLR at the ToA. Winter surface temperature shows a very pronounced increase since c. 2000 (Fig. 10.10). It is clear that MT to the Arctic has increased since then. Winter heat and moisture transport into the Arctic alters temperature, humidity, and cloud cover, as well as the radiative proper-

ties. Arctic import of heat and moisture during winter leads to cloud formation, which shifts the strongest radiative cooling from the surface to cloud tops, which are frequently warmer in winter due to temperature inversions. In the marginal sea ice zone, winter heat intrusions cause a temporary retreat of the ice margin, leading to enhanced heat loss by the ocean until the ice forms again (Woods & Caballero 2016). The advected heat that is not exported back to lower latitudes is distributed between increased OLR and increased downward longwave radiation. The enhanced downward radiation increases surface temperature, but due to the low thermal conductivity of ice, and since the ocean is always warmer than the atmosphere during winter, temperature inversions commonly result, often accompanied by humidity inversions, and the radiative cooling continues from the top of the inversion or the top of the clouds until the water vapor freezes and precipitates, restoring the original very cold condition (Fig. 10.8a).

Heat is more efficiently transported in winter to the Arctic at times when high pressure conditions prevail over the pole leading to a weak or split vortex. The warm air enters the central Arctic ascending over the cold air by isentropic lifting, pushing the cold air below outwards. Cold Arctic air masses then move over the mid–latitude

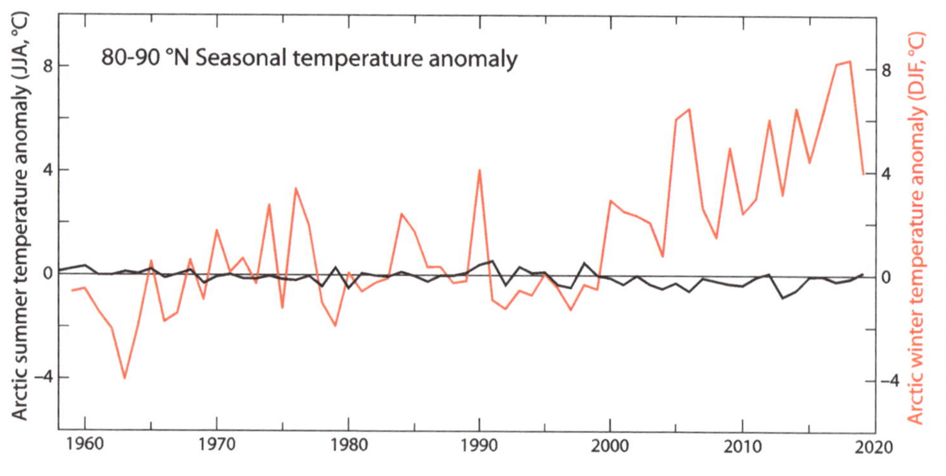

Fig. 10.10 Arctic seasonal temperature anomaly
Black curve, summer (June–August) mean temperature anomaly calculated from the operational atmosphere model at the European Center for Medium-range Weather Forecast (ECMWF) for the +80°N region. Red curve, the corresponding winter (December–February) mean temperature anomaly for the same region. Reference climate is ECMWF–ERA40 reanalysis model for 1958–2002. Data from the Danish Meteorological Institute.

continents producing anomalously cold temperatures and snow in a pattern known as warm Arctic/cold continents (WACC, Overland et al. 2011). Since Arctic amplification started, the frequency of mid-latitude cold winters has increased, something that models cannot explain (Cohen et al. 2020), but something similar took place between 1920–40 (Chen et al. 2018). Arctic amplification has turned out to be mainly a cold season phenomenon that started for reasons unknown to most climate scientists and models c. 2000. Arctic amplification is dependent on changes in MT.

10.4 El Niño/Southern Oscillation as part of the meridional transport system, modulated by the sun

ENSO has been explained either as a self-sustained and naturally oscillatory mode of the coupled ocean-atmosphere system or a stable mode where events are triggered by stochastic forcing (Wang et al. 2016). This indicates that ENSO is very misunderstood, as it constitutes a natural heat pump (Sun 2000), part of the global MT sys-

tem, modulated by external forcing. It is the coupled ocean-atmosphere system answer to a MT problem. Most solar energy enters the climate system in the tropical band, yet MT is very inefficient near the equator. The dry static heat transported poleward by the upper branch of the Hadley circulation is almost matched by the latent heat transported equatorward by its lower branch. On top of that, predominant surface winds at the Hadley cells are equatorward and then turn West at the equator. So, the wind-driven circulation does not direct heat poleward from the equator, instead it sloshes heat against the western margin of ocean basins, where western boundary currents direct the heat poleward and transfer large amounts of heat to the atmosphere (Yu & Weller 2007). ENSO is Earth's response to the reduced atmospheric poleward transport in the tropics. El Niño takes care of transporting the residual heat accumulated at the ocean surface and subsurface in the IPWP out of the tropics.

ENSO has three main modes, El Niño, La Niña, and Neutral, but, erroneously, only two of them are generally considered significant. However, a simple frequency analysis of the three of them separately reveals that the

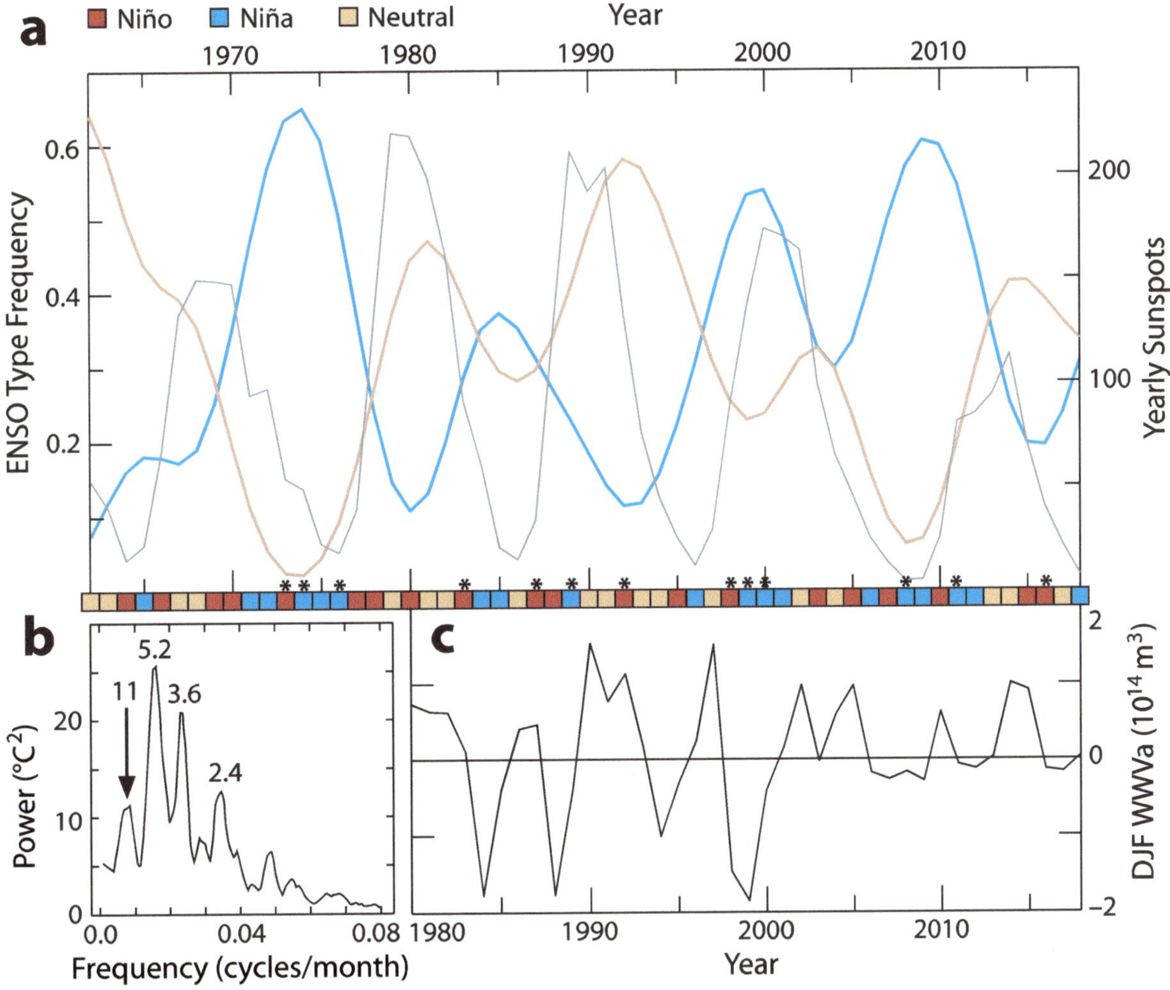

Fig. 10.11 ENSO modes and solar activity
a) Frequency of Niña years (blue thick line) and Neutral years (light brown thick line) in a 5-year centered moving average (gaussian filtered) between 1962–2018 showing almost perfect anti–correlation for the entire period. Small boxes are the ENSO mode classification after Domeisen et al. 2019, with dark red boxes for Niño years, and same color as curves for Niña and Neutral years. Asterisks mark strong Niño and Niña events with ≥1 °C anomaly in Oceanic Niño Index. Fine grey line is the number of yearly sunspots. **b)** Power spectrum of the 1900–2008 Niño-3.4 SST anomaly time series after Deser et al. 2010. An arrow marks the 11-year frequency peak that might correspond to the solar cycle effect. **c)** Dec–Feb average warm water volume anomaly above the 20 °C isotherm between 5°N–5°S, 120°E–80°W. Data from TAO Project Office of NOAA/PMEL.

opposite of La Niña is not El Niño, as is universally assumed, but Neutral. Los Niños typically take place every 2–3 years (range 1–4 years), so there are always 1–3 Niños in a 5-year period. Las Niñas and Neutral years are more variable, as there can be 0–4 of each in a 5-year period. Importantly, Los Niños (not shown in Fig. 10.11a) do not correlate significantly with Las Niñas or Neutrals, while Las Niñas and Neutrals display a very strong anti-correlation (Fig. 10.11a). This anti-correlation is a clear indication that the two basic states of ENSO are La Niña and Neutral. Tropical MT is weaker (less efficient) during Niña years (Sun 2000) when the warm water volume (WWV; Meinen & McPhaden 2000) accumulates. MT is stronger during Niño years when WWV decreases as shown in Fig.10.11c. Niños occur periodically (once every 3 years on average) and move heat out of the equatorial region and IPWP, reducing the WWV. During Neutral years tropical MT is intermediate and WWV changes less than during Los Niños or Las Niñas. The alternation between Niña-frequent and Neutral-frequent periods follows a quasi-decadal variability that is synchronized to the solar cycle (Fig. 10.11a), with Neutral years occurring more often during or after high solar activity years and Niñas predominating during or after low solar activity years. Niños occur with a different rhythm slightly disrupting the Niña-Neutral solar pattern. The synchronization to the solar cycle explains the ENSO frequency peak at 11-years (Fig. 10.11b). The explanation for the ENSO-solar synchronization is that it is part of the global MT system, and solar variability is one of several phenomena modulating MT (see Chap. 11).

Examination of ENSO conditions during the solar cycle further reveals the correlation between solar activity and sea-surface temperature in the equatorial Pacific. Analysis of ENSO sea-surface temperature (SST) data since 1950, when the Oceanic Niño Index (ONI) is available, performing an epoch analysis during the solar cycle (correcting for the variable duration of each solar cycle), allows the identification of five ENSO phases during the c. 11-year solar cycle (Fig. 10.12).

For the purpose of this analysis, the start and end of the solar cycle are defined by the monthly-smoothed international sunspot number (WDC–SILSO) changing from below to above 30 sunspots, a point well defined in every cycle. Temporal data (in months) is normalized as fractions of one solar cycle with variable duration (epoch analysis). Distribution of the ONI data (monthly, 3-month averaged; NOAA) in the same way allows the calculation of the average and standard deviation at each fraction of the solar cycle for the six and a half solar cycles since 1950, when ONI data becomes available.

Phase I starts when the peak in solar activity is on average reached (c. 2.5 years into the cycle as defined) and lasts around two and a half years during which positive ONI conditions are more probable, following the peak in solar activity. Phase II, of another two and a half years length, coincides with declining solar activity and is a highly variable period. Phase III, of around three years, coincides with the final decline in solar activity towards the solar minimum, and usually presents negative ONI conditions. Particularly frequent is a La Niña near the solar minimum. The last two phases are the ones with a

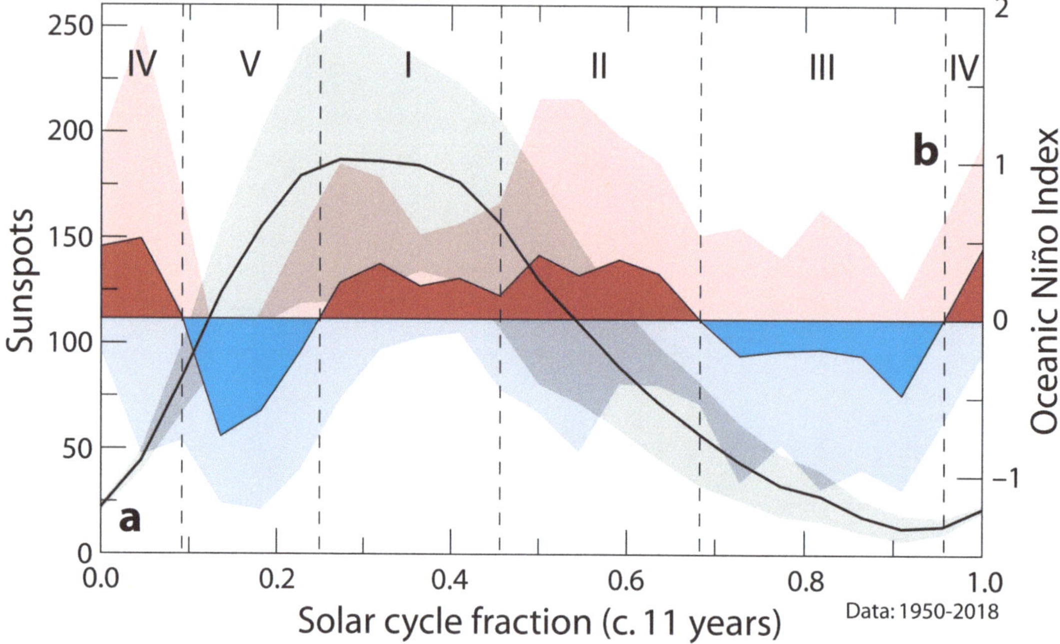

Fig. 10.12 Solar cycle–ENSO relationship
a) Black curve, left scale, epoch averaged monthly smoothed sunspot number, for six and a half solar cycles between 1950–2018, starting and ending the cycle epoch when the number of monthly smoothed sunspots changes from <30 to >30. Grey area corresponds to the standard deviation of the averages at each position. Data from WDC–SILSO, Royal Observatory of Belgium. **b)** Colored areas, right scale, Oceanic Niño Index data from NOAA between 1950–2018 distributed along the solar cycle as calculated for (a). Areas above zero represent average positive ONI conditions and areas below zero average negative ONI conditions. Light colored areas represent the standard deviation of the averages at each position. Five different phases can be identified from ENSO response to solar activity. Phase I, high solar activity and predominant positive ONI conditions. Phase II, declining solar activity and highly variable ENSO condition. Phase III, low solar activity and predominant negative ONI conditions. Phase IV, after the solar minimum, predominant El Niño conditions. Phase V, rapidly increasing solar activity and predominant La Niña conditions.

higher deviation from average. Phase IV, a short period of about 1.5 years, starts at the time of minimal solar activity, and generally displays positive ONI conditions that at times result in considerably strong Niños, like in 1998. Afterwards, phase V coincides with the period of rapidly rising solar activity, that reliably produces negative ONI conditions and frequent Niñas.

As we have seen (Fig. 10.11) solar modulation of ENSO is mainly due to a La Niña tendency to occur near the solar minimum (phases III and V), and it is statistically significant. We can examine Phase V that shows the larger negative ONI departure from average. To that end we average the ONI values for the months when the 13-month smoothed monthly total sunspot number (WDC–SILSO) is increasing between 35–80% of the total for solar cycles 19–24 (11/1955–11/1956, 7/1966–5/1967, 3/1978–2/1979, 3/1988–11/1988, 3/1998–8/1999, 12/2010–12/2011). The average ONI value for those 76 months is –0.649 (i.e. La Niña condition). A Monte Carlo experiment using the 816 12-contiguous monthly periods in the ONI dataset (1/1950–11/2018) and randomly drawing 100,000 groups of six and averaging them, only produces an equal or lower ONI value 0.7% of the time. The La Niña instance at 35–80% rising solar activity has a 99.3% probability of not being due to chance. ENSO is modulated by solar activity.

The evidence for a solar modulation of ENSO has been obscured because unlike La Niña or Neutral, El Niño frequency has a lower correlation to solar activity. Despite this difficulty, several authors have noticed and reported the association of La Niña, or more generally ENSO with the solar cycle (Anderson 1990; Landscheidt 2000; White & Liu 2008; Leamon et al. 2021; Lin et al. 2021). Haam and Tung (2012) tried to explain this as a spurious correlation between two autocorrelated series, but Wang et al. (2020) using information flow analysis found a causal link between solar activity and ENSO. Additionally, it is known that the quasi-biennial oscillation (QBO), ENSO and solar activity affect stratospheric transport (Rao et al. 2019) and the PV. The QBO modulates the effect of solar activity (Labitzke & van Loon 1988), solar activity modulates the QBO (Salby & Callaghan 2000), and ENSO modulates the QBO (Taguchi 2010). Solar activity, the QBO and ENSO are three inter–related factors affecting stratospheric transport (Rao et al. 2019), and a solar modulation of ENSO does not appear far-fetched.

The solar effect on ENSO is generally unrecognized, despite strong evidence and a significant bibliography, mainly because it is unexplained and does not emerge from models. A recent review on ENSO complexity by 45 prominent ENSO experts (Timmermann et al. 2018) fails to mention any solar effect and does not recognize that ENSO is the consequence of the tremendous need to transport heat outside the equatorial region in the absence of sufficient atmospheric transport.

If ENSO is an equatorial heat pump (Sun 2000) that oscillates between La Niña and Neutral conditions, with El Niño periodically pumping the accumulated residual heat out of the region, we would find an explanation for some of the most puzzling aspects of El Niño. The long-term frequency and magnitude of Niños would depend on the rate of accumulation of residual heat, i.e. on the long-term changes in MT. It is known from proxies that Los Niños and heat export from the IPWP were enhanced during the Little Ice Age and reduced during the Medieval Warm Period and the Holocene Climatic Optimum (Moy et al. 2002; Perner et al. 2018). Los Niños would become more abundant when the planet cools—because an enhanced MT is how it cools.

Another puzzling aspect of El Niño is that it comes in flavors that denote geographical variability (Central versus East Pacific Niños). There is evidence that there are long-term changes in their predominance during the Holocene (Karamperidou et al. 2015) and during the past decades (Fedorov et al. 2020). These El Niño changes probably reflect changes in MT that also have a strong geographical variability between the Pacific and Atlantic MT pathways, since ENSO and MT variability modes changed simultaneously at the 1997–98 Climatic Shift (see Chap. 11). ENSO events are locked to the annual cycle because ENSO belongs to the MT system, and MT is strongest towards the winter North Pole. The relationship between ENSO and the IPWP annual cycles has been termed the C–mode and has been shown to be modulated by the Atlantic Multidecadal Oscillation (AMO; Geng et al. 2020). During a positive AMO phase, El Niño events are distinctly weaker than those in an AMO negative phase. It is not that AMO modulation determines decadal shifts in ENSO properties, as the authors suggest. It is that both AMO and ENSO depend on MT changes in absolute strength and relative strength between the Pacific and Atlantic pathways.

10.5 The Sun, QBO and ENSO modulation of stratosphere–troposphere coupling

The QBO is a most remarkable atmospheric phenomenon and determines, along with ENSO, seasonal and inter-annual weather variability. In the equatorial stratosphere, strong zonal winds circle the Earth. They originate at an altitude of 10 hPa (c. 35 km) and migrate downward at c. 1 km/month until they dissipate at the base of the stratosphere at 80 hPa (c. 20 km). As the new zonal wind belt originates to replace the downward migrating previous one, it moves in an opposite direction, alternating easterly and westerly winds (Baldwin et al. 2001). The QBO is usually defined by wind speed at 30 hPa, where winds in one direction will start and increase in strength, and then decline and be replaced by winds moving in the opposite direction. The easterly and westerly phases of the QBO alternate every 22–34 months with an average of 28 months, but the periodicity is tuned to the yearly cycle, so the phase reversal occurs preferentially during boreal late spring. The angular momentum signature of the QBO, rather than having only a single spectral frequency peak at c. 28 months, includes two smaller spectral peaks at the annual and biannual frequencies.

In a breakthrough at the time, Lindzen and Holton (1968) proposed, and it was later demonstrated, that convection-originated, vertically-propagated gravity waves provided the necessary wave forcing (momentum) for QBO generation and maintenance. Current understanding is that equatorially trapped Kelvin waves provide the westerly momentum and Rossby-gravity waves provide easterly momentum to produce the QBO oscillation. The QBO is a tropical phenomenon that affects the global stratosphere through the modulation of winds, temperature, extra–tropical waves, meridional wind circulation, the transport of chemical constituents, and the distribution of ozone.

One of the most puzzling aspects of the QBO is that it also modulates the NH PV, a persistent, large-scale, mid–troposphere to stratosphere, low pressure winter zone that when strong contains a large mass of very cold, dense Arctic air, and when weak and disorganized allows masses of cold Arctic air to push equatorward, causing sudden temperature drops in ample regions of the NH. In 1980, Holton and Tan published a seminal work linking the QBO to the global atmospheric circulation. One of their findings, known since as the Holton–Tan effect, was a significant correlation between the westerly phase of the QBO and low zonal mean geopotential at the North Pole, when planetary waves were present. Lower geopotential means a more organized PV and lower temperatures. Holton and Tan had to introduce the planetary waves condition because at certain times the correlation broke down. Karin Labitzke (1987) noticed that the vortex–QBO correlation broke down at times during west winds, but only when the number of sunspots was near its maximum. She decided to segregate the data on stratospheric polar temperatures according to QBO phase. The very low correlation between solar activity and polar temperatures, when all the data is considered, becomes very high using the segregated data (Fig. 10.13). When she published her finding James Hol-

ton said: *"Superficially, I can't find anything wrong with it, but there is absolutely no physical basis, and that bothers me. These people have the highest correlation I've seen, but if I were a betting man, I would bet against it."* (Kerr 1987). Holton would have lost his bet, but it shows how observations are accepted or questioned as a function of how well they fit the dominant hypothesis.

The data indicates that stratospheric North Pole temperature correlation to solar forcing depends on QBO state. During QBOe (easterly) years stratospheric polar temperature is lowest when solar activity is highest, and highest when solar activity is lowest. The exact opposite occurs during QBOw (westerly) years (Fig. 10.13a). The lowest QBOe and highest QBOw temperatures are similar, the largest difference occurs during low solar activity years. The average North Pole 30 hPa winter temperature difference between QBOe and QBOw years under low solar activity conditions is an astounding 20 °C (Fig. 10.13). Winter North Pole stratospheric temperature reflects PV conditions that strongly influence the NH troposphere. The temperature is also affected by strong El Niño conditions (Fig. 10.13b).

The effect of solar activity on NH winter sea-level pressure, surface air temperature, 700-mb temperature and geopotential height showed surprisingly high correlations over extensive areas of the NH with a marked tendency for showing opposite patterns for each phase of the QBO (van

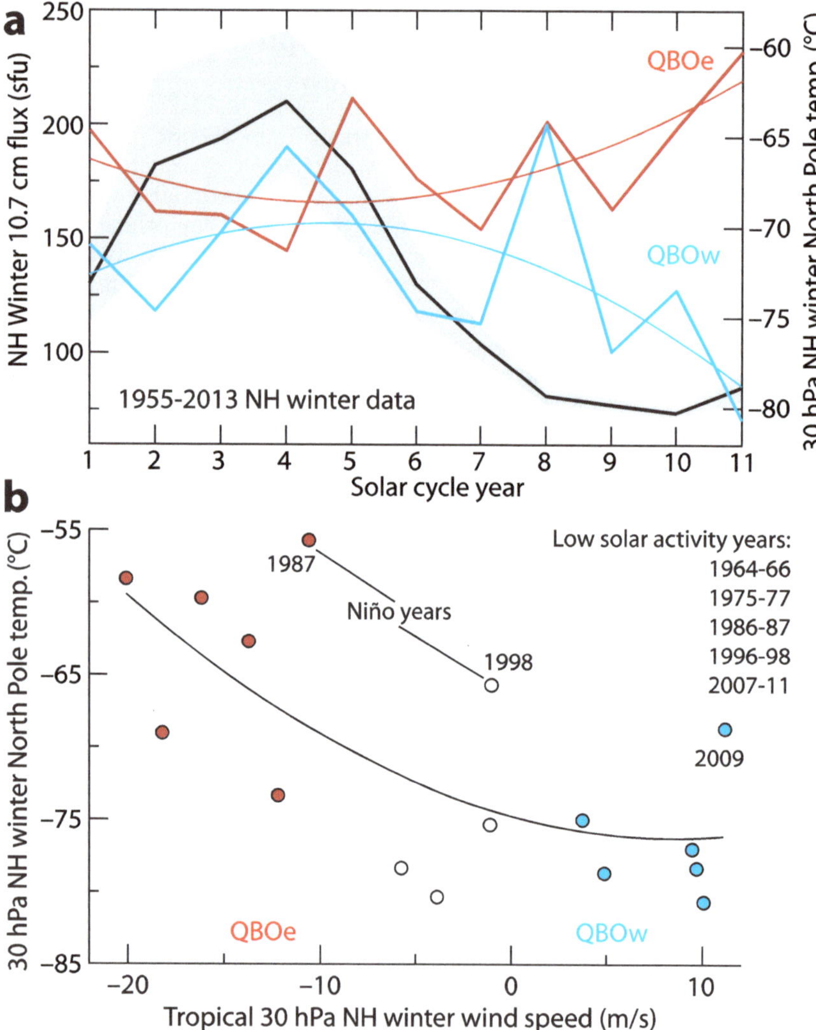

Fig. 10.13 Effect of solar activity on winter North Pole stratospheric temperature a) Black curve and light grey area, winter (DJF) 10.7 cm flux average and standard deviation between Dec. 1955 and Feb. 2013, a proxy for solar activity, adjusted to an 11-year solar cycle. Solar cycle 21 has no data for year 11, and solar cycle 23 has two points for year 10 and two for year 11. Colored curves correspond to winter temperature at 30 hPa (stratosphere) over the North Pole calculated as the average of the three more centered values among DJFM monthly average temperatures (outlier discarded) and plotted according to the position in the 11-year solar cycle. Dark-red thick curve is the temperature for winters when the QBO presented average DJF values lower than −5.8 ms^{-1} (negative values denote easterly wind) corresponding to QBOe (easterly). Dark-red thin curve is the quadratic regression. Light-blue thick curve is the temperature for winters when the QBO presented average DJF values higher than 1.1 ms^{-1} (positive values denote westerly wind) corresponding to QBOw (westerly). Light-blue thin curve is the quadratic regression. **b)** Scatter plot of 30 hPa winter North Pole temperature, determined as in (a) versus tropical 30 hPa winter wind speed for years with very low solar activity, corresponding to years 9 to 11 in the solar cycle as defined in (a), and indicated in the graph. Dark-red-filled dots are QBOe values used for the same color curve in (a). Light-blue-filled dots are QBOw values used for the same color curve in (a). Black thin curve is the quadratic regression. Strong El Niño years are indicated. Data on North Pole stratospheric temperature from the Institute of Meteorology at the Freie Universität Berlin. Data on 10.7 cm solar flux from the Royal Observatory of Belgium STAFF viewer.

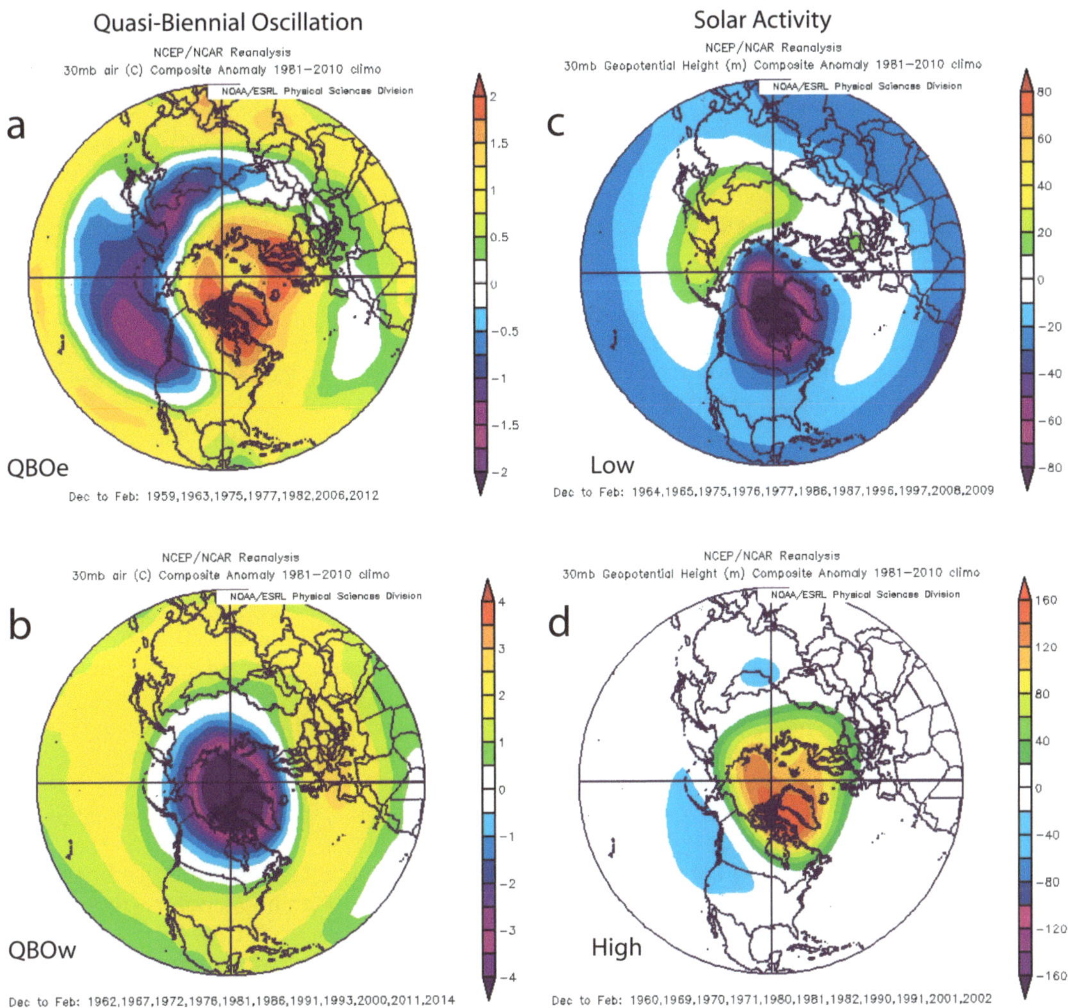

Fig. 10.14 The effect of QBO phase and solar activity on Northern Hemisphere winter stratospheric temperature and geopotential height

a) NCEP/NCAR composite of December–February 30 hPa temperature anomaly (°C, 1981–2010 baseline) for seven QBO easterly years. The situation corresponds to a disorganized polar vortex with more frequent cold Arctic surface air incursions at lower latitudes. **b)** Same as in (a) for eleven QBO westerly years. A well-organized polar vortex keeps Arctic air trapped underneath. **c)** Composite of December–February 30 hPa geopotential height anomaly (m, 1981–2010 baseline) for eleven solar minimum years. **d)** Composite of December–February 30 hPa geopotential height anomaly (m, 1981–2010 baseline) for eleven solar maximum years. Figure obtained from NOAA/ESRL reanalysis tool, Boulder, CO.

Loon & Labitzke 1988). The unusually high correlation between solar activity and sea-level pressure or surface temperature appear to explain an important fraction of the total interannual variability in the winter circulation (Peixoto & Oort 1992). The data segregation, pioneered by Labitzke, has been used very successfully to establish the relationship between phenomena that present phases, like the QBO, solar variability, and ENSO, and their effects on the PV, SSWs, the Pacific Decadal Oscillation and the NAO. Weather and climate models have problems reproducing a realistic QBO. For example, only 4 of more than 30 models used for the AR5 IPCC report had any sort of QBO. However, reanalysis readily displays the statistical association between the QBO phase and solar activity with stratospheric temperature and geopotential height (Fig. 10.14). Geopotential height is the actual height of a pressure surface above mean sea-level, and it is related to the density of the air below. A low geopotential height indicates the presence of cold dense air masses below, while a high geopotential height indicates the opposite.

ENSO has been suspected to affect the winter northern stratospheric PV since the 1980s, with El Niño years resulting in a more perturbed and warmer PV than La Niña years (see Fig. 10.13b), but the statistics were difficult to disentangle, since El Niño winters tend to coincide with QBOe, and not enough years of data were available to separate the three factors affecting the winter PV: QBO, solar cycle, and ENSO phases. However, the effect was seen in models, with El Niño years having double the probability of producing SSW than La Niña years. After confirming the statistical significance of the QBO and solar cycle effects on the PV, Camp & Tung (2007) were intrigued that it wasn't necessary to stratify the data by ENSO phase to see their effect. This was possible only if ENSO effect was small or if the perturbation from ENSO takes a spatial form that is almost orthogonal to the spatial form of both QBO and solar cycle perturbations. The latter turned out to be correct with El Niño producing a wide latitude of warming from mid-latitudes to the pole, while the QBO and solar cycle warm the polar stratosphere by approximately the same magnitude, but their spatial pattern is more confined to the polar region (Camp & Tung 2007). This difference suggests the ENSO effect is also strong on tropospheric/ocean MT, while the QBO and solar cycle effect is more directly on stratospheric MT.

The current understanding, supported by observations, reanalysis, and modeling, is that the winter polar stratosphere is warmed by the momentum transfer and pressure changes caused by different kinds of gravity waves that originate from convection and weather phenomena in the troposphere and propagate vertically and can affect the polar annular modes weakening them. SSWs are the extreme result of planetary waves breaking, and they can be triggered by unforced variability, QBOe, the solar cycle, or El Niño. QBO, solar activity, and ENSO act as gatekeepers by determining the propagation properties of the stratosphere to planetary waves. Combinations of these three factors during winter cause a constructive or destructive interference with the vertical planetary waves, and in the first case the waves are deflected poleward transmitting momentum and energy to the stratospheric PV where they break, causing heating that might result in extreme cases in SSW. Changes in solar UV activity determine stratospheric tropical ozone levels and heating creating temperature and wind speed anomalies. These anomalies are sometimes maintained and enhanced by their interaction with planetary waves when the propagation properties of the stratosphere allow it during certain winters, and they migrate poleward in the NH and downward to the polar troposphere. At the surface they determine the state of the dominant mode of variability, the Arctic Oscillation, then extend their influence to the North Atlantic Oscillation.

The stratosphere–troposphere coupling has been one of the main surprises about the atmosphere of the past decades (Baldwin et al. 2019). The Arctic Oscillation (AO, or Northern Annular Mode NAM) is strongly connected to the stratosphere. Sea level pressure, surface temperature and wind patterns respond to changes in the stratosphere with a similar pattern to the AO, with a slight delay of about two months. Also, Atlantic and Pacific jet streams and storm tracks shift systematically in response to stratospheric variability. Extreme winter and spring weather events in Europe and North America are frequently associated with changes in the stratosphere. Weather forecasts for these regions on monthly or seasonal timeframes improve greatly when stratospheric conditions are taken into account (Baldwin et al. 2019; Domeisen et al. 2020). Solar activity induced changes in stratospheric ozone, temperature, and propagation properties to planetary waves, have a way of affecting surface weather and climate through this stratosphere–troposphere coupling, and there is ample evidence of this top–down indirect mechanism from both observations and models (Baldwin et al. 2019).

While the easterly and westerly phases of the QBO have an opposite influence in modulating the effect of solar activity, it is the QBOe in combination with low solar activity that shows the largest departure from average conditions (Fig. 10.13). In QBOe winters, during low solar activity, the polar geopotential height is higher, polar stratospheric temperatures are higher, SSW occurs earlier, the PV is more frequently weaker and disorganized, the polar jet stream forms meanders that extend into lower latitudes, there is a higher frequency of blocking days, and the AO and NAO tend to be in their negative phase (Roy 2014; Hall et al. 2015). This results in winters that are colder in Northern Hemisphere mid–high latitudes. Meteorologists have learned that NH winters with QBOe and low solar activity (like 2017–18 winter) tend to be cold and with more snow, especially in non-La Niña years, unless a recent stratospheric-reaching volcanic eruption interferes and produces a warmer winter.

Given the complexity of the solar signal transmission through this indirect top–down pathway, that is both conditional and seasonal, we do not yet have a good quantitative understanding of the mechanism. A further complication comes from the inability of most models to include or realistically reproduce stratospheric phenomena. However, the qualitative knowledge is solidly grounded in observation, reanalysis, and modeling (Baldwind & Dunkerton 2005; Gray et al. 2010; Baldwin et al. 2019). During solar grand minima, solar activity gets stuck in low mode, and the frequency of cold winters in the NH multiplies, although warm winters can still occur, especially during QBOw phases. A significant weakening of the PV and prevailing winter AO negative conditions is accompanied by a general decrease in mid-latitude winter temperatures and an increase in snow precipitation, that result in glacier advances. These were the conditions observed during the Maunder period of the Little Ice Age that have been reproduced in models with clear solar attribution (Shindell et al. 2001).

As we have seen, there are several pathways by which the solar signal is transmitted to the atmosphere–ocean coupled system and amplified through several feedback mechanisms, affecting climate (Roy 2014; Fig. 10.15). Even though solar changes are small, the amplification energy is provided by the climate system. The pathways are complex, involving some of the lesser known climate phenomena, and act in a phase-, seasonal-, and latitudinal-dependent way, often with opposite results. To the combined effect of ENSO, the QBO and solar activity described in Fig. 10.15 we must add the modulation of the solar effect by the QBO and ENSO (Fig. 10.13), the modulation of ENSO by solar activity (Figs. 10.11 & 10.12), and the modulation of the QBO by ENSO (Taguchi 2010), and by solar activity (Salby & Callaghan 2000). This is why the solar variability signal, whose effect can be seen so clearly in paleoclimatic records, is so hard to see in real time. The existence of more than one pathway also multiplies the signal, and further enhancement is attained through the different lags that allow an accumulation of the effect.

10.6 The meridional transport of momentum

The transport of heat and moisture from the equator and tropics toward the middle and high latitudes is linked to the transport of angular momentum, that is exchanged between the solid Earth–ocean and the atmosphere. In low latitudes surface winds are easterly and flow in the opposite direction to the rotation of the Earth, so the atmosphere gains momentum through friction with the solid Earth–ocean that reduces its speed of rotation, while in middle latitudes surface winds are westerly and the atmosphere loses momentum to the solid Earth–ocean that increases its speed of rotation, so a poleward atmospheric

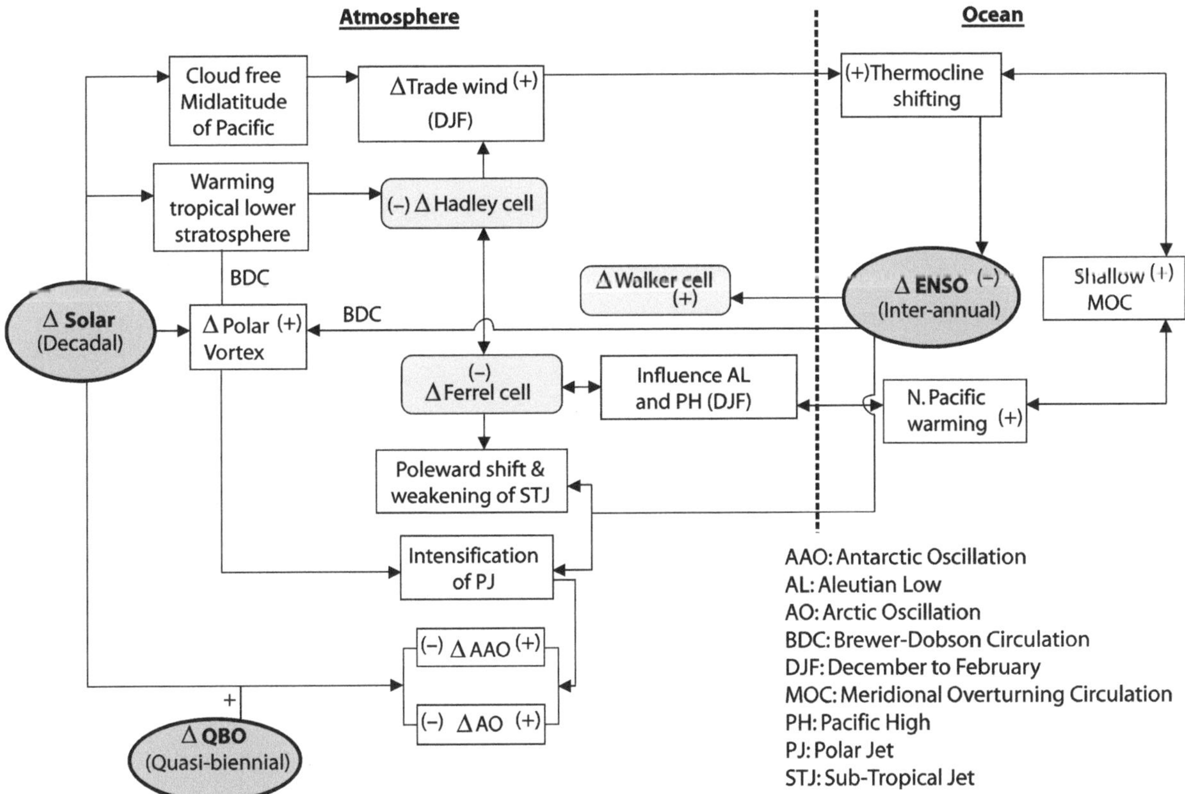

Fig. 10.15 The combined influences of ENSO, QBO and solar activity on the atmosphere–ocean coupled circulation, as a flow chart centered on the solar role
The three major factors are shown with oval outlines, whereas, the major circulations, responsible for modulating their effect are shown by rounded rectangles. The climatic effects are shown in boxes, with the direction of change shown by + (for increase) or − (for decrease). After Roy (2014). See the reference for additional information.

flux of angular momentum is required to conserve momentum and maintain the speed of rotation. MT not only transports heat and moisture poleward, but also momentum (Fig. 10.16).

Changes in the atmospheric angular momentum (AAM) must be balanced by changes in the speed of rotation of the solid Earth–ocean to preserve momentum, and they are mostly due to the seasonal changes in the zonal wind circulation. Zonal wind circulation is stronger in winter when more angular momentum resides in the atmosphere due to a deeper LTG, so the Earth rotates faster in January and July, and slower in April and October, when zonal circulation is weaker. Small changes in the speed of rotation of the Earth result in micro-second changes in the length-of-day (ΔLOD), the difference between the duration of the day and 86,400 Standard International seconds. LOD has been measured daily down to a 20 μs precision by interferometry since 1962. Seasonal variation in ΔLOD has been known for decades to reflect changes in zonal circulation (Lambeck & Cazenave

Fig. 10.16 Meridional transport of energy (left) and angular momentum (right) implied by the observed state of the atmosphere
In the energy budget there is a net radiative gain in the tropics and a net loss at high latitudes; to balance the energy budget at each latitude, a poleward energy flux is implied. In the angular momentum budget the atmosphere gains angular momentum in low latitudes due to easterly surface winds and loses it in middle latitudes due to westerly surface winds. A poleward atmospheric flux of angular momentum is implied. Meridional transport of energy and momentum is known to be modulated by ENSO, the QBO and solar activity. After Marshall & Plumb 2008.

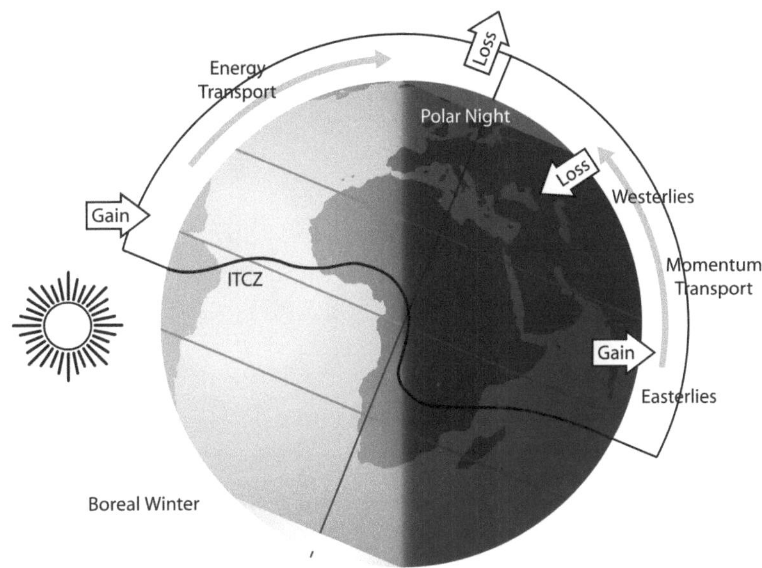

1973) and, therefore, in MT. The biennial component of ΔLOD reflects changes in the QBO (Lambeck & Hopgood 1981), while the 3–4 year component matches the ENSO signal (Haas & Scherneck 2004). All of them reflect changes in the strength of westerly winds associated with changes in the intensity of MT. During El Niño a tropospheric westerly anomaly induces a strong AAM increase driving a slowing of the planet. During the 2015–16 winter season, El Niño produced a ΔLOD excursion reaching 0.81 ms in January 2016.

The Sun, QBO and ENSO constitute three factors coupling the tropical stratosphere to the PV and the polar troposphere, regulating heat and moisture transport to the winter pole. Since they affect the zonal wind circulation it is not surprising to see they also affect the speed of rotation. But while the role of ENSO and the QBO in changing the AAM and ΔLOD is widely known and reported, the role of the Sun remains largely ignored. The existence of an 11-yr solar signal in ΔLOD was reported in 1962 by Danjon, rediscovered in 1969 by Stoyko, and again in 1980 by Currie (Lambeck 1980; Barlyaeva et al. 2014), indicating how little attention it has attracted. From the beginning it was proposed that the effect must be due to solar induced changes in the global atmospheric circulation (Challinor 1971), something that is confirmed by reanalysis (Weng 2012). A viable mechanism was proposed by Hines (1974) based on differential propagation of atmospheric waves. It is surprising that so little effort has been directed into clarify this promising solar–climate connection in a time when so much research has been directed into climate change. The solar activity signal in LOD is most apparent in the semi–annual variation in LOD (Le Mouël et al. 2010; Barlyaeva et al. 2014). This component is mainly due to the 6-month out of phase variation in zonal wind intensity caused by the difference in insolation between hemispheres. It is therefore linked to a fundamental feature of the climate system: the latitudinal distribution and transport of energy and momentum.

The semi–annual oscillation in ΔLOD that is under solar modulation has the following characteristics: From November to January the Earth accelerates to c. 0.2 ms-day (ΔLOD changes by –0.2 ms). Then it decelerates by nearly the same amount by April. Afterwards it accelerates to c. 1 ms-day by July (ΔLOD change of – 1 ms), before decelerating back to the initial value by the next November. The average amplitude is c. 0.35 ms, but

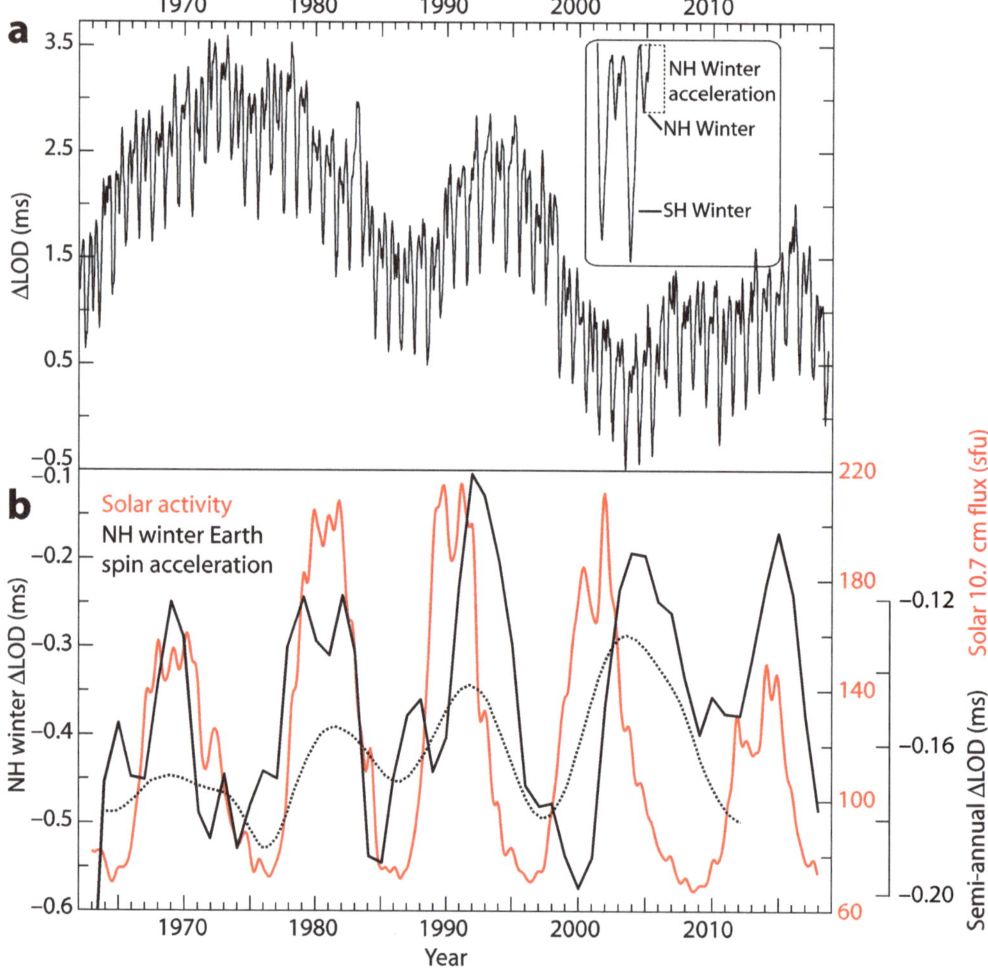

Fig. 10.17 Modulation of the semi-annual LOD variation by the solar 11-year Schwabe cycle
a) Monthly ΔLOD for the 1962–2018 period. The inset shows two years of data with four semi–annual components corresponding to northern (NH) and Southern Hemisphere (SH) winters. **b)** Black curve, left scale, 3-point smoothed amplitude of the NH winter change in ΔLOD from weekly data after 31-day smoothing. Lower values indicate a larger change in the Earth's rotation speed. Red curve, right scale, solar activity as determined by 10.7 cm flux (solar flux units, gaussian smoothed). Dotted curve, right scale, Fast Fourier Transform with a 4-yr window of the time derivative 0.5-yr component of LOD, 30-month smoothed, after Barlyaeva et al. 2014.

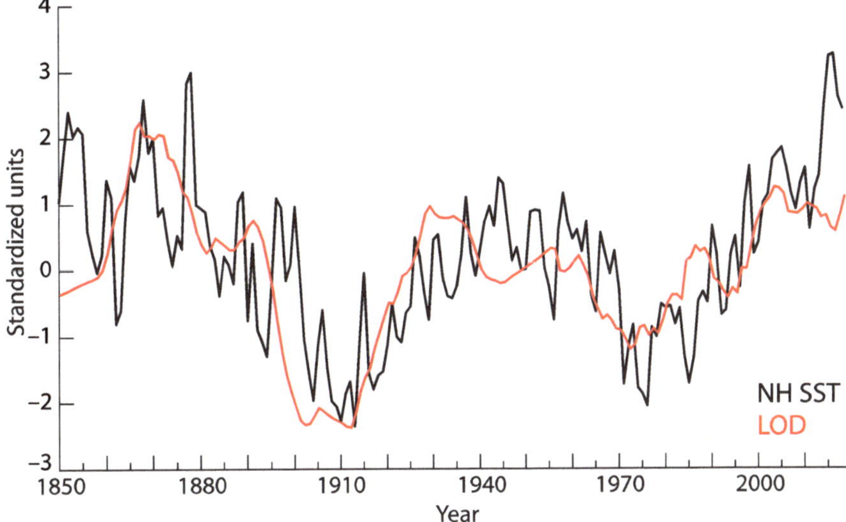

Fig. 10.18. Earth rotation and sea surface temperature anticorrelation
Black line, detrended yearly values of Northern Hemisphere SST, from HadSST3. Red line, inverted, detrended yearly values of ΔLOD. After. Mazzarella (2013).

the NH winter component is much smaller than the SH winter component (Fig. 10.17a, inset). The annual hemispheric temperature curves (Fig. 10.2) explain the annual and semi–annual components of LOD. The annual component is large because temperature contrasts are larger in winter, producing bigger changes in extra–tropical winds. The Earth rotates more slowly during NH winter, causing the day to be 1 ms longer than during SH winter, because more angular momentum resides in the atmosphere (higher AAM). Besides the asymmetric distribution in landmass between hemispheres, the NH in winter is the coldest hemisphere during the year, at a time when TSI is maximal due to the Earth being closer to the sun, creating the greatest temperature contrast. The semi-annual component is smaller and symmetric as the Earth goes through the equinoxes and is caused mainly by changes in tropical winds, with a significant tidal contribution.

Although the Earth rotation responds to multiple factors that alter the mass and momentum distribution with respect to the axis from the inner core to the stratosphere, it has been demonstrated that for periods of time between 14 days and 4 years changes in the AAM of the troposphere and stratosphere account for over 90% of the changes in LOD (Rosen & Salstein 1985). The Earth rotation is a sensitive, high precision gauge for changes that alter the global atmospheric zonal-mean zonal circulation. As the meridional circulation takes place at the expense of the zonal circulation, changes in ΔLOD can be used to study changes in MT.

The NH winter trough in ΔLOD might take place in Dec–Feb (DJF). By subtracting from the lowest weekly value for that period the highest weekly value from the prior 16 weeks, a measure of Earth's acceleration from ON to DJF, between 0 and −0.9 ms, is obtained for every year. The result is shown smoothed in Fig. 10.17b compared to solar activity. It displays an 11-year periodicity that agrees well in phase and amplitude with solar activity, although in some oscillations the Earth acceleration appears lagged, which is not surprising since LOD is also affected by the QBO and ENSO. While El Niño slows down the rotation (increases LOD) as does high solar activity, La Niña reduces LOD as does low solar activity. The correlation between LOD and solar activity is negative. Periods of low solar activity coincide with bigger NH winter acceleration

(ΔLOD decrease), while periods of high solar activity see almost no change in the rotation speed. So, the NH winter atmospheric circulation suffers more profound changes when solar activity is low than when it is high, confirming the effect seen in polar stratospheric temperature (Fig. 10.13a).

The link between changes in ΔLOD, changes in AAM, the strength of the MT and solar variability is very straightforward, and necessarily must go in the direction "solar → atmosphere → rotation." The momentum of the Earth system is conserved at the scales involved and it is not possible that changes in the speed of rotation of the Earth affect solar activity. Even before the effect of solar activity on the rotation of the planet had been considered, a relationship between multidecadal changes in ΔLOD and changes in climate had been proposed. In 1976, Lambeck and Cazenave reported on the similarity between the trends of numerous climate indices for the past two centuries and changes in ΔLOD, in particular surface temperature and pressure, were related to wind strength. They concluded that periods of increasing zonal winds correlate with an acceleration of the Earth while periods of decreasing zonal circulation correlate with a deceleration of the Earth. They found a lag of 5–10 years in the climatic indices. Their result has been reproduced multiple times, and an example is shown with SST and ΔLOD (Fig. 10.18; Mazzarella 2013). Lambeck and Cazenave (1976) ended with an interesting prediction, as the article was written after several decades of decreasing global temperature:

"if the hypothesis is accepted then the continuing deceleration of [the speed of rotation] for the last 10 yr suggests that the present period of decreasing average global temperature will continue for at least another 5–10 yr. Perhaps a slight comfort in this gloomy trend is that in 1972 the LOD showed a sharp positive acceleration that has persisted until the present..."

As they suggested, 4 years after the 1972 sharp decrease in ΔLOD took place, current global warming started, coinciding with the publication of the article. So, the gloomy cooling trend was transformed into—what to many is—an even gloomier warming trend.

The close correlation between SST and the AAM (and LOD) has been known for a long time. The correlation is explained as due to ocean–atmospheric coupling; where upwelling and downwelling depend on wind strength, and atmospheric pressure correlates with SST. Salstein (2015), one of the foremost experts in AAM, explains that the atmosphere has been simulated by a large number of models that are driven solely by the temperature of the underlying ocean surface. Based on these models, AAM has been calculated since the late 19th century from available SST data and checked against LOD estimations based on lunar occultation measurements. Multidecadal LOD shifts took place around 1870, 1910, 1935 and 1975 (Fig. 10.17). Another possible shift may have taken place around 2000 (see Chap. 11). They probably reflect multidecadal changes in ocean–atmospheric conditions reflecting changes in MT.

10.7 Conclusions

10a. The climate of the planet is characterized by its meridional transport of energy along the latitudinal gradient in temperature, its most fundamental property.

10b. Meridional transport is poleward, carried out mainly by the atmosphere over the ocean basins and stronger in winter.

10c. The Arctic in winter is the main heat sink for the planet. Part of the energy transported during the warm season is stored in the melting of ice and snow, and is lost through radiative cooling during the cold season after refreezing.

10d. Meridional transport to the Arctic in winter has a stratospheric component, and a tropospheric component that in a large part relies on discrete extreme Arctic intrusion events. Both components depend on the wave transmitting properties of the atmosphere, controlled by the thermal wind balance that determines the strength of zonal winds.

10e. Arctic amplification resulting in faster warming is the consequence of an increase in meridional transport, that leads to increased ice melting during the summer and increased surface temperature during the winter.

10f. ENSO is a part of the global meridional transport system extracting heat from the equatorial band. Pacing of La Niña and Neutral years is set by the solar cycle, while El Niño responds to the buildup of residual subsurface heat.

10g. The solar cycle, quasi-biennial oscillation, and ENSO, determine the status of the ozone layer in the winter Northern Hemisphere stratosphere and through it, the status of the polar vortex and winter meridional transport.

10h. The meridional transport of energy is linked to the meridional transport of momentum. Winter changes in the amount of atmospheric momentum alter the speed of Earth's rotation and are modulated by solar activity.

References

Alekseev G, Kuzmina S, Bobylev L et al (2019) Impact of atmospheric heat and moisture transport on the Arctic warming. International Journal of Climatology 39 (8) 3582–3592

Anderson RY (1990) Solar-cycle modulations of ENSO: a possible source of climatic change. In: Betancourt JL and MacKay AM (eds) Proceedings of the Sixth Annual Pacific Climate (PACLIM) Workshop, March 5–8, 1989: California Department of Water Resources, Interagency Ecological Studies Program Technical Report 23 77–81

Baldwin MP, Birner T, Brasseur G et al (2019) 100 years of progress in understanding the stratosphere and mesosphere. Meteorological Monographs 59 27–1

Baldwin MP & Dunkerton TJ (2005) The solar cycle and stratosphere–troposphere dynamical coupling. Journal of atmospheric and solar-terrestrial physics. 67 (1) 71–82

Baldwin MP, Gray LJ, Dunkerton TJ et al (2001) The quasi-biennial oscillation. Reviews of Geophysics 39 (2) 179–229

Barlyaeva T, Bard E & Abarca–del–Rio R (2014) Rotation of the Earth, solar activity and cosmic ray intensity. Annales Geophysicae 32 (7) 761–771

Beerling DJ & Royer DL (2011) Convergent Cenozoic CO_2 history. Nature Geoscience 4 (7) 418–420

Camp CD & Tung KK (2007) Stratospheric polar warming by ENSO in winter: A statistical study. Geophysical Research Letters 34 (4)

Carlson B, Lacis A, Colose C et al (2019) Spectral signature of the Biosphere: NISTAR finds it in our solar system from the Lagrangian L-1 point. Geophysical Research Letters 46 (17–18) 10679–10686

Challinor RA (1971) Variations in the rate of rotation of the Earth. Science 172 (3987) 1022–1025

Chen L, Francis J & Hanna E (2018) The "Warm-Arctic/Cold-continents" pattern during 1901–2010. International Journal of Climatology 38 (14) 5245–5254

Cohen J, Zhang X, Francis J et al (2020) Divergent consensuses on Arctic amplification influence on midlatitude severe winter weather. Nature Climate Change 10 (1) 20–29

Crick F (1988) What Mad Pursuit: A Personal View of Scientific Discovery. Basic Books, New York

Danish Meteorological Institute (2021) http://ocean.dmi.dk/arctic/meant80n_anomaly.uk.php Accessed 19 Nov 2021

Deser C, Alexander MA, Xie SP & Phillips AS (2010) Sea surface temperature variability: Patterns and mechanisms. Annual review of marine science 2 115–143

Domeisen DI, Garfinkel CI & Butler AH (2019) The teleconnection of El Niño Southern Oscillation to the stratosphere. Reviews of Geophysics 57 (1) 5–47

Domeisen DI, Butler AH, Charlton-Perez AJ et al (2020) The role of the stratosphere in subseasonal to seasonal prediction: 2. Predictability arising from stratosphere-troposphere coupling. Journal of Geophysical Research: Atmospheres 125 (2) e2019JD030923

Fedorov AV, Hu S, Wittenberg AT, Levine AF & Deser C (2020) ENSO Low-Frequency Modulation and Mean State Interactions. In: McPhaden MJ, Santoso A and Cai W (eds). El Niño Southern Oscillation in a Changing Climate. AGU Geophysical Monograph 253. John Wiley & Sons 173–198

Geng X, Zhang W, Jin FF et al (2020) Modulation of the relationship between ENSO and its combination mode by the Atlantic Multidecadal Oscillation. Journal of Climate 33 (11) 4679–4695

Gollan G & Greatbatch RJ (2017) The relationship between Northern Hemisphere winter blocking and tropical modes of variability. Journal of Climate 30 (22) 9321–9337

Gray LJ, Beer J, Geller M et al (2010) Solar influences on climate. Reviews of Geophysics 48 (4) RG4001

Haam E & Tung KK (2012) Statistics of solar cycle–La Niña connection: Correlation of two autocorrelated time series. Journal of the atmospheric sciences 69 (10) 2934–2939

Haas R & Scherneck HG (2004) The IVS Analysis Center at the Onsala Space Observatory. In: Vandenberg NR and Baver KD (eds) International VLBI Service for Geodesy and Astrometry, 2003 Annual Report by NASA/TP-2004-212254

Hall R, Erdélyi R, Hanna E et al (2015) Drivers of North Atlantic polar front jet stream variability. International Journal of Climatology 35 (8) 1697–1720

Hartmann DL (2016) Global Physical Climatology (2nd ed). Elsevier

Hines CO (1974) A possible mechanism for the production of sun–weather correlations. Journal of the Atmospheric Sciences 31 (2) 589–591

Holton JR & Tan HC (1980) The influence of the equatorial quasi–biennial oscillation on the global circulation at 50 mb. Journal of Atmospheric Sciences 37 (10) 2200–2208

Huber M & Caballero R (2011) The early Eocene equable climate problem revisited. Climate of the Past 7 (2) 603–633

Jones PD, New M, Parker DE et al (1999) Surface air temperature and its changes over the past 150 years. Reviews of Geophysics 37 (2) 173–199

Kang SM, Seager R, Frierson DM & Liu X. (2015) Croll revisited: Why is the northern hemisphere warmer than the southern hemisphere?. Climate Dynamics 44 (5) 1457–1472

Karamperidou C, Di Nezio PN, Timmermann A et al (2015) The response of ENSO flavors to mid-Holocene climate: implications for proxy interpretation. Paleoceanography 30 (5) 527–547

Kaspi Y & Schneider T (2013) The role of stationary eddies in shaping midlatitude storm tracks. J Atmos Sci 70 (8) 2596–2613

Kerr RA (1987) Sunspot–weather correlation found. Science 238 (4826) 479–481

Labitzke K (1987) Sunspots, the QBO, and the stratospheric temperature in the north polar region. Geophysical Research Letters 14 (5) 535–537

Labitzke K & van Loon H (1988) Associations between the 11-year solar cycle, the QBO and the atmosphere. Part I: the troposphere and stratosphere in the northern hemisphere in winter. Journal of Atmospheric and Terrestrial Physics 50 (3) 197–206

Lambeck K (1980) Changes in length-of-day and atmospheric circulation. Nature 286 (5769) 104–105

Lambeck K & Cazenave A (1973) The Earth's rotation and atmospheric circulation—I Seasonal variations. Geophysical Journal International 32 (1) 79–93

Lambeck K & Cazenave A (1976) Long term variations in the length of day and climatic change. Geophysical Journal International 46 (3) 555–573

Lambeck K & Hopgood P (1981) The Earth's rotation and atmospheric circulation, from 1963 to 1973. Geophysical Journal International 64 (1) 67–89

Landscheidt T (2000) Solar forcing of El Niño and La Niña. In: Wilson A (ed) The solar cycle and terrestrial climate. Proceedings of the 1st Solar and Space Weather Euroconference. ESA, Noordwijk, vol 463 p 135–140

Le Mouël JL, Blanter E, Shnirman M & Courtillot V (2010) Solar forcing of the semi-annual variation of length-of-day. Geophysical Research Letters 37 (15) L15307

Leamon RJ, McIntosh SW & Marsh DR (2021) Termination of solar cycles and correlated tropospheric variability. Earth and Space Science 8 (4) e2020EA001223

Liang M, Czaja A, Graversen R & Tailleux R (2018) Poleward energy transport: Is the standard definition physically relevant at all time scales?. Climate dynamics 50 (5) 1785–1797

Lin YF, Yu JY, Wu CR & Zheng F (2021) The Footprint of the 11-Year Solar Cycle in Northeastern Pacific SSTs and Its Influence on the Central Pacific El Niño. Geophysical Research Letters 48 (5) e2020GL091369

Lindzen RS & Holton JR (1968) A theory of the quasi–biennial oscillation. Journal of the Atmospheric Sciences 25 (6) 1095–1107

Marshall J & Plumb RA (2008) Atmosphere, Ocean and Climate Dynamics: An Introductory Text. International Geophysics Series vol 93. Academic Press

Mazzarella A (2013) Time-integrated North Atlantic Oscillation as a proxy for climatic change. Natural Science 5 (1A) 149–155

McIntyre ME & Palmer TN (1984) The 'surf zone' in the stratosphere. Journal of atmospheric and terrestrial physics 46 (9) 825–849

Meinen CS & McPhaden MJ (2000) Observations of warm water volume changes in the equatorial Pacific and their relationship to El Niño and La Niña. Journal of Climate 13 (20) 3551–3559

Mewes D & Jacobi C (2019) Heat transport pathways into the Arctic and their connections to surface air temperatures. Atmospheric Chemistry and Physics 19 (6) 3927–3937

Moy CM, Seltzer GO, Rodbell DT & Anderson DM (2002) Variability of El Niño/Southern Oscillation activity at millennial timescales during the Holocene epoch. Nature 420 (6912) 162–165

Nakamura N & Huang CS (2018) Atmospheric blocking as a traffic jam in the jet stream. Science 361 (6397) 42–47

Overland JE, Wood KR & Wang M (2011) Warm Arctic—cold continents: climate impacts of the newly open Arctic Sea. Polar Research 30 (1) 15787

Papritz L & Dunn–Sigouin E (2020) What configuration of the atmospheric circulation drives extreme net and total moisture transport into the Arctic. Geophysical Research Letters 47 (17) e2020GL089769

Peixoto JP & Oort AH (1992) Physics of climate. American Institute of Physics. New York

Perner K, Moros M, De Deckker P et al (2018) Heat export from the tropics drives mid to late Holocene palaeoceanographic changes offshore southern Australia. Quaternary Science Reviews 180 96–110

Randall DA (2015) An Introduction to the Global Circulation of the Atmosphere. Princeton University Press

Rao J, Yu Y, Guo D, Shi C et al (2019) Evaluating the Brewer–Dobson circulation and its responses to ENSO, QBO, and the solar cycle in different reanalyses. Earth and Planetary Physics 3 (2) 166–181

Rosen RD & Salstein DA (1985) Contribution of stratospheric winds to annual and semiannual fluctuations in atmospheric angular momentum and the length of day. Journal of Geophysical Research: Atmospheres 90 (D5) 8033–8041

Roy I (2014) The role of the Sun in atmosphere–ocean coupling. International Journal of Climatology 34 (3) 655–677

Salby M & Callaghan P (2000) Connection between the solar cycle and the QBO: The missing link. Journal of Climate 13 (2) 328–338

Salstein DA (2015) Angular Momentum of the Atmosphere. In: North GR, Pyle J and Zhang F (eds) Encyclopedia of Atmospheric Sciences 2nd ed. Elsevier Vol 2 p 43–50

Scotese CR, Song H, Mills BJ & van der Meer DG (2021) Phanerozoic paleotemperatures: The earth's changing climate during the last 540 million years. Earth-Science Reviews 103503

Shaw TA, Baldwin M, Barnes EA et al (2016) Storm track processes and the opposing influences of climate change. Nature Geoscience 9 (9) 656–664

Sherwood SC & Huber M (2010) An adaptability limit to climate change due to heat stress. Proceedings of the National Academy of Sciences 107 (21) 9552–9555

Shindell DT, Schmidt GA, Mann ME et al (2001) Solar forcing of regional climate change during the Maunder Minimum. Science 294 (5549) 2149–2152

Steinthorsdottir M, Vajda V, Pole M & Holdgate G (2019) Moderate levels of Eocene p CO_2 indicated by Southern Hemisphere fossil plant stomata. Geology 47 (10) 914–918

Sud YC, Walker GK & Lau KM (1999) Mechanisms regulating sea-surface temperatures and deep convection in the tropics. Geophysical Research Letters 26 (8) 1019–1022

Sun DZ (2000) The heat sources and sinks of the 1986–87 El Niño. Journal of climate 13 (20) 3533–3550

Taguchi M (2010) Observed connection of the stratospheric quasi-biennial oscillation with El Niño–Southern Oscillation in radiosonde data. Journal of Geophysical Research: Atmospheres 115 (D18)

Timmermann A, An SI, Kug JS et al (2018) El Niño–southern oscillation complexity. Nature 559 (7715) 535–545

van Loon H & Labitzke K (1988) Association between the 11-year solar cycle, the QBO, and the atmosphere. Part II: Surface and 700 mb in the Northern Hemisphere in winter. Journal of Climate 1 (9) 905–920

Wang C, Deser C, Yu J–Y et al (2016) El Niño and Southern Oscillation (ENSO): A Review. In: Glynn PW, Manzello DP and Enochs IC (eds) Coral reefs of the Eastern Tropical Pacific: persistence and loss in a dynamic environment. Springer

Wang G, Zhao C, Zhang M et al (2020) The causality from solar irradiation to ocean heat content detected via multi-scale Liang–Kleeman information flow. Scientific Reports 10 (1) 1–9

Wang X & Key JR (2005) Arctic surface, cloud, and radiation properties based on the AVHRR Polar Pathfinder dataset. Part I: Spatial and temporal characteristics. Journal of Climate 18 (14) 2558–2574

Wazneh H, Gachon P, Laprise R et al (2021) Atmospheric blocking events in the North Atlantic: trends and links to climate anomalies and teleconnections. Climate Dynamics 56 (7) 2199–2221

Weng H (2012) Impacts of multi-scale solar activity on climate. Part I: Atmospheric circulation patterns and climate extremes. Advances in Atmospheric Sciences 29 (4) 867–886

White WB & Liu Z (2008) Non-linear alignment of El Niño to the 11-yr solar cycle. Geophysical Research Letters 35 (19) L19607

Woods C, Caballero R & Svensson G (2013) Large-scale circulation associated with moisture intrusions into the Arctic during winter. Geophysical Research Letters 40 (17) 4717–4721

Woods C & Caballero R (2016) The role of moist intrusions in winter Arctic warming and sea ice decline. Journal of Climate 29 (12) 4473–4485

Wunsch C (2005) The total meridional heat flux and its oceanic and atmospheric partition. Journal of climate 18 (21) 4374–4380

Yang H, Li Q, Wang K et al (2015) Decomposing the meridional heat transport in the climate system. Climate Dynamics 44 (9–10) 2751–2768

Yu L & Weller RA (2007) Objectively analyzed air–sea heat fluxes for the global ice-free oceans (1981–2005). Bulletin of the American Meteorological Society 88 (4) 527–540

Yu Y & Ren R (2019) Understanding the variation of stratosphere–troposphere coupling during stratospheric northern annular mode events from a mass circulation perspective. Climate Dynamics 53 (9) 5141–5164

MERIDIONAL TRANSPORT AND SOLAR VARIABILITY ROLE IN CLIMATE CHANGE

"A new scientific truth does not triumph by convincing its opponents and making them see the light, but rather because its opponents eventually die, and a new generation grows up that is familiar with it."
Max Planck (1906). Known as the Planck Principle

11.1 Introduction

Meridional transport (MT) of energy is the most fundamental property of the climate system. Its role in maintaining the stability of the Earth climate has long been recognized (Yang et al. 2015). In the previous chapter we saw that it is modulated by several complex factors, and among them, strikingly, is solar activity. It is assumed that MT does not contribute to climate change, since external natural drivers and internal climate variability have an estimated zero net contribution to the observed warming from 1850–2019 (IPCC AR6 SMP 2021). In fact, MT is not even considered a climate driver. It is surprising that changes in the most fundamental property of a stable climate system, that at present cannot be properly measured (Wunsch 2005), are disregarded as a cause for climate change when there is ample evidence of its involvement in the effects of both external solar and volcanic drivers and internal multidecadal variability. In stark contrast, paleo-climatologists consider that the climates of the distant past can be described in a first approach by their latitudinal temperature gradient (LTG) from the equator to the poles, the primary driver of MT (Scotese et al. 2021). Different climates of the past are thus characterized by different MT intensities. The assumption that climate changes are responsible for MT changes ignores the clear possibility that MT variability is a driver of those climate changes. An obvious clue that we are neglecting the importance of MT in climate change is that our current theory of climate cannot explain the low gradient paradox (Huber & Caballero 2011) of the early Eocene equable climate (see Fig. 10.1), when both poles had a temperate climate, causing a very flat LTG that could only drive a very reduced MT. As with many paradoxes, the low gradient paradox is easily solved by changing the frame of reference. The climate was equable because there was a reduced MT. The more energy transported to the winter pole, a gigantic cooling radiator, the more energy is emitted to space by the largest heat sink at the top of the atmosphere (ToA). This framework-change places MT at the center of climate change at all timescales, including modern global warming.

The planet today is much colder than in the early Eocene because a lot more energy is being transported to the winter pole and, since the increase has taken place for millions of years, the ocean has cooled to the bottom and the poles have accumulated a huge amount of ice. The winter poles have very low greenhouse gas (GHG) concentrations, since they have the lowest absolute humidity of the planet, and receive less solar radiation. This results in a loss of energy to space through radiative cooling proportional to the amount of energy transported there by the atmosphere (see Sect. 10.3). The polar vortex (PV) limits winter MT to the polar caps by restricting atmospheric circulation and can be considered a proxy for MT. The southern PV is very strong, resulting in a weak modulation and a lower occurrence of strong perturbations (sudden stratospheric warmings; Ayarzagüena et al. 2021). The northern PV is weaker and susceptible to a stronger modulation by the several factors that affect MT, including the quasi-biennial oscillation (QBO), El Niño/Southern Oscillation (ENSO) and solar activity (see Chap. 10). In this chapter the climatic effects of externally and internally driven changes in MT are reviewed and a hypothesis on long-term modulation of MT by changes in solar activity is presented.

11.2 Volcanic effects on meridional transport

The effect of volcanic eruptions on the global climate is through the sulfate aerosols produced by vast amounts of sulfur dioxide injected into the stratosphere during strong explosive volcanic eruptions. Sulfate aerosols are washed from the troposphere in a matter of weeks, but their decay in the stratosphere is much slower, taking 1–2 years. Since there is little inter-hemispheric transport in the lower stratosphere (see Fig. 10.5), the eruption location determines if there is a global effect (tropical eruptions) or a mainly hemispheric effect (extra-tropical eruptions).

The climate is affected by eruptions through radiative, chemical and dynamic changes. Radiative effects result from sulfate aerosol presence in the lower stratosphere that reflect and scatter incoming shortwave radiation, causing an increase of 3–5 W/m^2 in reflected radiation (RSR), however they also absorb solar near-infrared radiation (IR) and terrestrial IR radiation, so the maximum cooling of the system is 2–3 W/m^2 (Stenchikov 2021). This perturbation of the Earth's radiative balance can only last while there are significant volcanic sulfate aerosols in the stratosphere (Fig. 11.1). Chemical effects in the stratosphere are also caused by the volcanic injection of SO$_2$. The most important one is believed to be the decrease in ozone from a

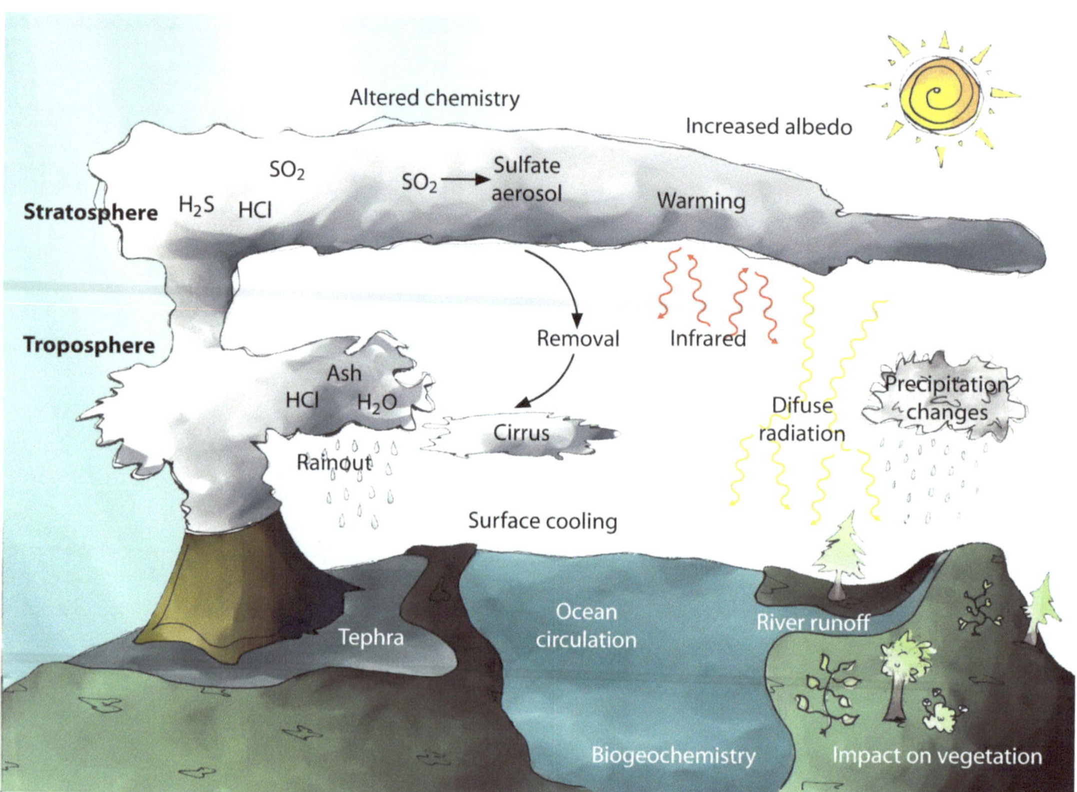

Fig. 11.1 Schematic overview of the climate effects after a volcanic eruption with large stratospheric sulfate injection After Timmreck (2012). Artist: Lucía M. Lagunas.

combination of transport perturbations, radiative perturbations on ozone photochemistry, and heterogeneous chemistry that is dependent on the presence of anthropogenic chlorine in the stratosphere. The drop in ozone levels leads to an increase in solar UV radiation reaching the surface. The chemical effect is also short lived and ozone levels recover when sulfate aerosols decay. The dynamic effects of a volcanic eruption result from the change in surface and lower stratosphere temperature patterns and gradients that affect MT.

Strong tropical eruptions have the biggest climatic effect on a global scale. Incoming solar radiation is higher at the tropics, and stratospheric aerosols have a bigger reflection effect there. As a result, surface cooling is stronger at lower latitudes, reducing the LTG and weakening the Hadley circulation and subtropical jets. The cooling also results in a decrease in atmospheric water vapor enhancing the temperature reduction (Soden et al. 2002). Due to stronger cooling over land than over ocean, the Asian summer monsoon becomes weaker. At the lower stratosphere, radiative effects cause the inverse temperature pattern, with a stronger tropical warming that peaks at 30°N, due to absorption of near IR radiation at the top of the aerosol cloud and of upwelling IR at the bottom (Robock 2000). The stratospheric LTG is enhanced even more by the decrease in ozone that reduces solar UV warming, as the ozone anomaly caused by tropical volcanic eruptions is larger at the NH high latitudes (Stenchikov et al. 2002).

Dynamic effects result from perturbations in MT, as indicated by their strongest impact being during the boreal winter, the period of strongest global transport (see Chap. 10). MT is strongly dependent on tropospheric and stratospheric LTGs, and they are altered in opposite directions by strong tropical eruptions. The strengthening of the lower stratosphere LTG results in an increase in the zonal-mean zonal wind anomaly and a reduction in the Brewer–Dobson circulation (BDC) shallow branch transport. An El Niño-like response is triggered at the equatorial Pacific that redirects the transport (Swingedouw et al. 2017), followed by La Niña-like condition the following year (Sun et al. 2018). In the NH, the strengthened stratospheric LTG results in a stronger PV causing a positive North Atlantic Oscillation (NAO) response during the first two winters (Guðlaugsdóttir et al. 2019), followed by negative NAO in the third year. The positive NAO and strengthened PV result in a cold Arctic/warm continents pattern that produces winter warming in northern Eurasia and North America, with colder conditions at the Mediterranean and the Middle East. Higher land/ocean temperature contrast during volcanic winters at mid-latitudes results in an increase in planetary wave number 1 activity and a decrease in number 2 activity (Graf et al. 2007) but, due to the altered zonal wind state, waves are reflected instead of breaking, and so the PV remains strong through the winter (Bittner et al. 2016). The resulting slowdown in stratospheric transport is manifested in a marked increase in stratospheric air mean age for 1–2 years (Diallo et al. 2017).

The changes in the LTG induced by stratospheric sulfate aerosols result in a different effect of extra-tropical volcanic eruptions versus tropical ones. NH eruptions still cause a Niño-like response, but they cannot strengthen the stratospheric LTG and zonal winds as much as tropical eruptions, and as a result the PV is weakened by the increased wave activity and a negative NAO is the result (Sjolte et al. 2021). In the NH, extra-tropical eruptions have a seasonal dependency as summer eruptions increase the stratospheric aerosol optical depth less than winter eruptions (Toohey et al. 2019). This is another indication that the effects of volcanic eruptions on climate depend on MT, which is stronger in winter. SH eruptions have a

Fig. 11.2 Regional and hemispheric temperature effect from volcanic eruptions
a) Gray bars, right scale, Northern Hemisphere temperature change for the years around eight strong eruption years. After Self et al. (1981). Black and red curves, left scale, inferred temperature change from tree-ring maximum latewood density records from Northern Europe (black curve) and Central Europe (red curve) for the years around 34 large volcanic eruptions (VEI index ≥5) within the AD 1111–1976 period. After Esper et al. (2013). **b)** 30–90°N mean annual decadally averaged temperature proxy reconstruction (Hegerl et al. 2007), from −10 to +60 years of the indicated volcanic eruptions (red curves), with highest global forcing (Sigl et al. 2015) of the past 1400 years. The 1230 and 1783 AD eruptions were not included due to another strong eruption taking place within the 60-year post-eruption period. Temperatures were detrended and expressed as anomaly to their 70-year period average. The thick black curve is the average.

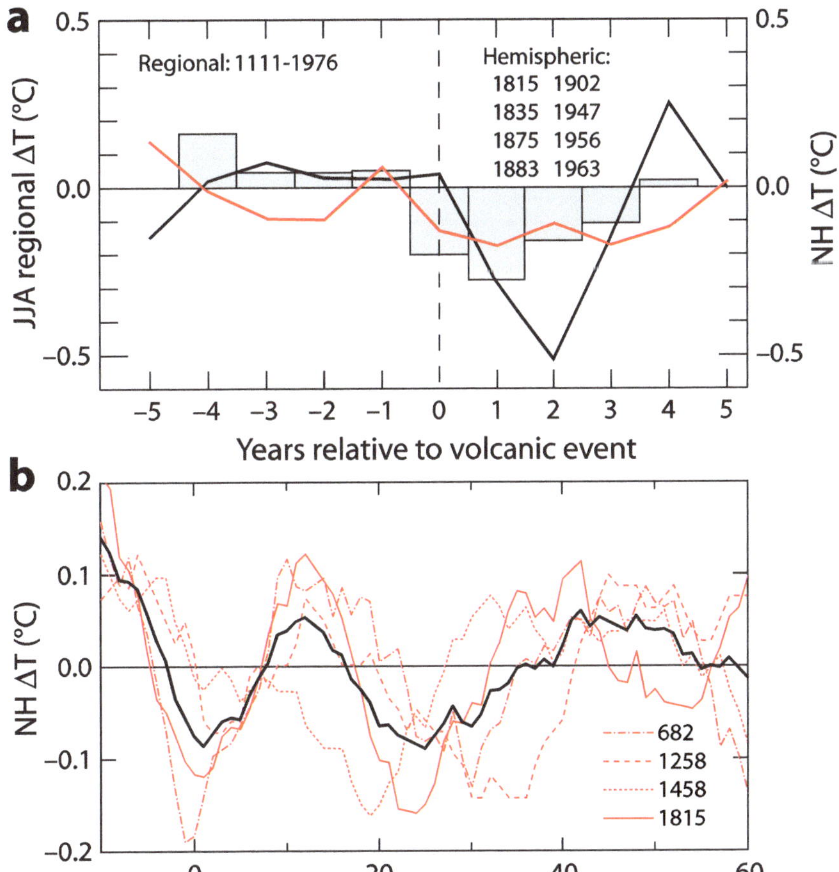

much smaller effect because the southern PV is stronger and MT is not as affected. The lack of intense effect on MT is also noticeable in that SH eruptions do not induce an El Niño-like response, instead they cause sea surface cooling and La Niña conditions (Liu et al. 2018).

The available evidence from the Pinatubo and El Chichón eruptions supports a short-term effect from volcanoes on temperatures, only lasting a few years. Instrumental temperature records from 19th and 20th century volcanic eruptions indicate a temperature decrease of 0.2–0.3 °C for 3–4 years, as do NH and European proxies (Self et al. 1981; Esper et al. 2013; Fig. 11.2a).

In theory, the substantial energy lost by the climate system due to the increased RSR by volcanic stratospheric aerosols cannot be recovered or compensated for, and should result in long-term cooling. This is not observed, as surface temperatures recover in a few years. Models, instead, simulate a centennial cooling of the entire ocean, inferring that the missing energy must come from enhanced surface flux (Brönnimann et al. 2019; Stenchikov 2021). This is referred as the ocean memory. It should be noted that models do not reproduce volcanic effects very well, like the El Niño response or the strong PV. They prescribe too much surface cooling compared to observations (Toohey et al. 2014; Swingedouw et al. 2017). Instead, volcanic aerosols result in two opposite effects on the energy balance. They increase RSR, reducing the energy gain, and they reduce MT towards the winter pole, reducing the polar energy loss. The 1991 Pinatubo eruption resulted in a decrease in outgoing longwave radiation (OLR) from the Arctic of c. 2 W/m² for about 5 years (see Sect. 11.4 below), so the energy loss by the climate system from

a volcanic eruption is lower than generally believed. This decrease in energy loss helps explain the rapid temperature recovery after stratospheric aerosols decay, reducing the need for a significant entire-ocean centennial memory to volcanic eruptions that has not been observed since the Pinatubo eruption. Thus, a more likely explanation, a significant change in MT, is not being contemplated in the climate models.

It is clear that there is a delayed response to volcanic forcing after about 1–3 decades. Hegerl et al. (2007) NH proxy temperature reconstruction is able to display it, because it is annually resolved and calibrated to avoid the loss of low-frequency variance. The four volcanic eruptions of the past 1400 years with the highest global forcing (Sigl et al. 2015) and not followed by another very strong volcanic eruption during the next 60 years, occurred in 682, 1258, 1458, and 1815 AD. Figure 11.2b shows the reconstructed temperature from −10 to +60 years of each eruption, after detrending and expressed as anomalies to the detrended average. The reconstruction reduces interannual variability while preserving low-frequency variability with decadal smoothing (Hegerl et al. 2007), and the smoothing displays volcanic induced cooling prior to the eruption year. Despite these eruptions being among the strongest in the past two millennia, the cooling is neither very intense, nor is it prolonged for more than a few years. This agrees with historic instrumental observations (Fig. 11.2a bars; Self et al. 1981), and regional proxy reconstructions for the same period (Fig. 11.2a lines; Esper et al. 2013). However, a second multi-year cooling of similar magnitude is observed with a delay of c. 20 years (1458 and 1815 eruptions) to c. 30 years (682 and 1258 erup-

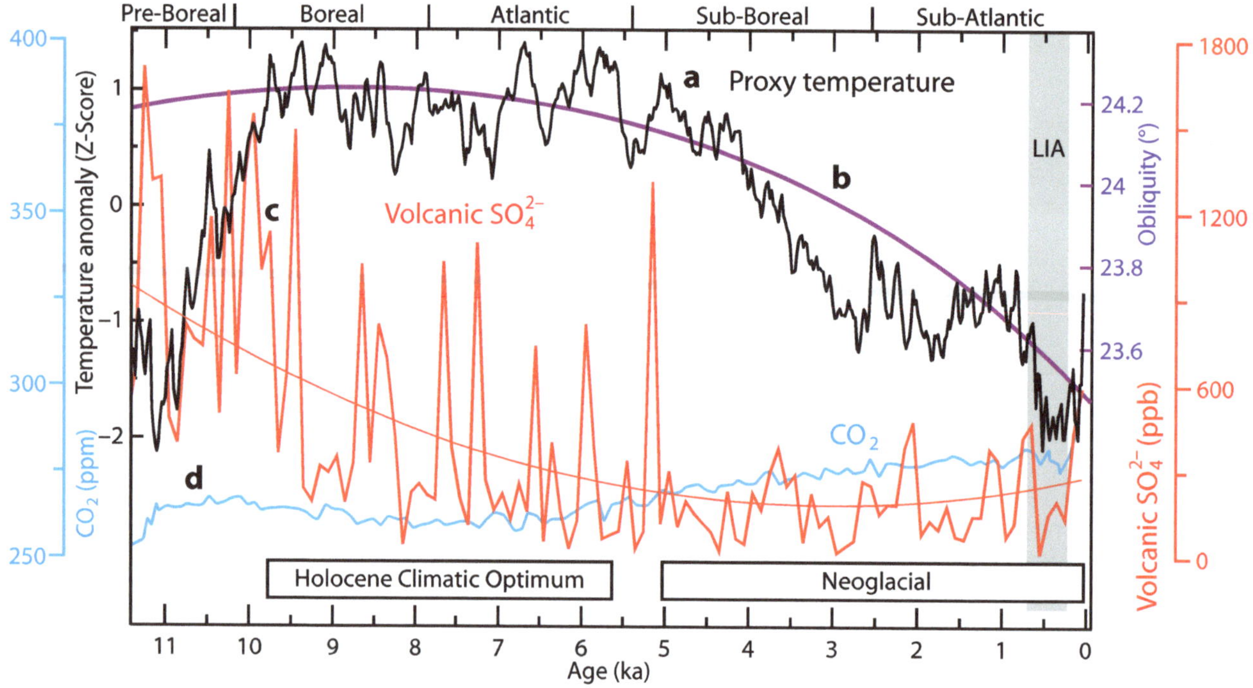

Fig. 11.3 Volcanic activity during the Holocene
a) Black curve, global temperature reconstruction from 73 proxies (See Chap. 4) expressed as distance to the average in standard deviations (Z-score). **b)** Purple curve, Earth's axis obliquity in degrees. **c)** Red curve, Holocene volcanic sulfate in the GISP2 ice core in parts per billion summed for each century in BP scale (rightmost point is 0–99 or 1851–1950), with quadratic trendline (thin red line). Data from Zielinski et al. 1996. **d)** Light blue curve, CO2 levels as measured in Epica Dome C (Antarctica) ice core. Data from Monnin et al. 2004. Grey bar, Little Ice Age (LIA).

tions). The recovery from this second cooling is also fast. This delayed response must come from the ocean-atmosphere coupled system as radiative and chemical effects from volcanic aerosols have long disappeared. A proposed explanation involves a delayed oceanic response by the Atlantic Meridional Overturning Circulation that excites a well-known 20-year oscillation that would, thus, have an origin in the supposed ocean memory to volcanic eruptions (Swingedouw et al. 2015). MT is very likely involved in this delayed effect, as multidecadal oscillations are a feature of the MT system (see below). If MT is involved in this delayed surface cooling, it would mean that MT is reduced initially by aerosol radiative effects causing a stronger vortex, but MT is increased 20–30 years after by a strengthened multidecadal oscillation of the stadium-wave (see below). The temporal variability of the delayed response fits the non-stationary nature of this oscillation that cannot be properly reproduced by models.

Proxy records (Fig. 11.2; Moberg et al. 2005; Hegerl et al. 2007; see also Fig. 6.12) show that the cooling caused by the strongest volcanic eruptions of the past 1500 years only caused a temporary cooling that was fully recovered in a few decades at most. In every case the underlying temperature trend, that was decreasing for the 1258 and 1458 eruptions but increasing for the 682 and 1815 eruptions, continued unaltered. This makes it difficult to explain the Little Ice Age (LIA) in terms of volcanic activity. Tree-ring analysis shows a similar effect for LIA volcanoes and instrumental-era volcanoes (Self et al. 1981; Esper et al. 2013; Fig. 11.2a). Most authors, accept that the LIA was probably caused by a combination of solar and volcanic forcings, and the discussion is about which was more important. Since both affect MT, one possibility is that they act synergistically to produce a larger combined

cooling. A causal link between altered MT and the LIA is suggested by a prolonged period of warm Arctic/cold continents (WACC) pattern, indicative of a predominantly weak PV, from c. 1400 to 1850 AD (Porter et al. 2019).

The idea that volcanoes could cause long-term climate cooling was already suggested in 1940, and in the early 1970s there was speculation that volcanoes could be a cause for glaciations. But by the late-1970s it was clear that the effect of volcanoes on climate was limited. As Rampino et al. reported in 1979: *"The eruption of Tambora in 1815, one of the largest eruptions during the past few thousand years, is associated with a hemispheric temperature decrease of only 0.5° to 1°C for 2 to 3 years. In this case, average global temperatures had already been decreasing since 1810 and then rose again in the 1820s."*

Rampino et al. (1979) observed that major historical eruptions associated with cooling were taking place after decadal length temperature decreases had been initiated, so they asked if rapid climate change could cause volcanic eruptions, by means of stress changes on the earth's crust, through loading and unloading of ice and water masses and through axial and spin-rate changes. This question was left unanswered and forgotten once scientific focus changed from cooling to warming in the 1980s. However, through better data collection the question reemerged at the turn of the century. Zielinski et al. published in 1996 the GISP2 ice-core record of 110,000 years of explosive volcanism, showing that the transitions between colder periods (stadials) and warmer periods (interstadials) displayed increased volcanism. The largest and most abundant volcanic signals over the past 110,000 years occur between 17,000 and 7,000 BP, during the deglaciation and warmest part of the Holocene (Zielinski et al. 1996; Fig. 11.3). The LIA, the coldest period in the Holocene, is completely unremarkable in volcanic terms in the GISP2 record. If there is any correlation between volcanic activity and temperature during the Holocene it is a positive one. The hypothesis that the LIA could be due to high volcanic activity cannot be sustained with the available evidence.

The following year McGuire et al. (1997) published a correlation between sea-level changes and the frequency of explosive volcanism in the Mediterranean. And Glazner et al. (1999) published under the suggestive title "Fire or ice," that the anti-correlation between volcanism and glaciation extended for the past 800,000 years in Californian volcanoes, located 300 km from the sea. This was also confirmed for continental volcanism in France and Germany by Nowell et al. (2006). Then research turned to the mechanisms. Jellinek et al. (2004) showed that Californian volcanoes were responding to ice changes in the obliquity (41-kyr) frequency with a time lag. The lag was related to glacial unloading, and was a few kyr for silicic volcanism but 10 kyr for basaltic volcanism. They advanced the hypothesis that the decrease in pressure from ice melting stimulated dike formation leading to more explosive volcanism. But this is not the only hypothesis, since volcanoes are varied and different types might respond to different stimuli. Caldera-forming volcanoes from the glaciated island arc of Kamchatka seem to respond to glacial loading and show peak activity during glacial maxima (Geyer & Bindeman 2011).

Kutterolf et al. (2013) conducted the most comprehensive analysis to date of the temporal frequency of volcanism in the Ring of Fire, where most of the world's largest volcanic eruptions have occurred in the recent past. They built an extensive dataset of 408 tephra (volcanic ejecta) layer dates for the past million years from multiple coring sites down-stratospheric-wind from volcanic areas along the Pacific Ring of Fire. Frequency analysis of volcanic eruption activity shows a significant peak at the 41-kyr period with non-significant peaks also at the 23-, 82-, and 100-kyr Milankovitch frequencies. The 41-kyr volcanic frequency peak coincides in period with the $\delta^{18}O$ change from benthic cores. The result indicates that volcanic activity is linked mainly to the 41-kyr obliquity orbital frequency. This frequency is linked to the determination of glaciations and deglaciations in the glacial cycle, while the 100-kyr eccentricity orbital frequency is related to the global amount of ice during glacial periods (see Sect. 2.7 and Fig. 2.12). The result suggests, in agreement with the GISP2 record, that rapid deglaciations linked to the obliquity cycle lead to periods of enhanced volcanic activity.

Ice cores (volcanic sulfate deposition) and volcanic (tephra layers) datasets independently show that the highest level of volcanic activity for the past 100,000 years took place between 13 and 7 kyr BP, with a rate 2–6 times above background level (Huybers & Langmuir 2009; Fig. 11.3). Part of the lag with respect to the start of the deglaciation at c. 18 kyr might be due to the melting starting at the Eastern Laurentide and Antarctic ice sheets that do not affect volcanic regions. After 7 kyr BP, once the ice sheets melted, volcanic activity declined and returned to background, glacial-comparable levels. During the Neoglaciation global cooling volcanic activity remained subdued (Fig. 11.3), so the evidence supports that during the Holocene volcanism has not been a cause for long-term cooling. Huybers and Langmuir (2009) have analyzed CO_2

output by volcanoes at deglaciation, considering that the increase in subaerial volcanism due to melting must have been compensated in part by a decrease in non–ridge associated submarine volcanism due to the increasing load caused by sea-level rise. Their modeling indicates that half of the CO_2 increase (c. 40 ppm) during deglaciation may be due to the volcanic response to climate change, constituting an important feedback factor to glacial termination.

Volcanic activity has no detectable climatic effect in the Late Pleistocene and Holocene. Volcanic eruptions have a short-term effect, and do not significantly affect climate. Their only possible climatic effect is a hypothetical (but reasonable) feedback effect at glacial terminations through their significant contribution to CO_2 increase. However, climate change has a strong effect on volcanism. The effect of obliquity-linked ice unloading, the corresponding sea-level increase, and isostatic adjustment on volcanic activity is detected in ice cores and volcanic records. Due to the asymmetric nature of ice changes, as melting is very rapid, but ice build-up is very slow, the effect of ice unloading has been well established, but that of ice loading has not. Anecdotal evidence since the 1970s suggests that ice loading might also increase volcanic activity, as an increase in eruptions has been observed after decades-long cooling periods (like the LIA), however, the evidence for this is not significant.

11.3 The circa 65-year oscillation and the stadium-wave hypothesis

The existence of a multidecadal mode of climate variation was first detected by Folland et al. (1984) in global sea-surface temperature (SST) and night marine air temperature records, and later correlated to precipitation records in the Sahel (Folland et al. 1986). This multidecadal oscillation was isolated by Schlesinger and Ramankutty (1994) in the global mean instrumental temperature record, as a 65–70-year NH periodicity, and attributed to internal variability of the coupled ocean-atmosphere system. It was termed the Atlantic Multidecadal Oscillation (AMO) by Kerr (2000). In the following years the c. 65-year oscillation was observed in North Atlantic sea level pressure and winds (Kushnir 1994), North Pacific and North American temperature (Minobe 1997), length of day and core angular momentum (Hide et al. 2000), fish populations (Mantua et al. 1997; Klyashtorin 2001), Arctic temperature and sea ice extent (Polyakov et al. 2004), the relative frequency of ENSO events (Verdon & Franks 2006), and global mean sea level (Jevrejeva et al. 2008). Most of these records display an additional c. 20-year periodicity that is also apparent in Greenland $\delta^{18}O$ ice core data (Chylek et al. 2012). This periodicity is most apparent at mid-latitudes subsurface temperatures, while the c. 65-year oscillation is most apparent at high latitudes deep water salinity levels (Frankcombe et al. 2010). This difference precludes a direct harmonic relation between them. Additionally, proxies indicate the c. 20-year periodicity was more intense during the LIA, while the c. 65-year oscillation appears more intensely in 20th century records, and might not have been present, at least with that periodicity, during the LIA (Gray et al. 2004).

It is generally believed that the c. 65-year oscillation originates from internal ocean-atmosphere variability,

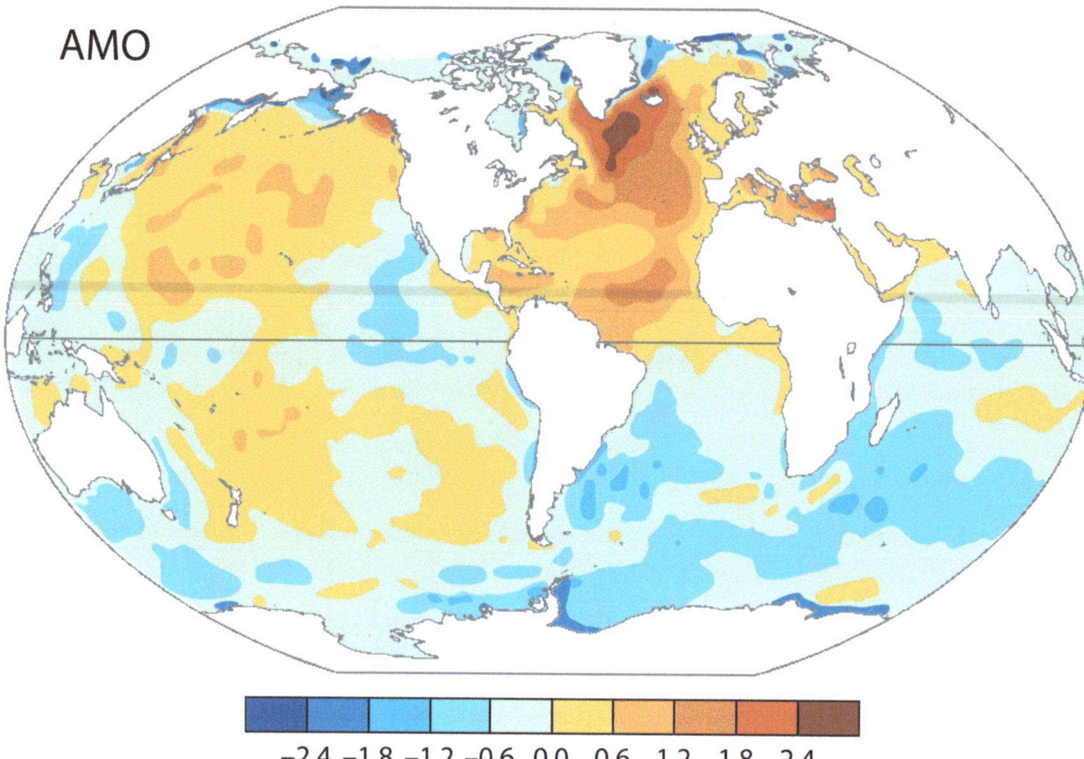

Fig. 11.4 Atlantic multidecadal oscillation spatial pattern Unitless (°C/°C) regression pattern of monthly SST anomalies (HadISST 1870–2008), after subtracting the global mean anomaly from the North Atlantic SST anomaly. It displays the °C of SST change per °C of AMO index. Besides displaying the AMO pattern, it shows that AMO is linked to the global surface MT system that extracts heat from the tropics in the main ocean basins. After Deser et al. 2010.

rather than being externally forced or randomly generated (Dai et al. 2015). An alternative explanation is that the c. 65-year oscillation reflects global MT system variability. The oscillation mostly affects the two ocean basins that communicate directly with both poles, particularly from the equator (ENSO) to the NH high latitudes (AMO and Pacific Decadal Oscillation, PDO), and it affects the rotation of the Earth through changes in the angular momentum of the atmosphere (Hide et al. 2000; Klyashtorin & Lyubushin 2007). The connection between ENSO and MT has already been explored in Section 10.4; a connection between ENSO and AMO also exists (Wang et al. 2020); and AMO is essentially a manifestation of the surface heat transport intensity through the Atlantic basin. As heat is continuously being introduced at the tropics at a near-constant rate, a warm AMO indicates a heat jam in the North Atlantic, perhaps due to a reduced ocean-atmosphere flux caused by a predominantly zonal wind pattern at mid-latitudes. The spatial pattern of the AMO, obtained by regression of North Atlantic SST anomalies after subtracting the global SST anomalies, reveals that the AMO is the Atlantic portion of a global MT system that moves heat poleward. The global system also includes the Pacific and Indian basins (Fig. 11.4). It shows that the NH SST oscillation of the AMO is phase-locked with other global SST oscillations, reflecting coordinated changes in the global MT system.

Decadal and multidecadal external solar and tidal forcings alter MT strength over different oceanic basins, and the c. 65-year oscillation could be an emergent temporal resonance from the intrinsic delays in the oceanic and atmospheric heat transmission. In this regard, the Bjerknes hypothesis establishes that hemispheric anomalies in the transport of heat by the atmosphere and the oceans should be of equal magnitude and opposite sign (Bjerknes compensation). The Bjerknes compensation has not been measured in nature due to our inability to properly sample

heat transport, but interestingly the Bergen climate model, when reproducing the Bjerknes compensation under constant forcing, generates a 60–80-year periodicity reminiscent of the AMO (Outten & Esau 2017).

The c. 65-year oscillation is important for multidecadal climate predictions. Divine and Dick (2006), in their study of the historical variability of sea ice edge position in the Nordic Seas, correctly identified the effect of the c. 65-year oscillation over any putative anthropogenic effect and ended with the conclusion that *"during decades to come, as the negative phase of the thermohaline circulation evolves, the retreat of ice cover may change to an expansion."* It must have taken courage to predict a sea ice expansion in 2006, when essentially everybody else was predicting a sea ice collapse, yet since 2007 Arctic sea ice has been showing a, still non-significant, modest growth in September extent that contrasts with the previous strong decline.

In her dissertation, Marcia Wyatt (2012) proposed a hypothesis on the dynamic transfer of a climate signal between the different ocean basins, Arctic sea ice, and the atmosphere via the c. 65-year oscillation, that she termed "the stadium-wave." The hypothesis neatly links all the different manifestations of the c. 65-year oscillation, accounting for their lags, and produces a complete set of "quasi-predictions" that should be good for as long as the oscillation maintains its periodicity (Fig. 11.5; Wyatt & Curry 2014). Wyatt & Curry 2014, is one of the few articles (with Divine & Dick 2006) that correctly predicted the current pause in Arctic sea ice melting at a time when the data suggested the opposite: *"this [sea-ice decline] trend should reverse... Rebound in West Ice Extent, followed by Arctic Seas of Siberia should occur after the estimated 2006 minimum of West Ice Extent and maximum of AMO"* (Wyatt & Curry 2014).

One consequence of the c. 65-year oscillation is the existence of climate shifts, the change from one climate

Fig. 11.5 The c. 65-year oscillation and the stadium-wave hypothesis

a) Simplified stadium-wave wheel cartoon showing a 60-year cycle from 1976 to 2036. Red color indicates the high warming phase, and blue color the low warming/cooling phase. AMO, Atlantic Multidecadal Oscillation. AO, Arctic Oscillation. NAO, North Atlantic Oscillation. PDO, Pacific Decadal Oscillation. LOD, Length of Day. NHT, Northern Hemisphere temperature. After. Wyatt & Curry (2014). **b)** Length of Day in milliseconds, daily data (grey) and long-term average smoothed (black). Data from IERS EOP. **c)** Atlantic Multidecadal Oscillation in °C detrended, monthly data unsmoothed (grey) and long-term average smoothed (black). Data from NOAA. **d)** Northern Hemisphere temperature in °C anomaly (1961–90 baseline), monthly data (grey) and long-term average smoothed (black). HadCRUT4 dataset from UK Met Office, Hadley Climate Research Unit. **e)** Arctic sea-ice extent in million km², September data (grey) and long term average smoothed (black). Data for 1962–1978 from Cea Pirón & Cano Pasalodos (2016). Data for 1979–2017 from NSIDC. **f)** Sea Level rate of change in mm/yr. Average of Church & White (2011), Ray & Douglas (2011), and Jevrejeva et al. (2014), after Dangendorf et al. (2017). Orange and blue bars are inflection points when a phase might have changed. A decrease in sea level rate is anticipated by the hypothesis.

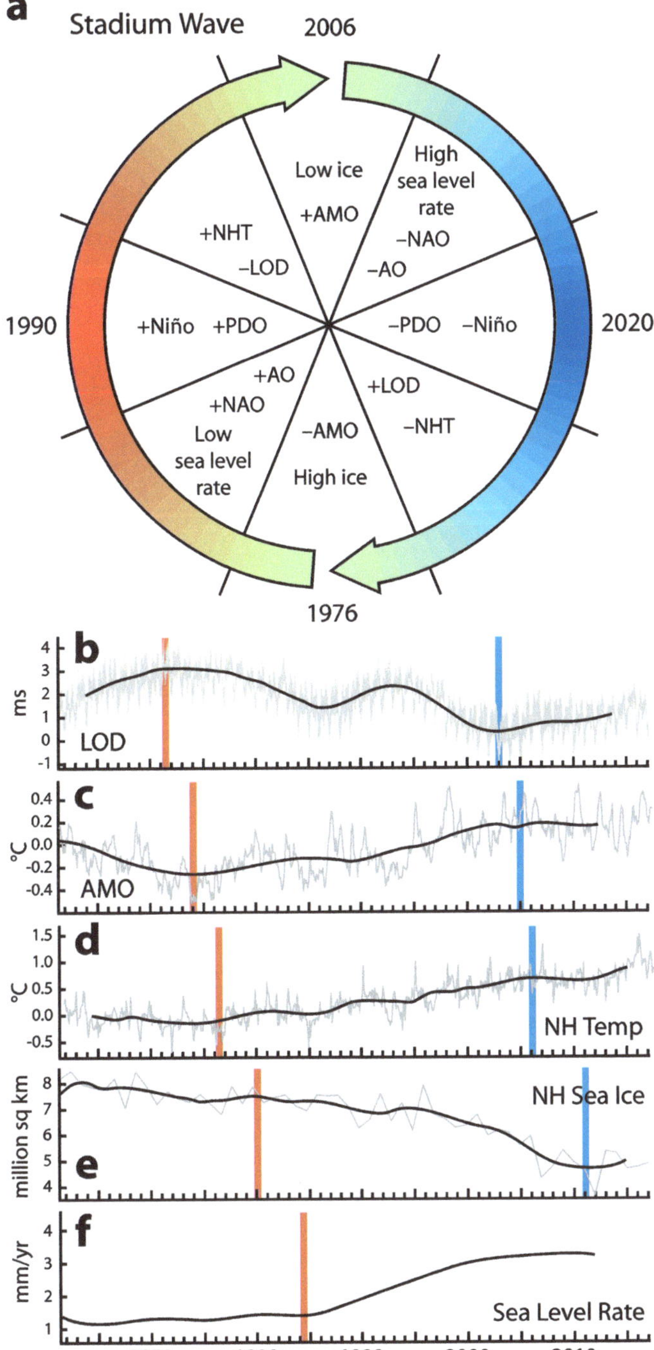

regime to a different one in a short period of a few years, with important climatic and ecological repercussions (Chavez et al. 2003). The PDO and AMO reflect the efficiency of MT through the respective ocean basins. Climatic shifts tend to occur when the oscillation alters the relative MT intensity between both basins and/or the strength of the global MT system.

This global MT system is the complex result of the geographically determined coupled atmosphere-ocean circulation in a rotating planet with its axis tilted in relation to the ecliptic, that receives most of its energy at the tropical band. As the transport intensity varies through time and space, authors focus their attention in describing its regional variability and talk about teleconnections and atmospheric bridges to try to explain what in essence are manifestations of a single very complex process (Fig. 11.6). The importance of MT for the planet's climate cannot be overstated and multidecadal changes in MT are an important and overlooked factor in climate change, as it has been erroneously assumed that over time they average to zero.

Another consequence of the c. 65-year oscillation is that c. 30 year warming and cooling phases and their associated effects on pressure, winds, precipitation, sea ice, and sea levels, should be properly accounted for by any global climate theory and models. This is clearly not the case for the leading CO_2 hypothesis of climate change and the models that support it, since the last cooling phase of the oscillation was assigned to anthropogenic aerosols, and the last warming phase to anthropogenic GHGs, leading to a completely unexpected, albeit predictable, warming pause 30 years later. Dai et al. (2015) show a step in the right direction. After deriving what they assume is the externally forced change in temperature from a large number of model simulations, they subtract it from observations. Almost all the interdecadal deviations left are explained by just two empirical orthogonal functions, one resembling the PDO, and the other similar to the AMO with a strong high latitude NH component. Inadvertently they are detecting the interdecadal changes in MT, and particularly the strong ENSO effect at the Pacific Equator, as well as

the strong MT effect on temperatures over the Arctic in winter. The problem is that any MT transport effect is assigned to external forcing (i.e. anthropogenic GHG increase) by their approach.

In theory, an internal climatic oscillation contributes only to short term changes because over several periods the effect should average to zero. However, the c. 65-year oscillation has a period long enough to have made an important contribution to Modern Global Warming. According to Chylek et al. (2014) one third of the post-1975 global warming is due to the positive phase of the AMO, and models overestimate GHG warming but compensate for it by overestimating aerosol cooling. However, the IPCC does not consider that internal variability has made a significant contribution to climate change between 1951–2010 (Fig. 9.12b). An alternative view is that a combina-

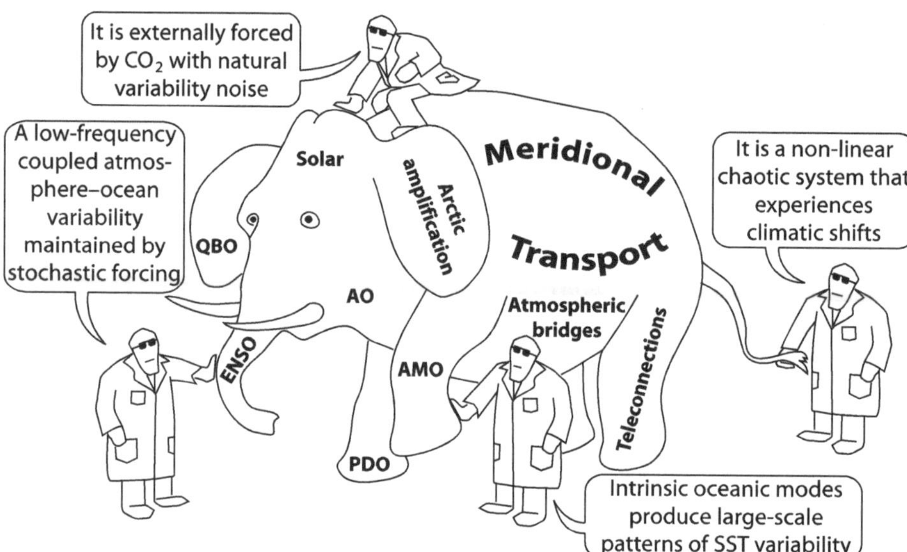

Fig. 11.6 Meridional transport is the overlooked climate factor
Meridional transport is both the elephant in the room that everybody ignores as an explaining factor for climate change, and the elephant from the Indian tale that blind people describe as a different animal when touching different parts of it.

tion of solar activity and a 65-year oscillation, if allowed an unconstrained contribution, can explain a great part of the changes in the global temperature rate of change during the 20th century (see Fig. 12.15), with residual changes attributable to the CO₂ increase and volcanic activity. That view requires that our current estimate of climate sensitivity to the different forcings is erroneous, a possibility supported by dynamical systems identification (de Larminat 2016).

The coincidence during the 20th century of two warming periods in the c. 65-year oscillation, and the Modern Solar Maximum (MSM, see Sect. 11.6 below) means that natural forcing and internal variability could have made a significant contribution to observed warming. This contribution could help explain the early 20th century warming in the absence of significant emissions, and the mid-century cooling despite increasing emissions. The natural contribution to the observed warming should come at the expense of considerably reducing the anthropogenic contribution.

11.4 The Climatic Shift of 1997–98

Between 1995 and 2005 a big shift took place in climate. The changes that took place became evident c. 2000 and were of a global scale. Solar activity changed from high at solar cycle (SC) 22 to low at SC24 (Fig. 11.7a). A predominantly Niño frequency pattern in ENSO turned into predominantly Niña, while warm water volume at the equator decreased in variability (Fig. 11.7b). Global (60°N-S) stratospheric water vapor decreased, and the tropical tropopause cooled (Fig. 11.7c; Randel & Park 2019). Global warming entered a reduced rate mode that became known as the hiatus, or "pause" (Fig. 11.7d; Fyfe et al. 2016). The tropics expanded as the Hadley cells increased their extent and intensity (Fig. 11.7e; Nguyen et al. 2013). The atmospheric angular momentum decreased causing the speed of rotation of the Earth to increase reducing the length of the day (Fig. 11.7f). The Earth's energy imbalance, the incoming solar radiation minus the total outgoing radiation, started to decrease, as shown by the ocean heat content (OHC) time derivative change in trend (Fig. 11.7g; Dewitte et al. 2019). Low cloud cover decreased (Fig. 11.7h; Veretenenko & Ogurtsov 2016;

Dübal & Vahrenholt 2021), while the albedo anomaly reached its lowest point in 1997 and started increasing (Goode & Pallé 2007), due to increasing high and middle altitude cloud cover.

The climatic shift of 1997–98 (97CS) was soon recognized (Chavez et al. 2003) as the opposite to the climatic regime shift of 1976–77 that was first described by Ebbesmeyer et al. (1991) and linked by Graham (1994) to a change in winter NH circulation. A westward shift in atmosphere-ocean variability in the tropical Pacific took place at the 97CS, characterized by a decrease of ENSO variability that coincides with a suppression of subsurface ocean temperature variability (Fig. 11.7b) and a weakening of atmosphere-ocean coupling in the tropical Pacific. The shift manifested as more central Pacific versus eastern Pacific El Niño events, and a frequency increase in ENSO, linked to a westward shift of the location of the wind-SST interaction region (Li et al. 2019).

The most conspicuous manifestations of the 97CS took place in the Arctic. Models predict that polar warming should be more intense, and tropical warming less intense, than the global average in what is termed polar amplification. Since it was evident that Antarctica, other than the western peninsula, was not warming (Zhu et al. 2021), the focus was placed on Arctic amplification. By 1995 Arctic amplification had not been clearly observed despite two decades of intense global warming (Curry et al. 1996). The situation was about to change and that year the climate of the Arctic started to shift. The cold season (October to April) >80°N surface temperature increased greatly, while summer (JJA) temperature remained unchanged (Fig. 11.7i; Danish Meteorological Institute 2021). It was accompanied by a change in most Arctic climate variables. Arctic sea-ice underwent an accelerated rate of decline (Fig. 11.7j), that by 2007 raised fears of an Arctic summer free of sea-ice in a near future. The Beaufort gyre stopped alternating its circulation regime between cyclonic and anticyclonic every few years as it had been doing since at least 1946, as the Arctic Ocean Oscillation index reflects (Fig. 11.7k; Proshutinsky et al. 2015). As a consequence, a large amount of freshwater started accumulating in the Arctic Ocean. Other changes affected Arctic winter cloud cover, that decreased from 1980–1998 (Scheweiger 2004; Wang & Key 2005) and have increased since 2005 (Wang

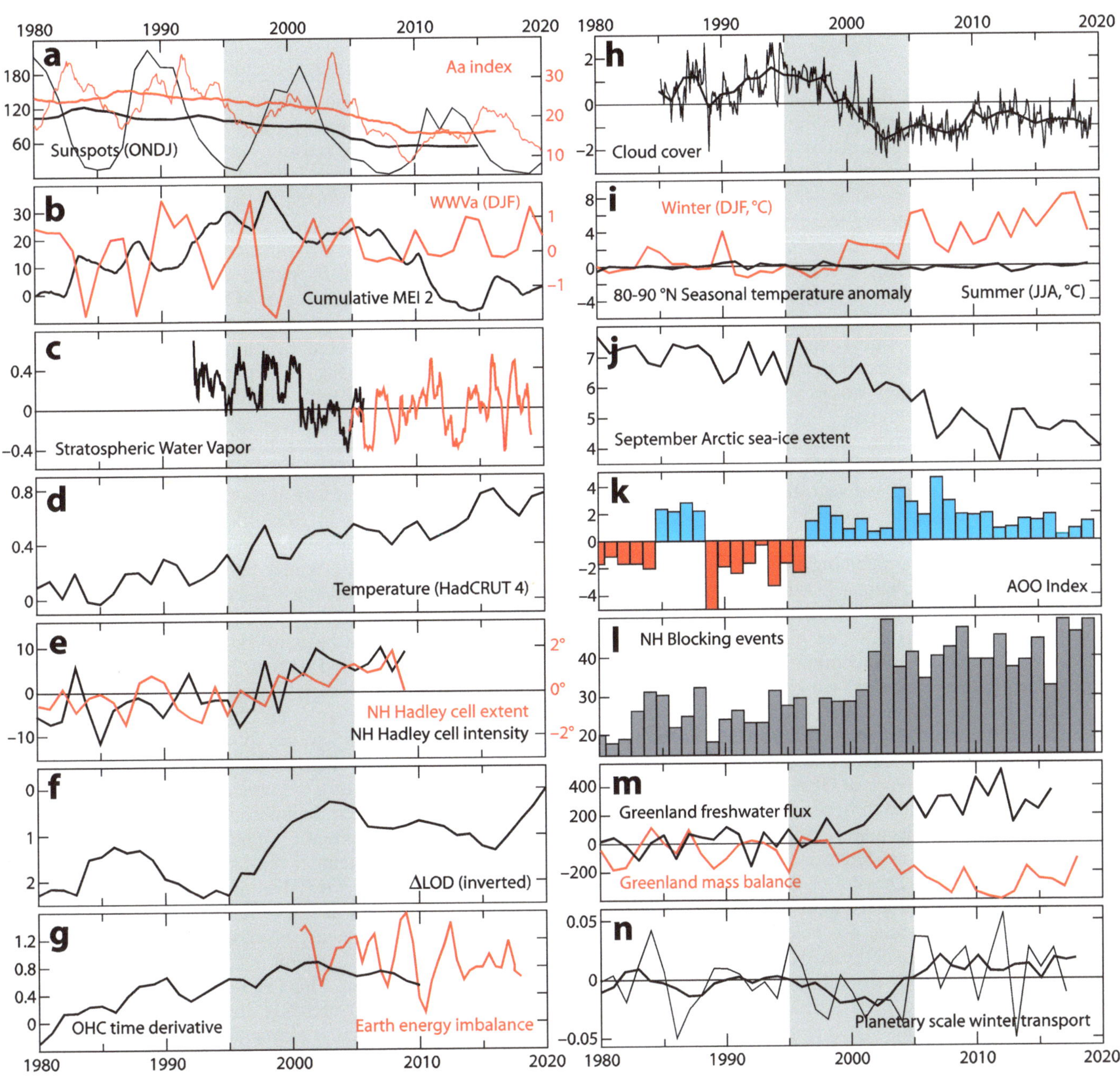

Fig. 11.7 Manifestations of the big climatic shift of 1997–98

Nearly simultaneous changes in climate related phenomena took place globally and in the Arctic between 1995 and 2005. **a)** Oct–Jan sunspots (thin black line) and 11-yr average Oct–Jan sunspots (thick black line). Solar activity decreased from high (108 sunspots 1980–1995) to low (54 sunspots 2005–2015). Data from WDC–SILSO. Antipodal amplitude (Aa) geomagnetic index 13-month average (red thin line) and 11-year average (red thick line) measuring magnetic disturbances caused mainly by the solar wind. Data from ISGI. **b)** Cumulative multivariate ENSO index v.2 changed from increasing to decreasing in 1998, indicating a shift in ENSO pattern. Data from NOAA. The change was also reflected in a strong reduction in the warm water volume anomaly variability at the equator (5°N–5°S, 120°E–80°W above 20 °C), where after 2000 negative values of −1 are no longer reached. Data in 10^{14} m^3 from TAO Project Office of NOAA/PMEL. **c)** Stratospheric water vapor monthly anomaly at 60°N–S, 17.5 km height, from solar occultation data (black line), and microwave sounder data (red line) in ppmv. The 2001 shift is confirmed by radiosonde tropopause temperature measurements. After Randel & Park 2019. **d)** Global surface average temperature anomaly in °C displaying the 1998–2013 pause in warming. From MetOffice HadCRUT 4.6 annual data. **e)** Ensemble mean annual Hadley cell intensity anomaly (in % from the mean) for the NH from eight reanalyses (black line), and ensemble mean annual-mean Hadley cell edge anomaly (in ° latitude) for the NH from eight reanalyses (red line). From Nguyen et al. 2013. **f)** Average annual change in length-of-day (ΔLOD) in ms, inverted (up is a shortening in LOD due to Earth spin acceleration). Data from IERS LOD C04 IAU2000A. **g)** Yearly increase of the 10-year running mean of the ocean heat content (black line), and annual mean Earth energy imbalance obtained as the difference between the incoming solar radiation and the total outgoing radiation (red line). Both in W/m^2. After Dewitte et al. 2019. **h)** Monthly (thin line) and yearly (thick line) 90°S–90°N cloud cover anomaly (%). Data from EUMETSAT CM SAF dataset, after Dübal & Vahrenholt (2021). **i)** 80–90°N summer (JJA, black line), and winter (DJF, red line) temperature anomaly (°C). Data from Danish Meteorological Institute. **j)** September average Arctic sea ice extent (10^6 km^2). Data from NSIDC. **k)** The Arctic Ocean Oscillation (AOO) index that reflects the alternation between sea-ice and ocean anticyclonic circulation (blue bars) and cyclonic circulation (red bars). After Proshutinsky et al. 2015. **l)** Number of NH blocking events per year. After Lupo 2020. **m)** Greenland freshwater flux (black line, km^3). After Dukhovskoy et al. 2019. Greenland ice-sheet mass balance (red line, Gt). After Mouginot et al. 2019. **n)** Winter (DJF) latent energy transport across 70°N by planetary scale waves, in PW. Thin line, annual; thick line, 5-year moving average. After Rydsaa et al. 2021.

et al. 2021), enhanced moisture transport (Nygård et al. 2020), greater atmospheric blocking frequency (Fig. 11.7l; Lupo 2020), augmented Greenland freshwater flux (Fig. 11.7m; Dukhovskoy et al. 2019), increased negative Greenland mass balance (Fig. 11.7m; Mouginot et al. 2019), augmented planetary-scale winter transport (Fig. 20n; Rydsaa et al. 2021), enhanced downwelling longwave radiation flux (DLR; Wang et al. 2021), and a greater OLR flux (Fig. 11.8) at the ToA, to cite some of the changes.

NOAA OLR satellite data (Liebmann & Smith 1996) shows that starting in mid-1996, seasonal mean Arctic OLR anomaly initiated a great change that lasted until late-2005 (Fig. 11.8, grey box). Seasonal variability (7-month average OLR anomaly; Fig 11.8, thin grey line) decreased markedly, while the monthly mean OLR at the ToA increased by 6 W/m². But the increase was not equally distributed along the year, the cold months (Nov–Apr) increased their monthly mean OLR by 8 W/m², while the summer months (JJA) increased it by only 3 W/m². The 97CS has created a much bigger winter energy deficit in the Arctic. Usually, changes in OLR and RSR are negatively correlated, so they partly cancel each other (Dewitte et al. 2019), but this is not true of the Arctic in winter, when there is little or no incoming SR. Planetary TOR is largest during the NH winter (see Fig. 10.2; Carlson et al. 2019) and became larger at the 97CS. Cao et al. (2020) have shown that the increase in Arctic OLR at the 97CS was associated with the expansion of the Hadley cell causing a simultaneous increase in OLR at the subtropics. Coincident with the increase in Arctic OLR, the Earth energy imbalance started to decrease, and the ocean started to accumulate less heat, as determined by the OHC time-derivative (Fig. 11.7f; Dewitte et al. 2019). At the same time global warming entered a hiatus (Fyfe et al. 2016).

The 97CS in the Arctic was celebrated as further confirmation of the CO₂ hypothesis of global warming with renewed calls to end CO₂ emissions (IPCC: AR5 Synthesis report 2014). However, there was no proper explanation for the changes that took place at that time (and not be-fore) in the Arctic and globally. While the Arctic summer climate was affected, the most prominent changes took place during the cold season. The changes in stratospheric water vapor (Randel & Park 2019) or the Beaufort gyre (Proshutinsky et al. 2015) remain unexplained, and the decrease in the rate of global warming received multiple explanations (which is equivalent to none), but without explanation of why it happened at that time. In contrast, a reorganization and increase of MT is supported by most, if not all, the phenomena involved, providing a powerful explanation. Randel et al. (2006) show that the decrease in stratospheric vapor (Fig. 11.7c) and cooling of the tropical tropopause, together with simultaneous changes in ozone are consistent with an increase in the mean tropical upwelling of the BDC, the stratospheric MT.

The start date of the 1995–2005 climate shift is difficult to determine, and it is further obscured by the great El Niño that took place in 1997–98. However, many of the altered trends in climatic parameters, like albedo (Goode & Pallé 2007), Arctic winter temperature (Fig. 11.7i), or changes in Greenland mass balance and freshwater flux (Fig. 11.7m; Dukhovskoy et al. 2019; Mouginot et al. 2019) can be traced back to 1997. The climatic shift of 1976–77 also coincided with an El Niño that started in the early summer of 1976, it is thus appropriate to date the most recent climatic shift to 1997–98. A few years after the 97CS the climate appears to have reduced the accelerated rate of climate change that it displayed in the 1995–2005 period, suggesting the climate has settled into a new regime. Arctic sea-ice decline has decreased greatly. The minimum summer extent took place in 2012, and in 2021 there was more September sea-ice extent than in 2007, 14 years before. Arctic OLR has also stabilized in a higher level and global low cloud cover in a lower level. Arctic extreme moisture events, stratospheric water vapor, and atmospheric blocking events all appear to have reduced their rate of increase. The 1976–77 climate shift also involved changes in winter NH circulation (Graham 1994), and atmospheric angular momentum (Marcus et al. 2011).

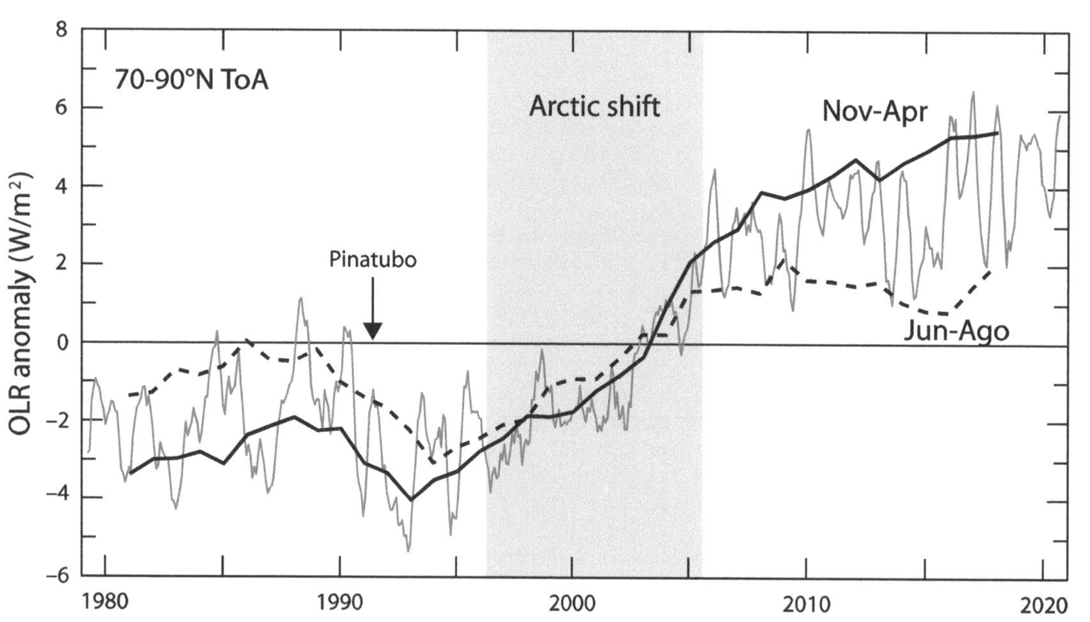

Fig. 11.8 Arctic region outgoing longwave radiation change Thin grey line, 7-month average of the monthly mean OLR anomaly in W/m² from interpolated OLR NOAA dataset. Thick black line, 5-year average of the cold season (Nov–Apr) mean. Thick black dashed line, 5-year average of the summer (JJA) mean. Grey box highlights the Arctic shift in OLR between mid-1996 and late 2005. The time of the Pinatubo eruption is indicated. Data from KNMI explorer (field=noaa_olr).

A less evident solar activity change and its correlation with stratospheric temperature changes has been proposed as forcing for the 1976–77 climate shift (Powell & Xu 2011).

The climate undergoes multidecadal shifts that cannot be explained, reproduced, or predicted by climate models. These climate shifts not only affect the internal variability of the climate system, but also the Earth radiative properties and thus they constitute a source for global climate change that is not accounted for in the IPCC list of natural and anthropogenic forcings that have contributed to observed surface temperature change over the period 1951–2010 (IPCC AR5 Synthesis report 2014). All the described climatic changes of the 97CS (Figs. 11.7 & 11.8), and those shared by the 1976–77 climate shift, have in common that they are either affected by changes in MT towards the winter pole, or they cause changes in it. Winter MT is a fundamental property of the climate system and at the center of the observed changes. It has the potential to be a major climate determinant with an importance even greater than changes in non-condensing GHGs.

11.5 Meridional transport modulation of global climate

The Holocene prior to 1950 underwent episodes of profound and abrupt climate change that significantly altered the vegetation of entire ecozones at times of very modest GHG changes. This Holocene variability was reviewed in Chap. 4. Additionally, in the distant past (e.g. the early Eocene, see Fig 10.1) the planet experienced at times warm, equable climate conditions, with reduced LTG and seasonality, characterized by the absence of year-round freezing conditions at the poles, when CO_2 levels were perhaps only twice current levels. Current climate models cannot reproduce those changes and conditions without resorting to unrealistic settings. Thus, besides GHGs and volcanic eruptions, other poorly identified climate change mechanisms must be at work. The transport of energy from the equator to the poles is a good candidate. The LTG is a defining feature of the different global climates the Earth has experienced during the Phanerozoic Eon, and energy is transported from the equator to the poles along, and as a function of, that gradient. The riddle that makes climate models choke on the equable climate problem is known as the equable climate paradox. Energy is moved to the poles along the LTG to keep them warmer than they would be otherwise, yet the warmer they are the less energy should move along the gradient and the colder they should be. The answer to the riddle is counterintuitive. The more energy that is moved to the poles in winter, the more energy the planet loses and the colder it becomes. The equable climate poles were kept warm through the winter not because of more heat transported there, but because the most abundant GHG in the planet, water vapor, must have created a permanent fog and cloud cover at the poles in winter that greatly reduced heat loss.

Since the 97CS, more energy is being transported to the Arctic in winter (Fig. 11.7i–n), and more energy is being lost at the Arctic ToA (Figs. 10.4 & 11.8). The result is a reduction in the Earth's energy imbalance and OHC increase (Fig. 11.7g; Dewitte et al. 2019) and a pause in global warming (Fig. 11.7d; Fyfe et al. 2016), temporarily interrupted by the great 2015 El Niño (Lean 2018). There

is so far no indication that the current period of increased MT and reduced warming has come to an end, and if past is precedent, the current climate pattern could last until mid-2030s. It is then evident that part of the warming that took place in the 1976–97 period, erroneously attributed to anthropogenic factors (see Fig. 9.12), was due to a reduction in MT. Also, Arctic amplification is a completely misunderstood phenomenon. Arctic amplification, especially in winter, results in planetary cooling, not warming. The more winter Arctic amplification, the less the planet will warm.

The great complexity of the ocean-atmosphere coupled global circulation with all its modes of variability, oscillations, teleconnections, and modulations, is just the manifestation of a single underlying cause, the transport of energy from its climate system entry point to its exit point. Mass (including water) is transported as a consequence of energy transport, directly or indirectly. Every phenomenon in the climate system is powered by the energy received from the sun as it moves towards its exit point. Due to the insolation gradient the exit point is on average at a higher latitude than the entry point, so the defining aspect of the climate system is the MT of energy. Every aspect of the ocean-atmosphere coupled global circulation must be analyzed in terms of its effect on energy MT. ENSO is a heat pump extracting energy from the equatorial Pacific (Sun 2000). The AMO is a readout of Atlantic surface MT, its positive phase indicating reduced MT resulting in heat accumulation. Teleconnections can be explained as MT changes in strength through different pathways, as prevailing temperature and pressure gradients change.

The general belief that horizontal transport of energy has a net zero effect, and the net radiation flux at ToA should not be affected by it, does not bear close scrutiny. If we had a region in the planet where incoming solar energy is zero or very low, and where its atmospheric GHG content is much lower, then transporting more energy there could only result in more energy loss at that ToA without compensating gain elsewhere. The energy flux of the planet would be altered by that change in transport. And we have not one but two of these regions at the poles, resulting in having one in cold season during most of the year. MT towards the North Pole is more important than towards the South Pole for geographic reasons (see Chap. 10). There is ample evidence that Arctic winter MT increased at the 97CS (see Sect. 11.4). The increased MT resulted in an increased energy loss at the ToA (Fig. 11.8), that affected the Earth's energy imbalance (Fig. 11.7g; Dewitte et al. 2019). Changes in MT constitute a climate forcing that is not accounted for in any of the climate change assessment reports published by the IPCC to date.

MT can be divided in two components. The BDC through the stratosphere, that has a shallow branch in the lower stratosphere and a deep branch in the mid-upper stratosphere. The other is through the troposphere with atmospheric and oceanic contribution (Fig. 11.9). Both components are variably coupled in time and space (Kidston et al. 2015). At the equatorial zone there is coupling through deep-convection and the ascending branch of the BDC (Collimore et al. 2003), and at high latitudes through the PV. The downward coupling in the mid-latitudes is complex and variable by longitude (Elsbury et al. 2021). The coupling is mainly exerted by changes in stratospheric temperature gradients and the response of the thermal

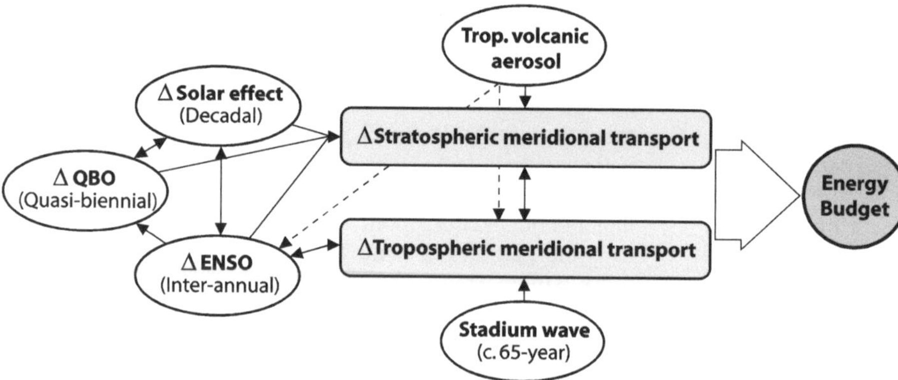

Fig. 11.9 Meridional transport diagram
In light-grey rounded rectangles the two components of meridional transport, with their known modulators in white ovals. Black arrows indicate coupling or modulation. Dashed arrows, the indirect effect from tropical volcanic eruptions on tropospheric meridional transport and ENSO. The effect on the energy budget of the Earth's climatic system is implied by the changes in energy transfer from a GHG-rich tropical region to a GHG-poor polar region. See text for references.

wind balance, which affects the strength of the mean zonal circulation, and the position and strength of tropospheric jets, eddies, and storm tracks (Kidston et al. 2015). The upward coupling depends on changes in convection and wave generation. As a consequence, the coupling is stronger in winter when temperature contrasts and wave generation at the troposphere are higher, and temperature gradients at the stratosphere are deeper.

Stratospheric MT is modulated by factors that alter its LTG (ozone, solar activity, and volcanic aerosols), or the zonal wind strength (QBO), as they determine the level of planetary waves transmission that powers the stratospheric transport. ENSO is part of the tropospheric MT and is determined by its conditions, but it is also a modulator of stratospheric transport, affecting the strength of the BDC (Domeisen et al. 2019), and thus participates in the stratosphere-troposphere MT coupling. Whether the QBO has an effect on ENSO is not known, but all other interactions between these three modulators (solar effect, QBO and ENSO) of stratospheric MT have been documented (see Sect. 10.5; Labitzke 1987; Calvo & Marsh 2011; Salby & Callaghan 2000; Taguchi 2010). The stadium-wave represents the coordinated sequential change affecting interconnected parts of the tropospheric MT (Wyatt & Curry 2014). It constitutes a strong multidecadal oscillation in MT. Whether it is due to internal variability or externally forced, or paced, is currently unknown. The importance of this multidecadal oscillation on climate variability cannot be overstated.

In figure 11.10 the global effect of the stadium-wave on MT since 1900 is analyzed. The 1912–2008 period has been subjectively divided into three phases of 32 years. Although the analyzed MT phenomena do not shift simultaneously (hence the name stadium-wave), the phases so defined describe periods of alternating prevailing conditions in MT well. The northern annular mode or Arctic oscillation (AO; Fig. 11.10a grey line), is the leading mode of extratropical circulation variability in the NH (Thompson & Wallace 2000). It is weaker than its SH counterpart, leading to a debate about its physical reality (Sun & Tan 2013). To act as a North-South seesaw of atmospheric mass exchange between the Arctic and mid-latitudes (a true annular mode), there must be a correlation between its three centers of action—the Arctic, Atlantic and Pacific sectors. The Arctic–Atlantic correlation is strong and constitutes the NAO. The Arctic–Pacific linkage is weaker, but it was found that the Aleutian Low and the Icelandic Low since the mid-1970s display a negative correlation from one winter to the next, that is stronger by late winter (Honda & Nakamura 2001). This Aleutian–Icelandic see-

saw appears to depend on the propagation of stationary waves and varies in strength with the changes in PV strength (Sun & Tan 2013). By calculating the Jan–Feb cumulative AO (Fig. 11.10a grey line) we can see that until c. 1940 positive AO values (i.e. strong vortex conditions) prevailed, but in the 1940–1980s period negative AO values were preponderant, only to change back afterwards. The Aleutian–Icelandic seesaw confirms the changes in PV strength with its 25-year moving correlation (Li et al. 2018; Fig. 11.10a black line).

The strength of the PV negatively reflects the magnitude of the mass and heat exchange between the Arctic and mid-latitudes. Over the Atlantic sector, the AMO measures SST anomalies. As is to be expected, positive AMO values reflect warm water accumulation in the North Atlantic under reduced MT and strong PV conditions (Fig. 11.10b black line). The NAO, which is part of the AO, represents the sea-level pressure dipole over the Atlantic, and when analyzing its detrended and cumulative value, it shows a great similarity with the AO, with some influence from AMO SSTs (Fig. 11.10b grey line). The prolonged NAO trends that last several decades are unexplainable with current models, that clearly do not reflect multidecadal MT regimes. Models consider NAO indices white noise without serial correlation (Eade et al. 2021). Without properly representing MT, climate models cannot explain climate change. Over the Pacific sector, the PDO also reflects SST anomalies. A positive PDO shows accumulation of warm water over the equatorial and eastern side of the Pacific. This is indicative of reduced MT, which moves heat out of the equator and towards the western Pacific boundary so the Kuroshio current can move it northward. The detrended cumulative PDO values (Fig. 11.10c) show that the phases of increased or decreased Pacific MT roughly coincide with those of the Atlantic. Climatic and ecological shifts in the Pacific identified in 1925, 1946, 1976 and 1997 (Mantua & Hare 2002) coincide with times when the PDO shifts from predominantly positive to negative or back (Fig. 11.10c black dots).

The MT is carried out by the meridional circulation and increases in atmospheric MT imply decreases in zonal circulation, since more meridional circulation transports more energy and momentum poleward. The atmospheric circulation index is a cumulative representation of the yearly anomaly in zonal (E–W) versus meridional (N–S) air-mass transfer in Eurasia (Klyashtorin & Lyubushin 2007). It shows that multidecadal periods, when the NH PV has been stronger and MT over the Atlantic and Pacific sectors has been lower (grey areas in Fig. 11.10), have been characterized by prevalent anomalies of zonal-type

Fig. 11.10 Multidecadal climate variability and meridional transport
a) Black line, Aleutian Low–Icelandic Low seesaw 25-year moving correlation as a proxy for polar vortex strength. After Li et al. 2018. Grey line, cumulative winter (Dec–Feb average) Arctic Oscillation index. 1899–2002 AO index data from DW Thompson. Dept. of Atmos. Sci. CSU. Thompson & Wallace 2000. **b)** Black line, 4.5-year average of the Atlantic Multidecadal Oscillation index. Data from NOAA unsmoothed from the Kaplan SST V2. Grey line, cumulative 1870–2020 detrended cold season (Nov–Apr average) North Atlantic Oscillation index. Data from CRU, U. East Anglia. Jones et al. 1997. **c)** Cumulative PDO. 1870–2018 detrended annual average cumulative PDO index from HadISST 1.1. Data from NOAA. Black dots mark the years 1925, 1946, 1976 and 1997 when PDO regime shifts took place (Mantua & Hare 2002; see Sect. 11.4). **d)** Black line, zonal atmospheric circulation index, cumulative anomaly. After Klyashtorin & Lyubushin 2007. Grey line, 1900–2020 inverted detrended annual ΔLOD. Data in ms from IERS. **e)** Detrended 1895–2015 annual global surface average temperature, 10-year averaged. Data from Met Office HadCRUT 4.6. **f)** Dashed line, 8.2–16.6 years band-pass of the monthly mean total sunspot number. Data from WDC–SILSO. Grey line, 6.6–11 years band-pass of the monthly AMO index. Black line, inverted 20-year running correlation of the band-pass sunspot and AMO data. Black dots as in c, showing their position with respect to solar minima.

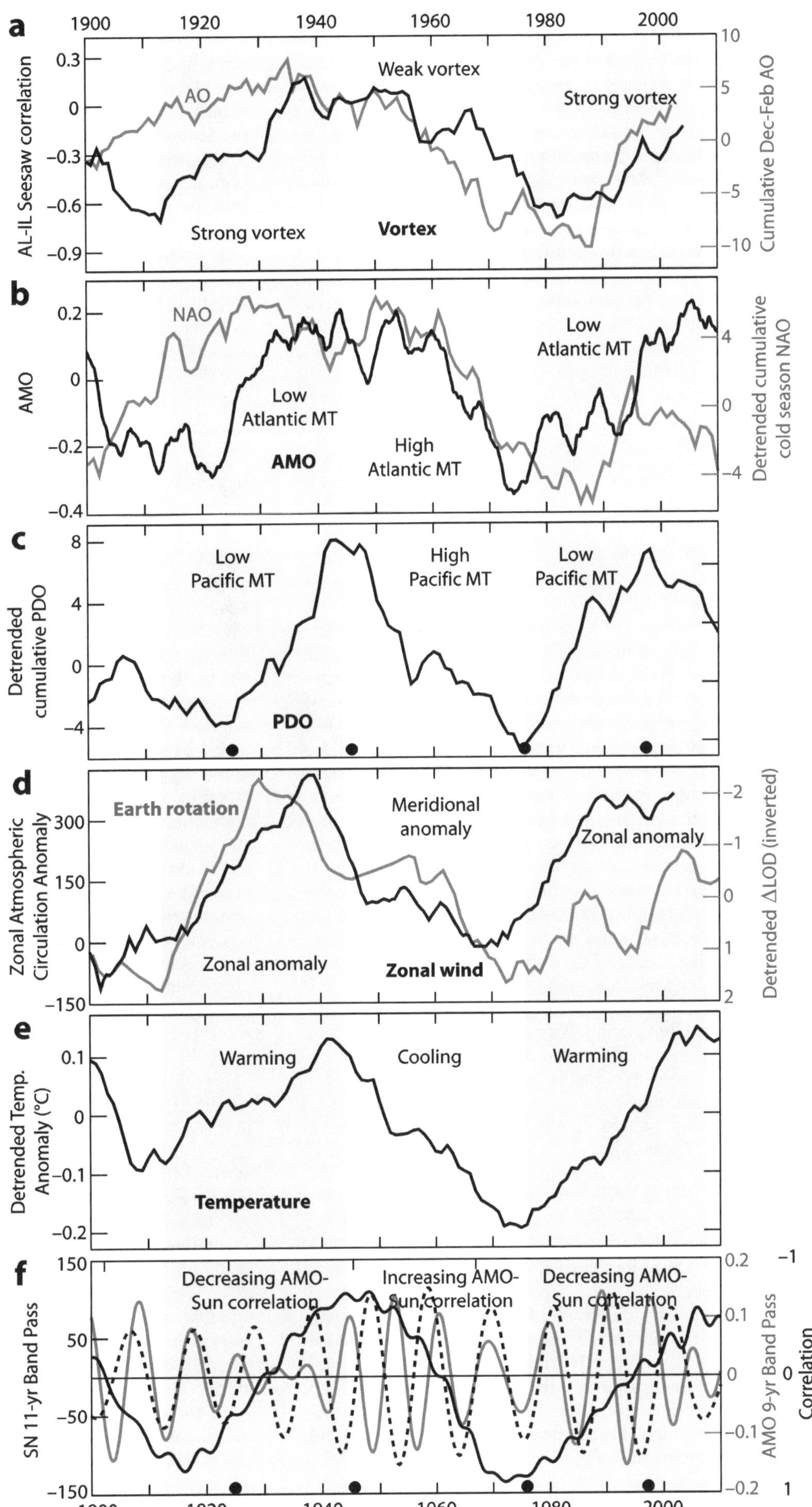

(Fig. 11.10d black line), while periods of weaker PV and higher MT present predominant anomalies of meridional-type, as should be expected. The effect that these persistent changes in predominant atmospheric circulation patterns have on the transfer of momentum between the atmosphere and the solid Earth–ocean cause changes in Earth's rotation speed, measured as changes in the length of day (see Sect. 10.6). Periods of increasing zonal circulation correlate with an acceleration of the Earth and a decrease in ΔLOD while periods of decreasing zonal circulation correlate with a deceleration of the Earth and an increase in ΔLOD (Lambeck & Cazenave 1976; Fig. 11.10d grey line). The rotation of the Earth is also affected by tidal and core–mantle interaction forces, changes in atmospheric momentum are the main seasonal and interannual factor. Changes in ΔLOD support the existence of multidecadal periods of reduced MT and increased zonal circulation (grey areas in Fig. 11.10), alternating with periods of increased MT and reduced zonal circulation. Changes in the rate of rotation of the Earth integrate global changes in atmospheric circulation that support the global effect of MT changes.

Multidecadal changes in MT are the cause of the multidecadal oscillation known as the stadium-wave, because SST changes in the AMO and PDO are a response to changes in the global atmospheric circulation. Sea surface transfer of energy and moisture to the atmosphere are highest at ocean basin's extra-tropical western boundaries (Yu & Weller 2007). From there, poleward energy transport is carried out almost exclusively by the meridional circulation of the atmosphere (Fasullo & Trenberth 2008). A reduction in atmospheric meridional circulation and the corresponding increase in zonal circulation mean less poleward energy transport, and since annual incoming energy is near constant and ocean heat transport is only partially dependent on wind-driven circulation, more heat accumulates at each latitudinal band, but particularly at mid-latitudes. The heat accumulation on land and in the sea surface as a result of the reduction in MT, produces the stadium-wave effects and an increase in the global temperature. When the global average surface temperature anomaly is detrended, periods of reduced (increased) MT correspond to warming (cooling) with respect to the trend (Fig. 11.10e). Considering this evidence, the general belief (incorporated into models) that the 1940–1975 hiatus was due to an increase in aerosols, and the 1975–2000 warming is due to anthropogenic factors should be re-evaluated.

The causes behind the multidecadal stadium-wave changes in MT are unknown. The c. 65-year oscillation is non-stationary. Proxy reconstructions indicate that the AMO had a shorter periodicity and less power during the LIA and a longer periodicity and more power during the Medieval Warm Period (MWP; Chylek et al. 2012; Wang et al. 2017). A modulation of ENSO by solar activity was shown in Sect. 10.4, and a modulation of the changes in Earth rotation, and therefore of changes in atmospheric momentum, by solar activity was shown in Sect. 10.5. One possibility is that internal variability and external solar forcing are responsible for the current periodicity and strength of the stadium-wave. Alternatively internal variability in MT might be responding to the warming trend imposed by anthropogenic and natural causes, mainly the increase in solar activity associated to the MSM. Interestingly, the four multidecadal climatic shifts recognized in

the Pacific (Mantua & Hare 2002) took place at or right after a solar minimum (Figs. 11.10c & f, dots and dashed line), and the two grey areas and middle white area in figure 11.10 representing alternating MT regimes span three solar cycles between solar minima. It has been shown that the Holton–Tan effect, that relates the tropical QBO phase to the strength of the PV through planetary wave propagation, is stronger at solar minima (Labitzke et al. 2006), and that the Holton–Tan effect weakened substantially during the 1977–1997 period of reduced MT (Lu et al. 2008). This implies that during winter at solar minima the stratospheric tropical-polar coupling, and the stratospheric-tropospheric coupling are stronger, and they might constitute an appropriate time for a coordinated shift in MT strength that takes effect during the ensuing solar cycle. We should see if future MT and climate shifts also take place at or immediately following solar minima.

If solar minima are the times when MT shifts occur, one interesting correlation may provide an explanation for the cause of the c. 65-year oscillation pacing. The AMO has a 9.1-year strong frequency peak that is also found in the PDO (Muller et al. 2013). This frequency is clearly discernible in a 4.5-year averaged AMO index (Fig. 11.10b black curve). The origin of this clear AMO frequency variability has not been adequately researched, but Scafetta (2010) suggests that it has a lunisolar tidal origin. The difference in frequency between this reported 9.1-yr tidal cycle and the 11-yr solar cycle is such that they change from correlated to anti-correlated (i.e. constructive to destructive interference) with a periodicity that not only matches the AMO, but is exactly synchronized to it (compare black curves in Fig. 11.10b & f). One can speculate that a constructive or destructive interference between the effect of oceanic and atmospheric tides on the tropospheric component of MT and the effect of the solar cycle on the stratospheric component of MT might result in the periodical change in MT strength that produces the observed climatic shifts. In support of this hypothesis two intrinsic components of c. 4.5 and 11 years are found in the Fourier analysis of the daily NAO autocorrelation series (Álvarez–Ramírez et al. 2011). The 11-year component is phase synchronized to the solar cycle except during solar minima, indicating that NAO predictability increases with solar activity, and became strongly anti-correlated during the 1997 solar minimum, when the 97CS took place. A c. 65-yr climate oscillation that depends on solar activity would explain both the changes in intensity and periodicity over the last centuries as solar activity has been changing. Its 20th century intensity and periodicity is the result of the MSM and should change if solar activity changes over several decades.

It can be argued that when a detrended quasi-periodic oscillation in the climate system displays little amplitude variation from one period to the next, over time-multiples of its period should average to zero. Similarly, other factors known to affect MT, like the QBO and ENSO average to zero in similar or shorter timeframes. However, AMO reconstructions show that its values and amplitude have increased greatly over the last two cycles, since about 1850 (Moore et al. 2017). This change in the c. 65-year oscillation suggests that MT is important in Modern Global Warming, since it coincides with the strong melting of glaciers and increase in sea-level rise that started around 1850 and precedes the strong increase in CO_2 emissions after 1945 (Boden et al. 2009). There is an external factor that affects MT and does not average to zero even in very long timeframes. It is solar activity, that is known to present centennial and millennial cycles (see Chap. 8). There has been a long-standing scientific debate about whether there is an important effect of solar activity on climate. Sunspot records show that the average number of sunspots increased by 24% from the 1700–1843 to the 1844–1996 period (Fig. 11.11). Solar variability is clearly involved in MT variability (see Sect. 10.4 to 10.6). The effect that solar variability has on MT, and the effect that MT has on the planet's energy imbalance (Figs. 11.7g & 11.8) settles the controversy on the solar activity effect on climate.

11.6 The search for a controversial sun-climate connection

Sunspots, known from antiquity, were first observed with a telescope in 1610. Variable stars were discovered in 1638. However, it was not until 1801 when William Herschel was the first to publish a connection between sunspots, the hypothesized variable nature of the Sun, and climate: *"I am now much inclined to believe, that openings (sunspots) with great shallows, ridges, nodules, and corrugations, instead of small indentations, may lead us to expect a copious emission of heat, and therefore mild seasons."* (Herschel 1801). He then initiated 220 years of controversy over the role of solar variability in climate change when in his article he proceeded to link the relative abundance of sunspots to the price of wheat in England, as a climate proxy; a result now seen with skepticism. In 1843 Heinrich Schwabe published his observation that changes in the number of sunspots followed on average an 11-year period. In the following decades a multitude of reports were published correlating sunspots to all sorts of weather and climate variables. Poor statistics and the inconsistency of the reported correlations, together with a

lack of mechanism, led to the dismissal of the sun-climate connection hypotheses after the 1930s.

During the first half of the 20th century the solar output was believed to be constant. Instrumental measurements at the surface were too variable and imprecise to detect a significant change in solar output until satellite measurements started in the 1970s. Nevertheless, solar physicists were convinced from the 1950s onward that the sun was a variable star. Together with a revision of the sunspot record and the acceptance of the radiocarbon record as a proxy for past solar activity prior to sunspot records, solar variability led to a revival of the sun-climate connection hypotheses in the mid-1970s.

Several authors connected the coldest periods of the LIA with the Maunder and Spörer Grand Solar Minima (Eddy 1976). The revival of the sun-climate connection was short-lived, after the satellite data showed that Total Solar Irradiance (TSI) varied only by 0.1 % over the solar cycle, and the climate effects proved to be elusive once more and difficult to explain. By the mid-1980s this area of research was again discredited, after the confirmation of global warming and its explanation by the CO_2 hypothesis. Sun-climate connection supporters had trouble explaining a significant effect from such a small change in TSI without a great sensitivity to the small solar changes. Besides, these changes did not correlate well with surface temperature, and appeared to run opposite between 1980–2000 (solar cycles SC21–23) when temperatures increased, and solar activity decreased. It is a mistake to assume that changes in TSI are the only way solar activity affects climate; and in fact, solar activity during the 1980–2000 period was above or at the 300 year average (Fig. 11.11).

If anything has been made clear in these past 220 years is that the sun-climate connection, that paleoclimatic evidence strongly supports, cannot be a simple linear regression or it would have been discovered long ago. And insufficient evidence for a complex relationship should not lead to rejection or dismissal, but to more research. Frie-

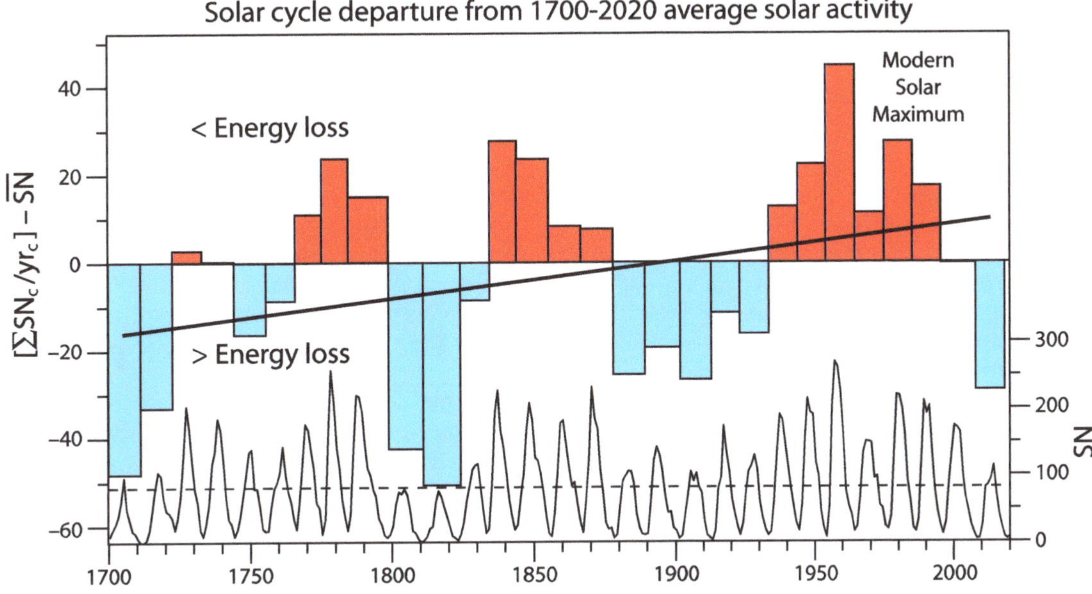

Fig. 11.11 Solar cycle departure from 1700–2020 average solar activity
Top, the 320-yr average yearly sunspot number is subtracted from the sum of the yearly sunspot number for all the years of each cycle divided by the number of years in the cycle. Bar width is proportional to cycle duration. Thick line is the linear trend. The Modern Solar Maximum (1935–2005), during which no solar cycle was below average, is indicated. **Bottom**, yearly mean total sunspot number. Dashed line the 1700–2020 average. Data from WDC–SILSO.

drich von Hayek, in his 1974 Nobel lecture "The Pretense of Knowledge," warned that *"there may thus well exist better 'scientific' evidence for a false theory, which will be accepted because it is more 'scientific', than for a valid explanation, which is rejected because there is no sufficient quantitative evidence for it."*

The best argument in favor of an important sun-climate connection is the one uncovered by Roger Bray (1971): the correspondence between past solar activity and past climate at the centennial and millennial timescales is just too good to ignore. A correspondence that builds up as one studies Paleoclimatology with an open mind, and that has taken authors like Rohling et al. (2002) to say: *"In view of these findings, we call for an in-depth multidisciplinary assessment of the potential for solar modulation of climate on centennial scales."* Magny et al. (2013) affirm: *"On a centennial scale, the successive climatic events which punctuated the entire Holocene in the central Mediterranean coincided with cooling events associated with deglacial outbursts in the North Atlantic area and decreases in solar activity during the interval 11700–7000 cal BP, and to a possible combination of NAO-type circulation and solar forcing since ca. 7000 cal BP onwards."* Hu et al. (2003) state: *"Our results imply that small variations in solar irradiance induced pronounced cyclic changes in northern high-latitude environments. They also provide evidence that centennial-scale shifts in the Holocene climate were similar between the subpolar regions of the North Atlantic and North Pacific, possibly because of Sun-ocean-climate linkages."* Just those three articles have 50 authors who are among the most respected in paleoclimatology. The sun-climate connection is a common understanding among many paleoclimatologists, just not discussed and often ignored outside the subfield to avoid supporting a hypothesis that competes with the dominant one. General circulation climate models just have no room for a sun-climate connection (see Fig. 9.12c).

After early reports of a relationship between the solar cycle and the QBO (Berson & Kulkarni 1968), it was unfortunate that just when the global warming-CO_2 hypothesis became politically prominent, the first solid evidence of the sun-climate connection was discovered by Karin Labitzke (1987). She cracked a 186-year old nut. In the words of Richard Kerr (1987), senior writer for Science magazine: *"It suddenly occurred to her to plot on one graph the winter time temperature at about 23 kilometers over the North Pole during west equatorial winds against sunspot number. Perhaps, she reasoned, solar activity could only make itself felt under one set of conditions... When Labitzke plotted only pole temperatures from the QBO's west phase, a strong correlation became clear. The more sunspots, the warmer the winter time temperature, due to vortex breakdowns and the subsequent intrusion of warmer air."* In 1974 Colin Hines had proposed a solar-climate mechanism that would act at middle and high latitudes in winter, based on anomalous absorption or reflection of planetary waves. Karin Labitzke's discovery that the PV is under solar control depending on QBO, ENSO and volcanic aerosol conditions, provided the tropospheric effect that Hines' mechanism required. Labitzke's 1987 article has been cited many times, but she passed away in 2015 without receiving the recognition she deserved for demonstrating the first solid sun-climate connection after 186 years of continuous attempts. She was unlucky that

the dominant hypothesis does not have a place for her finding. Thirty more years of data have confirmed Labitzke's finding (see Fig. 10.13).

The change in TSI from solar cycle minimum to maximum (c. 0.17 W m^{-2} globally averaged; Gray et al. 2010) is too small to even account for the c. 0.14 °C surface warming observed for the solar cycle, and some regional responses are clearly above what should be expected based on energy changes alone, so it is generally accepted that there must exist some amplification mechanism, and two of them have been confirmed by observations and to a certain degree reproduced in models. The more direct one involves the amplification of the solar cycle signal after it heats the surface and, thus, has been termed the "bottom-up" mechanism (Fig. 11.12a). It would act in the tropical region strengthening the Hadley and Walker circulations through increased evaporation, moisture transport towards the Intertropical Convergence Zone and precipitation there, reducing tropical cloud cover and acting as a positive feedback producing more upwelling of cold water in the eastern Pacific in a La Niña-like pattern, further reducing cloud cover and allowing stronger surface heating (Meehl et al. 2009).

A more indirect mechanism acts through differential production of stratospheric ozone at the tropics (c. 3%), and ozone warming by the larger variation in UV radiation (c. 7%) during the solar cycle and has been termed the "top down" mechanism (Haigh 1996; Fig.11.12b). The signal starts as a large temperature change in the tropical stratosphere (c. 1.5 K) that results in anomalous zonal winds in the subtropics to maintain thermal wind balance. This affects the propagation of planetary waves in winter and then the signal follows two routes. The polar route affects the strength of the PV, propagating down to the troposphere and the surface by inducing NAO anomalies. The tropical route would affect the strength of the ascending branch of the BDC, that determines the temperature of the tropical tropopause through the adiabatic cooling of ascending air (Gray et al. 2010). Multiple mechanisms have been proposed that couple the stratospheric changes to tropospheric changes, in many cases affecting the strength of the Walker and Hadley circulation.

Although the "bottom-up" and "top-down" mechanisms together could be responsible for the observed climate changes associated to the 11-yr solar cycle, the result is a forcing too small to produce large climate changes unless it is maintained for a long time or there are other mechanisms not properly accounted for that could alter the energy flux at the ToA.

11.7 The Winter Gatekeeper hypothesis

From the solar effects on climate reviewed in sections 5.11, 6.9, and 8.7, two crucial asymmetries can be deduced, one from solar activity, and the other one from its effect on climate. The first is that solar variability is asymmetrical. Strongly active solar cycles still have low activity years, while weakly active cycles lack high activity years. During multi-decadal periods of persistently elevated solar activity, like the 70-year long MSM (1935–2004), the elevated solar activity is interrupted every 5.5 years by a solar minimum. Even though Schwabe minima during a prolonged maximum are shorter and have more activity than minima outside it, they still limit how high

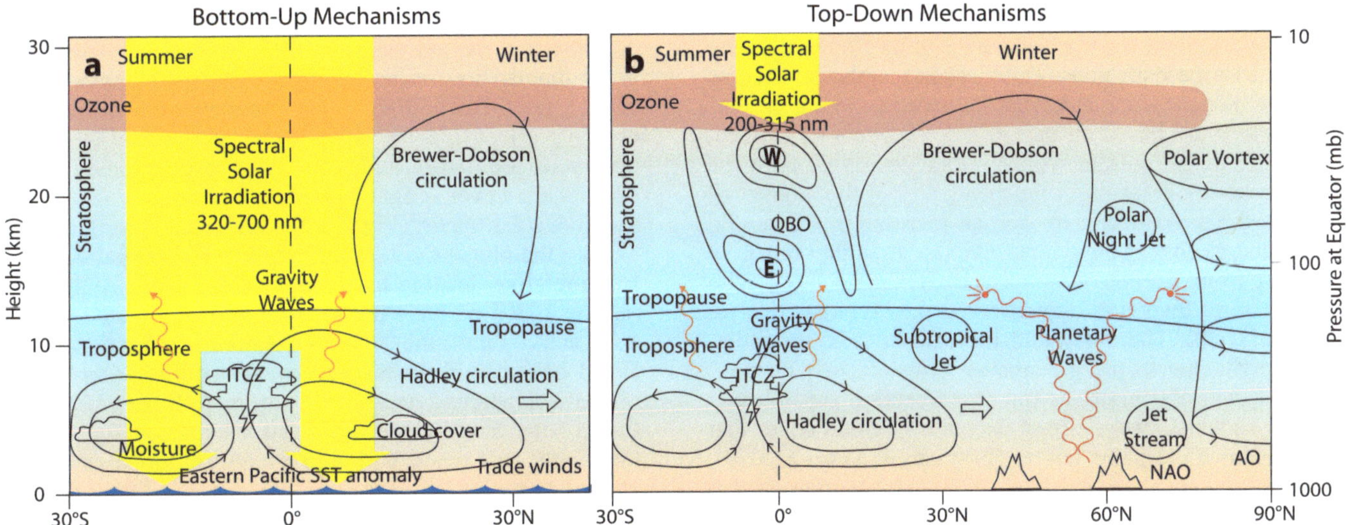

Fig. 11.12 Summary of proposed amplification mechanisms for solar variability effects on climate
a) Bottom-up mechanisms. Only the tropical Pacific during the boreal winter is represented. In less clouded subtropical areas, peak solar activity increases evaporation. The enhanced moisture is transported by trade winds to convergence zones increasing precipitation, and strengthening Walker (not shown) and Hadley circulations. Intensified trade winds increase equatorial ocean upwelling reducing equatorial SST and driving a cloud reduction, through enhanced subsidence, that acts as a positive feedback similar to La Niña conditions, allowing more solar radiation to reach the surface. The Hadley circulation expands poleward increasing the area of the tropics. After the solar peak, the eastern equatorial surface transitions to higher SSTs a couple of years later. The effect is stronger during the boreal winter.
b) Top-down mechanisms. The NH and the tropical SH during boreal winter are represented. UV solar irradiance, variable with the solar cycle, is responsible for the ozone layer and its temperature gradients. Different types of tropical waves (orange) originating from convection, are responsible for the creation and maintenance of the quasi-biennial oscillation (QBO), that together with the Brewer–Dobson circulation is responsible for the poleward transport of ozone. The position of the subtropical Jet Stream is determined by the Hadley circulation, while the strength and position of the Jet Stream and the Polar Night Jet depend on the strength of the polar vortex. Depending on stratospheric conditions, planetary-scale Rossby waves (red) can be deflected during the winter, causing stratospheric warming and a weakening of the polar vortex. The polar vortex determines the winter state of the Arctic Oscillation (AO), which strongly influences the North Atlantic Oscillation (NAO). Solar activity level, through its effect on stratospheric conditions, influences Northern Hemisphere winter weather far more than its small change in irradiance suggests. The ITCZ, Intertropical Convergence Zone, is the climatic equator.

solar activity can be above the long-term average. By contrast, the fall in solar activity during a grand solar minimum is larger, as the 11-yr Schwabe cycle becomes undetectable and there are no 11-yr solar maxima in sunspots. According to Eddy (1976), during the Maunder Minimum a decade could pass between two sunspot observations, indicating that solar activity becomes low for the entire duration of the SGM, and peaks in activity are sometimes insufficient to produce noticeable sunspots. The opposite to the 70-year long Maunder Minimum (1645–1715) is the 70-year long MSM (1935–2004), but due to solar asymmetry the first is readily recognized as a SGM, while the second arouses controversy. This asymmetry is intuitively understood when we realize that we have identified and named multiple SGM from the past, but not a single grand maximum, and even the observed MSM is disputed (Muscheler et al. 2005). If we define solar grand maxima as a symmetrical deviation from long-term average solar activity towards the opposite extreme from a SGM, they do not exist. The monthly average sunspot number for the 1749–2021 period is 81.6 sunspots per month (WDC–SILSO sunspot data). During the MSM the average was 108.5, only a 33% increase. During the Maunder Minimum the decrease must have been close to a 100%. The asymmetry in solar variability, implies that activity can decrease from the average a lot more than it can increase. This has very important repercussions for the solar effect on climate and explains why the paleoclimate record readily registers the abrupt climatic effect of prolonged low

solar activity (see Sect. 4.8 & table 4.1), while the climatic effect of prolonged high solar activity is not so obvious. In addition, as reviewed in section 8.7, the evidence suggests that reduced solar activity induced by a low in one of the solar cycles cannot be compensated by a high in other cycles. Cyclical solar activity changes would then be subtractive among cycles and not additive, introducing an additional low bias asymmetry in solar activity variability over time.

The second asymmetry pertains to the climatic effect of persistently altered solar activity. The IPCC only recognizes the effect of linear changes in TSI. As changes in TSI are very small (0.1% over the solar cycle), and average to nearly zero over the medium-term, the IPCC (see Fig. 9.12) and climate models see no significant role for solar forcing in climate change. However, as we have seen in Chap. 10, there is a non-linear solar activity effect that acts during cold seasons modifying the Earth's rotation speed. These changes reflect differences in the activation of the meridional atmospheric circulation responsible for the amount of heat transported to the poles during the cold season, when the insolation and temperature latitudinal gradients are steepest. This cold-season specific effect, tied to the strength of the PV, is seen in climate reanalysis and observations that indicate that it affects atmospheric and oceanic phenomena, including the AO and NAO, blocking events frequency, zonal wind strength, the subpolar gyre strength, and the North Atlantic winter storm track. Unlike the effect of changes in TSI that is year-round and global, this non-linear cold season-specific effect must result in important changes in the amount of heat directed to the dark pole. This heat is mainly destined to exit the planet, as ice has great thermal insulation properties (igloos are based on that principle). Most of the heat delivered to the pole in winter is radiated as OLR in the long clear-sky polar night independently of the amount of CO_2 in the atmosphere. In fact, an increase in CO_2 could favor energy loss at the winter pole due to higher radiative cooling from warmer than surface CO_2 molecules under temperature inversion conditions (van Wijngaarden & Happer 2020). The delay and increase in effective radiation height due to the greenhouse effect (GHE) matters little in the polar night, as the heat retention capability is greatly reduced by the strong decrease in cloud cover that accompanies the polar winter (Eastman & Warren 2010), and the low absolute humidity of the winter polar atmosphere.

The winter asymmetric effect of solar activity on climate establishes solar variability as the most important long-term gatekeeper of the great amount of heat that leaves the planet at the poles every cold season, the main heat sink for the planet (see Sect. 10.3; Fig. 10.4). This role is termed here "the Winter Gatekeeper hypothesis" (WGK-h), and could conceivably dominate over GHE changes, that produce a relatively small forcing over the entire planet, if the energy involved is sufficiently large during long periods of persistently altered solar activity, like the LIA and MSM.

The WGK-h (Fig. 11.13) states that the level of solar activity is one of several factors that determine zonal winds strength and thus the propagation properties of the winter atmosphere to planetary waves. Poleward and upward wave propagation controls PV strength, that is the main modulator of heat and moisture MT towards the winter pole. Winters of high (low) solar activity promote stronger (weaker) zonal circulation reducing (increasing) MT, leading to a colder (warmer) Arctic winter, warmer (colder) mid-latitudes winter, warmer (colder) tropical band due to reduced (increased) BDC upwelling, and lower (higher) energy loss at the winter pole. The difference in energy loss at the winter pole is large enough to greatly affect the climate of the planet when solar activity is consistently high or low for several consecutive solar cycles (i.e. decades).

The annual effect of solar activity on winter MT is variable, because ENSO, the QBO, stratospheric volcanic aerosols, and other climatic conditions also affect the strength of zonal winds, and thus the climatic result on a given year can be opposite to the one expected from solar activity alone. As MT depends on atmospheric and oceanic transport, it responds not only to the stratospheric signal that involves solar activity, but also on a tropospheric one that involves the ocean (Fig. 11.13). This double variability leads to the inconsistency of solar effects that have plagued solar-climate studies. Solar signal is part of a complex system that determines the strength of winter MT, but its long turnover rate (decadal to multidecadal) has a strong cumulative effect at times.

The WGK-h implies that the solar effect on climate has a profound bias towards cooling, as revealed by paleoclimatic studies. During multidecadal periods of high solar activity, like the MSM, over the course of a solar cycle a few years of high solar activity are followed by a few years of low solar activity, while during multidecadal periods of low solar activity, like the Maunder Minimum, all years display low solar activity. Although some winters, ENSO, the QBO and volcanic eruptions might induce the opposite effect to solar activity, their action is either temporary or oscillatory and tends to zero over time. Solar activity, however, might be reduced during several consecutive solar cycles (Fig. 11.11), and the resulting energy loss by the planet during most winters for several decades is cumulative, leading to global cooling. Since SC16 during the 1920s we have not seen a prolonged period of low solar activity until the 97CS. High solar activity during the MSM has very likely favored energy conservation by the planet contributing to global warming. Since 1997, reduced solar activity has promoted a larger energy loss at the winter pole, contributing to a hiatus in global warming known as the pause. If correct, the pause is likely to continue until solar activity becomes high again.

The mechanisms for the solar effect on climate are already known. Differential heating of ozone by UV, creating a temperature gradient in the stratosphere that affects zonal wind strength, and through planetary waves the PV, propagating to the troposphere through thermal wind balance and stratosphere-troposphere coupling (Lean 2017). However, the WGK-h proposes that the long-term climatic effect of solar variability is mediated through its effect on the MT of heat towards the winter pole, and that the stronger global climatic effects are due to cumulative enhanced energy loss at the winter pole during prolonged periods of low solar activity. The main role for solar variability in climate is to act as a winter gatekeeper, promoting energy conservation during years of high solar activity and allowing a higher energy loss during years of low solar activity. Geographic determinants establish that the solar energy gatekeeping role is more variant and can have a larger effect in the North Atlantic winter storm track than at the South polar cap.

The WGK-h explains the strong paleoclimatic effect of periods of prolonged low solar activity, like the LIA, and its alternation with warmer periods like the MWP or Modern Global Warming according to a c. 1000-yr solar cycle revealed by solar and climate proxies (Marchitto et al. 2010; see Fig. 8.3), known as the Eddy Cycle. It also explains why the North Atlantic region behaves as a climate variability hotspot. Many paleoclimatic features like Bond events, Dansgaard–Oeschger events, Heinrich events, the MWP or the LIA manifest primarily and more prominently in the North Atlantic region, that constitutes a preferred corridor for MT. Several of these features, like Bond events (Bond et al. 2001), the MWP and the LIA have been linked to concurrent long-term changes in solar activity by many authors. Changes in MT constitute perhaps the main climate change factor in sub-Milankovitch frequencies, and solar activity is one of its variability main determinants, but not the only one.

The WGK-h explains the WACC pattern linked to solar activity (Kobashi et al. 2015), when the Arctic warms and the northern continents cool during the winter due to changes in MT. It explains why Arctic sea-ice suffered a great reduction at the 97CS and not during the previous decades of global warming. It also explains Arctic amplification since 2000 as a cold season phenomenon, with little summer amplification as a manifestation of low solar activity. The hiatus in global warming is a reduction in the

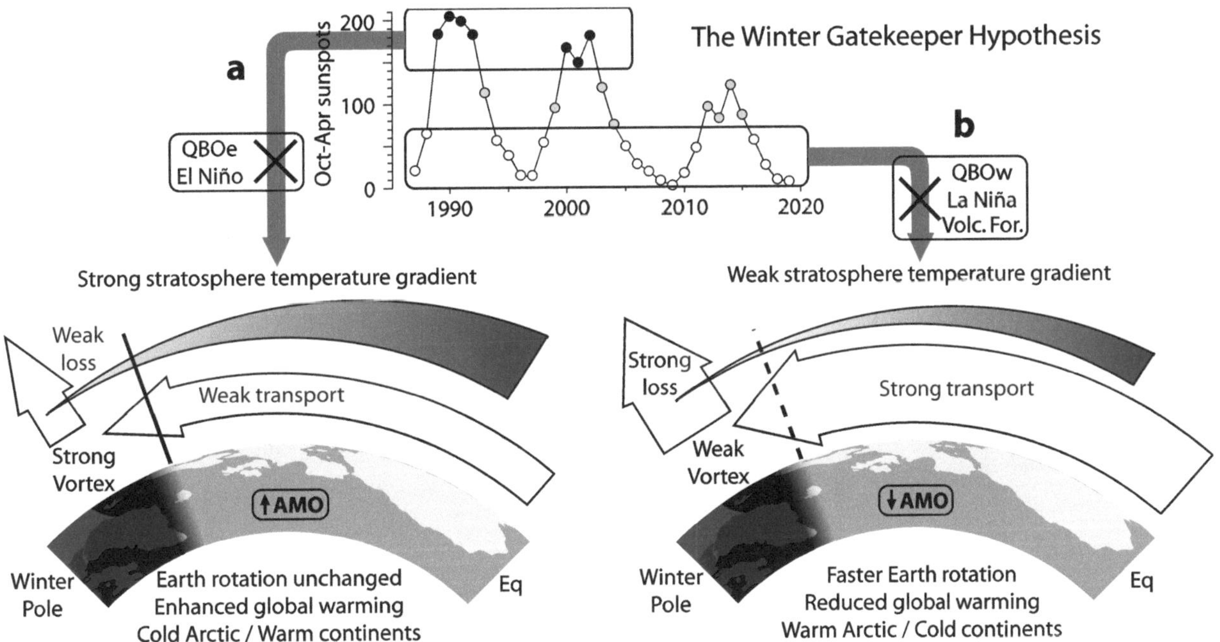

Fig. 11.13 The Winter Gatekeeper hypothesis of solar variability effect on climate
a) High solar activity winters promote a strong stratosphere latitudinal temperature gradient through increased ozone and enhanced ozone heating by higher UV radiation. The effect on zonal wind from thermal wind balance alters planetary wave propagation leading to a strong polar vortex that reduces meridional transport and heat loss at the winter pole. The effect on the stratosphere temperature gradient from high solar activity is opposed by easterly QBO and El Niño conditions, while tropospheric meridional transport is strongly affected by the c. 65-year oscillation, here represented over the Atlantic by the AMO, that denotes a weaker transport when it is changing to higher values (heat accumulation in the North Atlantic). The climatic effect is enhanced global warming and a cold Arctic/warm continents winter pattern. **b)** Low solar activity winters promote a weak stratosphere latitudinal temperature gradient due to lower UV radiation, leading to a weak polar vortex that increases meridional transport and heat loss at the winter pole. The effect on the stratosphere temperature gradient from low solar activity is opposed by westerly QBO, La Niña conditions and volcanic aerosol forcing. The tropospheric meridional transport is strong when the c. 65-year oscillation is in descending phase, and the AMO is changing to lower values (heat reduction in the North Atlantic). An increased transport manifests in a faster Earth rotation as zonal winds decrease, and less angular momentum resides in the atmosphere. The climatic effect is reduced global warming and a warm Arctic/cold continents winter pattern.

time derivative of global average surface temperature or the time derivative of OHC (Dewitte et al. 2019). The WGK-h explains how the reduction in solar activity contributed to the 97CS that caused the hiatus. And if the hiatus in global warming can be attributed at least in part to the reduction in solar activity, according to the WGK-h, then the 20th century warming must also have a high solar activity component during the MSM. The relative contributions to modern warming from solar activity and anthropogenic factors might be difficult to separate, but if high MT is capable of stopping or greatly reducing the rate of warming during the Pause, despite a constant increase in GHG forcing, it is reasonable to suppose that both might have contributed similarly to the previous warming.

The WGK-h explains why the semi-annual component of the changes in the Earth's speed of rotation, manifested as ΔLOD changes, responds to changes in solar activity (Le Mouël et al. 2010). It also explains why the multidecadal trend in ΔLOD changes correlates with climatic changes (Lambeck & Cazenave 1976; Mazzarella 2013). The WGK-h also explains the solar modulation of ENSO (see Sect. 10.4). As low solar activity promotes a stronger MT, it induces La Niña conditions at the equatorial Pacific, probably in response to a higher BDC upwelling through tropical stratosphere-troposphere coupling. This is the opposite effect to tropical volcanic eruptions that produce a weaker MT and stronger PV, inducing El Niño conditions in the equatorial Pacific probably through

a reduction in tropical upwelling by the opposite mechanism.

The evidence supporting the WGK-h is profuse. Paleoclimatic evidence supports a linkage between prolonged low solar activity and planetary cooling (see Chap. 5 & 8), particularly or more prominently in the North Atlantic region (Bond et al. 2001; Jiang et al. 2015). Low solar activity is also linked to a WACC pattern indicative of a more active MT during boreal winter (Kobashi et al. 2015; Porter et al. 2019). The linkage between the poleward transport of heat and the transport of momentum by the atmosphere manifests as changes in zonal wind circulation that affect the speed of rotation of the planet and are measured by ΔLOD. As the semi-annual component in ΔLOD varies according to solar activity (Le Mouël et al. 2010; Barlyaeva et al. 2014), MT towards the winter pole must respond also to solar activity. As required by the hypothesis, stratospheric planetary wave amplitude is modulated by solar activity (Powell & Xu 2011), with low solar activity resulting in increased planetary wave amplitude that should promote a stronger BDC and weaker PV.

The biennial oscillation (BO) of the PV (Fig. 11.14a) results from the solar cycle modulation of the bimodality in the QBO and its interaction with the strong polar annual variation (Baldwin & Dunkerton 1998; Salby & Callaghan 2006; Christiansen 2010). This oscillation changes the PV from a strong configuration one year to a weak configuration the next. At the 1976–77 climate shift, the bimodality in the QBO and the BO weakened resulting in a predominantly strong vortex phase (Fig. 11.14a; Christiansen 2010). At the 97CS the bimodality in the QBO and the BO changed again to a stronger-bimodality weaker-vortex phase. These changes coincide with a 1977–97 phase when the effect of the QBO on the strength of the PV by the Holton–Tan mechanism weakened considerably (Lu et al. 2008). The QBO at 50 hPa, and extratropical winds at 54°N and 10 hPa broke their correlation while becoming more predominantly westerly (positive) as shown by their cumulative value (Fig. 11.14b; Lu et al. 2008), effectively uncoupling the QBO and the PV for the period 1977–97. The stronger PV that coincided with the high activity SC21 and 22 resulted in the slight cooling trend in winter Arctic temperature (Fig. 11.14c, grey area), while the weaker PV that coincided with the medium activity of SC20 and 23 resulted in warming trends in the winter Arctic (Fig. 11.14c, white areas). The relationship between the strength of the PV and winter Arctic surface temperature is very clear.

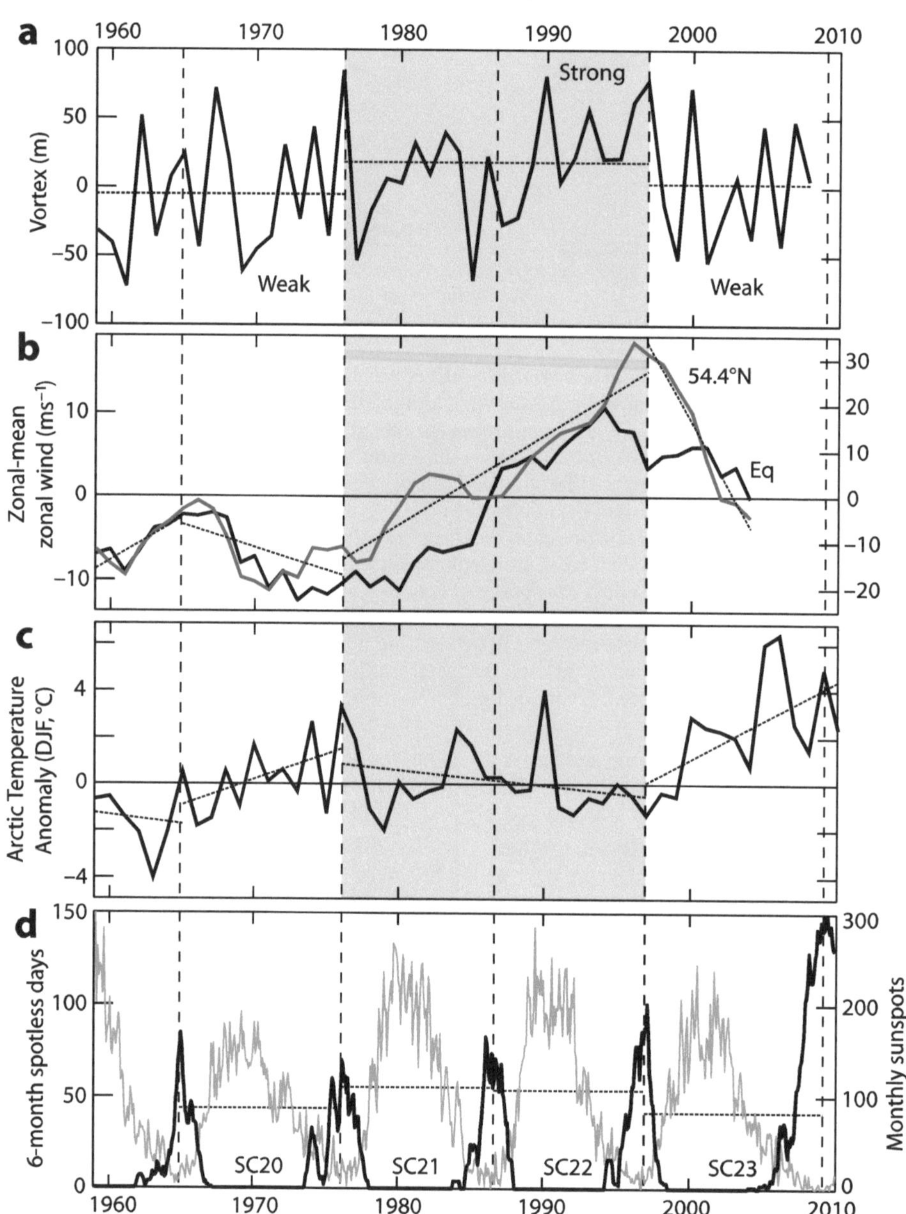

Fig. 11.14 Polar vortex, zonal wind, Arctic temperature and the solar cycle
a) October–March mean vortex at 20 hPa, as the leading principal component of the mean geopotential height north of 20°N empirical orthogonal function, from the NCEP/NCAR reanalysis dataset. Higher values denote a strong vortex for that winter. Circa 1976 a regime shift took place from a weak vortex displaying bimodality to a stronger vortex with unimodality. The opposite shift took place c. 1997. Dotted lines are average values for the periods separated by 1976 and 1997. After Christiansen 2010. **b)** Black line, cumulative of the 3-year averaged November–March average zonal-mean zonal wind at the equator at 50 hPa (right scale). Grey line, cumulative of the 3-year averaged November–March average zonal-mean zonal wind at 54.4°N at 10 hPa (left scale). Dotted lines, linear trends for the cumulative 54.4°N data for the periods 1959–65, 1965–76, 1976–97 and 1997–2004. Data from Lu et al. 2008. **c)** Winter (December–February) mean temperature anomaly calculated from the operational atmosphere model at the European Center for Medium-range Weather Forecast for the +80 °N region. Dotted lines, linear trends as in b) except the last period ending in 2010. Data from the Danish Meteorological Institute. **d)** Black line, sunspot spotless days in a running 6-month window. Grey line, monthly sunspots. Horizontal dotted lines are the average monthly number of sunspots for each solar cycle (SC). Vertical dashed lines mark the solar minima. Data from WDC–SILSO.

Fig. 11.15 Arctic winter temperature is solar modulated
Black curve, smoothed 10.7 cm solar flux as a proxy for solar activity. The third order polynomial least-squares fit has been calculated with all the data available since 1947 to reduce the border effect in the graphed data. Data from the Royal Observatory of Belgium STAFF viewer. Red curve, winter (December–February) mean temperature anomaly calculated from the operational atmosphere model at the European Center for Medium-range Weather Forecast for the +80 °N region, with a third order polynomial least-squares fit. Data from the Danish Meteorological Institute.

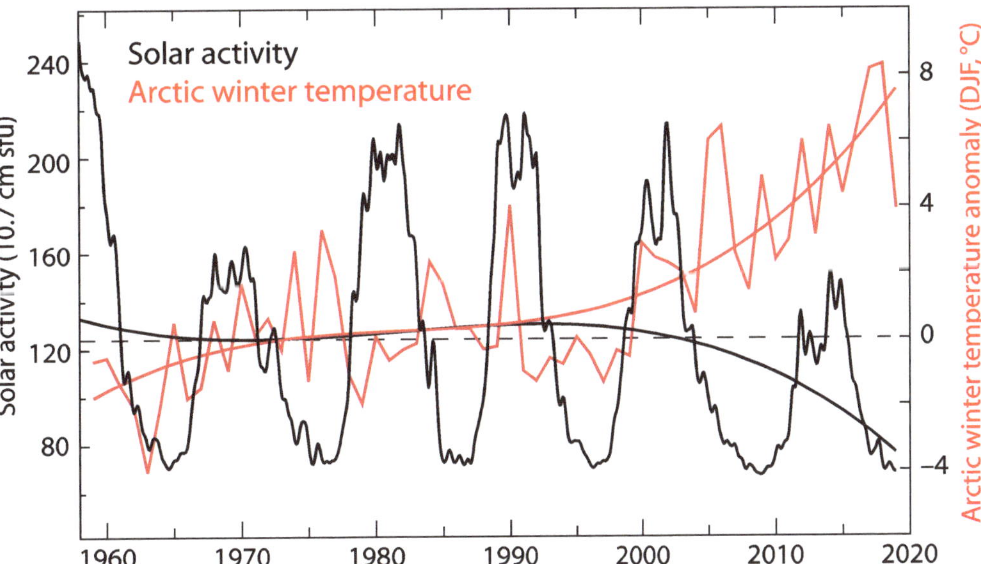

As required by the WGK-h, seasonal patterns of 80–90 °N temperature anomaly display very important changes over time (Fig. 11.7i). Arctic summer and winter temperature anomaly did not display any significant long-term deviation from average during the 1970–99 period, indicating a surprising lack of effect from the global warming experimented by most of the planet at the time, and in stark contrast to the polar amplification predicted by theory. Starting in 1997, Arctic summer temperature anomaly displays a small decrease of about half a degree (Fig. 11.7i), while Arctic winter temperature anomaly experiments a huge increase reaching +8 °C average during the 2017–18 winter (Fig. 11.15). The heat responsible for this winter temperature increase is transported to the Arctic from lower latitudes (see Sect. 10.3). It is paradoxical and contrary to the prevalent view, that Arctic warming was less pronounced during the rapid global warming period of the 1980s and 1990s and is more pronounced in coincidence with a period of reduced warming that has been termed the pause or hiatus in global warming. This apparent contradiction can be resolved if solar activity regulates the amount of heat directed to the poles during winter. According to this WGK-h, the increase in winter poleward heat transport responsible for the temperature increase in the Arctic in that season is contributed by the persistent decrease in solar activity since 2004. The negative correlation between long-term solar activity and Arctic winter temperature is clear (Fig. 11.15).

When solar activity is low the effect of the equatorial stratosphere on the PV (Holton–Tan effect) is stronger and the PV becomes anomalously weaker. Thus, at solar minimum the solar effect is maximum. The biggest positive deviations from trend in winter Arctic temperature usually take place during solar minima (Fig. 11.15). The climatic shifts of 1976 and 1997 taking place at the solar minimum constitutes evidence of the WGK-h. The 1925 shift also took place at the SC15–16 minimum, and the 1946 shift at the SC17–18 minimum (Fig. 11.10c & f; Mantua et al. 1997). Solar activity level between minima determines the level of equatorial-polar atmospheric coupling and the Arctic climate over that cycle (Fig. 11.14d). Since regime shifts in atmospheric circulation and climate appear to take place at solar minima, over the following years the activity of the solar maximum determines if a

shift takes place. If the activity is similar to the prior cycle there is no shift, if it is markedly different the shift starting at the solar minimum is confirmed. A predictable result is a higher frequency of climate phases that span two solar cycles, like the 1976–1997 period, providing an explanation for the repeated finding of a 22-year solar signal in climate proxies, like the bidecadal drought rhythm in the Western US (Cook et al. 1997), or tree-ring width in the Arctic (Ogurtsov et al. 2020) and Southern Chile (Rigozo et al. 2007).

Numerical model experiments on the impact of stratospheric zonal wind changes that reproduce solar signals in observations, when extended over 50 years of weak or strong stratospheric polar-night jet show that the persistent weakening of the PV leads to a very strong surface cooling over high and middle northern latitudes (Kodera et al. 2016). The cooling, consistent with LIA features, results from dynamical processes and increased albedo and can be explained by a change in the UV part of the solar spectrum, despite very little total energy change.

A strengthening of the lower branch of the BDC under low solar activity conditions was proposed by Kodera & Kuroda (2002) and reproduced in numerical experiments by Matthes et al. (2010). It has been observed in reanalysis (Rao et al. 2019), where the BDC shows a significant strengthening at the 97CS. Observationally, the shift involved an important cooling of the tropical tropopause proportional to the decrease in stratospheric water vapor measured at the time (Fig. 11.7c; Randel & Park 2019). The cooling at the tropopause is due to an enhanced adiabatic cooling from the increased tropical upwelling that results in a deeper dehydration of the air entering the stratosphere and a reduction in tropical ozone due to enhanced transport (Hood 2003; Gray et al. 2010). The changes in ozone amplify the solar signal. However, the pattern of solar induced changes in the tropical tropopause layer are complex and at solar minimum the tropopause cooling takes place over the eastern Pacific/Atlantic while at solar maximum there is cooling over the Indian Ocean/western Pacific (Krüger et al. 2008), indicating that the changes in the BDC are not only in intensity, but also in the location of the dominant upwelling locations. The solar-induced surface temperature pattern changes at these locations are opposite to the tropopause temperature changes, as there is surface warming in the eastern Pacific and cooling over the Indian and central Pacific oceans during low solar activity (Kodera et al. 2016). White et al. (2003), demonstrated that the upper ocean tropical diabatic heat storage budget experiments decadal anomalies phase-locked to the solar cycle, but they are almost an order of magnitude too strong to be caused by changes in solar irradiation. The evidence supports the idea that the increase in MT required by the WGK-h at times of low solar activity takes place in part through a strengthening of the BDC that increases the energy flux from the ocean to the troposphere, and then to the stratosphere, through increased atmospheric upwelling at certain regions producing many of the effects assigned to the "bottom up" mechanisms of solar effect. The regulation of the BDC by solar activity is also strongly modulated by the QBO phase (Matthes et al. 2010) since it depends on planetary wave propagation.

There are many consequences of the solar-induced changes in the Arctic. The WGK-h requires an increase in cold-season Arctic OLR when decadal solar activity decreases. This increase was observed at the 97CS (Fig. 11.8). The energy lost since 1997 at the poles must have contributed to the pause in global warming. At the same time the strong wintertime warming in the Arctic has little effect on the regional cryosphere, as Arctic winter temperature is c. 25 °C below freezing on average. Meanwhile, the modest summer temperature decrease has a stabilizing effect on summer sea-ice that displays a pause in extent change since 2007 (see Fig. 13.11). We have the apparently paradoxical result that the big increase in yearly averaged Arctic temperature, that is being publicized as evidence of hefty Arctic amplification, coincides with a pause in Arctic summer sea-ice extent change that might even lead to a modest increase in summer sea-ice over the present solar cycle (SC25, 2020–c. 2031). Unless the Arctic temperature increase is seasonally analyzed, it is difficult to understand what is happening, but then it becomes clear that the Arctic amplification is not actually an amplification of global warming. Arctic winter warming is a strong indication that the climatic effect of solar variability is being profoundly misunderstood, and the contribution from the MSM in solar activity to modern global warming is much larger than accounted for in IPCC reports and current climate models. It doesn't escape this author that if the WGK-h were accepted, the urgency to combat climate change would greatly diminish, something that is unlikely to happen soon. A clear prediction from this hypothesis is that Arctic winter temperature anomaly will start to decrease when a new active solar cycle takes place. This could happen with solar cycle 26, predicted to increase in activity c. 2032 (see Fig. 13.6).

11.8 An outline for planetary climatology

The planet has been in the Late Cenozoic Ice Age for the past 34 Ma because it is hemorrhaging heat at the winter pole from two gigantic cooling radiators. In the early Cenozoic, heat loss at the winter pole was limited by an intense cloud-, fog-, and water vapor-GHE during the polar night. For unknown reasons the climate started a transition from the late Triassic–early Cenozoic warm mode to the late Cenozoic cool mode (Frakes et al. 1992), in what appears to be a c. 150-Myr cycle (see Fig. 9.4). The water vapor reduction associated with the cooling is what drove the increasing heat loss at the winter pole. With the cooling the LTG increased, and the MT responded by transporting more heat to the winter pole, intensifying planetary heat loss. Through this positive-feedback the planet slowly cooled from the surface to the bottom of the ocean, driving a reduction in CO_2 levels. If it is indeed a 140-Myr cycle, an inverse process should take the planet out of the ice age over the next 50 Myr.

MT is the answer to why and how the climate changes at all timescales. The amount of energy the planet receives only changes significantly due to the annual revolution, as the Earth changes its distance from the sun, and due to eccentricity changes (95, 125 and 405 kyr cycle). There is speculation that it might change at longer timescales as the solar system moves around the galaxy center (see Sect. 9.3.2). At all other timescales between 1 year and the eccentricity cycle the amount of energy the planet receives does not vary significantly. The energy received at the ToA during the Last Glacial Maximum, 20 ka, and the Holocene Climatic Optimum, 9 ka, was essentially the same. The planet surface temperature difference, however, is thought to have been about 6 °C (Tierney et al. 2020). Neither albedo nor GHG changes can explain why and how the planet transitioned from the glacial maximum to the optimum. Both states are quite stable and cannot be very sensitive to small changes in albedo and GHGs that at times during the c. 90-kyr long glacial period manifested important variations (Ahn & Brook 2014). The explanation obviously lies in the different distribution of the near-constant incoming solar radiation over the geometry of the planet that changes its orientation towards the sun, known as Milankovitch forcing. The difference in energy at the equator from changes in Earth orientation is small, but grows with latitude, so the latitudinal insolation gradient is the foremost physical parameter that determines climate, and it does so by altering the LTG (Davis & Brewer 2009; see Fig. 2.17) that, in turn, strongly affects MT. In theory MT should not affect the energy balance of the planet, because it just moves energy within the system, but the polar regions during the cold season suffer a tremendous GHG deficit with respect to the global average, and OLR varies with MT. Forcing of LTG by insolation gradient changes explains the propagation of orbital signatures through the climate system caused by changes in MT. Climate models treat LTG as an emergent feature, overestimating the role of GHGs in driving polar amplification (Davis & Brewer 2009), neglecting the role of MT.

For multi-decadal to centennial timescales, MT is still the most important determinant for climate change, and larger than the increase in non-condensing GHGs. In addition to planetary LTG changes, MT responds to changes in the zonal-mean zonal wind strength. Most of the transport is carried out by the atmosphere through the meridional circulation and across the PV (Fig. 11.16). A decrease in zonal wind strength not only results in an increase in meridional wind strength, but also in a weakening of the PV through the propagation of planetary waves, increasing MT. Zonal wind strength is favored by a strong stratospheric LTG through thermal wind balance, and affected by tropospheric changes in ocean-atmosphere circulation. The ToA energy gain/loss ratio is largest at the tropical

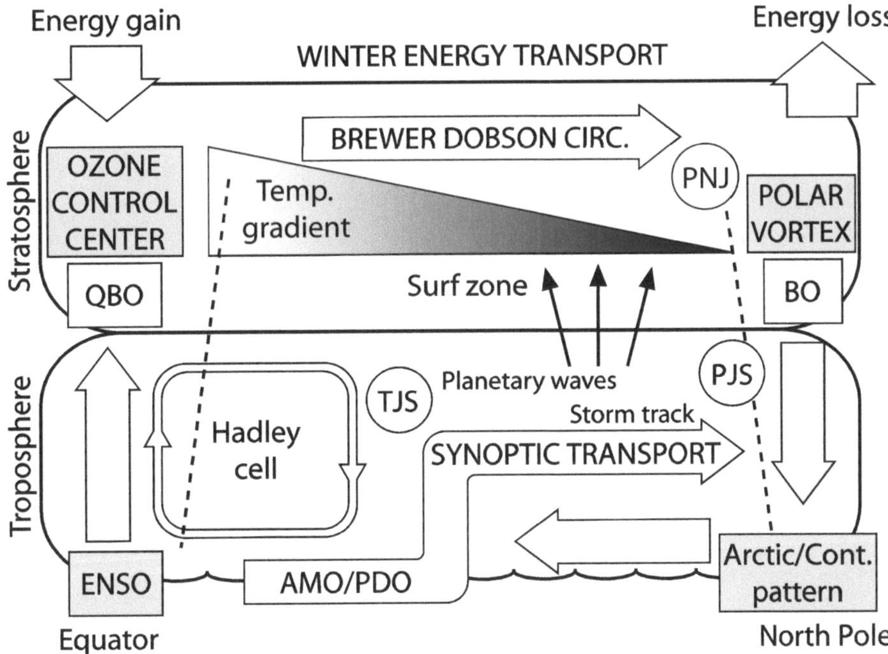

Fig. 11.16 Winter meridional transport outline
Energy gain/loss ratio at the ToA determines the maximal energy source at the tropical band and the maximal energy sink at the Arctic in winter. Incoming solar energy is distributed at the stratosphere and troposphere/surface where it is subjected to different transport modulation. Energy (white arrows) ascends from the surface to the stratosphere at the tropical pipe (left dashed line) and is transported towards the polar vortex (right dashed line) by the Brewer–Dobson circulation. Stratospheric transport is determined by UV heating at the tropical ozone layer, that establishes a temperature gradient affecting zonal wind strength through thermal wind balance, and by the QBO. This double control determines the behavior of planetary waves (black arrows) and determines if the polar vortex undergoes a biennial coupling with the QBO (BO). At the tropical ocean mixed-layer ENSO is the main energy distribution modulator. While the Hadley cell participates in energy transport and responds to its intensity by expanding or contracting, most energy transport in the tropics is done by the ocean. Changes in transport intensity result in the main modes of variability, the AMO and PDO. Outside the tropics most of the energy is transferred to the troposphere, where synoptic transport by eddies along storm tracks is responsible for the bulk of the transport to high latitudes. The strength of the polar vortex determines the high latitudes winter climate regime. A weak vortex promotes a warm Arctic/cold continents winter regime, where more energy enters the Arctic exchanged by cold air masses moving out. Jet streams (PJS, polar; TJS, tropical; PNJ, polar night) constitute the boundaries and limit transport.

bands and the energy is absorbed by the climate system mainly at two levels, the tropical ozone layer for the UV part of the spectrum and the ocean photic zone for the visible part (Fig. 11.16). 3.85 W/m² of 200–300 nm UV energy are absorbed at the stratosphere (Eddy et al. 2003; one fourth of 15.4 W/m²), constituting 5 % of the solar energy absorbed by the atmosphere, and 1.6 % of the total solar energy absorbed by the climate system (Wild et al. 2019). Tropical heat and moisture are transported towards the winter pole, where the ToA energy gain/loss ratio is most negative. The amount transported depends on the energy distribution from the ozone control center and from the equatorial ocean, where the most prominent distribution center is ENSO due to the large size of the Pacific basin. At the polar end, transport depends on the PV, whose strength is determined by the stratospheric LTG created at the ozone control center, and by planetary wave generation and transmission. A strong PV reduces the intensity of the entire transport system all the way to its origin at the tropical sea surface mixed layer, not only through a reduction in BDC upwelling, but also in synoptic heat transport by the atmosphere to the pole, that results in oceanic heat accumulation, inducing a positive phase at the internal modes of variability, AMO and PDO. PV strength also limits cold air export from the Arctic towards the continents, alternating between WACC and cold Arctic/warm continents winter conditions. This temperature pattern contrast is a reflection of tropospheric winter MT.

Today we know that what keeps the winter pole from being much colder is the MT. What we don't appear to know is that the same MT is responsible for the heat loss that keeps the planet in an Ice Age (Fig. 11.17). MT is the climate control knob, and it responds primarily to the LTG. This planet's lower troposphere has an overabundance of the GHG water vapor, which is only scarce in Antarctica and the Arctic in winter. Two factors limit the heat loss by the planet from the winter pole. Winter sea-ice limits the heat loss by the ocean. Less winter sea-ice means more cooling from increased heat loss, not more warming. The PV limits the heat loss by the atmosphere by keeping a strong low-pressure center above the winter pole with very cold air inside that limits heat-loss by radiative cooling. Arctic winter warming by warm moist air advection from mid-latitudes increases heat-loss by the planet. The GHG content in the Earth's atmosphere is lowest at the winter poles due to cold-air lowering the water vapor content, and reduced moisture flux through sea-ice. Average absolute humidity as low as 0.2 g/m³ can be recorded for the month of January on the Arctic surface, and it is 10 to 1000 times lower a few kilometers above the surface (Tomasi et al. 2020). Under polar nighttime conditions, the Arctic atmospheric increase in GHG content and temperature due to moisture and heat transport from mid-

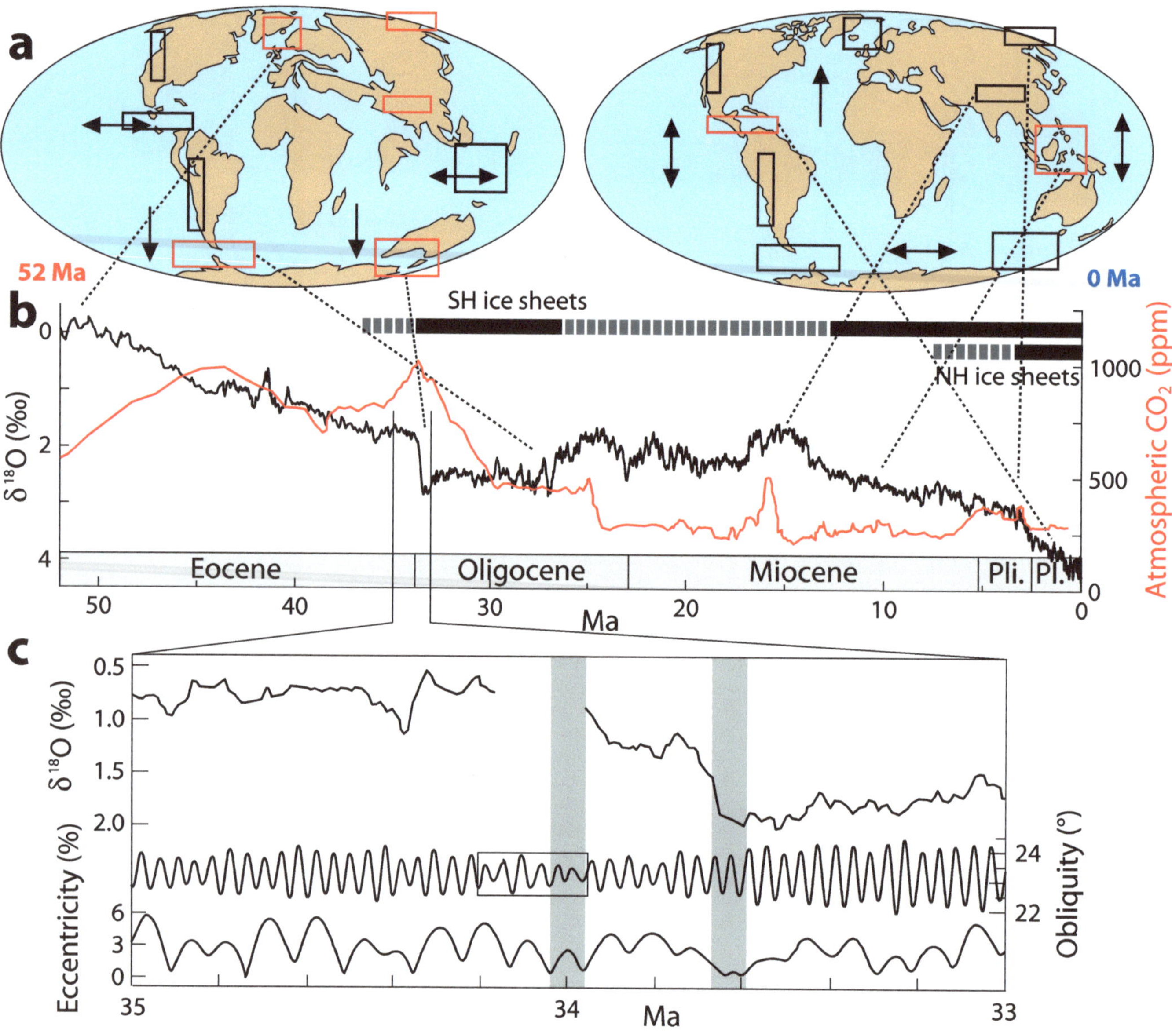

Fig. 11.17 Meridional transport as the main determinant for climate evolution

a) Left, mountain ranges and ocean gateways affecting meridional transport in the Cenozoic. Black boxes indicate active, well-developed geological features affecting meridional transport. Red boxes indicate features undergoing development. The Arctic Gateway began opening about 55 Ma. The Tasman Gateway opened between 36 and 30 Ma, while the Drake Passage opened either at 30 Ma or around 20 Ma. Vertical arrows indicate meridional transport (global cooling) is favored, and horizontal arrows zonal transport (global warming) is favored. **Right,** the world in the Pleistocene has developed significant geological features that favor meridional transport. The Himalayas reached modern elevation by about 15 Ma. The Indonesian Passage is still open, but significant restrictions occurred at about 11 Ma. The Bering Strait began its existence about 5.3 Ma, while the Panama Gateway completed closure around 3 Ma. After Lyle et al. 2008. Red boxes indicate geological changes affecting meridional transport. **b)** Black curve, global deep-sea δ¹⁸O data as a temperature and continental ice proxy. Full bar above represents ice volume >50% of present, and dashed bar ≤50%. After Zachos et al. 2001. Red curve, average CO₂ data after Beerling & Royer 2011. See Fig. 9.8 for details. **c)** High resolution δ¹⁸O changes in benthic foraminiferal calcite show that Antarctic glaciation took place faster than previously thought, in two steps. At the end of the Eocene, Antarctica had already developed several ice sheets in altitude. A tipping point was reached when low eccentricity (grey bar) promoted ice growth at a time when low obliquity amplitude (box) promoted summer ice survival, triggering Antarctic glaciation in just 80 kyr. The glaciation was completed 400 kyr later during another period of low eccentricity (grey bar). After Coxall et al. 2005. At the early Eocene (52 Ma) the world geography was very favorable to zonal circulation. There was a well-developed circumglobal seaway formed by the Tethys Sea, the Panama Gateway and the Indonesian Passage. After the Mid-Oligocene Glacial Interval c. 26 Ma, and until the end of the Mid-Miocene Climatic Optimum at c. 14 Ma (a 12 Myr interval) the planet entered a warm period that apparently nobody can explain. At the time CO₂ levels collapsed, according to proxies (Beerling and Royer 2011), from 450 to 200 ppm, and remained very low for the entire period except during the time of the Columbia River Flood Basalt flows. So, during the Late Oligocene to the Mid-Miocene warm period, CO₂ changes do not explain temperature changes. Recent research suggests most of this period was characterized by a strongly reduced LTG (Guitián et al. 2019), indicative of reduced MT. The climatic isolation of Antarctica by the Antarctic Circumpolar Current and the Southern Annular Mode must have hindered MT of heat from the tropics causing regional cooling, yet globally the planet warmed in response to reduced MT, and Antarctica entered a period when its ice-sheets waxed and waned following orbital changes, until the raising of the Himalayas set the planet again in a cooling trend. The present world geography is very favorable to meridional circulation, so the Late Cenozoic Ice Age should continue until that changes.

latitudes can only result in increased OLR and energy loss by the planet. Even though the moisture intrusions result in enhanced DLR from clouds and surface warming, upon return of clear-sky conditions that are frequent during the Arctic winter, the warmer surface should suffer increased radiative cooling, followed by a temperature inversion that results in radiative cooling from higher in the troposphere. From a whole climate system perspective, winter Arctic amplification due to enhanced MT must result in a net energy loss.

Since Arctic amplification during the past two decades has been mostly a winter phenomenon, the evidence for the implied energy loss comes from a simultaneous negative trend in the Earth energy imbalance and a decreasing trend of the OHC time derivative (Fig. 11.7g; Dewitte et al. 2019). The energy loss from the enhanced MT towards the Arctic in winter is the most parsimonious explanation for the reduced rate of surface warming known as the pause, or hiatus, that took place simultaneously.

MT variability affects the entire climate system and is affected by internal climate variability and external forcing changes. Solar variability is just one of several known modulators of MT (Fig. 11.9). On inter-annual to decadal time-scales the effect of solar variability is masked by the effect of the QBO, ENSO and volcanic aerosol short-term effects. On multidecadal timescales the solar effect is noticeable but confounded by volcanic multidecadal effects and the c. 65-yr stadium wave oscillation that at times can temporarily override the solar effect on MT (Veretenenko & Ogurtsov 2019). It is, however, on centennial to millennial timescales when solar variability becomes the undisputed MT modulator due to its long cycles (see Chap. 8), while at the multi-millennial timescale Milankovitch forcing effect on MT becomes dominant. Multiple paradoxes regarding the solar effect on climate can be resolved through the combined modulation of MT. It explains why the solar effect is almost unnoticeable over the 11-yr solar cycle yet extraordinarily strong in paleoclimatology. It explains how a very small change in TSI and a small change in UV can produce such huge effect as the LIA. And it explains the inconsistency of the solar influence on climate, that at times appears to produce opposing results when another MT modulator is overruling the solar effect.

11.9 Conclusions

11a. Volcanoes produce a short-term weather effect from radiative and meridional transport changes, and a long-term climate effect from delayed meridional transport changes. Climate change has an important effect on volcanic activity, that follows ice-volume changes associated with the glacial cycle, which is driven by Milankovitch forcing.

11b. A c. 65-year oscillation has been observed in multiple climate-related phenomena that can explain a significant part of climate variability. This stadium-wave of dynamic energy-transfer within the climate system is a manifestation of changes in meridional transport.

11c. The last climatic shift took place in 1997–98, because of changes in meridional transport, and resulted in more energy being lost at the Arctic, a reduction in the Earth's energy imbalance, and Arctic amplification. It proves that meridional transport variability is one of the main forcings of climate change.

11d. Meridional transport takes place through coupled stratospheric and tropospheric/ocean components undergoing complex regulation by multiple factors. Its changes are reflected in multiple climate variables, including the rate of global warming.

11e. Due to solar modulation of meridional transport, solar minima are periods of stronger tropical-polar stratosphere coupling, and when coordinated changes in meridional transport that produce climate shifts are more probable.

11f. The "Winter Gatekeeper" hypothesis proposes that solar activity is the main long-term regulator of the amount of heat transported to the pole in winter, that is later emitted to space.

11g. The "Winter Gatekeeper" hypothesis explains the strong paleoclimatic effect of periods of prolonged low solar activity, like the Little Ice Age, the North Atlantic region hotspot climate variability, the warm Arctic/cold continents pattern linked to low solar activity, Arctic winter amplification, the effect of solar activity on Earth's rotation, and solar ENSO modulation.

11h. Modern global warming is in part a consequence of high solar activity, and the pause in global warming is in part a consequence of low solar activity, and likely to continue until solar activity becomes high again.

11i. Meridional transport is the answer to why and how the climate changes at all timescales.

References

Ahn J & Brook EJ (2014) Siple Dome ice reveals two modes of millennial CO_2 change during the last ice age. Nature Communications 5 3723

Álvarez–Ramírez J, Echeverría JC & Rodríguez E (2011) Is the North Atlantic Oscillation modulated by solar and lunar cycles? Some evidences from Hurst autocorrelation analysis. Advances in Space Research 47 (4) 748–756

Ayarzagüena B, Baldwin MP, Birner T et al (2021) Sudden stratospheric warmings: a phenomenon with global effects. SPARC newsletter 57 8–11

Baldwin MP & Dunkerton TJ (1998) Biennial, quasi-biennial, and decadal oscillations of potential vorticity in the northern stratosphere. Journal of Geophysical Research: Atmospheres 103 (D4) 3919–3928

Barlyaeva T, Bard E & Abarca-del-Rio R (2014) Rotation of the Earth, solar activity and cosmic ray intensity. Annales Geophysicae 32 (7) 761–771

Beerling DJ & Royer DL (2011) Convergent Cenozoic CO2 history. Nature Geoscience 4 (7) 418–420

Berson FA & Kulkarni RN (1968) Sunspot cycle and the quasi-biennial stratospheric oscillation. Nature 217 (5134) 1133–1134

Bittner M, Timmreck C, Schmidt H et al (2016) The impact of wave-mean flow interaction on the Northern Hemisphere polar vortex after tropical volcanic eruptions. Journal of Geophysical Research: Atmospheres 121 (10) 5281–5297

Boden TA, Marland G & Andres RJ (2009) Global, regional, and national fossil-fuel CO_2 emissions. Carbon dioxide information analysis center, Oak ridge national laboratory, US department of energy. Doi 10.3334/CDIAC/00001_V2010

Bond G, Kromer B, Beer J et al (2001) Persistent solar influence on North Atlantic climate during the Holocene. Science 294 (5549) 2130–2136

Bray JR (1971) Solar-climate relationships in the post-Pleistocene. Science 171 (3977) 1242–1243

Brönnimann S, Franke J, Nussbaumer SU et al (2019) Last phase of the Little Ice Age forced by volcanic eruptions. Nature geoscience 12 (8) 650–656

Calvo N & Marsh DR (2011) The combined effects of ENSO and the 11 year solar cycle on the Northern Hemisphere polar stratosphere. Journal of Geophysical Research: Atmospheres 116 (D23)

Cao Y, Liang S & Yu M (2020) Observed low-frequency linkage between Northern Hemisphere tropical expansion and polar vortex weakening from 1979 to 2012. Atmospheric Research 243 105034

Carlson B, Lacis A, Colose C et al (2019) Spectral signature of the Biosphere: NISTAR finds it in our solar system from the Lagrangian L-1 point. Geophysical Research Letters 46 (17–18) 10679–10686

Cea Pirón MA & Cano Pasalodos JA (2016) Nueva serie de extensión del hielo marino ártico en septiembre entre 1935 y 2014. Revista de Climatología 16 1–19

Chavez FP, Ryan J, Lluch-Cota SE & Ñiquen M (2003) From anchovies to sardines and back: multidecadal change in the Pacific Ocean. Science 299 (5604) 217–221

Church JA & White NJ (2011) Sea-level rise from the late 19th to the early 21st century. Surveys in Geophysics 32 (4–5) 585–602

Christiansen B (2010) Stratospheric bimodality: Can the equatorial QBO explain the regime behavior of the NH winter vortex?. Journal of climate 23 (14) 3953–3966

Chylek P, Folland C, Frankcombe L et al (2012) Greenland ice core evidence for spatial and temporal variability of the Atlantic Multidecadal Oscillation. Geophysical Research Letters 39 (9)

Chylek P, Klett JD, Lesins G et al (2014) The Atlantic Multidecadal Oscillation as a dominant factor of oceanic influence on climate. Geophysical research letters 41 (5) 1689–1697

Collimore CC, Martin DW, Hitchman MH et al (2003) On the relationship between the QBO and tropical deep convection. Journal of climate 16 (15) 2552–2568

Cook ER, Meko DM & Stockton CW (1997) A new assessment of possible solar and lunar forcing of the bidecadal drought rhythm in the western United States. Journal of Climate 10 (6) 1343–1356

Coxall HK, Wilson PA, Pälike H et al (2005) Rapid stepwise onset of Antarctic glaciation and deeper calcite compensation in the Pacific Ocean. Nature 433 (7021) 53-57

Curry JA, Schramm JL, Rossow WB & Randall D. (1996) Overview of Arctic cloud and radiation characteristics. Journal of Climate 9 (8) 1731–1764

Dai A, Fyfe JC, Xie SP & Dai X (2015) Decadal modulation of global surface temperature by internal climate variability. Nature Climate Change 5 (6) 555–559

Dangendorf S, Marcos M, Wöppelmann G et al (2017) Reassessment of 20th century global mean sea level rise. Proceedings of the National Academy of Sciences 114 (23) 5946–5951

Danish Meteorological Institute (2021) http://ocean.dmi.dk/arctic/meant80n_anomaly.uk.php Accessed 19 Nov 2021.

Davis BA & Brewer S (2009) Orbital forcing and role of the latitudinal insolation/temperature gradient. Climate dynamics 32 (2–3) 143–165

de Larminat P (2016) Earth climate identification vs. anthropic global warming attribution. Annual Reviews in Control 42 114–125

Deser C, Alexander MA, Xie SP & Phillips AS (2010) Sea surface temperature variability: Patterns and mechanisms. Annual review of marine science 2 115–143

Dewitte S, Clerbaux N & Cornelis J (2019) Decadal changes of the reflected solar radiation and the earth energy imbalance. Remote Sensing 11 663

Diallo M, Ploeger F, Konopka P et al (2017) Significant contributions of volcanic aerosols to decadal changes in the stratospheric circulation. Geophysical research letters 44 (20) 10-780

Divine DV & Dick C (2006) Historical variability of sea ice edge position in the Nordic Seas. Journal of Geophysical Research: Oceans 111 C01001

Domeisen DI, Garfinkel CI & Butler AH (2019) The teleconnection of El Niño Southern Oscillation to the stratosphere. Reviews of Geophysics 57 (1) 5–47

Dübal HR & Vahrenholt F (2021) Radiative Energy Flux Variation from 2001–2020. Atmosphere 12 (10) 1297

Dukhovskoy DS, Yashayaev I, Proshutinsky A et al (2019) Role of Greenland freshwater anomaly in the recent freshening of the subpolar North Atlantic. Journal of Geophysical Research: Oceans 124 (5) 3333–3360

Eade R, Stephenson DB, Scaife AA & Smith DM (2021) Quantifying the rarity of extreme multi-decadal trends: how unusual was the late twentieth century trend in the North Atlantic Oscillation?. Climate Dynamics DOI 10.1007/s00382-021-05978-4

Eastman R & Warren SG (2010) Interannual variations of Arctic cloud types in relation to sea ice. Journal of Climate 23 (15) 4216–4232

Ebbesmeyer CC, Cayan DR, McLain DR et al (1991) 1976 step in the Pacific climate: forty environmental changes between 1968–1975 and 1977–1984. In: Betancourt JL and Tharp VL (eds) Proceedings of the Seventh Annual Pacific Climate (PACLIM) Workshop, April 1990: California Department of Water Resources, Interagency Ecological Studies Program Technical Report 26 115–126

Eddy JA (1976) The Maunder Minimum. Science 192 (4245) 1189–1202

Eddy JA, Bond GC, Bradley RS et al (2003) Living with a Star: New Opportunities in Sun-Climate Research. Report of the NASA LWS Sun-Climate Task Group

Elsbury D, Peings Y & Magnusdottir G (2021) Variation in the Holton–Tan effect by longitude. Quarterly Journal of the Royal Meteorological Society 147 (736) 1767–1787

Esper J, Schneider L, Krusic PJ et al. (2013) European summer temperature response to annually dated volcanic eruptions over the past nine centuries. Bulletin of volcanology 75 (7) 736

Fasullo JT & Trenberth KE (2008) The annual cycle of the energy budget. Part II: Meridional structures and poleward transports. Journal of Climate 21 (10) 2313–2325

Folland CK, Palmer TN & Parker DE (1986) Sahel rainfall and worldwide sea temperatures 1901–85. Nature 320 (6063) 602–607

Folland CK, Parker DE & Kates FE (1984) Worldwide marine temperature fluctuations 1856–1981. Nature 310 (5979) 670–673

Frakes LA, Francis JE & Syktus JI (1992) Climate modes of the Phanerozoic. Cambridge University Press, Cambridge

Frankcombe LM, Von Der Heydt A & Dijkstra HA (2010) North Atlantic multidecadal climate variability: an investigation of dominant time scales and processes. Journal of climate 23 (13) 3626–3638

Fyfe JC, Meehl GA, England MH et al (2016) Making sense of the early-2000s warming slowdown. Nature Climate Change 6 (3) 224–228

Geyer A & Bindeman I (2011) Glacial influence on caldera-forming eruptions. Journal of Volcanology and Geothermal Research 202 (1–2) 127–142

Glazner AF, Manley CR, Marron JS & Rojstaczer S (1999) Fire or ice: Anticorrelation of volcanism and glaciation in California over the past 800,000 years. Geophysical Research Letters 26 (12) 1759–1762

Goode PR & Pallé E (2007) Shortwave forcing of the Earth's climate: Modern and historical variations in the Sun's irradiance and the Earth's reflectance. Journal of Atmospheric and Solar-Terrestrial Physics 69 (13) 1556–1568

Graf HF, Li Q & Giorgetta MA (2007) Volcanic effects on climate: revisiting the mechanisms. Atmospheric Chemistry and Physics 7 (17) 4503–4511

Graham NE (1994) Decadal-scale climate variability in the tropical and North Pacific during the 1970s and 1980s: Observations and model results. Climate Dynamics 10 (3) 135–162

Gray LJ, Beer J, Geller M et al (2010) Solar influences on climate. Reviews of Geophysics 48 (4) RG4001

Gray ST, Graumlich LJ, Betancourt JL & Pederson GT (2004) A tree-ring based reconstruction of the Atlantic Multidecadal Oscillation since 1567 AD. Geophysical Research Letters 31 (12)

Guitián J, Phelps S, Polissar PJ et al (2019) Midlatitude temperature variations in the Oligocene to early Miocene. Paleoceanography and Paleoclimatology 34 (8) 1328–1343

Guðlaugsdóttir H, Sjolte J, Sveinbjörnsdóttir ÁE et al (2019) North Atlantic weather regimes in $\delta^{18}O$ of winter precipitation: isotopic fingerprint of the response in the atmospheric circulation after volcanic eruptions. Tellus B: Chemical and Physical Meteorology 71 (1) 1633848

Haigh JD (1996) The impact of solar variability on climate. Science 272 (5264) 981–984

Hegerl GC, Crowley TJ, Allen M et al (2007) Detection of human influence on a new, validated 1500-year temperature reconstruction. Journal of Climate 20 (4) 650–666

Herschel W (1801) Observations tending to investigate the nature of the Sun, in order to find the causes or symptoms of its variable emission of light and heat; with remarks on the use that may possibly be drawn from solar observations. Philosophical Transactions of the Royal Society of London 91 XIII 265–318

Hide R, Boggs DH & Dickey JO (2000) Angular momentum fluctuations within the Earth's liquid core and torsional oscillations of the core–mantle system. Geophysical Journal International 143 (3) 777–786

Hines CO (1974) A possible mechanism for the production of sun-weather correlations. Journal of the Atmospheric Sciences 31 (2) 589–591

Honda M & Nakamura H (2001) Interannual seesaw between the Aleutian and Icelandic lows. Part II: Its significance in the interannual variability over the wintertime Northern Hemisphere. Journal of Climate 14 (24) 4512–4529

Hood LL (2003) Thermal response of the tropical tropopause region to solar ultraviolet variations. Geophysical research letters 30 (23)

Hu FS, Kaufman D, Yoneji S et al (2003) Cyclic variation and solar forcing of Holocene climate in the Alaskan subarctic. Science 301 (5641) 1890–1893

Huber M & Caballero R (2011) The early Eocene equable climate problem revisited. Climate of the Past 7 (2) 603–633

Huybers P & Langmuir C (2009) Feedback between deglaciation, volcanism, and atmospheric CO_2. Earth and Planetary Science Letters 286 (3–4) 479–491

Intergovernmental Panel on Climate Change (2014) Summary for Policymakers. In: Core Writing Team, Pachauri RK & Meyer LA (eds) Climate Change 2014: Synthesis Report. Contribution of Working Groups I, II and III to the Fifth Assessment Report of the Intergovernmental Panel on Climate Change. IPCC, Geneva

Intergovernmental Panel on Climate Change (2021) Summary for Policymakers. In: Masson–Delmotte V, Zhai P, Pirani A et al (eds) Climate Change 2021: The Physical Science Basis. Contribution of Working Group I to the Sixth Assessment Report of the Intergovernmental Panel on Climate Change. Cambridge University Press. In Press

Jellinek AM, Manga M & Saar MO (2004) Did melting glaciers cause volcanic eruptions in eastern California? Probing the mechanics of dike formation. Journal of Geophysical Research: Solid Earth 109 (B9)

Jevrejeva S, Moore JC, Grinsted A & Woodworth PL (2008) Recent global sea level acceleration started over 200 years ago? Geophysical Research Letters 35 (8) L08715

Jevrejeva S, Moore JC, Grinsted A et al (2014) Trends and acceleration in global and regional sea levels since 1807. Global and Planetary Change 113 11–22

Jiang H, Muscheler R, Björck S et al (2015) Solar forcing of Holocene summer sea-surface temperatures in the northern North Atlantic. Geology 43 (3) 203–206

Jones PD, Jónsson T & Wheeler D (1997) Extension to the North Atlantic Oscillation using early instrumental pressure observations from Gibraltar and south-west Iceland. International Journal of Climatology: A Journal of the Royal Meteorological Society 17 (13) 1433–1450

Kerr RA (1987) Sunspot-weather correlation found. Science 238 (4826) 479–481

Kerr RA (2000) A North Atlantic climate pacemaker for the centuries. Science 288 (5473) 1984–1985

Kidston J, Scaife AA, Hardiman, SC et al (2015) Stratospheric influence on tropospheric jet streams, storm tracks and surface weather. Nature Geoscience 8 (6) 433–440

Klyashtorin LB (2001) Climate change and long-term fluctuations of commercial catches: the possibility of forecasting. FAO Fisheries Technical Paper 410. FAO, Rome

Klyashtorin LB & Lyubushin AA (2007) Cyclic climate changes and fish productivity. Moscow. Vniro publishing

Kobashi T, Box JE, Vinther BM et al (2015) Modern solar maximum forced late twentieth century Greenland cooling. Geophysical Research Letters 42 (14) 5992–5999

Kodera K & Kuroda Y (2002) Dynamical response to the solar cycle. Journal of Geophysical Research: Atmospheres 107 (D24) ACL-5

Kodera K, Thiéblemont R, Yukimoto S & Matthes K (2016) How can we understand the global distribution of the solar cycle signal on the Earth's surface?. Atmospheric Chemistry and Physics 16 (20) 12925–12944

Krüger K, Tegtmeier S & Rex M (2008) Long-term climatology of air mass transport through the Tropical Tropopause Layer (TTL) during NH winter. Atmospheric Chemistry and Physics 8 (4) 813–823

Kushnir Y (1994) Interdecadal variations in North Atlantic sea surface temperature and associated atmospheric conditions. Journal of Climate 7 (1) 141–157

Kutterolf S, Jegen M, Mitrovica JX et al (2013) A detection of Milankovitch frequencies in global volcanic activity. Geology 41 (2) 227–230

Labitzke K (1987) Sunspots, the QBO, and the stratospheric temperature in the north polar region. Geophysical Research Letters 14 (5) 535–537

Labitzke K, Kunze M & Brönnimann S (2006) Sunspots, the QBO and the stratosphere in the North Polar Region–20 years later. Meteorologische Zeitschrift 15 (3) 355–363

Lambeck K & Cazenave A (1976) Long term variations in the length of day and climatic change. Geophysical Journal International 46 (3) 555–573

Le Mouël JL, Blanter E, Shnirman M & Courtillot V (2010) Solar forcing of the semi-annual variation of length-of-day. Geophysical Research Letters 37 (15) L15307

Lean JL (2017) Sun-climate connections. In: Oxford Research Encyclopedia of Climate Science. DOI: 10.1093/acrefore/9780190228620.013.9

Lean JL (2018) Observation-based detection and attribution of 21st century climate change. Wiley Interdisciplinary Reviews: Climate Change 9 (2) e511

Li F, Orsolini YJ, Wang H et al (2018) Modulation of the Aleutian–Icelandic low seesaw and its surface impacts by the Atlantic multidecadal oscillation. Advances in Atmospheric Sciences 35 (1) 95–105

Li X, Hu ZZ & Becker E (2019) On the westward shift of tropical Pacific climate variability since 2000. Climate Dynamics 53 (5) 2905–2918

Liebmann B & Smith CA (1996) Description of a complete (interpolated) outgoing longwave radiation dataset. Bulletin of the American Meteorological Society 77 (6) 1275–1277

Liu F, Li J, Wang B et al (2018) Divergent El Niño responses to volcanic eruptions at different latitudes over the past millennium. Climate dynamics 50 (9) 3799–3812

Lu H, Baldwin MP, Gray LJ & Jarvis MJ (2008) Decadal-scale changes in the effect of the QBO on the northern stratospheric polar vortex. Journal of Geophysical Research: Atmospheres 113 (D10)

Lupo AR (2020) Atmospheric blocking events: a review. Annals of the New York Academy of Sciences. 1–20

Lyle M, Barron J, Bralower TJ et al (2008) Pacific Ocean and Cenozoic evolution of climate. Reviews of Geophysics 46 (2)

Magny M, Combourieu–Nebout N, De Beaulieu JL et al (2013) North–south palaeohydrological contrasts in the central Mediterranean during the Holocene: tentative synthesis and working hypotheses. Climate of the Past 9 (5) 2043–2071

Mantua NJ, Hare SR, Zhang Y et al (1997) A Pacific interdecadal climate oscillation with impacts on salmon production. Bulletin of the american Meteorological Society 78 (6) 1069–1080

Mantua NJ & Hare SR (2002) The Pacific decadal oscillation. Journal of oceanography 58 (1) 35–44

Marchitto TM, Muscheler R, Ortiz JD et al (2010) Dynamical response of the tropical Pacific Ocean to solar forcing during the early Holocene. Science 330 (6009) 1378–1381

Marcus SL, de Viron O & Dickey JO (2011) Abrupt atmospheric torque changes and their role in the 1976–1977 climate regime shift. Journal of Geophysical Research: Atmospheres 116 (D3)

Matthes K, Mars, DR, Garcia RR et al (2010) Role of the QBO in modulating the influence of the 11 year solar cycle on the atmosphere using constant forcings. Journal of Geophysical Research: Atmospheres 115 (D18)

Mazzarella A (2013) Time-integrated North Atlantic Oscillation as a proxy for climatic change. Natural Science 5 (1A) 149–155

McGuire WJ, Howarth RJ, Firth CR et al (1997) Correlation between rate of sea-level change and frequency of explosive volcanism in the Mediterranean. Nature 389 (6650) 473–476

Meehl GA, Arblaster JM, Matthes K et al (2009) Amplifying the Pacific climate system response to a small 11-year solar cycle forcing. Science 325(5944) 1114–1118

Minobe S (1997) A 50–70 year climatic oscillation over the North Pacific and North America. Geophysical Research Letters 24 (6) 683–686

Moberg A, Sonechkin DM, Holmgren K et al (2005) Highly variable Northern Hemisphere temperatures reconstructed from low-and high-resolution proxy data. Nature 433 (7026) 613–617

Moore GWK, Halfar J, Majeed H et al (2017) Amplification of the Atlantic Multidecadal Oscillation associated with the onset of the industrial-era warming. Scientific Reports 7 (1) 1–10

Mouginot J, Rignot E & Bjørk AA (2019) Forty-six years of Greenland Ice Sheet mass balance from 1972 to 2018. Proceedings of the national academy of sciences 116 (19) 9239–9244

Muller RA, Curry J, Groom D et al (2013) Decadal variations in the global atmospheric land temperatures. Journal of Geophysical Research: Atmospheres 118 (11) 5280–5286

Muscheler R, Joos F, Müller SA & Snowball I (2005) How unusual is today's solar activity?. Nature 436 (7050) E3-E4

Nguyen H, Evans A, Lucas C et al (2013) The Hadley circulation in reanalyses: Climatology, variability, and change. Journal of Climate 26 (10) 3357–3376

Nowell DA, Jones MC & Pyle DM (2006) Episodic quaternary volcanism in France and Germany. Journal of Quaternary Science 21 (6) 645–675

Nygård T, Naakka T & Vihma T (2020) Horizontal moisture transport dominates the regional moistening patterns in the Arctic. Journal of Climate 33 (16) 6793–6807

Ogurtsov M, Veretenenko SV, Helama S et al (2020) Assessing the signals of the Hale solar cycle in temperature proxy records from Northern Fennoscandia. Advances in Space Research 66 (9) 2113–2121

Outten S & Esau I (2017) Bjerknes compensation in the Bergen Climate Model. Climate Dynamics 49 (7–8) 2249–2260

Planck M (1906) Schriften des Vereins für Socialpolitik. Bände 116

Polyakov IV, Alekseev GV, Timokhov LA et al (2004) Variability of the intermediate Atlantic water of the Arctic Ocean over the last 100 years. Journal of climate 17 (23) 4485–4497

Porter SE, Mosley-Thompson E & Thompson LG (2019) Ice core $\delta^{18}O$ record linked to Western Arctic sea ice variability. Journal of Geophysical Research: Atmospheres 124 (20) 10784–10801

Powell Jr AM & Xu J (2011) Possible solar forcing of interannual and decadal stratospheric planetary wave variability in the Northern Hemisphere: An observational study. Journal of Atmospheric and Solar-Terrestrial Physics 73 (7–8) 825–838

Proshutinsky A, Dukhovskoy D, Timmermans ML et al (2015) Arctic circulation regimes. Philosophical Transactions of the Royal Society A: Mathematical, Physical and Engineering Sciences 373 (2052) 20140160

Rampino MR, Self S & Fairbridge RW (1979) Can rapid climatic change cause volcanic eruptions? Science 206 (4420) 826–829

Randel WJ, Wu F, Voemel H et al (2006) Decreases in stratospheric water vapor after 2001: Links to changes in the tropical tropopause and the Brewer-Dobson circulation. Journal of Geophysical Research: Atmospheres 111 (D12)

Randel W & Park M (2019) Diagnosing observed stratospheric water vapor relationships to the cold point tropical tropopause. Journal of Geophysical Research: Atmospheres 124 (13) 7018–7033

Rao J, Yu Y, Guo D, Shi C et al (2019) Evaluating the Brewer–Dobson circulation and its responses to ENSO, QBO, and the solar cycle in different reanalyses. Earth and Planetary Physics 3 (2) 166–181

Ray RD & Douglas BC (2011) Experiments in reconstructing twentieth-century sea levels. Progress in Oceanography 91 (4) 496–515

Rigozo NR, Nordemann DJR, Echer MS et al (2007) Solar activity imprints in tree ring width from Chile (1610–1991). Journal of atmospheric and solar-terrestrial physics 69 (9) 1049–1056

Robock A (2000) Volcanic eruptions and climate. Reviews of Geophysics 38 (2) 191–219

Rohling E, Mayewski P, Abu–Zied R et al (2002) Holocene atmosphere-ocean interactions: records from Greenland and the Aegean Sea. Climate Dynamics 18 (7) 587–593

Rydsaa JH, Graversen RG, Heiskanen TI & Stoll PJ (2021) Changes in atmospheric latent energy transport into the Arctic: Planetary versus synoptic scales. Quarterly Journal of the Royal Meteorological Society 147 2281–2292

Salby M & Callaghan P (2000) Connection between the solar cycle and the QBO: The missing link. Journal of Climate 13 (2) 328–338

Salby ML & Callaghan PF (2006) Relationship of the quasi-biennial oscillation to the stratospheric signature of the solar cycle. Journal of Geophysical Research: Atmospheres 111 (D6)

Scafetta N (2010) Empirical evidence for a celestial origin of the climate oscillations and its implications. Journal of Atmospheric and Solar-Terrestrial Physics 72 (13) 951–970

Schweiger AJ (2004) Changes in seasonal cloud cover over the Arctic seas from satellite and surface observations. Geophysical Research Letters 31 (12)

Schlesinger ME & Ramankutty N (1994) An oscillation in the global climate system of period 65–70 years. Nature 367 (6465) 723–726

Scotese CR, Song H, Mills BJ & van der Meer DG (2021) Phanerozoic paleotemperatures: The earth's changing climate during the last 540 million years. Earth-Science Reviews 103503

Self S, Rampino MR & Barbera JJ (1981) The possible effects of large 19th and 20th century volcanic eruptions on zonal and hemispheric surface temperatures. Journal of Volcanology and Geothermal Research 11 (1) 41–60

Sigl M, Winstrup M, McConnell JR et al (2015) Timing and climate forcing of volcanic eruptions for the past 2,500 years. Nature 523 (7562) 543–549

Sjolte J, Adolphi F, Guðlaugsdòttir H & Muscheler R (2021) Major Differences in Regional Climate Impact Between High- and Low-Latitude Volcanic Eruptions. Geophysical Research Letters 48 (8) e2020GL092017

Soden BJ, Wetherald RT, Stenchikov GL & Robock A (2002) Global cooling after the eruption of Mount Pinatubo: A test of climate feedback by water vapor. Science 296 (5568) 727–730

Stenchikov G (2021) The role of volcanic activity in climate and global changes. In: Letcher TM (ed) Climate change. Observed Impacts on Planet Earth, 3rd edn. Elsevier, Amsterdam, p 607–643

Stenchikov G, Robock A, Ramaswamy V et al (2002) Arctic Oscillation response to the 1991 Mount Pinatubo eruption: Effects of volcanic aerosols and ozone depletion. Journal of Geophysical Research: Atmospheres 107 (D24) ACL-28

Sun DZ (2000) The heat sources and sinks of the 1986–87 El Niño. Journal of climate 13 (20) 3533–3550

Sun J & Tan B (2013) Mechanism of the wintertime Aleutian low–Icelandic low seesaw. Geophysical Research Letters 40 (15) 4103–4108

Sun W, Liu J, Wang B et al (2018) A "La Niña-like" state occurring in the second year after large tropical volcanic eruptions during the past 1500 years. Climate Dynamics 1–15

Swingedouw D, Ortega P, Mignot J et al (2015) Bidecadal North Atlantic ocean circulation variability controlled by timing of volcanic eruptions. Nature communications 6 (1) 1–12

Swingedouw D, Mignot J, Ortega P et al (2017) Impact of explosive volcanic eruptions on the main climate variability modes. Global and Planetary Change 150 24–45

Taguchi M (2010) Observed connection of the stratospheric quasi-biennial oscillation with El Niño–Southern Oscillation in radiosonde data. Journal of Geophysical Research: Atmospheres 115 (D18)

Thompson DW & Wallace JM (2000) Annular modes in the extratropical circulation. Part I: Month-to-month variability. Journal of climate 13 (5) 1000–1016

Tierney JE, Zhu J, King J et al (2020) Glacial cooling and climate sensitivity revisited. Nature 584 (7822) 569–573

Timmreck C (2012) Modeling the climatic effects of large explosive volcanic eruptions. Wiley Interdisciplinary Reviews: Climate Change 3 (6) 545–564

Tomasi C, Petkov BH, Drofa O & Mazzola M (2020) Thermodynamics of the Arctic Atmosphere. In: Kokhanovsky A, Tomasi C (eds) Physics and Chemistry of the Arctic Atmosphere. Springer, Cham, p 53–152

Toohey M, Krüger K, Bittner M et al (2014) The impact of volcanic aerosol on the Northern Hemisphere stratospheric polar vortex: mechanisms and sensitivity to forcing structure. Atmospheric Chemistry and Physics 14 (23) 13063–13079

Toohey M, Krüger K, Schmidt H et al (2019) Disproportionately strong climate forcing from extratropical explosive volcanic eruptions. Nature Geoscience 12 (2) 100–107

van Wijngaarden WA & Happer W (2020) Dependence of Earth's Thermal Radiation on Five Most Abundant Greenhouse Gases. arXiv preprint arXiv:2006.03098

Verdon DC & Franks SW (2006) Long-term behaviour of ENSO: Interactions with the PDO over the past 400 years inferred from paleoclimate records. Geophysical Research Letters 33 (6) L06712

Veretenenko S & Ogurtsov M (2016) Cloud cover anomalies at middle latitudes: Links to troposphere dynamics and solar variability. Journal of Atmospheric and Solar-Terrestrial Physics 149 207–218

Veretenenko S & Ogurtsov M (2019) Manifestation and possible reasons of ~60-year oscillations in solar-atmospheric links. Advances in Space Research 64 (1) 104–116

von Hayek FA (1974) The Pretence of Knowledge – Prize Lecture to the memory of Alfred Nobel. In: Nobel Prize Outreach. https://www.nobelprize.org/prizes/economic-sciences/1974/hayek/lecture/ Accessed 20 Dec 2021

Wang J, Yang B, Ljungqvist FC et al (2017) Internal and external forcing of multidecadal Atlantic climate variability over the past 1,200 years. Nature Geoscience 10 (7) 512–517

Wang X & Key JR (2005) Arctic surface, cloud, and radiation properties based on the AVHRR Polar Pathfinder dataset. Part II: Recent Trends. Journal of Climate 18 (14) 2575–2593

Wang X, Liu J, Yang B et al (2021) Seasonal Trends in Clouds and Radiation over the Arctic Seas from Satellite Observations during 1982 to 2019. Remote Sensing 13 (16) 3201

Wang G, Zhao C, Zhang M et al (2020) The causality from solar irradiation to ocean heat content detected via multi-scale Liang–Kleeman information flow. Scientific Reports 10 (1) 1–9

White WB, Dettinger MD & Cayan DR (2003) Sources of global warming of the upper ocean on decadal period scales. Journal of Geophysical Research: Oceans 108 (C8)

Wild M, Hakuba MZ, Folini D et al (2019) The cloud-free global energy balance and inferred cloud radiative effects: an assessment based on direct observations and climate models. Climate Dynamics 52 (7) 4787–4812

Wunsch C (2005) The total meridional heat flux and its oceanic and atmospheric partition. Journal of climate 18 (21) 4374–4380

Wyatt MG (2012) A multidecadal climate signal propagating across the Northern Hemisphere through indices of a synchronized network. Dissertation, University of Colorado at Boulder

Wyatt MG & Curry JA (2014) Role for Eurasian Arctic shelf sea ice in a secularly varying hemispheric climate signal during the 20th century. Climate dynamics 42 (9–10) 2763–2782

Yang H, Li Q, Wang K et al (2015) Decomposing the meridional heat transport in the climate system. Climate Dynamics 44 (9–10) 2751–2768

Yu L & Weller RA (2007) Objectively analyzed air–sea heat fluxes for the global ice-free oceans (1981–2005). Bulletin of the American Meteorological Society 88 (4) 527–540

Zachos J, Pagani M, Sloan L et al (2001) Trends, rhythms, and aberrations in global climate 65 Ma to present. Science 292 (5517) 686–693

Zhu J, Xie A, Qin X et al (2021) An Assessment of ERA5 Reanalysis for Antarctic Near-Surface Air Temperature. Atmosphere 12 (2) 217

Zielinski GA, Mayewski PA, Meeker LD et al (1996) A 110,000-yr record of explosive volcanism from the GISP2 (Greenland) ice core. Quaternary Research 45 (2) 109–118

MODERN GLOBAL WARMING

"... is there any reason, really, to think that our modern science may not suffer from similar blunders? In fact, the more successful the fact, the more worrisome it may be. Really successful facts have a tendency to become impregnable to revision."
Stuart Firestein (2012)

12.1 Introduction

Modern Global Warming (MGW) is the change in climate that has been taking place from the coldest period of the Little Ice Age (LIA) to the present. It spans roughly 300 years. It is characterized by a preponderance of warming periods over cooling periods, resulting in the warming of the planet, expansion of tropical areas, cryosphere contraction, sea level rise (SLR), and a change in dominant weather and precipitation patterns. LIA's nadir appears to have been the late Maunder Minimum period of 1660–1715 (Luterbacher et al. 2001; Fig. 12.1). Afterwards, most of the eighteen century was warmer, but was followed by an intense cold relapse in 1790–1820, before the LIA finally ended around 1840. The LIA is the closest the planet has been in 12,000 years to returning to glacial conditions. All over the world most glaciers reached their maximum Holocene extent in the LIA (Solomina et al. 2015; see Fig. 4.12). But for the past 300 years, MGW has interrupted the Neoglacial cooling trend of the last five millennia. The last 70 years of MGW (25%) have seen considerable, and increasing, human-caused emissions of greenhouse gases (GHG). There is great concern than this and other human actions (deforestation, cattle raising, and changes in land use) might have an important impact over climate, precipitating a climate crisis. To some authors the climate crisis is already taking place (Pierrehumbert 2019).

In the previous chapters we have reviewed how the climate has been changing for the past 800,000 years, and with greater detail for the past 12,000 years. Climate change is the norm, and climate has never been stable for long. It is within this context of past climate change that MGW must be evaluated.

12.2 Modern Global Warming is consistent with Holocene climatic cycles

It is often said that MGW is unusual because it counters a Neoglacial cooling trend that has been ongoing for several millennia. However, this is a superficial observation. Several multi-centennial warming periods have taken place within the Neoglacial cooling trend. A warming period took place between 1250–850 yr BP (AD 700–1100), leading to the Medieval Warm Period (MWP; Hegerl et al. 2007; Fig. 12.1), and another one at 2800–2500 yr BP (850–550 BC), leading to the Roman Warm Period (RWP; Drake 2012).

Current models propose that the world would be cooling if it wasn't for the human influence on climate (Meehl

Fig. 12.1 Climate variability over the past 1500 years
Black curve, proxy reconstruction of 30–90°N mean annual decadally averaged temperature over land for the period AD 558–1960, plotted relative to its 1880–1960 mean. Gray shaded ranges give the 95% uncertainty bounds of decadal temperature estimates. After Hegerl et al. (2007). Red curve, AD 500–1925 low frequency component from an AD 1–1979 multi-proxy reconstruction of the Northern Hemisphere surface temperature, plotted relative to its 1880–1960 mean. After Moberg et al. (2005). Blue curve, AD 1500–2000 Northern Hemisphere ground surface temperature reconstruction from boreholes, plotted also relative to its 1880–1960 mean. After Pollack & Smerdon (2004). Main climatic periods are indicated by background color. Multi-centennial warming periods are indicated by horizontal continuous lines and vertical dashed lines.

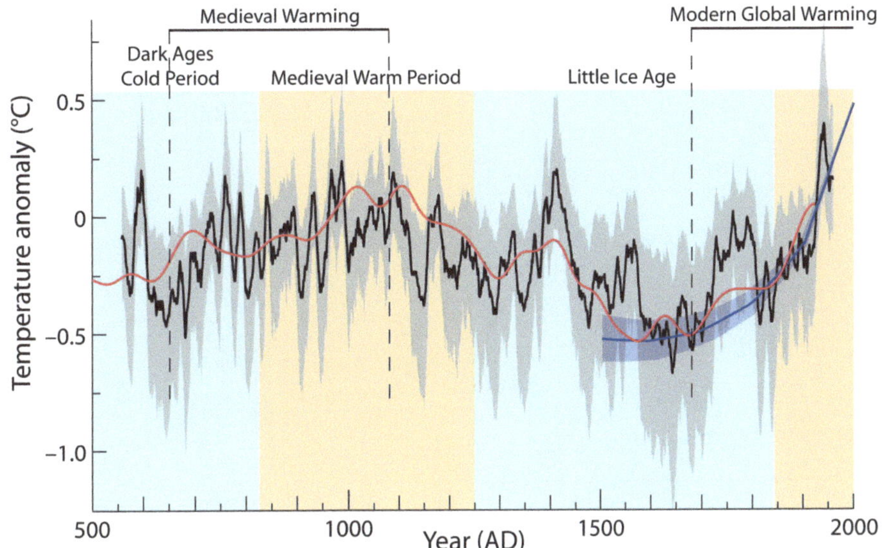

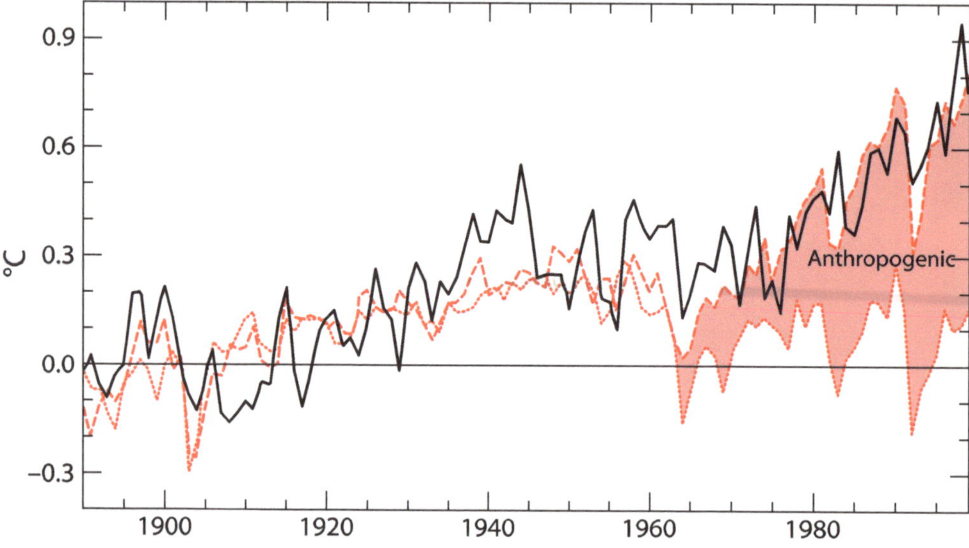

Fig. 12.2 Models simulate global cooling without anthropogenic forcing
Red dashed line, four-member ensemble mean for globally averaged surface air temperature anomalies (°C) for all forcings (volcano + solar + GHG + sulfate + ozone); the red dotted line is the ensemble mean for globally averaged temperature response to natural forcings only (volcano + solar). Department of Energy Parallel Climate Model, after Meehl et al. (2004). Red shaded area represents the anthropogenic effect, that according to the study is only significant after 1960. Black continuous line is the observations after Folland et al. (2001).

et al. 2004; Fig. 12.2). However, the proposition that the world should be cooling absent an anthropogenic effect, contradicts our knowledge of Holocene climate cycles. One of the main cycles is the c. 1000-yr Eddy cycle found in climate and solar activity proxy records of the Early and Late Holocene (see Chap. 8). The periodicity and phase of this cycle is maintained from Early to Late Holocene, and reflected in the Bond events of increased iceberg activity in the North Atlantic (see Fig. 8.2). The start of the Medieval Warming c. AD 700, and the start of the MGW at c. AD 1700 are separated by 1000 years. The peak of the MWP at c. AD 1100 and the trough of the LIA at c. AD 1600 are separated by 500 years (Fig. 12.1). Based on this cycle it can be projected that the period AD 1600–2100 should be a period of net warming, to be followed by a cooling period, AD 2100–2600, if the cycle maintains its beat (Fig. 12.3).

There is certainly a significant anthropogenic contribution to MGW, but it is clear that warming at this time is not unusual, and in fact, it is about what should be expected. The most logical conclusion is that natural warming is contributing to the observed warming. If models are not capable of simulating this natural warming, of millennial cyclic origin, then models must be wrong, and our knowledge of climate change is insufficient.

12.3 Modern Global Warming is within Holocene variability

How unusual is the warming observed during MGW? This is a very difficult question to answer. Temperature is an intrinsic intensive property that is changing during the course of a day at any point on the surface of the planet in an unpredictable direction and rate. If there is a global average temperature, we have no way of measuring it. However, we have devised methods of measuring temperature (or radiation) at different points on the surface (with huge areas unsampled) or in the atmosphere. A consistent mathematical treatment of this data gives a consistent value that we term average temperature, although it is not a temperature, but a conversion of intrinsic intensive measurements into an extrinsic extensive value using multiple assumptions.

However, the global average temperature concept is useful as the calculated value shows much less change over time than the measured values, and we term that change "anomaly," wrongly implying that it should be constant over time. The change in the global anomaly over the years shows a correlation to real physical and biological phenomena, like length of the growing season, extent of the cryosphere, and SLR, among others, and thus it is useful. However, two dangers should be avoided when dealing with the global temperature anomaly. The first is

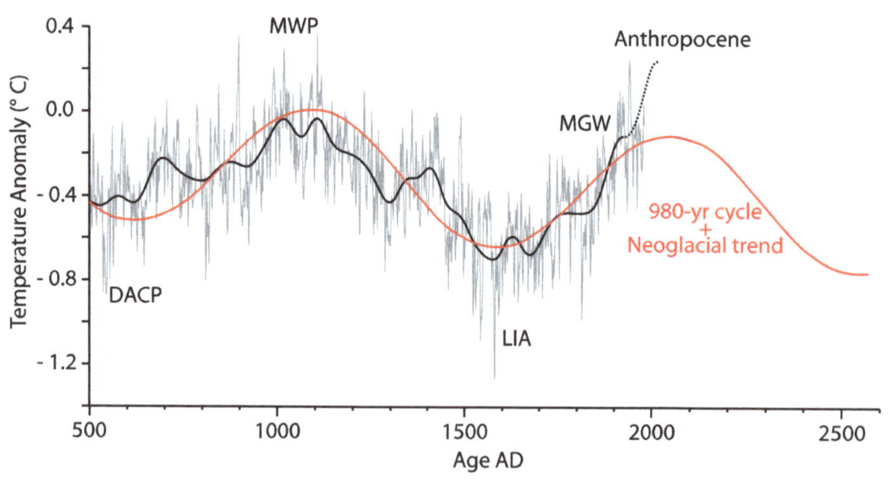

Fig. 12.3 Warming and cooling periods of the past 1500 years, fitted to known climate cyclic behavior
Moberg et al. (2005), reconstruction of Northern Hemisphere temperature anomaly for the period AD 500–1978 (thin grey curve), and its low frequency component (black curve). The 980-year Eddy cycle is shown by the red sinusoidal, with a declining Neoglacial trend of −0.2 °C/millennium. As Moberg's reconstruction ends in 1978, the dotted line represents the 1975–2000 warming, that is similar in magnitude to the 1910–1945 warming. DACP, Dark Ages Cold Period. MWP, Medieval Warm Period. LIA, Little Ice Age. MGW, Modern Global Warming. Peak natural warming is expected in AD 2050–2100.

using the same units for temperature as for the temperature anomaly. The degrees in the temperature anomaly are different from the degrees in temperature, since the connection to physical degrees is lost in the conversion from intrinsic to extrinsic. Many authors are unaware of this problem and attempt to compare proxy derived local temperatures to an instrumental-measured calculated global anomaly. Also, the precision given in a temperature anomaly is not a precision in measurement, but a precision in calculation. This is also important as the real uncertainty cannot be calculated, due to multiple assumptions in the process that are not properly evaluated. Another danger is that averaging changes in temperature ignores differences in enthalpy (changes in internal energy due to changes in heat content). Due to its low humidity, especially in winter, big changes in Arctic air temperature can take place with small changes in heat content. The weight that Arctic air temperatures should have in a global average is an unresolved question that is biasing instrumental temperature anomalies, relative to temperature proxies.

So, going back to the problem we are now calculating a global temperature with our chosen method, but with no way to relate it to anything similar from the past. Even our calculated anomaly becomes pure fiction (if it wasn't already) when moving into the 19th century. The way we estimate climate change from the past is through proxies. The relationship of proxies to temperature is convoluted. Some proxies respond to summer temperature changes, while others to winter or spring temperatures. Other factors, like rate of deposition, rate of upwelling, precipitation, cloud cover, storm frequency, or wind, might affect a proxy often without a clear possibility of correction, as the researcher might be unaware of the bias. The resolution of proxies cannot match the resolution of our measurements. The 2014–16 El Niño that increased our annual global anomaly by 0.4°C for a short period would not be resolved by most proxies. And proxies are always local in nature. That's why most serious scientists abstain from attempting to calculate past global temperature averages from collections of proxies, and avoid linking them to modern instrumental temperature anomalies. They are two very different things.

An independent means of assessing how unusual MGW is within the Holocene is through biology. Trees grow on the slope of mountains up to a certain altitude—the treeline—above which they are unable to survive. Temperature is the primary control on treeline forma-

tion and maintenance (Körner 2007) and consequently the treeline has been moving higher over the past century at 52% of the locations studied, while receding at only 1% (Harsch et al. 2009). The advances have taken place mainly in the extra-tropical NH, where more warming has been experienced, and particularly in locations where winter warming has been stronger. This aspect is also important as it indicates that winter tree survival might be a limiting factor for treeline altitude, and not only growth-season mean temperature.

There are many studies all over the NH showing that the treeline was much higher than present during the HCO. In the Italian Alps, it was 400 m higher than today from 8.4 and 4 ka (Badino et al. 2018), and 150–200 m higher between 9 and 2.5 ka in the Swiss Central Alps (Tinner & Theurillat 2003; Fig. 12.4). In the Pyrenees it was 400 m higher than the current treeline (Cunill et al. 2012). In the Swedish Scandes 600–700 m higher between 9.5–6.5 ka (Kullman 2017). In the British Columbia it was 235 m higher from 10.6 to 7.5 ka (Pisaric et al. 2003). In New Zealand's South Island, where mean annual temperatures were at least 1.5°C warmer than present in the early Holocene, treelines were lower however, suggesting shorter and cooler summers (McGlone et al. 2011).

Randin et al. (2013) showed that half of the 18 deciduous tree species they studied in Europe filled their thermal niche both at high latitude and high altitude, while 7 reached their latitudinal thermal limit, but not their elevational limit, where competition for space is stronger. We must take into account that present elevated CO_2 levels might give current trees an advantage over early Holocene trees, and that the NH-land is the region that has experienced the strongest recent climate warming. The difference in treeline altitude between now and the early Holocene imply that MGW is not unusual enough to have returned us to Holocene Climatic Optimum conditions. Therefore, present global warming is within Holocene variability. Reasoner and Tinner (2009) quantify the summer temperature difference in the Alps between now and the Holocene Optimum as: *"Assuming constant lapse rates of 0.7 °C / 100 m, it is possible to estimate the range of Holocene temperature oscillations in the Alps to 0.8–1.2 °C between 10,500 and 4,000 cal. yBP, when average (summer) temperatures were about 0.8–1.2 °C higher than today."*

The cryosphere confirms that present conditions are within Holocene variability, as globally glaciers reached

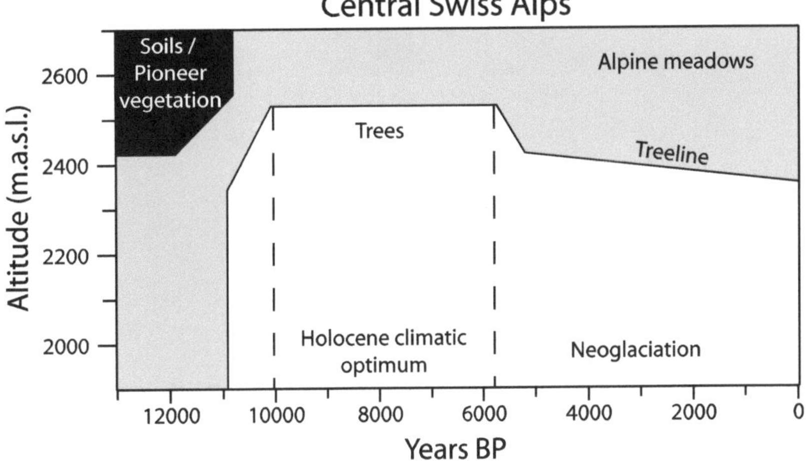

Fig. 12.4 Holocene treeline changes in the Alps Fluctuations of the treeline in the Swiss Central Alps during the Holocene. The limits of the vegetational zones are placed between the sites according to the presence of the respective vegetation type as inferred by macrofossil analysis. Altitude in meters above sea-level. After Tinner & Theurillat 2003. Current treeline in the Swiss Central Alps is 150–200 m below Holocene Climatic Optimum treeline limit.

their shortest extent at times between 10,000 and 5,000 years ago, when many glaciers that now exist were absent (Solomina et al. 2015). Arctic sea ice was also very much reduced during the Holocene Climatic Optimum compared to present day, and perhaps ice free (less than 1 million km^2) during the summers at some periods (Jakobsson et al. 2010; Stein et al. 2017).

MGW is not unusual by Holocene standards in its amplitude, duration, and timing. We cannot rule out that the magnitude of the warming, while not unusual for the Holocene, is unusual for the Neoglaciation that, after all, is characterized by a multi-millennial downward trend in temperature. If that is the case however it is very difficult to demonstrate because of the mentioned problems of comparing present and past temperatures. Circumstantial evidence supports that the RWP was warmer than present (Holzhauser et al. 2005), but the RWP was extraordinarily long, a millennium, so some of its effects might be due to the long time spent in a warm state not necessarily warmer than the present.

12.4 Modern Global Warming coincides with an increase in solar activity

Sunspots and cosmogenic isotopes indicate that solar activity has been increasing for the past 300 years. The average sunspot number (yearly data, WDC–SILSO) over the period 1700–2021 is 79 sunspots. There are three periods of more than two solar cycles ($\geq$ 30 years) that on average have significantly more than 79 sunspots, and constitute periods of higher solar activity. The first was the period of 1765–1795, formed by solar cycles 2–4. The second was longer, 1835–1875, formed by solar cycles 8–11. And the longest has been the period 1935–2004, defined by solar cycles 17–23, that has been termed the Modern Solar Maximum (MSM; Solanki et al. 2004; see also Fig. 11.11). All three coincide with periods of increasing temperatures according to proxies (Mohberg et al. 2005; Hegerl et al. 2007) and more recently instruments. And they are sepa-

rated by periods of low solar activity known as extended minima, constituting the Maunder, Dalton, and Gleissberg minima, that coincide with periods of decreasing temperature (Fig. 12.5).

There is no linear correlation between solar activity and temperature, however, periods of several solar cycles that deviate significantly from average tend to coincide with periods of warming and cooling. de Larminat (2016) has demonstrated that using dynamical systems identification the MWP and the LIA are better explained by a much higher climate sensitivity to solar forcing than proposed by IPCC. The obvious conclusion is that the MSM has had a bigger contribution to MGW than currently accepted by IPCC.

The MSM does not stand out for its level of solar activity, as solar cycles of similar activity have taken place in every century for which there are sunspot records. What is unusual in the MSM is its length. It lasts for seven solar cycles and seven decades when solar activity was significantly higher than average. Even solar cycle 20 was slightly above average. Knowing from cosmogenic records that solar activity was very low during the LIA, a similarly long period of high solar activity must have not taken place for at least 600 years. The extraordinary coincidence of a millennial-frequency warming period with a millennial-frequency period of high solar activity should raise all type of questions. Instead solar variability is assigned no role in MGW by IPCC (see Fig. 9.12).

12.5 Modern Global Warming displays an unusual cryosphere response

If MGW is not unusual by Holocene standards, it becomes important to inquire about the climatic response to the increased atmospheric CO_2 levels. Is there anything unusual about MGW? The answer is a clear yes. The cryosphere (with the exception of Antarctica) is showing a very unusual response to MGW. For the last two decades glaciologists have recognized that global glacier changes over the past 170 years are not cyclical and greatly exceed

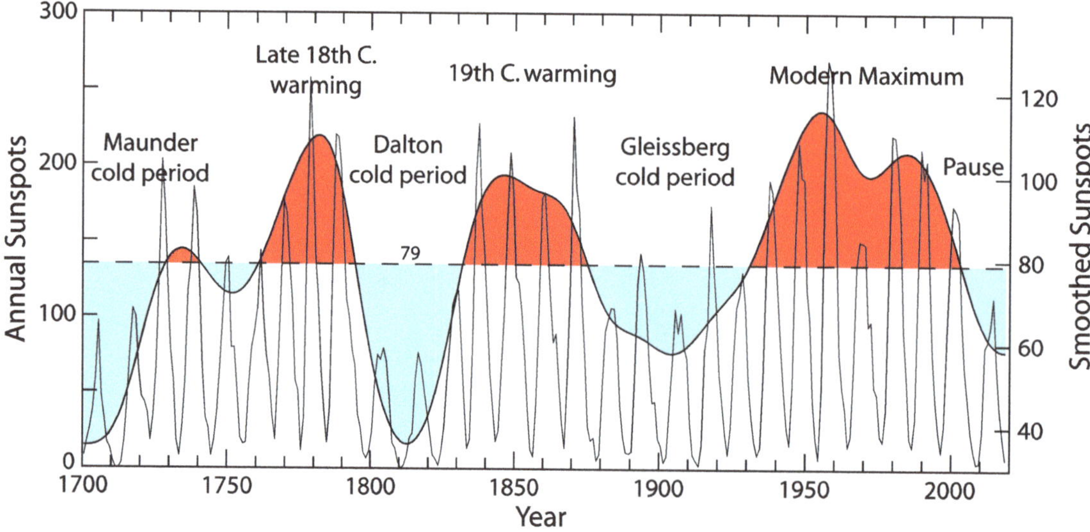

Fig. 12.5 Solar activity since 1700
Thin line, left scale, annual sunspots (WDC–SILSO). Thick line, right scale, gaussian smoothing of the annual sunspot record highlighting multi-cycle changes in solar activity. Dashed line, right scale, average number of annual sunspots for the 1700–2018 period. Red areas define periods of extended above average solar activity. Blue areas define periods of extended below average solar activity.

Fig. 12.6 Modern glacier retreat is not cyclical
a) Time-distance diagram for glacier extent on the Central Cumberland Peninsula (Baffin Island) through the Holocene. After Briner et al. (2009). **b)** Glacier fluctuations in the Himalaya and Karakoram up to 1980, defined by radiocarbon dating. After Owen (2009). **c)** Relative glacier extent fluctuations in western Canada during the Holocene. After Koch & Clague (2006). **d)** Combined equilibrium-line altitude (ELA) variations along the South–North coastal transect in Norway. The glacier growth index is obtained by adding standardized ELA estimates from southern and northern Norway. After Bakke et al. (2008).

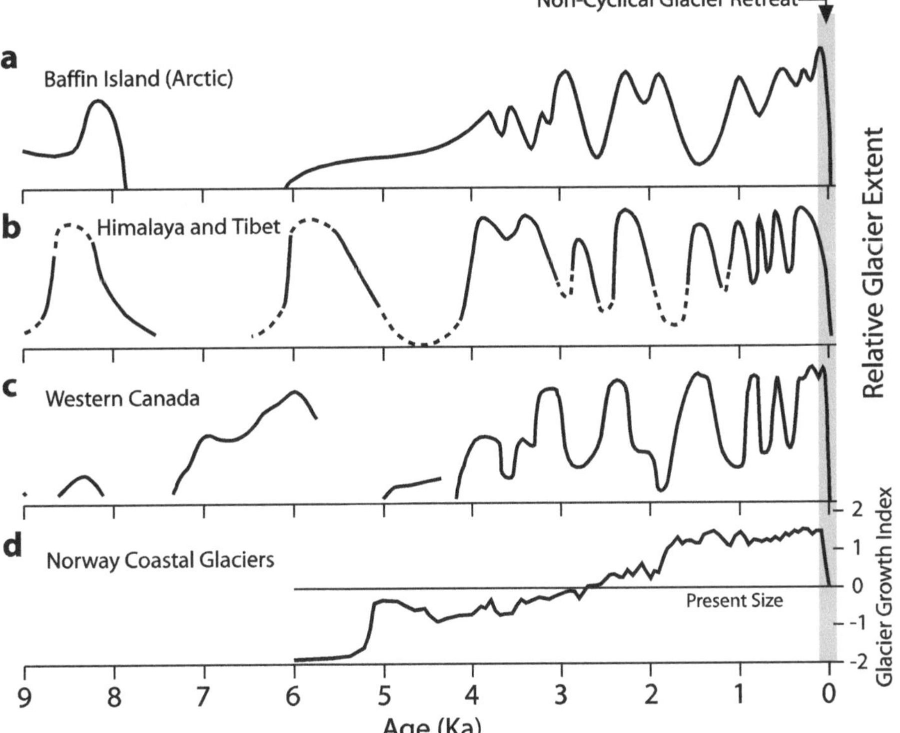

the range of the previously known periodic variations of glaciers (Solomina et al. 2008; Fig. 12.6). Koch et al. (2014), attest that the global scope and magnitude of glacier retreat likely exceed the natural variability of the climate system and cannot be explained by natural forcing alone. Goehring et al. (2012) state that after 5 kyr BP, the Rhône Glacier was larger than today, and its present extent therefore likely represents its smallest since the middle Holocene. Solomina et al. (2008) defend that Alpine glacier volumes have become smaller now than during at least the past 5000 years. And Bakke et al. (2008; Fig. 12.6d) have measured a retreat of maritime glaciers along western Scandinavia over the last century that is unprecedented in the entire Neoglacial period spanning the last 5200 years. Solomina et al. (2016) resume the global glacier situation: *"The current globally widespread glacier retreat is unusual in the context of the past two millennia and, indeed, for the whole Holocene. Contemporary glacier retreat breaks a long-term trend of increased glacier activity that dominated the past several millennia. The trend of glacier retreat is global, and the rate of this retreat has increased in the past few decades. The observed widespread glacial retreat in the past 100–150 years requires additional forcing outside the realm of natural changes for their explanation."*

Global glacier retreat is probably the only climate-associated phenomenon that shows a clear acceleration over the past decades. The World Glacier Monitoring Service, an organization participated by 32 countries, holds a dataset of 42,000 glacier front variations since 1600, that show that the rates of early 21st-century glacier mass loss are without precedent on a global scale, at least since 1850 (Zemp et al. 2015).

Unusual glacier retreat is confirmed by the loss of small permanent ice patches, also known as glacierets. These permanent ice patches have captured and preserved archeological organic remains during their long existence,

and the remains are now being released as they melt. This is the origin of the new subfield of 'ice patch' archeology, that has developed in three regions, North America, the Alps, and Norway. Some plant remains, like tree trunks, or the Quelccaya plants dated at c. 5200 BP (Thompson et al. 2006) are naturally occurring and their burial in ice is related only to climatic conditions, but archeological remains (Fig. 12.7b) reflect human activity and are thus more complex. Alpine findings are related to the use of mountain passes when conditions improved, and their dating shows asynchrony with North American findings, associated with summer hunting of caribou, that takes refuge from insects over ice patches. Thus, Alpine findings are more frequent from warm phases and at the beginning of cold phases, while North American findings are more frequent from cold phases when ice patches became more widespread. (Fig. 12.7c). The finding of North American ice-patch remains reveals a millennial periodicity coincident with the millennial Eddy solar cycle analyzed in section 8.2. Most organic remains, like leather, caribou dung, or corpses (Ötzi, dated at c. 5200 BP; Fig. 12.7a; see Sect. 6.6), are not preserved when exposed for even relatively short periods, and it is clear that they have remained continuously frozen since first buried in ice. Their present climate induced unburial is clear evidence that small permanent ice patches are experiencing a reduction not seen since the Mid-Holocene Transition. *"The ['ice patch' archeology] field is characterized by a sense of urgency about recovering and preserving both those occasional human remains melting from alpine ice and newly-exposed artifacts of rare and fragile organic technology"* (Reckin 2013).

Arctic sea-ice has displayed a different behavior to glaciers, with a very sharp reduction at the turn of the century (1996–2007), losing 30% of its summer extent in just a decade. This reduction is not outside Holocene variability, as multiple studies document a much lower Arctic sea

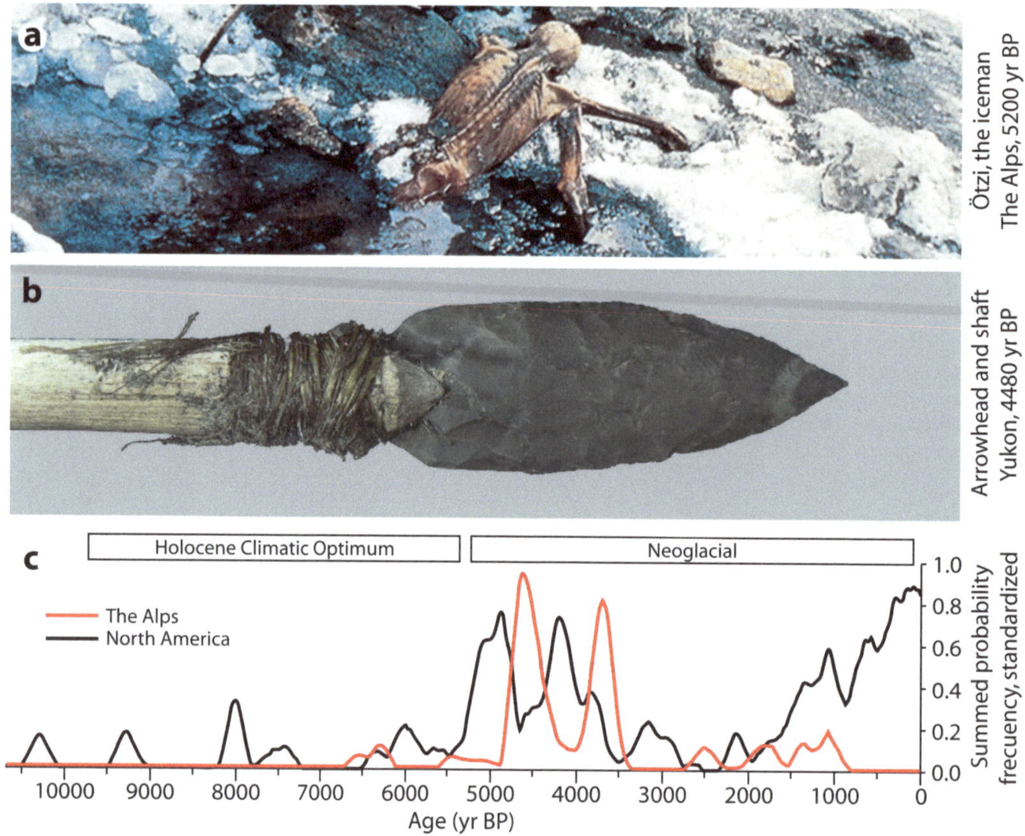

Fig. 12.7 Ice-patch archeology, evidence of non-cyclical cryosphere reduction
a) Ötzi, the alpine iceman, in situ before his removal from the site at Niederjoch, Italy, in 1991. Dated at c. 5200 BP. Image by Paul Hanny, © South Tyrol Museum of Archaeology – www.iceman.it reproduced with permission. **b)** Stone projectile point hafted with sinew to a wooden shaft radiocarbon dated to 4480 ± 60 BP, recovered from an ice patch in Yukon. © Government of Yukon, reproduced with permission. **c)** Summed probability of available radiocarbon dates for the Alps (red line) and North America (black line) ice patch archeological findings. They demonstrate the probability that a date from each collection will fall within a particular period and exclude typologically-dated material (most of which is Roman in age). They have been smoothed to a 200-year interval to remove extraneous noise and emphasize more general trends. Observe the millennial periodicity in North American findings that corresponds to the millennial Eddy solar cycle. After Reckin (2013).

ice extent between 9000 and 4000 BP (Belt et al. 2010; Jakobsson et al. 2010; Stranne et al. 2014; Stein et al. 2017). It has been hypothesized here (see Chap. 11) that the strong decrease in Arctic sea-ice in just a decade was partly due to a climate regime change induced by an abrupt intensification of meridional heat transport to the Arctic. The difficulty in reconstructing past sea-ice levels means that we don't have much confidence in how the present reduction in sea-ice extent compares to previous reductions during past warm periods. Nevertheless, an analysis of polar ice shelves (thick floating ice platforms), shows an unusual response in present polar ice retreat. Although ice shelves have collapsed and broken up at different times during the Holocene, this is the first time in the Holocene when a synchronous retreat in ice shelves from the Arctic and both sides of the Antarctic Peninsula is known to have occurred (Hodgson 2011).

Is this unusual, non-cyclic, cryosphere reduction due to the increase in temperature? Additional factors resulting from the contemporary industrial development complicate the interpretation. It is known that an increase in light-absorbing particles (black carbon, organic carbon, and dust) in snow and ice causes a decrease in albedo and accelerates melting (Bond et al. 2013). Decreased albedo also accelerates snow aging. When melting uncovers the underlying surface it triggers a strong snow-albedo feed-back that warms the surface and the air above it. The albedo effect from light-absorbing particles depends a lot more on their abundance than on temperature changes or CO_2 changes. Black carbon and organic carbon are components of soot resulting from the combustion of fossil fuels, biofuels, and burning of biomass, and have increased not only with the increase in fossil fuel use, but also with industrial development and population increase. While the evidence is strong that the unusual cryosphere reduction has an anthropogenic origin, the relative contribution from changes in temperature, GHGs, and light-absorbing particles cannot be properly assessed. The recent outsized cryosphere changes with respect to other global warming effects constitute evidence that the increase in light-absorbing particles from soot must be an important factor.

Antarctica is an exception to the global reduction of the cryosphere. The continent hasn't warmed for the past 200 years (Fig. 12.8), and it is currently debated if Antarctic melting is contributing to SLR and by how much (Zwally et al. 2015). Antarctic lack of climatic response to MGW and CO_2 increase is not well understood, and it might have to do with the exceptional conditions of the continent that make it unique in many aspects.

12.6 Extremely unusual CO₂ levels during the last quarter of Modern Global Warming

Another stark difference between MGW and previous Holocene warming periods is the great increase in CO_2 levels. Atmospheric CO_2 has been increasing since c. 1785, following the 18th century warming, but the rate of increase has been growing continuously due to anthropogenic emissions, reaching the highest values in 800,000 years by the first decades of the 20th century, and it is now fast approaching a doubling of the Late Pleistocene average value of 225 ppm (Fig. 12.8). It is absolutely clear that the increase in CO_2 levels is due to human emissions, as we have emitted double the amount that has ended up in the atmosphere, the rest being taken up by the oceans and biosphere. As a result land plants are showing an important increase in global leaf area (Zhu et al. 2016), also known as greening.

The Antarctic Plateau is the only place on Earth where we can measure CO_2 levels and proxy temperatures in a consistent manner for the past 800,000 years from ice cores. Much has been written about the close correlation between CO_2 and temperature over the Pleistocene (Fig. 12.8a). The change in CO_2 levels between glacial and interglacial periods, of only 70–90 ppm, is considered by most authors too small to drive the glacial cycle, although Shakun et al. (2012) defend that the CO_2 change at termi-nations can explain a large part of the temperature increase during deglaciations. But we can test the hypothesis because over the last 200 years CO_2 levels have increased by 125 ppm, an increase comparable to that of a glacial termination in terms of CO_2 forcing. Surprisingly, Antarctica shows absolutely no warming for the past 200 years (Schneider et al. 2006; Fig. 12.8b). The only place where we can measure both past temperature and past CO_2 levels with confidence shows no temperature response to the huge increase in CO_2 over for the last two centuries. This evidence supports that CO_2 has very little effect over Antarctic temperature, if any, and it cannot be responsible for the observed correlation over the past 800,000 years. It also raises doubts over the proposed role of CO_2 over glacial terminations and during MGW.

12.7 Relationship between CO₂ levels and temperature during Modern Global Warming

Physics shows that adding carbon dioxide leads to warming under laboratory conditions. It is generally assumed that a doubling of CO_2 should produce a direct forcing of 3.7 W/m² (IPCC–Third Assessment Report, Ramaswamy et al. 2001), that translates to a warming of 1 °C (by differentiating the Stefan–Boltzmann equation) to 1.2 °C (by models taking into account latitude and season). But that is a maximum value valid only if total energy outflow is the

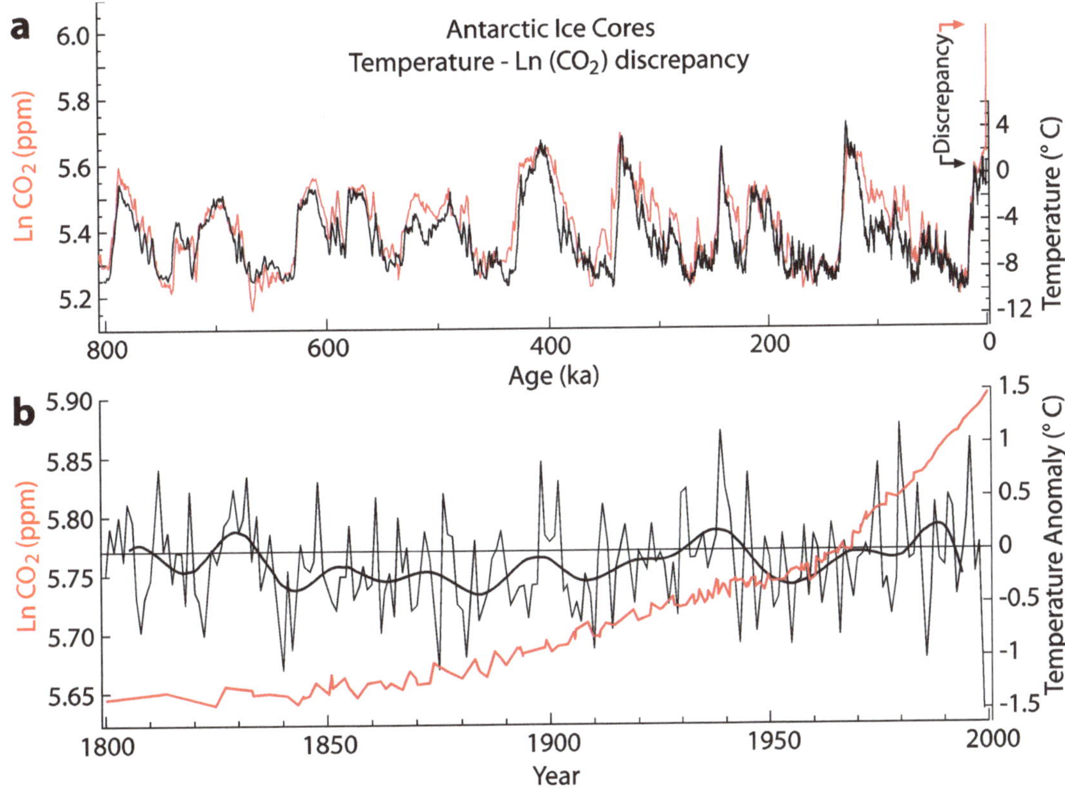

Fig. 12.8 Antarctic ice cores temperature–CO₂ discrepancy
a) Temperature curve (black) for the past 800,000 years from EPICA Dome C Ice Core 800KYr Deuterium data. After Jouzel et al. (2007). CO_2 curve (red) from Antarctic Ice Cores Revised 800 kyr CO_2 data (to 2001). After Bereiter et al. 2015; and from NOAA annual mean CO_2 data (2002–2017). Due to the logarithmic effect of CO_2 on temperature, the comparison is more appropriately done with the Ln(CO_2). The correlation shows a very big discrepancy over the last 200 years. **b)** CO_2 curve (red) as in (a). Temperature curve (black) for the past 200 years from 5 high resolution Antarctic ice cores. After Schneider et al. (2006). No temperature change is observed in response to the massive increase in CO_2.

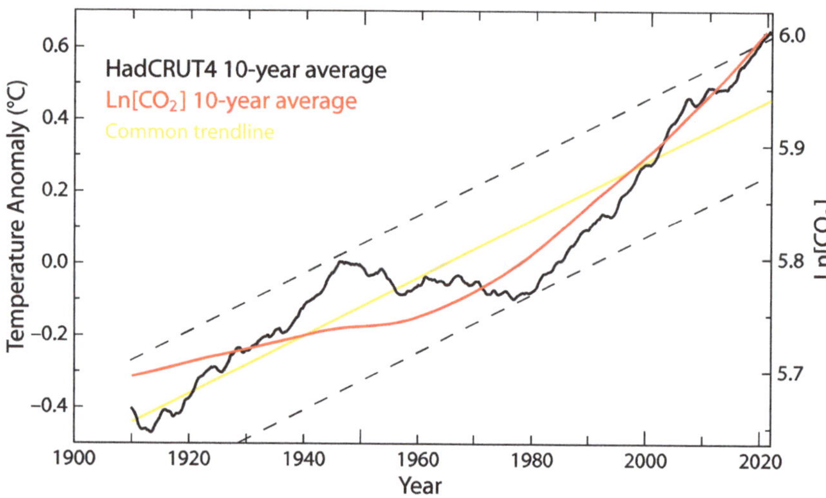

Fig. 12.9 The difference between temperature increase and CO₂ increase
Thick black curve, HadCRUT4 10-year trailing moving average surface temperature anomaly, relative to 1961–1990, from 1900 to 2021. Data from UK Met Office. Thick red curve, 10-year trailing moving average of the natural logarithm of the 1900–2021 annual atmospheric CO₂ concentration (ppm). CO₂ levels are from Law Dome (Etheridge et al. 1996) to 1958 and from NOAA afterwards. Thick orange, both curves have been plotted having the same trendline to avoid visual bias.

same as radiative outflow. As there is also conduction, convection, and evaporation, the final warming without feedbacks is probably less. Then we have the problem of feedbacks, which are unknown and can't be properly measured (see Sect. 9.5). For some of the feedbacks, like cloud cover we don't even know the sign of their contribution. And they are huge, a 1% change in albedo has a radiative effect of 3.4 W/m² (Farmer & Cook 2013), almost equivalent to a full doubling of CO₂. So, we cannot measure how much the Earth has warmed in response to the increase in CO₂ for the past 70 years, and how much for other causes.

Looking at borehole records and proxy reconstructions (Fig. 12.1), it becomes very clear that most of the acceleration in the rate of MGW took place between 1700 and 1950, when little human-caused GHGs were produced. The rate of warming has not changed much since 1950, despite the bulk of GHGs being emitted in these past 70 years. However, if the increase in global average temperature over the past 7 decades was mainly a consequence of the rapid increase in CO₂, the rate of temperature change should show dependence on the rate of change of the natural logarithm of CO₂ concentration. This is because the proposed link between CO₂ and temperature is

based on a molecular mechanism where every added molecule has slightly less effect than the previous. Even accounting for the logarithmic response of global average temperatures to CO₂, the curves for proposed cause and effect are diverging (Fig. 12.9). The global temperature anomaly between 1900 and 2021 is not significantly different from a linear trend. On the other hand, atmospheric CO₂ increase has been so fast over the 1958–2021 period that the rate of change of its logarithm displays a pronounced acceleration (Fig. 12.9).

The lack of MGW acceleration during the 20th–21st centuries can be more readily appreciated when looking at the change in warming rate (decadal trend change; Fig. 12.10). Over that period the warming rate has been oscillating between −0.2 and +0.4 °C/decade with an average of +0.16 °C/decade. Neither the warming rate maximum, nor the length of the warming periods have increased despite the huge increase in CO₂ levels. The expected warming effect of the additional CO₂ is not perceptible in warming rates. What can be seen in the warming rate record is that cooling periods have become less intense, from −0.4 °C/decade in the late 19th century, to −0.2 °C/decade in the mid-20th century, to zero in the 21st century pause. This decrease in cooling rate over time is a feature of MGW.

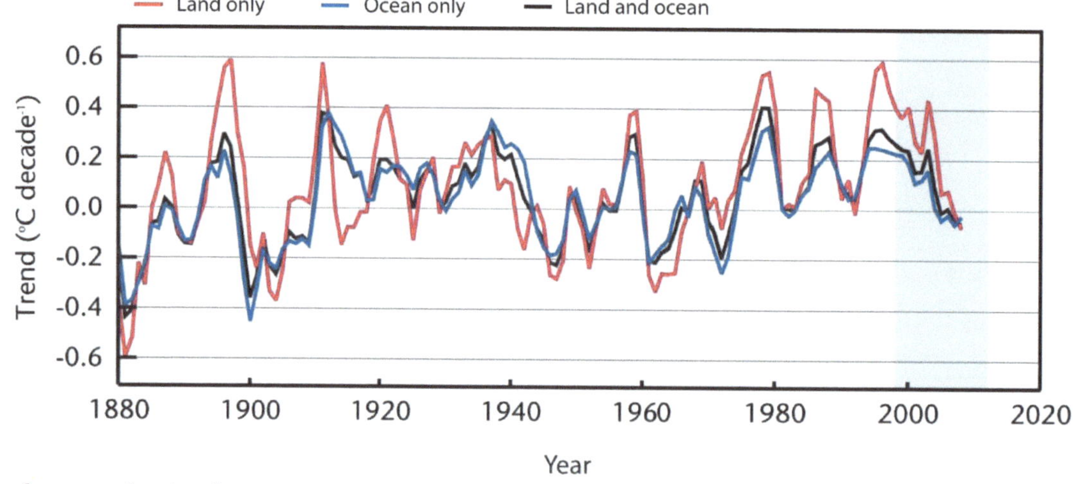

Fig. 12.10 Surface warming trend
Running nine-year trends in surface warming. Red line, land only. Blue line, ocean only. Black line, land and ocean combined. The recent slowdown in global warming is highlighted by the grey area. Reproduced from: UK Met Office 2013, Synopsis report CSc 02. The recent pause in global warming (2): What are the potential causes? Contains public sector information licensed under the Open Government Licence v2.0

Fig. 12.11 Atmospheric CO₂ and global surface temperature rate of change
a) Thin black curve, 12-month rate of change in atmospheric CO_2 in ppm/year. Thick black curve, gaussian smoothing. Red curve, quadratic regression to the 12-month rate of change in atmospheric CO_2. Mauna Loa CO_2 monthly mean data is from NOAA. **b)** Black curve, 1951-2021 centered averaged 181-month rate of change in Global Surface Average Temperature (GSAT) in °C/year. Red curve, quadratic regression to the averaged 181-month rate of change in GSAT. Circle indicates the pause in global warming, a 17-year period of no warming centered in 2006. GSAT monthly mean data is from Hadley Climate Research Unit, UK Met Office.

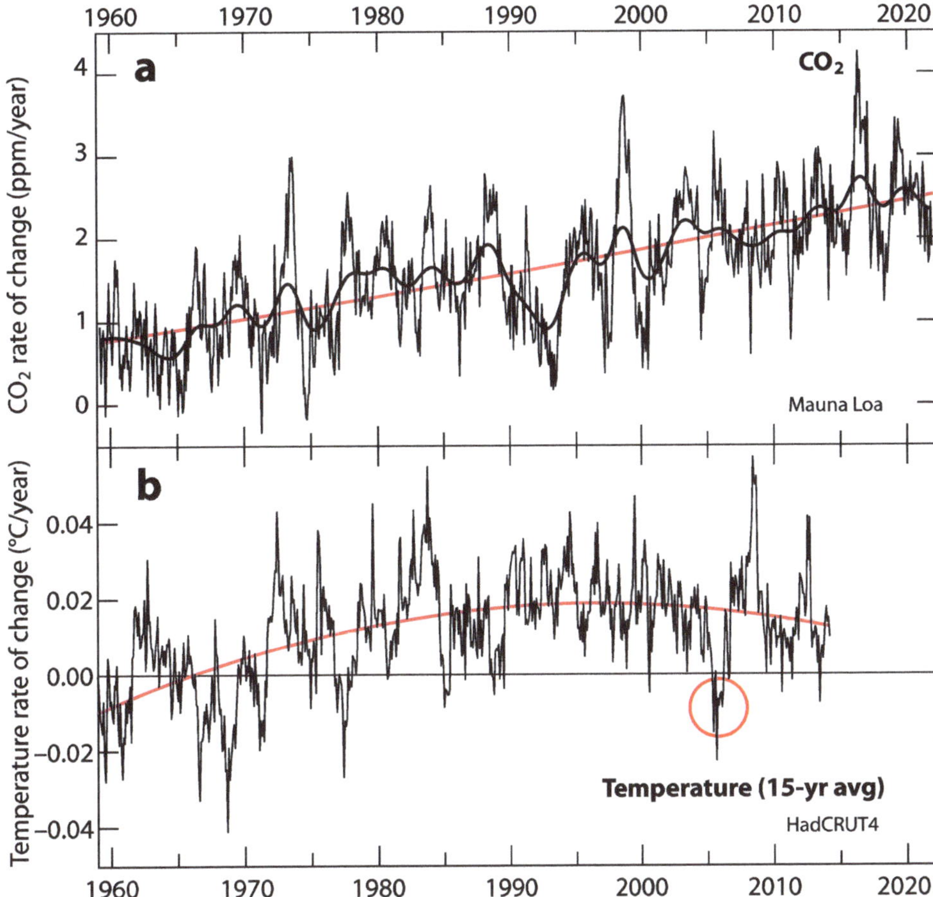

The world is warming because it cools less during cooling periods, not because it warms more during warming periods. The reasons for this are unclear, and not discussed often in the scientific literature. There is a coincidental reduction in periods of very low solar activity, that also usually coincide with cooling periods (see Sect. 12.10 below), but other factors cannot be ruled out, including an effect from increased CO_2 levels at reducing the severity of cooling periods, or a reduction in volcanic activity.

Figure 12.10 appears to indicate that global warming has decelerated over the past decades. This should not be possible, as according to theory the rapid increase in atmospheric CO_2 levels should be driving a long-term acceleration in the rate of warming. To research this issue in more detail the annual rates of increase in atmospheric CO_2 (Mauna Loa, from NOAA) and in global surface average temperature (HadCRUT 4, from UK Met Office) are compared in Fig. 12.11. CO_2 rate of change has been increasing almost linearly from 0.8 to 2.5 ppm/year between 1959 and 2021 (Fig. 12.11a). The acceleration in CO_2 increase is thus an almost constant c. 0.028 ppm/yr². The variation in global temperature rate of change is very noisy

Fig. 12.12 IPCC proposed contributions to observed surface temperature change over the period 1951–2010
IPCC assessed likely warming trends over the 1951–2010 period from well-mixed greenhouse gases, other anthropogenic forcings (including the cooling effect of aerosols and the effect of land use change), natural forcings and natural internal climate variability. The observed surface temperature change is shown in black, with the 5 to 95% uncertainty range due to observational uncertainty. The attributed warming ranges are based on observations combined with climate model simulations, in order to estimate the contribution of individual external forcings to observed warming. After IPCC (2014) Summary for policymakers.

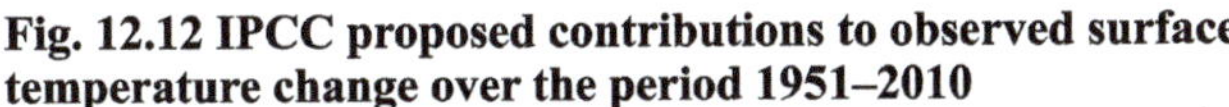

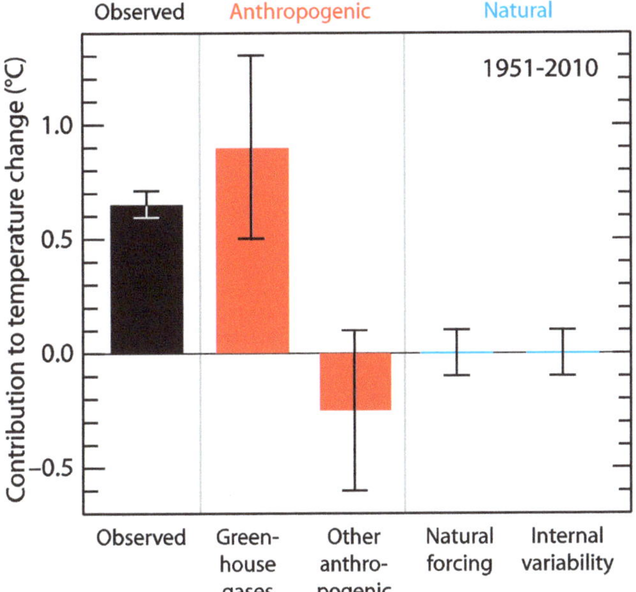

and has a very small trend. A 15-year centered moving average in the HadCRUT4 monthly temperature anomaly rate of change confirms that global warming reached maximum acceleration in the mid 90s and appears to have been decelerating over the last two decades (Fig. 12.11b). The recent scientific discussion over the existence of a global warming hiatus (Karl et al. 2015; Fyfe et al. 2016) is easily settled, as the hiatus can be mathematically defined as a 17.5 year period between February 1998 and July 2015 when the accumulated monthly rate of warming was below zero (HadCRUT4; Fig. 12.11b, circle). If the rate of global warming continues decreasing over the next decades we may expect more such periods without warming in the future. Some researchers are already warning about this possibility (Maher et al. 2020).

The lack of a corresponding MGW acceleration to the rapid increase in CO_2 over the past decades only has two possible explanations. The first is that the ongoing increase in the proposed anthropogenic forcing exactly matches in magnitude and time an ongoing decrease in natural forcing (Fig. 12.2). The second is that MGW responds more to natural causes, and only weakly to anthropogenic forcing. The first explanation constitutes an *"ad hoc"* match of hypothesis to evidence, requires an unrelated coincidence of decadal precision within a multi-century process (natural cooling should have started when we started our emissions), and it is disavowed by the IPCC, that considers natural forcing over the 1950–2010 period too small to have contributed to the observed temperature change in any direction (Fig. 12.12). That natural forcing has had no role over a 60-year period is difficult to believe.

The second explanation requires only an insufficient knowledge of the response of the climatic system to CO_2,

and an insufficient knowledge of natural forcings and climate feedbacks. That our knowledge is insufficient is clear and demonstrated every time the *"argumentum ad ignorantiam"* that "we don't know of anything else that could cause the observed warming" is used. New research into solar variability mechanisms (see Sects. 10.4 to 10.6 & Chap. 11) has produced hypotheses that indicate that solar forcing is probably not adequately represented in models. The response from the hydrological cycle to the warming constitutes another area of great uncertainty.

12.8 Uniform variation in sea level during Modern Global Warming

SLR is one of the main consequences of MGW as it is driven mainly by the addition of water from melting of the cryosphere, and thermal expansion from the warming oceans (steric SLR). A sea level reconstruction since the 18th century using tide gauge records (Jevrejeva et al. 2008; Fig. 12.13a) shows that SLR has been a feature of MGW for over two centuries. The central estimate on 20th-century average SLR is c. 1.6 mm/yr (1.2–1.9 mm/yr range), and the acceleration is usually estimated at c. 0.01 mm/yr² (Church & White 2011; Jevrejeva et al. 2014; Hogarth 2014; Fig. 12.13b). SLR displays a 60-year oscillation, like many other climatic manifestations (see Sect. 11.3). The recent period of satellite altimetry (since 1993) coincides with the crest of the oscillation, and thus shows a higher rate of SLR, c. 3 mm/yr, but no acceleration, to the surprise of some authors (Fasullo et al. 2016). If the 60-year oscillation continues affecting SLR, over the next couple of decades we should expect a deceleration of SLR rates towards c. 2 mm/yr.

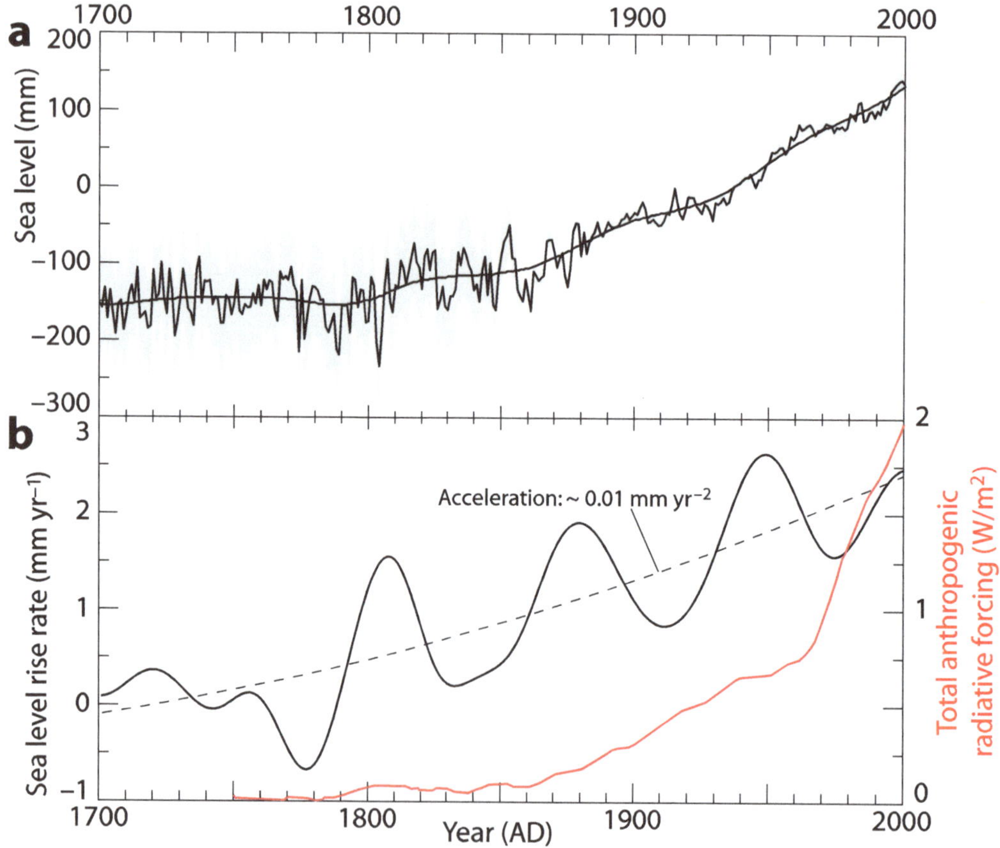

Fig. 12.13 Sea level acceleration started over 200 years ago
a) Time series of yearly global sea level calculated from 1023 tide gauge records corrected for local datum changes and glacial isostatic adjustment. Time variable trend detected by Monte–Carlo-Singular Spectrum Analysis with 30-year windows. Grey shading represents the standard errors. After Jevrejeva et al. (2008). **b)** Black line, left scale, Gaussian smoothing of the yearly rate of change in sea level, displaying periodical multidecadal variability. Dashed line, left scale, quadratic regression to sea level rate of change over time, corresponding to a sea level rise acceleration of c. 0.01 mm/yr². Red line, right scale, IPCC calculated total anthropogenic radiative forcing. After Jevrejeva et al. (2008), and Myhre et al. (2013).

As was the case with temperature, SLR precedes the big increase in emissions, and does not respond perceptibly to the great increase in anthropogenic forcing after 1960. Figure 12.13b displays the long term average SLR acceleration as a dashed line, and the increase in anthropogenic forcing (IPCC–AR5, Myhre et al. 2013) with a red line. The evidence shows that the big increase in anthropogenic forcing, has not provoked any perceptible effect on SLR acceleration. The belief that a decrease in our emissions should affect the rate of SLR has no basis in the evidence. A projection of the observed SLR and acceleration for the past 120 years gives a value of c. 300 mm more in 2100 than in 2021.

Cryosphere melting is considered the main factor driving SLR, followed by ocean temperature increase. SLR displays a small acceleration of c. 0.01 mm/yr^2 over the past two centuries (Fig. 12.13), while global temperature shows a linear increase over the past century, and the ocean is warming a lot less than the surface. It has been recently estimated from changes in atmospheric noble gases, that the ocean has warmed +0.1 °C for the past 50 years (Bereiter et al. 2018). The best candidate for causing the observed SLR acceleration is therefore the observed increase in cryosphere melting since c. 1850.

12.9 Modern Global Warming and the CO₂ hypothesis

The CO$_2$ hypothesis proposes that changes in atmospheric CO$_2$ levels are the main driver of Earth temperature changes (Lacis et al. 2010). It is based on the spectral absorption and radiation properties of certain gases, of which water vapor is by far the most abundant, and CO$_2$ is a distant second. Water vapor levels are locally determined and highly variable due to condensation. CO$_2$ levels are global, as it is a well-mixed gas that does not condense, and before industrialization it changed very slowly over time from natural causes. CO$_2$ hypothesis considers that water vapor changes are not the driving factor, but a feedback, proposing without clear evidence that the relevant causal relationship is CO$_2$ → temperature → water vapor. Past

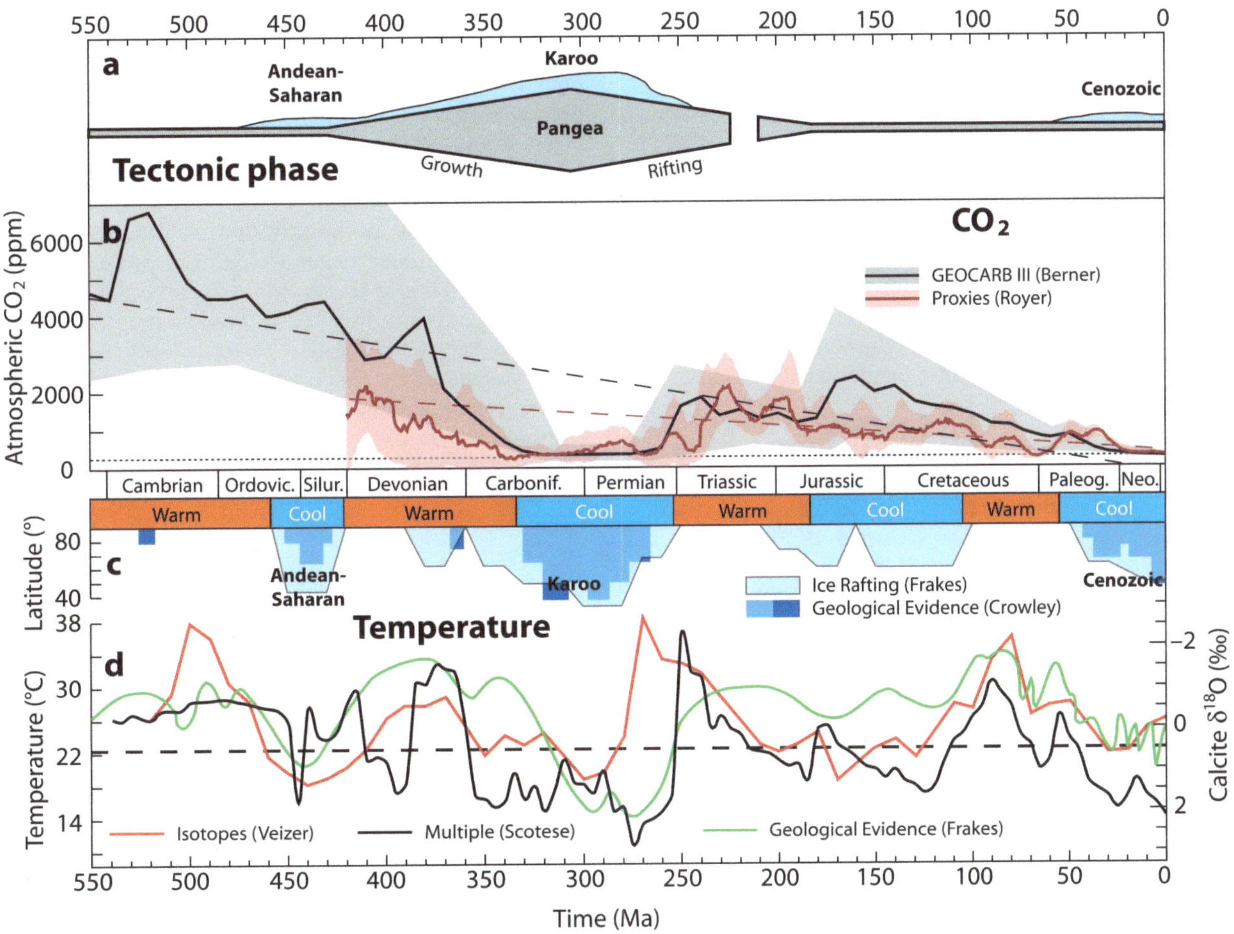

Fig. 12.14 Phanerozoic Eon conditions do not support the CO₂ hypothesis
a) Schematic representation of glacio-epochs during the past 550 million years in Earth history, and their relationship to phases of supercontinent assembly and break up. Glaciations are indicated and represented by the blue area above the scheme. After Eyles (2008). **b)** Phanerozoic atmospheric CO₂ levels in ppm from GEOCARB III model (black curve with grey estimate of error; after Berner & Kothavala 2001), Multi-proxy reconstruction (dark red curve with pink 95% confidence interval, after Foster et al. 2017). **c)** Alternating warm (orange) and cool (blue) modes, after Frakes et al. (1992), and glacial indicators, measured by the latitude reached from one of the poles. Ice rafting evidence (ice blue areas; after Frakes & Francis 1988), and direct geological evidence for glaciation (azure and blue areas; after Crowley 1998). **d)** Phanerozoic temperature reconstructions: Mean global temperature estimate (no scale) based on sedimentology and paleoecology (green curve; after Frakes 1979, and Frakes et al. 1992). Mean global temperature reconstruction based on geological evidence and isotopic studies (black curve, left scale; after Scotese 2018). Tropical marine shallow water calcite δ¹⁸O levels detrended (red curve, right scale; after Veizer et al. 2000).

water vapor levels cannot be determined, but in the distant past, cold periods of the planet (ice ages) were associated to lower CO_2 levels than warm periods, and this is the supporting evidence offered by proponents of the CO_2 hypothesis. But the interpretation of this evidence, is far from straightforward, as changes in temperature also lead to changes in CO_2. Gas solubility in the ocean is dependent on temperature, and there are huge ocean carbon stores that can affect atmospheric levels as temperature changes. In well resolved records, changes in temperature generally precede changes in CO_2 by hundreds to thousands of years. Another problem with the hypothesis is that it is generally accepted that a progressive decrease in CO_2 levels has taken place for the past 550 million years (the Phanerozoic Eon), from c. 5000 ppm in the Cambrian to c. 225 ppm in the Late Pleistocene (Berner & Kothavala 2001; Fig. 12.14b). This decrease does not appear to have produced a progressive decrease in temperature, that displays a cyclical range-bound oscillation (Frakes et al. 1992; Scotese 2018; Veizer et al. 2000; Fig. 12.14d), alternating between icehouse and hothouse conditions over the entire Phanerozoic.

The CO_2 hypothesis is not new, and can be traced to Arrhenius (1896), however it did not become the dominant hypothesis to explain temperature changes until the last warming phase of MGW started in the late 1970s, and temperature and CO_2 were both increasing. In the 20th century, while MGW was taking place, humanity embarked in the ultimate experiment to determine the validity of the CO_2 hypothesis and set about to burn huge fossil fuel natural stores while industrializing, to raise CO_2 levels beyond what the world has had in perhaps millions of years. After 70 years with CO_2 levels increasing faster than ever recorded, and above any previously recorded level for the Pleistocene, it is time to analyze the results.

1.   The world has continued warming as before. The warming during the 1975–1998 (or 1975–2009) period is not statistically significantly different from the warming during the 1910–1940 period (Jones 2010)
2.   The temperature increase since 1950 shows no discernible acceleration (Fig. 12.11b) and can be fitted to a linear increase (Fig. 12.9). The logarithm of the CO_2 increase, however, displays a very clear acceleration (Fig. 12.9). A linear relation between supposed cause and effect cannot be established
3.   Sea level has continued rising as before. Its acceleration is not responding perceptibly to the increase in anthropogenic forcing (Fig. 12.13)
4.   The cryosphere shows a non-cyclical retreat in glacier extent with evidence of acceleration (Fig. 12.6; Zemp et al. 2015). The reduction of the size of ice shelves is also unusual. We cannot distinguish if the cryosphere is responding mainly to the CO_2 increase, the temperature increase, or to the increase in light-absorbing particles

Despite CO_2 levels that are almost double the Late Pleistocene average, the climatic response is subdued, still within Holocene variability, below the Holocene Climatic Optimum and below warmer interglacials

Lack of support for the CO_2 hypothesis from Antarctic ice cores (Fig. 12.8), and from results 1–3 has forced the proponents of the hypothesis to make numerous new unsupported assumptions. They assume that all warming since 1950 is anthropogenic in nature (IPCC–AR5, Myhre et al. 2013; Fig. 12.12). That past recorded temperatures must be cooler than previously thought (Karl et al. 2015). That the oceans (Chen & Tung 2014), and volcanic eruptions (Fasullo et al. 2016), are delaying the surface warming and SLR. And essentially concluding that more time is required to observe the warming and SLR acceleration. All these might be true, but the simplest explanation (Occam's favorite) is that an important part of the warming is due to natural causes, and CO_2 only has a weak effect on temperature. If after 70 years of extremely unusual CO_2 levels, a lot more time is required to see substantive effects, then the hypothesis needs to be changed. As proposed it does not call for long delays, due to the near instantaneous effect of the atmospheric response to more CO_2. The CO_2 hypothesis is at its core an atmospheric-driven hypothesis of climate. There is a significant possibility however that the climate is actually ocean-driven, directly forced by the Sun and changes in meridional transport, and mediated by H_2O changes of state.

If the high sensitivity of the cryosphere was due mainly to the CO_2 increase, that would actually be an argument for a reduced sensitivity from the rest of the planet. The air above the cryosphere is the coldest of the planet, as it is not warmed much from below, and therefore it has the lowest humidity of the planet. The ratio of CO_2 to water vapor in the air above the cryosphere is the highest and the one that changes the most with the increase in CO_2. There is the possibility that air dryness, and its low capacity to produce water vapor in response to warming, might contribute to the cryosphere particular sensitivity to CO_2-induced warming, but it would imply the rest of the planet is less sensitive. If CO_2 sensitivity is highest over the cryosphere (except Antarctica), and lower over the rest of the planet, this points to a negative feedback by H_2O response, in its three states, to temperature changes. Antarctica doesn't show increased sensitivity because it has not been warming through the entire MGW, regardless of CO_2.

There are multiple possible H_2O temperature regulatory mechanisms, and the proposition that H_2O only acts as a fast-positive feedback to CO_2 changes is too simplistic. The huge water mass in Earth's oceans and its slow mixing, add a great thermal inertia that resists temperature changes. Atmospheric humidity determines how changes in energy translate into changes in temperature, as humid air has a higher heat capacity and responds to the same energy change with a lower temperature change than dry air. Atmospheric humidity responds very fast to temperature changes through evaporation and condensation. This mechanism is proportional to water availability, and works better above the oceans than over land, and very little over the cryosphere, inversely correlating to MGW temperature changes, that are highest in the Arctic (polar amplification), and lower over the oceans than over land. To that we must add other region-specific temperature-regulating mechanisms by H_2O. Deep convection is a tropical atmospheric phenomenon that takes place when the surface of the tropical ocean reaches 26–30 °C. The ocean flips from absorbing energy to releasing it, and convection takes the energy very high in the troposphere, cooling the ocean (Sud et al. 1999) and effectively limiting its maximum temperature. Polar sea-ice is a negative feedback that releases heat when it forms in the autumn, then absorbs heat, when it melts in spring, and it acts as an insulator preventing ocean heat loss during winter. Ice-albedo effect is a positive feedback, in that a decrease in ice reduces albedo, driving further ice loss. But ice-albedo feedback is ameliorated because ice extent moves opposite to sunlight (maximum ice coincides with minimum albedo when it is darker), and by the high inclination of the Sun's rays at polar latitudes, making water more reflective. So the albedo effect is not driving Arctic sea ice melting as demonstrated by the 13-year pause in summer Arctic sea-ice extent loss (2007–2021), after losing 30% the previous 10-year period.

Due to its huge thermal inertia, changes in its three states, cloud condensation, humidity regulation, and effective saturation of IR absorption, H_2O is a good candidate to explain the observed resistance of planetary temperatures to increasing CO_2 forcing. Only in the cryosphere, where humidity is very low and sublimation a very ineffective change of state, CO_2 increase, and light-absorbing particles increase, are driving a non-cyclical melting that affects SLR.

12.10 Modern Global Warming attribution

Climate science has an important problem of confidence regarding the cause of the recent global warming, and it is due to the Early Twentieth Century Warming (ETCW), the most pronounced warming in the instrumental climate record prior to the recent warming. Unless the ETCW can be satisfactorily explained we cannot have confidence on our recent warming attribution. The ETCW, that took place between 1910 and 1945, was of comparable magnitude (0.5 °C versus 0.6 °C) to the Late Twentieth Century Warming (LTCW) between 1975–2000, yet CO_2 emissions were several times lower during the ETCW. Between 1910–1945 atmospheric CO_2 increased on average 0.5 ppm/year, while between 1975–2000 it increased on average 1.75 ppm/year, 3.5 times faster. Recent studies (Hegerl et al. 2018) suggest that known factors at the accepted climate sensitivity values can only explain half of the warming at most. As with anything that we can't properly explain, they label the period as anomalous in a subtle attempt to dismiss its relevance. The aerosol cooling wildcard has been used to support the 1945–1975 cooling so it

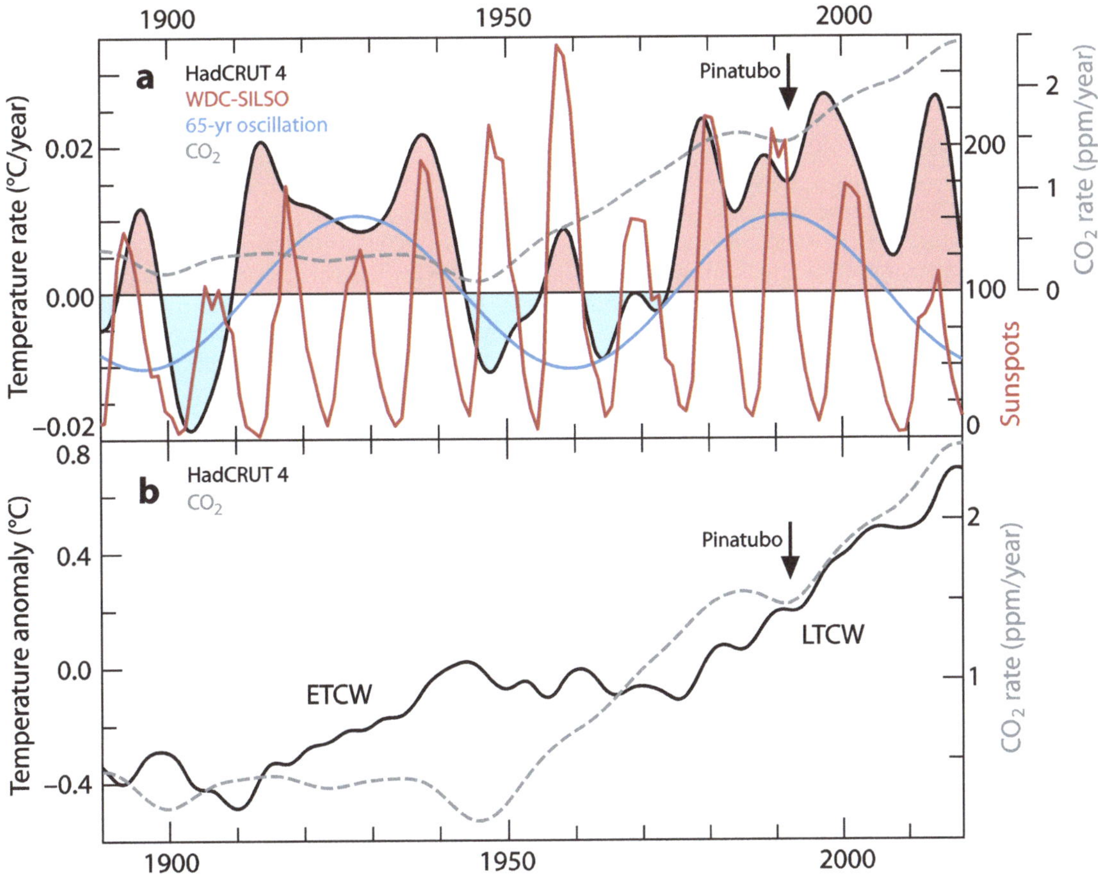

Fig. 12.15 Modern Global Warming attribution
a) Black line, left scale, gaussian smoothing of the 12-month temperature change between January 1890 and March 2021 to reveal decadal changes. Data from HadCRUT 4.6 global monthly surface temperature dataset, UK Met Office. Areas above or below the zero line are proportional to the amount of warming or cooling observed respectively. Dark red line, right inner scale, annual sunspot number. Data from WDC–SILSO, Royal Observatory of Belgium. Medium grey dashed line, right outer scale, gaussian smoothed yearly CO_2 increase. Data to 1959 from Law Dome after Etheridge et al. (1996). Data since 1959 from Mauna Loa, NOAA. Light blue sinusoid represents the 65-yr oscillation after Schlesinger & Ramankutty (1994), extrapolated to 2021 and with arbitrary amplitude. **b)** Black line, left scale, gaussian smoothing of the 13-month averaged global surface temperature anomaly with 1961–1990 baseline. Data from HadCRUT 4.6 global monthly surface temperature dataset, UK Met Office. Medium grey dashed line, right scale, gaussian smoothed yearly CO_2 increase as in (a). Pinatubo volcanic eruption is indicated.

should promote cooling also during the ETCW, making it more difficult to explain such a strong warming for a small increase in GHGs and a level of solar activity that was actually lower than the 20th century average. During 1910–1945 the average monthly sunspot number was 70.7, while during 1900–2000 it was 90.9 (WDC–SILSO monthly dataset). If solar activity, volcanic activity, GHGs, and aerosols are woefully unable to explain the ETCW, what caused it?

The known factor that is left after natural and anthropogenic forcings have been ruled out is internal variability due to changes in meridional transport (see Chap. 11). The problem is that the IPCC assigns no role to internal variability in the 1951–2010 period (IPCC 2014 SPM; Fig. 12.12). So if the ETCW is actually explainable in terms of internal variability, then the causes for the LTCW have been misdiagnosed. Already Schlesinger and Ramankutty (1994) identified that *"the rapid rise in global-mean temperature between about 1908 and 1946, and the subsequent reversal of this warming until about 1965, were the result of the oscillation in [the North Atlantic, North America, Eurasia, and Africa regions]."* Research on the ETCW in the Arctic also identifies interdecadal oceanic variability as the main factor contributing to the observed warming (Tokinaga et al. 2017).

To understand what has caused 20th century warming, both early and late, and the cooling periods prior and between them, one has to look not at the temperature, but at its rate of change. As in dynamics, where forces act on the speed of a body, not on its position, climate forcings act by changing the temperature rate of change (the derivative of temperature over time). Temperature rate data displays a low signal to noise ratio, and requires averaging or smoothing to be evaluated. A gaussian smoothing that reduces variability to the decadal rate is displayed in Fig. 12.15a. The zero line divides the temperature rate into periods when it is positive and warming takes place, increasing global temperature, and periods when it is negative and cooling takes place, decreasing global temperature. Analyzed in this way, the temperature rate shows correlation with solar activity, strongly modulated by the 65-yr oceanic oscillation (Fig. 12.15a). No doubt, part of this effect is due to the solar modulation of El Niño/Southern Oscillation (see Sect. 10.4).

The decadal variability in the temperature rate shows clear correspondence with the decadal variability in solar activity, while the interdecadal variability shows clear correspondence with a 65-yr sinusoidal oscillation representing the interdecadal oscillation with the period and phase identified by Schlesinger and Ramankutty (1994). Volcanic forcing shows only a clear effect for the 1991 Pinatubo eruption (Fig. 12.15, arrows) that briefly interrupted the increase in atmospheric CO_2, and temporarily reduced the rate of warming, affecting its correlation with solar activity during solar cycles 22 and 23. The effect of the increase in CO_2, perhaps compensated by the increase in aerosols, only becomes obvious after the Pinatubo eruption when the area below the temperature trend, that represents the amount of warming, is bigger than it should correspond to solar activity and internal variability alone.

The 1910 cold period, according to this attribution interpretation, was due to the coincidence of low solar activity and the low in the 65-yr oscillation. The 1910–1945 warming was cooperative between increasing solar

activity and the high in the 65-yr oscillation. The 1945–1975 period saw less cooling because the low in the 65-yr oscillation coincided with very high solar activity that countered it. The 1975–2005 warming period is due mainly to high solar activity and the high in the 65-yr oscillation, with a significant contribution from the CO_2 increase.

The unusual warming of the 20th century is mainly due to the coincidence of two highs in the 65-yr oscillation within the century, and an unusually high level of solar activity that reduced the mid-20th century cooling and increased the LTCW. If we interpret the role of the interdecadal oscillation as a redistribution of heat from the lows to the highs of the oscillation, then this interpretation of MGW attribution assigns 10–20% of the 20th century temperature increase to GHGs, and 80–90% to solar activity. The MSM is a period of 70 years with above average solar activity. It is the longest such period in the 320-yr sunspot record. Since we know from proxies that solar activity was very low most of the time during the LIA, we can affirm that the Modern Maximum represents the longest period of high solar activity in the last 600 years. Almost a millennial type of event. That the Modern Maximum coincides with the biggest period of warming in 600 years should not be considered as unrelated coincidence.

A test of this interpretation is about to take place. Extrapolation of the forcings represented in Fig. 12.15a indicates we have entered, for the first time since 1900, a period when low solar activity coincides with a low in the 65-yr oscillation. If both together account for most of the observed temperature rate of change the result can only be a negative rate (i.e. cooling), that might have begun in February 2016. If the increase in GHGs is responsible for most of the warming observed since 1951 and solar activity and internal variability have played a very modest role at best, as the IPCC claims (Fig. 12.12), then more warming is inevitable for as long as CO_2 keeps increasing.

12.11 Conclusions

12a. Modern Global Warming is one of several multi-centennial warming periods that have taken place in the last 3000 years.

12b. Holocene climate cycles project that the period AD 1600–2100 should be a period of warming.

12c. A consilience of evidence supports the postulate that Modern Global Warming is within Holocene variability.

12d. Modern Global Warming displays an unusual non-cyclical cryosphere retreat. The contraction appears to have undone most of the Neoglacial advance.

12e. The last quarter (70 yr) of Modern Global Warming is characterized by extremely unusual and fast rising, very high CO_2 levels, higher than at any time during the Late Pleistocene. This increase in CO_2 is human caused.

12f. The increase in temperature over the past 120 years shows no perceptible acceleration, and contrasts with the accelerating CO_2 forcing.

12g. Sea level has been increasing for the past 200 years, and its modest acceleration for over a century shows no perceptible response for the last decades to strongly accelerating anthropogenic forcing.

12h. The available evidence can only support a higher sensitivity to increased CO_2 in the cryosphere, limited by its sensitivity to the increase in light-absorbing particles. Both are driving unusual melting and a small long-term sea level rise acceleration. The rest of the planet shows a lower sensitivity, indicating a negative feedback by H_2O, that prevents CO_2 from having the same effect elsewhere.

12i. Global warming attribution interpretations based on temperature rate of change point to solar activity as the main cause of Modern Global Warming, with a significant contribution from GHGs only in the latter part of it.

References

Arrhenius S (1896) On the influence of carbonic acid in the air upon the temperature of the ground Philosophical Magazine 41 237–276

Badino F, Ravazzi C, Valle F et al (2018) 8800 years of high-altitude vegetation and climate history at the Rutor Glacier forefield, Italian Alps. Evidence of middle Holocene timberline rise and glacier contraction. Quaternary Science Reviews 185 41–68

Bakke J, Lie Ø, Dahl SO et al (2008) Strength and spatial patterns of the Holocene wintertime westerlies in the NE Atlantic region. Global and Planetary Change 60 (1–2) 28–41

Belt ST, Vare LL, Massé G et al (2010) Striking similarities in temporal changes to spring sea ice occurrence across the central Canadian Arctic Archipelago over the last 7000 years Quaternary Science Reviews 29 (25–26) 3489–3504

Bereiter B, Eggleston S, Schmitt J et al (2015) Revision of the EPICA Dome C CO_2 record from 800 to 600 kyr before present. Geophysical Research Letters 42 (2) 542–549

Bereiter B, Shackleton S, Baggenstos D et al (2018) Mean global ocean temperatures during the last glacial transition. Nature 553 (7686) 39–44

Berner RA & Kothavala Z (2001) GEOCARB III: a revised model of atmospheric CO_2 over Phanerozoic time. American Journal of Science 301 (2) 182–204

Bond TC, Doherty SJ, Fahey DW et al (2013) Bounding the role of black carbon in the climate system: A scientific assessment. Journal of Geophysical Research: Atmospheres 118 (11) 5380–5552

Briner JP, Davis PT & Miller GH (2009) Latest Pleistocene and Holocene glaciation of Baffin Island Arctic Canada: key patterns and chronologies. Quaternary Science Reviews 28 (21–22) 2075–2087

Chen X & Tung KK (2014) Varying planetary heat sink led to global-warming slowdown and acceleration. Science 345 (6199) 897–903

Church JA & White NJ (2011) Sea-level rise from the late 19th to the early 21st century. Surveys in Geophysics 32 (4–5) 585–602

Crowley TJ (1998) Significance of tectonic boundary conditions for paleoclimate simulations. In: Crowley TJ and Burke K (eds) Tectonic boundary conditions for climate reconstructions. Oxford monographs on geology and geophysics 39. Oxford University Press, New York, p 3–20

Cunill R, Soriano JM, Bal MC et al (2012) Holocene treeline changes on the south slope of the Pyrenees: a pedoanthracological analysis. Vegetation history and archaeobotany 21 (4–5) 373–384

de Larminat P (2016) Earth climate identification vs. anthropic global warming attribution. Annual Reviews in Control 42 114–125

Drake BL (2012) The influence of climatic change on the Late Bronze Age Collapse and the Greek Dark Ages. Journal of Archaeological Science 39 (6) 1862–1870

Etheridge DM, Steele LP, Langenfelds RL et al (1996) Natural and anthropogenic changes in atmospheric CO_2 over the last 1000 years from air in Antarctic ice and firn. Journal of Geophysical Research: Atmospheres 101 (D2) 4115–4128

Eyles N (2008) Glacio-epochs and the supercontinent cycle after ~ 30 Ga: tectonic boundary conditions for glaciation. Palaeogeography, Palaeoclimatology, Palaeoecology 258 (1) 89–129

Farmer GT & Cook J (2013) Earth's albedo, radiative forcing and climate change. In: Climate Change Science: A Modern Synthesis, Springer, Dordrecht p 217–229

Fasullo JT, Nerem RS & Hamlington B (2016) Is the detection of accelerated sea level rise imminent? Scientific reports 6 31245

Firestein S (2012) Ignorance. How It Drives Science. Oxford University Press, New York

Folland CK, Rayner NA, Brown SJ et al (2001) Global temperature change and its uncertainties since 1861. Geophysical Research Letters 28 (13) 2621–2624

Foster GL, Royer DL & Lunt DJ (2017) Future climate forcing potentially without precedent in the last 420 million years. Nature Communications 8 p14845

Frakes LA (1979) Climates throughout geologic time. Elsevier, Amsterdam

Frakes LA & Francis JE (1988) A guide to Phanerozoic cold polar climates from high-latitude ice-rafting in the Cretaceous. Nature 333 (6173) 547–549

Frakes LA, Francis JE & Syktus JI (1992) Climate modes of the Phanerozoic. Cambridge University Press, Cambridge

Fyfe JC, Meehl GA, England MH et al (2016) Making sense of the early-2000s warming slowdown. Nature Climate Change 6 (3) 224–228

Goehring BM, Vacco DA, Alley RB & Schaefer JM (2012) Holocene dynamics of the Rhone Glacier Switzerland deduced from ice flow models and cosmogenic nuclides. Earth and Planetary Science Letters 351 27–35

Harsch MA, Hulme PE, McGlone MS & Duncan RP (2009) Are treelines advancing? A global meta-analysis of treeline response to climate warming. Ecology letters 12 (10) 1040–1049

Hegerl GC, Crowley TJ, Allen M et al (2007) Detection of human influence on a new validated 1500-year temperature reconstruction. Journal of Climate 20 (4) 650–666

Hegerl GC, Brönnimann S, Schurer A & Cowan T (2018) The early 20th century warming: Anomalies causes and consequences. Wiley Interdisciplinary Reviews: Climate Change 9 (4) e522

Hodgson DA (2011) First synchronous retreat of ice shelves marks a new phase of polar deglaciation. Proceedings of the National Academy of Sciences 108 (47) 18859–18860

Hogarth P (2014) Preliminary analysis of acceleration of sea level rise through the twentieth century using extended tide gauge data sets (August 2014). Journal of Geophysical Research: Oceans 119 (11) 7645–7659

Holzhauser H, Magny M & Zumbuühl HJ (2005) Glacier and lake-level variations in west-central Europe over the last 3500 years. The Holocene 15 (6) 789–801

Intergovernmental Panel on Climate Change (2014) Summary for Policymakers. In: Core Writing Team, Pachauri RK & Meyer LA (eds) Climate Change 2014: Synthesis Report. Contribution of Working Groups I, II and III to the Fifth Assessment Report of the Intergovernmental Panel on Climate Change. IPCC, Geneva, p 4–6

Jakobsson M, Long A, Ingólfsson Ó et al (2010) New insights on Arctic Quaternary climate variability from palaeo-records and numerical modelling. Quaternary Science Reviews 29 (25–26) 3349–3358

Jevrejeva S, Moore JC, Grinsted A & Woodworth PL (2008) Recent global sea level acceleration started over 200 years ago? Geophysical Research Letters 35 (8) L08715

Jevrejeva S, Moore JC, Grinsted A et al (2014) Trends and acceleration in global and regional sea levels since 1807. Global and Planetary Change 113 11–22

Jones P (2010) In: Harrabin R (writer) Q&A: Professor Phil Jones. BBC News, February 13. http://news.bbc.co.uk/2/hi/85 11670.stm Accessed 04 Mar 2019

Jouzel J, Masson–Delmotte V, Cattani O et al (2007) Orbital and millennial Antarctic climate variability over the past 800,000 years. Science 317 (5839) 793–796

Karl TR, Arguez A, Huang B et al (2015) Possible artifacts of data biases in the recent global surface warming hiatus. Science 348 (6242) 1469–1472

Koch J & Clague JJ (2006) Are insolation and sunspot activity the primary drivers of global Holocene glacier fluctuations? Pages Newsletter 14 (3) 20–21

Koch J, Clague JJ & Osborn G (2014) Alpine glaciers and permanent ice and snow patches in western Canada approach their smallest sizes since the mid-Holocene consistent with global trends. The Holocene 24 (12) 1639–1648

Körner C (2007) The use of 'altitude' in ecological research. Trends in ecology & evolution 22 (11) 569–574

Kullman L (2017) Further details on holocene treeline, glacier/ice patch and climate history in Swedish Lapland. International Journal of Research in Geography 3 (4) 61–69

Lacis AA, Schmidt GA, Rind D & Ruedy RA (2010) Atmospheric CO_2: Principal control knob governing Earth's temperature. Science 330 (6002) 356–359

Luterbacher J, Werner JP, Smerdon JE et al (2016) European summer temperatures since Roman times. Environmental Research Letters 11 (2) 024001

Maher N, Lehner F & Marotzke J (2020) Quantifying the role of internal variability in the temperature we expect to observe in the coming decades. Environmental Research Letters 15 (5) 054014

McGlone MS, Hall GM & Wilmshurst JM (2011) Seasonality in the early Holocene: Extending fossil-based estimates with a forest ecosystem process model. The Holocene 21 (4) 517–526

Meehl GA, Washington WM, Ammann CM et al (2004) Combinations of natural and anthropogenic forcings in twentieth-century climate. Journal of Climate 17 (19) 3721–3727

Moberg A, Sonechkin DM, Holmgren K et al (2005) Highly variable Northern Hemisphere temperatures reconstructed from low-and high-resolution proxy data. Nature 433 (7026) 613–617

Myhre GD, Shindell F–M, Bréon W et al (2013) Anthropogenic and Natural Radiative Forcing. In: Stocker TF, Qin D, Plattner G–K et al (eds) Climate Change 2013: The Physical Science Basis. Contribution of Working Group I to the Fifth Assessment Report of the Intergovernmental Panel on Climate Change. Cambridge University Press, Cambridge p 659–740

Owen LA (2009) Latest Pleistocene and Holocene glacier fluctuations in the Himalaya and Tibet. Quaternary Science Reviews 28 (21–22) 2150–2164

Pierrehumbert R (2019) There is no Plan B for dealing with the climate crisis. Bulletin of the Atomic Scientists 75 (5) 215–221

Pisaric MF, Holt C, Szeicz JM et al (2003) Holocene treeline dynamics in the mountains of northeastern British Columbia Canada inferred from fossil pollen and stomata. The Holocene 13 (2) 161–173

Pollack HN & Smerdon JE (2004) Borehole climate reconstructions: Spatial structure and hemispheric averages. Journal of Geophysical Research: Atmospheres 109 D11

Ramaswamy V, Boucher O, Haigh J et al (2001) Radiative Forcing of Climate Change. In: Houghton JT, Ding Y, Griggs DJ et al (eds) Climate change 2001: the scientific basis. Contribution of working group I to the Third Assessment Report of the Intergovernmental Panel on Climate Change. Cambridge University Press, Cambridge, p 349–416

Randin CF, Paulsen J, Vitasse Y et al (2013) Do the elevational limits of deciduous tree species match their thermal latitudinal limits?. Global Ecology and Biogeography 22 (8) 913–923

Reasoner MA & Tinner W (2009) Holocene treeline fluctuations. In: Gornitz V (ed) Encyclopedia of Paleoclimatology and Ancient Environments. Springer, Dordrecht, p 442–446

Reckin R (2013) Ice patch archaeology in global perspective: archaeological discoveries from alpine ice patches worldwide and their relationship with paleoclimates. Journal of World Prehistory 26 (4) 323–385

Schlesinger ME & Ramankutty N (1994) An oscillation in the global climate system of period 65–70 years. Nature 367 (6465) 723–726

Schneider DP, Steig EJ, van Ommen TD et al (2006) Antarctic temperatures over the past two centuries from ice cores. Geophysical Research Letters 33 (16)

Scotese CR (2018) Phanerozoic Temperatures: Tropical Mean Annual Temperature (TMAT), Polar Mean Annual Temperature (PMAT) and Global Mean Annual Temperature (GMAT) for the last 540 million Earth's Temperature. History Research Workshop, Smithsonian National Museum of Natural History, Washington DC 30–31 Mar 2018 https://www.researchgate.ne t/publication/324017003_Phanerozoic_Temperatures_Tropical _Mean_Annual_Temperature_TMAT_Polar_Mean_Annual_T emperature_PMAT_and_Global_Mean_Annual_Temperature _GMAT_for_the_last_540_million_years Accessed 13 Jun 2022

Shakun JD, Clark PU, He F et al (2012) Global warming preceded by increasing carbon dioxide concentrations during the last deglaciation. Nature 484 (7392) 49–54

Solanki SK, Usoskin IG, Kromer B et al (2004) Unusual activity of the Sun during recent decades compared to the previous 11,000 years. Nature 431 (7012) 1084–1087

Solomina O, Haeberli W, Kull C & Wiles G (2008) Historical and Holocene glacier-climate variations: General concepts and overview. Global and Planetary Change 60 (1) 1–9

Solomina ON, Bradley RS, Hodgson DA et al (2015) Holocene glacier fluctuations. Quaternary Science Reviews 111 9–34

Solomina ON, Bradley RS, Jomelli V et al (2016) Glacier fluctuations during the past 2000 years. Quaternary Science Reviews 149 61–90

Stein R, Fahl K, Schade I et al (2017) Holocene variability in sea ice cover primary production and Pacific-Water inflow and climate change in the Chukchi and East Siberian Seas (Arctic Ocean). Journal of Quaternary Science 32 (3) 362–379

Stranne C, Jakobsson M & Björk G (2014) Arctic Ocean perennial sea ice breakdown during the Early Holocene Insolation Maximum. Quaternary Science Reviews 92 123–132

Sud YC, Walker GK & Lau KM (1999) Mechanisms regulating sea-surface temperatures and deep convection in the tropics. Geophysical Research Letters 26 (8) 1019–1022

Thompson LG, Mosley–Thompson E, Brecher H et al (2006) Abrupt tropical climate change: Past and present. Proceedings of the National Academy of Sciences 103 (28) 10536–10543

Tinner W & Theurillat JP (2003) Uppermost limit, extent, and fluctuations of the timberline and treeline ecocline in the Swiss Central Alps during the past 11,500 years. Arctic, Antarctic, and Alpine Research 35 (2) 158–169

Tokinaga H, Xie SP & Mukougawa H (2017) Early 20th-century Arctic warming intensified by Pacific and Atlantic multidecadal variability. Proceedings of the National Academy of Sciences 114 (24) 6227–6232

Veizer J, Godderis Y & François LM (2000) Evidence for decoupling of atmospheric CO_2 and global climate during the Phanerozoic eon. Nature 408 (6813) 698–701

Zemp M, Frey H, Gärtner–Roer I et al (2015) Historically unprecedented global glacier decline in the early 21st century. Journal of Glaciology 61 (228) 745–762

Zhu Z, Piao S, Myneni RB et al (2016) Greening of the Earth and its drivers. Nature Climate Change 6 (8) 791–795

Zwally HJ, Li J, Robbins JW et al (2015) Mass gains of the Antarctic ice sheet exceed losses. Journal of Glaciology 61 (230) 1019–1036

13

21ST CENTURY CLIMATE CHANGE

"The future is unknowable, but the past should give us hope."
Winston S. Churchill (1958)

13.1 Introduction

Karl Popper's falsifiability criterion for science requires that hypotheses not only explain known evidence, but also must be testable by evidence still unknown. However, a problem arises when any failed ex-ante prediction made by a hypothesis, can be post-hoc explained in multiple ways leaving the hypothesis nearly intact. An example is the pause in global warming that took place between 1998–2014, while accelerated warming was the CO_2-hypothesis outstanding prediction for the 21st century (IPCC 1990 SPM). The pause was explained in multiple ways (see: Nature Climate Change vol. 4, issue 3, 2014, and Nature 2014 Focus: Recent slowdown in global warming), which is the same as saying it was not adequately explained. To meet Karl Popper's scientific criterion, the CO_2 hypothesis of climate change must make predictions that cannot be post-hoc explained when they fail. When demanding urgent action on CO_2 emissions, the predictions that can falsify the hypothesis are being made for a period ending in 2100, 80 years away. Its falsifiability is being moved forward in time until it no longer matters for present policy decisions.

As with any other activity, forecasting has been the subject of systematic studies, and three of the foremost experts in forecasting principles have established the golden rule of forecasting: *"be conservative by adhering to cumulative knowledge about the situation and about forecasting methods"* (Armstrong et al. 2015). Research has shown that ignoring the guidelines deduced from the golden rule greatly increases forecasting error. However, climate forecasting is dominated by radical predictions, many of which are absurd, yet they are given disproportionate positive attention. Two of the authors (Green & Armstrong 2007) analyzed the IPCC-Fourth Assessment Report, concluding that its forecasts were not the outcome of scientific procedures, but *"the opinions of scientists transformed by mathematics and obscured by complex writing,"* and warned that research on forecasting has shown that experts' predictions are not useful in situations involving uncertainty and complexity. Previous research by Philip E. Tetlock had already demonstrated that expert forecasting is usually worse than basic extrapolation algorithms, and that there is a perverse inverse relationship between fame and accuracy in forecasting (Tetlock 2005).

With that knowledge we can attempt to conservatively forecast future climate change for the next decades. The starting premise for the conservative forecast presented here, derived from past and present climate change evidence, is that greenhouse gases (mainly CO_2), solar variability, and oceanic oscillations, are all significant climate variables in the centennial timeframe considered. It is important to remark that forecasts only consider a very limited number of variables and the rest is assumed as invariant. This necessary simplification means that with increasing time the chance of a forecast being correct decreases even if the variables considered were correctly projected. The future is, after all, unknowable.

13.2 Changes in CO_2 emissions and atmospheric levels

Atmospheric CO_2 levels are very likely to continue increasing for the next 30 years, even if changes in the rate of emissions take place. Due to the large size of natural stores, sinks, and sources, the trend in atmospheric CO_2 levels responds slowly to changes in emissions. The 7% decrease in emissions caused by the societal response to the COVID-19 pandemic in 2020 (Le Quéré et al. 2021) is, as expected, undetectable in atmospheric levels a year later. A conservative forecast by extrapolating the increase in CO_2 values since 1960 gives 498 ppm of CO_2 by 2050.

Regarding CO_2 emissions, the failure of past projections shows how difficult it is to forecast future emissions. It is very easy to extrapolate fossil fuel consumption that has experienced continuous growth for over a century, but several factors are very likely to have a significant impact in fossil fuel production for the 2022–2050 period, making a simple extrapolation a non-realistic forecast. The UN population forecast shows that there is a profound and inevitable demographic change taking place (United Nations DESA 2017). The UN medium-variant projection shows every region except Africa reaching peak population by 2050. The aged >60 population is the fastest growing group and by 2050 all regions of the world except Africa will have at least a quarter of its population above age 60. Population demographics suggests a growing negative pressure on per capita energy use, whose increase has been driven in the 21st century by the growing Asian middle class. For countries with high old-age dependency ratio (number of people above 64-years old per 100 people in working age), what it is observed is a decrease in per capita energy use with time (Fig. 13.1). China's former one-child policy is going to turn its demographic dividend into a demographic burden very fast. China's working popula-

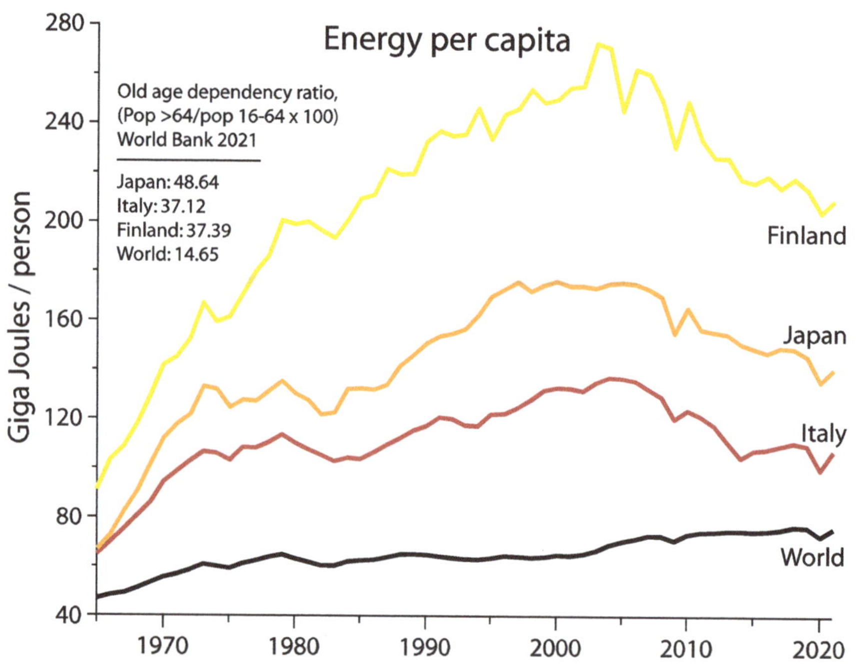

Fig. 13.1 Declining energy per capita for countries with aged population
Primary energy consumption in Giga Joules per person, for the world (black line) and the three countries with highest old-age dependency ratio (number of people above 64-years old per 100 people in working age. Italy, dark red; Japan, orange; Finland, yellow). Japan's population has been declining since 2010, Italy's population since 2014, while Finland's population is still growing. Source of data: BP Statistical Energy Review (2022), and World Bank.

tion has already peaked in 2010, and by 2040 it should have an old-age dependency ratio similar to Japan.

Besides an increasing fraction of older people, population decrease should also reduce the total energy demand. From the supply side, coal production is showing an unexpected lack of growth, and oil production is generally expected to peak within the 2022–2050 period for a variety of reasons, including reducing energy return on energy invested (EROEI, manifested in increasing costs of production), energy transition mainly to natural gas, but also to alternative energy sources, and active global policies to reduce oil and coal burning.

Because of these and other economic factors, our CO_2 emissions have been growing more slowly for the past few years, having already reached its lowest 5-year average value since there are records (Fig. 13.2, purple line). Our emissions are growing now at a rate like RCP 4.5 (Fig. 13.2, black line). If the present trend slowing continues, a decrease in CO_2 emissions should start before 2050 and should continue for the foreseeable future, unless demographic trends change. The slowing in emissions rate started years before the Paris Agreement, so a recovery from COVID-19 pandemic effects is unlikely to change the sign of that trend.

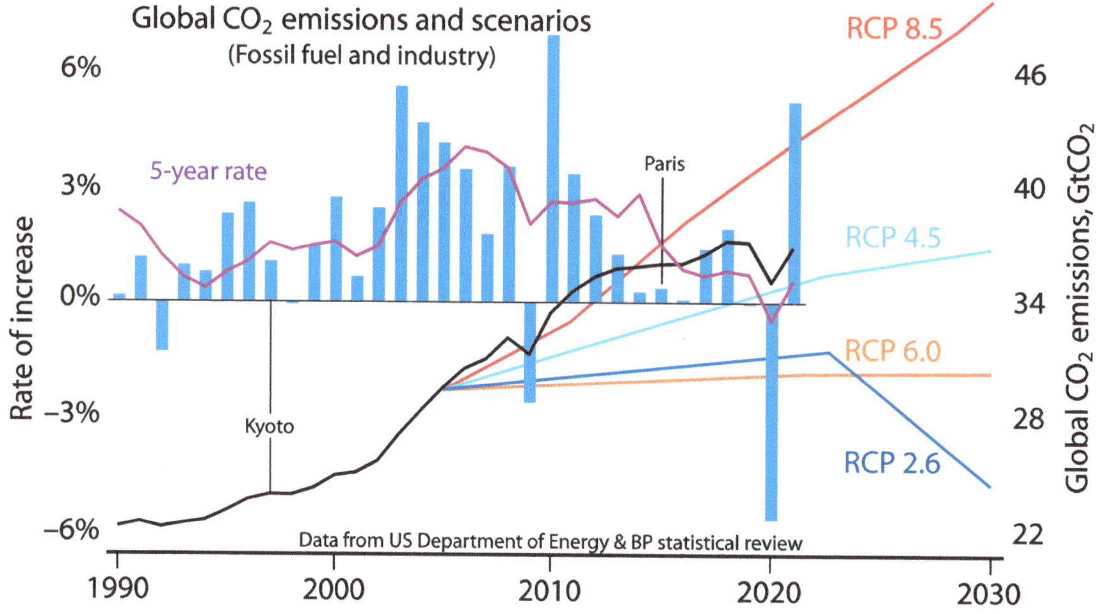

Fig. 13.2 Global CO₂ emissions are decelerating
Yearly percentage rate of increase in global CO_2 emissions from fossil fuels and industry (blue bars, left scale), and its 5-year average (purple line, left scale). Global CO_2 emissions from fossil fuels and industry, in gigatons of CO_2 (black line, right scale). CO_2 emissions considered by the four IPCC AR5 emissions scenarios in the four representative concentration pathways (colored lines, right scale). Since 2011 our emissions have been growing at a similar rate to RCP 4.5. Source: Gilfillan et al. (2019). Data for 2017–21 from BP Statistical Energy Review (2022), adding cement and flaring contribution estimates. RCP data from IPCC AR5.

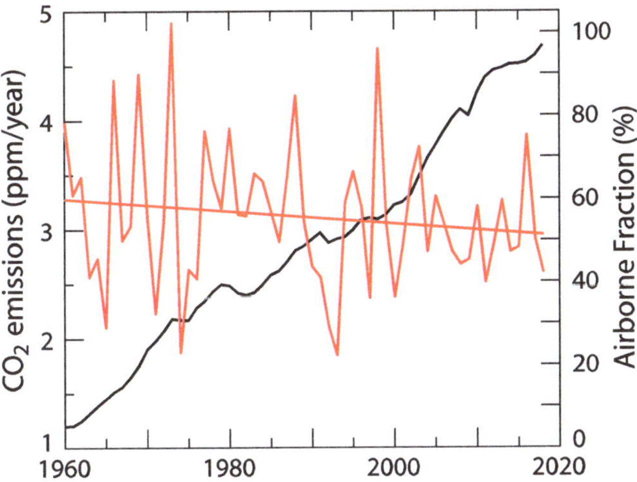

Fig. 13.3 The decreasing airborne fraction
The airborne fraction (red) is the fraction of our CO_2 emissions (black, in ppm of CO_2 equivalent) that remains in the atmosphere each year. The declining trend shows how over time a smaller part of our much larger emissions remains in the atmosphere. After Hansen, et al. (2013). Atmospheric CO_2 data from Mauna Loa, NOAA. Emissions in Gt carbon from Le Quéré et al. (2018), were converted to CO_2 by multiplying by 3.664, and afterwards to ppm by dividing by 7.8.

What would happen to atmospheric CO_2 levels under a likely decreasing emissions scenario from 2050? This scenario is similar to RCP 4.5 that shows stabilizing atmospheric CO_2 at c. 500 ppm (van Vuuren et al. 2011). However, carbon sinks have been a considerable source of positive surprises to climate researchers. First, it was the "missing sink" (Schindler 1999), since it could not be explained where the emitted CO_2 that did not remain in the atmosphere was going. Environmentalists were slow to accept that the biosphere was expanding and greening in response to increasing CO_2 and warming, despite the opposite effect being well-documented during glacial periods. Then climate scientists became worried that the land (Canadell et al. 2007) and ocean (Schuster & Watson 2007) carbon sinks were saturating. However, the opposite has been found, and sinks are actually increasing their rate of uptake (Keenan et al. 2016). If in the 1960s they were taking up c. 40% of our CO_2 emissions, they are now taking up c. 50% of our much larger current emissions (Fig. 13.3; Hansen et al. 2013).

The reason why sinks are taking up more CO_2 from the atmosphere is that we are farther from equilibrium. Since atmospheric CO_2 changed very slowly before anthropogenic emissions from fossil fuels, it can be assumed that sinks (K) and sources (S) were at equilibrium at 280 ppm ($\Delta K = \Delta S$). Due to warming the oceans release c. 16 ppm/°C, so current equilibrium is c. 290 ppm. Since the current level (c. 400 ppm) is above equilibrium level, sinks are larger than sources ($\Delta K > \Delta S$), and the farther we are from equilibrium, the larger the difference between sinks and sources ($\Delta K–\Delta S$). If we stabilize emissions (E) near present levels, as current trend suggests, the difference between sinks and sources will continue increasing until it matches emissions ($\Delta K–\Delta S = E$), reaching a new equilibrium for constant emissions. Since we are c. 120 ppm above equilibrium and sinks are absorbing 50% of our emissions ($\Delta K–\Delta S = 0.5E$), it can be calculated that for constant current emissions the new equilibrium lies at 240 ppm (120/0.5) above the present equilibrium value of 290 ppm, or 530 ppm.

Given constant emissions at present levels, atmospheric CO_2 should increase with a logarithmically decreasing rate towards 530 ppm, at which point sinks should match sources plus emissions ($\Delta K = \Delta S + E$). One of the biggest mistakes in the climate change debate is assuming that we need zero emissions to stabilize CO_2 levels. Deep ocean carbon stores are so large that carbon sinks can be considered unlimited in terms of anthropogenic emissions. The planet has dealt with much higher perturbations of atmospheric CO_2 levels in the past, as supported by the large $\delta^{13}C$ excursions associated with the formation of large igneous provinces that formed over tens of thousands of years. The IPCC hypothesis predicts that under constant emissions there should be a constant increase in atmospheric CO_2 levels and the airborne fraction of anthropogenic CO_2 should increase as sinks saturate. However, if we stabilize our CO_2 emissions, after just 10 years it should become apparent that the airborne fraction of fossil fuel CO_2 is decreasing and the rate of increase in total atmospheric CO_2 is slowing down. Once more we are poised for another positive surprise by carbon sinks. Fossil fuel CO_2 emissions are growing more slowly now (Fig. 13.2) and there is the possibility that they will decrease in a few decades. Once our emissions decrease, atmospheric CO_2 will start slowly decreasing, as sinks and sources equilibrate to our decreasing emissions.

13.3 Fossil fuel changes

The biggest part of our CO_2 emissions is due to the burning of fossil fuels, and a cursory analysis of future fossil fuel production is required for a more accurate forecasting of future CO_2 levels. Coal production reached a maximum in 2013 and shows a flat trend for the past 10 years (BP Statistical Energy Review 2022; Fig. 13.4). The decline in coal production has been completely unexpected. It is probably helped by an increasing trend in coal plant retirements reaching 30 GW/year (Shearer et al. 2017), as many coal plants are very old, and by the effect of the COVID-19 pandemic. There are plans to increase the number and capacity of coal plants worldwide and coal production should increase again, but coal plant implementation rate has been low lately (37% for the period 2010–2016), with most plant projects halted, cancelled, or shelved. From January 2016 to January 2017 the amount of coal power capacity in pre-construction planning decreased from 1,090 to 570 GW (Shearer et al. 2017). A higher future coal production is possible, particularly with the energy crisis caused by the Russo-Ukrainian War of 2022, because coal reserves are abundant and coal is cheaper than other fuels, but as coal is being increasingly replaced by gas and other energy sources, it is unlikely that coal production will return to the vigorous growth of the 2002–2011 period. The unexpected drop in coal production from 2013 to 2020 increases the chances that peak coal production could take place several decades earlier than forecasted. ExxonMobil (2018) Outlook for Energy places Peak Coal in 2025. We cannot discard the possibility that Peak Coal has already taken place.

Oil is the least abundant fossil fuel. Several signs indicate we are approaching the end of oil growth (Peak

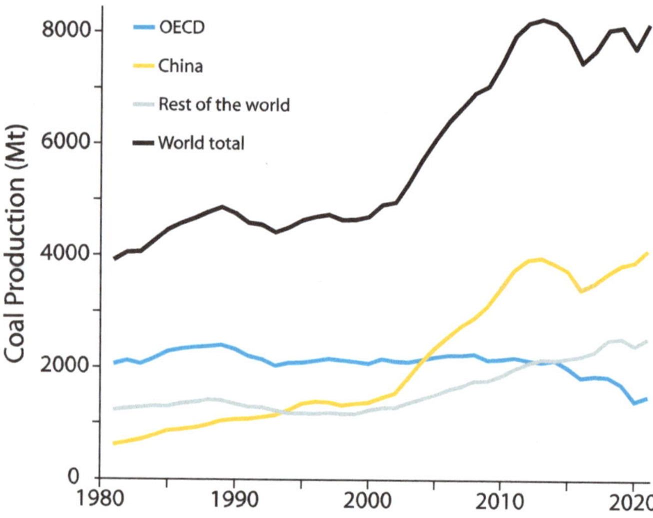

Fig. 13.4 Lack of growth in coal production since 2013
Coal production 1981–2021 in million tonnes for the world (black line), China (orange line), OECD (blue line), and the rest of the world (grey line). Source of data: BP Statistical Energy Review (2022).

Oil). Oil is categorized by its density (specific gravity), and with time the proportion of light oil from tight formations, liquid condensate from natural gas, and heavy oil, has been growing at the expense of more desirable intermediate density oil. It is also clear to anybody that we would not have to resort to fracturing shale rocks with high pressured water to obtain low-producing wells that decline by 75% in just three years, if we could still get sufficient oil by more conventional methods.

The impression that peak oil is approaching is confirmed by analyzing the oil growth curve. For the past 30 years oil growth has been declining from c. 2% to c. 1% (Fig. 13.5). This is a period when oil production has not been constrained, and more oil could have been produced if more demand for it existed. The decline can be attrib-

uted to a multitude of factors, including global economy growth rate, increasing oil use efficiency, economic changes that reduce energy intensity, demographic changes, and active policies to reduce oil consumption. If this long-term trend continues, peak oil is expected to be reached c. 2065 when oil growth should cease, but linear trend extension is a poor way of forecasting. While it is hard to imagine realistic scenarios that would invert a decreasing 37-year long trend in oil growth, there are several scenarios that might accelerate Peak Oil. Obtaining oil from more difficult geologic formations leads to a higher cost oil that, to be sustainable over the long term, must be adequately reflected in oil price, and should promote oil substitution. The net energy yield of our global oil operations is decreasing, becoming a less efficient, less competitive process. Concerns over CO_2 emissions are also driving oil substitution with policies that for example support the increase in electric vehicles.

The conservative forecast for oil production proposed here is in stark contrast to every single official projection by the International Energy Agency, the U.S. Energy Information Administration, British Petroleum Statistical Review of World Energy, or ExxonMobil Outlook for Energy, as none of them projects a decline in world oil production for the next decades. Therefore, it is fair to ask if it is really conservative to predict a Peak Oil within the next three decades. After all the business-as-usual projection has proven superior so far despite repeated claims of impending Peak Oil in the past. Each one must decide about that, and the author has exposed the reasons that lead him to believe Peak Oil is a conservative prediction. The author also believes no official projection will ever anticipate a decline in production as they cannot afford to be wrong on that. They will only see a decline after a decline is taking place, and therefore have no predictive value. The lack of growth in coal production since 2013 is an example. It was never predicted before it took place, but it is predicted in some scenarios afterwards. The last peak in oil production (crude + condensate) was established in November 2018 and predates the COVID-crisis by more than a year. This oil peak is likely to remain in place for several years to come, a situation that has no precedent. Oil upstream capital expenditure has been insufficient since 2015 and new oil discoveries are at a multi-decadal low. On top of this situation the COVID-19 pandemic response has been very negative for oil capital expenditure and production, and the situation is likely to continue because the COVID-induced global economic crisis, compounded by the Ukraine war and sanctions consequences, is going to be severe and multi-year.

If Peak Oil does take place before 2050 (or has taken place in 2018, as believed by this author), the lack of oil growth will have to be compensated by other energy sources, if global energy consumption is to continue growing unaffected. While our response to Peak Oil might be to increase our coal consumption, it is reasonable to assume that we will also substitute it by gas and other energy sources, leading to a decrease in CO_2 emissions. A likely response will also be a reduction in energy consumption per capita that several countries are already showing (Fig. 13.1). Climate change scenarios that consider GHG emissions must factor in the almost inevitable reduction in our CO_2 emissions during the 21st century to avoid being unrealistic.

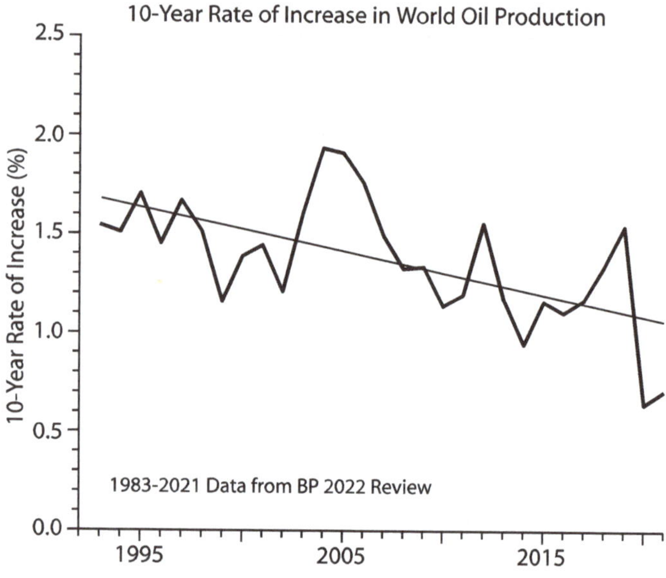

Fig. 13.5 Decrease in the rate of change of world oil production
Ten-year average of the percentage change in oil production between 1983 and 2021 with its linear trend. Source of data: BP Statistical Energy Review (2022).

13.4 Changes in solar activity

Forecasting solar activity has proven difficult. There is currently no known mechanism that can explain long-term solar variability, and accurate prediction beyond the next cycle has not been demonstrated so far. Cycle forecasting takes advantage of the presence of repeating features like solar extended minima every c. 100 years (centennial lows) identified from past activity, despite not knowing what causes them. Humans were capable of predicting seasons thousands of years before they could explain them.

Solar activity has been increasing for the past 300 years according to sunspot observations and solar proxies (Fig. 13.6a, trendline). The Feynman centennial solar cycle (Feynman & Ruzmaikin 2014; see Sect. 8.6) defines three oscillations (F1–F3, Fig. 13.6a) delimited by the lowest sunspot minima every c. 100 years. The last oscillation, F3, has the highest average sunspot number (93.4 SN/year). This period, and particularly between 1935–2004, has been termed the Modern Solar Maximum (MSM; Kobashi et al. 2015). We are currently in a centennial extended minimum in solar activity between F3 and F4, that should be named the Clilverd solar minimum since Clilverd et al. (2006) were capable of predicting it.

The cycle-based forecast indicates it should affect mainly solar cycle (SC) 24 and SC25 with increasing solar activity afterwards. SC25 has already been forecasted to be intermediate between SC24 and SC20 by the reliable polar fields method (Svalgaard 2018), supporting cycle forecasting.

A cycle-based forecast for 2020–2050 must consider the centennial periodicity. Analogues for SC25–27 are SC6–8, and SC15–17. Also, since long-term solar activity is increasing within the millennial periodicity towards its 2050–2100 maximum, SC25–27 should have higher activity than SC15–17, as that period had higher activity than SC6–8 (Fig. 13.6). The forecast therefore is:

- SC25 should be slightly above SC24, but below SC23
- SC26 should also be above SC24, and similar to SC23
- SC27 should be similar to SC22

From 2022–2035 solar activity is forecasted to be below average. From 2035–2055 activity should increase inversely to the 1980–2000 decline (Fig. 13.6). SC29 should have somewhat reduced solar activity due to the pentadecadal periodicity that also affected SC20 and SC10. From 2080 solar activity should decrease due to the

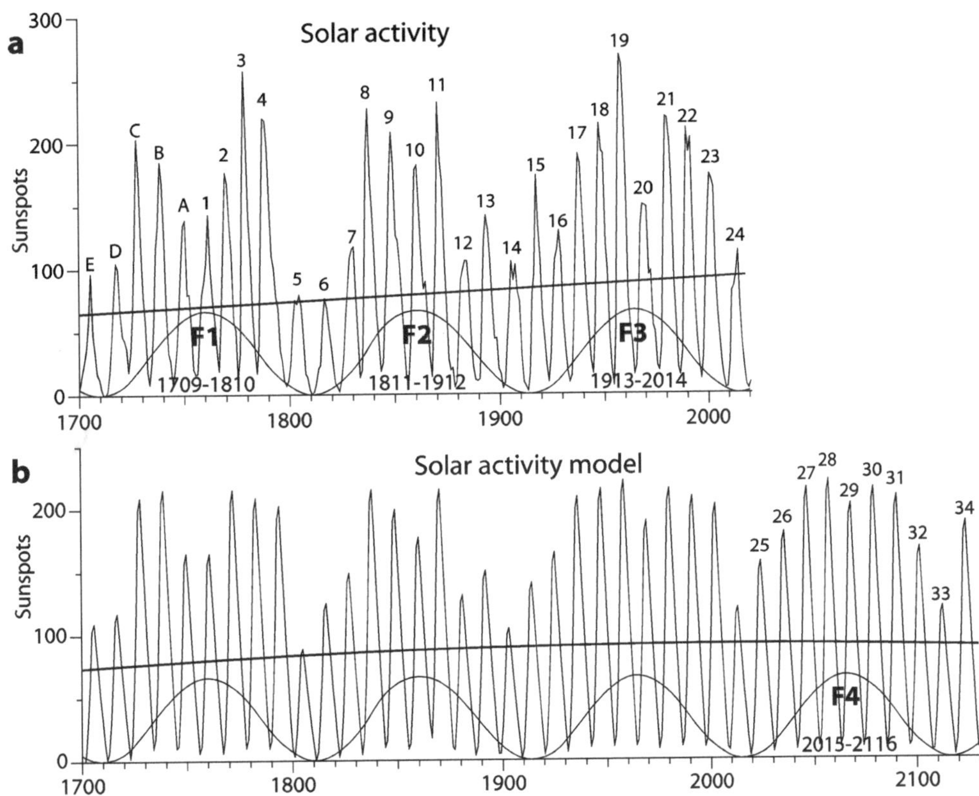

Fig. 13.6 Sunspot forecasting based on solar activity cycles
a) International annual sunspot number 1700–2020 in black, with rising linear trend. The centennial Feynman periodicity represented as a sinusoidal curve with minima at the times of lowest sunspot numbers, defining the centennial periods F1–F3, with their span indicated by the dates below. F3 shows the highest solar activity of the three. F2 was affected by the presence of a bicentennial (de Vries) cycle low at SC12–13. Source of data: WDC–SILSO, Royal Observatory of Belgium, Brussels. b) Solar model built on the spectral properties of solar activity from cosmogenic records and sunspot records. The model assumes default maximum activity for each cycle that is then lowered by the distance to the lows of the five cycles considered: 2500-yr, 1000-yr, 210-yr, 100-yr and 50-yr cycles. Cycles dates and periods deduced from past activity are projected into the future, producing a solar activity forecast for 2022–2130. F4 is set to coincide with a peak in the millennial Eddy cycle identified from Holocene solar proxy records, and likely to have more sunspots than F3 despite another de Vries cycle low expected for SC31–32. SC1, SC10, SC20, and SC29 constitute lows in the pentadecadal solar periodicity, that reduces sunspot numbers at the peak of the centennial periodicity.

bicentennial (de Vries) periodicity that affects solar activity a few decades in advance of the centennial minima

In summary, 21st century solar activity should be similar or a little higher than 20th century solar activity, due to being at the peak of the millennial Eddy solar cycle. This level of solar activity corresponds to the Holocene highest 25%, and no doubt is contributing to the present warm period. Lower than average solar activity should only take place in the 2006–2035 period. The Sun should promote warming during the 2035–2100 period but should reach maximal millennial activity during the 2050–2080 period.

13.5 A mid-21st century solar grand minimum is highly improbable

A mid-21st century solar grand minimum (21stC-SGM) prediction breaks the golden rule of forecasting, as it is a non-conservative prediction. Surprisingly a high number of well-known authors that question the CO_2 hypothesis have embraced the 21stC-SGM hypothesis. In the famous for the wrong reasons first issue of the Pattern Recognition in Physics journal, N.–A. Mörner and eighteen more authors signed a letter (Mörner et al. 2013), stating a conclusion and two implications that challenged IPCC interpretation of climate change and triggered the termination of the journal by its owners. Of interest here is the second implication that served as a basis to doubt IPCC claims: *"Several papers have addressed the question about the evolution of climate during the 21st century. Obviously, we are on our way into a grand solar minimum."*

There is no convincing evidence for a 21stC-SGM, that would justify such a non-conservative forecast. Abdussamatov (2013) was probably the first to write about this issue in Russian in 2007, but he mistakes the current centennial low for a bicentennial one, and ignores the modulating effect of the 2500-yr Bray solar cycle on the bicentennial (de Vries) cycle amplitude. But the consensus dissolves upon scrutiny, as many of those authors have not published on the issue, and Scafetta (2014), and Charvátová & Hejda (2014) actually do not predict a 21stC-SGM, but a tame, run-of-the-mill, centennial low. Charvátová foresees nearly identical activity for SC24–26 as for SC12–14, very far from SGM low values. Mörner, the first signatory, was so convinced of the coming 21stC-SGM that he did not present any evidence in his numerous articles about the issue (Mörner 2015). Salvador (2013), and Shepherd et al. (2014) rely on different models for their 21stC-SGM prediction, the first a tidal torque model, and the second a solar dynamo model. Salvador's model disputes the approaching millennial peak in solar activity, with a projection of 160 years of very low solar activity, while the dynamo model from Zharkova's group doesn't adequately hindcast past solar activity. Together with Steinhilber & Beer (2013) evidence-based forecast, they all have the problem that the very reliable polar fields method has already forecasted a SC25 with more activity than SC24 (Svaalgard 2018), contradicting their proposed continuous decline in solar activity towards the predicted SGM.

A conservative forecast that a SGM is not going to take place doesn't need supporting evidence, as the Sun only expends c. 17% of its time in SGM conditions (Usoskin et al. 2007), so the chances are skewed against it. However, an analysis of what is known about SGM builds a strong case against the 21stC-SGM hypothesis. 30 SGM have been identified during the 11,700-year Holocene based on a very high rate of cosmogenic isotopes production (Usoskin et al. 2007; Inceoglu et al. 2015; Usoskin et al. 2016; Fig. 13.7). The average is one SGM every 400 years, but SGM show a tendency to cluster. 17 SGM

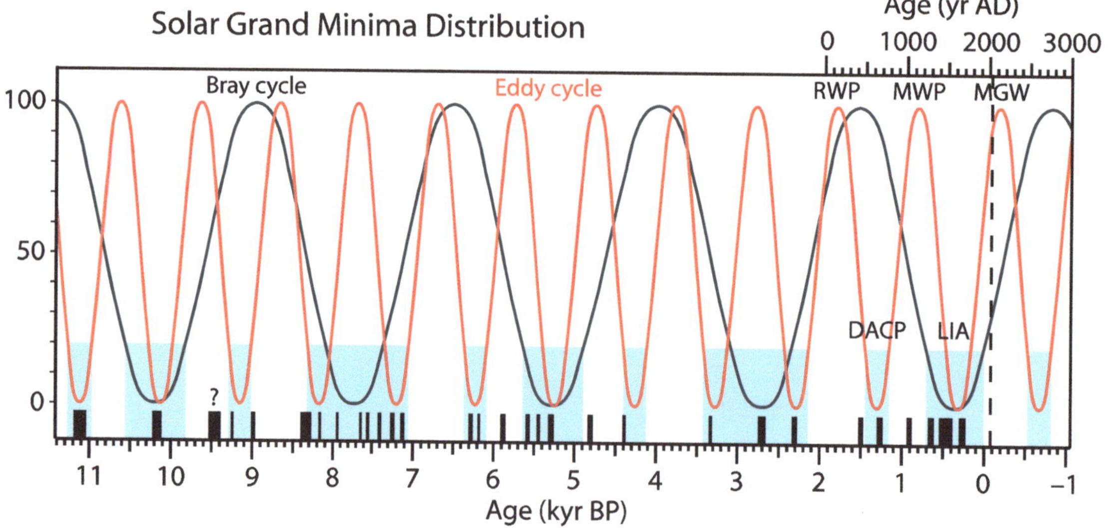

Fig. 13.7 Solar Grand Minima distribution during the Holocene
Thirty Solar Grand Minima (SGM) from solar proxy records during the Holocene, identified in the literature, are indicated by black boxes of thickness proportional to their duration. The 2500-yr Bray solar cycle (black sinusoidal) and the 1000-yr Eddy cycle (red sinusoidal) identified in solar proxy records are displayed at their proposed time-evolution that best matches both solar activity and climate changes consistent with their periodicity. Periods where any of the cycles is at its lowest 20% of their relative sinusoidal function are marked in blue, and comprise 54% of the Holocene. Present position is indicated by a dashed line. SGM show a bias towards clustering at the blue areas. Question mark, a questioned SGM that could correspond to an increase in cosmic rays from a supernova (see Sect. 8.7). RWP: Roman Warm Period. DACP: Dark Ages Cold Period. MWP: Medieval Warm Period. LIA: Little Ice Age. MGW: Modern Global Warming. Source for SGM data: Usoskin et al. (2007); Inceoglu et al. (2015); Usoskin et al. (2016).

(57%) are at around two centuries of another SGM, and 7 clusters of 2 or more SGM can be recognized. That is why the de Vries bicentennial cycle is so important for SGM as it is a very favored spacing. Usoskin et al. (2016) have shown that SGM have a statistically significant tendency to cluster at the lows of the 2500-yr Bray solar cycle, challenging random-probability based analyses (Lockwood 2010). If we also consider the 1000-yr Eddy solar cycle, we can see that 26 SGM (87%) fall at or right next to the periods when one of these two cycles is at its lowest 20% (colored areas in Fig. 13.7, 54% of time). The conclusion is clear, SGM tend to occur nine out of ten times when solar activity is at its lowest coinciding with the lows of the 2500 and 1000-yr solar cycles. In periods like the present, outside the lows of both solar cycles, the Sun expends only 7.5% of its time in a SGM, and with a frequency of c. 1 SGM in 1000 years. Forecasting a SGM for the mid-21st century is really a low-probability non-conservative proposition. The probability for a next SGM should become high again c. AD 2450.

13.6 Changes in global surface average temperature anomaly

Over periods of a few years climate variability appears to be dominated by El Niño/Southern Oscillation (ENSO) variability (Fig. 13.8), that so far has resisted forecasting attempts. The failure of the 2014 and 2017 El Niño forecasts (ECMWF 2021) shows the difficulty of forecasting ENSO and global temperatures even for a few months. The February 2017 El Niño forecast (ECMWF 2021) failed because it did not take into account solar activity control of ENSO (see Sect. 10.4). February 2017 belongs already to the late decline in activity towards the end of the solar cycle that corresponds to phase III in Fig. 10.12, when a La Niña is more probable than an El Niño.

The successful 2020 La Niña prediction (Vinós 2019) anticipated a continuation of the decline in global surface average temperature (GSAT) observed since February 2016. From mid-2019 the PDO returned to the negative values that prevailed during the pause (1999–2014). The combined effect of ENSO, oceanic oscillations, and low solar forcing from the Clilverd extended solar minimum suggest that the lack of significant global warming should continue perhaps until the late 2020s or early 2030s establishing a new pause. GSAT (HadCRUT4.6) is already in 2021 lower than 95% of CMIP5 models (RCP4.5 emissions scenario; Fig. 13.9) and CMIP6 models project even more warming (Meehl et al. 2020). A new pause would make the model-observations disparity even more striking, as the GSAT appears to be increasing in a linear fashion since 1950, while models project an accelerating path as prescribed by the CO_2 hypothesis.

The effect of multiple solar cycles with lower than average activity on temperature is not well determined, but previous similar periods, known as solar extended minima, coincide with cool periods in the early decades of the 18th century (Maunder minimum), 19th century (Dalton minimum), and 20th century (Gleissberg minimum). A lag of c. 10–20 years has been found between the decrease in solar activity and its effect on tree-ring growth and ice core temperatures in several reconstructions (Eichler et al. 2009; Breitenmoser et al. 2012; Anchukaitis et al. 2017). A

longer lag has been found on the maximum effect of solar activity reduction on the increase in heat transport, causing cooling in low latitudes and warming in high latitudes (Kobashi et al. 2015). The present solar extended minimum, named here the Clilverd minimum, includes SC24 and most likely SC25. A conservative forecast on the effect of the Clilverd minimum on temperatures indicates no additional global warming before 2035, and perhaps even a slight cooling. This forecast is also supported by the position of the 65-year oscillation in global temperature (see Fig. 12.15), that also indicates no warming for the first 3 decades of the 21st century.

It is remarkable that a knowledge-based conservative forecasting for the next 15 years as the one presented here, agrees so well with the naive no-change forecast proposed by Green and Armstrong in 2007, that has been superior to the IPCC early forecast. It is important to emphasize that although very variable in the short term in different places, Earth temperature is extraordinarily constant over the long term. 0.2 °C is a small variation in temperature at temporal scales of less than one year, but a significant variation on yearly averages, an important variation in decadal temperature scales, and a huge variation in millennial scales. The Neoglacial trend that has driven glacier expansions all over the globe, culminating at the LIA, was just –0.2 °C/millennium or less over the past five millennia (–0.38 °C/millennium in Greenland; Kobashi et al. 2015), due to Milankovitch forcing. The planet lost about one degree average from the Holocene Climatic Optimum to the second millennium AD, and this amount caused considerable glacier expansion and biome changes, reducing tropical forests and expanding the tundra. Higher temperature changes are observed on shorter timeframes, but they also have lower and shorter effects. From July 2013 to February 2016 the global surface average temperature anomaly increased by 0.4 °C but has since lost most of it (Fig. 13.9). Multidecadal to centennial temperature forecasting, to be conservative, must strongly constrain the amount of temperature change that it allows. That is the reason why

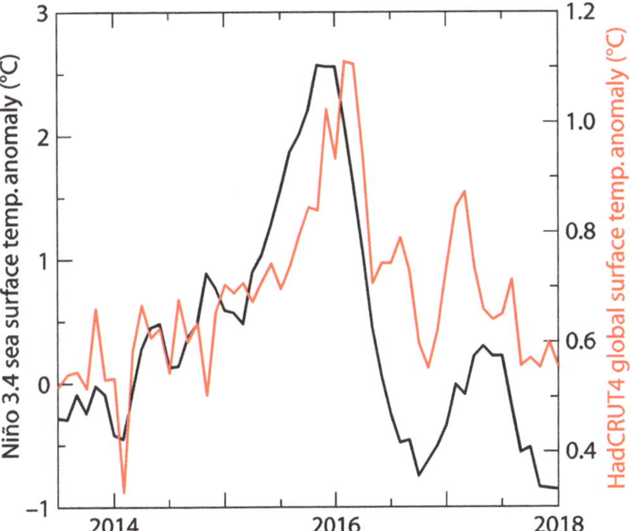

Fig. 13.8 ENSO-Global temperature relation
June 2013–January 2018 Niño 3.4 region sea surface temperature anomaly (black line, left scale), and monthly global surface average temperature anomaly (red line, right scale). Data from Australian Bureau of Meteorology and UK HadCRUT4.6 dataset.

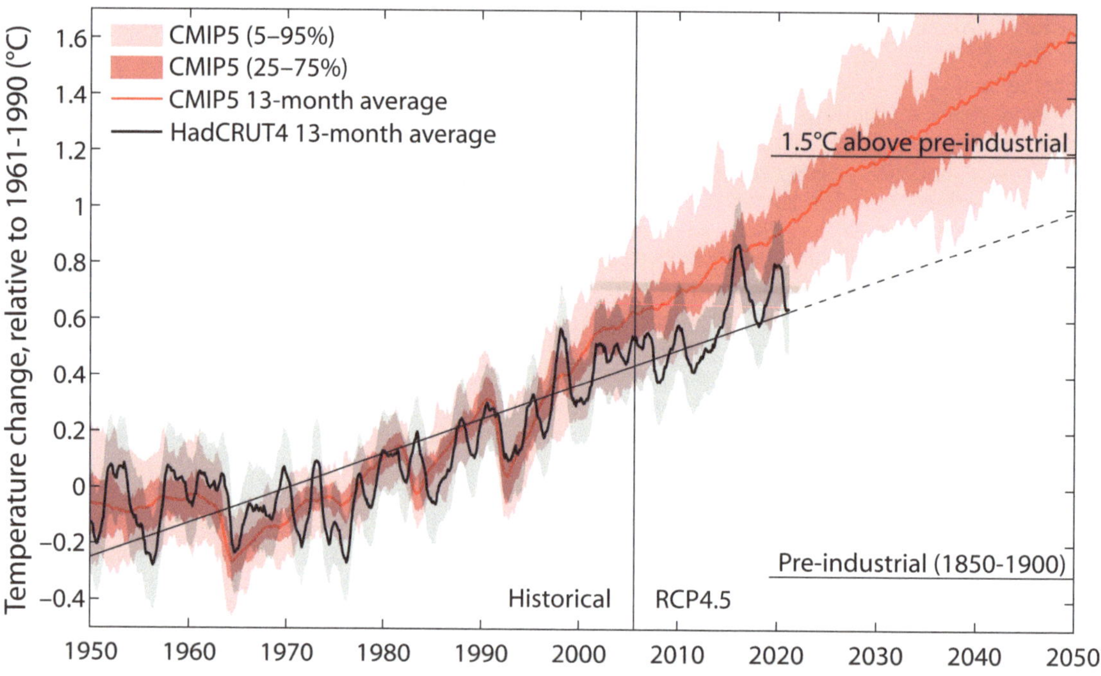

Fig. 13.9 Global temperature change 1950–2021: comparing observations to models
Black curve, global surface temperature anomaly (°C; monthly HadCRUT 4.6 13-month averaged) with its linear trend (thin continuous line), and 95% confidence interval (grey area). Data from UK Met Office. Red curve, Coupled Model Intercomparison Project Phase 5 (CMIP5) multi-model mean temperature anomaly 1950–2050 under RCP 4.5 conditions (13-month averaged) with 25–75% (medium red area), and 5–95% values (light red area) for the 42 models used. Data from KNMI climate explorer, Trouet & van Oldenborgh (2013). CMIP5 models were initialized in 2006 (vertical line) and reproducing historical climate to that point was a prerequisite.

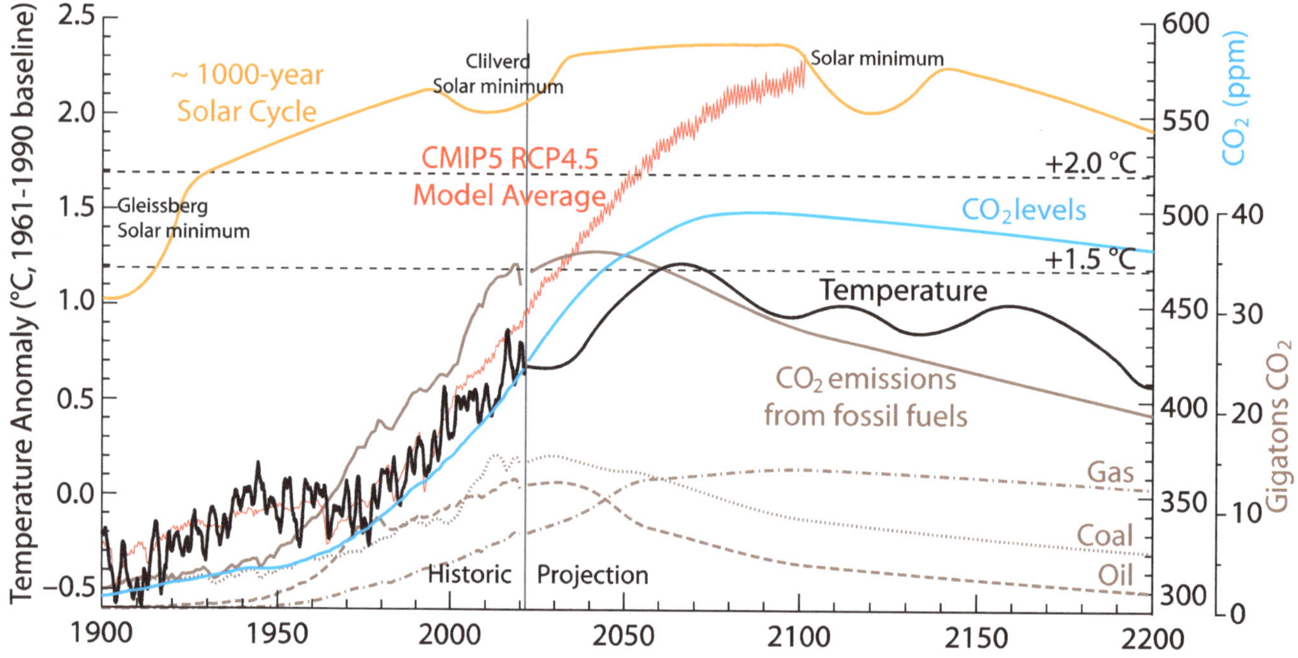

Fig. 13.10 Conservative temperature, CO_2 level, and emissions forecast to AD 2200
CO_2 emissions from fossil fuels forecast (brown continuous line, right outer scale) based on a peak in oil consumption by 2018–35 (dashed brown line), a second peak in coal consumption before 2050 (dotted brown line) and increasing gas consumption to 2100 (dash-dotted brown line), producing a peak in CO_2 emissions from fossil fuels at c. 35 Gtons by 2050. Historical CO_2 emissions from fossil fuels are from Gilfilland et al. (2019), archived at CDIAC (AppState) and updated with data from BP Statistical Energy Review (2022). Atmospheric CO_2 levels (blue line, right inner scale) should therefore stabilize at c. 500 ppm by 2080 before starting to decrease slowly as sinks remove more than is added. Historical atmospheric CO_2 levels are from Law Dome (after Etheridge et al. 1996) to 1958 and from NOAA afterwards. An idealized millennial cycle in solar activity is represented by the orange line, peaking at 2050–2080, temporarily reduced by centennial and bicentennial lows indicated with their names. CMIP5 model-mean temperature anomaly (red line) projects reaching +1.5 °C above pre-industrial by the 2030s, and +2.0 °C by the 2050s. Temperature anomaly should stabilize until the 2030s, increasing afterwards and peaking c. the 2070s at c. +1.5 °C above pre-industrial, due to CO_2, solar activity forecasts, and oceanic oscillations. Afterwards temperature anomaly could enter a slightly declining undulating plateau as both CO_2 and solar activity slowly decline. Historic temperature anomaly is from UK MetOffice HadCRUT 4.6 global surface monthly dataset (13-month averaged).

Green and Armstrong no-change forecast was better than IPCC 0.3 °C/decade forecast. Frequent claims that we are on course to +2 °C above pre-industrial average temperature by 2100, included in some official climate scenarios, require sustained long-term warming rates above what it has been observed over the past seven decades, and thus are highly unlikely to be correct in any case.

For the past 120 years, the global temperature anomaly has been changing by a long-term trend and a 65-year oscillation that have not been affected much by the increase in CO_2. The long term temperature trend is c. +0.12 °C/decade in global temperature anomaly, while the 65-year oscillation departs from trend by c. ± 0.2 °C (Fig. 13.9). After 2035 a likely increase in solar activity, and an expected change in phase in the 65-year oscillation, suggest a forecast for resumed warming during the c. 2035–2065 period. Therefore, for the entire period 2018–2065 we should expect a continuation of the observed linear increase in temperature since 1950 (Fig. 13.10) of c. 0.12°C/decade. Any small deviation from this linear trend should be towards lower values if emissions decrease or if solar activity is lower than expected. An increase in any of them beyond what is calculated is unlikely as they are already considered a high scenario.

Of the two most important external forcings contributing to Modern Global Warming, stabilizing emissions suggest that maximum CO_2 levels could be reached by c. 2075, while maximum solar forcing is forecasted for the 2050–2080 period. If those forecasts are correct, it follows that maximum temperature should be reached in the 2050–2100 period at c. +1.5 °C above pre-industrial values, and remain essentially stable, with variability due to natural oscillations and a small declining trend, for the rest of the present warm period (Fig. 13.10). This warm period, currently unofficially known as the Anthropocene, could last 2–3 centuries more, until c. AD 2250–2350, if the Medieval Warm Period is a good analog.

This conservative forecast of essentially no change in temperature for the next 200 years, allowing for ± 0.5 °C multidecadal changes, rests on the following assumptions:

- A continuation of the CO_2 emissions stabilizing trend observed over the past 10 years with a small declining trend of 0.3%/year after 2050

- An increase in solar activity peaking c. 2080 and a decrease afterwards
- Unsaturating carbon sinks for the period and amounts considered
- A trend to equilibrate carbon sinks and sources plus emissions at an airborne fraction close to zero

The forecast does not depend on any change in policies or heroic reductions in emissions. Policies already being implemented, limitations in fossil fuel availability, and natural demographic changes set the course for future reductions in emissions. Faster reductions should not affect the forecast very much, as atmospheric levels should react slowly to them, and the effect of CO_2 on temperature appears to be lower than estimated by the IPCC (see Sects. 12.7 & 12.10).

13.7 Consequences for Arctic sea ice

The 30% decline in Arctic sea-ice extent that took place between 1995 and 2007 led to numerous radical forecasts, predicting in some cases a summer ice-free Arctic by 2016 (Maslowski et al. 2012) due mainly to albedo feedback leading sea-ice into a "death spiral" (Serreze 2008). Of course, radical forecasts are seldom true, and the albedo effect on sea-ice has turned out to be lower than expected, because summer Arctic sea ice extent has refused to decline any further for the past 14 years. Green and Armstrong (2007) are proven correct in their assessment that experts' predictions are not useful in situations involving uncertainty and complexity, when biases tend to go unchecked.

A knowledge-based Arctic sea ice forecast must take into account the known 60 and 20-year periodicities in sea ice (Polyakov et al. 2004; Divine & Dick 2006; Wyatt & Curry 2014) probably responsible for the present Arctic summer melting pause. These oscillations are likely to produce no change to slight growth in Arctic summer ice until c. 2035, when significant melting is more likely to renew. A conservative forecast is that Arctic summer sea ice will decrease at a slower rate for the period 2022–2050. By 2050 there should still be close to 4 million km^2 of summer sea ice in the Arctic (Fig. 13.11; table 13.1). A return to a warming, melting phase around 2040 might

Table 13.1 Twenty-first century climate projections
The 2021 values for some climate indexes are given in the first data column. The rest of the columns give the projected or assumed corresponding values for 2050 (light blue background) and 2100 (darker blue background), for this author conservative prediction and IPCC CMIP5 RCP4.5 emissions scenario. The biggest differences are for sea-level rise and temperature increase, where the conservative prediction is closer to observed changes for the past decades. (*) indicates 2020 data.

		Projection			
		Conservative		IPCC RCP 4.5	
Year	2021	2050	2100	2050	2100
Atmospheric CO2 (ppm)	416	480–490	~ 500	490	540
CO2 emissions (Gt/year)	35*	35	28	42	15
Temperature anomaly 1961–90 baseline (°C)	0.64	1.0	0.9	1.6	2.2
Temperature anomaly above pre-industrial (°C)	0.95	1.3	1.2	1.9	2.5
Arctic sea ice extent September (million km^2)	4.92	3.8	2.5	2.6	1.4
Sea level rise (mm from 2000)	77	170	340	225	525

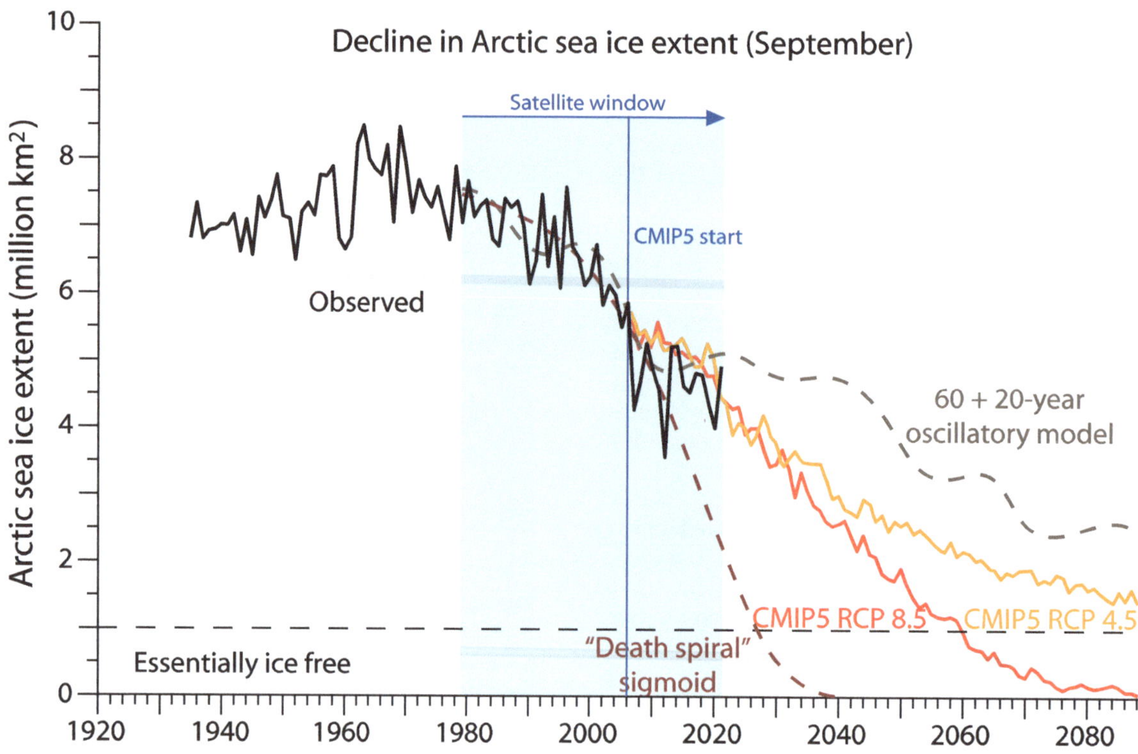

Fig. 13.11 Projected Arctic sea ice decline
Model simulations (continuous colored curves), and observations (black curve) of Arctic sea ice extent for September (1935–2090). Colored curves for RCP scenarios are model averages (CMIP5). After Walsh et al. (2014). Brown dashed curve is a model based on the known 60 and 20-year periodicities in Arctic sea ice. Black continuous curve is NSIDC September Arctic sea ice extent for the satellite window (1979–2021), while 1935–1978 September Arctic sea ice extent data is from Cea Pirón & Cano Pasalodos (2016) reconstruction. Dark red dashed curve is a sigmoid survival curve fitted to 1979–2012 data with ice-free conditions near 2030, following the Arctic sea-ice death spiral proposed by Mark Serreze (2008). The conservative projection (brown dashed curve), explains the pause in Arctic sea-ice melting since 2007 and suggests over 2 million km² of Arctic sea ice remaining by summer 2100.

further reduce Arctic sea ice that could be down to c. 2.5 million km² of summer sea ice (table 13.1) by 2100. With such low levels it cannot be ruled out that some summer might see an ice-free condition (< 1 million km²). This forecast is not too far from the IPCC RCP 4.5 projection (Fig. 13.11; table 13.1), probably because the cryosphere (except Antarctica) is showing a strong response to the increase in temperature and soot (light absorbing particles) levels.

The conservative forecast however is in stark contrast to the many alarmist projections from polar scientists that believe Arctic sea ice is past a tipping point and only accelerated rates of melting are possible now. Those projections that see an Arctic free of ice every summer before 2100 are very likely to be wrong. Lack of significant melting progress for the next decade and a half might clarify the issue.

13.8 Consequences for sea-level rise

In 2007 the IPCC made public its Fourth Assessment Report (AR4). Among AR4 emissions scenarios was B1, that contemplates slow growth in CO_2 emissions to 2050 followed by moderate decrease in emissions afterwards. This scenario is the one that best agrees with the conservative projection outlined above, and projects a 300 mm increase in sea levels for 2000–2100 (central estimate; Fig. 13.12). Seven years later the IPCC published its Fifth Assessment Report (AR5), and among the new scenarios

RCP 4.5 is the most similar to B1. However, the IPCC sea-level model is now a lot more aggressive and projects 525 mm for similar emissions (table 2). Such a strong upward revision responded to claims that models used in the 4th report substantially underestimated the observed past sea-level rise, although no acceleration has been observed since 1993. Despite the 60% increase due to a change of assumptions, the IPCC was severely criticized for producing estimates of sea-level rise that were too conservative. To provide a view that satisfied the consensus, Horton et al. (2014) conducted an expert elicitation (poll) on sea-level rise among authors of articles related to sea-level rise. Although they were only asked for a low and high scenarios, a mean projection can be obtained by averaging both (Fig. 13.12). This intermediate scenario derived from Horton et al. (2014) projects a rise of c. 800 mm for 2000–2100. In 2017 NOAA published their updated global sea-level rise scenarios where the intermediate scenario, that is most consistent with RCP 4.5, forecasts one meter of sea-level rise for 2000–2100 (Sweet et al. 2017; Fig. 13.12). Surprisingly, and despite lack of acceleration in sea-level rise since 1993 (28 years), projections are becoming significantly more pessimistic with time. It reminds of Mark Twain's famous quote: *"There is something fascinating about science. One gets such wholesale returns of conjecture out of such a trifling investment of fact"* (Life in the Mississippi, 1883).

Past sea-level increases for the last 70 years have taken place under rapidly increasing emissions. However,

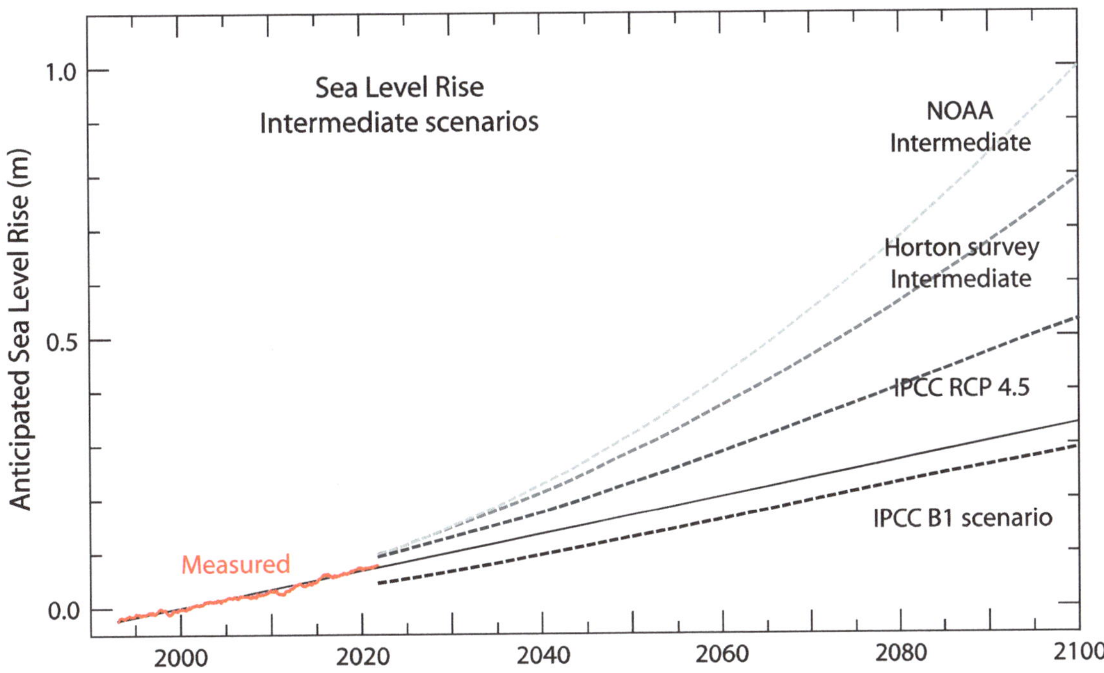

Fig. 13.12 Sea-level rise intermediate scenarios for 2100
Red continuous curve, sea-level rise measured since 1993 and zeroed in 2000. Data from NASA. Dashed curves, sea-level rise projections for the 2000–2100 period under intermediate emissions scenarios from different sources. 2007 IPCC AR4 B1 scenario (dashed black); 2014 IPCC AR5 RCP 4.5 scenario (dashed dark grey); Horton et al. (2014), survey intermediate scenario (average of the high and low scenarios; dashed medium grey); 2017 NOAA intermediate scenario (dashed light grey, Sweet et al. 2017). Black line, 1993-2021 linear trend extrapolated to 2100. All models were started at zero in 2000 and are shown since 2021.

only a small 0.01 mm/yr^2 acceleration has been detected by most researchers (Church & White 2011; Jevrejeva et al. 2014; Hogarth 2014). The added uncertainty in future CO_2 levels and emissions suggests a simple linear extrapolation might be sufficient for projecting future sea levels, that appear to be increasing quite constantly despite changing atmospheric CO_2 and global temperature. Such a forecast would see a 170 mm increase every 50 years, for a total 340 mm increase in the 21st century (Fig. 13.12; table 13.1). This increase is too small to constitute a problem on a global basis but might add to the problem of local sea-level rise in areas where land subsidence or lack of sufficient sedimentation are going to require adapting measures.

13.9 Other climate change consequences for the 21st century

A conservative forecast is that most extreme weather phenomena should continue occurring in an unpredictable manner without a significant change in frequency. Storm data from the last 6500 years shows clearly that storms increase in frequency and strength with cooling, and decrease with warming (see Fig. 7.7c). The reasons are that warming reduces heat transport due to a decrease in the latitudinal temperature gradient, and that the atmospheric heat engine has a reduced ability to generate work due to an increase in the power required by the intensification of the hydrological cycle (Laliberté et al. 2015). The association of weaker storm strength to warming is supported by a global wind stilling during the 1980–2000 rapid warming period that reversed course with the pause (Zeng et al. 2019). Records (Maue 2011) and models (Sugi et al. 2015)

also show a reduction in tropical cyclone activity with global warming.

The only extreme weather phenomenon that is credibly projected to increase is the frequency and intensity of heat waves. However, the change could be smaller than anticipated as Modern Global Warming is having more effect on minimum, rather than on maximum, temperatures producing generally warmer winters. This has an added benefit because a 68-authors global study of 5 million non-optimal temperature associated deaths found that 90% of them were cold-related (Zhao et al. 2021). From a societal point of view, adaptation to increased heat-waves requires cheap, abundant energy.

The effect over the biosphere is more difficult to forecast, as it has shown very high adaptability through much bigger climate changes in the past. If we accept that the world in 2021 was c. 0.95 °C above the pre-industrial average, the conservative forecast indicates it might only increase a further 0.55 °C before stabilizing. Therefore, we might have already seen over 60% of the total expected warming. The negative effects strictly from climate change over the biosphere are very limited, while the positive effects are abundant and profound. Most biomes, but particularly semi-arid ones, have responded to warming and CO_2 increase through an increase in leaf area (also known as greening; Zhu et al. 2016). These three factors (temperature, CO_2, and greening) appear to have caused an increase in global terrestrial net primary production of 12% between 1961–2010 (Li et al. 2017). The effect of this increased energy flux through ecosystems is beneficial to nearly all species. The net effect of the warming and increased CO_2 is clearly positive for the biosphere. It is reasonable to think that too much of a good thing should reach a point when the net effect starts being negative, but there is no evidence that we are close to that point or that it should be reached within the 21st century.

The loss of Arctic sea ice has been proposed to be a clear risk to polar bears, and the species was included in the US endangered species list solely on those grounds. However polar bears might not be very sensitive to summer ice reductions as their ice-dependent hunting takes place in spring and is negatively affected by too much ice. The species has survived very reduced or even absent summer Arctic sea ice during the Holocene Climatic Optimum and the past warmer interglacial. The main danger to polar bears has historically been human hunting, and since the international hunting limitation by the Oslo agreement of 1973, polar bear population estimates have been increasing, apparently unaffected by the loss of 30% of summer Arctic sea ice in the 1995–2007 period (Crockford 2020). At present there is no evidence that polar bears are threatened during the 21st century from climate change, even if the projected summer ice loss in the Arctic takes place (Fig. 13.11).

Regarding other possible consequences, our knowledge is too limited to say much. Claims of sinking nations, hordes of climate refugees, and a new normal every time there is an extreme weather event, are wildly exaggerated and agenda-driven. The highest return for our limited resources is very likely to come from adaptation policies, and no-regrets policies. Policies to prevent or reduce climate change are destined to be highly ineffectual given the strong natural component of climate change, as the past demonstrates.

13.10 Projections

13a. Human CO_2 emissions are stabilizing. Peak coal and oil, and current trends make a decrease in emissions very likely before 2050. Atmospheric CO_2 levels should reach 500 ppm but might stabilize soon afterwards.

13b. According to solar cycles, solar activity should increase after the present extended solar minimum, and 21st century solar activity should be as high or higher than 20th century. A mid-21st century solar grand minimum is highly improbable.

13c. Global warming might stall or slightly reverse for the period 2000–2035. Cyclic factors suggest renewed warming for the 2035–2065 period at a similar rate to the last half of the 20th century. Afterwards global warming could end, with temperatures stabilized around +1.5 °C above pre-industrial, and a very slow decline for the last part of 21st century and beyond.

13d. The present summer Arctic sea ice melting pause might continue until c. 2035. Renewed melting is probable afterwards, but it is unlikely that the Arctic summer will become consistently ice free even by 2100.

13e. The rate of sea-level rise can be conservatively projected to a 340 mm increase by 2100 over 2000 levels. Most rates published are extremely non-conservative and very unlikely to take place.

13f. Climate change should remain subdued and net positive for the biosphere for the 21st century. Adaptation is likely to be the best strategy, as it has always been.

References

Abdussamatov HI (2013) Grand minimum of the total solar irradiance leads to the little ice age. Journal of Geology & Geosciences 2 (2) 113

Anchukaitis KJ, Wilson R, Briffa KR et al (2017) Last millennium Northern Hemisphere summer temperatures from tree rings: Part II spatially resolved reconstructions. Quaternary Science Reviews 163 1–22

Armstrong JS, Green KC & Graefe A (2015) Golden rule of forecasting: Be conservative. Journal of Business Research 68 (8) 1717–1731

BP statistical review of world energy (2022) 71st ed. London. https://www.bp.com/en/global/corporate/energy-economics/statistical-review-of-world-energy.html Accessed 30 Jun 2022

Breitenmoser P, Beer J, Brönnimann S et al (2012) Solar and volcanic fingerprints in tree-ring chronologies over the past 2000 years. Palaeogeography, Palaeoclimatology, Palaeoecology 313 127–139

Canadell JG, Pataki DE, Gifford R et al (2007) Saturation of the terrestrial carbon sink. In: Canadell JG, Pataki DE and Pitelka LF (eds) Terrestrial ecosystems in a changing world. Springer, Berlin, p 59–78

Cea Pirón MA & Cano Pasalodos JA (2016) Nueva serie de extensión del hielo marino ártico en septiembre entre 1935 y 2014. Revista de Climatología 16 1–19

Charvátová I & Hejda P (2014) Responses of the basic cycle of 178.7 and 2402 yr in solar-terrestrial phenomena during Holocene. Pattern Recognition in Physics 2 21–26

Church JA & White NJ (2011) Sea-level rise from the late 19th to the early 21st century. Surveys in Geophysics 32 (4–5) 585–602

Churchill WS (1958). A History of the English-Speaking Peoples, 1st ed. Vol. IV 387. Cassell, London

Clilverd MA, Clarke E, Ulich T et al (2006) Predicting solar cycle 24 and beyond. Space weather 4 (9) 1–7

Crockford SJ (2020) The state of the polar bear 2019. GWPF Report 39. https://www.thegwpf.org/content/uploads/2020/02/StatePB2019.pdf Accessed 01 Jul 2022

Divine DV & Dick C (2006) Historical variability of sea ice edge position in the Nordic Seas. Journal of Geophysical Research: Oceans 111 C01001

Eichler A, Olivier S, Henderson K et al (2009) Temperature response in the Altai region lags solar forcing. Geophysical Research Letters 36 (1) L01808

Etheridge DM, Steele LP, Langenfelds RL et al (1996) Natural and anthropogenic changes in atmospheric CO_2 over the last 1000 years from air in Antarctic ice and firn. Journal of Geophysical Research: Atmospheres 101 (D2) 4115–4128

European Center for Medium-range Weather Forecast (2021) https://apps.ecmwf.int/webapps/opencharts/products/seasonal_system5_nino_plumes?base_time=201702010000&nino_area=Nino3-4 Accessed 23 Dec 2021

ExxonMobil (2018) Outlook for Energy: A view to 2040. http://cdn.exxonmobil.com/~/media/global/files/outlook-for-energy/2018/2018-outlook-for-energy.pdf Accessed 03 Mar 2019

Feynman J & Ruzmaikin A (2014) The Centennial Gleissberg Cycle and its association with extended minima. Journal of Geophysical Research: Space Physics 119 (8) 6027–6041

Gilfillan D, Marland G, Boden T & Andres R (2019) Global, Regional, and National Fossil-Fuel CO_2 Emissions. Carbon Dioxide Information Analysis Center at Appalachian State University. https://energy.appstate.edu/research/work-areas/cdiac-appstate Accessed 12 Aug 2020

Green KC & Armstrong JS (2007) Global warming: Forecasts by scientists versus scientific forecasts. Energy & Environment 18 (7) 997–1021

Hansen J, Kharecha P & Sato M (2013) Climate forcing growth rates: doubling down on our Faustian bargain. Environmental Research Letters 8 (1) 011006

Hogarth P (2014) Preliminary analysis of acceleration of sea level rise through the twentieth century using extended tide gauge data sets (August 2014). Journal of Geophysical Research: Oceans 119 (11) 7645–7659

Horton BP, Rahmstorf S, Engelhart SE & Kemp AC (2014) Expert assessment of sea-level rise by AD 2100 and AD 2300. Quaternary Science Reviews 84 1–6

Inceoglu F, Simoniello R, Knudsen MF et al (2015) Grand solar minima and maxima deduced from ^{10}Be and ^{14}C: magnetic dynamo configuration and polarity reversal. Astronomy & Astrophysics 577 A20

Intergovernmental Panel on Climate Change (1990) Executive Summary. In: Houghton JT, Jenkins GJ & Ephraums JJ (eds) Climate Change. The IPCC Scientific Assessment. Cambridge University Press, Cambridge, p XI

Jevrejeva S, Moore JC, Grinsted A et al (2014) Trends and acceleration in global and regional sea levels since 1807. Global and Planetary Change 113 11–22

Keenan TF, Prentice IC, Canadell JG et al (2016) Recent pause in the growth rate of atmospheric CO_2 due to enhanced terrestrial carbon uptake. Nature Communications 7 13428

Kobashi T, Box JE, Vinther BM et al (2015) Modern solar maximum forced late twentieth century Greenland cooling. Geophysical Research Letters 42 (14) 5992–5999

Laliberté F, Zika J, Mudryk L et al (2015) Constrained work output of the moist atmospheric heat engine in a warming climate. Science 347 (6221) 540–543

Le Quéré C, Andrew RM, Friedlingstein P et al (2018) Global carbon budget 2018. Earth System Science Data 10 2141–2194

Le Quéré C, Peters GP, Friedlingstein P et al (2021) Fossil CO_2 emissions in the post-COVID-19 era. Nature Climate Change 11 (3) 197–199

Li P, Peng C, Wang M et al (2017) Quantification of the response of global terrestrial net primary production to multifactor global change. Ecological Indicators 76 245–255

Lockwood M (2010) Solar change and climate: an update in the light of the current exceptional solar minimum. Proceedings of the Royal Society A: Mathematical Physical and Engineering Sciences 466 (2114) 303–329

Maslowski W, Clement Kinney J, Higgins M & Roberts A (2012) The future of Arctic sea ice. Annual Review of Earth and Planetary Sciences 40 625–654

Maue RN (2011) Recent historically low global tropical cyclone activity. Geophysical Research Letters 38 (14)

Meehl GA, Senior CA, Eyring V et al (2020) Context for interpreting equilibrium climate sensitivity and transient climate response from the CMIP6 Earth system models. Science Advances 6 (26) eaba1981

Mörner N–A (2015) The approaching new grand solar minimum and little ice age climate conditions. Natural Science 7 510–518

Mörner N–A, Tattersall R, Solheim JE et al (2013) General conclusions regarding the planetary–solar–terrestrial interaction. Pattern Recognition in Physics 1 205–206

Nature (2014) Focus: Recent slowdown in global warming. https://www.nature.com/collections/sthnxgntvp Accessed 03 Mar 2019

Polyakov IV, Alekseev GV, Timokhov LA et al (2004) Variability of the intermediate Atlantic water of the Arctic Ocean over the last 100 years. Journal of climate 17 (23) 4485–4497

Salvador RJ (2013) A mathematical model of the sunspot cycle for the past 1000 yr. Pattern Recognition in Physics 1 117–122

Scafetta N (2014) The complex planetary synchronization structure of the solar system. Pattern Recognition in Physics 2 (1) 1–19

Schindler DW (1999) Carbon cycling: The mysterious missing sink. Nature 398 (6723) 105–106

Schuster U & Watson AJ (2007) A variable and decreasing sink for atmospheric CO_2 in the North Atlantic. Journal of Geophysical Research: Oceans 112 C11006

Serreze M (2008) In: Reuters (News agency) Arctic ice second-lowest ever; polar bears affected. Reuters Market News, 27 Aug 2008. https://www.reuters.com/article/us-arctic-ice-polar-bears-idUSN2745499020080827 Accessed 03 Mar 2019

Shearer C, Ghio N, Myllyvirta L et al (2017) Boom and Bust 2017. Tracking the Global Coal Plant Pipeline. CoalSwarm/Sierra Club/Greenpeace Report. http://endcoal.org/wp-content/uploads/2017/03/BoomBust2017-English-Final.pdf Accessed 03 Mar 2019

Shepherd SJ, Zharkov SI & Zharkova VV (2014) Prediction of solar activity from solar background magnetic field variations in cycles 21–23. The Astrophysical Journal 795 (1) 46

Steinhilber F & Beer J (2013) Prediction of solar activity for the next 500 years. Journal of Geophysical Research: Space Physics 118 (5) 1861–1867

Sugi M, Yoshida K & Murakami H (2015) More tropical cyclones in a cooler climate?. Geophysical Research Letters 42 (16) 6780–6784

Svalgaard L (2018) Prediction of Solar Cycle 25. In: SORCE 2018 Sun–Climate Symposium. Lake Arrowhead, California, 19–23 Mar 2018. https://www.leif.org/research/Prediction-of-SC25.pdf Accessed 03 Mar 2019

Sweet WV, Kopp RE, Weaver CP et al (2017) Global and regional sea level rise scenarios for the United States. NOAA Technical Report NOS CO-OPS 083

Tetlock PE (2005) Expert Political Judgment: How Good Is It? How Can We Know? Princeton University Press, Princeton

Trouet V & van Oldenborgh GJ (2013) KNMI Climate Explorer: a web-based research tool for high-resolution paleoclimatology Tree-Ring Research 69 (1) 3–13

United Nations, Department of Economic and Social Affairs, Population Division (2017) World Population Prospects: The 2017 Revision, Key Findings and Advance Tables. ESA/P/WP/248

Usoskin IG, Gallet Y, Lopes F et al (2016) Solar activity during the Holocene: the Hallstatt cycle and its consequence for grand minima and maxima. Astronomy & Astrophysics 587 A150

Usoskin IG, Solanki SK & Kovaltsov GA (2007) Grand minima and maxima of solar activity: new observational constraints. Astronomy & Astrophysics 471 (1) 301–309

van Vuuren DP, Edmonds J, Kainuma M et al (2011) The representative concentration pathways: an overview. Climatic change 109 (1–2) 5–31

Vinós J (2019) ENSO predictions based on solar activity. https://judithcurry.com/2019/09/01/enso-predictions-based-on-solar-activity/ Accessed 23 Dec 2021

Walsh J, Wuebbles D, Hayhoe K et al (2014) Our changing climate. In: Melillo JM, Richmond TC and Yohe GW (eds) Climate Change Impacts in the United States: The Third National Climate Assessment. US Global Change Research Program, p 19–67

Wyatt MG & Curry JA (2014) Role for Eurasian Arctic shelf sea ice in a secularly varying hemispheric climate signal during the 20th century. Climate dynamics 42 (9–10) 2763–2782

Zeng Z, Ziegler AD, Searchinger T et al (2019) A reversal in global terrestrial stilling and its implications for wind energy production. Nature Climate Change 9 (12) 979–985

Zhao Q, Guo Y, Ye T et al (2021) Global, regional, and national burden of mortality associated with non-optimal ambient temperatures from 2000 to 2019: a three-stage modelling study. The Lancet Planetary Health 5 (7) e415–e425

Zhu Z, Piao S, Myneni RB et al (2016) Greening of the Earth and its drivers. Nature Climate Change 6 (8) 791–795

14

THE NEXT GLACIATION

"A global deterioration of climate, by order of magnitude larger than any hitherto experienced by civilized mankind is a very real possibility and indeed may be due very soon."

George Kukla and Robert Matthews (1972)

14.1 Introduction

The expected timeframe for the next glaciation is so far in the future that traditionally it has only attracted academic interest. There was a small peak of popular interest in the early 1970s at the end of the mid-20th century cooling period. In January 1972, geologists George Kukla and Robert Matthews organized a meeting on the end of the present interglacial, and afterwards they wrote president Richard Nixon calling for federal action on the observed climate deterioration that had the potential to lead to the next glaciation. Ironically, concerns over the end of the interglacial led to the creation of NOAA's Climate Analysis Center in 1979 (Reeves & Gemmill 2004), that would substantially contribute to global warming research. Some popular magazines reported about a coming ice age at times of harsh winter weather during the early 1970s.

Current academic consensus is that a return to glacial conditions is not possible under any realistic condition for tens of thousands of years, and the IPCC expresses virtual certainty that a new glacial inception is not possible for the next 50 kyr if CO_2 levels remain above 300 ppm (IPCC–AR5, Masson–Delmotte et al. 2013, p 435). This claim expressed on so certain terms is in stark contrast with the lack of precedent for any interglacial spanning over two obliquity oscillations.

14.2 Interglacial evolution

Each interglacial is different. They all have different astronomical signatures, different initial conditions, different evolution, and are subjected to non-linear chaotic climate unpredictability. But they all take place during a single obliquity oscillation. 104 marine isotope stages (MIS) have been identified over the Pleistocene's 2.6 Ma, half of them (Fig. 14.1b, odd numbers) corresponding to warm periods. On average there is one every 50,000 years, almost corresponding to the obliquity frequency (41 kyr). The match is not exact because some obliquity oscillations have failed to produce an interglacial.

For the last 800 kyr, after the Mid-Pleistocene Transition (1.5–0.7 Ma; Fig. 14.1), the planet has become so cold, and the ice-sheets grown so large, that to produce an interglacial outside the periods of high eccentricity requires the simultaneous concert of high obliquity, high northern summer insolation, and very large unstable ice sheets. This has had the effect of spacing interglacials from an obliquity-linked 41-kyr cycle to its multiples 82 or

123 kyr (see Fig. 2.11). A side effect is that after one or more obliquity oscillations without an interglacial the planet gets colder, and when an interglacial is finally produced, it reaches a warmer state. The climate of the planet has become more unstable in the Middle and Late Pleistocene, rapidly transitioning from more extreme cold to more extreme warm, and back, contributing to numerous species extinctions, and perhaps to the evolution of our species (Fig. 14.1c).

The majority of interglacials of the past 800 kyr are the product of very similar orbital and ice-volume conditions and present a common pattern (Fig. 14.2). The Holocene interglacial is the result of similar conditions, and belongs to this group. Nearly all exceptions can be explained in terms of particular orbital and ice-volume conditions that do not apply to the Holocene (see Chap. 2).

Antarctica leads the deglaciation over the Northern Hemisphere and reaches its highest temperature at the obliquity peak when the Laurentide and Fennoscandian ice-sheets are not completely melted yet. The asynchrony between a Southern Hemisphere cooling from declining obliquity and low summer insolation, and a Northern Hemisphere warming from ice-sheets melting and high summer insolation results in a global optimum that has a different span depending on latitude. Interglacial temperature decline presents a delay with respect to obliquity decline of 5,500–8,000 years (see Fig. 2.11), observed since the late Pliocene (Donders et al. 2018). This delay has been attributed to a lag in the ice volume change with respect to its rate of change (Huybers 2009). Ocean thermal inertia could also contribute to the lag. Once northern summer insolation is declining at its fastest rate, the interglacial enters a phase of slowly declining temperature (c. −0.2°C/millennium) that in the Holocene has been termed the Neoglaciation. Despite temperature decline and modest glacier expansion, sea levels are quite stable over this period, as there is no significant ice-sheet build up. The Holocene has been clearly at this stage since c. 5000 yr ago, until Modern Global Warming.

When northern summer insolation becomes low, and obliquity is at its fastest rate of decline, the interglacial reaches glacial inception. This tipping point appears to take place during a global Little Ice Age (LIA)-type cold period when due to the start of ice-sheet build up, sea-level starts dropping. The intensification of ice-albedo and vegetation feedbacks result in a point of no return. Regardless of insolation changes, once glacial inception takes place, the glaciation will continue through advances and retreats

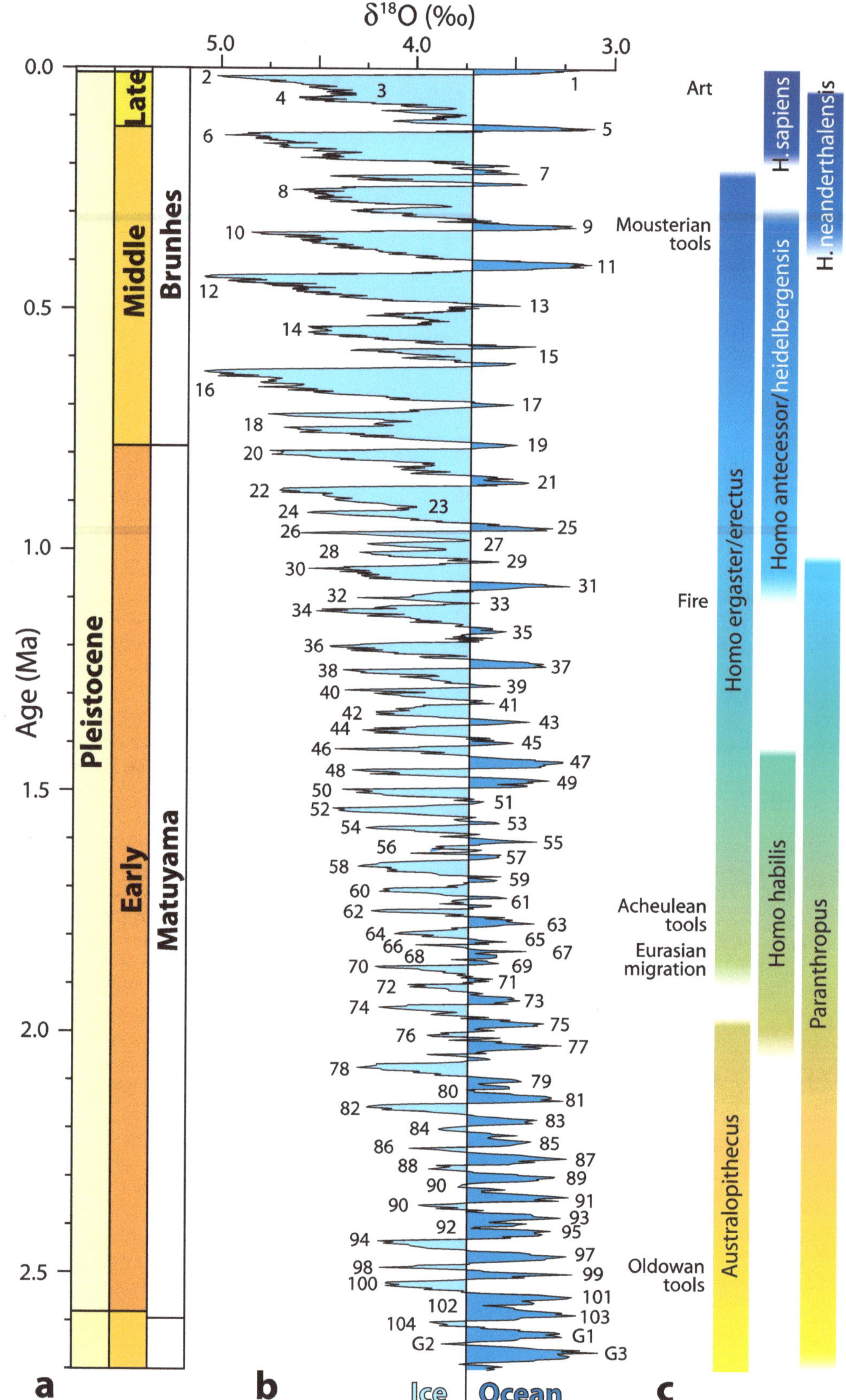

Fig. 14.1 The Pleistocene climatic madhouse

a) Pleistocene subdivisions according to the International Commission on Stratigraphy. **b)** LR04 stack of 57 benthic cores record of δ¹⁸O, an ice and temperature proxy, subdivided between ocean (dark blue), and ice (light blue) at the level that identifies the generally accepted interglacials. δ¹⁸O in essence measures the distribution of water between ice and oceans. Numbers correspond to marine isotope stages (after Lisiecki & Raymo 2005). **c)** Pleistocene hominins showing the main species of the genus *Homo*, and the *Australophitecus* and *Paranthropus* genera, as well as the approximate time for some of the main cultural advances. No species distinction is made between *ergaster* and *erectus*, and between *antecessor* and *heildebergensis*. No inference is made regarding evolutionary relationship.

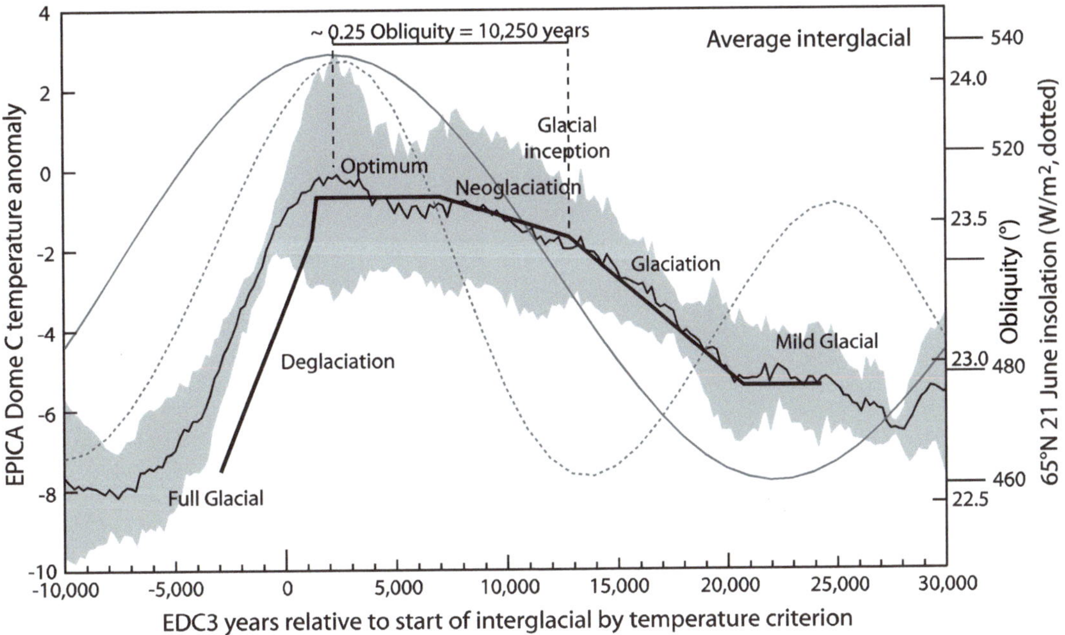

Fig. 14.2 Average of six of the past ten interglacials
An average interglacial (black curve and 1σ grey bands) was constructed from EPICA Dome C deuterium data for interglacials MIS 5e, 7c–a, 9e, 15a, 15c and 19c, after aligning them at the zero date, corresponding to 787.0, 624.4, 579.6, 335.5, 214.7, and 131.4 ka in EDC3 dates. The obliquity average for all of them (grey sinusoid continuous line) and the insolation average at 65°N on 21st June for all but MIS 7c–a (grey dotted line) were also averaged from the alignment. The thick line represents the different global substages of a typical interglacial. Antarctica leads in deglaciation, with the rest of the planet lagging (Jouzel, et al. 2007; Laskar et al. 2004).

in a relaxation-type dynamic until the conditions are met for a new interglacial. During the past 2.3 million years no interglacial has been able to continue from one obliquity oscillation to the next. Low obliquity conditions have always led to the end of the interglacial.

Despite their very different temporal scale, the similarities between Antarctic interglacial records and Greenland Dansgaard–Oeschger oscillations (Fig. 14.3; see also Chap. 3) suggest similar dynamics are at play. Both have been proposed to be astronomically paced (Milanković 1920; Rahmstorf 2003). The warming phase is explosive, fed by a fast ice-melting feedback, producing an early peak. It appears to constitute an excitable system from a stable glacial state. A slow declining phase from the peak towards an inflection point (unstable point, Fig. 14.3) suggests a quasi-stable warm phase as the warming conditions wane. At the inflection point the intensification of the slow ice-accumulation feedback accelerates the cooling phase increasing the climatic instability and producing a faster relaxation towards the stable state. Fast-slow dynamics acting on excitation cycles have been discussed as an explanation for both Dansgaard–Oeschger events and interglacials (Crucifix 2012), in which the cold stable and

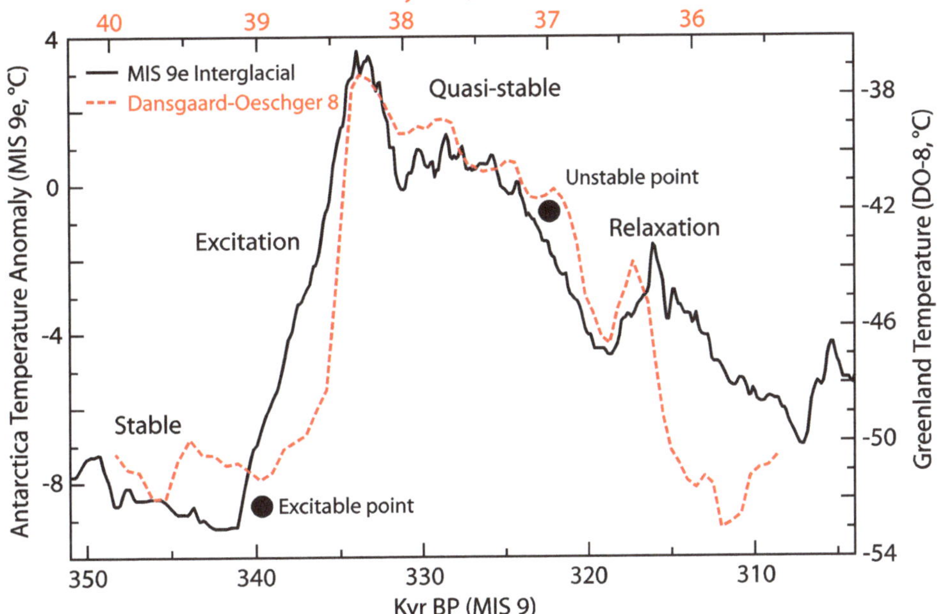

Fig. 14.3 Comparison of MIS 9e interglacial and Daansgard–Oechsger event 8
With a different time-scale, MIS 9e (black line, EPICA) and DO–8 (red dashed line, GISP2) present a fast (excitation) transition from an excitable point, and a slow (relaxation) transition from an unstable point, between a stable cold state and a quasi–stable warm state. Data from Jouzel, et al. (2007) and Alley (2000).

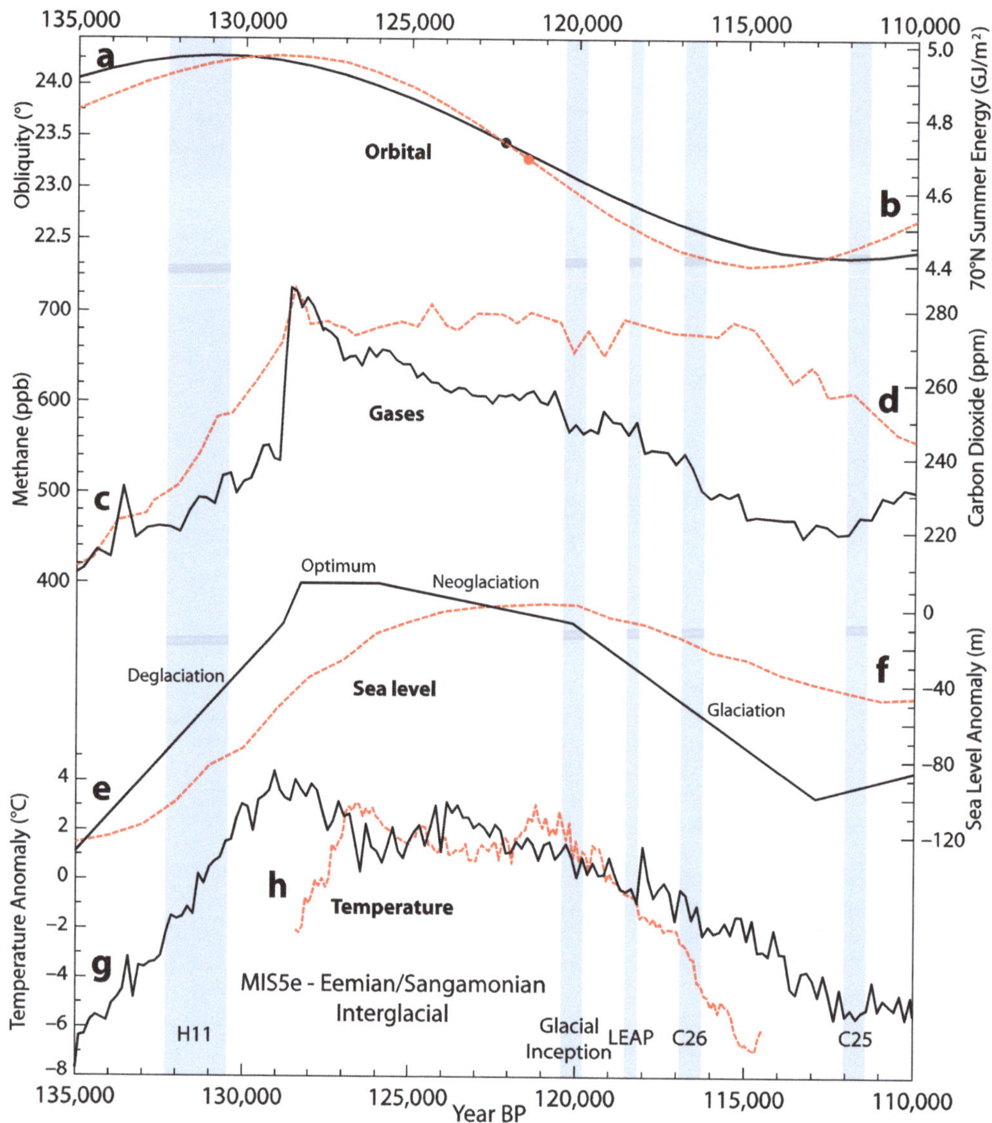

Fig. 14.4 The Eemian interglacial and its transition to the Weichselian glaciation
a) Obliquity configuration (black continuous line; after Laskar et al. 2004). **b)** Summer energy at 70°N (red dashed line; after Huybers 2006). Dots indicate current values. **c)** Methane levels (black continuous line; after Loulergue et al. 2008). **d)** Carbon dioxide levels (red dashed line; after Bereiter et al. 2015). **e)** Interglacial temperature profile (black continuous line, no scale). **f)** Sea level (red dashed line; after Spratt & Lisiecki 2016). **g)** Antarctic temperature anomaly (black continuous line; after Jouzel et al. 2007). **h)** Greenland temperature anomaly (red dashed line; after NEEM Community Members 2013). H11: Heinrich event 11. LEAP: Late Eemian aridity pulse. C25 and C26: North Atlantic cold events 25 and 26.

warm quasi-stable states constitute the different branches of a slow manifold in an excitable dynamic system.

The last interglacial is usually referred as the Eemian, its stratigraphic name in Western Europe after the Dutch river Eem. The Eemian stratum is dated between 126–115 ka in Northern Europe and 126–110 ka in Southern Europe. MIS 5e is dated between 132–115 ka. Weichselian (also Wisconsinan or Würm) glacial inception takes place earlier, at 120 ka, when both Antarctica and Greenland initiate their cooling at a time when CO_2 shows a temporary decline, and methane and global sea level start declining (Spratt & Lisiecki 2016; Fig. 14.4). At this time North Atlantic subpolar waters become colder and Northern European boreal forests start retreating (Govin et al. 2015). Climate becomes more unstable and at 118.6 ka the Late Eemian Aridity Pulse (Sirocko et al. 2005; Fig. 14.4), a 470-year cold and dry period, takes place in Central Europe. Only 1500 years later the C26 cold period is rec-

ognized in Greenland records (Govin et al. 2015). Finally, at 115 ka CO_2 levels start decreasing, with over 5,000 years delay to glacial inception. Temperate forest retreat reaches Southern Europe. Methane levels bottom at 113 ka, and CO_2 at 109 ka. Between 108 and 106 ka glacial conditions are established in most records (Govin et al. 2015). The first large iceberg discharge takes place at 107 ka, indicating well-developed ice sheets. While a deglaciation usually takes about 5,000 years, a glaciation requires almost 15,000 years.

14.3 Studying the future by looking at the past

How glaciations (or ice ages as they are popularly known) were discovered, and the 140-year debate that took place between CO_2 supporters and astronomical cycles proponents to explain them, was described in chapter 9 (see Sect. 9.2). The scientific debate was settled in favor of the latter ones. The evidence is so strong for the orbital-driven Pleistocene glacial oscillations, that it has become widely accepted as an exception to the CO_2 hypothesis of climate. No alternative to Milankovitch Theory has successfully explained why CO_2 levels oscillate at Milankovitch frequencies. Milankovitch is a problem for the hypothesis that CO_2 is climate's main control knob, as it doesn't leave much room for CO_2 to explain the climate of the past 3 Myr, when very large oscillations have taken place at quasi-regular intervals. An explanation offered is that very low CO_2 levels during the Pleistocene allowed glacial pacing by Milankovitch orbital changes. However, this explanation, that sounds very much like a "no true Scotsman" statement to explain an exception, is unconvincing, as precisely the lower the CO_2 levels the more climatic effect its changes must have. But since temperature and CO_2 levels follow Milankovitch oscillations, an unresolved question is how much of the temperature change is caused by the CO_2 change. Central to this question is the climate sensitivity to CO_2 at equilibrium (ECS). At one extreme, Carolyn Snyder (2016), attributes all temperature changes to CO_2 changes, estimating an Earth system sensitivity of

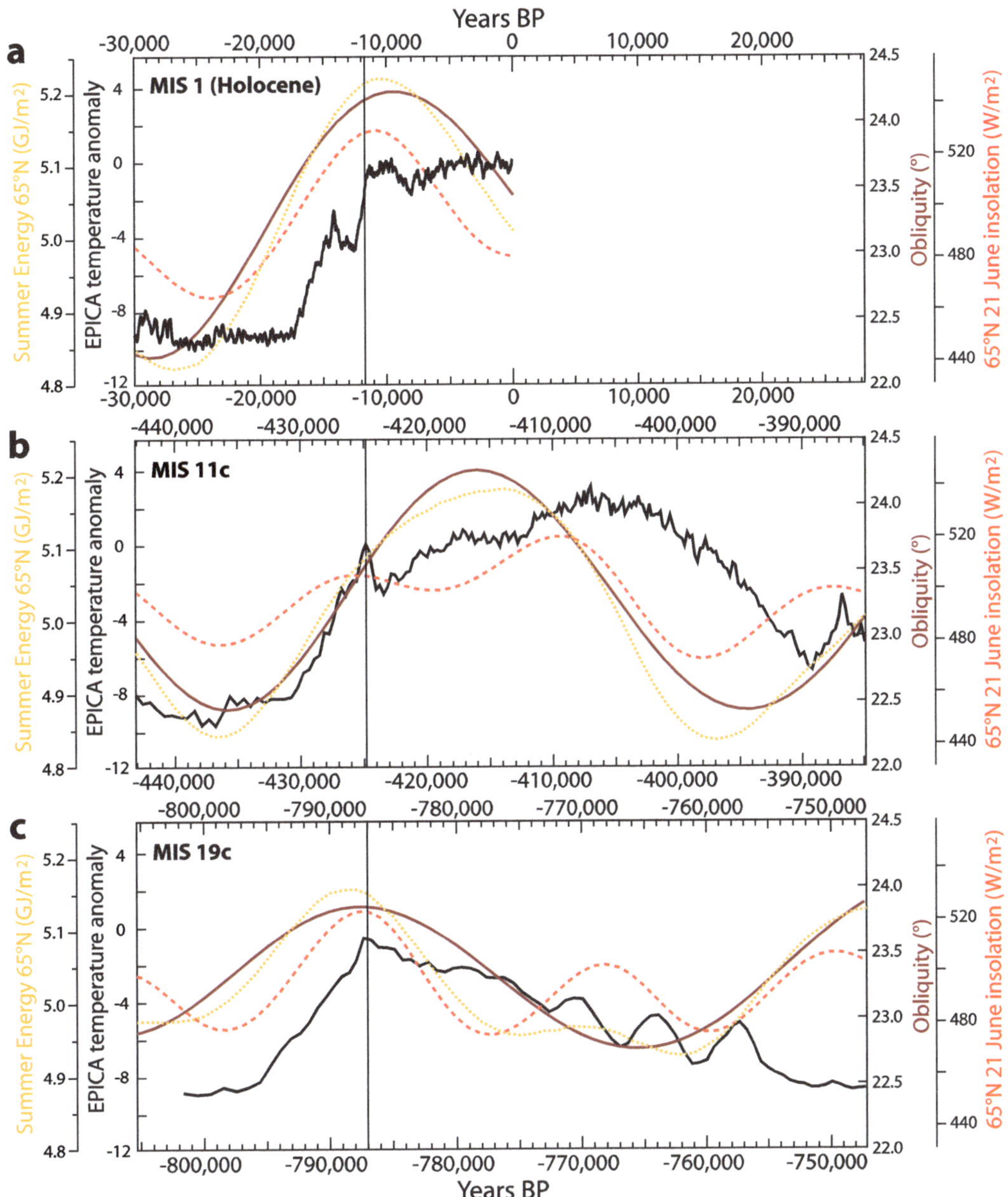

Fig. 14.5 Low eccentricity interglacials of the past 800 ka
EPICA Dome C deuterium proxy expressed as temperature anomaly for **a)** MIS 1, **b)** MIS 11c, and **c)** MIS 19c, aligned at their interglacial start (vertical line). MIS 1 and MIS 19c have a similar orbital configuration with obliquity (dark red continuous line) rising slightly ahead of northern summer insolation (red dashed line) producing a peak in summer energy (orange dotted line) at the start of the interglacial. For MIS 11c northern summer insolation, started increasing from a relatively high level several kyr earlier than obliquity, and continued being high for two oscillations resulting in a very wide summer energy peak. When obliquity, northern summer insolation, and summer energy decreased for MIS 11c, there was a longer than usual delay until temperature fell. Despite its unusual length due to its unusual orbital configuration, MIS 11c could not continue through an obliquity minimum. The proximity of its obliquity minimum makes the Holocene orbital configuration unfavorable for a long interglacial. MIS 1 temperature data averaged with a 200-yr running window. After Jouzel et al. (2007); Laskar et al. (2004); Huybers (2006).

9°C warming per doubling of CO_2 over millennia (disputed for ignoring orbital forcing by Schmidt et al. 2017). At the other extreme, with an ECS of 1.5°C (Lewis & Curry 2018), CO_2 contribution to the glacial-interglacial temperature change would be relatively minor (c. 15%).

Chapter 2 reviewed Milankovitch Theory of the glacial cycle showing how glacials and interglacials result from the interplay between the 41-kyr obliquity cycle and a 100-kyr ice-volume cycle. A clear consequence derived from evidence is that once an interglacial takes place and most extrapolar ice is melted, the 100-kyr ice-volume cycle plays no role and the end of the interglacial is decided solely on obliquity's terms. High summer insolation at high northern latitudes might delay the inevitable, but obliquity always has the last word on finishing interglacials. Obliquities below 23° do not support interglacials, and their end becomes only a question of when. The question of the end of the present interglacial has become a question of what happens when obliquity gets below 23° (in 3,500 years), and to answer that question scientists can choose between looking at the past for adequate analogues analyzing their differences with the present interglacial, or use models and run them far into the future to see what they say. The main problem with this second approach is that models must be programmed with all the relevant functions correctly, or otherwise their output is useless. And as we have seen (Chap. 2), to study the end of the present interglacial the relevant orbital parameters are summer energy (or obliquity) and eccentricity. By contrast, almost every study dealing with the issue has used the erroneous 65°N summer insolation parameter, and that mistake is reflected in every model.

14.4 MIS 11c is a poor Holocene analog

Most authors use MIS 11c interglacial as analog to the Holocene, because of similarly low eccentricity. However, precession and obliquity do not align in the same way for MIS 11c as for the Holocene (see Sect. 2.10, interglacials of atypical duration). MIS 11c is an anomalous interglacial in terms of duration, and clearly this is not due to low eccentricity as MIS 19c had similarly low eccentricity and a standard duration (Fig. 14.5).

During the Middle to Late Pleistocene only a combination of high obliquity, high ice volume, and high 65°N summer insolation provides enough summer energy and ice-melting positive feedbacks to terminate the glacial period. Afterwards (as in MIS 19c, Fig. 14.5c), the fall in summer energy brings about glacial inception after a delay of several thousand years. In the case of MIS 11c, an unlikely coincidence of several factors working together resulted in the longest interglacial of the past million years. First, 65°N summer insolation was relatively high for many thousands of years before glacial termination, above 480 W/m² (today's value). This, together with extraordinarily high ice-volume instability (the largest amount of ice in the entire Pleistocene), could have "primed" the interglacial, that started unusually soon after summer energy began increasing, without the usual wait to obliquity increase. The obliquity peak took place right in the middle of two insolation peaks. It is extremely unusual that such combination would produce an interglacial in the Middle to Late Pleistocene, but in the case of MIS 11c the 65°N summer insolation at the minimum between the two insolation peaks never falls below 500 W/m², a value higher than today's. As a result, summer energy shows a very broad peak of 20,000 years, versus the usual 10–12,000 years (Fig. 14.5b, dotted line). When summer energy was declining, came the second peak in 65°N summer insolation, increasing the temperature and producing a very warm interglacial, particularly so late. When finally, the orbital conditions were adequate for glacial inception, the interglacial had lasted almost 25,000 years, almost double the usual duration. The high amount of heat gained during such a long interglacial probably caused the delay between falling summer energy and glacial inception to be of 10,000 years instead of the usual 6,000. MIS 11c ended up lasting close to 35,000 years. However, despite such a long interglacial expanding two precession peaks, it could not extend from one obliquity oscillation to the next. Once glacial inception is reached there is no turning back even under rising obliquity and summer energy.

MIS 19c is a better analog in terms of eccentricity, obliquity, and precession. The main difference is that low values in obliquity and insolation during MIS 19c, and relatively low ice-volume prior to it, resulted in a cool, short interglacial (Fig. 14.5c). The values of obliquity, insolation, and summer energy were lower at MIS 19c glacial inception than they are currently.

14.5 The long interglacial hypothesis

Intermediate complexity (simplified) climate models have been used since the early 1990s to explore glacial conditions under elevated CO_2, and Loutre and Berger (2000) tried to specifically address the question of the end of the present interglacial. They found that at CO_2 concentrations of 210 ppm the ice-sheets would form at 15 kyr AP (after present), while no ice-sheets form for 130 kyr AP with CO_2 levels of 250 ppm. The 65°N summer insolation changes for the next 50 kyr are so small that they find that with CO_2 levels reproducing Vostok records (a decrease from 296 to 184 ppm over the next 114 kyr), the present interglacial should last c. 50,000 years more. Since the only long interglacial of the Middle to Late Pleistocene is MIS 11c, it was confirmed as a suitable analog because it had similarly low 65°N summer insolation changes and relatively high CO_2 levels (as a warm interglacial).

The main conclusions from Loutre and Berger (2000), were that no insolation threshold will be crossed within the next 40 kyr, and that a glacial inception within the next 50 kyr would require unnaturally low CO_2 levels. This modeling result was confirmed multiple times (Cochelin et al. 2006; Mysak 2008). Archer and Ganopolski (2005) painted a more drastic scenario, with the release of 5,000 GtC (545 GtC were released between 1870–2014) preventing glaciation for the next half million years. Ganopolski et al. (2016) determined a 65°N summer insolation/CO_2 threshold for glacial inception from model realizations. With such a threshold, not even the pre-industrial CO_2 level of 280 ppm could produce an interglacial at present, and present cumulative carbon emissions already preclude a glacial inception over the next 50 kyr.

The long Holocene interglacial hypothesis has become an axiom that is seldom questioned in the scientific literature and fully endorsed by the IPCC. However, this axiom is based exclusively on model studies that rest on three assumptions that have not been demonstrated:

1a. Glacial inception has depended in the past on 65°N summer insolation

2a. Climate has a high sensitivity to CO_2 levels. Models produce an average of 3°C/doubling of CO_2

3a. CO_2 levels remain elevated for tens of thousands of years after a pulse

If any or all of the following alternative possibilities were to be correct:

1b. Glacial inception depends on summer energy (obliquity)

2b. Climate has a low to medium sensitivity to CO_2, c. 1.5°C/doubling of CO_2

3b. CO_2 levels artificially elevated after a pulse can go back down close to baseline over a few centuries

…the prediction would be completely different. And we already know that 1a is wrong. We have already shown evidence that glacial inception is driven by the fall in obliquity, well reflected in summer energy, and not by 65°N summer insolation. Models don't appear to be as sensitive to obliquity changes as climate and it probably comes from their inability to reproduce the changes in the latitudinal temperature gradient and the resulting changes in meridional transport of energy. Another thing models ignore is that although low eccentricity promotes small changes in 65°N summer insolation due to a nearly circular orbit, it also clearly promotes ice-volume build up during low summer energy periods. The 100 and 400-kyr eccentricity cycles are associated with high ice-volume at times of low eccentricity (see Fig. 2.12b), and present eccentricity is very low and will decrease over the next 26 kyr, reaching a value of almost zero (0.002). MIS 11c, despite being a very long and very warm interglacial with high CO_2 levels, was followed by 60 kyr of small 65°N summer insolation changes, analogous to what awaits in the future. During that period the planet managed to accumulate more ice than during the equivalent period after MIS 5e, when much higher insolation changes and higher eccentricity occurred. Models do not appropriately reflect the 100-kyr ice cycle paleoclimatologists have recognized for the past five decades.

Not all authors have accepted the unproven model assumptions at face value. Tzedakis et al. (2012) circumvented the problem of CO_2 sensitivity by trying to determine the natural length of the Holocene under 245 ppm CO_2 using the MIS 19c analogy. If CO_2 sensitivity turns out to be low their scenario could be realistic. Their study concludes that with 245 ppm CO_2 the end of the current interglacial would occur within the next 1500 years.

Vettoretti and Peltier (2004, 2011) have questioned the soundness of the assumption that glacial inception depends on insolation and CO_2. They studied separately the effect of the different components of orbital changes, finding that a low obliquity value is most important in determining the strength of the inception process, followed in order of importance by the magnitude of the eccentricity-precession forcing. They also find that areas of perennial snow cover are much more sensitive to the insolation regime than to GHG concentrations. They conclude with a glacial inception at the next obliquity minimum in 10 Kyr

in the absence of modern anthropogenic forcing. These results contradict some of the assumptions of the long interglacial hypothesis.

14.6 The fat tail of anthropogenic CO_2 adjustment time

Despite intense research, knowledge of the carbon cycle is still very inadequate. Particularly net carbon fluxes between different reservoirs have a big uncertainty, due in great measure to large differences in regional measurements (Ballantyne et al. 2015; Fig. 14.6). Also sink behavior has been a source of significant surprises in the past (Schindler 1999), due to the unexpected fast growth in net global carbon uptake by ocean and biosphere sinks (Ballantyne et al. 2015). Additionally, models indicate that the fraction of our emissions that remains in the atmosphere (airborne fraction) should increase over time, but we have evidence of the opposite (Keenan et al. 2016). Our imperfect knowledge of the carbon cycle is built into Earth system models that from different emissions scenarios produce the atmospheric CO_2 concentrations that general circulation models (climate models) use as input. Millar et al. (2017) have shown that the Earth systems models are working incorrectly, and they contribute to current models running too hot.

Given the problems delimitating and predicting the evolution of the carbon cycle over the past several decades it is surprising that the IPCC would write: *"The removal of all the human-emitted CO_2 from the atmosphere by natural processes will take a few hundred thousand years (high confidence) … we assessed that about 15 to 40% of CO_2 emitted until 2100 will remain in the atmosphere longer than 1000 years"* (IPCC–AR5, Ciais et al. 2013, pg. 472).

So, we were unable to predict a few decades ago that over 50% of our fast-growing CO_2 emissions would disappear from the atmosphere without any time delay, or that the fraction removed could actually increase despite exponentially increasing emissions, yet we have high confidence that 15–40% will remain in the atmosphere 1000 years from now. Clearly, we hugely underestimated the carbon sinks capacity to deal with our emissions, so we cannot have high confidence in distant future predictions. The problem is that we are dealing with a situation without precedent and so the answers that we can obtain from sci-

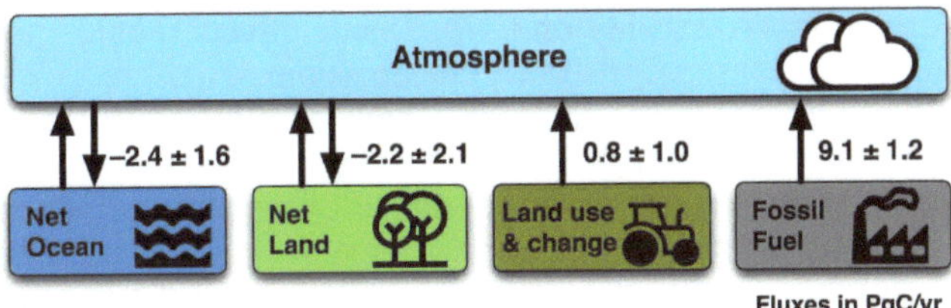

Fig. 14.6 Diagram of the global carbon budget atmospheric fluxes
Major net fluxes of carbon to the atmospheric reservoir of CO_2 are from fossil fuel emissions and land use/land conversion. Net land and net ocean uptake of carbon from the reservoir of atmospheric CO_2 is the residual of much larger fluxes in both senses. Flux of carbon indicated in petagrams of C per year. Error estimates for each flux in 2010 are expressed as ± 2σ. Error estimates are of the same magnitude as the fluxes except for fossil fuels. After Ballantyne et al. (2015).

ence carry a huge uncertainty that cannot be properly constrained by evidence. The Paleocene–Eocene Thermal Maximum is usually cited as precedent; however, its isotopic carbon excursion took place so long ago that our poor knowledge of its source, amount, and release timescale precludes any meaningful estimation of its decay.

IPCC confidence comes essentially from David Archer's studies, that since 1997 has become the authority of reference. It is clear that the unburial and release of huge carbon fossil stores constitutes a long-term perturbation of the carbon cycle. Since carbon only permanently exits the carbon cycle very slowly through calcium carbonate sea-bottom burial, and silicate rock weathering, the different compartments of the carbon cycle will have to deal with excess carbon for a very long time and this should necessarily lead to an equilibration between compartments at higher levels than prior to the perturbation. At present the complexity of the effects involved is being studied with box-modeling, but every step in the process requires taking assumptions. It is assumed that the land biosphere, that is currently a sink due to an increase in photosynthesis over respiration, should reach equilibrium within decades after the end of anthropogenic emissions and then become a net source as atmospheric levels decrease. The reduction in atmospheric CO_2 is then assumed to occur mainly through ocean uptake on a timescale of centuries, driven by changes in oceanic chemistry and ocean mixing. It is assumed that as more carbon dioxide dissolves in the ocean, it will compromise the ocean's buffering capacity and that ocean acidification will increase the Revelle factor (dissolved CO_2 to dissolved inorganic carbon). This is then expected to reduce the efficiency of the ocean carbon sink until it stops taking CO_2 after about 1000 years, when 14–30% of the maximum level reached remains in the atmosphere (Archer & Ganopolski 2005). Higher temperature is also expected to contribute to a decrease in the ocean carbon sink efficiency.

Archer's (2005) worst case scenario involves anthropogenic emissions of 1600 GtC by 2100 (545 GtC emitted 1870–2014) and increasing afterwards. Up to 1000 GtC should be contributed by a reversal of the land biosphere and soils sinks, and the rest to 5000 GtC total contributed by permafrost and marine methane clathrate deposits. A more realistic scenario considering fossil fuel supply-side constrains and extrapolating observed warming leads to only 500–1000 GtC addition at most. But this amount disregards any effort to reduce emissions. It is to be expected that a higher certainty on CO_2 climatic effects should lead to more intense efforts to curtail emissions.

It is impossible to have a high confidence that 14–30% of the carbon emitted will remain in the atmosphere 1000 years from now. That number comes from a set of assumptions made using a poor understanding of the carbon cycle, and it could be much lower. Those models are unable to reproduce or explain the significant increase of 20 ppm in CO_2 that took place between 6,000 and 600 yr BP. Initializing the models at 6,000 yr BP doesn't produce the pre-industrial CO_2 levels of 280 ppm, unless ad hoc assumptions are introduced, indicating models cannot be trusted to project atmospheric CO_2 levels thousands of years into the future.

The National Research Council set in 2008 the Committee on the Importance of Deep-Time Geologic Records for Understanding Climate Change. This body produced a 2011 report: Understanding Earth's Deep Past: Lessons for Our Climate Future (National Research Council 2011). This expert committee fully acknowledges the uncertainty present in CO_2 adjustment time estimates from box-models:

"Although box-model calculations should not be considered definitive, they do suggest that the fossil-fuel perturbation may interfere with the natural glacial–interglacial oscillation driven by predictable changes in Earth's orbit, perhaps forestalling the onset of the next Northern Hemisphere 'ice age' by tens of thousands of years. A more convincing exposition of the central question of "how long" requires more comprehensive models. Scientific confidence in those models will only be high if they can be evaluated against observation. The historical record, and even the expanse of the Quaternary climate record, contains nothing comparable."

The proposed fat-tail of anthropogenic CO_2 adjustment time should be taken as a possible scenario if certain assumptions are correct, and not as what is expected to happen.

14.7 Glacial inception in the Holocene

Glacial inception is the transition from interglacial climate to glaciation, a process characterized by ice-sheet build up and falling sea-levels. However, there is no unambiguous definition of glacial inception that allows it to be placed at a specific point in time for each interglacial. In the excitation/relaxation dynamic model of glaciations discussed in chapter 2 (see Sect. 2.9, Fig. 2.15), and illustrated in Fig. 14.3, glacial inception can be understood as an irreversible commitment from a quasi-stable interglacial state into a relaxation process towards a stable glacial state, taking place once the conditions that made the interglacial possible have disappeared, and once the downward drift in temperature allows the boundary crossing at the commitment point (Fig. 14.3, unstable point).

A point of inflection can be observed in the Antarctic proxy temperature record of past interglacials. In each case the slowly declining temperature of the late interglacial suddenly accelerates into a terminal decline towards glacial conditions (Fig. 14.7). In the case of the Eemian interglacial, this inflection point takes place at 120 ka, coinciding with the determination of glacial inception by different criteria (Fig. 14.4), confirming the date. It can be assumed that the inflection point in the cooling rate corresponds to glacial inception in all cases, and can be explained as a point when the intensification of positive feedbacks (like ice-albedo, vegetation changes, or changes in oceanic currents), leads to a steepening of the latitudinal temperature gradient, an intensification of meridional transport of heat towards the poles, and the consequent accelerated cooling of the planet into glaciation.

The start of an interglacial is also lacking a formal definition. In the case of the Holocene the start is formally placed 11,700 years ago (Walker et al. 2009). At that time EPICA Dome C deuterium proxy temperature record shows no anomaly with respect to current value (0°C anomaly). For a consistent comparison we can define the start of every interglacial at the time they first reach 0°C anomaly in the EPICA Dome C record. For cooler interglacials that didn't reach 0°C anomaly, picking a lower temperature would lead to overestimating their length, as a

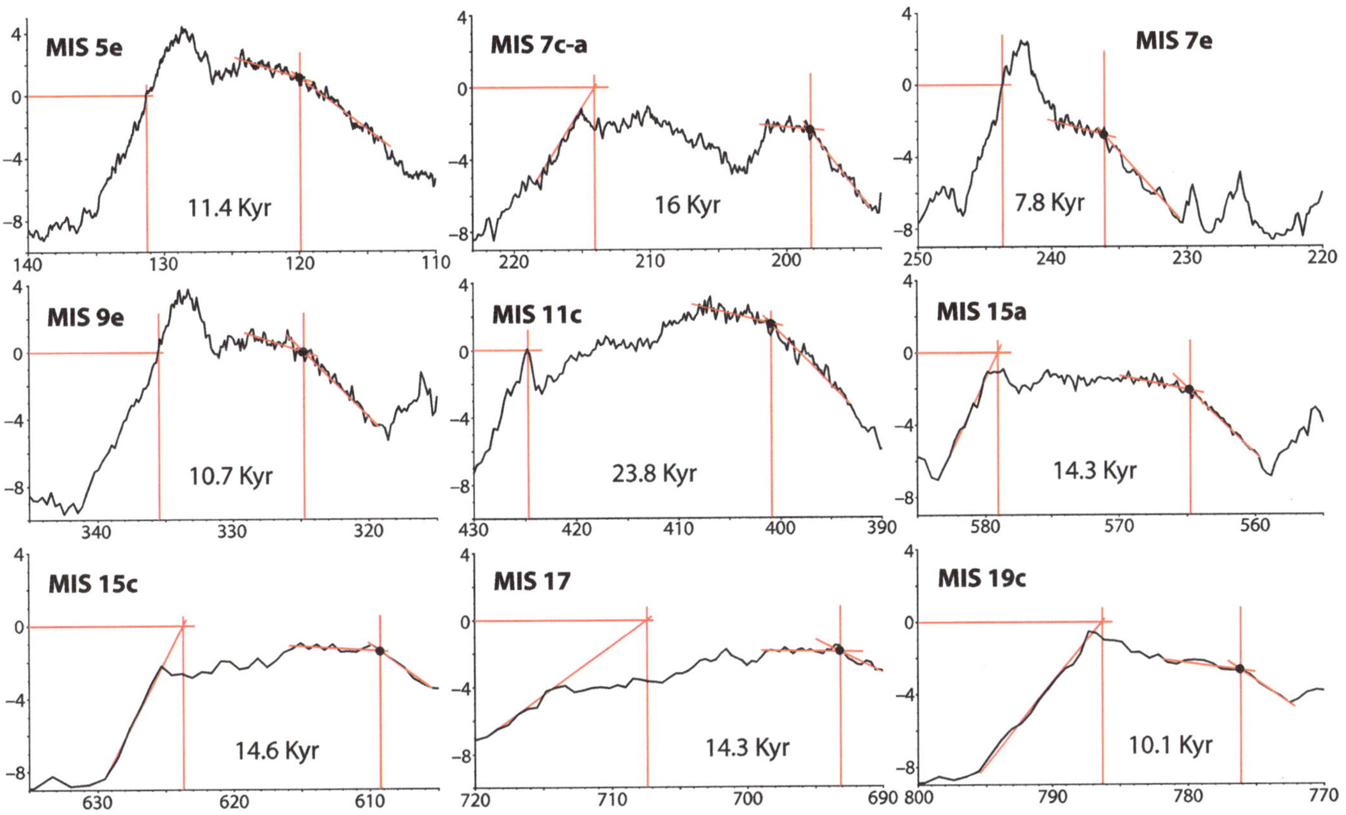

Fig. 14.7 Interglacial length normalization
The start of an interglacial is defined, as for the Holocene, by the time EPICA Dome C temperature anomaly reaches 0°C, or by extrapolating the rate of warming to the 0°C value. The end of an interglacial is defined at the inflection point where EPICA Dome C temperature anomaly increases its rate of cooling towards glacial values.

lower temperature is reached earlier. A more correct choice, in the author's opinion, is to extrapolate the warming trend to the point where it would have reached 0°C anomaly, picking that time as the normalized start of the interglacial (Fig. 14.7; table 14.1). This choice avoids the significant bias of overestimating the length of cool interglacials. MIS 13a cannot be normalized and it is not analyzed under the criteria chosen here.

Normalized in this way interglacials are between 10 and 16 kyr in length, with an average of 13 kyr, with two exceptions: MIS 7e and MIS 11c. Orbital configuration explains MIS 7e and MIS 11c anomalous length (Fig. 14.8, see also Fig. 2.16). A consistent rule is that all interglacials end when obliquity is low. No interglacial of the past 800 ka has gone beyond 15.5 kyr from the obliquity maximum (Fig. 14.8a). Since MIS 7e had a late start with respect to the obliquity cycle it became a very short interglacial. Since MIS 11c started early in the obliquity cycle due to its unusual precessional insolation, it became a very long interglacial. Low eccentricity allows long interglacials when other conditions are present, but it does not cause them to be long.

The other orbital rule is that interglacials of the past 800 ka start when the combination of obliquity and precessional insolation is high enough (high summer energy). Precessional insolation is irrelevant for glacial inception, as three interglacials (MIS 7c–a, MIS 11c, and MIS 17) were capable of surviving through an insolation minimum, yet they suffered glacial inception close to the next maximum, when obliquity dropped. A typical interglacial (Fig. 14.8, line between circles) starts 2000 years before obliq-

uity maximum, and 1000 years before insolation maximum, and lasts 13,000 years. So far, the Holocene is extraordinarily close to a typical interglacial in astronomical terms and length.

Orbital configuration alone can explain when interglacials start and end, while changes in CO_2 levels cannot. Interglacial temperature is inversely correlated to the amount of ice-volume before deglaciation starts (see Fig.

Interglacial	Start	End	Length
MIS 1	11,700		
MIS 5e	131,400	120,000	11,400
MIS 7c–a	(−214,200)	198,200	16,000
MIS 7e	243,800	236,000	7,800
MIS 9e	335,500	324,800	10,700
MIS 11c	424,800	401,000	23,800
MIS 15a	(−579,100)	564,800	14,300
MIS 15c	(−623,800)	609,200	14,600
MIS 17	(−707,400)	693,100	14,300
MIS 19c	(−786,300)	776,200	10,100

Table 14.1 Normalized interglacial length
Dates in yr BP for the start, end, and length, of normalized interglacials. Dates between parenthesis are extrapolated from the rate of warming. These dates and lengths are used to compare interglacial orbital conditions for the rest of the chapter.

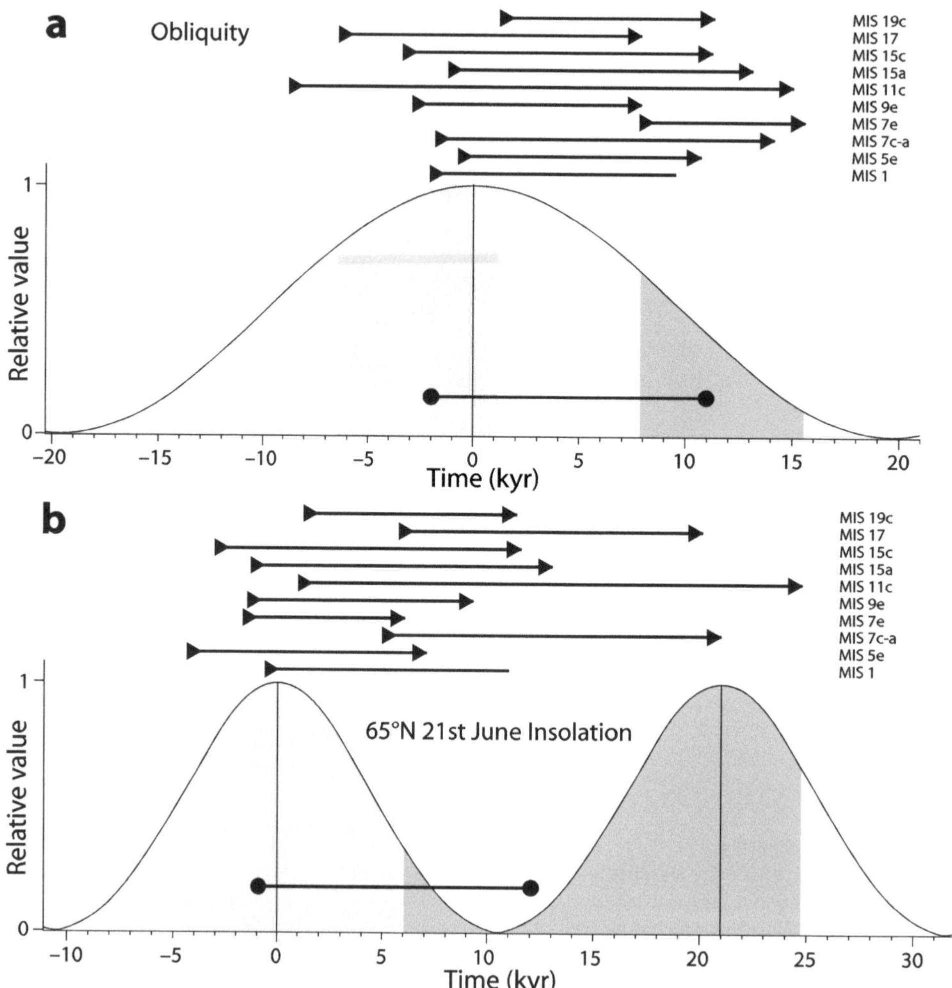

Fig. 14.8 Interglacial orbital configuration
a) Interglacial start and end dates (triangles in arrows) relative to the obliquity maximum. Light grey area indicates interglacial start for all interglacials except MIS 7e and MIS 11c that had an anomalous length due to starting too late and too early respectively in the obliquity cycle. Dark grey area indicates interglacial end for all interglacials. Circles indicate start and end of a typical interglacial with average 13 kyr length. Interglacials start when obliquity is high and end when obliquity is low. **b)** Interglacial start and end dates (triangles in arrows) relative to northern summer insolation maximum. Light grey area indicates interglacial start for all interglacials. Dark grey area indicates interglacial end for all interglacials. Circles indicate start and end of a typical interglacial with average 13 kyr length. Interglacials start when insolation is high but can end at any time in the insolation cycle.

2.15; warmer interglacials correspond to previous higher ice-volume). Interglacial temperature is also directly correlated to CO_2 (see Fig. 12.8). And ice-volume is inversely correlated to eccentricity (see Figs. 2.12 & 2.15). As it is difficult to explain why CO_2 levels would inversely correlate to prior ice-volume, the most likely explanation is that CO_2 levels are a consequence of temperature levels, not a cause (eccentricity → ice-volume → temperature → CO_2). Eemian glacial inception and the next 5000 years of cooling took place under stable 270 ppm CO_2 levels, indicating that glacial inception is responding to orbital changes, not CO_2 changes. Despite this evidence IPCC expresses virtual certainty that a new glacial inception is not possible for the next 50 kyr if CO_2 levels remain above 300 ppm (IPCC–AR5, Masson–Delmotte et al. 2013, pg. 435). Ice core measurements indicate CO_2 levels at past glacial inceptions have always been below 300 ppm, but there is simply no evidence indicating how high CO_2 levels must be to stop a glacial inception, if that is even possible.

It is widely known that there is a delay between the astronomical signal and the geological evidence of climate change. This delay, in the case of obliquity is c. 6000 years (Huybers 2009; Donders et al. 2018). The logical conclusion is that the astronomical threshold for glacial inception is crossed c. 6000 years before it takes place. This inference is supported by the presence at the end of interglacials of a period of declining temperatures before reaching the inflection point that indicates glacial inception has taken place (Fig. 14.7). In the Holocene this period is termed Neoglaciation, and it is also observed between 126–120 ka in the Eemian (Fig. 14.4).

Analysis of the orbital conditions that produce a glacial inception requires examining them 6000 years before the inflection point, in the cooling rate at the end of the interglacial. Glacial inception does not take place at 65°N, but at 70°N, where ice sheets start to grow (Birch et al. 2017). Examination of 70°N summer energy (at 250 W/m² threshold) 6000 years before glacial inception (Fig. 14.9, diamonds) reveals a threshold at 4.96 GJ/m² when the glacial inception orbital "decision" has already been taken for all previous interglacials (Fig. 14.9, continuous line). A maximum summer energy value prior to glacial inception

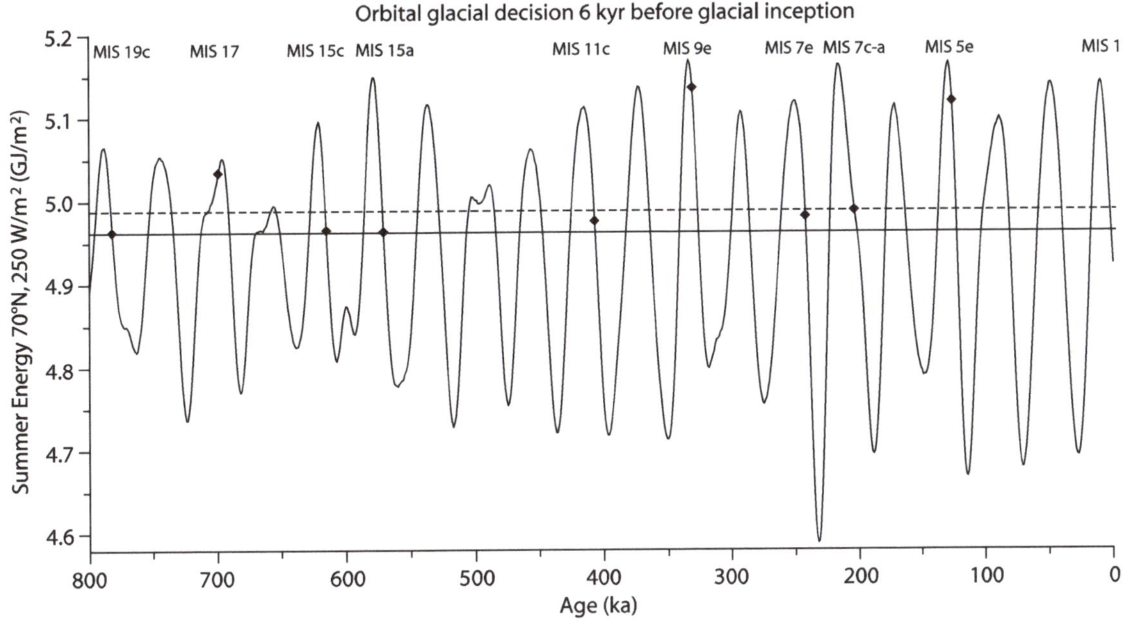

Fig. 14.9 Orbital decision to end an interglacial
Summer energy at 70°N with a 250 W/m² threshold for the past 800 ka. Diamonds mark the position 6 kyr before glacial inception as observed (Fig. 14.7) in the EPICA Dome C temperature proxy record for each interglacial except MIS 13. Continuous line marks the lowest value observed (4.96 GJ/m²). Dashed line marks the highest value for six of the nine interglacials considered. The Holocene (MIS 1) is already below the lower threshold value as the curve ends at the present.

"decision" by six of the nine interglacials is apparent at 4.99 GJ/m² (Fig. 14.9, dashed line). This narrow window of just 0.03 GJ/m² constitutes the range where most interglacials commit to glacial inception 6000 years later, while some interglacials appear to take the decision earlier. Two of the exceptions, MIS 9e and MIS 5e, have in common a well below average duration (10.7 and 11.4 kyr), and being very warm very early, as the only interglacials to reach +3.7 °C in the entire 800 kyr EPICA Dome C ice core (Fig. 14.7; Jouzel et al. 2007). It is possible that being very warm, very early, sets an interglacial on an early cooling trend. By contrast, the other exception, MIS 17, is the coolest interglacial of the past 800 kyr, as it is the only one that didn't reach −1.5 °C in EPICA data, and is still considered an interglacial. It is not unreasonable to assume that temperature affects interglacial duration, and thus the

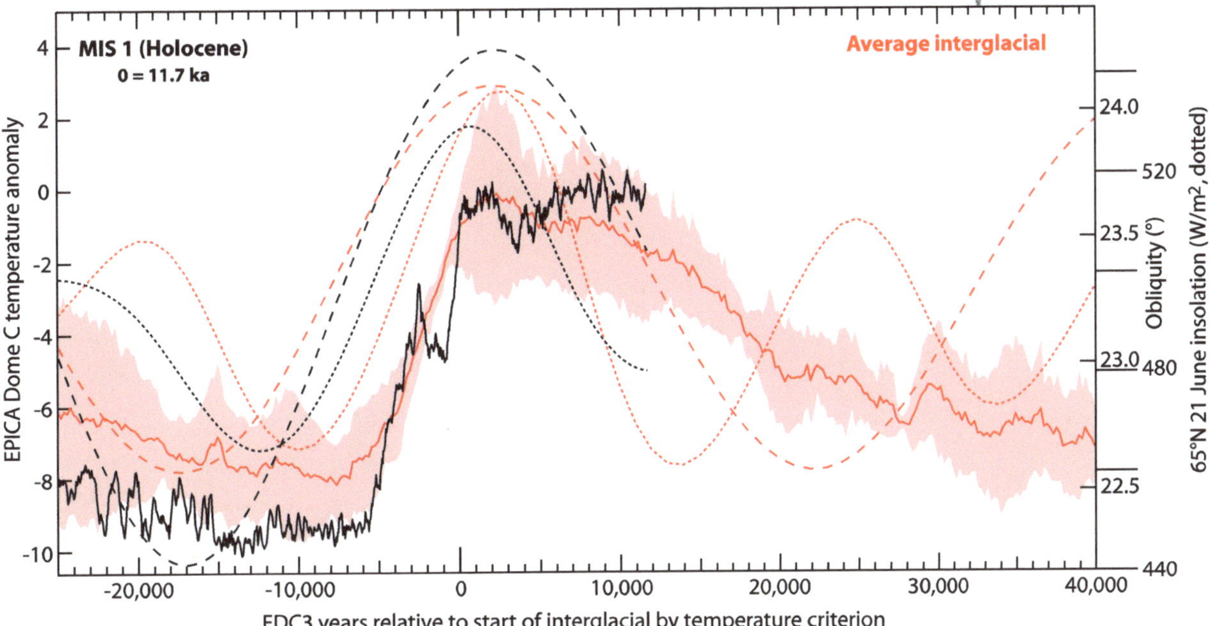

Fig. 14.10 The Holocene is a typical interglacial
Holocene temperature profile (black solid curve, from EPICA Dome C data), obliquity evolution (black dashed curve), and 65°N summer insolation evolution (black dotted curve), do not show a significant deviation from the respective values of an average interglacial (red curves and 1σ pink bands) from MIS 5e, 7c–a, 9e, 15a, 15c and 19c, aligned at 787.0, 624.4, 579.6, 335.5, 214.7, and 131.4 ka in EDC3 dates. Interglacial average summer insolation curve does not include MIS 7c–a values. The Holocene was aligned at 11.7 ka. Data after Jouzel, et al. (2007); Laskar et al. (2004). The Holocene data closeness to that of most interglacials does not provide support for hypotheses of a long interglacial on astronomical grounds, and suggests that statistically the Holocene is close to reaching the beginning of its end.

time to reach glacial inception. But for interglacials that deviate less from the average temperature, the summer insolation energetic window 6000 years before glacial inception is surprisingly narrow. The Holocene is within one standard deviation of the average interglacial temperature (Fig. 14.10).

The upper limit of 4.99 GJ/m² was crossed by the Holocene 2400 years ago, and the 4.96 GJ/m² limit was crossed 1400 years ago, so the orbital decision to end the Holocene was likely taken between those two dates. By orbital considerations alone the Holocene should undergo glacial inception in 3600–4600 years, but it could have as little as 1000 years left if the start of the Neoglaciation 5000 years ago corresponds to the Holocene's orbital commitment towards glacial inception. The average duration of Holocene-like interglacials is 13,800 years. That length would place the end of the Holocene, 2000 years from now, right at the center of the obliquity range for the end of every interglacial, 12 kyr after the obliquity maximum is reached (Fig. 14.8a, dark grey area), slightly over one fourth of the obliquity cycle. The orbital high probability time frame for glacial inception is 1500–4500 years in the future. But between 1500–2500 years from now, there should be a period when two consecutive lows in the Eddy solar cycle separated by a low in the Bray solar cycle are expected, defining a period similar to 8.4–7.1 ka when eight solar grand minima took place in rapid succession (see Fig. 13.7). This period coincided with one of the hardest climatic periods in the Holocene so far, marked by the succession of the 8.2, 7.7, and 7.2-ka ACEs (see Fig. 6.4), coinciding with very important changes in human societies (see Sects. 6.4 & 6.5). 1500–2500 years from now define a solar high probability time frame for glacial inception within the orbital window of time.

The analysis of the leading orbital decision to the lagging glacial inception, 6 kyr later, provides possible answers to some questions, like why the Holocene did not end at the LIA. The Neoglacial Period started at 5.2 ka, and it is likely that the LIA took place too early, and the interglacial was too young then for glacial inception. The intense cooling accompanied by glacier extension, and a reduction from 280 to 270 ppm CO_2 (Eemian's glacial inception level), indicates it was probably a close call, that resulted in a temperature rebound afterwards. Similarly, Ruddiman's "Early anthropogenic hypothesis" (Ruddiman 2007), that states that a glaciation was prevented by early agricultural release of greenhouse gases, is unnecessary. With or without human intervention the Holocene should not have ended yet, as it has not reached the situation when similar interglacials enter glacial inception. The question of the GHG increase during the Middle to Late Holocene remains controversial, but it is unrelated to the length of the present interglacial.

14.8 The next glaciation

Without human intervention the next glaciation should start in just 1500–4500 years. The question that we cannot answer with any degree of certainty is how high CO_2 levels would have to be to prevent glacial inception. Summer energy is going to be very low for the next 20,000 years and that should require sufficiently elevated CO_2 levels for that long. Alternatively technology could develop to a point when it is possible for humankind to prevent the next glaciation. Those are questions that cannot be answered, but we can make reasonable inferences from what we do know.

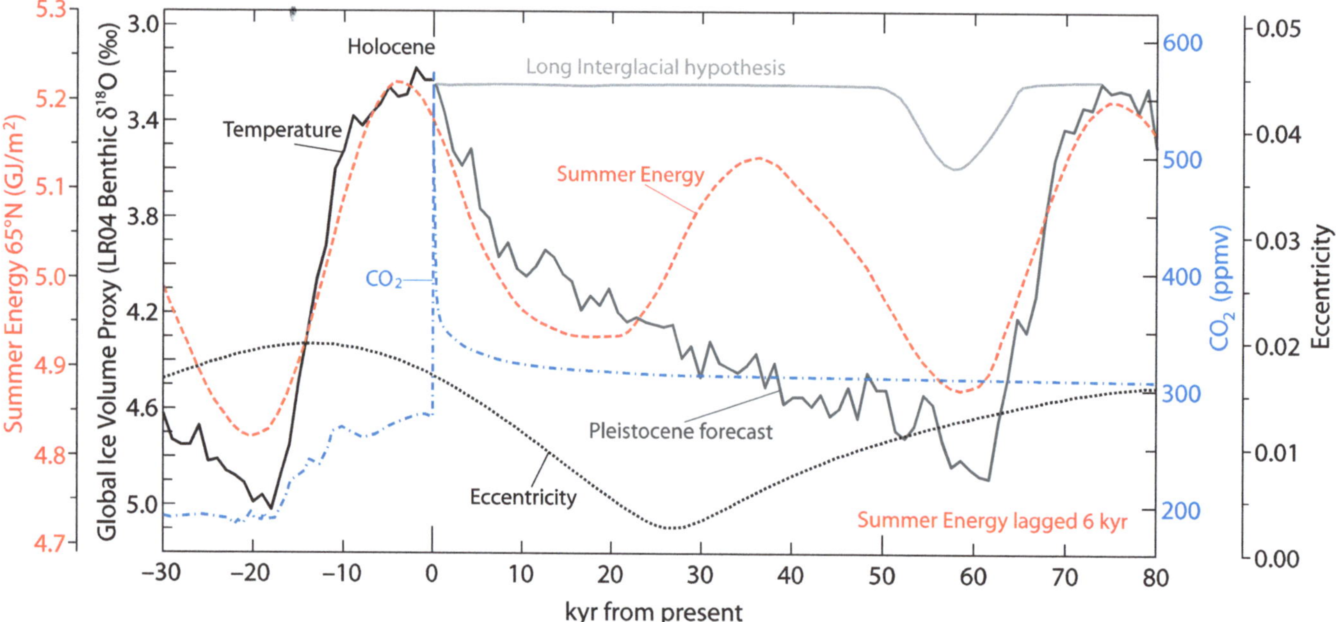

Fig. 14.11 Future climate forecasts for the next 80 kyr
LR04 benthic stack global ice volume proxy for the past 30 kyr (black curve; after Lisiecki & Raymo 2005). Orbital eccentricity for the past 30 kyr and future 80 kyr (black fine dotted curve; after Laskar 2004). Summer energy at 65°N with 275 W/m² threshold for the past 30 kyr and the future 80 kyr lagged by 6000 years (red dashed curve; after Huybers 2006). Past CO_2 levels from ice cores and modeled long term CO_2 concentration evolution after a 1250 GtC pulse (blue dash-dotted curve; after Archer 2005). Future global ice volume forecast considering only orbital conditions by reproducing ice volume change after MIS 11c, 402–322 Kyr BP, a period with similar orbital evolution (dark grey curve). Future global ice volume modeled after a 1000 GtC pulse, representing the long interglacial hypothesis (light grey curve; after Ganopolski et al. 2016).

Over the past 150 years it is calculated that we have produced 545 GtC leading to an increase in atmospheric CO_2 of 125 ppm. Estimates from reputable sources place fossil fuel peak production in a few decades (see Sect. 13.3). Even if fossil carbon is more abundant, we might already have extracted one-third to half of what we will extract over the next centuries. Supply constraints should limit our emissions even in the unlikely case that we don't limit them ourselves. If these estimates are correct, peak atmospheric CO_2 levels should not go much above 550 ppm (Fig. 14.11, dash-dotted line). A rapid decline in CO_2 should follow as oceans and the biosphere absorb most of it, but according to current models about 320 ppm should remain for a very long time (Archer 2005). If this assumption is correct, the question is if 320 ppm of CO_2 could stop a glaciation, as the IPCC claims with virtual certainty. We know the Eemian entered glaciation with 270 ppm, 50 ppm below future estimated levels. An opposite result from just 50 ppm difference would require a very high climate sensitivity to CO_2. The 100 ppm increase between 1959 and 2021 has been accompanied by climatic variability within interglacial range, as Holocene Climatic Optimum conditions have not been reproduced (see Sect. 12.3). Achieving changes of an interglacial-glacial scale might require a much larger amount of GHG than available.

When glacial inception takes place, ice caps at Baffin Island (Fig. 14.12) and the Canadian Arctic Archipelago will start to grow, initiating the Laurentide ice sheet (Birch et al. 2017), and ice caps should also grow initially over West Siberian islands. Glaciers in Norway should grow to the sea and start releasing icebergs due to increased winter precipitation. The Fennoscandian ice sheet growth, however, should be delayed by an intensification of the Atlantic Meridional Overturning Circulation, that is expected to bring more heat towards the Nordic seas (Born et al.

2010). In 10,000 years large ice sheets should have developed causing sea level to fall by 30–40 m. Current eccentricity is very low and is going to continue decreasing to almost zero over the next 26 kyr (Fig. 14.11, dotted line). While low eccentricity prevents insolation from going too low, it also prevents it from going too high, so the ice accumulated over low summer energy periods doesn't melt significantly during periods of higher summer energy. The result is that low eccentricity promotes a faster ice-sheet growth.

The next high summer energy period in 35 kyr cannot result in a new interglacial. Obliquity and insolation oscillations are misaligned during that period (not shown), and despite a rapid ice-sheet growth ice-volume should still be too low, so every necessary condition for a Middle to Late Pleistocene interglacial is missing in 35 kyr. Quite the contrary, the low eccentricity should cause the ice-volume to grow through the summer energy peak as it happened during MIS 10 glacial period under similarly low eccentricity (Fig. 14.11, dark grey line). On the positive side, the rapid ice-volume accumulation over the next 60 kyr should create the right conditions for a new interglacial in 70 kyr, when a correct orbital alignment and sufficient ice-volume should produce the next interglacial.

If Berger and Loutre (2002), Archer (2005), and Ganopolski et al. (2016) are correct, and the residual CO_2 in the atmosphere will allow, for the first time in two million years, the survival of an interglacial through an obliquity minimum, then the Holocene should last for at least 50 kyr more (Fig. 14.11, light grey line).

Milankovitch forcing is a very powerful force when acting over millennia. With the help of the appropriate feedbacks it puts the planet into very cold glacial periods and then melts the ice sheets into warm interglacials. It is very difficult for scientists and people in general, living during a multi-centennial warming period characterized by

Fig. 14.12 Baffin Island ice caps
The Barnes and Penny ice caps in Baffin Island (Canada) are the last remnants of the Laurentide ice sheet of the Wisconsinan glaciation. They are projected to disappear in 300 years if Modern Global Warming continues intensifying (Gilbert et al. 2017). Instead, together with the Canadian Arctic Archipelago and West Siberian islands, they might constitute the starting places for the next glaciation. Image composite from several NASA worldview satellite images for Aug–Sep 2010–2014.

a strong increase in GHG to imagine that in the long run Milankovitch forcing might win. The Romans that for many centuries lived in a warm world characterized by technical progress could not imagine that their mighty empire could fall amid a cooling and worsening climate and terrible plagues into a millennial dark age of lost knowledge and declining civilization. A new glacial period would constitute humankind's biggest test and clearly has the potential to constitute its worst catastrophe. The world is so populated that an important cooling would place us in overshooting conditions. The precautionary principle requires that we start preparing for that possibility over the next decades and centuries while we are in a warm optimum, as cooling periods are rife with troubles. For the past 2 million years, when obliquity declined enough, a glacial period always followed. Obliquity is declining fast, and we should not place too much confidence in computer models that tell us this time will be different.

14.9 Conclusions

14a. Typical interglacials last on average 13 kyr from their start to glacial inception, when their temperature decrease significantly accelerates towards glacial levels. The Holocene is astronomically, and according to its Antarctica temperature evolution, a typical interglacial.

14b. The glacial cycle fits a model of a stable glacial state that reaches an excitable point where fast excitation (rapid warming) takes it to a quasi-stable interglacial state that slowly degrades to an unstable point (glacial inception) where a slow relaxation takes it back to the glacial state. This dynamic system is defined as a fast-slow excitable system around a two-branch slow manifold.

14c. Due to the c. 6000-yr lag between orbital forcing and ice-volume effect, the orbital threshold for glacial inception is crossed several millennia before glacial inception takes place. Analysis of the past 800 kyr indicates the orbital threshold to terminate the Holocene was crossed 1400–2400 years ago.

14d. In the absence of sufficient anthropogenic forcing, glacial inception might take place in 1500–4500 years as determined by orbital parameters, average interglacial length, Neoglaciation length, and solar variability periodicities.

14e. The long interglacial hypothesis rests on the wrong astronomical parameter, high equilibrium climate sensitivity to CO_2, and uncertain model predictions of very long-term CO_2 decay rates. The virtual certainty by the IPCC that a glaciation is not possible for the next 50 kyr if CO_2 levels remain above 300 ppm is unsupported by evidence.

References

Alley RB (2000) The Younger Dryas cold interval as viewed from central Greenland. Quaternary Science Reviews 19 (1–5) 213–226

Archer D (2005) Fate of fossil fuel CO_2 in geologic time. Journal of geophysical research: Oceans 110 C9

Archer D & Ganopolski A (2005) A movable trigger: Fossil fuel CO_2 and the onset of the next glaciation. Geochemistry Geophysics Geosystems 6 (5)

Ballantyne AP, Andres R, Houghton R, et al (2015) Audit of the global carbon budget: estimate errors and their impact on uptake uncertainty. Biogeosciences 12 (8) 2565–2584

Bereiter B, Eggleston S, Schmitt J et al (2015) Revision of the EPICA Dome C CO_2 record from 800 to 600 kyr before present. Geophysical Research Letters 42 (2) 542–549

Berger A & Loutre MF (2002) An exceptionally long interglacial ahead? Science 297 (5585) 1287–1288

Birch L, Cronin T & Tziperman E (2017) Glacial inception on Baffin Island: the role of insolation meteorology and topography. Journal of Climate 30 (11) 4047–4064

Born A, Kageyama M & Nisancioglu KH (2010) Warm Nordic Seas delayed glacial inception in Scandinavia. Climate of the Past 6 (6) 817–826

Bosmans JHC, Hilgen FJ, Tuenter E & Lourens LJ (2015) Obliquity forcing of low-latitude climate. Climate of the Past 11 (10) 1335–1346

Ciais P, Sabine C, Bala G et al (2013) Carbon and other biogeochemical cycles. In: Stocker TF, Qin D, Plattner G–K et al (eds) Climate Change 2013: The Physical Science Basis. Contribution of Working Group I to the Fifth Assessment Report of the Intergovernmental Panel on Climate Change. Cambridge University Press, Cambridge, p 465–570

Cochelin ASB, Mysak LA & Wang Z (2006) Simulation of long-term future climate changes with the green McGill paleoclimate model: The next glacial inception. Climatic Change 79 (3–4) 381–401

Crucifix M (2012) Oscillators and relaxation phenomena in Pleistocene climate theory. Philosophical Transactions of the Royal Society A: Mathematical, Physical and Engineering Sciences 370 (1962) 1140–1165

Donders T, Van Helmond NA, Verreussel R et al (2018) Land-sea coupling of early Pleistocene glacial cycles in the southern North Sea exhibit dominant Northern Hemisphere forcing. Climate of the Past 14 (3) 397–411

Ganopolski A, Winkelmann R & Schellnhuber HJ (2016) Critical insolation–CO_2 relation for diagnosing past and future glacial inception. Nature 529 (7585) 200–203

Gilbert A, Flowers GE, Miller GH et al (2017) The projected demise of Barnes Ice Cap: Evidence of an unusually warm 21st century Arctic. Geophysical Research Letters 44 (6) 2810–2816

Govin A, Capron E, Tzedakis PC et al (2015) Sequence of events from the onset to the demise of the Last Interglacial: Evaluating strengths and limitations of chronologies used in climatic archives. Quaternary Science Reviews 129 1–36

Huybers P (2006) Early Pleistocene glacial cycles and the integrated summer insolation forcing. Science 313 (5786) 508–511

Huybers PJ (2009) Pleistocene glacial variability as a chaotic response to obliquity forcing. Climate of the Past 5 481–488

Jouzel J, Masson–Delmotte V, Cattani O et al (2007) Orbital and millennial Antarctic climate variability over the past 800,000 years. Science 317 (5839) 793–796

Keenan TF, Prentice IC, Canadell JG et al (2016) Recent pause in the growth rate of atmospheric CO_2 due to enhanced terrestrial carbon uptake. Nature Communications 7 13428

Kukla G & Matthews R (1972). Letter to the US President. As reported in: Reeves RW & Gemmill D. Origins of a 'diagnostics climate center' 29th Annual Climate Diagnostics & Prediction Workshop, Madison, WI. Oct 20 2004. https://www.cpc.ncep.noaa.gov/products/outreach/proceedings/cdw29_proceedings/Reeves.pdf Accessed 04 Jun 2022

Laskar J, Robutel P, Joutel F et al (2004) A long-term numerical solution for the insolation quantities of the Earth. Astronomy & Astrophysics 428 (1) 261–285

Lewis N & Curry J (2018) The impact of recent forcing and ocean heat uptake data on estimates of climate sensitivity. Journal of Climate 31 (15) 6051–6071

Lisiecki LE & Raymo ME (2005) A Pliocene-Pleistocene stack of 57 globally distributed benthic δ^{18}O records. Paleoceanography 20 1

Loulergue L, Schilt A, Spahni R et al (2008) Orbital and millennial-scale features of atmospheric CH_4 over the past 800,000 years. Nature 453 (7193) 383–386

Loutre MF & Berger A (2000) Future climatic changes: Are we entering an exceptionally long interglacial? Climatic Change 46 (1–2) 61–90

Masson–Delmotte V, Schulz M, Abe–Ouchi A et al (2013) Information from paleoclimate archives. In: Stocker TF, Qin D, Plattner G–K et al (eds) Climate Change 2013: The Physical Science Basis. Contribution of Working Group I to the Fifth Assessment Report of the Intergovernmental Panel on Climate Change. Cambridge University Press, Cambridge, p 383–464

Milanković M (1920) Théorie mathématique des phénomènes thermiques produits par la radiation solaire. Académie Yougoslave des Sciences et des Arts de Zagreb. Gauthier Villars, Paris

Millar RJ, Fuglestvedt JS, Friedlingstein P et al (2017) Emission budgets and pathways consistent with limiting warming to 1.5 °C. Nature Geoscience 10 (10) 741–747

Mysak LA (2008) Glacial inceptions: Past and future. Atmosphere-Ocean 46 (3) 317–341

National Research Council (2011) Understanding Earth's deep past: Lessons for our climate future. The National Academies Press, Washington DC

NEEM Community Members (2013) Eemian interglacial reconstructed from a Greenland folded ice core. Nature 493 (7433) 489–494

Rahmstorf S (2003) Timing of abrupt climate change: A precise clock. Geophysical Research Letters 30 (10) 1510–1514

Reeves RW & Gemmill D (2004) Origins of a 'diagnostics climate center.' In: Proceedings of NOAA's 29th Annual Climate Diagnostics and Prediction Workshop, 18–22 Oct 2004, Madison, WI. https://origin.cpc.ncep.noaa.gov/products/outreach/proceedings/cdw29_proceedings/Reeves.pdf Accessed 04 Mar 2019

Ruddiman WF (2007) The early anthropogenic hypothesis: Challenges and responses. Reviews of Geophysics 45 (4) RG4001

Schindler DW (1999) Carbon cycling: The mysterious missing sink. Nature 398 (6723) 105–106

Schmidt GA, Severinghaus J, Abe–Ouchi A et al (2017) Overestimate of committed warming. Nature 547 (7662) E16

Sirocko F, Seelos K, Schaber K et al (2005) A late Eemian aridity pulse in central Europe during the last glacial inception. Nature 436 (7052) 833–836

Snyder CW (2016) Evolution of global temperature over the past two million years. Nature 538 (7624) 226–228

Spratt RM & Lisiecki LE (2016) A Late Pleistocene sea level stack. Climate of the Past 12 (4) 1079–1092

Tzedakis PC, Channell JE, Hodell DA et al (2012) Determining the natural length of the current interglacial. Nature Geoscience 5 (2) 138–141

Vettoretti G & Peltier WR (2004) Sensitivity of glacial inception to orbital and greenhouse gas climate forcing. Quaternary Science Reviews 23 (3–4) 499–519

Vettoretti G & Peltier WR (2011) The impact of insolation greenhouse gas forcing and ocean circulation changes on glacial inception. The Holocene 21 (5) 803–817

Walker M, Johnsen S, Rasmussen SO et al (2009) Formal definition and dating of the GSSP (Global Stratotype Section and Point) for the base of the Holocene using the Greenland NGRIP ice core, and selected auxiliary records Journal of Quaternary Science 24 (1) 3–17

PEER REVIEWS

Kotkin J (2019) Age of Amnesia. Quillette. https://quillette.com/2019/07/15/age-of-amnesia/ Accessed 12 Jun 2022

Anonymous reviewer 1

The explanations that "Modern Global Warming coincides with an increase in solar activity" and that the contributions of CO_2 levels are seriously overestimated by a politically-charged IPCC will be very helpful to climate change deniers. They will also be used as argument that there is no need for action to reduce Green House Gases now (not only CO_2, but all others). This is a logical following to the conclusions of this book, and even if it was not the author's intention, I think the arguments for and against these activities and plans should have been presented.

I started reading the book with much interest, as the author is a respected biologist addressing a new field of research and examining the current consensus. Fresh insights, and reevaluation of what is taken for granted, are always welcome. I was however quickly taken aback by the regular criticism of the IPCC reports, the picking of some evidence or literature but not all options, and the general conclusion that *"Modern Global Warming coincides with an increase in solar activity"*, with additional misleading statements like *"Sea level has been increasing for the past 200 years, and its modest acceleration for over a century shows no perceptible response for the last decades to strongly accelerating anthropogenic forcing."* (conclusion 11g of Chapter 11) or the *"forecast for a stabilization of the current warming"*.

Chapter 1 includes analyses of a formal statement by the Geological Society of London in 2010, and casts doubts about how it links climate change and CO_2 levels. This has been superseded by other documents, based on work done since, in particular a UK House of Commons Science and Technology Committee Inquiry on Public Understanding of Climate Change in 2013, and a joint statement in December 2015. With 24 of the UK's foremost academic institutions, the Geological Society published a joint Climate Communiqué calling on governments to take immediate action to avert the serious risks posed by climate change. This 1-page communiqué unambiguously starts with the statement "The scientific evidence is now overwhelming that the climate is warming and that human activity is largely responsible for this change through emissions of greenhouse gases". This communiqué is quoted later, but immediately denigrated (although it doesn't focus on CO_2 only, but on greenhouse gases in general, including methane).

Chapter 2 starts with a good presentation at the base of the key elements of the glacial cycle, but it soon includes "redrawing" of data published in other articles, and it is often hard to understand how it was "redrawn". Why not present the data directly from these articles? The use of point data (e.g. Epica Dome C) should be contrasted with other measurements from other parts of the world, as used in the basis of the Geological Society statements the author objects to. Most key figures supporting the arguments of the chapter are "©Author, 2019. All rights reserved" but they have not been peer-reviewed by experts or published: one would have to dig through the original articles, synthesise all their data and then assess how well these graphs are supported by the original data. Chapter 2 ends with a good summary of the author's conclusions, but I cannot agree with all of them until I have seen the original data. Overall, the figures with "©Author, 2019. All rights reserved" comprise between 46% (Chapter 8) and 89% (Chapter 2) of all figures. This is positive, as it shows the strong engagement of the author in making the information accessible, but this is negative, as none of these figures have been peer-reviewed, and it is not clear how they use the data published by others (without checking each single article).

Simple but regular criticism of the articles written by experts in the field, even if backed by some detailed explanations of point measurements, cannot detract from the fact the other articles were peer-reviewed and published. Chapter 3 follows the same pattern, questioning peer-reviewed papers and redrawing data. We need to be persuaded by more than selective use of evidence. In other places, sweeping statements like *"Ötzi, the iceman from Tyrol [...] was a testimony to both the changing climate of the Mid-Holocene Transition and its devastating effect on human societies"* also need to be backed up by references. The mention in Chapter 12 that *"At present there is no evidence that polar bears are threatened during the 21st century from climate change"* does not match the evidence either. And one might argue that polar bears are far from the only animals in the Arctic: all other species are affected by climate change, and poleward species migration is observed everywhere in response to warming waters (e.g. Copernicus Marine Service Ocean State Report, 2018).

The conclusions of Chapter 8 are that a large number of solar cycles are superposing to give current observations of climate change: a 1,000-year solar cycle (Eddy cycle), at least between 11,500 – 4,500 BP and 2,000 – 0 BP; a 210-year solar cycle (de Vries cycle); a 100-year (Feynman cycle); a pentadecadal solar cycle; an 11-year solar cycle (Schwabe cycle). This starts to look like the epicycles used in the Middle Ages to explain planetary orbits before a better explanation was found. There are also regular references to the (yet) unsubstantiated theory that the position of the Earth in the galaxy, or changes in galactic cosmic rays, are an element of this variability.

Section 9.7 is a long charge against the IPCC, and it does not refer to the enormous amount of measurements, models and literature supporting the different IPCC reports, or points to the equally important body of literature arguing about some points (but then later refuted). Some of the references in favour of the author's argument, e.g. Curry and Webster (2011), are presented without mention they were published in a Forum as part of a wider debate, and strongly disputed as inaccurate by Hergler et al. (2011) in the same journal (for completeness, the authors' response should be cited too, as it is part of the controversy).

Chapter 12 presents forecasts for "fossil fuel changes" (Section 12.3) and uses only 3 references, 2 from fossil fuel energy companies (BP Energy Review 2018 and ExxonMobil), possibly counter-balanced by 1 reference to a report by environmental group EndCoal. I teach a final-year undergraduate unit and we always encourage students tackling this issue to look beyond these reports, which might present vested interests or partial views, and also consider peer-reviewed articles and publications, across a broader spectrum. One of conclusions (12c) is that "Global warming might stall or slightly reverse for the period 2000-2035", and 19 years into the 35-year period, is not supported by facts.

In conclusion, the subject is interesting, and a reanalysis of the scientific consensus is always welcome. But I do not think the manuscript, in its present stage, should be published. The references need to include a broader spectrum of evidence and debate; the figures need to explain much more how the original data was redrawn or adapted, and the controversies need to be revisited to better address why there has been a consensus building on specific issues. In particular, I cannot agree with the *"suspicion that anthropogenic forcing of climate change has been seriously overestimated"*, the assertion that *"the importance of solar variations in controlling the Earth's climate"* and the *"forecast for a stabilization of the current warming"*. Most of the references are a bit old, pre 2011. At the time much was being written about how climate drove societal changes. Cyclic pattern as drivers is too simple. The 8.2 event is now not seen as important as the 4.2 ka event, which is not really mentioned, and much has been written on the latter in the last 3-4 years.

Author's reply to reviewer 1

Although I am well aware that climate has become a politically charged issue, the manuscript intentionally avoids entering into political argumentations. What Bernhard Url, Executive Director of the European Food Safety Authority, said about glyphosate is perfectly valid here: *"It is the role of science to provide informed opinion, but what society decides to do is a debate that must stay in the political arena."*

Whether glyphosate should be allowed or measures to reduce greenhouse gases should be taken is a political decision that does not correspond to science but to society. I disagree that arguments for and against these activities and plans should have been presented, because they are out of the scope of a science book. The contribution of CO_2 levels to modern warming is an open question despite IPCC reports, as demonstrated by the inability to constrain the value of climate sensitivity to CO_2 in 40 years.

If what I have seen in climate science from my reexamination of the evidence was well represented by the IPCC reports it would be unnecessary for me to write this book. The IPCC is a political body, but their reports are a compendium of climate science. As such the only general criticism that is possible is about what has been left out of the reports. Most of the science included in the IPCC report is correct to the best of our knowledge. My manuscript frequently cites the IPCC reports, but includes other articles that are not included in the IPCC reports. As it is the case for the IPCC reports, it is not possible to present all the literature, particularly when contradictory, and the IPCC fifth report from the first working group has >1300 pages. I have reflected a minimal part of the bibliography studied for space considerations, but I'll be happy to include and discuss the evidence from other works if the reviewer can be more specific on what he is missing. The main point of the book is that it examines the consensus and the evidence that supports it and finds it lacking. This is very important because we know from history that consensus is no guarantee for being right. Alfred Wegener based his 1912 continental drift theory on abundant evidence, and yet the geologist's consensus dismissed it for 50 years. During this time scientists were able to support it or attack it in conferences, books and articles. That my manuscript does not support the consensus should not be an impediment to its publication as it is in the best interest of science that anybody is allowed to present their case.

I am sorry the reviewer is taken aback by the statement that *"Modern Global Warming coincides with an increase in solar activity."* The modern solar maximum is defended both by important solar physicists like Ilya Usoskin, Sami Solanki and Jürg Beer, as well as very important climatologists like Bo Vinther, Takuro Kobashi, Thomas Blunier and James WC White. In this last article after analyzing 2100 years of Greenland temperature records they conclude: *"high solar activity during the modern solar maximum (approximately 1950s–1980s) resulted in a cooling over Greenland and surrounding subpolar North Atlantic."* This peer-reviewed article is restricted to subpolar latitudes, but it is non-controversial that if high solar activity cooled those latitudes it could very well have been at the expense of warming lower latitudes.

The modern solar maximum is evident by running a 70-year average on the monthly sunspot database (1749-2019) available at SILSO (Fig. R1).

It defines the longest period in the sunspot record (1935-2005) with above average solar activity. It is a fact that most of the unusual warming observed in what has been termed global warming has taken place within this period, so while the attribution of a significant part of global warming to high solar activity is controversial (and should be debated), the coincidence of a unique period of warming within at least the past 600 years with a unique period of above average solar activity within at least the past 600 years is a fact according to the available evidence.

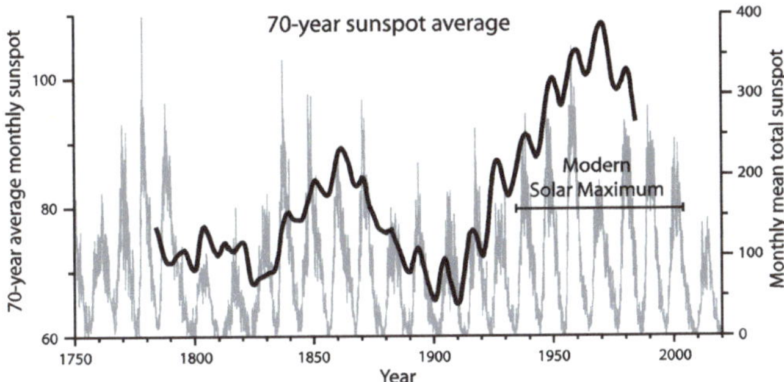

Fig. R.1 The modern solar maximum

I don't agree that the statement *"Sea level has been increasing for the past 200 years, and its modest acceleration for over a century shows no perceptible response for the last decades to strongly accelerating anthropogenic forcing,"* is misleading. The increase in sea levels for the past 200 years is well established in the literature according to data and models. See:

Jevrejeva, S., Moore, J.C., Grinsted, A. and Woodworth, P.L., 2008. Recent global sea level acceleration started over 200 years ago?. Geophysical Research Letters, 35(8).

Grinsted, A., Moore, J.C. and Jevrejeva, S., 2010. Reconstructing sea level from paleo and projected temperatures 200 to 2100 AD. Climate Dynamics, 34(4), pp.461-472.

The lack of significant acceleration in sea level rise so far is a known issue:

Fasullo, J.T., Nerem, R.S. and Hamlington, B., 2016. Is the detection of accelerated sea level rise imminent?. Scientific reports, 6, p.31245.

It does not appear to be imminent, as the increase for the past 6 years since the article was published (02/2016-02/2022) has been of 15.8 mm according to NASA, or 2.63 mm/year, well below the average since 1993, and more supportive of my thesis of a sea level raise slowdown due to the "stadium wave" (see Sect. 11.6, Fig. 11.5).

The strongly accelerating anthropogenic forcing has been shown multiple times, for example in the IPCC AR5:

Myhre, G., et al., 2013: Anthropogenic and Natural Radiative Forcing. In: Climate Change 2013: The Physical Science Basis. Contribution of Working Group I to the Fifth Assessment Report of the Intergovernmental Panel on Climate Change.

And in its figure 8.18.1.

Then, if everything the phrase says is correct, why should it be misleading. What I believe is misleading is to lead people to believe that a reduction in CO_2 emissions should reduce the rate of sea level increase, as there is no evidence for that.

Finally I accept that I might be wrong on my "forecast for a stabilization of the current warming," but that is how hypotheses are tested. Predictions are an important part of science, particularly when they help differentiate between competing hypotheses. The idea that there is only one explanation for modern global warming, and that only more warming must be expected does not foster scientific debate on alternative explanations. If alternative predictions are not allowed, is it science? For the first time in 70 years, solar activity is below its average since 2005, while CO_2 continues increasing. It is a unique chance to explore the merits of this alternative explanation beyond the diffi-

culties in identifying its mechanisms of action. The scientific literature should explore it as it is an opportunity to clarify the issue even for scientists that believe greenhouse gases are solely responsible for all the warming.

The introduction (chapter 1) has been completely rewritten, and the references to the Geological Society of London statements have been eliminated.

Redrawing of figures becomes a necessity when the rights for a figure reuse have to be paid by the author, and in many cases a new figure has to be made to clearly show the point being made, as for example in the interglacials alignment. Such alignments are being published here for the first time and they are very informative to define an average interglacial and to examine their orbital cause (figures 2.9, 2.10, and 2.16). Epica Dome C data is preferred for its good resolution, as ice core data has an age model far superior to all other proxies, but the LR04 stack of benthic cores is from all over the world's oceans and shows essentially the same thing. It is used in figures 2.3, 2.4, 2.7, 2.13, and 2.15. Here is a superposition of both (Fig. R2). The level of agreement is extraordinarily good, and any small dating difference should in principle be resolved in favor of ice core data due to its superior age model and because benthic data is orbitally tuned, a process not exempt from controversies.

The source for all the data is from publicly available official repositories like NOAA paleoclimate data repository:

In the cases where the original data has been processed (averaged, filtered, …) it is indicated in the figure caption so reproducing any figure should be a quick process taking minutes. If the referee so wishes I can write a detailed report on how every figure was made including original figures from publications if available. This is too much detail for the book as the figure captions are already very long. It would help if the referee would provide a list of the figures of interest, as the book has 180 figures and more than half of them have been made by me.

Criticism is essential to the scientific process as articles defending opposing hypotheses are published all the time, being mutually exclusive. Different hypotheses are proposed based on the same evidence and scientists energetically defend opposite points of view and theories in the scientific literature. In particular chapter 3 deals with Dansgaard-Oeschger events, a subject where great controversies have taken place over their causes and periodicity. Over time more evidence and better analysis settle these disputes but that is not the role of peer-review that should try to avoid gatekeeping valid scientific debate by preselecting favored or consensus views. A general comment that the manuscript is criticizing peer-review published

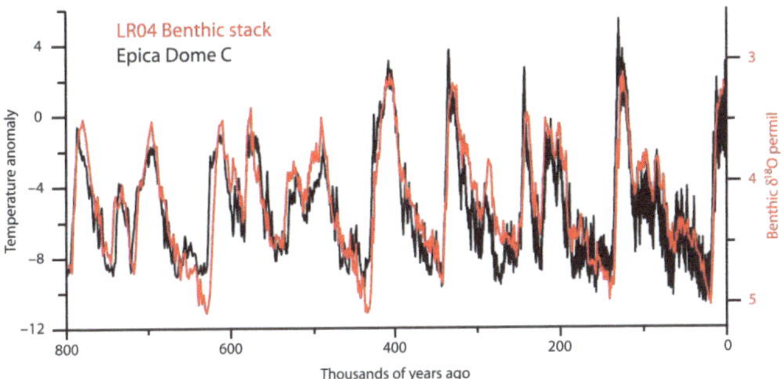

Fig. R.2. Epica and LR04 comparison

articles is difficult to address without pointing to each case to be able to support the specific criticism.

The statement about Ötzi can be supported on the following reference:

Magny, M. & Haas, J.N., 2004. A major widespread climatic change around 5300 cal. yr BP at the time of the Alpine Iceman. Journal of Quaternary Science, 19 (5), pp.423-430.

This reference is already included in the manuscript, but not referenced in this paragraph. The manuscript has been modified to include it. From the reference:

"Finally, although the onset of Neoglaciation favoured snow accumulation and glacier advance in the Tyrolean Alps and, as a result, the excellent preservation of the Alpine Iceman, the changes in palaeoenvironmental conditions induced by this mid-Holocene climate reversal may have led, via a more or less complex causal network, to substantial perturbations within human societies, as suggested by changes in cultural development in central Europe (Arbogast et al., 1996; Berglund, 2003; Magny, 2004), strong variations in human settlement patterns (Vernet and Faure, 2000) and by a rapid development of hierarchical societies in the overpopulated Nile valley and Mesopotamia (Sirocko et al., 1993) as well as in South America (Sandweiss et al., 2001)."

More evidence for the abrupt climate change at the time Ötzi died is discussed by David Bressan.

Regarding polar bears their number increased after the 1970s agreement to limit their hunting. The Polar Bear Specialist Group and Regehr et al., 2016 give the following central estimates for polar bears:

    1993: 25,000
    1997: 24,500
    2001: 23,250
    2005: 22,500
    2015: 26,000

Regehr, E.V., Laidre, K.L., Akçakaya, H.R., Amstrup, S.C., Atwood, T.C., Lunn, N.J., Obbard, M., Stern, H., Thiemann, G.W. and Wiig, Ø., 2016. Conservation status of polar bears (Ursus maritimus) in relation to projected sea-ice declines. Biology Letters, 12(12), p.20160556.

So for the past 25 years and despite the decrease in Arctic sea-ice the polar bear population has been stable, and Inuit reports indicate it might be even growing:

Wong, P.B., Dyck, M.G., Hunters, A., Hunters, I., Hunters, M. and Murphy, R.W., 2017. Inuit perspectives of polar bear research: lessons for community-based collaborations. Polar Record, 53 (3), pp. 257-270.

"Across the north, Inuit still report recent increases in polar bear abundance and the ability of polar bears to adapt to rapidly changing environments (Keith 2005; Tyrell 2006; Dowsley 2007; Kotierk 2010; Arviat Hunters and Trappers 2011; Kotierk 2012)."

It is clear that there are other animals besides polar bears and all are affected by climate change. It is outside the scope of the manuscript to analyze the situation and responses to climate change by different species. The polar bear is a special case as it was included in the Red List solely for climate concerns over the loss of sea-ice, and despite 25,000 individuals being a respectable number for such a large predator from a very restricted environment. It was therefore singled out due to climate concerns and

there is no evidence so far that their population is in decline.

Cycles have long been recognized as a feature of climate. Evidence for periodicities in proxy records led to the confirmation of Milankovitch theory of orbital changes being responsible for the most important climatic changes of the Pleistocene. It is a complication that solar activity presents so many quasi-cycles, but they are real. The 11, 50, 100 and 200-year periodicities are observable in sunspot records, while the 200, 1000, and 2500-year periodicities are observable in cosmogenic isotope records. The current extended solar minimum (Solar cycles 24-25) was predicted on the basis of the 100-year periodicity (Clilverd et al. 2006). Some of these periodicities appear in climate records as attested in the bibliography presented. In particular the 200-year periodicity in tree-rings, and the 1000-year periodicity in monsoon proxy records, methane records, and iceberg activity proxy records. I find surprising that scientists that have no problem in accepting the 11-year cycle in sunspots and the 41-kyr and 100-kyr cycles in glaciations have so much problem accepting the existence of cycles in between. There is not much support in the manuscript for the climatic effect of galactic cosmic rays, as I don't find the evidence convincing. The galactic position of the Solar System hypothesis is mentioned, but I also say that it is controversial and not well constrained by evidence, so it is hardly an endorsement.

Section 9.7 does not globally criticize IPCC reports, as this section is quite specific about climate change attribution, that is assessed in AR5 chapter 10 "Detection and Attribution of Climate Change: from Global to Regional." This chapter explains how attribution is done in page 872: *"In general, a component of an observed change is attributed to a specific causal factor if the observations can be shown to be consistent with results from a process-based model that includes the causal factor in question, and inconsistent with an alternate, otherwise identical, model that excludes this factor."* The problem with this approach is that to be acceptable, models have to reproduce the warming and they do so based on our incomplete knowledge, and therefore their use to prove attribution constitutes circular reasoning as they cannot be used as evidence of attribution when their coding already presupposes it. This is explained in section 9.7 and in my opinion is a valid criticism of attribution studies reflected in IPCC's AR5. If models are wrong …, [and since they can't reproduce the climate of the past they must be wrong, see for example:

Liu, Z., Zhu, J., Rosenthal, Y., Zhang, X., Otto-Bliesner, B.L., Timmermann, A., Smith, R.S., Lohmann, G., Zheng, W. and Timm, O.E., 2014. The Holocene temperature conundrum. Proceedings of the National Academy of Sciences, 111(34), pp.E3501-E3505.

Or the Early Eocene equable climate problem.]… then the uncertainty reflected in IPCC reports is incorrect and probably much higher. This is something that should be openly debated.

The paragraph about Curry and Webster 2011 has been changed to reflect the controversy. Thank you for pointing it out. It now reads:

"Curry and Webster (2011) essay on climate uncertainty claimed that it is too big to be simplified, as the IPCC does, or hidden as in the media portraits of global

warming. Their claim was contested by a group of IPCC authors (Hegerl et al. 2011, followed by Curry and Webster reply). While many scientists fear that uncertainty recognition will lead to inaction and furiously combat anyone raising the issue as a "merchant of doubt" (Oreskes and Conway 2010), Curry and Webster propose ideas for dealing with uncertainty, and convincingly argue that trust is more important than certainty for public confidence (Curry and Webster 2011)."

Regarding possible vested sources in chapter 12 fossil fuel changes, the following academic reference has been added to the text:

Wang, J., Feng, L., Tang, X., Bentley, Y. and Höök, M., 2017. The implications of fossil fuel supply constraints on climate change projections: A supply-side analysis. Futures, 86, pp.58-72.

They reach a similar conclusion to mine:

"Climate projections calculated in this paper indicate that the future atmospheric CO_2 concentration will not exceed 610 ppm in this century; and that the increase in global surface temperature will be lower than 2.6 °C compared to pre-industrial level even if there is a significant increase in the production of non-conventional fossil fuels. Our results indicate therefore that the IPCC's climate projections overestimate the upper-bound of climate change."

Similar views are expressed by:

Berg, P. and Boland, A., 2014. Analysis of ultimate fossil fuel reserves and associated CO2 emissions in IPCC scenarios. Natural resources research, 23(1), pp.141-158.

Höök, M. and Tang, X., 2013. Depletion of fossil fuels and anthropogenic climate change—A review. Energy Policy, 52, pp.797-809.

Höök, M., Sivertsson, A. and Aleklett, K., 2010. Validity of the fossil fuel production outlooks in the IPCC Emission Scenarios. Natural Resources Research, 19(2), pp.63-81.

The conclusion that *"Global warming might stall or slightly reverse for the period 2000-2035"* is consistent with the evidence that the only period of warming since 2002 was due to the big El Niño between 2014 and 2016. Since 2016 the global surface average temperature has been declining. I expected further declines when the increase in solar activity with the new solar cycle coincides with La Niña conditions as has happened in the last six solar cycles (70 years). This has also taken place. A decline of 0.2-0.3°C over the 2020–2025 period is possible and in accordance with the science presented in the manuscript. That would leave the global surface average temperature unchanged since 2002 and for over 20 years and is consistent with an analogous situation 138 years ago that shows a good agreement for the past 20 years, indicating climate conditions are perfectly capable of making this prediction correct (Fig. R3). As the prediction looks risky, particularly if one accepts that IPCC models do a good job at representing climate, it is a good test of the alternative scientific explanation discussed in the manuscript.

Certainly the number of references can be expanded and other points of view better represented. Since the manuscript was already quite long and IPCC views are presented in nearly every book published on the subject I considered it was more novel and interesting to present a more critical view that is supported by many paleoclimate studies and scientific authors. Although I don't mind giving great detail about the making of the figures, a much

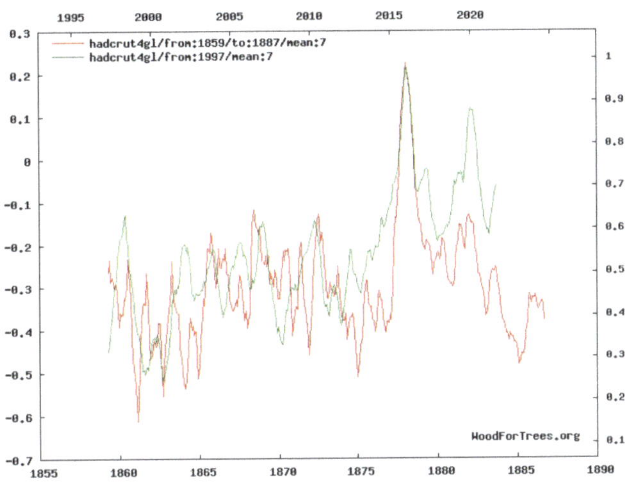

Fig. R.3. Fractal comparison of the great Niños of 1876 and 2016

1870s fractal analog (red, left and bottom scales) of the recent temperature record (green, right and top scales) shows that a 0.2-0.3 °C cooling over the years following a great Niño would not be unprecedented. Data from woodfortrees.org.

more detailed explanation in the figure captions would likely be detrimental to most readers. As an alternative I offer to provide that information to the reviewer so their correctness can be checked.

In science, disagreement is normal and productive. Different hypotheses are tested and reinforced or discarded. The book provides evidence from the published literature of the importance of natural mechanisms of climate change. Since the view of IPCC is that *"The best estimate of the human-induced contribution to warming is similar to the observed warming over this period,"* it is clear that a significant contribution by natural factors implies a significant overestimation of anthropogenic factors. And it is clear that models run hot (as shown in figure 12.10), predicting more warming than is taking place and supporting that point.

Although we don't understand the mechanisms by which solar activity affects climate, despite a great amount of scientific literature on the issue (a hypothesis is presented in the manuscript), the correlation between climate proxies and solar proxies supports a very important role, particularly on the multi-decadal to millennial timescales. This position is defended by prominent climate scientists as in:

E. Rohling, P. Mayewski, R. Abu-Zied, J. Casford, and A. Hayes, 2002. Holocene atmosphere-ocean interactions: records from Greenland and the Aegean Sea. Climate Dynamics 18 7 587-593

"In view of these findings, we call for an in-depth multi-disciplinary assessment of the potential for solar modulation of climate on centennial scales."

This disagreement exists in the scientific literature, it is a valid position in light of the evidence, and it is something that should be openly debated.

The stabilization of global warming in the first third of the 21st century is a clear possibility. A hiatus in global warming was identified prior to the 2014-16 El Niño, and Hansen et al. 2018 in "Global Temperature in 2017" affirm that: *"Therefore, because of the combination of the strong*

2016 El Niño and the phase of the solar cycle, it is plausible, if not likely, that the next 10 years of global temperature change will leave an impression of a 'global warming hiatus'."

The possibility of two warming hiatuses separated by a strong El Niño is defensible.

I must disagree that the references are old. The manuscript includes 758 unique references, of which 627 were published since 2000. The age distribution of the references since 1950 is represented in figure R4. The peak year is 2013, which is reasonable since the manuscript was started in 2016. The number of annual references is above the 21st century average for the 2010-2018, years indicating the author's effort to keep the bibliography up to date during the writing process. It usually takes 2-3 years for most scientific articles to show their true impact through the citations in their colleagues' articles. Nevertheless, more references can be added and particularly any references missed by the reviewer.

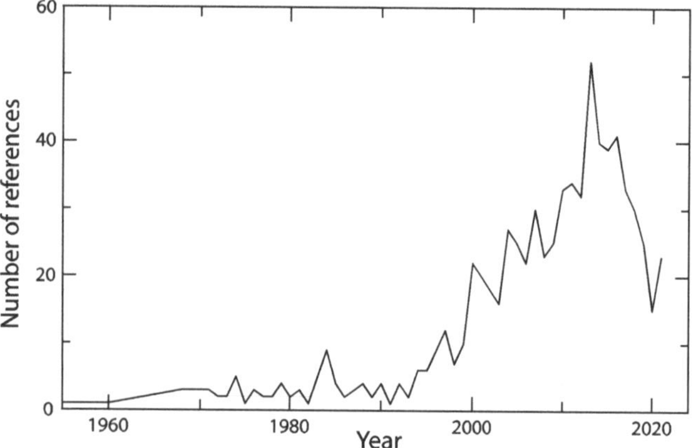

Fig. R.4. Age distribution of references in the book

Reviewer 2

"Climate of the Past, Present and Future", by Javier Vinós, presents a comprehensive description of our climate's long-term variability, together with an extensive analysis of its plausible forcings. Data and numerous publications are presented and critically analyzed which study the climate from thousands and even millions of years ago. The analysis and results shown are based on studies by many authors, mainly experts on the subject and from various orientations regarding climate change cause attribution.

The book consists of 14 Chapters. The first is an Introduction that effectively invites to start the "journey" of reading it, even for those with "biased" positions regarding the main responsible for our climate's long-term trend.

Chapters 2, 3, 4, 5, 7, and 8 analyze mainly long-term periodicities detected in solar activity and in characteristics of the Earth's orbit, among others, which could explain some aspects of the climate variability, not only based on statistical analysis but on physical foundations as well.

Chapter 6 analyzes the effect of abrupt changes in climate on society in the past, associating them with solar forcing and highlighting how little its role is recognized. The latter, according to Vinos, may induce overestimating other forcing factors, such as anthropogenic activity, for example.

Chapter 9 deals with greenhouse gases and climate, going through various aspects and showing successes and failures in their association. I highlight here the conclusion regarding climate uncertainties, which given their magnitude really prevents us yet from determining with full accuracy the roles of both, anthropogenic and natural forcings.

Chapters 10 and 11 deal with the meridional transport of energy and its relation to solar variability. This transport, as indicated and highlighted by the author, is a fundamental property that determines key aspects of the climate and its variability at all scales.

The last three Chapters deal with the present climate and predictions, ending with a discussion of the next glacial period. The significant role of the greenhouse effect is recognized, even though limited to a short period of time in paleoclimatic terms. The role of the Sun would seem to prevail when the research is extended to millions of years and even on scales of centuries.

All Chapters are accompanied by figures supporting many of the arguments. It is also noteworthy the synthesis provided by the conclusions at the end of each Chapter summarizing the ideas expressed by the author throughout each of them.

Some points are expressed sarcastically, and it seems that the author does not agree with the more general ideas about the preponderant responsibility of anthropogenic activity in forcing the latest climate long-term changes. However, in my opinion, his ultimate intention is clear: to warn of the role of natural forces, especially the Sun, in changing the climate of our planet, which affects us all equally.

The book's topic is too broad and dates back a long time, in addition to being under scrutiny for the last few decades. Even so, this work reflects a good selection of points to study in depth before finding a single origin to our climate trend, which is certainly shared among more than one forcing.

Dr. Ana G. Elías
Laboratorio de Ionosfera, Atmósfera Neutra y Magnetosfera (LIANM)
FACET, Universidad Nacional de Tucumán, Argentina

Reviewer 3

During a period of overwhelming concerns about the potential impacts of human induced climate change, Javier Vinós' carefully reasoned book *Climate of the Past, Present and Future, A Scientific Debate*, is an important and well-timed contribution to the discussion.

The book title accurately describes the book's very substantial content.

The climate problem is tackled with an emphasis on the role of solar variability, arguing that not every variation in climate can be explained by variations in CO_2 concentration. The book commences with a detailed description of the influences of orbital variability on climate. The Milankovitch theory is explained in some detail, noting the impacts and roles of obliquity, eccentricity, and precession and their relative roles in past climate variations. He also introduces additional feedbacks involving, for example, ice volume loading. Finally, he clarifies the role of CO_2

concentrations in the termination of glacial events, noting the complexity of Earth's climate.

Vinós also describes some problems with Milankovitch theory, using as an example the apparent interhemispheric synchronicity of glacial and interglacial events. For example, during the last glaciation the Antarctic winter sea-ice sheet extended 10-15° closer to the equator and Australia, New Guinea and Chile all contained glaciers The usual explanation is that CO_2 changes induced by insolation effects in the Northern Hemisphere cause the ocean absorption of more CO_2 than a warmer ocean and vice versa. Since Earth's atmosphere is well mixed, the changes in greenhouse effect become one-signed, globally producing interhemispheric synchronicity. Yet, based on ice-core CO_2 records, Vinós argues that the decreases in CO_2 concentration cannot completely explain the symmetry, arguing further that other feedbacks besides greenhouse effects must be involved. I emphasize this aspect because of evidence of interhemispheric synchronicity in the top-of-the-atmosphere radiation balance and in precipitation appear to exist in the present climate state. This suggests an overall constraint, possibly dynamic, on global climate.

The third chapter provides a detailed accounting of rapid climate variability associated within the Dansgaard-Oeschger cycles (D-O). Between 85-12 kYrs BP Greenland ice cores show the rapid changes between cold (stadial) and warm (interstadial) events. The author discusses several hypotheses that may explain the D-O cycles. Importantly, he concludes CO_2 variability occurring between stadial and interstadial periods cannot account completely for the rapidity and magnitude of the events, but that these events are tied to changes in ocean structure and ice extent in the northern Atlantic Ocean. He believes they are associated with an oscillating Atlantic Meridional Overturning Circulation (AMOC). This is a key chapter as the author raises the question that if the D-O transitions are not related to CO_2 concentrations, can similar transitions explain, at least partially, recent climate variations?

Chapter 4 describes climate variability during the Holocene, which occupies the period between the end of the last glaciation to the present. Climate variability within this period has had a profound influence on the waxing and waning of advanced societies across the planet (see also Chapter 6). A detailed account of temperature, CO_2, and methane variations over the Holocene using data from Greenland ice cores shows a major warming starting at the end of the last glacial period until about 9 kyr BP (+3°C), followed by a relatively warm period around 9 kyr BP, with the first part being relatively dry (10.5-7.8 kyr BP) and the second being relatively moist (until 5 kyr BP). The period 9 kyr to 5 kyr BP is termed the Holocene climatic optimum. preceding a distinct cooling (-3°C) until about 200 years ago, referred to as the Neoglacial period. The proxy climate records are displayed graphically in Figure 4.5 along with changes in methane (inversely related to temperature) and CO_2 . The figure also shows a synthesis of temperature variations from a suite of climate model simulations. What is surprising is the abruptness of the start of the Neoglacial Period (5 kYr BP), which is evident from various proxy data. Of particular interest is the absence of the grand cooling associated with the Neoglacial in the model results (Figure 4.5e). Vinós notes that the rise in CO_2 through the Neoglacial Period did not coincide

with any warming, perhaps reflecting overwhelming obliquity effects. But an important conclusion is that the Holocene proxy temperature trends are opposite to what one would expect from increasing CO_2.

Whereas Vinós ties the longer-term climate variability in the Holocene with variations in the planet's obliquity, he also associates more rapid changes with the 2500 Bray solar cycle (Chapter 5) for which he lists proxy data support for the cycle. It is hypothesized that the Bray cycle alters the structure of the stratosphere and the changes are dynamically transmitted downwards, causing tropospheric reorganization ("top-down" influence) whilst insolation variability impacts the surface ("bottom up"), as described by Joanna Haigh in 1996. Vinós notes Haigh's modelling work[1], in which she found pathways for solar influences on climate via stratospheric chemistry. Thus, it is not surprising that models without an interactive upper atmosphere that includes chemistry may minimize or underestimate the influences of solar variability. The Haigh model with stratospheric chemistry perhaps represents a base level of complexity required for simulation of climate variability associated with solar influences. With the absence of upper atmosphere chemistry, simpler models may have led to an overemphasis on the role of CO_2 in climate change. An important conclusion is that one should be cautious about dismissing solar influences on climate until we have models that more accurately replicate past climate.

The importance of understanding past climate as a basis for understanding current climate change is underlined by noting the influence of climate change on past human societies. Chapter 6 is especially interesting as it explains low solar activity within Bray cycle during the Holocene resulted in rapidly occurring global cooling, which are related to periods of societal crisis. Vinos hypothesizes how the cycle impacts global circulation through changes in major circulation patterns. It is also noted that climate crises, severe as they were, produced methods of adaptation and eventually societal advances. The two chapters that follow contain details of climate variability seemingly associated with solar variations.

The overall role of greenhouse gas concentrations in climate variability is the focus of Chapter 9. Early in Earth's history global temperatures were very high even though the radiation from the Sun was considerably lower. This is referred to as the "period of the faint sun". Only CO_2 concentrations of >5000 ppm could explain high global temperatures during a prolonged period of low solar radiation. From this early peak, concentrations decreased to order 500 ppm with lower values occurring during periods of glaciation. The meridional transport of heat critical to the global energy balance on all time scales, which is discussed at length in Chapter 10 and its modulation by CO_2 variability and variability of solar forcing, is described in Chapter 11. The atmosphere transports heat meridionally mainly by large transient eddies in the middle and high latitudes. The ocean carries heat mainly by the large-scale horizontal gyres, and by the meridional over-

---

[1] Haigh, J.D., 1994: The role of stratospheric ozone in modulating the solar radiative forcing of climate. Nature 370, 544.

Haigh, J. D., 1996: The impact of solar variability on climate, Science, 272, 5264.

turning cells driven by the sinking and spreading of cold water.

Vinós attempts to tie ocean and atmospheric heat transports with solar activity using the "stadium wave" hypothesis[2] that synchronizes ocean, atmospheric and sea-ice indices with an approximate 65 yr period. One of the interesting factors is that the phases of the stadium wave coincide with periods of more rapid recent warming 1910-1940 and 1970-2000 and may explain the lull in surface temperature change 1945-1970, a period when there was a major increase in greenhouse gas emissions. Of some importance, the stadium wave hypothesis offers a possible explanation for the nonlinearity of recent climate and greenhouse gas emissions relationships.

Climate change since 1900 is discussed in detail in Chapter 13, including Vinós' speculation of climate change through the 21st century. He argues that CO_2 emissions are expected to level off mid-century based on two major factors: projected changes in global demographics and increases in carbon sinks. The United Nations estimates that populations will tend to level off by the end of the 21st century. In fact, the annual global population growth peaked at 2.1% in 1968, has since dropped to 1.1%, and may drop even further to 0.1% by 2100, which would be a growth rate not seen since pre-industrial revolution days. All regions show a similar decrease in growth rate except for Africa. Furthermore, populations tend to become older, and energy use per capita has fallen significantly from a peak at the beginning of the 20th century. World energy use per capita has continued to increase largely driven by African demographics. However, sinks of CO_2 over land areas have increased in efficiency because of the "greening" of the planet. Vinós points out estimates that in 1960, 40% of CO_2 emissions were taken up by land plants compared to the current 50% uptake of what are now much larger emission levels. Vinós hypothesizes that with the aging of the global population and eventually its decrease and the increased efficiency of carbon sinks that warming will stabilize at about 1.5°C above preindustrial levels.

Finally, Vinós looks to the "far" future and asks whether another glacial period is imminent. Concerns were expressed by many climatologists in the 1970s that the decrease of global temperatures in the 1950-70 period may be harbingers of an imminent ice age. The decrease in global temperature occurred during an acceleration in carbon emissions following WWII. Of course, an ice age did not occur and has not occurred since. However, Vinós expects the future interglacial, resulting from orbital effects, to be rather long delayed due to the "fat tail" of anthropogenic CO_2 recovery or the exaggerated time for the current concentration to fall.

The text is well complemented by carefully constructed figures with excellent captions. It includes extensive references and glossary. The book emphasizes the importance of solar variability more so than those who believe that the "CO_2 knob" controls climate. The author

presents a balanced view of climate change noting the importance of orbital changes of the Earth, changes in greenhouse gas concentrations, both natural and anthropogenic, solar variability and the impact of each of these factors on the dynamical transport of energy towards the poles.

Overall, the book is very readable. Who will be interested in this work? My thoughts are that this book should be read by anyone interested in climate variability, but especially graduate students and young investigators. It would make an excellent text for a graduate course.

Dr. Peter J. Webster
Emeritus Professor, School of Earth, and Atmospheric Sciences.
Georgia Institute of Technology.

Reviewer 4

"Climate of the Past, Present & Future; A scientific debate" is a powerful discourse on climate change from the onset of 800,000 years ago to the future of next ice age. This book excels at showcasing the historical chronicles of climate change, its effects, picturing our very future in this scenario and in the discussion about the ramifications of climate change for humanity.

Over the past 20+ years, the public has been informed on various theories and hypothesis about the causes/effects of rising temperatures due to Carbon dioxide emissions. We've seen plenty of definite evidence of climate change at an unexpected higher rate.

The book calls attention towards the natural causes of climate change as well as the anthropogenic factors. It also depicts the major player in keeping the earth warm; Greenhouse effect because it keeps some of the planet's heat that would otherwise escape from the atmosphere out to space. In fact, without the greenhouse effect the Earth's average global temperature would be much colder and life on Earth as we know it would not be possible

There's a huge number of differences between past climate change and modern climate change, both in terms of causes and in terms of the potential impacts, but by knowing about the past changes and their attributions we can understand causes of climate change, and responses of the earth's systems to both rapid and gradual change. Past environments basically act as experiments that we can study, and use to constrain models for the future. The author successfully conveyed this message to the readers.

It further explores the current scientific consensus about fluctuations in solar activity and their account for a significant portion of our global warming. Many thanks for the wise author's hard work on giving us such an excellent knowledge of the climate change.

Dr. Shican Qiu
Depatment of Geophysics, College of the Geology Engineering and Geomatics
Chang'an University, Xi'an, 710054, China

---

[2] Wyatt, M.G., and J.A. Curry (2014). Role for Eurasian Arctic shelf sea ice in a secularly varying hemispheric climate signal during the 20th century, Clim. Dyn., 42, 2763–2782.
    Kravstov, S., Wyatt, M.G, Curry, J.A. and Tsonis, A.A. (2014). Two contrasting views of multidecadal climate variability in the twentieth century. Geophys. Res. Lettrs., 41 (19), 6881-6888.

¹⁴C: Unstable isotope of carbon having an atomic weight of 14, and a half-life of c. 5,700 years. It is produced by the effect of cosmic and solar high-energy radiation on atmospheric nitrogen. It is used for radiocarbon dating going back some 40 ka, and as a proxy for past solar activity. Its production is affected by solar magnetic activity and geomagnetic changes.

– A –

aa index: The antipodal amplitude geomagnetic index is a measure of the disturbance level of the Earth's magnetic field based on magnetometer observations at two nearly antipodal stations in the UK and Australia. It is the oldest geomagnetic index as it has been reconstructed back to 1868.

Abrupt climate change: A climate change characterized by a persistent alteration in one or more climate variables at a rate that is higher than what is observed 80% of the time, leading to a different climate state that can last decades or longer.

Abrupt climatic event (ACE): A period of centuries that displays significantly altered climate variables on a global or hemispheric scale, as a result of abrupt climate change, constituting a different climate state.

Advection: The transfer of some property of the atmosphere or ocean, like heat, moisture or salinity, by predominantly horizontal mass motions of water or air. In meteorology and oceanography it is the horizontal equivalent of the predominantly vertical convection.

African Humid Period: Each period when northern summer insolation is sufficiently high to allow the West African Monsoon reach the Sahara region. The annual precipitation over the Sahara is then sufficient to allow a savanna-type of environment instead of a desert. The last African humid period took place between 12.4–5.5 kyr BP.

Albedo: Is the fraction (percentage) of solar radiation that is reflected from a surface. Atmospheric albedo due to cloud cover is the largest contributor to Earth's albedo. Surface albedo is highest from ice and lowest generally from the ocean.

Alkenone: Long-chain unsaturated methyl and ethyl n-ketone produced by some phytoplankton species. They are stable enough to have been detected in 120-million-year old marine sediments. Some are used as a proxy for past sea-surface temperature because their degree of saturation is related to the temperature at the time they were produced.

Allerød Oscillation: An interstadial between the Older Dryas and the Younger Dryas in the North Atlantic region, now identified with phases a to c3 of Greenland Interstadial 1, between 13,900–12,900 BP. It was interrupted by a cold relapse known as the Intra-Allerød Cold Period (IACP, phase b of GI–1).

Altithermal: The middle part of a tripartite division of the Holocene based on palynology (pollen science) proposed by Ernst Antevs in 1948 and published as Neothermal Climatic Sequence. It means period of high temperature and corresponds to the Holocene Climatic Optimum.

Anathermal: The earliest part of a tripartite division of the Holocene based on palynology (pollen science) proposed by Ernst Antevs in 1948 and published as Neothermal Climatic Sequence. It means period of rising temperature.

Angular momentum: A vector quantity determined by the rotational momentum of a rotating body or system, equivalent to the product of the angular velocity of the body or system and its moment of inertia with respect to the rotation axis. The direction of the vector is the rotation axis.

Anomaly: Referred to temperature indicates a scale, usually in Kelvin or Celsius degrees, where the zero value has been placed at the average temperature of a certain period of time, usually 30 years. The name is unfortunate, as it suggests that temperature changes are anomalous.

Antarctic Isotope Maxima (AIM): Millennial scale periods of warming in Antarctica characterized by an increase in $\delta^{18}O$ and δD isotope ratios in ice cores.

Anthropogenic Global Warming (AGW): Global warming resulting from past and present human activities.

Aphelion: The point in an orbit which is farthest from the sun. For the Earth it happens currently around the 5th of July.

Apogee: The point of greatest distance in the orbit of any satellite of the Earth.

Apsidal precession: Also known as orbital precession, it is the precession (gradual rotation) of the line connecting the apsides, the points closest and farthest to the main body. The apsidal precession period for the Earth is 113 kyr, and participates in Milankovitch orbital frequencies that constitute a climate forcing. The apsidal precession period for the Moon is 8.85 years.

Arctic Oscillation (AO): Also known as Northern Hemisphere Annular Mode, the AO is a mode of climate variability that affects the winds circulating counterclockwise around the Arctic. A positive AO is characterized by strong winds, a ring-like jet stream, low surface pressure in the Arctic, and cold masses of air confined to polar regions. A negative AO is characterized by weaker winds, a meandering jet stream, high surface pressure in the Arctic, and cold masses of air penetrating non-polar latitudes. The AO index is calculated by comparing the 20–90°N 1000 mBar geopotential height field to its main mode of variability for the 1979–2000 period.

Atlantic Meridional Overturning Circulation (AMOC): A system of surface and deep oceanic currents in the Atlantic Ocean responsible for the transport of heat, salt, carbon and nutrients. Surface currents transport heat and moisture northward from the tropics, while deep cold currents return the salt southward. Regions of overturning at both ends link both subsystems.

Atlantic Multidecadal Oscillation (AMO): A recurrent mode of climate variability in the North Atlantic associated to changes in sea surface temperature, changes in North American, European and North African precipitation, and the intensity of North Atlantic hurricanes. It is characterized by alternating phases of 20–40 years with a c. 0.6 °C amplitude in sea surface temperature.

Axial precession: The slow and continuous change in the orientation of an astronomical body's axis of rotation. The Earth's axial precession period is close to 25,800 years, and participates in Milankovitch orbital frequencies that constitute a climate forcing.

– B –

Bjerknes compensation: The proposition by Jacob Bjerknes in 1964 that variability in latitudinal heat transport by the ocean is largely compensated by variability of the opposite sign in latitudinal heat transport by the atmosphere. Although not formally demonstrated due to difficulties in measuring heat transport by the ocean, it is generally accepted.

Blytt–Sernander sequence: A Northern European climatic sequence based upon pollen and plant remains in Danish peat bogs by Axel Blytt and Rutger Sernander. It was built and refined during the first half of the 20th century when it was popular and shown to occur more widely. It fell in disuse in the 1970s.

Bølling–Allerød Period: An interstadial between the Oldest Dryas and the Younger Dryas in the North Atlantic region, now identified with Greenland Interstadial 1, between 14,700–12,900 BP. It includes the Bølling Oscillation, the Older Dryas, and the Allerød Oscillation.

Bølling Oscillation: An interstadial between the Oldest Dryas and the Older Dryas in the North Atlantic region, now identified with phase e of Greenland Interstadial 1, between 14,700–14,100 BP.

Bond cycle: An oceanic sedimentary pattern from the glacial period discovered by Gerard Bond and Rusty Lotti in 1995 that linked the occurrence of 3–4 Dansgaard–Oeschger events between two Heinrich events.

Bond event: Periods of increased ice-rafting activity in the North Atlantic during the Holocene identified by Gerard Bond in 1997. Their number and periodicity have been the subject of controversy, but their climatic importance is not, as they correlate to precipitation changes at both sides of the Atlantic, periods of weakness in the Asian Monsoon, and aridity events in the Middle East.

BP: Before Present. Label to indicate a number of years before 1950 in the Gregorian calendar. This scale is common in Geology, and 1950 was selected because radiocarbon dating was developed in the early 1950s and used to date materials prior to that date.

Bray solar cycle: A solar activity periodicity of c. 2500-year period first described by Roger Bray in 1968 and linked to a climate periodicity of the same period and

phase.

– C –

C-mode: A combination mode of climate variability resulting from the nonlinear interaction of the western Pacific warm pool annual cycle and El Niño/Southern Oscillation variability. The C-mode exhibits profound impacts on tropical climate, ENSO phase locking, and inter-hemispheric sea level variability in the Pacific.

Cal (years): Calibrated years, also calendar years. Dating obtained from converting radiocarbon years to calendar years.

Climate change: A change in climate identified by statistically significant changes in its climatological variables that persist for an extended period, typically decades or longer. By this definition climate is always changing.

Climate sensitivity: See equilibrium climate sensitivity.

Climate: General pattern of weather conditions for an area. Climate is statistically defined in terms of the mean and variability of relevant climatological variables over a period of time ranging from months to thousands or millions of years.

CMIP: Coupled Model Intercomparison Project. A collaborative framework designed to improve knowledge for global coupled ocean–atmosphere general circulation models. Organized in 1995 by the Working Group on Coupled Modelling of the World Climate Research Programme. Its most recently completed phase of the project (2014–2020) is phase 6.

CO$_2$ hypothesis: Hypothesis proposing that the amount of CO$_2$ in Earth's atmosphere is the main factor governing the temperature of its surface, and that changes in CO$_2$ levels caused most large climate changes in the past and are responsible for present global warming.

Conduction: Thermal conduction is the transfer of heat between particles through collisions. The flow of energy is spontaneous from a hotter to a colder body and its rate depends on the temperature gradient and the properties of the conductive medium.

Convection: The transfer of some property of the atmosphere or ocean, like heat, moisture or salinity, by predominantly vertical mass motions of water or air. In meteorology and oceanography it is the vertical equivalent of the predominantly horizontal advection.

Cosmogenic isotopes: Unstable isotopes produced naturally by the effect of cosmic radiation on Earth's atmosphere. The main ones are ^{14}C and ^{10}Be.

– D –

Dansgaard–Oeschger event (D–O): A glacial-period abrupt climate event centered in the North Atlantic–Nordic Seas region characterized by abrupt warming, measured at 7–13 °C in Greenland ice cores, over a period of seven decades, followed by a slower return to glacial conditions over several centuries to a few millennia. Their effect is hemispheric and they are tied to Antarctic isotope maxima to produce a global climate feature that is well registered in global methane levels.

Dark Ages Cold Period (DACP): A climate interval after the Roman Warm Period and before the Medieval

Warm Period characterized by cooling. It is usually dated c. AD 400–900.

de Vries solar cycle: A solar activity periodicity of c. 210-year period named after Hessel de Vries, who in 1958 identified two maxima in ^{14}C production near AD 1500 and 1700.

Demographic dividend: The increase in economic productivity due to the increase in the number of people in the workforce relative to the number of dependents. It is the positive economic impact of a shift in a population's age structure towards a larger share from the working-age population.

Dryas octopetala: an Arctic–alpine flowering plant in the family *Rosaceae* that has given name to the last three cold periods before the Holocene.

– E –

Early Holocene: The first part after dividing the Holocene in three periods of similar length. It previously had a variable span depending on the area and proxy studied, but in 2018 the International Union of Geological Sciences established its correspondence to the Greenlandian Stage between 11,700–8,326 B2K.

Early Twentieth Century Warming (ETCW): The period of global warming between 1910 and 1945, that was of comparable magnitude (0.5 °C versus 0.6 °C) to the Late Twentieth Century Warming (LTCW) between 1975–2000, despite a much lower increase in atmospheric CO_2 levels.

Earth System Model: A model that incorporates biogeochemistry and the carbon cycle, and from an emissions pathway produces a model of resulting CO_2 atmospheric levels.

Easterlies: Prevailing pattern of surface winds from the east toward the west. At the Hadley cell they are known as trade winds, and at the Polar cell as polar easterlies.

Eccentricity: In astrodynamics eccentricity is a measure of the ellipticity of an orbit, with a value between zero for a circular orbit and one for a parabola.

Eddy: Fluid current with a different direction to the general flow. They are responsible for most of the energy and angular momentum transfer within the fluid. The size and number of eddies is a measure of turbulence. Examples of atmospheric eddies are hurricanes, cyclones and anticyclones, and Rossby waves. Oceanic eddies are responsible for upwelling and downwelling.

Eddy solar cycle: A solar activity periodicity of c. 1000-year period named after John A. Eddy, who described it in 1976.

Eemian: A stratigraphic period in Western Europe named after the Dutch river Eem. The Eemian stratum is dated between 126–115 ka in Northern Europe and 126–110 ka in Southern Europe. It has also become a popular name for the last interglacial before the Holocene.

El Niño/Southern Oscillation (ENSO): Irregularly periodic oscillation of 2–5 years in sea surface temperatures and predominant winds strength over the tropical eastern Pacific Ocean, that affects the weather of much of the world.

El Niño: The warm phase of the El Niño Southern Oscillation (ENSO) associated to warm surface waters in the Central–East Pacific Ocean and a weakening or reversal of the easterly trade winds.

Enthalpy: Enthalpy comprises a system's internal energy, plus the amount of work required to establish its volume and pressure. It is the preferred expression of energy changes in many scientific disciplines as changes in enthalpy equal the energy transfer through heat and work when there is no transfer of matter at constant pressure.

Equable climate problem: Refers to the inability by climate models to reproduce past hothouse climates of the Earth (e.g. Early Eocene, Cretaceous), characterized by reduced equator to pole temperature difference, warm polar regions with a reduced seasonality and ice free conditions at both poles, without resorting to unrealistic greenhouse gas concentrations or altered physical parameters.

Equilibrium Climate Sensitivity (ECS): The amount of warming produced by a doubling of atmospheric CO_2 levels after the oceans have had time to equilibrate.

Equilibrium line altitude (ELA): The line in a glacier that separates the zone of ice accumulation from the zone of ice ablation. The movement of this line determines if the glacier is growing or shrinking.

EROEI: Energy return on energy invested. The amount of energy obtained after discounting the amount of energy expended in obtaining it. A measure of the efficiency in energy production.

– F –

Feedback: A feedback occurs when part of the output from a system is added or subtracted to the input modifying the result. Amplifying feedbacks are positive and dampening feedbacks negative. Systems dominated by negative feedbacks are inherently stable and systems where positive feedbacks predominate are unstable.

Fennoscandian ice sheet: The ice sheet that forms over Scandinavia and surrounding areas of Eurasia during glacial periods.

Ferrel Cell: Part of the atmospheric circulation pattern proposed by William Ferrel in 1856 to explain prevailing wind patterns between latitudes 35–60° in both hemispheres. Part of the ascending air at 60° diverges at high altitude westward and towards the equator meeting an opposite circulation from the Hadley cell at 30° latitude. There it subsides and strengthens the high pressure ridges beneath. Then the air flows eastward and northward near the surface. The Ferrel cell is driven by the existence of the Hadley and Polar cells, as it lacks a powerful heat source or sink. Due to this it is a weaker cell with more mixed winds, and its characteristic surface winds are called prevailing westerlies. The Ferrel cell is not a very good representation of reality as at 10-km height strong westerlies are usually found.

Feynman solar cycle: A solar activity periodicity of c. 100-year period named after Joan Feynman, who described it in 1982 based on solar wind and geomagnetic activity.

Forcing: Any process or perturbation that drives climate change.

– G –

Ga: Giga anni (10^9 years), a unit of time to indicate age in billions of years from the present, taking 1950 as the reference present date.

General circulation model (GCM): Numerical models representing physical processes in the atmosphere, ocean, cryosphere and land surface.

Geopotential height: It is the actual height of a pressure surface above mean sea-level, and is related to the density of the air below. A low geopotential height indicates the presence of cold dense air masses below, while a high geopotential height indicates the opposite. It is measured in meters relative to a given pressure. On a weather map height contours connect points of equal geopotential height.

Glacial cycle: The alternation of glacial and interglacial periods during the Pleistocene according to Milankovitch orbital frequencies.

Glacial inception: The transition from an interglacial to a glacial period. The establishment of glacial conditions after an interglacial is a very slow process that can take 15 kyr or more. It is generally considered that glacial inception starts when non-polar ice sheets start to grow and sea levels show a marked decline.

Glacial period: An interval of time within an ice age when surface temperature is several degrees lower than present, and polar and mountain ice sheets are much more extensive, covering large parts of the northern hemisphere.

Glacial termination: A period of time of around 5–10 kyr when the transition from a glacial period to an interglacial takes place. They are usually dated at the midpoint, defined as the time when the sea level rise reaches 50% of its change.

Great Oxygen Crisis: The Great Oxygen Crisis, or Great Oxygenation Event, was the biologically induced appearance of oxygen in Earth's atmosphere at the beginning of the Proterozoic eon, 2.45 Ga. The oxygen produced by Cyanobacteria was first absorbed by iron and other elements producing thousands of new minerals. When it started to accumulate in the oceans and atmosphere it resulted in a mass extinction of the, until then, dominant anaerobic life forms. It also reacted with atmospheric methane, a greenhouse gas, greatly reducing its concentration and likely triggering the Huronian glaciation, the longest and harshest ice age known in Earth's history.

Greenhouse effect (GHE): It is the difference between the temperature at which a planet must emit infrared radiation to balance the absorbed solar radiation and the temperature at its surface. It is primarily due to atmospheric greenhouse gases that absorb and emit infrared radiation. Due to the GHE, the Earth is 33°C warmer than it would be with an atmosphere transparent to infrared radiation, or no atmosphere.

Greenhouse gas (GHG): A gas that absorbs and emits energy within the thermal infrared part of the spectrum. The main GHGs in Earth's atmosphere are water vapor (H_2Ov), carbon dioxide (CO_2), methane (CH_4), nitrous oxide (N_2O), ozone (O_3), chlorofluorocarbons (CFCs), and hydrofluorocarbons (HFCs). In climatology the term might refer only to non-condensing greenhouse gases, excluding water vapor.

Greenhouse theory: Theory describing how the surface temperature of a planet, with an atmosphere containing greenhouse gases, is determined by the balance between the absorbed solar radiation and the emitted infrared radiation. Due to the presence of greenhouse gases, infrared radiation emission to space takes place mostly from the atmosphere instead of the surface, and the temperature of the surface becomes warmer. Changes in the amount of greenhouse gases cause an imbalance between absorbed and emitted energy, due to a change in the height of infrared radiation emission. The balance is restored by a change in the surface and atmosphere temperature, causing a change in climate.

Greenland Ice Core Chronology 2005 (GICC05): A stratigraphical timescale for Greenland ice cores developed from NGRIP, GRIP, and DYE–3 ice cores through annual layer counting that reaches back to 60,202 b2k. It was further extended with a flow model mainly based on NGRIP to 104 ka b2k.

Greenland ice core project (GRIP): A multinational European research project that between 1990–1992 drilled an ice core at the Greenland ice sheet summit. It also refers to the ice core produced.

Greenland Ice Sheet Project 2 (GISP2): A 1998–1993 drilling project at the Greenland ice sheet summit, and the ice core that it produced.

Greenlandian stage: The first stratigraphic stage of the Holocene, between 11,700–8,326 B2K.

– H –

HadCRUT: A global surface temperature dataset produced by the Hadley Centre of the UK Met Office and the Climatic Research Unit of the University of East Anglia. Current version is HadCRUT5.

Hadley Cell: Part of the atmospheric circulation pattern proposed by George Hadley in 1735 to explain prevailing wind patterns near the equator (trade winds). High insolation in the equatorial band causes warm air to rise. At high altitude the warm air moves poleward and deviates eastward by the Coriolis force. Upon reaching 30° latitude the air sinks and closes the loop by moving equatorward and westward at the surface creating the trade winds (easterlies).

Hale solar cycle: A solar magnetic cycle of c. 22 years discovered by George Ellery Hale in 1919, that comprises two 11-year sunspot cycles, during which the polarity of the solar magnetic polar fields and the polarity of the sunspot pairs reverses twice.

Halocline: A water body layer that displays a strong vertical salinity gradient, separating two zones that display more moderate salinity changes. Due to salinity effect on water density, the halocline plays an important role in water vertical stratification, strongly limiting vertical exchange of mass, heat, salt, ions, and nutrients.

Heinrich event (HE): Recurrent climate event during the glacial period responsible for the deposition of a very prominent layer of ice-rafted debris at the bottom of the North Atlantic Ocean. They were discovered by Hartmut Heinrich in 1988. Heinrich events constitute periods of very intense iceberg activity indicative of a massive collapse at the Atlantic borders of the ice sheets. They coincide with colder periods in Greenland and warming periods in Antarctica. There are six of them between 60 and 17 ka, with the Younger Dryas possibly constituting the most recent one.

Hiatus: In climatology, each period of time during the instrumental era of temperature measurements (since 1850) when little or no warming has taken place. Hiatuses appear to be the low phase of a c. 65-yr periodicity. The first hiatus took place between 1879 and 1909. The second hiatus happened between 1944 and 1974. The third hiatus started around 1998 and is still ongoing.

Holocene: The current interglacial and geological epoch. The base of the Holocene has been stratigraphically defined at 11,700 B2K (11,650 BP) by the International Union of Geological Sciences.

Holocene Climatic Optimum (HCO): A period within the Holocene when the highest global average temperatures were reached. Although its temporal span varied between different regions, globally it can be considered to have taken place roughly between 9600 and 5500 BP.

Holocene Thermal Maximum: An alternative name for the Holocene Climatic Optimum.

Holton–Tan Effect: A phenomenon in which the strength of northern stratospheric winter polar vortex synchronizes with the equatorial quasi-biennial oscillation. The vortex becomes stronger and colder when the QBO is in its westerly phase and weaker and warmer when it is in its easterly phase.

Hypsithermal: An alternative name for the Holocene Climatic Optimum.

– I –

Ice Age: Any geological period of the Earth history characterized by the presence of large continental ice sheets. We are currently in the Quaternary Ice Age, as both Greenland and Antarctica are covered by ice sheets. Within an ice age there are alternating colder glacial periods, or stadials, and warmer interglacials, or interstadials. Historically and popularly the term ice age is used as a synonym for glacial periods, generating confusion.

Ice-rafted debris (IRD): Deposits of material, usually of petrological origin, found at the bottom of a water body that have been transported by ice. During the Pleistocene the transport of sediments by icebergs to areas of high iceberg melting rates was a primary mechanism of sediment transport. The mineralogical nature of the sediments allows to trace the regional origin of the icebergs responsible for the transport.

Indo–Pacific Warm Pool (IPWP): A large area ($>30 \times 10^6$ km^2) in the tropical western Pacific Ocean and eastern Indian Ocean of c. 7% of the planet surface that is permanently above 28 °C. The high temperature causes deep convection, producing clouds that rise to 15 km high, and result in significant atmospheric circulation effects. It is an important component of the global climate system.

Infrared radiation (IR): Radiation with a wavelength between 0.75–1000 μm. The relevant IR for climate is the thermal infrared, between 3–15 μm.

Interferometry: A technique based on the superposition of electromagnetic waves in order to extract information from their interference. In radio astronomy, very-long-baseline interferometry allows the precise timing of the arrival of a signal from a very distant source with atomic clocks at faraway radiotelescopes for imaging remote cosmic radio sources and astrometry. When used in reverse it allows the precise measurement of changes in Earth's rotation speed and the mapping of tectonic movements.

Intergovernmental Panel on Climate Change (IPCC): The United Nations body in charge of producing reports assessing the published science on climate change.

Interstadial: A warm interval within a climatological time sequence.

Intertropical Convergence Zone (ITCZ): Is the climatic equator of the planet, the area encircling the Earth where the northeast and southeast trade winds converge, producing what is known by sailors as the calms. It is formed by high tropical insolation driving warm, moist air convection, forming the ascending branch of the Hadley cell. As the air ascends it cools down forming a band of clouds and thunderstorms that encircle the globe near the Equator. The ITCZ location varies through the seasons, moving North from January to July and South from July to January, following the band of maximal solar flux. Tropical monsoons are linked to the position of the ITCZ, and long-term changes in its position due to changes in insolation derived from precessional and obliquity changes have a very important effect on paleoclimate evolution.

– L –

La Niña: The cold phase of the El Niño Southern Oscillation (ENSO) associated to cold surface waters in the Central–East Pacific Ocean and a strengthening of the easterly trade winds.

Last Glacial Maximum (LGM): The time during the last glacial period when ice sheets were at their greatest extent. It is defined based on sea level being 125 meters below current level between 26.5 and 19.0 ka.

Late Cenozoic Ice Age: The present ice age, that started at 33.9 Ma at the Eocene–Oligocene Boundary whith the beginning of Antarctic glaciation. It spans the second half of the Cenozoic Era.

Late Holocene: The last part after dividing the Holocene in three periods of similar length. It previously had a variable span depending on the area and proxy studied, but in 2018 the International Union of Geological Sciences established its correspondence to the Meghalayan Stage from 4,250 B2K to the present.

Late Twentieth Century Warming (LTCW): The period of global warming between 1975 and 2000, that was of comparable magnitude (0.6 °C versus 0.5 °C) to the Early Twentieth Century Warming (ETCW) between 1910–1945, despite a much higher increase in atmospheric CO$_2$ levels.

Latitudinal insolation gradient (LIG): The gradient determined by the angle of incidence of solar radiation in the amount of energy received from the sun in a period of time at the planet's surface (for example, kWh/m^2/day), that changes with latitude. This gradient acts on the climate system through differential solar heating, which determines the Earth's latitudinal temperature gradient that drives the atmospheric and ocean circulation and creates the different climatic zones. The LIG changes with the seasons and on longer timescales with changes in obliquity and precession.

Latitudinal temperature gradient (LTG): The surface temperature gradient, determined mainly by differential solar heating, that changes with latitude, and by the efficiency of the heat transport from the tropics to the poles. The LTG drives the atmospheric and ocean circulation and creates the different climatic zones. The LTG changes with the seasons and on longer timescales with changes in obliquity and precession, but unlike the LIG it also changes when there is a latitudinal change in surface temperatures, like with Arctic warming amplification. The LTG is a central property of the planet's climate system.

Laurentide ice sheet: The ice sheet that forms over North America during glacial periods.

Length of day (LOD): A measure of day-length fluctuations determined by the difference between the astronomically determined duration of the day and 86,400 International System seconds.

Linear Pottery culture (LBK): Linearbandkeramik (German), the first Central European Neolithic culture between 7550–6950 BP.

Little Ice Age (LIA): A climate interval after the Medieval Warm Period characterized by cooling and mountain glaciers expansion. There is no agreement on the LIA timespan. In this book the LIA is considered to start after the 1257 Samalas eruption and end after the effects of the 1835 Cosigüina eruption were over, spanning 1258–1840.

Low gradient paradox: The physical paradox posed by equable climates with warm poles requiring enhanced meridional fluxes of heat to sustain mild high latitude temperatures while keeping low latitudes from becoming exceedingly warm and the turbulence theory standpoint that the meridional heat flux is proportional to the meridional temperature gradient.

Lunisolar: Caused by both the sun and the moon.

– M –

Medieval Climatic Anomaly: A synonym for the Medieval Warm Period.

Medieval Warm Period (MWP): A climate interval after the Dark Ages Cold Period and before the Little Ice Age characterized by warming and mountain glaciers contraction. It is usually dated c. AD 950–1250.

Medithermal: The latest part of a tripartite division of the Holocene based on palynology (pollen science) proposed by Ernst Antevs in 1948 and published as Neothermal Climatic Sequence. It means period of intermediate temperature, and corresponds to the Neoglaciation.

Meghalayan stage: The last stratigraphic stage of the Holocene, from 4,250 B2K to the present.

Meltwater pulse: An acceleration in sea-level rise resulting from outbursts of glacial meltwater and/or greatly enhanced iceberg production during periods of ice-sheet collapse. Rates of sea-level rise during meltwater pulses may have been of 35–60 mm/year and persisted for a few centuries. The increased addition of cold fresh water to the ocean may have had important climatic effects, affecting also oceanic currents and vertical stratification of water.

Mid-Holocene Transition (MHT): A period of time between c. 6,000 and 4,800 BP that marks the end of the Holocene Climatic Optimum and the beginning of the Neoglacial Period, when a complete reorganization of the Earth's climate took place. The principal cause of this global climatic shift was the redistribution of solar energy as the northern summer insolation decrease reached its maximum rate. It produced a southward shift of the Intertropical Convergence Zone causing increased dryness and desertification at around 30°N latitude in South America, Africa and Asia. It also led to an increase in the amplitude of the El Niño Southern Oscillation.

Mid-Pleistocene Transition (MPT): A period of time at the transition from the Early to Middle Pleistocene, from about 1.25 to 0.7 Ma, characterized by a reduction in the frequency of interglacials and the emergence of a c. 100-kyr periodicity in $\delta^{18}O$ records. Once the MPT ended, the Pleistocene cooling and increase in global ice levels ended.

Middle Holocene: The second part after dividing the Holocene in three periods of similar length. It previously had a variable span depending on the area and proxy studied, but in 2018 the International Union of Geological Sciences established its correspondence to the Northgrippian Stage between 8,326–4,250 B2K.

Milankovitch Theory: The theory proposed by Milutin Milanković in 1920 to explain the alternation of interglacial and glacial periods during the Pleistocene as a result of long-period changes in the orbit of the Earth caused by the gravitational pull of the sun, the moon and the planets. It was demonstrated in 1976 that Pleistocene climate proxies follow the orbital frequencies proposed by Milanković.

Modern Global Warming (MGW): The period of warming of c. 300 years from the bottom of the Little Ice Age between 1650–1700 to the present.

Modern Solar Maximum (MSM): The period 1935–2004, defined by solar cycles 17–23, constituting the longest period in the sunspot record of above average decadal solar activity.

– N –

NCEI PDO index: The Pacific Decadal Oscillation index produced by the National Centers for Environmental Information from the National Oceanic and Atmospheric Administration sea-surface temperature dataset ERSST. The NCEI PDO index closely follows the Mantua PDO index.

Neoglacial Period: The period of the Holocene between c. 5200–400 BP characterized by an increase in global glacier advances and a decrease in global temperature. It is thought to have been driven by the decrease in Earth's obliquity and in northern summer insolation due to precession.

Neoglaciation: The increasing trend in global glacier advances after the Holocene Climatic Optimum identified and named by François Matthes in the 1940s.

Nodal precession: It is the precession of the orbital plane of a satellite around the rotational axis of the main body. In the case of the Moon it is the time it takes the ascending node to move through 360° relative to the vernal equinox. It is about 18.6 years.

North Atlantic Current (NAC): A current originating between the Grand Banks and Mid Atlantic Ridge with waters from the Gulf Stream and the Slope Water Current, characterized by warm temperature and high salinity. It mixes with northern cold polar water from the sub-polar gyre and branches out at the Irminger, Norway, and Canary currents.

North Atlantic Deep Water (NADW): A cold (2.2–3.5 °C), high salinity, high oxygen, low nutrients, layer that occupies the 1500–3500 depth level through the length of the Atlantic. Its sources are the Nordic Seas, the Labrador Sea, and the Mediterranean Sea. It flows south over time, being part of the thermohaline circulation, closing the loop of the Atlantic circulation.

North Atlantic Oscillation (NAO): A north–south dipole of atmospheric pressure mode of variability over the North Atlantic that displays prominent climatic teleconnections. One center of the dipole is located over Greenland and the other center of opposite sign is located in the Central North Atlantic between 35–40°N. The North Atlantic Oscillation (NAO) index is constructed from the pressure difference between the Icelandic Low and the Azores High. The oscillation switches between a positive mode with strong Icelandic Low and Azores High and a negative mode with weak Icelandic Low and Azores High. Strong NAO positive phases display above-average temperature in the Eastern US and Northern Europe and below-average temperature in Greenland and often in Southern Europe and the Middle East. They are also associated with above-average winter precipitation over Northern Europe and Scandinavia, and below-average winter precipitation over Southern and Central Europe. Opposite patterns of temperature and precipitation anomalies are typically observed during strong NAO negative phases.

North Greenland Eemian Ice Drilling Project (NEEM): An international project that between 2007–2012 succeeded in obtaining an ice-core from North–West Greenland that contained ice from the Eemian interglacial and from the previous glacial period.

North Greenland Ice Core Project (NGRIP): A multinational European drilling project that between 1999–2004 obtained a core reaching 105 ka from Central Greenland.

Northgrippian Stage: The second stratigraphic stage of the Holocene, between 8,326–4,250 B2K.

– O –

Obliquity: The angle between the earth's orbital plane (ecliptic) and equatorial plane, also called axial tilt. It can change between 22.1° and 24.5°, and currently is equal to 23°26′ (23.44°) and decreasing. It is the main Milankovitch parameter for orbital forcing of climate, responsible for the spacing and occurrence of interglacials.

Oceanic Niño index (ONI): NOAA's El Niño/Southern Oscillation index based on sea-surface temperature in the Niño 3.4 region (5°N–5°S, 120–170°W).

Old-age dependency ratio: The population ages 65-plus divided by the population ages 16–64. A measure of the economic burden of an ageing population.

Older Dryas: A colder period in the North Atlantic region of c. 200 years duration at c. 14,000 BP within Greenland Interstadial 1 (Bølling–Allerød stadial), corresponding to GI-1d, palynologically characterized by the abundance of *Dryas octopetala* leaves in Scandinavian lake sediments and peat bogs.

Oldest Dryas: The cold period prior to the start of the Bølling Oscillation at 14,700 BP. It is not well defined and largely in disuse, coinciding with Heinrich event 1 (16.8–14.7 kyr BP) after the Last Glacial Maximum. It is characterized by the abundance of *Dryas octopetala* leaves in Scandinavian lake sediments and peat bogs.

Orbital precession: See apsidal precession.

Orbital tuning: The process of adjusting the time scale of a geologic or climate record so that the observed fluctuations correspond to calculated orbital changes. Overtuning can result if the features in the record do not correspond to the chosen orbital changes, leading to circular reasoning.

– P –

Pacific Decadal Oscillation (PDO): A climatic mode of variability in the North Pacific with ample teleconnections. It is defined as the leading pattern of sea surface temperature anomalies in the North Pacific basin. It is strongly influenced by ENSO and represents a long-term envelop of ENSO variability. Its phases can last decades and when positive present negative SST anomalies in central and western North Pacific and positive SST anomalies in the eastern North Pacific and the opposite when positive. A weak mirror image of these anomalies occur across the South Pacific.

Paleocene–Eocene Thermal Maximum (PETM): A very large carbon isotope $\delta^{13}C$ excursion that took place approximately at 55.5 Ma and was contemporaneous with a significant warm and oxygen isotope anomaly. The size and ^{13}C depletion of the excursion has made a large release of methane clathrates the favored explanation.

Palynology: The study of plant pollen, spores and certain microscopic plankton organisms (collectively termed palynomorphs) in both living and fossil form.

Pause: See hiatus.

Pentadecadal solar cycle: A statistically weak reduction in solar activity every c. 50 years that is supported by cosmogenic ^{10}Be frequency analysis in the annually resolved DYE–3 Antarctic ice core.

Perigee: The point of least distance in the orbit of any satellite of the Earth.

Perihelion: The point in an orbit which is closest to the sun. For the Earth it happens currently around the 4th of January.

Petrological tracer: A mineral sediment whose origin can be traced to geological formations within a certain region.

Pleistocene Glaciation: Also known as Quaternary Glaciation, it comprises the last 2.59 Myr within the Late Cenozoic Ice Age when ice-sheets in the Northern Hemisphere began to grow outside Greenland at intervals. The 2.59 Ma boundary was chosen as a convenient point within a period of global cooling with major cooling phases between 2.8–2.4 Ma and cold evidence at mid-latitudes at the time of the Isthmus of Panama closure.

Polar Cell: Part of the atmospheric circulation pattern. Very cold air at high altitude in polar regions sinks creating a high pressure area and moving equatorward and westward at surface (polar easterlies) towards the 60° parallel, where it meets opposite warmer more humid winds from the Ferrel cell. The air ascends and diverges, with part of it moving polarward and eastward at high altitude to close the loop.

Polar circulation index (PCI): A statistical parameter introduced by Paul Mayewski in 1994 to measure changes in the relative size and intensity of the atmospheric circulation system, that transported air masses to Greenland, from the main empirical orthogonal function from ice-core chemical data.

Polar see-saw: A hypothesis that provides an explanation for why temperature changes in the two polar regions were out of phase at certain times during the last glacial period. The Atlantic Thermohaline Circulation is proposed to be responsible for variable heat transport leading to the polar alternative warming or cooling.

Polar vortex: A large region of low pressure cold air rotating cyclonically (clockwise in the Southern Hemisphere, counter-clockwise in the Northern Hemisphere) around both poles that manifests both in the troposphere and stratosphere. The stratospheric polar vortex is an Autumn-Spring phenomenon, while the tropospheric polar vortex usually persists, albeit weakened, during the Summer.

Precession: In a rotating body or system, precession is the comparatively slow (with respect to the rotation speed) change in the orientation of the rotating axis. Earth's axial precession is responsible for the slow displacement of the equinoxes (and seasons) along its orbit, with very important climatic repercussions, constituting one of Milankovitch orbital forcings. Earth's orbit around the sun also has a rotation axis that presents precession (apsidal precession), modifying the frequencies of the precession of the equinoxes.

– Q –

Quasi-Biennial Oscillation (QBO): Is a quasi-periodic oscillation in the strong stratospheric winds that circle the planet high above the equator, descending about 1 km per month. The new belt that develops above the old one has an opposite orientation. At a given height (measured at 30 hPa) westerlies and easterlies alternate every c. 14 months. The amplitude of the easterly phase (QBOe, negative values of wind speed) is about twice as strong as that of the westerly phase (QBOw, positive values of wind speed), and lasts a little longer, but climatically low speed easterly winds (−5–0 m/s) behave as westerly winds. The QBO has important repercussions on the northern hemisphere climate, particularly during winters, affecting the strength of the Polar Vortex and the Jet Stream.

Quaternary: The current and most recent of the three periods of the Cenozoic Era that spans the last 2.59 Myr and is divided into two epochs: the Pleistocene (2.59 Ma to 11.7 ka) and the Holocene (11.7 ka to today).

Quaternary Glaciation: See Pleistocene Glaciation.

– R –

Radiative forcing (RF): The net change in the energy balance of the Earth system due to some imposed perturbation.

Representative concentration pathway (RCP): A greenhouse gas concentration trajectory adopted by the IPCC for its Assessment Reports for the purpose of modelling climate change.

Revelle effect: The low CO_2 increase in the ocean of about 1/10 that results from a CO_2 increase in the atmosphere. It is due to the buffering effect of dissolved inorganic carbon species that regulates the amount of CO_2 in water according to the Revelle factor. The Revelle effect contributes to the increase in atmospheric CO_2 levels.

Revelle factor: The ratio of instantaneous change of CO_2 to the change in dissolved inorganic carbon species that determines the buffering capacity of ocean water. It is inversely correlated to the amount of dissolved inorganic carbon, to the buffering capacity of ocean water, and to the capacity of the ocean to take up increased atmospheric CO_2. The Revelle factor has different values in different parts of the oceans.

Roman Warm Period (RWP): A very long climate interval after the 2.8 Ka event and before the Dark Ages Cold Period characterized by warming and mountain glaciers contraction. Some authors date it at 2500–1600 BP (550 BC – AD 350), while others restrict it to 250 BC – AD 350. Historical and climatic evidence suggests the Roman Warm Period could have been as warm or warmer than the present.

Rossby wave: also planetary wave, is a kind of inertial wave generated in rotating planets due to differences in the Coriolis effect with latitude. Atmospheric Rossby waves are huge meanders in high-altitude winds with wavelengths of several hundreds of kilometers. Oceanic Rossby waves are much smaller and generally associated to the thermocline.

– S –

Sapropel: Dark-coloured sediments of marine, estuarine, or lacustrine deposition that are rich in organic matter derived from aquatic plants and animals. Mediterranean sapropels reflect monsoonal variations caused by Milankovitch orbital changes.

Schwabe solar cycle: A solar activity periodicity of c. 11-year period named after Samuel Heinrich Schwabe, who described it in 1843. It is traditionally measured by counting the number of sunspots in the sun.

Seasonality: The variation between the seasons. In paleoclimatology the difference between the seasons has changed over time with changes in precessional-linked insolation. Currently northern hemisphere winters are warmer and summers cooler than they were during the Early Holocene, displaying a decrease in seasonality over time.

Stadial: A cold interval within a climatological time sequence.

Stadium wave hypothesis: The hypothesis proposed by Marcia Glaze Wyatt in 2012 of a multidecadally varying climate signal that propagates across the Northern Hemisphere within a network sequence of synchronized ocean, atmosphere, and sea-ice indices. All indices vary at the same timescale of c. 64 years peak-to-peak throughout 20th century with one index leading the next in a consistently ordered lead-lag fashion.

Stefan–Boltzmann equation: Equation describing the Stefan–Boltzmann law that relates the energy radiated by a body to its temperature.

Stratigraphy: A branch of geology dedicated to the study of rock layers of sedimentary and volcanic origin, their successions and interpretation in terms of a general time scale. It provides a basis for historical geology.

Suess effect: The depletion of the heavy carbon isotopes ^{14}C and ^{13}C due to the burning of fossil fuels. Due to their biological origin fossil fuels are low in the stable

[13]C that is selected against by photosynthetic organisms, and because of their ancient origin they are low in ^{14}C due to radioactive decay.

Summary for policymakers (SPM): Summary of the IPCC reports intended to aid policymakers. The form is approved line by line by governments.

Syzygy: A straight-line configuration of three or more celestial bodies in a gravitational system.

– T –

Tephra: All pyroclastic solid materials, such as ash, dust, cinders, or blocks, that are ejected into the air during a volcanic eruption.

Thermocline: Thin layer in a large body of fluid that separates a zone with increased temperature mixing from a zone with reduced temperature mixing, resulting in a more rapid temperature rate of change than above and below.

Thermohaline circulation (THC): Also called the Great Ocean Conveyor Belt. The main component of general oceanic circulation below the surface and at depth controlled by differences in temperature and salinity. Cold, salty water sinks mainly in the polar regions and outside the Mediterranean and spreads slowly into the rest of the oceans before resurfacing at multiple places as cold, less saline, rich in nutrients water that is warmed and increases in salinity as it moves towards the sinking areas. In the process it transports huge amounts of heat, nutrients and ions over vast distances. It is responsible for part of the heat transport from the equator towards the poles, and it links both poles by the polar see-saw.

Torque: A measure of the force that causes an object to rotate acquiring angular acceleration. It is equal to the product of the magnitude of the force by the distance from its point of application to the axis of rotation.

Total solar irradiance (TSI): The total amount of solar radiation in W/m^2 received outside the Earth's atmosphere on a surface normal to the incident radiation, and at the Earth's mean distance from the Sun. It can only be measured reliably from satellites and the record extends back only to 1978. The solar cycle variation of TSI is of the order of 0.1%.

Treeline: The edge of the habitat, at high altitudes or high latitudes, beyond which trees cannot grow.

Tuning: See orbital tuning.

Turbidite: A sedimentary deposit consisting of material that has moved down the steep slope at the edge of a continental shelf by gravity-induced turbidity currents.

– V –

Varve: A sedimentary layer at the bottom of a water body deposited within a year, that may present two layers of alternately finer and coarser silt or clay, reflecting differences in seasonal sedimentation within the year.

– W –

Westerlies: Dominant pattern of surface winds from the west toward the east. At the Ferrel cell they are known as prevailing westerlies.

– Y –

Younger Dryas (YD): The last Greenland stadial, GS–1 between 12,900–11,700 BP, before the Holocene. It took place during glacial termination causing a reversal in climate conditions at the North Atlantic region with worldwide effects. It is considered by some researchers as a Heinrich event (HE) caused by the collapse of ice-sheets' oceanic margins, accompanied by ice-rafted deposition in the North Atlantic bottom, and ending in the last abrupt Dansgaard–Oeschger event as other HEs. It is characterized by the abundance of *Dryas octopetala* leaves in Scandinavian lake sediments and peat bogs.